Animal Behavior

with Commentaries

EDITORS

Lynne D. Houck and Lee C. Drickamer

THE UNIVERSITY OF CHICAGO PRESS
Chicago & London

Lynne D. Houck is a research associate (associate
professor) in ecology and evolution at the
University of Chicago. Lee C. Drickamer is
professor of zoology at Southern Illinois
University at Carbondale.

The University of Chicago Press, Chicago 60637
The University of Chicago Press, Ltd., London

ISBN: 0-226-35456-3 (cloth)
        0-226-35457-1 (paper)

Library of Congress Cataloging-in-Publication Data

Foundations of animal behavior : classic papers with commentaries /
    Lynne D. Houck and Lee C. Drickamer, editors.
        p.   cm.
    "Published in association with the Animal Behavior Society."
    Includes bibliographical references.
    1. Animal behavior.   I. Houck, Lynne D.   II. Drickamer, Lee C.
III. Animal Behavior Society.
QL751.6.F66   1996
591.51—dc20   96-33734

⊗ The paper used in this publication meets
the minimum requirements of the American
National Standard for Information Sciences—
Permanence of Paper for Printed Library
Materials, ANSI Z39.48-1984.

Contents

Preface   xv

Introduction
Lynne D. Houck and Lee C. Drickamer   1

PART ONE

Historical Origins
The Making of a Science
Bennett G. Galef, Jr.   5

1
Charles Darwin (1872)
Instinct

The Origin of Species, chapter 8.
13

2
George C. Romanes (1882)
Introduction

Animal Intelligence. London: Kegan, Paul, Trench, pp. 360–66
38

3

E. L. Thorndike (1898)

Introduction

> Animal intelligence; an experimental study of the associative processes in animals.
> Psychological Review Monograph Supplement 2(8):1–109
>
> 45

4

J. B. Watson (1924)

What is behaviorism?

> *Behaviorism.* New York: W. W. Norton, pp. 6–19
>
> 55

PART TWO

The Emerging Science
Defining the Goals, Approaches, and Methods
Lee C. Drickamer and Charles T. Snowdon   71

5

Daniel S. Lehrman (1953)

A critique of Konrad Lorenz's theory of instinctive behavior

> The Quarterly Review of Biology 28:337–63
>
> 87

6

Niko Tinbergen (1963)

On aims and methods of ethology

> Zeitschrift für Tierpsychologie 20:410–33
>
> 114

7

George W. Barlow (1968)

Ethological units of behavior

> In The Central Nervous System and Fish Behavior, pp. 217–32, D. Ingle, ed. Chicago:
> University of Chicago Press
>
> 138

8

Peter J. B. Slater (1973)

Describing sequences of behavior

> In Perspectives in Ethology, vol. 1, pp. 131–53, P. P. G. Bateson and P. H. Klopfer, eds. New York: Plenum Press
>
> 154

9

Jeanne Altmann (1974)

Observational study of behavior: sampling methods

> Behaviour 48:227–65
>
> 177

10

Thelma E. Rowell (1967)

A quantitative comparison of the behaviour of a wild and caged baboon group

> Animal Behaviour 15:499–509
>
> 218

11

G. P. Baerends (1958)

Comparative methods and the concept of homology in the study of behaviour

> Archives neerlandische Zoologie (supplement 1)
>
> 229

12

William Hodos and C. B. G. Campbell (1969)

Scala naturae: Why there is no theory in comparative psychology

> Psychological Review 76:337–50
>
> 246

PART THREE

Development and Learning
Genetic Influences, Learning, and Instinct
Charles T. Snowdon   261

13
Wallace Craig (1918)
Appetites and aversions as constituents of instincts
Biological Bulletin 34:91–107
276

14
Peter Marler and Miwako Tamura (1964)
Culturally transmitted patterns of vocal behavior in sparrows
Science 146:1483–86
293

15
Gilbert Gottlieb (1968)
Prenatal behavior of birds
The Quarterly Review of Biology 43:148–74
297

16
Margaret Bastock (1956)
A gene mutation which changes a behavior pattern
Evolution 10:421–39
324

17
David R. Bentley and Ronald R. Hoy (1972)
Genetic control of the neuronal network generating cricket song patterns
Animal Behavior 20:478–92
343

18

Jane Van Zandt Brower (1958)

Experimental studies of mimicry in some North American Butterflies, Part 1:
The monarch, Danaus plexippus, *and viceroy,* Limenitis archippus,
archippus

        Evolution 12:32–47

        358

19

John Garcia and Robert A. Koelling (1966)

Relation of cue to consequence in avoidance learning

        Psychonomic Science 4:123–24

        374

20

Harry F. Harlow and Robert R. Zimmermann (1959)

Affectional responses in the infant monkey

        Science 130:421–32

        376

PART FOUR

Neural and Hormonal Mechanisms of Behavior
Physiological Causes and Consequences
Elizabeth Adkins-Regan   389

21

Niko Tinbergen (1950)

The hierarchical organization of nervous mechanisms underlying instinctive
behaviour

        Symposium of the Society for Experimental Biology4:305–12

        406

22

E. von Holst (1954)

Relations between the central nervous system and the peripheral organs

        British Journal of Animal Behavior 2:89–94

        414

23

Jürgen Aschoff (1960)

Exogenous and endogenous components in circadian rhythms

Cold Spring Harbor Symposia on Quantitative Biology 25:11–28

420

24

Martha K. McClintock (1971)

Menstrual synchrony and suppression

Nature 229:244–45

438

25

Daniel S. Lehrman (1965)

Interaction between internal and external environments in the regulation of the reproductive cycle of the ring dove

In Sex and Behavior, F. A. Beach, ed., 344–80. New York: Wiley

440

26

Frank A. Beach and A. Marie Holz-Tucker (1949)

Effects of different concentrations of androgen upon sexual behavior in castrated male rats

Journal of Comparative and Physiological Psychology 42:433–53

466

27

Charles H. Phoenix, Robert W. Goy, Arnold A. Gerrall and William C. Young (1959)

Organizing action of prenatally administered testosterone propionate on the tissues mediating mating behavior in the female guinea pig

Endocrinology 65:369–82

487

28

Vincent G. Dethier and Dietrich Bodenstein (1958)

Hunger in the blowfly

    Zeitschrift für Tierpsychologie 15:129–40

    501

29

R. A. Hinde (1960)

Energy models of motivation

    Symposia of the Society for Experimental Biology 14:199–213

    513

PART FIVE

Sensory Processes, Orientation, and Communication
Biology of the Umwelt

Fred C. Dyer and H. Jane Brockmann   529

30

Karl von Frisch and Martin Lindauer (1956)

The "language" and orientation of the honey bee

    Annual Review of Entomology 1:45–48

    539

31

Donald R. Griffin and Robert Galambos (1941)

The sensory basis of obstacle avoidance by flying bats

    Journal of Experimental Zoology 86:481–506

    553

32

Kenneth D. Roeder and Asher E. Treat (1961)

The detection and evasion of bats by moths

    American Scientist 49:135–48

    579

33

Arthur D. Hasler (1960)

Guideposts of migrating fishes

Science 132:785–92

594

34

Edward O. Wilson and William H. Bossert (1963)

Chemical communication among animals

Recent Progress in Hormone Research 19:673–716

602

35

Stephen T. Emlen (1969)

Bird migration: influence of physiological state upon celestial orientation

Science 165:716–18

646

36

Peter Marler (1961)

The logical analysis of animal communication

Journal of Theoretical Biology 1:295–317

649

PART SIX

Evolution of Behavior

Approaches to Studying Behavioral Change

Stevan J. Arnold and H. Jane Brockmann   673

37

Konrad Lorenz (1971)

Comparative studies of the motor patterns of Anatinae (1941), translated by Robert Martin, in Studies in Animal and Human Behavior, vol. 2, pp. 14–18, 106–14.

Cambridge: Harvard University Press

683

38

Howard E. Evans (1962)

The evolution of prey-carrying mechanisms in wasps

> Evolution 16:468–483
> 697

39

Esther Cullen (1957)

Adaptations in the kittiwake to cliff-nesting

> Ibis 99:275–302
> 713

40

Niko Tinbergen, G. J. Broekhuysen, F. Feekes, J. C. W. Houghton, H. Kruuk, and E. Szulc (1962)

Egg shell removal by the black-headed gull, Larus ridibundus L.; a behaviour component of camouflage

> Behaviour 19:74–117, pp. 74–85 and 109–17
> 741

41

William D. Hamilton (1964)

The genetical evolution of social behaviour, I

> Journal of Theoretical Biology 7:1–16
> 764

42

Gordon H. Orians (1969)

On the evolution of mating systems in birds and mammals

> American Naturalist 103:589–603
> 780

43

Robert L. Trivers (1972)

Parental investment and sexual selection

> In Sexual Selection and the Descent of Man, 1871–1971, B. Campbell, ed.,
> pp. 136–79. Chicago: Aldine
> 795

44

John Maynard Smith and George R. Price (1973)

The logic of animal conflict

Nature 246:15–18

839

List of Contributors   843

Preface

This volume was inspired by the simple observation that students of animal behavior are highly rewarded by reading original papers in the field. The excitement of discovering original works provides a feeling of being connected to the field in a way that reading textbooks cannot offer.

Obtaining access to original work, however, can be a time-consuming matter. In teaching a course, for example, the process involves identifying and selecting appropriate papers, obtaining clean copies from files or libraries, duplicating each paper and then compiling the total collection for distribution to the students. This process was made more difficult by recent efforts to enforce the copyright laws, which require written permission for duplicating each paper, and for renewing this permission each time the course is taught. An alternative is a single volume of reprinted articles that provides students with easy access to original readings.

The impetus to put together this particular volume was crystallized by the appearance in 1991 of a book entitled *Foundations of Ecology* (edited by Leslie A. Real and James H. Brown [University of Chicago Press]). The focus of this book was a historical perspective on research in

ecology, and it included reprints of papers widely acknowledged to have had a major influence on work in the field. The style of that *Foundations* volume prompted us to consider a similar treatment for the field of animal behavior. Indeed, we have retained this historical view, as well as a similar format featuring commentaries on each group of reprinted papers.

We began by forming an editorial board, inviting six active researchers in animal behavior to join us in selecting papers and providing commentaries. Our board was comprised of Elizabeth Adkins-Regan (Cornell University), Stevan J. Arnold (University of Chicago), H. Jane Brockmann (University of Florida), Fred C. Dyer (Michigan State University), Bennett G. Galef, Jr. (McMaster University), and Charles T. Snowdon (University of Wisconsin). The range of research experience and expertise among us spans topics that includes social influences on feeding and reproductive behavior, the hormonal basis of gender-specific behavior, evolutionary stable reproductive strategies, and the application of quantitative genetics to theoretical models of sexual selection. Of perhaps greater importance was our shared interest in earlier historical approaches—

both conceptual and empirical—and the impact of key works on current research paradigms.

At our first meeting, each of us brought a restricted list of "must include" papers. Our initial goal was to eliminate duplicates and quickly select a final roster of 40 to 50 papers. Combining our individual choices, however, yielded a truly shocking compilation of more than 200 papers. Moreover, each of these papers was considered to be a pivotal contribution in its time. Brief thoughts of persuading our editor to produce a five-volume set were replaced soon enough by a sense of reality. We set about the demanding task of selecting a subset of particular papers from our cherished—but extensive—original list.

The elimination of nearly 80 percent of our original "must include" choices required painful and far from unanimous decisions. The final selection of papers to be reprinted, however, has been based on a strong consensus of the editors and board members. The ultimate culling in part reflects works that have strongly affected our personal approaches to behavioral research. Another influence on the final list was the goal of representing broadly different research approaches, from neuroethology to theoretical models of behavioral evolution. Reducing the number of included papers also was accomplished by concentrating primarily on the three decades ending with the early 1970s. The era immediately following World War II was a prolific time during which innovative thinking combined with technological advances to change the nature of animal behavior research. Our focus on this period excludes much

excellent work begun in the 1970s, in behavioral ecology to name but one area. Currently, these more recent papers are readily available in libraries, and we trust that these contributions will be included in future collections.

We often take for granted the current knowledge afforded by earlier behavioral research. One aim in creating this volume has been to bring beginning students—and their teachers—in direct contact with the work of ethologists whose research helped shape the field as it is today. Looking back to gain an appreciation of innovative contributions made in earlier times can reveal how major changes in ways of thinking about behavior engendered creative research. Perhaps a more important lesson is that the study of animal behavior is a dynamic field, one that welcomes contributions from students of all ages.

In terms of contributions, the editors and members of the editorial board unanimously agreed that all royalties from the sale of this book will be donated to the Animal Behavior Society for support of student research. We are pleased that this work is being sponsored by the Animal Behavior Society.

In preparing this volume, we have benefited greatly from many discussions with our colleagues. We particularly thank the following for their suggestions and advice: Stuart Altmann, Brian Charlesworth, John Dumbacher, Steven Emlen, Mart Gross, Robert Johnston, Sarah Lenington, Shaibal Mitra, Stephen Pruett-Jones, and Thomas Seeley.

Lynne D. Houck
Lee C. Drickamer

Introduction

In 1973, the field of animal behavior was formally acknowledged when three ethologists, Konrad Lorenz, Niko Tinbergen, and Karl von Frisch, were awarded the Nobel Prize. This prestigious award drew attention to ethology as a significant field of scientific research, an evaluation supported by many excellent studies published during the three decades prior to the Nobel award. Indeed, this period witnessed major conceptual, theoretical, and empirical contributions to behavioral research. What were the major advances during this time? How did conceptual innovations provide a legacy that informs current research in animal behavior? The nature of these contributions are the subject of this volume.

Our proximate goals in creating this book were twofold. Our first goal was to identify a group of papers, each of which has had a major impact on subsequent behavioral research. The selection of articles deliberately spans a range of research interests within the field of animal behavior. Our second goal was to include introductory commentaries that provide a historical context for the reprinted articles. These commentaries reveal how and why each paper had a critical impact on the conduct of behavioral research. Ultimately, we hope that this collection stimulates an enthusiastic appreciation for the evolving trajectories of behavioral studies.

This volume is divided into six parts, each beginning with a commentary that introduces the papers reprinted in that section. Most of the papers (39 of 44) reflect work published between the early 1940s and the 1970s. The exceptions to this emphasis are the four papers that constitute Part 1. These papers, published between 1872 and 1924, provide a brief history of the early emergence of behavioral studies as a scientific field. We take particular pleasure in beginning with excerpts from Charles Darwin, thus grounding our understanding of animal behavior within an evolutionary perspective. Three other papers included in Part 1 illustrate extremes in early thinking on how behavior should be studied. These extremes range from an initial approach endorsing anecdotes and anthropomorphism to the subsequent advocacy of rigorous scientific methods.

By moving from an early historical period to work published primarily after 1940, we deliberately have omitted behavioral work conducted in the earlier decades of the twentieth century. Adherence to a strong scientific approach characterized the studies that helped to distinguish ethology as a separate field.

1

Critical papers during this important period include work by Charles O. Whitman, Oskar Heinroth, and Jacob von Uexkull. Many of these papers have been reprinted in a thoughtful volume by Burghardt (1985) and, with the single exception of a paper by Wallace Craig, were not included here. Other work from this period, however, is considered by Bekoff and Jamieson (1990).

In Part 2, the focus is on methods and approaches that have made an impact on subsequent behavioral research. Contributions by Tinbergen and Lehrman provide overviews that helped bring realistic goals in sight of researchers. Other approaches emphasize the proper use of sampling methods, the value of comparing behavior in the field vs. in the laboratory, and the advantages of casting behavioral elements in terms of units and sequences.

In the remaining portion of the volume, Parts 3 to 6, we consider contributions within broadly defined areas of research: developmental influences on instinct and learning, neuroendocrine mechanisms that support behavior, sensory processes that affect communication and orientation, and approaches to studying behavioral change in an evolutionary context.

The papers discussed in Part 3 have provided critical insights into studies of development and learning. Our current appreciation for the interconnectedness of learning and instinct echoes the early contribution by Wallace Craig on appetites and aversions. Questions of nature-nurture effects also were addressed by elegant demonstrations of behaviors having strong components of learning (song dialects in sparrows) or innate expression (courtship in *Drosophila*). The belief that behaviors present at birth necessarily were innate was shaken by work showing that learning and sensation could occur

even during the prenatal period. Moreover, Harry Harlow and R. Zimmermann demonstrated that, in addition to food, mothers provide critical emotional resources, a concept that has greatly affected child care.

Neuroendocrine mechanisms and their influence on behavior are the subject of Part 4. The technological advances of current research in neural and hormonal studies make it difficult to envision a research climate dominated by the paradigm of stimulus-reflex actions. Yet is it within this context that Aschoff convinced skeptics that endogenous clocks existed. Other studies demonstrated that cyclic behavior could be entrained by conspecifics (McClintock), and that social stimulation could effect physiological changes (Lehrman), including hormonal responses. During this period of landmark research in endocrinology, the field of neuroethology was being established. Models of neural action distinguished between active and passive actions (von Holst), and provided a hierarchical scheme (Tinbergen) that could account for complex sequences of behavior better than stimulus-reflex models. These few examples reflect phenomenal changes in our understanding of neural and hormonal support of behavior.

The papers collected in Part 5 also concentrate on mechanistic studies but emphasize the nature of an animal's sensory world. The survival value of behavior is of concern, in addition to the physiological basis for that behavior. Here we consider bats that use sonar to navigate with accuracy in total darkness (Griffin and Galambos), moths that are able to sense and avoid the sonar pulses of their predatory bats (Roeder and Treat), and salmon that use odor cues to return precisely to their natal stream (Hasler). Also in this section are reprints of pioneering work on the language of bees (von Frisch and Lindauer) and the use of pheromones to communicate (Wilson and Bossert). These discoveries of animal abilities were impressive contributions in their time, and they continue to fascinate and inspire.

In Part 6, the focus on survival value continues, but with an evolutionary interpretation instead of a proximate description. How did behaviors change over evolutionary time, and what was the evolutionary function of these particular behaviors? Two researchers who pioneered studies of behavioral change used phylogenetic histories to focus on displays in ducks and their relatives (Lorenz) and in wasps (Evans). Each of these studies was exemplary in considering a large group of behaviorally diverse species. Other work focused on adaptation and survivorship in studying the evolutionary function of particular behaviors. Cullen's comparisons of related gull species were noteworthy as among the first to consider how different environments might influence behavioral change. In using a different approach to understanding adaptation, Tinbergen et al. devised experiments that directly tested whether egg-shell removal affected chick survivorship in a population of gulls. New concepts of how evolutionary change could occur were introduced by theoretical models that made behavioral predictions (Hamilton, Orians, Maynard Smith and Price, and Trivers). Compared with early historical attempts to understand behavior by using anecdotal stories, these theoretical models represent a great distance traveled.

The papers reprinted here provided crucial stepping-stones that have supported and informed current research. Just as these works contributed to change in their time, the field of animal behavior continues to evolve. Within the last decade, the use of molecular techniques has transformed our capabilities. Studies of parentage now can be based on DNA fingerprinting, and measures of mating success can be directly evaluated. At a different level, the use of molecular data (along with increased computational power) can provide another dimension for evaluating phylogenetic relatedness and hence inferences of behavioral evolution. These changes are exciting in their own right, but particularly in their promise of enhancing behavioral studies that affect conservation efforts. The field of animal behavior embraces a range of perspectives, from ecology and evolution to physiological mechanisms and the molecular nature of gene action. Perhaps our greatest challenge will be to combine perspectives from past and present work in order to focus these broadly based research capabilities on studies of behavioral biodiversity and species conservation.

Lynne D. Houck
Chicago, Illinois

Lee C. Drickamer
Carbondale, Illinois

Literature Cited

Burghardt, G. M. 1985. *Foundations of Comparative Ethology*. New York: Van Nostrand Reinhold. 441 pp.

Bekoff, M., and D. Jamieson. 1990. *Interpretation and Explanation in the Study of Animal Behavior*. Volumes 1 (505 pp.) and 2 (465 pp). Boulder: Westview Press.

1 Historical Origins

The Making of a Science

Bennett G. Galef, Jr.

Humans have been carefully observing the behavior of other animals since before the dawn of recorded history. Extraordinary paintings of horses, mammoths, and reindeer found deep in the caves around Cro-Magnon in France testify to the sophisticated powers of observation and depiction of our Magdalenian forebears who lived and hunted in the valley of the Dordogne some 30,000 years ago. Elegant bas reliefs and paintings of birds, fish, and mammals, common in 5,000-year-old tombs of Egyptian royalty, demonstrate that animal watching did not end with the development of agriculture. Similarly, written descriptions from the fourth century B.C.E. of phenomena as diverse as the role of experience in the development of nightingale song, the copulatory postures of spine-covered hedgehogs, and brood parasitism by European cuckoos (Aristotle, 1970) show that interest in the behavior of animals survived the birth of Western civilization.

Such fragmentary evidence of early interest in animal behavior tantalizes as it informs, for we know nothing of the methods used in ancient times for observation or experiment and little of the motives of those studying animals in centuries long past. Only in relatively modern times has description of animal behavior been coupled with discussion of how and why the behavior of animals should be observed and analyzed. Such discussions, central to the emergence of a coherent science of animal behavior, became a focus of intellectual activity during the nineteenth century in Great Britain. Consequently, the present review begins in earnest in nineteenth-century England, where early development of a science of animal behavior took place.

Victorian England

During the decades from 1789 (date of publication of Gilbert White's *Natural History of Selborne*) to 1859 (date of publication of Charles Darwin's *Origin of Species*), the study of natural history was pursued with steadily increasing enthusiasm by a broad cross section of British society, "from the aristocrats who competed in turning their parks over to elands, beavers and kangaroos, to the artisans who hoarded their pennies to buy the *Entomologist's Weekly Intelligencer*" (Barber 1980). A Wardian case (terrarium) or Warrington case (aquarium) could be found in the parlor of almost every middle-class Victorian home, and biological supply houses sprang up across England to provide specimens and equipment for local amateur naturalists. By mid-century, run-of-the-mill natural history texts, like J. G. Wood's *Common Objects of the Country* (1858), could sell 100,000 copies in a single week (Barber 1980).

One reason for the exceptional popularity of natural history as a pastime of the Victorians lay in acceptance by gentlefolk of observation of nature as a form of "rational amusement," in contrast to "vulgar amusements" like reading novels or attending the theater. To qualify as rational amusement, an activity had to be, at least superficially, either useful or morally uplifting, preferably both. The study of natural history had the good fortune to be perceived as not only scientific, and therefore educational, but also as reverent and inspirational.

In the pre-Darwinian worldview, each animal and plant was seen as perfectly designed for its particular role in nature as the result of intelligent acts by an all-knowing and all-powerful Creator. Consequently, studying the structure and function of living things was seen as a means of learning about the God who designed them. In the words of Alexander Pope, one of the leading poets of the era, it was possible to look "through Nature up to Nature's God." By studying the beauty and perfection of God's creatures, one could find evidence not only of God's existence but also of His perfection and beauty (Paley 1802). Unfortunately, the necessity of seeing reflections only of God's beauty, perfection, and beneficence in every living thing interfered seriously with both objective description and unbiased analysis of the natural world.

Given the Victorians' professed educational and religious motivations for studying nature, one might expect Victorian books of natural history to be scholarly and serious. They were not. The texts tried to make animals and plants as entertaining and accessible as possible. They were filled with amusing tales about thieving magpies, of elephants who never forgot, of wise old owls and clever foxes. Even sophisticated scientists saw nothing wrong with treating these stories as the raw data of animal behavior. Darwin (1871, p. 449), in a chapter of his *Descent of Man* devoted to "mental powers," relates the following story, one of a number of unlikely tales. "At the Cape of Good Hope an officer had often plagued a certain baboon, and the animal, seeing him approaching one Sunday for parade, poured water into a hole and hastily made some thick mud, which he skillfully dashed over the officer as he passed by. . . . For long afterwards the baboon rejoiced and triumphed whenever he saw his victim."

Writers of books of natural history used their considerable talents to give "seemingly dry disquisitions and animals of the lowest type, by little touches of pathos and humour, that living and personal interest, to bestow which is generally the function of the poet" (Kingsley, 1855, p. 7). Exercise of poetic license, rather than a striving for accuracy, guided the pen of the Victorian naturalist-writer much as the desire to amuse shapes the scripts of many nature films seen on television today. Success lay in attributing human motives and human-like intelligence to the actions of every bird, bee, and Barbary ape. Scientific journals were as open to anecdote (informal observations of behavior) and to anthropomorphism (attributions of human mental experiences to animals) as were lay publications. The distinguished British journal *Nature* (1883, p. 580) published a letter from a Mr. Oswald Fitch describing a domestic cat that took some of its own dinner of fish bones "from the house to the garden, and, being followed, was seen to have placed them in front of a miserably thin and evidently hungry stranger cat who was devouring them; not satisfied with that, our cat returned, procured a fresh supply, and repeated its charitable offer. . . . This act of benevolence over, our cat . . . ate its own dinner off the remainder of the bones." Such anthropomorphic interpretation of anecdote was the life's blood of the successful Victorian naturalist.

It is against this background of anecdote, anthropomorphism, and inference of mind from behavior that Darwin's (1872) chapter on instinct, reprinted here from the sixth edition of his *Origin of Species*, must be viewed. In general, Darwin's manuscript is surprisingly modern in tone. Darwin's uncommon common sense serves him well in excluding the obviously exaggerated stories that encumbered the work of many of his predecessors and successors.

Darwin focuses on behavioral phenomena (brood parasitism in birds, slave-making in ants, the behavior of sterile castes of social insects, etc.) many of which still concern us today. Further, he is generally unwilling to attribute human-like emotional responses to animals (though he lapses occasionally, for example, in describing *Formica flava* [a species of ant] as "courageous" and *F. sanguinea* [another species of ant] as on occasion "terrified" and at other times "taking heart"). Darwin also joins the more cynical of his contemporaries in questioning the accuracy of the observations reported by his correspondents: He undertakes his own observations of milking of aphids by ants and the behavior of ant "slaves" and their "masters." He conducts his own experiments on cell-making by honey bees.

In the summary of his chapter, Darwin uses evidence of imperfection in instincts (for example, the occasional decision by European cuckoos to lay their "dull and pale-coloured" eggs among the "bright greenish-blue" eggs of the Hedge-warbler or the tendency of south American shiny cowbirds [*Molothrus bonariensis*] to start, but never complete, their own nests and to lay many eggs on bare ground, where they surely perish) to support his view that instincts have a natural origin. For, if perfection of design is the hallmark of the acts of an omnipotent Creator, then imperfection is evidence of a natural, rather than a supernatural origin of instinctive behaviors.

Darwin's main goal in the chapter is to suggest ways in which natural selection might have acted gradually to shape rudimentary forms of instinctive behaviors into the sophisticated instincts that so impressed his contemporaries.

Darwin's methods are, in large part, those of objective description and comparison between species. His conclusion is that perfection of instinct is a "consequence of one general law leading to the advancement of all organic beings,— namely, multiply, vary, let the strongest live and the weakest die."

If Darwin's successors had built on the firm foundation laid down by Darwin in his chapter on instinct, the subsequent history of animal behavior might have been relatively straightforward. Unfortunately, in some later work, as for example in the *Descent of Man* (as quoted above), Darwin (1871) lapses into the language of his contemporaries, relating stories of the "deceit," "artful revenge," "loyalty," etc., of a variety of animals. Equally unfortunately, Darwin chose as protégé and intellectual heir in matters behavioral a young biologist, George Romanes, who adopted wholeheartedly a poetic rather than an objective approach to describing and interpreting animal behavior.

During his brief career in science, Romanes published three lengthy volumes on the comparative study of animal mind. The short passage extracted and reproduced here is from *Animal Intelligence*, a volume Romanes (1882, p. vi) described in its preface as a "text-book of the facts of Comparative Psychology." It proved to be a compendium of occasionally informative description or experiment, numerous wildly implausible anecdotes, and consistently anthropomorphic interpretation of both. The contrast with Darwin's striving for accuracy and objectivity in his chapter on instinct is evident.

The description, in the opening paragraphs of the passage, of rats exhibiting "wise caution," "boldness," "courage," "wonderful cunning," "unselfishness," "affection for human beings," "wonderful intelligence," etc., is no idle literary conceit. As Romanes (1882, p. 1) explained in his introduction to *Animal Intelligence*, "in our objective analysis of other or foreign minds . . . all of our knowledge of their operations is derived, as it were through the medium of ambassadors—these ambassadors being the activity of the organism. . . . Starting with what I know subjectively of my own individual mind, and the activities which in my own organism they prompt, I proceed by analogy to infer from the observable activities of other organisms what are the mental operations that underlie them." Animal behavior was but an ambassador of the real object of interest, animal mind, which could be studied only by introspection and analogy.

As the passages reproduced here make clear, Romanes was no skeptic; he lacked the sense of what was likely or possible that had led a contemporary, Douglas Spalding (1873, p. 291), to note wryly "the many extraordinary and exceptional feats of dogs and other animals, which seem to be constantly falling under the observation of everybody except the few that are interested in these matters." A similar skepticism led Darwin to undertake the experiments on cell-building in honey bees described in his chapter on instinct.

Romanes's use of unlikely anecdote and of introspection, as well as his attribution of human-like mental processes and emotional states to animals, were consistent with the approach of his contemporaries writing popular natural-history texts. But introspection, anecdote, and anthropomorphism all interfered with development of animal behavior as a scientific enterprise. Each had to be purged from discussions of animal behavior before the field could mature as a science. The purge was not long in coming.

The Turn of the Century

It is somewhat ironic that Romanes chose as a successor (as Darwin had chosen Romanes) C. L. Morgan, who was to write that it was necessary "always to look narrowly at every anecdote of animal intelligence and emotion, and to endeavor to distinguish observed fact from observer's inference" (Morgan, 1890, p. vii–ix). Indeed, it was Morgan who was the first major figure in nineteenth-century biology to question systematically the anthropomorphic analyses of his contemporaries.

Surely the most important sentence Morgan ever wrote, possibly the most important single sentence in the history of the study of animal behavior, appeared in various forms scattered through his (1894) *An Introduction to Comparative Psychology*. The most commonly cited version of what Morgan (1894, p. 53) was to call his "canon," is as follows: "In no case may we interpret an action as the outcome of the exercise of a higher psychical faculty, if it can be interpreted as the outcome of one which stands lower in the psychological scale." It is but one of a series of attacks on the poetic Victorian approach to the interpretation of the behavior of animals that took place late in the nineteenth and early in the twentieth century.

If taken literally, Morgan's canon is not very useful without some description of the "psychological scale" of what constitutes relatively simple and relatively complex psychological processes. Such a scale is difficult to provide, though Morgan, like Spencer (1855) before him, tried. The importance of Morgan's canon resulted not so much from its literal interpretation as from its appeal to the law of parsimony in explaining behavior. Simple explanations were to be preferred to complex ones and anthropomorphic attributions of human-like intellectual and emotional capacities to animals were no longer to be accepted uncritically.

The usefulness of Morgan's canon in analyses of behavior was forcefully documented by cases such as Pfungst's 1907 study of the performance of Clever Hans, a horse widely reported in the popular press to be able to read numbers and words, to spell, and to make arithmetic calculations. Hans consistently answered correctly questions posed by strangers (by tapping or pointing with his hoof) even when his trainer, Herr von Osten, was not present.

A student, Oskar Pfungst, was asked by the director of the Psychological Institute of the University of Berlin to determine whether Hans really could read and do arithmetic. By using the ingenious method of mixing trials on which questioners knew and did not know the answers to the questions they were asking, Pfungst was able to show that Hans could answer questions reliably only when his examiners knew the answers to the questions that they asked.

Pfungst's observations revealed that Hans had learned to respond to small, involuntary movements made by questioners when the correct answer was arrived at, and Pfungst concluded that Hans could neither read nor write nor perform calculations with numbers. Pfungst's analysis had the important result of promoting a more critical attitude toward those who attributed human-like mental powers to animals (Pfungst 1965).

Anthropomorphism was not the only aspect of the popular Victorian approach to the study of animal behavior to come under attack at the turn of the century. The pages from Thorndike's (1898) *Ani-*

mal Intelligence reprinted here bear witness to the growing unpopularity of the use of anecdotal evidence in discussions of animal behavior. Thorndike's great contributions to the development of the general field of animal behavior lay in his explicit insistence on study of behavior in a systematic, quantitative manner and on his development of experimental methods for use in behavioral investigations. Thorndike used relatively large numbers of subjects in his studies, made an effort to standardize testing conditions, measured performance quantitatively and compared systematically the performance of subjects given different training regimens. His work was innovative and important, especially given that it formed the empirical basis for the first doctoral thesis in psychology that reported the results of research with animals.

Thorndike happened to work in the laboratory on problems of animal learning that in subsequent decades became a central part of research carried out in the laboratories of experimental animal psychologists. However, Thorndike's message was the same regardless of the locale in which behavior was observed or the type of behavior under study; the days of informal, unsystematic observation of behavior had come to an end.

Thorndike was not the first to carry out and carefully describe experimental studies with animals. Some twenty-five years earlier, Spalding (1873) had published an elegant series of experiments on the development of behavior in newly hatched chickens and had discovered the phenomenon that was later to be called "imprinting." At about the same time, Fabre, working in France, had initiated his classic experimental work on insects (Fabre 1879–82). However, it was Thorndike who explicitly rejected the prevailing anecdotalism and provided an explicit rationale for a quantitative approach to the study of behavior. It was Thorndike's work at the turn of the century that shaped the study of behavior for decades to come.

Ten years later, when Margaret Washburn published her highly influential textbook, *The Animal Mind* (1908; it was to go through four editions, the last published in the late 1930s), she rejected absolutely what she called "the method of anecdote" and adopted instead "the method of experiment." In Washburn's words (1908, p. 11) "the ideal method for the study of a higher animal involves patient observation upon a specimen known from birth, watched in its ordinary behavior and environment and occasionally experimented upon with proper control of the conditions without frightening it or otherwise rendering it abnormal."

In spite of the relative sophistication of her views on method, Washburn (1908, p. 11) like Romanes and Morgan before her, felt that the reason for studying the behavior of animals was to find out about their minds, and that "all psychic interpretation must be based on the analogy of human performance. . . . whether we will or no, we must be anthropomorphic in the notions we form of what takes place in the mind of an animal."

Rejection of attempts to study animal mind or animal consciousness using introspective methods is the subject of the fourth and final article reprinted in the present section. J. B. Watson, its author and father of the "behaviorist" approach to the study of psychology, was concerned with developing scientific methods for studying humans. The chapter reprinted here, from Watson's 1924 text, *Behaviorism*, reflects that interest. However, the behaviorist philosophy that Watson first introduced in 1913 obviously also was applicable to the study of animals.

Watson's (1913, p. 158) views were

presented clearly and forcefully in the opening paragraph of his classic article, "Psychology as the behaviorist views it," quoted below. "Psychology as the behaviorist views it is a purely objective experimental branch of natural science. Its theoretical goal is the prediction and control of behavior. Introspection forms no essential part of its methods."

Watson argued that lack of progress in psychology, as compared with other sciences, was due to the dependence of psychologists on subjective data based on private introspections because such introspections could be neither proven nor disproven in the same way as could the empirical claims in other sciences where data were open to public assessment. Watson believed that subjective data based on introspection were without value. He insisted that psychologists should focus instead on the study of behavior. Watson's behaviorism had a profound influence on the future development of ethology as well as of comparative psychology. For decades after Watson, behaviorists, those trained in biology as well as those trained in psychology, maintained that questions about the mind and consciousness of animals were unanswerable in principle and that a true science of behavior must be concerned only with observable events: the stimuli impinging on organisms and organisms' responses to those stimuli.

Although the view of the world that follows from such an approach is limited in important ways, it was a significant improvement over the introspectionism and anthropomorphism that had dominated the study of behavior during preceding decades. The rejection of anthropomorphism, introspectionism, and anecdotalism that occurred at the turn of the century set the stage for the exponential growth in scientific study of animal behavior that is still in progress today.

While the great advances in knowledge described in the remainder of this volume occurred largely in the decades after the Second World War, the major intellectual battles that made such advances possible had been fought and won in the preceding generation. An objective, scientific study of animal behavior required rejection of the introspective, anecdotal, and anthropomorphic methodologies that characterized the field in the nineteenth century.

The Present

In recent years, there has been a resurgence, under the label of "cognitive ethology" (Griffin 1976, 1992), of interest in using anthropomorphism, introspectionism, and anecdotalism in the study of the behavior of animals. Thinking about how animals might think has generated many interesting experiments and has led to discovery of previously unsuspected abilities in animals. Informal observations of the behavior of animals in uncontrolled settings have similarly generated some new and exciting hypotheses concerning the behavioral capacities of animals. It has become clear that even hypotheses about behavior that are based on the unsatisfactory methods of the nineteenth century can be useful, if predictions deduced from those hypotheses are stated in testable form and are then tested using systematic, quantitative methodologies of the sort first advocated by Thorndike.

The total rejection of anthropomorphism, anecdotalism, and introspectionism that developed at the turn of the century was a necessary antidote to prevailing excess. However, there is nothing inherently counterproductive either in trying to imagine how an animal might view the world or in using unsystematic observations as clues to understanding the abilities of other organisms, if (and it is an important "if") it is recognized that such activities are a means of generating hypotheses, not of testing them. Anecdotes, introspection, and projection onto animals of human-like mental and emotional states impede development of animal behavior as a scientific discipline only if, as happened in the nineteenth century, such activities replace systematic analyses of observable behaviors in testing hypotheses.

Literature Cited

Aristotle. 1970. *Historia Animalium*. Vol. 2. Cambridge, Mass.: Harvard University Press.

Barber, L. 1980. *The Heyday of Natural History*. Garden City: Doubleday.

Darwin, C. 1859. *On the Origin of Species by Means of Natural Selection*. London: Murray.

Darwin, C. 1871. *The Descent of Man and Selection in Relation to Sex*. London: Murray.

Darwin, C. 1872. *The Origin of Species*. 6th Edition. London: Murray.

Fabre, J. H. 1879–82. *Souvenirs Entomologiques*. Paris: Delagrave.

Fitch, O. 1883. Benevolence in animals. *Nature*, 27, 580.

Griffin, D. R. 1976. *The Question of Animal Awareness*. New York: Rockefeller University Press.

Griffin, D. R. 1992. *Animal Minds*. Chicago: University of Chicago Press.

Kingsley, C. 1855. *Glaucus; or the Wonders of the Shore*. Cambridge: Macmillan.

Morgan, C. L. 1890. *Animal Life and Intelligence*. London: Edward Arnold.

Morgan, C. L. 1894. *An Introduction to Comparative Psychology*. London: Scott.

Paley, W. 1802. *Natural Theology, or Evidences of the Existence and Attributes of the Deity*. London: Wilkes & Taylor.

Pfungst, O. 1965. *Clever Hans: The Horse of Mr. von Osten*. New York: Holt.

Romanes, G. J. 1882. *Animal Intelligence*. London: Kegan, Paul, Trench.

Spencer, H. 1855. *Principles of Psychology*. London: Longman.

Spalding, D. A. 1873. Instinct with original observations on young animals. *Macmillan's Magazine*, 27, 282–93.

Thorndike, E. L. 1898. Animals' intelligence: An experimental study of the associative processes in animals. *Psychological Review Monograph Supplements*, vol. 2, no. 4. 1–109.

Washburn, M. F. 1908. *The Animal Mind*. New York: Macmillan.

Watson, J. B. 1913. Psychology as the behaviorist views it. *Psychological Review*, 20, 158–77.

Watson, J. B. 1924. *Behaviorism*. New York: Norton.

White, G. 1789. *The Natural History and Antiquities of Selborne*. London: T. Bentley.

Wood, J. G. 1858. *Common Objects of the Country*. London: G. Routledge.

CHAPTER VIII

INSTINCT

Instincts comparable with habits, but different in their origin—Instincts graduated —Aphides and ants—Instincts variable—Domestic instincts, their origin— Natural instincts of the cuckoo, molothrus, ostrich, and parasitic bees—Slave-making ants—Hive-bee, its cell-making instinct—Changes of instinct and structure not necessarily simultaneous—Difficulties of the theory of the Natural Selection of instincts—Neuter or sterile insects—Summary.

MANY instincts are so wonderful that their development will probably appear to the reader a difficulty sufficient to overthrow my whole theory. I may here premise that I have nothing to do with the origin of the mental powers, any more than I have with that of life itself. We are concerned only with the diversities of instinct and of the other mental faculties in animals of the same class.

I will not attempt any definition of instinct. It would be easy to show that several distinct mental actions are commonly embraced by this term; but every one understands what is meant, when it is said that instinct impels the cuckoo to migrate and to lay her eggs in other birds' nests. An action, which we ourselves require experience to enable us to perform, when performed by an animal, more especially by a very young one, without experience, and when performed by many individuals in the same way, without their knowing for what purpose it is performed, is usually said to be instinctive. But I could show that none of these characters are universal. A little dose of judgment or reason, as Pierre Huber expresses it, often comes into play, even with animals low in the scale of nature.

Frederick Cuvier and several of the older metaphysicians have compared instinct with habit. This comparison gives, I think, an accurate notion of the frame of mind under which an instinctive action is performed, but not necessarily of its origin. How unconsciously many habitual actions are performed, indeed not rarely in direct opposition to our conscious will! yet they may be modified by the will or reason. Habits easily become associated with other habits, with certain periods of time, and states of the body. When once acquired, they often remain constant throughout life. Several other points of resemblance between instincts and habits could be pointed out. As in repeating a well-known song, so in instincts, one action follows another by a sort of rhythm; if a person be interrupted in a song, or in repeating anything by rote, he is generally forced to go back to recover the habitual train of thought; so P. Huber found it was with a caterpillar, which makes a very complicated hammock; for if he took a caterpillar which had completed its hammock up to, say, the sixth stage of construction, and put it into a hammock completed up only to the third stage, the caterpillar simply re-performed the fourth, fifth, and sixth stages of construction. If, however, a caterpillar were taken out of a hammock made up, for instance, to the third stage, and were put into one finished up to the sixth stage, so that much of its work was already done for it, far from deriving any benefit from this, it was much embarrassed, and in order to complete its hammock, seemed forced to start from the third stage, where it had left off, and thus tried to complete the already finished work.

If we suppose any habitual action to become inherited—and it can be shown that this does sometimes happen—then the resemblance between what originally was a habit and an instinct becomes so close as not to be distinguished. If Mozart, instead of playing the pianoforte at three years old with wonderfully little practice, had played a tune with no practice at all, he might truly be said to have done so instinctively. But it would be a serious error to suppose that the greater number of instincts have been acquired by habit in one generation, and then transmitted by inheritance to succeeding generations. It can be clearly shown that the most wonderful instincts with which we are acquainted, namely, those of the hive-bee and of many ants, could not possibly have been acquired by habit.

It will be universally admitted that instincts are as important as corporeal structures for the welfare of each species, under its present conditions of life. Under changed conditions of life, it is at least possible that slight modifications of instinct might be profitable to a species; and if it can be shown that instincts do vary ever so little, then I can see no difficulty in natural selection preserving and continually accumulating variations of instinct to any extent that was profitable. It is thus, as I believe, that all the most complex and wonderful instincts have originated. As modifications of corporeal structure arise from, and are increased by, use or habit, and are diminished or lost by disuse, so I do not doubt it has been with instincts. But I believe that the effects of habit are in many cases of subordinate importance to the effects of the natural selection of what may be called spontaneous variations of instincts;—that is of variations produced by the same unknown causes which produce slight deviations of bodily structure.

No complex instinct can possibly be produced through natural selection, except by the slow and gradual accumulation of numerous slight, yet profitable, variations. Hence, as in the case of corporeal structures, we ought to find in nature, not the actual transitional gradations by which each complex instinct has been acquired—for these could be found only in the lineal ancestors of each species—but we ought to find in the collateral lines of descent some evidence of such gradations; or we ought at least to be able to show that gradations of some kind are possible; and this we certainly can do. I have been surprised to find, making allowance for the instincts of animals having been but little observed except in Europe and North America, and for on instinct being known amongst extinct species, how very generally gradations, leading to the most complex instincts, can be discovered. Changes of instinct may sometimes be facilitated by the same species having different instincts at different periods of life, or at different seasons of the year, or when placed under different circumstances, &c.; in which case either the one or the other instinct might be preserved by natural selection. And such instances of diversity of instinct in the same species can be shown to occur in nature.

Again, as in the case of corporeal structure, and conformably to my theory, the instinct of each species is good for itself, but has never, as far as we can judge, been produced for the exclusive good of others. One of the strongest instances of an animal apparently performing an action for the sole good of another, with which I am acquainted, is that of aphides voluntarily yielding, as was first observed by Huber, their sweet excretion to ants: that they do so voluntarily, the following facts show. I removed all the ants from a group of about a dozen aphides on a dock-plant, and prevented their attendance during several hours. After this interval, I felt sure that the aphides would want to excrete. I watched them for some time through a lens, but not one excreted; I then tickled and stroked them with a hair in the same manner, as well as I could, as the ants do with their antennæ; but not one excreted. Afterwards I allowed an ant to visit them, and it immediately seemed, by its eager way of running about, to be well aware what a rich flock it had discovered; it then began to play with its antennæ on the abdomen first of one aphis and then of another; and each, as soon as it felt the antennæ, immediately lifted up its abdomen and excreted a limpid drop of sweet juice, which was eagerly devoured by the ant. Even the quite young aphides behaved in this manner; showing that the action was instinctive, and not the result of experience. It is certain, from the observations of Huber, that the aphids show no dislike to the ants: if the latter be not present they are at last compelled to eject their excretion. But as the excretion is extremely viscid, it is no doubt a convenience to the aphides to have it removed; therefore probably they do not excrete solely for the good of the ants. Although there is no evidence that any animal performs an action for the exclusive good of another species, yet each tries to take advantage of the instincts of others, as each takes advantage of the weaker bodily structure of other species. So again certain instincts cannot be considered as absolutely perfect; but as details on this and other such points are not indispensable, they may be here passed over.

As some degree of variation in instincts under a state of nature, and the inheritance of such variations, are indispensable for the action of natural selection, as many instances as possible ought to be given; but want of space prevents me. I can only assert that instincts certainly do vary—for instance, the migratory instinct, both in extent and direction, and in its total loss. So it is with the nests of birds, which vary partly in dependence on the situations chosen, and on the nature and temperature of the country inhabited, but often from causes wholly unknown to us: Audubon has given several remarkable cases of differences in the nests of the same species in the northern and southern United States. Why, it has been asked, if instinct be variable, has it not granted to the bee "the ability to use some other material when wax was deficient"? But what other natural material could bees use? They will work, as I have seen, with wax hardened with vermilion or softened with lard. Andrew Knight observed that his bees, instead of laboriously collecting propolis, used a cement of wax and turpentine, with which he had covered decorticated trees. It has lately been shown that bees, instead of searching for pollen, will gladly use a very different substance, namely oatmeal. Fear of any particular enemy is certainly an instinctive quality, as may be seen in nestling birds, though it is strengthened by experience, and by the sight of fear of the same enemy in other animals. The fear of man is slowly acquired, as I have elsewhere shown, by the various animals which inhabit desert islands; and we see an instance of this even in England, in the greater wildness of all our large birds in comparison with our small birds; for the large birds have been most persecuted by man. We may safely attribute the greater wildness of our large birds to this cause; for in uninhabited islands large birds are not more fearful than small; and the magpie, so wary in England, is tame in Norway, as is the hooded crow in Egypt.

That the mental qualities of animals of the same kind, born in a state of nature, vary much, could be shown by many facts. Several cases could also be adduced of occasional and strange habits in wild animals, which, if advantageous to the species, might have given rise, through natural selection, to new instincts. But I am well aware that these general statements, without the facts in detail, will produce but a feeble effect on the reader's mind. I can only repeat my assurance, that I do not speak without good evidence.

Inherited Changes of Habit or Instinct in Domesticated Animals

The possiblity, or even probability, of inherited variations of instinct in a state of nature will be strengthened by briefly considering a few cases under domestication. We shall thus be enabled to see the part which habit and the selection of so-called spontaneous variations have played in modifying the mental qualities of our domestic animals. It is notorious how much domestic animals vary in their mental qualities. With cats, for instance, one naturally takes to catching rats, and another mice, and these tendencies are known to be inherited. One cat, according to Mr. St. John, always brought home game-birds, another hares or rabbits, and another hunted on marshy ground and almost nightly caught woodcocks or snipes. A number of curious and authentic instances could be given of various shades of disposition and of taste, and likewise of the oddest tricks, associated with certain frames of mind or periods of time, being inherited. But let us look to the familiar case of the breeds of the dogs: it cannot be doubted that young pointers (I have myself seen a striking instance) will sometimes point and even back other dogs the very first time that they are taken out; retrieving is certainly in some degree inherited by retriev-ers; and a tendency to run round, instead of at, a flock of sheep, by shep-herd dogs. I cannot see that these actions, performed without experience by the young, and in nearly the same manner by each individual, per-formed with eager delight by each breed, and without the end being known—for the young pointer can no more know that he points to aid his master, than the white butterfly knows why she lays her eggs on the leaf of the cabbage—I cannot see that these actions differ essentially from true instincts. If we were to behold one kind of wolf, when young and without any training, as soon as it scented its prey, stand motionless like a statue, and then slowly crawl forward with a peculiar gait; and another kind of wolf rushing round, instead of at, a herd of deer, and driving them to a distant point, we should assuredly call these actions instinctive. Do-mestic instincts, as they may be called, are certainly far less fixed than natural instincts; but they have been acted on by far less rigorous selec-tion, and have been transmitted for an incomparably shorter period, under less fixed conditions of life.

How strongly these domestic instincts, habits, and dispositions are in-herited, and how curiously they become mingled, is well shown when dif-ferent breeds of dogs are crossed. Thus it is known that a cross with a bull-dog has affected for many generations the courage and obstinacy of greyhounds; and a cross with a greyhound has given to a whole family of shepherd-dogs a tendency to hunt hares. These domestic instincts, when thus tested by crossing, resemble natural instincts, which in a like manner become curiously blended together, and for a long period exhibit traces of the instincts of either parent: for example, Le Roy describes a dog, whose great-grandfather was a wolf, and this dog showed a trace of its wild pa-rentage only in one way, by not coming in a straight line to his master, when called.

Domestic instincts are sometimes spoken of as actions which have be-come inherited solely from long-continued and compulsory habit; but this is not true. No one would ever have thought of teaching, or probably could have taught, the tumbler-pigeon to tumble,—an action which, as I have witnessed, is performed by young birds, that have never seen a pi-geon tumble. We may believe that some one pigeon showed a slight ten-dency to this strange habit, and that the long-continued selection of the best individuals in successive generations made tumblers what they now are; and near Glasgow there are house-tumblers, as I hear from Mr. Brent, which cannot fly eighteen inches high without going head over heels. It may be doubted whether any one would have thought of training a dog to point, had not some one dog naturally shown a tendency in this line; and this is known occasionally to happen, as I once saw, in a pure terrier: the act of pointing is probably, as many have thought, only the exaggerated pause of an animal preparing to spring on its prey. When the first tendency to point was once displayed, methodical selection and the inherited effects of compulsory training in each successive generation would soon complete the work; and unconscious selection is still in progress, as each man tries to procure, without intending to improve the breed, dogs which stand and hunt best. On the other hand, habit alone in some cases has sufficed; hardly any animal is more difficult to tame than the young of the wild rabbit; scarcely any animal is tamer than the young of the tame rabbit; but I can hardly suppose that domestic rabbits have often been selected for tameness alone; so that we must attribute at least the greater part of the inherited change from extreme wildness to extreme tameness, to habit and long-continued close confinement.

Natural instincts are lost under domestication: a remarkable instance of this is seen in those breeds of fowls which very rarely or never become "broody," that is, never wish to sit on their eggs. Familiarity alone prevents our seeing how largely and how permanently the minds of our domestic animals have been modified. It is scarcely possible to doubt that the love of man has become instinctive in the dog. All wolves, foxes, jackals, and species of the cat genus, when kept tame, are most eager to attack poultry, sheep, and pigs; and this tendency has been found incurable in dogs which have been brought home as puppies from countries such as Tierra del Fuego and Australia, where the savages do not keep these domestic animals. How rarely, on the other hand, do our civilised dogs, even when quite young, require to be taught not to attack poultry, sheep, and pigs! No doubt they occasionally do make an attack, and are then beaten; and if not cured, they are destroyed; so that habit and some degree of selection have probably concurred in civilising by inheritance our dogs. On the other hand, young chickens have lost, wholly by habit, that fear of the dog and cat which no doubt was originally instinctive in them; for I am informed by Captain Hutton that the young chickens of the parent-stock, the Gallus bankiva, when reared in India under a hen, are at first excessively wild. So it is with young pheasants reared in England under a hen. It is not that chickens have lost all fear, but fear only of dogs and cats, for if the hen gives the danger-chuckle, they will run (more especially young turkeys) from under her, and conceal themselves in the surrounding grass or thickets; and this is evidently done for the instinctive purpose of allowing as we see in wild ground-birds, their mother to fly away. But this instinct retained by our chickens has become useless under domestication, for the mother-hen has almost lost by disuse the power of flight.

Hence, we may conclude, that under domestication instincts have been acquired, and natural instincts have been lost, partly by habit, and partly by man selecting and accumulating, during successive generations, peculiar mental habits and actions, which at first appeared from what we must in our ignorance call an accident. In some cases compulsory habit alone has sufficed to produce inherited mental changes; in other cases, compulsory habit has done nothing, and all has been the result of selection, pursued both methodically and unconsciously: but in most cases habit and selection have probably concurred.

Special Instincts

We shall, perhaps, best understand how instincts in a state of nature have become modified by selection by considering a few cases. I will select only three,—namely, the instinct which leads the cuckoo to lay her eggs in other birds' nests; the slave-making instinct of certain ants; and the cell-making power of the hive-bee. These two latter instincts have generally and justly been ranked by naturalists as the most wonderful of all known instincts.

Instincts of the Cuckoo.—It is supposed by some naturalists that the more immediate cause of the instinct of the cuckoo is, that she lays her eggs, not daily, but at intervals of two or three days; so that, if she were to make her own nest and sit on her own eggs, those first laid would have to be left for some time unincubated, or there would be eggs and young birds of different ages in the same nest. If this were the case, the process of laying and hatching might be inconveniently long, more especially as she migrates at a very early period; and the first hatched young would probably have to be fed by the male alone. But the American cuckoo is in this predicament; for she makes her own nest, and has eggs and young successively hatched, all at the same time. It has been both asserted and denied that the American cuckoo occasionally lays her eggs in other birds' nests; but I have lately heard from Dr. Merrell, of Iowa, that he once found in Illinois a young cuckoo together with a young jay in the nest of a Blue jay (Garrulus cristatus); and as both were nearly full feathered, there could be no mistake in their identification. I could also give several instances of various birds which have been known occasionally to lay their eggs in other birds' nests. Now let us suppose that the ancient progenitor of our European cuckoo had the habits of the American cuckoo, and that she occasionally laid an egg in another birds' nest. If the old bird profited by this occasional habit through being enabled to migrate earlier or through any other cause; or if the young were made more vigorous by advantage being taken of the mistaken instinct of another species than when reared by their own mother, encumbered as she could hardly fail to be by having eggs and young of different ages at the same time; then the old birds or the fostered young would gain an advantage. And analogy would lead us to believe, that the young thus reared would be apt to follow by inheritance the occasional and aberrant habit of their mother, and in their turn would be apt to lay their eggs in other birds' nests, and thus be more successful in rearing their young. By a continued process of this nature, I believe that the strange instinct of our cuckoo has been generated. It has, also, recently been ascertained on sufficient evidence, by Adolf Müller, that the cuckoo occasionally lays her eggs on the bare ground, sits on them, and feeds her young. This rare event is probably a case of reversion to the long-lost, aboriginal instinct of nidification.

It has been objected that I have not noticed other related instincts and adaptations of structure in the cuckoo, which are spoken of as necessarily co-ordinated. But in all cases, speculation on an instinct known to us only in a single species, is useless, for we have hitherto had no facts to guide us. Until recently the instincts of the European and of the non-parasitic American cuckoo alone were known; now, owing to Mr. Ramsay's observations, we have learnt something about three Australian species, which lay their eggs in other birds' nests. The chief points to be referred to are three: first, that the common cuckoo, with rare exceptions, lays only one egg in a nest, so that the large and voracious young bird receives ample food. Secondly, that the eggs are remarkably small, not exceeding those of the skylark,—a bird about one-fourth as large as the cuckoo. That the small size of the egg is a real cause of adaptation we may infer from the fact of the non-parasitic American cuckoo laying full-sized eggs. Thirdly, that the young cuckoo, soon after birth, has the instinct, the strength, and a properly shaped back for ejecting its foster-brothers, which then perish from cold and hunger. This has been boldly called a beneficent arrangement, in order that the young cuckoo may get sufficient food, and that its foster-brothers may perish before they had acquired much feeling!

Turning now to the Australian species; though these birds generally lay only one egg in a nest, it is not rare to find two and even three eggs in the same nest. In the Bronze cuckoo the eggs vary greatly in size, from eight to ten times in length. Now if it had been of an advantage to this species to have laid eggs even smaller than those now laid, so as to have deceived certain foster-parents, or, as is more probable, to have been hatched within a shorter period (for it is asserted that there is a relation between the size of eggs and the period of their incubation), then there is no difficulty in believing that a race or species might have been formed which would have laid smaller and smaller eggs; for these would have been more safely hatched and reared. Mr. Ramsay remarks that two of the Australian cuckoos, when they lay their eggs in an open nest, manifest a decided preference for nests containing eggs similar in colour to their own. The European species apparently manifests some tendency towards a similar instinct, but not rarely departs from it, as is shown by her laying her dull and pale-coloured eggs in the nest of the Hedge-warbler with bright greenish-blue eggs. Had our cuckoo invariably displayed the above instinct, it would assuredly have been added to those which it is assumed must all have been acquired together. The eggs of the Australian Bronze cuckoo vary, according to Mr. Ramsay, to an extraordinary degree in colour; so that in this respect, as well as in size, natural selection might have secured and fixed any advantageous variation.

In the case of the European cuckoo, the offspring of the foster-parents are commonly ejected from the nest within three days after the cuckoo is hatched; and as the latter at this age is in a most helpless condition, Mr. Gould was formerly inclined to believe that the act of ejection was performed by the foster-parents themselves. But he has now received a trustworthy account of a young cuckoo which was actually seen, whilst still blind and not able even to hold up its own head, in the act of ejecting its foster-brothers. One of these was replaced in the nest by the observer, and was again thrown out. With respect to the means by which this strange and odious instinct was acquired, if it were of great importance for the young cuckoo, as is probably the case, to receive as much food as possible soon after birth, I can see no special difficulty in its having gradually acquired, during successive generations, the blind desire, the strength, and structure necessary for the work of ejection; for those young cuckoos which had such habits and structure best developed would be the most securely reared. The first step towards the acquisition of the proper instinct might have been more unintentional restlessness on the part of the young bird, when somewhat advanced in age and strength; the habit having been afterwards improved, and transmitted to an earlier age. I can see no more difficulty in this, than in the unhatched young of other birds acquiring the instinct to break through their own shells;—or than in young snakes acquiring in their upper jaws, as Owen has remarked, a transitory sharp tooth for cutting through the tough egg-shell. For if each part is liable to individual variations at all ages, and the variations tend to be inherited at a corresponding or earlier age,—propositions which cannot be disputed,—then the instincts and structure of the young could be slowly modified as surely as those of the adult; and both cases must stand or fall together with the whole theory of natural selection.

Some species of Molothrus, a widely distinct genus of American birds, allied to our starlings, have parasitic habits like those of the cuckoo; and the species present an interesting gradation in the perfection of their instincts. The sexes of Molothrus badius are stated by an excellent observer, Mr. Hudson, sometimes to live promiscuously together in flocks, and sometimes to pair. They either build a nest of their own, or seize on one belonging to some other bird, occasionally throwing out the nestlings of the stranger. They either lay their eggs in the nest thus appropriated, or oddly enough build one for themselves on the top of it. They usually sit on their own eggs and rear their own young; but Mr. Hudson says it is probable that they are occasionally parasitic, for he has seen the young of this species following old birds of a distinct kind and clamouring to be fed by them. The parasitic habits of another species of Molothrus, the M. bonariensis, are much more highly developed than those of the last, but are still far from perfect. This bird, as far as it is known, invariably lays its eggs in the nests of strangers; but it is remarkable that several together sometimes commence to build an irregular untidy nest of their own, placed in singularly ill-adapted situations, as on the leaves of a large thistle. They never, however, as far as Mr. Hudson has ascertained, complete a nest for themselves. They often lay so many eggs—from fifteen to twenty—in the same foster-nest, that few or none can possibly be hatched. They have, moreover, the extraordinary habit of pecking holes in the eggs, whether of their own species or of their foster-parents, which they find in the appropriated nests. They drop also many eggs on the bare ground, which are thus wasted. A third species, the M. pecoris of North America, has acquired instincts as perfect as those of the cuckoo, for it never lays more than one egg in a foster-nest, so that the young bird is securely reared. Mr. Hudson is a strong disbeliever in evolution, but he appears to have been so much struck by the imperfect instincts of the Molothrus bonariensis that he quotes my words, and asks, "Must we consider these habits, not as especially endowed or created instincts, but as small consequences of one general law, namely, transition?"

Various birds. as has already been remarked, occasionally lay their eggs in the nests of other birds. This habit is not very uncommon with the Gallinaceæ, and throws some light on the singular instinct of the ostrich. In this family several hen-birds unite and lay first a few eggs in one nest and then in another; and these are hatched by the males. This instinct may probably be accounted for by the fact of the hens laying a large number of eggs, but, as with the cuckoo, at intervals of two or three days. The instinct, however, of the American ostrich, as in the case of the Molothrus bonariensis, has not as yet been perfected; for a surprising number of eggs lie strewed over the plains, so that in one day's hunting I picked up no less than twenty lost and wasted eggs.

Many bees are parasitic, and regularly lay their eggs in the nests of other kinds of bees. This case is more remarkable than that of the cuckoo; for these bees have not only had their instincts but their structure modified in accordance with their parasitic habits; for they do not possess the pollen-collecting apparatus which would have been indispensable if they had stored up food for their own young. Some species of Sphegidæ (wasplike insects) are likewise parasitic; and M. Fabre has lately shown good reason for believing that, although the Tachytes nigra generally makes its own burrow and stores it with paralysed prey for its own larvæ, yet that, when this insect finds a burrow already made and stored by another sphex, it takes advantage of the prize and becomes for the occasion parasitic. In this case, as with that of the Molothrus or cuckoo, I can see no difficulty in natural selection making an occasional habit permanent, if of advantage to the species, and if the insect whose nest and stored food are feloniously appropriated, be not thus exterminated.

Slave-making instinct.—This remarkable instinct was first discovered in the Formica (Polyerges) rufescens by Pierre Huber, a better observer even than his celebrated father. This ant is absolutely dependent on its slaves; without their aid, the species would certainly become extinct in a single year. The males and fertile female do no work of any kind, and the workers or sterile females, though most energetic and courageous in capturing slaves, do no other work. They are incapable of making their own nests, or of feeding their own larvæ. When the old nest is found inconvenient, and they have to migrate, it is the slaves which determine the migration, and actually carry their masters in their jaws. So utterly helpless are the masters, that when Huber shut up thirty of them without a slave, but with plenty of the food which they like best, and with their own larvæ and pupæ to stimulate them to work, they did nothing; they could not even feed themselves, and many perished of hunger. Huber then introduced a single slave (F. fusca), and she instantly set to work, fed and saved the survivors; made some cells and tended the larvæ, and put all to rights. What can be more extraordinary than these well-ascertained facts? If we had not known of any other slave-making ant, it would have been hopeless to speculate how so wonderful an instinct could have been perfected.

Another species, Formica sanguinea, was likewise first discovered by P. Huber to be a slave-making ant. This species is found in the southern parts of England, and its habits have been attended to by Mr. F. Smith, of the British Museum, to whom I am much indebted for information on this and other subjects. Although fully trusting to the statements of Huber and Mr. Smith, I tried to approach the subject in a sceptical frame of mind, as any one may well be excused for doubting the existence of so extraordinary an instinct as that of making slaves. Hence, I will give the observations which I made in some little detail. I opened fourteen nests of F. sanguinea, and found a few slaves in all. Males and fertile females of the slave species (F. fusca) are found only in their own proper communities, and have never been observed in the nests of F. sanguinea. The slaves are black and not above half the size of their red masters, so that the contrast in their appearance is great. When the nest is slightly disturbed, the slaves occasionally come out, and like their masters are much agitated and defend the nest: when the nest is much disturbed, and the larvæ and pupæ are exposed, the slaves work energetically together with their masters in carrying them away to a place of safety. Hence, it is clear, that the slaves feel quite at home. During the months of June and July, on three successive years, I watched for many hours several nests in Surrey and Sussex, and never saw a slave either leave or enter a nest. As, during these months, the slaves are very few in number, I thought that they might behave differently when more numerous; but Mr. Smith informs me that he has watched the nests at various hours during May, June, and August, both in Surrey and Hampshire, and has never seen the slaves, though present in large numbers in August, either leave or enter the nest. Hence he considers them as strictly household slaves. The masters, on the other hand, may be constantly seen bringing in materials for the nest, and food of all kinds. During the year 1860, however, in the month of July, I came across a community with an unusually large stock of slaves, and I observed a few slaves mingled with their masters leaving the nest, and marching along the same road to a tall Scotch-fir-tree, twenty-five yards distant, which they ascended together, probably in search of aphides or cocci. According to Huber, who had ample opportunities for observation, the slaves in Switzerland habitually work with their masters in making the nest, and they alone open and close the doors in the morning and evening; and, as Huber expressly states, their principal office is to search for aphides. This difference in the usual habits of the masters and slaves in the two countries, probably depends merely on the slaves being captured in greater numbers in Switzerland than in England.

One day I fortunately witnessed a migration of F. sanguinea from one nest to another, and it was a most interesting spectacle to behold the masters carefully carrying their slaves in their jaws instead of being carried

SLAVE-MAKING INSTINCT   195

by them, as in the case of F. rufescens. Another day my attention was
struck by about a score of the slave-makers haunting the same spot, and
evidently not in search of food; they approached and were vigorously re-
pulsed by an independent community of the slave-species (F. fusca);
sometimes as many as three of these ants clinging to the legs of the slave-
making F. sanguinea. The latter ruthlessly killed their small opponents,
and carried their dead bodies as food to their nest, twenty-nine yards distant; but they were prevented from getting any pupæ to rear as slaves.
I then dug up a small parcel of the pupæ of F. fusca from another nest,
and put them down on a bare spot near the place of combat; they were
eagerly seized and carried off by the tyrants, who perhaps fancied that,
after all, they had been victorious in their late combat.

At the same time I laid on the same place a small parcel of the pupæ
of another species, F. flava, with a few of these little yellow ants still
clinging to the fragments of their nest. This species is sometimes, though
rarely, made into slaves, as has been described by Mr. Smith. Although
so small a species, it is very courageous, and I have seen it ferociously attack other ants. In one instance I found to my surprise an independent
community of F. flava under a stone beneath a nest of the slave-making
F. sanguinea; and when I had accidentally disturbed both nests, the little
ants attacked their big neighbours with surprising courage. Now I was
curious to ascertain whether F. sanguinea could distinguish the pupæ of
F. fusca, which they habitually make into slaves, from those of the little
and furious F. flava, which they rarely capture, and it was evident that
they did at once distinguish them; for we have seen that they eagerly and
instantly seized the pupæ of F. fusca, whereas they were much terrified
when they came across the pupæ, or even the earth from the nest, of F.
flava, and quickly ran away; but in about a quarter of an hour, shortly
after all the little yellow ants had crawled away, they took heart and carried off the pupæ.

One evening I visited another community of F. sanguinea, and found a
number of these ants returning home and entering their nests, carrying
the dead bodies of F. fusca (showing that it was not a migration) and
numerous pupæ. I traced a long file of ants burthened with booty, for
about forty yards back, to a very thick clump of heath, whence I saw the
last individual of F. sanguinea emerge, carrying a pupa; but I was not
able to find the desolated nest in the thick heath. The nest, however, must
have been close at hand, for two or three individuals of F. fusca were rushing about in the greatest agitation, and one was perched motionless with
its own pupa in its mouth on the top of a spray of heath, an image of de-
spair over its ravaged home.

Such are the facts, though they did not need confirmation by me, in regard to the wonderful instinct of making slaves. Let it be observed what a
contrast the instinctive habits of F. sanguinea present with those of the
continental F. rufescens. The latter does not build its own nest, does not
determine its own migrations, does not collect food for itself or its young,
and cannot even feed itself: it is absolutely dependent on its numerous

slaves. Formica sanguinea, on the other hand, possesses much fewer slaves, and in the early part of the summer extremely few: the masters determine when and where a new nest shall be formed, and when they migrate, the masters carry the slaves. Both in Switzerland and England the slaves seem to have the exclusive care of the larvæ, and the masters alone go on slave-making expeditions. In Switzerland the slaves and masters work together, making and bringing materials for the nest; both, but chiefly the slaves, tend, and milk, as it may be called their aphides; and thus both collect food for the community. In England the masters alone usually leave the nest to collect building materials and food for themselves, their slaves and larvæ. So that the masters in this country receive much less service from their slaves than they do in Switzerland.

By what steps the instinct of F. sanguinea originated I will not pretend to conjecture. But as ants which are not slave-makers will, as I have seen, carry off the pupæ of other species, if scattered near their nests, it is possible that such pupæ originally stored as food might become developed; and the foreign ants thus unintentionally reared would then follow their proper instincts, and do what work they could. If their presence proved useful to the species which had seized them—if it were more advantageous to this species to capture workers than to procreate them—the habit of collecting pupæ, originally for food, might by natural selection be strengthened and rendered permanent for the very different purpose of raising slaves. When the instinct was once acquired, if carried out to a much less extent even than in our British F. sanguinea, which, as we have seen, is less aided by its slaves than the same species in Switzerland, natural selection might increase and modify the instinct—always supposing each modification to be of use to the species—until an ant was formed as abjectly dependent on its slaves as is the Formica rufescens.

Cell-making instinct of the Hive-Bee.—I will not here enter on minute details on this subject, but will merely give an outline of the conclusions at which I have arrived. He must be a dull man who can examine the exquisite structure of a comb, so beautifully adapted to its end, without enthusiastic admiration. We hear from mathematicians that bees have practically solved a recondite problem, and have made their cells of the proper shape to hold the greatest possible amount of honey, with the least possible consumption of precious wax in their construction. It has been remarked that a skilful workman with fitting tools and measures, would find it very difficult to make cells of wax of the true form, though this is effected by a crowd of bees working in a dark hive. Granting whatever instincts you please, it seems at first quite inconceivable how they can make all the necessary angles and planes, or even perceive when they are correctly made. But the difficulty is not nearly so great as it at first appears: all this beautiful work can be shown, I think, to follow from a few simple instincts.

I was led to investigate this subject by Mr. Waterhouse, who has shown that the form of the cell stands in close relation to the presence of adjoining cells; and the following view may, perhaps, be considered only as a modification of this theory. Let us look to the great principle of gradation, and see whether Nature does not reveal to us her method of work. At one end of a short series we have humble-bees, which use their old cocoons to hold honey, sometimes adding to them short tubes of wax, and likewise making separate and very irregular rounded cells of wax. At the other end of the series we have the cells of the hive-bee, placed in a double layer: each cell, as is well known, is an hexagonal prism, with the basal edges of its six sides bevelled so as to join an inverted pyramid, of three rhombs. These rhombs have certain angles, and the three which form the pyramid-al base of a single cell on one side of the comb enter into the composition of the bases of three adjoining cells on the opposite side. In the series be-tween the extreme perfection of the cells of the hive-bee and the simplicity of those of the humble-bee we have the cells of the Mexican Melipona do-mestica, carefully described and figured by Pierre Huber. The Melipona itself is intermediate in structure between the hive and humble bee, but more nearly related to the latter; it forms a nearly regular waxen comb of cylindrical cells, in which the young are hatched, and, in addition, some large cells of wax for holding honey. These latter cells are nearly spherical and of nearly equal sizes, and are aggregated into an irregular mass. But the important point to notice is, that these cells are always made at that degree of nearness to each other that they would have intersected or brok-en into each other if the spheres had been completed; but this is never permitted, the bees building perfectly flat walls of wax between the spheres which thus tend to intersect. Hence, each cell consists of an outer spherical portion, and of two, three, or more flat surfaces, according as the cell adjoins two, three, or more other cells. When one cell rests on three other cells, which, from the spheres being nearly of the same size, is very frequently and necessarily the case, the three flat surfaces are united into a pyramid; and this pyramid, as Huber has remarked, is manifestly a gross imitation of the three-sided pyramidal base of the cell of the hive-bee. As in the cells of the hive-bee, so here, the three plane surfaces in any one cell necessarily enter into the construction of three adjoining cells. It is obvious that the Melipona saves wax, and what is more important, la-bour, by this manner of building; for the flat walls between the adjoin-ing cells are not double, but are of the same thickness as the outer spher-ical portions, and yet each flat portion forms a part of two cells.

Reflecting on this case, it occurred to me that if the Melipona had made its spheres at some given distance from each other, and had made them of equal sizes and had arranged them symmetrically in a double layer, the resulting structure would have been as perfect as the comb of the hive-bee. Accordingly I wrote to Professor Miller of Cambridge, and this geometer has kindly read over the following statement, drawn up from his informa-tion, and tells me that it is strictly correct:—

If a number of equal squares be described with their centres placed in two parallel layers; with the centre of each sphere at the distance of ra-dius $\times \sqrt{2}$, or radius $\times 1.41421$ (or at some lesser distance), from the centres of the six surrounding spheres in the same layer; and at the same distance from the centres of the adjoining spheres in the other and parellel layer; then, if planes of intersection between the several spheres in both layers be formed, there will result a double layer of hexagonal prisms united together by pyramidal bases formed of three rhombs; and the rhombs and the sides of the hexagonal prisms will have every angle identically the same with the best measurements which have been made of the cells of the hive-bee. But I hear from Prof. Wyman, who has made numerous careful measurements, that the accuracy of the workmanship of the bee has been greatly exaggerated; so much so, that whatever the typical form of the cell may be, it is rarely, if ever, realised.

Hence we may safely conclude that, if we could slightly modify the instincts already possessed by the Melipona, and in themselves not very wonderful, this bee would make a structure as wonderfully perfect as that of the hive-bee. We must suppose that Melipona to have the power of forming her cells truly spherical, and of equal sizes; and this would not be very surprising, seeing that she already does so to a certain extent, and seeing what perfectly cylindrical burrows many insects make in wood, apparently by turning round on a fixed point. We must supposse the Melipona to arrange her cells in level layers, as she already does her cylindrical cells; and we must further suppose, and this is the greatest difficulty, that she can somehow judge accurately at what distance to stand from her fellow-labourers when several are making their spheres; but she is already so far enabled to judge of distance, that she always describes her spheres so as to intersect to a certain extent; and then she unites the points of intersection by perfectly flat surfaces. By such modifications of instincts which in themselves are not very wonderful,—hardly more wonderful than those which guide a bird to make its nest,—I believe that the hive-bee has acquired, through natural selection, her inimitable architectural powers.

But this theory can be tested by experiment. Following the example of Mr. Tegetmeier, I separated two combs, and put between them a long, thick, rectangular strip of wax: the bees instantly began to excavate minute circular pits in it; and as they deepened these little pits, they made them wider and wider until they were converted into shallow basins, appearing to the eye perfectly true or parts of a sphere, and of about the diameter of a cell. It was most interesting to observe that, wherever several bees had begun to excavate these basins near together, they had begun their work at such a distance from each other, that by the time the basins had acquired the above-stated width (*i. e.* about the width of an ordinary cell), and were in depth about one-sixth of the diameter of the sphere of which they formed a part, the rims of the basins intersected or broke into each other. As soon as this occurred, the bees ceased to excavate, and began to build up flat walls of wax on the lines of intersection between the basins, so that each hexagonal prism was built upon the scalloped edge of a smooth basin, instead of on the straight edges of a three-sided pyramid as in the case of ordinary cells.

I then put into the hive, instead of a thick, rectangular piece of wax, a thin and narrow, knife-edged ridge, coloured with vermilion. The bees instantly began on both sides to excavate little basins near to each other, in the same way as before; but the ridge of wax was so thin, that the bottoms of the basins, if they had been excavated to the same depth as in the former experiment, would have broken into each other from the opposite sides. The bees, however, did not suffer this to happen, and they stopped their excavations in due time; so that the basins, as soon as they had been a little deepened, came to have flat bases; and these flat bases, formed by thin little plates of the vermilion wax left ungnawed, were situated, as far as the eye could judge, exactly along the planes of imaginary intersection between the basins on the opposite sides of the ridge of wax. In some parts, only small portions, in other parts, large portions of a rhombic plate were thus left between the opposed basins, but the work, from the unnatural state of things, had not been neatly performed. The bees must have worked at very nearly the same rate in circularly gnawing away and deepening the basins on both sides of the ridge of vermilion wax, in order to have thus succeeded in leaving flat plates between the basins, by stopping work at the planes of intersection.

Considering how flexible thin wax is, I do not see that there is any difficulty in the bees, whilst at work on the two sides of a strip of wax, perceiving when they have gnawed the wax away to the proper thinness, and then stopping their work. In ordinary combs it has appeared to me that the bees do not always succeed in working at exactly the same rate from the opposite sides; for I have noticed half-completed rhombs at the base of a just commenced cell, which were slightly concave on one side, where I suppose that the bees had excavated too quickly, and convex on the opposed side where the bees had worked less quickly. In one well-marked instance, I put the comb back into the hive, and allowed the bees to go on working for a short time, and again examined the cell, and I found that the rhombic plate had been completed, and had become *perfectly flat:* it was absolutely impossible, from the extreme thinness of the little plate, that they could have effected this by gnawing away the convex side; and I suspect that the bees in such cases stand on opposite sides and push and bend the ductile and warm wax (which as I have tried is easily done) into its proper intermediate plane, and thus flatten it.

From the experiment of the ridge of vermillion wax we can see that, if the bees were to build for themselves a thin wall of wax, they could make their cells of the proper shape, by standing at the proper distance from each other, by excavating at the same rate, and by endeavouring to make equal spherical hollows, but never allowing the spheres to break into each other. Now bees, as may be clearly seen by examining the edge of a growing comb, do make a rough, circumferential wall or rim all round the comb; and they gnaw this away from the opposite sides, always working circularly as they deepen each cell. They do not make the whole three-sided pyramidal base of any one cell at the same time, but only that one rhombic plate which stands on the extreme growing margin, or the two plates, as the case may be; and they never complete the upper edges of the rhombic plates, until the hexagonal walls are commenced. Some of these statements differ from those made by the justly celebrated elder Huber, but I am convinced of their accuracy; and if I had space, I would show that they are conformable with my theory.

Huber's statement that the very first cell is excavated out of a little parallel-sided wall of wax, is not, as far as I have seen, strictly correct; the first commencement having always been a little hood of wax; but I will not here enter on details. We see how important a part excavation plays in the construction of the cells; but it would be a great error to suppose that the bees cannot build up a rough wall of wax in the proper position—that is, along the plane of intersection between two adjoining spheres. I have several specimens showing clearly that they can do this. Even in the rude circumferential rim or wall of wax round a growing comb, flexures may sometimes be observed, corresponding in position to the planes of the rhombic basal plates of future cells. But the rough wall of wax has in every case to be finished off, by being largely gnawed away on both sides. The manner in which the bees build is curious; they always make the first rough wall from ten to twenty times thicker than the excessively thin finished wall of the cell, which will ultimately be left. We shall understand how they work, by supposing masons first to pile up a broad ridge of cement, and then to begin cutting it away equally on both sides near the ground, till a smooth, very thin wall is left in the middle; the masons always piling up the cut-away cement, and adding fresh cement on the summit of the ridge. We shall thus have a thin wall steadily growing upward but always crowned by a gigantic coping. From all the cells, both those just commenced and those completed, being thus crowned by a strong coping of wax, the bees can cluster and crawl over the comb without injuring the delicate hexagonal walls. These walls, as Professor Miller has kindly ascertained for me, vary greatly in thickness; being, on an average of twelve measurements made near the border of the comb, $\frac{1}{352}$ of an inch in thickness; whereas the basal rhomboidal plates are thicker, nearly in the proportion of three to two, having a mean thickness, from twenty-one measurements, of $\frac{1}{229}$ of an inch. By the above singular manner of building, strength is continually given to the comb, with the utmost ultimate economy of wax.

It seems at first to add to the difficulty of understanding how the cells are made, that a multitude of bees all work together; one bee after working a short time at one cell going to another, so that, as Huber has stated a score of individuals work even at the commencement of the first cell. I was able practically to show this fact, by covering the edges of the hexagonal walls of a single cell, or the extreme margin of the circumferential rim of a growing comb, with an extremely thin layer of melted vermilion wax; and I invariably found that the colour was most delicately diffused by the bees—as delicately as a painter could have done it with his brush —by atoms of the coloured wax having been taken from the spot on which it had been placed, and worked into the growing edges of the cells all round. The work of construction seems to be a sort of balance struck between many bees, all instinctively standing at the same relative distance from each other, all trying to sweep equal spheres, and then building up, or leaving ungnawed, the planes of intersection between these spheres. It was really curious to note in cases of difficulty, as when two pieces of comb met at an angle, how often the bees would pull down and rebuild in different ways the same cell, sometimes recurring to a shape which they had at first rejected.

When bees have a place on which they can stand in their proper positions for working,—for instance, on a slip of wood, placed directly under the middle of a comb growing downwards, so that the comb has to be built over one face of the slip—in this case the bees can lay the foundations of one wall of a new hexagon, in its strictly proper place, projecting beyond the other completed cells. It suffices that the bees should be enabled to stand at their proper relative distances from each other and from the walls of the last completed cells, and then, by striking imaginary spheres, they can build up a wall intermediate between two adjoining spheres; but, as far as I have seen, they never gnaw away and finish off the angles of a cell till a large part both of that cell and of the adjoining cells has been built. This capacity in bees of laying down under certain circumstances a rough wall in its proper place between two just-commenced cells, is important, as it bears on a fact, which seems at first subversive of the foregoing theory; namely, that the cells on the extreme margin of wasp-combs are sometimes strictly hexagonal; but I have not space here to enter on this subject. Nor does there seem to me any great difficulty in a single insect (as in the case of a queen-wasp) making hexagonal cells, if she were to work alternately on the inside and outside of two or three cells commenced at the same time, always standing at the proper relative distance from the parts of the cells just begun, sweeping spheres or cylinders, and building up intermediate planes.

As natural selection acts only by the accumulation of slight modifications of structure or instinct, each profitable to the individual under its conditions of life, it may reasonably be asked, how a long and graduated succession of modified architectural instincts, all tending towards the present perfect plan of construction, could have profited the progenitors of the hive-bee? I think the answer is not difficult: cells constructed like those of the bee or the wasp gain in strength, and save much in labour and space, and in the materials of which they are constructed. With respect to the formation of wax, it is known that bees are often hard pressed to get sufficient nectar, and I am informed by Mr. Tegetmeier that it has been experimentally proved that from twelve to fifteen pounds of dry sugar are consumed by a hive of bees for the secretion of a pound of wax; so that a prodigious quantity of fluid nectar must be collected and consumed by the bees in a hive for the secretion of the wax necessary for the construction of their combs. Moreover, many bees have to remain idle for many days during the process of secretion. A large store of honey is indispensable to support a large stock of bees during the winter; and the security of the hive is known mainly to depend on a large number of bees being support-ed. Hence the saving of wax by largely saving honey and the time con-sumed in collecting the honey must be an important element of success to any family of bees. Of course the success of the species may be dependent on the number of its enemies, or parasites, or on quite distinct causes, and so be altogether independent of the quantity of honey which the bees can collect. But let us suppose that this latter circumstance determined, as it probably often has determined, whether a bee allied to our humble-bees could exist in large numbers in any country; and let us further suppose that the community lived through the winter, and consequently required a store of honey: there can in this case be no doubt that it would be an ad-vantage to our imaginary humble-bee if a slight modification in her in-stincts led her to make her waxen cells near together, so as to intersect a little; for a wall in common even to two adjoining cells would save some little labour and wax. Hence it would continually be more and more ad-vantageous to our humble-bees, if they were to make their cells more and more regular, nearer together, and aggregated into a mass, like the cells of the Melipona; for in this case a large part of the bounding surface of each cell would serve to bound the adjoining cells, and much labour and wax would be saved. Again, from the same cause, it would be advantageous to the Melipona, if she were to make her cells closer together, and more reg-ular in every way than at present; for then, as we have seen, the spherical surfaces would wholly disappear and be replaced by plane surfaces; and the Melipona would make a comb as perfect as that of the hive-bee. Be-yond this stage of perfection in architecture, natural selection could not lead; for the comb of the hive-bee, as far as we can see, is absolutely per-fect in economising labour and wax.

Thus, as I believe, the most wonderful of all known instincts, that of the hive-bee, can be explained by natural selection having taken advant-age of numerous, successive, slight modifications of simpler instincts; nat-ural selection having, by slow degrees, more and more perfectly led the bees to sweep equal spheres at a given distance from each other in a double layer, and to build up and excavate the wax along the planes of in-tersection; the bees, of course, no more knowing that they swept their spheres at one particular distance from each other, than they know what are the several angles of the hexagonal prisms and of the basal rhombic plates; the motive power of the process of natural selection having been the construction of cells of due strength and of the proper size and shape for the larvæ, this being effected with the greatest possible economy of labour and wax; that individual swarm which thus made the best cells with least labour, and least waste of honey in the secretion of wax, having succeeded best, and having transmitted their newly-acquired economical instincts to new swarms, which in their turn will have had the best chance of succeeding in the struggle for existence.

OBJECTIONS TO THEORY   ∞ʒ

Objections to the Theory of Natural Selection as applied to Instincts: Neuter and Sterile Insects

It has been objected to the foregoing view of the origin of instincts that "the variations of structure and of instinct must have been simultaneous and accurately adjusted to each other, as a modification in the one without an immediate corresponding change in the other would have been fatal." The force of this objection rests entirely on the assumption that the changes in the instincts and structure are abrupt. To take as an illustration the case of the larger titmouse (Parus major) alluded to in a previous chapter; this bird often holds the seeds of the yew between its feet on a branch, and hammers with its beak till it gets at the kernel. Now what special difficulty would there be in natural selection preserving all the slight individual variations in the shape of the beak, which were better and better adapted to break open the seeds, until a beak was formed, as well constructed for this purpose as that of the nuthatch, at the same time that habit, or compulsion, or spontaneous variations of taste, led the bird to become more and more of a seed-eater? In this case the beak is supposed to be slowly modified by natural selection, subsequently to, but in accordance with, slowly changing habits or taste; but let the feet of the titmouse vary and grow larger from correlation with the beak, or from any other unknown cause, and it is not improbable that such larger feet would lead the bird to climb more and more until it acquired the remarkable climbing instinct and power of the nuthatch. In this case a gradual change of structure is supposed to lead to changed instinctive habits. To take one more case: few instincts are more remarkable than that which leads the swift of the Eastern Islands to make its nest wholly of inspissated saliva. Some birds build their nests of mud, believed to be moistened with saliva; and one of the swifts of North America makes its nest (as I have seen) of sticks agglutinated with saliva, and even with flakes of this substance. Is it then very improbable that the natural selection of individual swifts, which secreted more and more saliva, should at last produce a species with instincts leading it to neglect other materials, and to make its nest exclusively of inspissated saliva? And so in other cases. It must, however, be admitted that in many instances we cannot conjecture whether it was instinct or structure which first varied.

No doubt many instincts of very difficult explanation could be opposed to the theory of natural selection—cases, in which we cannot see how an instinct could have originated; cases, in which no intermediate gradations are known to exist; cases of instincts of such trifling importance, that they could hardly have been acted on by natural selection; cases of instincts almost identically the same in animals so remote in the scale of nature, that we cannot account for their similarity by inheritance from a common progenitor, and consequently must believe that they were independently acquired through natural selection. I will not here enter on these several cases, but will confine myself to one special difficulty, which at first appeared to me insuperable, and actually fatal to the whole theory. I allude to the neuters or sterile females in insect-communities; for these neuters often differ widely in instinct and in structure from both the males and fertile females, and yet, from being sterile, they cannot propagate their kind.

The subject well deserves to be discussed at great length, but I will here take only a single case, that of working or sterile ants. How the workers have been rendered sterile is a difficulty; but not much greater than that of any other striking modification of structure; for it can be shown that some insects and other articulate animals in a state of nature occasionally become sterile; and if such insects had been social, and it had been profitable to the community that a number should have been annually born capable of work, but incapable of procreation, I can see no especial difficulty in this having been effected through natural selection. But I must pass over this preliminary difficulty. The great difficulty lies in the working ants differing widely from both the males and the fertile females in structure, as in the shape of the thorax, and in being destitute of wings and sometimes of eyes, and in instinct. As far as instinct alone is concerned, the wonderful difference in this respect between the workers and the perfect females, would have been better exemplified by the hive-bee. If a working ant or other neuter insect had been an ordinary animal, I should have unhesitatingly assumed that all its characters had been slowly acquired through natural selection; namely, by individuals having been born with slight profitable modifications, which were inherited by the offspring; and that these again varied and again were selected, and so onwards. But with the working ant we have an insect differing greatly from its parents, yet absolutely sterile; so that it could never have transmitted successively acquired modifications of structure or instinct to its progeny. It may well be asked how is it possible to reconcile this case with the theory of natural selection?

First, let it be remembered that we have innumerable instances, both in our domestic productions and in those in a state of nature, of all sorts of differences of inherited structure which are correlated with certain ages, and with either sex. We have differences correlated not only with one sex, but with that short period when the reproductive system is active, as in the nuptial plumage of many birds, and in the hooked jaws of the male salmon. We have even slight differences in the horns of different breeds of cattle in relation to an artificially imperfect state of the male sex; for oxen of certain breeds have longer horns than the oxen of other breeds, relatively to the length of the horns in both the bulls and cows of these same breeds. Hence I can see no great difficulty in any character becoming correlated with the sterile condition of certain members of insect-communities: the difficulty lies in understanding how such correlated modifications of structure could have been slowly accumulated by natural selection.

This difficulty, though appearing insuperable, is lessened, or, as I believe, disappears, when it is remembered that selection may be applied to the family, as well as to the individual, and may thus gain the desired end.

Breeders of cattle wish the flesh and fat to be well marbled together: an animal thus characterised has been slaughtered, but the breeder has gone with confidence to the same stock and has succeeded. Such faith may be placed in the power of selection, that a breed of cattle, always yielding oxen with extraordinarily long horns, could, it is probable, be formed by carefully watching which individual bulls and cows, when matched, produced oxen with the longest horns; and yet no ox would ever have propagated its kind. Here is a better and real illustration: according to M. Verlot, some varieties of the double annual Stock from having been long and carefully selected to the right degree, always produce a large proportion of seedlings bearing double and quite sterile flowers; but they likewise yield some single and fertile plants. These latter, by which alone the variety can be propagated, may be compared with the fertile male and female ants, and the double sterile plants with the neuters of the same community. As with the varieties of the stock, so with social insects, selection has been applied to the family, and not to the individual, for the sake of gaining a serviceable end. Hence we may conclude that slight modifications of structure or of instinct, correlated with the sterile condition of certain members of the community, have proved advantageous: consequently the fertile males and females have flourished, and transmitted to their fertile offspring a tendency to produce sterile members with the same modifications. This process must have been repeated many times, until that prodigious amount of difference between the fertile and sterile females of the same species has been produced, which we see in many social insects.

But we have not as yet touched on the acme of the difficulty; namely, the fact that the neuters of several ants differ, not only from the fertile females and males, but from each other, sometimes to an almost incredible degree, and are thus divided into two or even three castes. The castes, moreover, do not commonly graduate into each other, but are perfectly well defined; being as distinct from each other as are any two species of the same genus, or rather as any two genera of the same family. Thus in Eciton, there are working and soldier neuters, with jaws and instincts extraordinarily different: in Cryptocerus, the workers of one caste alone carry a wonderful sort of shield on their heads, the use of which is quite unknown: in the Mexican Myrmecocystus, the workers of one caste never leave the nest; they are fed by the workers of another caste, and they have an enormously developed abdomen which secretes a sort of honey, supplying the place of that excreted by the aphides, or the domestic cattle as they may be called, which our European ants guard and imprison.

It will indeed be thought that I have an overweening confidence in the principle of natural selection, when I do not admit that such wonderful and well-established facts at once annihilate the theory. In the simpler case of neuter insects all of one caste, which, as I believe, have been rendered different from the fertile males and females through natural selection we may conclude from the analogy of ordinary variations, that the successive, slight, profitable modifications did not first arise in all the neuters in the same nest, but in some few alone; and that by the survival of the

communities with females which produced most neuters having the advantageous modifications, all the neuters ultimately came to be thus characterised. According to this view we ought occasionally to find in the same
nest neuter insects, presenting gradations of structure; and this we do
find, even not rarely, considering how few neuter insects out of Europe
have been carefully examined. Mr. F. Smith has shown that the neuters of
several British ants differ surprisingly from each other in size and sometimes in colour; and that the extreme forms can be linked together by individuals taken out of the same nest: I have myself compared perfect
gradations of this kind. It sometimes happens that the larger or the smaller sized workers are the most numerous; or that both large and small are
numerous, whilst those of an intermediate size are scanty in numbers.
Formica flava has larger and smaller workers, with some few of intermediate size; and, in this species, as Mr. F. Smith has observed, the larger
workers have simple eyes (ocelli), which though small can be plainly distinguished, whereas the smaller workers have their ocelli rudimentary.
Having carefully dissected several specimens of these workers, I can af-
firm that the eyes are far more rudimentary in the smaller workers than
can be accounted for merely by their proportionally lesser size; and I fully
believe, though I dare not assert so positively, that the workers of intermediate size have their ocelli in an exactly intermediate condition. So that
here we have two bodies of sterile workers in the same nest, differing not
only in size, but in their organs of vision, yet connected by some few members in an intermediate condition. I may digress by adding, that if the
smaller workers had been the most useful to the community, and those
males and females had been continually selected, which produced more
and more of the smaller workers, until all the workers were in this condition; we should then have had a species of ant with neuters in nearly the
same condition as those of Myrmica. For the workers of Myrmica have
not even rudiments of ocelli, though the male and female ants of this genus have well-developed ocelli.

I may give one other case: so confidently did I expect occasionally to
find gradations of important structures between the different castes of
neuters in the same species, that I gladly availed myself of Mr. F. Smith's
offer of numerous specimens from the same nest of the driver ant (Anom-
ma) of West Africa. The reader will perhaps best appreciate the amount
of difference in these workers, by my giving not the actual measurements,
but a strictly accurate illustration: the difference was the same as if we
were to see a set of workmen building a house, of whom many were five
feet four inches high, and many sixteen feet high; but we must in addition
suppose that the larger workmen had heads four instead of three times as
big as those of the smaller men, and jaws nearly five times as big. The
jaws, moreover, of the working ants of the several sizes differed wonderfully in shape, and in the form and number of the teeth. But the important fact for us is, that, though the workers can be grouped into castes of
different size, yet they graduate insensibly into each other, as does the
widely-different structure of their jaws. I speak confidently on this latter

point, as Sir J. Lubbock made drawings for me, with the camera lucida, of the jaws which I dissected from the workers of the several sizes. Mr. Bates, in his interesting 'Naturalist on the Amazons,' has described analogous cases.

With these facts before me, I believe that natural selection, by acting on the fertile ants or parents, could form a species which should regularly produce neuters, all of large size with one form of jaw, or all of small size with widely different jaws; or lastly, and this is the greatest difficulty, one set of workers of one size and structure, and simultaneously another set of workers of a different size and structure;—a graduated series having first been formed, as in the case of the driver ant, and then the extreme forms having been produced in greater and greater numbers, through the survival of the parents which generated them, until none with an intermediate structure were produced.

An analogous explanation has been given by Mr. Wallace, of the equally complex case, of certain Malayan Butterflies regularly appearing under two or even three distinct female forms; and by Fritz Müller, of certain Brazilian crustaceans likewise appearing under two widely distinct male forms. But this subject need not here be discussed.

I have now explained how, as I believe, the wonderful fact of two distinctly defined castes of sterile workers existing in the same nest, both widely different from each other and from their parents, has originated. We can see how useful their production may have been to a social community of ants, on the same principle that the division of labour is useful to civilised man. Ants, however, work by inherited instincts and by inherited organs or tools, whilst man works by acquired knowledge and manufactured instruments. But I must confess, that, with all my faith in natural selection, I should never have anticipated that this principle could have been efficient in so high a degree, had not the case of these neuter insects led me to this conclusion. I have, therefore, discussed this case, at some little but wholly insufficient length, in order to show the power of natural selection, and likewise because this is by far the most serious special difficulty which my theory has encountered. The case, also, is very interesting, as it proves that with animals, as with plants, any amount of modification may be effected by the accumulation of numerous, slight, spontaneous variations, which are in any way profitable, without exercise or habit having been brought into play. For peculiar habits confined to the workers or sterile females, however long they might be followed, could not possibly affect the males and fertile females, which alone leave descendants. I am surprised that no one has hitherto advanced this demonstrative case of neuter insects, against the well-known doctrine of inherited habit, as advanced by Lamarck.

Summary

I have endeavoured in this chapter briefly to show that the mental qualities of our domestic animals vary, and that the variations are inherited. Still more briefly I have attempted to show that instincts vary slight-ly in a state of nature. No one will dispute that instincts are of the highest importance to each animal. Therefore there is no real difficulty, under changing conditions of life, in natural selection accumulating to any ex-tent slight modifications of instinct which are in any way useful. In many cases habit or use and disuse have probably come into play. I do not pre-tend that the facts given in this chapter strengthen in any great degree my theory; but none of the cases of difficulty, to the best of my judgment, annihilate it. On the other hand, the fact that instincts are not always absolutely perfect and are liable to mistakes:—that no instinct can be shown to have been produced for the good of other animals, though ani-mals take advantage of the instincts of others;—that the canon in natural history, of "Natura non facit saltum," is applicable to instincts as well as to corporeal structure, and is plainly explicable on the foregoing views, but is otherwise inexplicable,—all tend to corroborate the theory of nat-ural selection.

This theory is also strengthened by some few other facts in regard to instincts; as by that common case of closely allied, but distinct species, when inhabiting distant parts of the world and living under considerably different conditions of life, yet often retaining nearly the same instincts. For instance, we can understand, on the principle of inheritance, how it is that the thrush of tropical South America lines its nest with mud, in the same peculiar manner as does our British thrush; how it is that the Horn-bills of Africa and India have the same extraordinary instinct of plaster-ing up and imprisoning the females in a hole in a tree, with only a small hole left in the plaster through which the males feed them and their young when hatched; how it is that the male wrens (Troglodytes) of North America build "cock-nests," to roost in, like the males of our Kitty-wrens,—a habit wholly unlike that of any other known bird. Final-ly, it may not be a logical deduction, but to my imagination it is far more satisfactory to look at such instincts as the young cuckoo ejecting its fos-ter-brothers,—ants making slaves,—the larvæ of ichneumonidæ feeding within the live bodies of caterpillars,—not as specially endowed or cre-ated instincts, but as small consequences of one general law leading to the advancement of all organic beings,—namely, multiply, vary, let the strongest live and the weakest die.

saluted; *a ceremony which he never performed but once again upon a similar occasion.* Finding him extremely tractable, I made it my custom to carry him always after breakfast into the garden. . . . I had not long habituated him to this taste of liberty before he began to be impatient for the return of the time when he might enjoy it. He would invite me to the garden by drumming upon my knee, and by a look of such expression as it was not possible to misinterpret. If this rhetoric did not immediately succeed, *he would take the skirt of my coat between his teeth and pull it with all his force.* He seemed to be happier in human society than when shut up with his natural companions.

Rats and Mice.

Rats are well known to be highly intelligent animals. Unlike the hare or rabbit, their shyness seems to proceed from a wise caution rather than from timidity ; for, when circumstances require, their boldness and courage in combat is surprising. Moreover, they never seem to lose their presence of mind ; for, however great their danger, they seem always ready to take advantage of any favouring circumstances that may arise. Thus, when matched with so formidable an opponent as a ferret in a closed room, they have been known to display wonderful cunning in taking advantage of the light—keeping close under the window so as to throw the glare into the eyes of the enemy, darting forwards time after time to deliver a bite, and then as often retiring to their vantage-ground.[1] But the emotions of rats do not appear to be of an entirely selfish character. There are so many accounts in the anecdote books of blind rats being led about by their seeing companions, that it is difficult to discredit an observation so frequently confirmed.[2] Moreover, rats have been frequently known to assist one another in defending themselves from dangerous enemies. Several observations of this kind are recorded by the trustworthy writer Mr. Rodwell, in his somewhat elaborate work upon this animal.

[1] See Watson's *Reasoning Power in Animals*, and *Quarterly Review*, c. i., p. 135.
[2] See especially Jesse, *Gleanings*, &c., iii., p. 206 ; and *Quarterly Review*, c. i., p. 135.

Again, as showing affection for human beings, I may
quote the following :—' The mouse which had been tamed
by Baron Trench in his prison having been taken from
him, watched at the door and crept in when it was
opened; being removed again, it refused all food, and
died in three days.'[1]

With regard to general intelligence, every one knows
the extraordinary wariness of rats in relation to traps,
which is only equalled in the animal kingdom by that of
the fox and the wolverine. It has frequently been regarded
as a wonderful display of intelligence on the part of rats
that while gnawing through the woodwork of a ship, they
always stop before they completely perforate the side;
but, as Mr. Jesse suggests, this is probably due to their
distaste of the salt water. No such disparaging explanation,
however, is possible in some other instances of the display
of rat-intelligence. Thus, the manner in which they
transport eggs to their burrows has been too frequently
observed to admit of doubt. Rodwell gives a case in
which a number of eggs were carried from the top of a
house to the bottom by two rats devoting themselves to
each egg, and alternately passing it down to each other at
every step of the staircase.[2] Dr. Carpenter also received
from an eye-witness a similar account of another instance.[3]
According to the article in the *Quarterly Review*, already
mentioned, rats will not only convey eggs from the top of
the house to the bottom, but from bottom to top. ' The
male rat places himself on his fore-paws, with his head
downwards, and raising up his hind legs and catching
the egg between them, pushes it up to the female, who
stands on the step above, and secures it with her fore-paws
till he jumps up to her; and this process is repeated from
step to step till the top is reached.'

'The captain of a merchantman,' says Mr. Jesse,
' trading to the port of Boston, in Lincolnshire, had constantly missed eggs from his sea stock. He suspected
that he was robbed by his crew, but not being able to dis-

[1] Thompson, *Passions of Animals*, p. 368.
[2] *The Rat, its Natural History*, p. 102.
[3] Mrs. Lee, *Anecdotes of Animals*, p. 264.

362   ANIMAL INTELLIGENCE.

cover the thief, he was determined to watch his store-room. Accordingly, having laid in a fresh stock of eggs, he seated himself at night in a situation that commanded a view of his eggs. To his great astonishment he saw a number of rats approach; they formed a line from his egg baskets to their hole, and handed the eggs from one to another in their fore-paws.' [1]

Another device to which rats resort for the procuring of food is mentioned in all the anecdote books, and it seemed so interesting that I tried some direct experiments upon the subject. I shall first state the alleged facts in the words of Watson :—

As to oil, rats have been known to get oil out of a narrow-necked bottle in the following way :—One of them would place himself, on some convenient support, by the side of the bottle, and then, dipping his tail into the oil, would give it to another to lick. In this act there is something more than what we call instinct; there is reason and understanding. [2]

Jesse also gives the following account :—

A box containing some bottles of Florence oil was placed in a store-room which was seldom opened; the box had no lid to it. On going to the room one day for one of the bottles, the owner found that the pieces of bladder and cotton at the mouth of each bottle had disappeared, and that much of the contents of the bottles had been consumed. The circumstance having excited suspicion, a few bottles were refilled with oil, and the mouths of them secured as before. Next morning the coverings of the bottles had been removed, and some of the oil was gone. However, upon watching the room, which was done through a little window, some rats were seen to get into the box, and insert their tails into the necks of the bottles, and then with-drawing them, they licked off the oil which adhered to them. [3]

Lastly, Rodwell gives another case similar in all essential respects, save that the rat licked its own tail instead of presenting it to a companion.

The experiment whereby I tested the truth of these

---

[1] Jesse, *Gleanings*, &c., ii., p. 281.
[2] *Reasoning Power in Animals*, p. 293.
[3] *Loc. cit.*

statements was a very simple one. I recorded it in 'Nature' as follows :—

It is, I believe, pretty generally supposed that rats and mice use their tails for feeding purposes when the food to be eaten is contained in vessels too narrow to admit the entire body of the animal. I am not aware, however, that the truth of this supposition has ever been actually tested by any trustworthy person, and so think the following simple experiments are worth publishing. Having obtained a couple of tall-shaped preserve bottles with rather short and narrow necks, I filled them to within three inches of the top with red currant jelly which had only half stiffened. I covered the bottles with bladder in the ordinary way, and then stood them in a place infested by rats. Next morning the bladder covering each of the bottles had a small hole gnawed through it, and the level of the jelly was reduced in both bottles to the same extent. Now, as this extent corresponded to about the length of a rat's tail if inserted at the hole in the bladder, and as this hole was not much more than just large enough to admit the root of this organ, I do not see that any further evidence is required to prove the manner in which the rats obtained the jelly, viz., by repeatedly introducing their tails into the viscid matter, and as repeatedly licking them clean. However, to put the question beyond doubt, I refilled the bottles to the extent of half an inch above the jelly level left by the rats, and having placed a circle of moist paper upon each of the jelly surfaces, covered the bottles with bladder as before. I now left the bottles in a place where there were no rats or mice, until a good crop of mould had grown upon one of the moistened pieces of paper. The bottle containing this crop of mould I then transferred to the place where the rats were numerous. Next morning the bladder had again been eaten through at one edge, and upon the mould there were numerous and distinct tracings of the rats' tails, resembling marks made with the top of a pen-holder. These tracings were evidently caused by the animals sweeping their tails about in a fruitless endeavour to find a hole in the circle of paper which covered the jelly.

With regard to mice, the Rev. W. North, rector of Ashdown, in Essex, placed a pot of honey in a closet, in which a quantity of plaster rubbish had been left by builders. The mice piled up the plaster in the form of a heap against the sides of the pot, in order to constitute an

inclined plane whereby to reach the rim. A quantity of
the rubbish had also been thrown into the pot, with the
effect of raising the level of the honey that remained to
near the rim of the pot ; but, of course, the latter fact may
have been due to accident, and not to design.[1] This is a
case in which mal-observation does not seem to have been
likely.

Powelsen, a writer on Iceland, has related an account
of the intelligence displayed by the mice of that country,
which has given rise to a difference of competent opinion,
and which perhaps can hardly yet be said to have been
definitely settled. What Powelsen said is that the mice
collect in parties of from six to ten, select a flat piece of
dried cow-dung, pile berries or other food upon it, then
with united strength drag it to the edge of any stream
they wish to cross, launch it, embark, and range themselves round the central heap of provisions with their heads
joined over it, and their tails hanging in the water,
perhaps serving as rudders. Pennant afterwards gave
credit to this account, observing that in a country where
berries were scarce, the mice were compelled to cross
streams for distant forages.[2] Dr. Hooker, however, in his
' Tour in Iceland,' concludes that the account is a pure
fabrication. Dr. Henderson, therefore, determined on
trying to arrive at the truth of the matter, with the following result :—' I made a point of inquiring of different
individuals as to the reality of the account, and am happy
in being able to say that it is now established as an important fact in natural history by the testimony of two eye-
witnesses of unquestionable veracity, the clergyman of
Briamslaek, and Madame Benedictson of Stickesholm, both
of whom assured me that they had seen the expedition
performed repeatedly. Madame Benedictson, in particular, recollected having spent a whole afternoon, in her
younger days, at the margin of a small lake on which
these skilful navigators had embarked, and amusing herself
and her companions by driving them away from the sides
of the lake as they approached them. I was also informed

---

[1] Jesse, *Gleanings*, iii., p. 176.
[2] *Introduction to Arctic Zoology*, p. 70.

that they make use of dried mushrooms as sacks, in which they convey their provisions to the river, and thence to their homes.'[1]

Before leaving the mice and rats I may say a few words upon certain mouse- and rat-like animals which scarcely require a separate section for their consideration. Of the harvesting mouse Gilbert White says :—

One of their nests I procured this autumn, most artificially plaited and composed of blades of wheat, perfectly round, and about the size of a cricket-ball, with the aperture so ingeniously closed that there was no discovering to what part it belonged. It was so compact and well filled that it would roll across the table without being discomposed, though it contained eight little mice that were naked and blind. As the nest was perfectly full, how could the dam come at her litter respectively, so as to administer a teat to each? Perhaps she opens different places for that purpose, adjusting them again when the business is over; but she could not possibly be contained herself in the ball with the young ones, which, moreover, would be daily increasing in size. This wonderful procreant cradle, an elegant instance of the efforts of instinct, was found in a wheat-field, suspended on the head of a thistle.

Pallas has described the provident habits of the so-called ' rat-hare ' (*Lagomys*), which lays up a store of grass, or rather hay, for winter consumption. These animals, which occur in the Altai Mountains, live in holes or crevices of rock. About the middle of the month of August they collect grass, and spread it out to dry into hay. In September they form heaps or stacks of the hay, which may be as much as six feet high, and eight feet in diameter. It is stored in their chosen hole or crevice, protected from the rain.

The following is quoted from Thompson's ' Passions of Animals,' pp. 235–6 :—

The life of the harvester rat is divided between eating and fighting. It seems to have no other passion than that of rage, which induces it to attack every animal that comes in its way, without in the least attending to the superior strength of its enemy. Ignorant of the art of saving itself by flight, rather

[1] Dr. Henderson, *Journal of a Residence in Iceland in* 1814 *and* 1815, vol. ii., p. 187.

366   ANIMAL INTELLIGENCE.

than yield, it will allow itself to be beaten to pieces with a stick.
If it seizes a man's hand, it must be killed before it will quit its
hold. The magnitude of the horse terrifies it as little as the
address of the dog, which last is fond of hunting it.   When a
harvester perceives a dog at a distance, it begins by emptying
its cheek-pouches, if they happen to be filled with grain ; it then
blows them up so prodigiously, that the size of the head and
neck greatly exceeds that of the rest of the body.   It rears itself
upon its hind legs, and thus darts upon the enemy. If it catches
hold, it never quits it but with the loss of its life ; but the dog
generally seizes it behind, and strangles it.   This ferocious disposition prevents it from being at peace with any animal whatever.   It even makes war against its own species.   When two
harvesters meet, they never fail to attack each other, and the
stronger always devours the weaker.   A combat between a
male and a female commonly lasts longer than between two
males.   They begin by pursuing and biting each other, then
each of them retires aside, as if to take breath.   After a short
interval they renew the combat, and continue to fight till one of
them falls.   The vanquished uniformly serves as a repast to the
conqueror.

If we contrast the fearless disposition of the harvester
with the timidity of the hare or rabbit, we observe that in
respect of emotions, no less than in that of intelligence,
the order Rodentia comprises the utmost extremes.

The so-called ' prairie-dog' is a kind of small rodent,
which makes burrows in the ground, and a slight elevation above it.   The animals being social in their habits,
their warrens are called ' dog-towns.'   Prof. Jillson, Ph.D.,
kept a pair in confinement (see ' American Naturalist,'
vol. v., pp. 24–29), and found them to be intelligent and
highly affectionate animals.   These burrows he found to
contain a ' granary,' or chambers set apart for the reception
of stored food.   With regard to the association said to
exist between this animal and the owl and rattle-snake,
Prof. Jillson says, ' I have seen many dog-towns, with
owls and dogs standing on contiguous, and in some cases
on the same mound, but never saw a snake in the vicinity.'
The popular notion that the owl acts the part of sentry to
the dog requires, to say the least, confirmation.

ANIMAL INTELLIGENCE; AN EXPERIMENTAL STUDY OF THE ASSOCIATIVE PROCESSES IN ANIMALS.

This monograph is an attempt at an explanation of the nature of the process of association in the animal mind. Inasmuch as there have been no extended researches of a character similar to the present one either in subject-matter or experimental method, it is necessary to explain briefly its standpoint.

Our knowledge of the mental life of animals equals in the main our knowledge of their sense-powers, of their instincts or reactions performed without experience, and of their reactions which are built up by experience. Confining our attention to the latter we find it the opinion of the better observers and analysts that these reactions can all be explained by the ordinary associative processes without aid from abstract, conceptual, inferential thinking. These associative processes then, as present in animals' minds and as displayed in their acts, are my subject-matter. Any one familiar in even a general way with the literature of comparative psychology will recall that this part of the field has received faulty and unsuccessful treatment. The careful, minute, and solid knowledge of the sense-organs of animals finds no counterpart in the realm of associations and habits. We do not know how delicate or how complex or how permanent are the possible associations of any given group of animals. And although one would be rash who said that our present equipment of facts about instincts was sufficient or that our theories about it were surely sound, yet our notion of what occurs when a chick grabs a worm are luminous and infallible compared to our notion of what happens when a kitten runs into the house at the familiar call. The reason that they have satisfied us as well as they have is just that they are so vague. We say that the kitten associates the sound 'kitty kitty' with the experience of nice milk to drink, which does very well for a common-sense answer. It also suffices as a rebuke to those who would have the kitten ratiocinate about the matter, but it fails to tell what real mental content is present. Does the kitten feel *" sound of call, memory-image of milk in a saucer in the kitchen, thought of running into the house, a feeling, finally, of ' I will run in'? "* Does he perhaps feel only the sound of the bell and an impulse to run in, similar in quality to the impulses which make a tennis player run to and fro when playing? The word association may cover a multitude of essentially different processes, and when a writer attributes anything that an animal may do to association his statement has only the negative value of eliminating reasoning on the one hand and instinct on the other. His position is like that of a zoölogist who should to-day class an animal among the ' worms.' To give to the word a positive value and several definite possibilities of meaning is one aim of this investigation.

The importance to comparative psychology in general of a more scientific account of the association-process in animals is evident. Apart from the desirability of knowing all the facts we can, of whatever sort, there is the especial consideration that these associations and consequent habits have an immediate import for biological science. In the higher animals the bodily life and preservative acts are largely directed by these associations. They, and not instinct, make the animal use the best feeding grounds, sleep in the same lair, avoid new dangers and profit by new changes in nature. Their higher development in mammals is a chief factor in the supremacy of that group. This, however, is a minor consideration. The main purpose of the study of the animal mind is to learn the development of mental life down through the phylum, to trace in particular the origin of human faculty. In relation to this chief purpose of comparative psychology the associative processes assume a rôle predominant over that of sense-powers or instinct, for in a study of the associative processes lies the solution of the problem. Sense-powers and instincts have changed by addition and supersedence, but the cognitive side of consciousness has changed not only in quantity but also in quality. Somehow

out of these associative processes have arisen human conscious-
nesses with their sciences and arts and religions. The association of ideas proper, imagination, memory, abstraction, generalization, judgment, inference, have here their source. And in
the metamorphosis the instincts, impulses, emotions and sense-
impressions have been transformed out of their old natures.
For the origin and development of human faculty we must look
to these processes of association in lower animals. Not only
then does this department need treatment more, but promises to
repay the worker better.

Although no work done in this field is enough like the
present investigation to require an account of its results, the
method hitherto in use invites comparison by its contrast and,
as I believe, by its faults. In the first place, most of the books
do not give us a psychology, but rather a *eulogy*, of animals.
They have all been about animal *intelligence*, never about
animal *stupidity*. Though a writer derides the notion that
animals have reason, he hastens to add that they have marvellous capacity of forming associations, and is likely to refer to
the fact that human beings only rarely reason anything out,
that their trains of ideas are ruled mostly by association, as if,
in this latter, animals were on a par with them. The history
of books on animals' minds thus furnishes an illustration of the
well-nigh universal tendency in human nature to find the marvelous wherever it can. We wonder that the stars are so big
and so far apart, that the microbes are so small and so thick
together, and for much the same reason wonder at the things
animals do. They used to be wonderful because of the
mysterious, God-given faculty of instinct, which could almost
remove mountains. More lately they have been wondered at
because of their marvellous mental powers in profiting by experience. Now imagine an astronomer tremendously eager to
prove the stars as big as possible, or a bacteriologist whose
great scientific desire is to demonstrate the microbes to be very,
very little! Yet there has been a similar eagerness on the part
of many recent writers on animal psychology to praise the
abilities of animals. It cannot help leading to partiality in deductions from facts and more especially in the choice of facts

for investigation.   How can scientists who write like lawyers, defending animals against the charge of having no power of rationality, be at the same time impartial judges on the bench? Unfortunately the real work in this field has been done in this spirit.   The level-headed thinkers who might have won valuable results have contented themselves with arguing against the theories of the eulogists.   They have not made investigations of their own.

   In the second place the facts have generally been derived from anecdotes.   Now quite apart from such pedantry as insists that a man's word about a scientific fact is worthless unless he is a trained scientist, there are really in this field special objections to the acceptance of the testimony about animals' intelligent acts which one gets from anecdotes.   Such testimony is by no means on a par with testimony about the size of a fish or the migration of birds, etc.   For here one has to deal not merely with ignorant or inaccurate testimony, but also with prejudiced testimony.   Human folk are as a matter of fact eager to find intelligence in animals.   They like to.   And when the animal observed is a pet belonging to them or their friends, or when the story is one that has been told as a story to entertain, further complications are introduced.   Nor is this all.   Besides commonly misstating what fact sthey report, they report only such facts as show the animal at his best.   Dogs get lost hundreds of times and no one ever notices it or sends an account of it to a scientific magazine.   But let one find his way from Brooklyn to Yonkers and the fact immediately becomes a circulating anecdote.   Thousands of cats on thousands of occasions sit helplessly yowling, and no one takes thought of it or writes to his friend, the professor; but let one cat claw at the knob of a door supposedly as a signal to be let out, and straightway this cat becomes the representative of the cat-mind in all the books.   The unconscious distortion of the facts is almost harmless compared to the unconscious neglect of an animal's mental life until it verges on the unusual and marvelous.   It is as if some denizen of a planet where communication was by thought-transference, who was surveying humankind and reporting their psychology, should be oblivious to all our

ANIMAL INTELLIGENCE.   5

inter-communication save such as the psychical-research society has noted. If he should further misinterpret the cases of mere coincidence of thoughts as facts comparable to telepathic communication, he would not be more wrong than some of the animal psychologists. In short, the anecdotes give really the *abnormal* or *super-normal* psychology of animals.

Further, it must be confessed that these vices have been only ameliorated, not obliterated, when the observation is first-hand, is made by the psychologist himself. For as men of the utmost scientific skill have failed to prove good observers in the field of spiritualistic phenomena,[1] so biologists and psychologists before the pet terrier or hunted fox often become like Samson shorn. They, too, have looked for the intelligent and unusual and neglected the stupid and normal.

Finally, in all cases, whether of direct observation or report by good observers or bad, there have been three other defects. Only a single case is studied, and so the results are not necessarily true of the type; the observation is not repeated, nor are the conditions perfectly regulated; the previous history of the animal in question is not known. Such observations may tell us, if the observer is perfectly reliable, that a certain thing takes place, but they cannot assure us that it will take place universally among the animals of that species, or universally with the same animal. Nor can the influence of previous experience be estimated. All this refers to means of getting knowledge about what animals *do*. The next question is, " What do they *feel?*" Previous work has not furnished an answer or the material for an answer to this more important question. Nothing but carefully designed, crucial experiments can. In abandoning the old method one ought to seek above all to replace it by one which will not only tell more accurately *what they do*, and give the much-needed information *how they do it*, but also inform us *what they feel* while they act.

To remedy these defects experiment must be substituted for

---

[1] I do not mean that scientists have been too credulous with regard to spiritualism, but am referring to the cases where ten or twenty scientists have been sent to observe some trick-performance by a spiritualistic ' medium,' and have all been absolutely confident that they understood the secret of its performance, *each of them giving a totally different explanation.*

6   *E. L. THORNDIKE.*

observation and the collection of anecdotes. Thus you immediately get rid of several of them. You can repeat the conditions at will, so as to see whether or not the animal's behavior is due to mere coincidence. A number of animals can be subjected to the same test, so as to attain typical results. The animal may be put in situations where its conduct is especially instructive. After considerable preliminary observation of animals' behavior under various conditions, I chose for my general method one which, simple as it is, possesses several other marked advantages besides those which accompany experiment of any sort. It was merely to put animals when hungry in enclosures from which they could escape by some simple act, such as pulling at a loop of cord, pressing a lever, or stepping on a platform. (A detailed description of these boxes and pens will be given later.) The animal was put in the enclosure, food was left outside in sight, and his actions observed. Besides recording his general behavior, special notice was taken of how he succeeded in doing the necessary act (in case he did succeed), and a record was kept of the time that he was in the box before performing the successful pull, or clawing, or bite. This was repeated until the animal had formed a perfect association between the sense-impression of the interior of that box and the impulse leading to the successfnl movement. When the association was thus perfect, the time taken to escape was, of course, practically con stant and very short.

If, on the other hand, after a certain time the animal did not succeed, he was taken out, but *not fed*. If, after a sufficient number of trials, he failed to get out, the case was recorded as one of complete failure. Enough different sorts of methods of escape were tried to make it fairly sure that association in general, not association of a particular sort of impulse, was being studied. Enough animals were taken with each box or pen to make it sure that the results were not due to individual peculiarities. None of the animals used had any previous acquaintance with any of the mechanical contrivances by which the doors were opened. So far as possible the animals were kept in a uniform state of hunger, which was practically utter hunger. That is, no cat or dog was experimented on when the experiment involved any important question of fact or theory, unless I was sure that his motive was of the standard strength. With chicks this is not practicable, on account of their delicacy. But with them dislike of loneliness acts as a uniform motive to get back to the other chicks. Cats (or rather kittens), dogs and chicks were the subjects of the experiments. All were apparently in excellent health, save an occasional chick.

By this method of experimentation the animals are put in situations which call into activity their mental functions and permit them to be carefully observed. One may, by following it, observe personally more intelligent acts than are included in any anecdotal collection. And this actual vision of animals in the act of using their minds is far more fruitful than any amount of histories of what animals have done without the history of how they did it. But besides affording this opportunity for purposeful and systematic observation, our method is valuable because it frees the animal from any influence of the observer. The animal's behavior is quite independent of any factors save its own hunger, the mechanism of the box it is in, the food outside, and such general matters as fatigue, indisposition, etc. Therefore the work done by one investigator may be repeated and verified or modified by another. No personal factor is present save in the observation and interpretation. Again, our method gives some very important results which are quite uninfluenced by *any* personal factor in any way. The curves showing the progress of the formation of associations, which are obtained from the records of the times taken by the animal in successive trials, are facts which may be obtained by any observer who can tell time. They are absolute, and whatever can be deduced from them is sure. So also the question of whether an animal does or does not form a certain association requires for an answer no higher qualification in the observer than a pair of eyes. The literature of animal psychology shows so uniformly and often so sadly the influence of the personal equation that any method which can partially eliminate it deserves a trial.

Furthermore, although the associations formed are such as could not have been previously experienced or provided for by heredity, they are still not too remote from the animal's ordinary

8   *E. L. THORNDIKE.*

course of life. They mean simply the connection of a certain act
with a certain situation and resultant pleasure, and this general
type of association is found throughout the animal's life normally.
The muscular movements required are all such as might often be
required of the animal.   And yet it will be noted that the acts required are nearly enough like the acts of the anecdotes to enable
one to compare the results of experiment by this method with
the work of the anecdote school.   Finally, it may be noticed
that the method lends itself readily to experiments on imitation.

We may now start in with the description of the apparatus
and of the behavior of the animals.[1]

DESCRIPTION OF APPARATUS.

The shape and general apparatus of the boxes which were
used for the cats is shown by the accompaning drawing of box
K.   Unless special figures are given, it should be understood

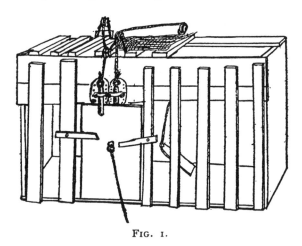

FIG. I.

that each box is approximately 20 inches long by 15 broad by
12 high.   Except where mention is made to the contrary, the
door was pulled open by a weight attached to a string which ran
over a pulley and was fastened to the door, just as soon as the

[1] The experiments now to be described were for the most part made in the
Psychological Laboratory of Columbia University during the year '97–'98, but a
few of them were made in connection with a general preliminary investigation
of animal psychology undertaken at Harvard University in the previous year.

animal loosened the bolt or bar which held it. Especial care was taken not to have the widest openings between the bars at all near the lever, or wire-loop, or what not, which governed the bolt on the door. For the animal instinctively attacks the large openings first, and if the mechanism which governs the opening of the door is situated near one of them the animal's task is rendered easier. You do not then get the association process so free from the helping hand of instinct as you do if you make the box without reference to the position of the mechanism to be set up within it. These various mechanisms are so simple that a verbal description will suffice in most cases. The facts which the reader should note are the nature of the movement which the cat had to make, the nature of the object at which the movement was directed, and the position of the object in the box. In some special cases attention will also be called to the force required. In general, however, that was very slight (20 to 100 grams if applied directly). The various boxes will be designated by capital letters.

A. A string attached to the bolt which held the door ran up over a pulley on the front edge of the box, and was tied to a wire loop (2½ inches diameter) hanging 6 inches above the floor in front center of box. Clawing or biting it, or rubbing against it even, if in a certain way, opened the door. We may call this box A '*O at front.*'

B. A string attached to the bolt ran up over a pulley on the front edge of the door, then across the box to another pulley screwed into the inside of the back of the box 1¼ inches below the top, and passing over it ended in a wire loop (3 inches in diameter) 6 inches above the floor in back center of box. Force applied to the loop or *to the string* as it ran across the top of the box between two bars would open the door. We may call B '*O at back.*'

B1. In B1 the string ran outside the box, coming down through a hole at the back, and was therefore inaccessible and invisible from within. Only by pulling the loop could the door be opened. B1 may be called '*O at back 2nd.*'

C. A door of the usual position and size (as in Fig. 1) was kept closed by a wooden button 3½ inches long, ⅞ inch wide,

½ inch thick. This turned on a nail driven into the box ½ inch above the middle of the top edge of the door. The door would fall inward as soon as the button was turned from its vertical to a horizontal position. A pull of 125 grams would do this if applied sideways at the lowest point of the button 2¼ inches below its pivot. The cats usually clawed the button round by downward pressure on its top edge, which was 1¼ inches above the nail. Then, of course, more force was necessary. C may be called '*Button*.'

D. The door was in the extreme right of the front. A string fastened to the bolt which held it ran up over a pulley on the top edge and back to the top edge of the back side of the box (3 inches in from the right side) and was there firmly fastened. The top of the box was of wire screening and arched over the string ¾ inch above it along its entire length. A slight pull on the string anywhere opened the door. This box was 20 × 16, but a space 7 × 16 was partitioned off at the left by a wire screen. D may be called '*String*.'

D1 was the same box as B, but had the string fastened firmly at the back instead of running over a pulley and ending in a wire loop. We may call it '*String 2nd*.'

E. A string ran from the bolt holding the door up over a pulley and down to the floor outside the box, where it was fastened 2 inches in front of the box and 1½ inches to the left of the door (looking from the inside). By poking a paw out between the bars and pulling this string inward the door would be opened. We may call E '*String outside*.'

In F the string was not fastened to the floor but ended in a loop 2½ inches in diameter which could be clawed down so as to open the door. Unless the pull was in just the right direction, the string was likely to catch on the pulley. This loop hung 3 inches above the floor, and 1¾ inches in front of the box. We may call F '*String outside unfastened*.'

G was a box 29 × 20½ × 22½, with a door 29 × 12 hinged on the left side of the box (looking from within), and kept closed by an ordinary thumb-latch placed 15 inches from the floor. The remainder of the front of the box was closed in by wooden bars. The door was a wooden frame covered with screening.

I. What is Behaviorism?

THE OLD AND NEW PSYCHOLOGY CONTRASTED

Two opposed points of view are still dominant in American psychological thinking—introspective or subjective psychology, and behaviorism or objective psychology.[1] Until the advent of behaviorism in 1912, introspective psychology completely dominated American university psychological life.

The conspicuous leaders of introspective psychology in the first decade of the twentieth century were E. B. Titchener of Cornell and William James of Harvard. The death of James in 1910 and the death of Titchener in 1927 left introspective psychology without emotional leadership. Although Titchener's psychology differed in many points from that of William James, their fundamental assumptions were the same. In the first place, both were of German

---

[1] In the last few decades there have been two other more or less prominent but temporary points of view—the so-called functional psychology of Dewey, Angell and Judd and the Gestalt Psychologie of Wertheimer, Koffka and Köhler. In my opinion both of these points of view are, as it were, illegitimate children of introspective psychology. Functional psychology, which one rarely hears of now, owed its vogue to considerable patter about the physiologically adaptive functions of the mind. The mind with them is a kind of adjusting "guardian angel." The philosophy behind it smacks very much of the good old Christian philosophy of Berkeley (interaction or control of the body by the deity).

Gestalt psychology makes its patter about "configurational response (really inborn!)." As a psychological theory it cannot gain very much headway. It is as obscure as Kant's treatment of imagination, which it resembles quite a little. The kernel of truth behind it has been very much better and more clearly expressed by William James in his *Principles* in the chapters on Sensation and Perception. Those chapters could be read with profit by the sponsors of Gestalt. Gestalt is still a part of introspective psychology. Incidentally a bit of collateral reading for any student who works on Gestalt is Hobhouse's *Mind in Evolution*.

6　　　　　BEHAVIORISM

dropped from his scientific vocabulary all subjective terms such as sensation, perception, image, desire, purpose, and even thinking and emotion as they were subjectively defined.

THE BEHAVIORIST'S PLATFORM

The behaviorist asks: Why don't we make what we can *observe* the real field of psychology? Let us limit ourselves to things that can be observed, and formulate laws concerning only those things. Now what can we observe? We can observe *behavior—what the organism does or says*. And let us point out at once: that *saying* is doing—that is, *behaving*. Speaking overtly or to ourselves (thinking) is just as objective a type of behavior as baseball.

The rule, or measuring rod, which the behaviorist puts in front of him always is: Can I describe this bit of behavior I see in terms of "stimulus and response"? By stimulus we mean any object in the general environment or any change in the tissues themselves due to the physiological condition of the animal, such as the change we get when we keep an animal from sex activity, when we keep it from feeding, when we keep it from building a nest. By response we mean anything the animal does—such as turning toward or away from a light, jumping at a sound, and more highly organized activities such as building a skyscraper, drawing plans, having babies, writing books, and the like.

SOME SPECIFIC PROBLEMS OF THE BEHAVIORISTS

You will find, then, the behaviorist working like any other scientist. His sole object is to gather facts about behavior—verify his data—subject them both to logic and to mathematics (the tools of every scientist). He brings the new-born individual *into his experimental nursery* and begins to set problems: What is the baby doing now? What is the stimulus that makes him behave this way? He finds that the stimulus of tickling the cheek brings the response of turning the mouth to the side stimulated. The stimulus

WHAT IS BEHAVIORISM? 7

of the nipple brings out the sucking response. The stimulus of a rod placed on the palm of the hand brings closure of the hand and the suspension of the whole body by that hand and arm if the rod is raised. Stimulating the infant with a rapidly moving shadow across the eye will not produce blinking until the individual is sixty-five days of age. Stimulating the infant with an apple or stick of candy or any other object will not call out attempts at reaching until the baby is around 120 days of age. Stimulating a properly brought up infant at any age with snakes, fish, darkness, burning paper, birds, cats, dogs, monkeys, will not bring out that type of response which we call "fear" (which to be objective we might call reaction "X") which is a catching of the breath, a stiffening of the whole body, a turning away of the body from the source of stimulation, a running or crawling away from it. (See page 152.)

On the other hand, there are just two things which will call out a fear response, namely, a loud sound, and loss of support.

Now the behaviorist finds from observing children brought up *outside of his nursery* that hundreds of these objects will call out fear responses. Consequently, the scientific question arises: If at birth only two stimuli will call out fear, how do all these other things ever finally come to call it out? Please note that the question is not a speculative one. It can be answered by experiments, and the experiments can be reproduced and the same findings can be had in every other laboratory if the original observation is sound. Convince yourself of this by making a simple test.

If you will take a snake, mouse or dog and show it to a baby who has never seen these objects or been frightened in other ways, he begins to manipulate it, poking at this, that or the other part. Do this for ten days until you are logically certain that the child will always go toward the dog and never run away from it (positive reaction) and that it does not call out a fear response at any time. In contrast to this, pick up a steel bar and strike upon it loudly behind the infant's head. Immediately the fear response is called forth. Now try this: At the instant you show him the animal and just as he begins to reach for it, strike the steel bar behind his head. Repeat the experiment three or four times. A new and important change is apparent. The animal now calls out the same response as the steel bar, namely a fear response. We call this, in behavioristic psychology, the *conditioned emotional response*—a form of *conditioned reflex.*

Our studies of conditioned reflexes make it easy for us to account for the child's fear of the dog on a thoroughly natural science basis without lugging in consciousness or any other so-called mental process. A dog comes toward the child rapidly, jumps upon him, pushes him down and at the same time barks loudly. Oftentimes one such combined stimulation is all that is necessary to make the baby run away from the dog the moment it comes within his range of vision.

There are many other types of conditioned emotional responses, such as those connected with *love,* where the mother by petting the child, rocking it, stimulating its sex organs in bathing, and the like, calls out the embrace, gurgling and crowing as an unlearned original response. Soon this response becomes conditioned. The mere sight of the mother calls out the same kind of response as actual bodily contacts. In *rage* we get a similar set of facts. The stimulus of holding the infant's moving members brings out the original unlearned response we call rage. Soon the mere sight of a nurse who handles a child badly throws the child into a fit. Thus we see how relatively simple our emotional responses are in the beginning and how terribly complicated home life soon makes them.

The behaviorist has his problems with the adult as well. What methods shall we use systematically to condition the adult? For example, to teach him business habits, scientific habits? Both manual habits (technique and skill) and laryngeal habits (habits of speech and thought) must be formed and tied together before the task of learning is complete. After these work habits are formed, what system of changing stimuli shall we surround him with in order to keep his level of efficiency high and constantly rising?

In addition to vocational habits, there comes the problem of his emotional life. How much of it is carried over from childhood? What part of it interferes with his present adjustment? How can we make him lose this part of it; that is, uncondition him where unconditioning is necessary, and condition him where conditioning is necessary? Indeed we know all too little about the amount and kind of emotional or, better, visceral habits (by this term we mean that our stomach, intestines, breathing, and circulation become conditioned—form habits) that should be formed. We do know that they are formed in large numbers and that they are important.

Probably more adults in this universe of ours suffer vicissitudes in family life and in business activities because of poor and insufficient visceral habits than through the lack of technique and skill in manual and verbal accomplishments. One of the large problems in big organizations today is that of personality adjustments. The young men and young women entering business organizations have plenty of skill to do their work but they fail because they do not know how to get along with other people.

DOES THIS BEHAVIORISTIC APPROACH LEAVE ANYTHING OUT OF PSYCHOLOGY?

After so brief a survey of the behavioristic approach to the problems of psychology, one is inclined to say: "Why, yes, it is worth while to study human behavior in this way, but the study of behavior is not the whole of psychology. It leaves out too much. Don't I have sensations, perceptions, conceptions? Do I not forget things and remember things, imagine things, have visual images and auditory images of things I once have seen and heard? Can I not see and hear things that I have never seen or heard in nature? Can I not be attentive or inattentive? Can I not will to do a thing or will not to do it, as the case may be? Do not certain things arouse pleas-

ure in me, and others displeasure? Behaviorism is trying to rob us
of everything we have believed in since earliest childhood."

Having been brought up on introspective psychology, as most of
us have, you naturally ask these questions and you will find it
hard to put away the old terminology and begin to formulate
your psychological life in terms of behaviorism. Behaviorism is new
wine and it will not go into old bottles. It is advisable for the time
being to allay your natural antagonism and accept the behavioristic
platform at least until you get more deeply into it. Later you will
find that you have progressed so far with behaviorism that the
questions you now raise will answer themselves in a perfectly satisfactory natural science way. Let me hasten to add that if the behaviorist were to ask you what you mean by the subjective terms
you have been in the habit of using he could soon make you
tongue-tied with contradictions. He could even convince you that
you do not know what you mean by them. You have been using
them uncritically as a part of your social and literary tradition.

TO UNDERSTAND BEHAVIORISM BEGIN TO OBSERVE PEOPLE

This is the fundamental starting point of behaviorism. You will
soon find that instead of self-observation being the easiest and most
natural way of studying psychology, it is an impossible one; you
can observe in yourselves only the most elementary forms of response. You will find, on the other hand, that when you begin to
study what your neighbor is doing, you will rapidly become pro-
ficient in giving a reason for his behavior and in setting situations
(presenting stimuli) that will make him behave in a predictable
manner.

DEFINITION OF BEHAVIORISM

Definitions are not as popular today as they once were. The definition of any one science, physics, for example, would necessarily
include the definition of all other sciences. And the same is true

of behaviorism. About all that we can do in the way of defining a science at the present time is to mark a ring around that part of the whole of natural science that we claim particularly as our own.

Behaviorism, as you have already grasped from our preliminary discussion, is, then, a natural science that takes the whole field of human adjustments as its own. Its closest scientific companion is physiology. Indeed you may wonder, as we proceed, whether behaviorism can be differentiated from that science. It is different from physiology only in the grouping of its problems, not in fundamentals or in central viewpoint. Physiology is particularly interested in the functioning of parts of the animal—for example, its digestive system, the circulatory system, the nervous system, the excretory systems, the mechanics of neural and muscular response. Behaviorism, on the other hand, while it is intensely interested in all of the functioning of these parts, is intrinsically interested in what the whole animal will do from morning to night and from night to morning.

The interest of the behaviorist in man's doings is more than the interest of the spectator—he wants to control man's reactions as physical scientists want to control and manipulate other natural phenomena. It is the business of behavioristic psychology to be able to predict and to control human activity. To do this it must gather scientific data by experimental methods. Only then can the trained behaviorist predict, given the stimulus, what reaction will take place; or, given the reaction, state what the situation or stimulus is that has caused the reaction.

Let us look for a moment more closely at the two terms—stimulus and response.

WHAT IS A STIMULUS?

If I suddenly flash a strong light in your eye, your pupil will contract rapidly. If I were suddenly to shut off all light in the room in which you are sitting, the pupil would begin to widen. If a pistol

shot were suddenly fired behind you you would jump and possibly turn your head around. If hydrogen sulphide were suddenly
released in your sitting room you would begin to hold your nose
and possibly even seek to leave the room. If I suddenly made the
room very warm, you would begin to unbutton your coat and
perspire. If I suddenly made it cold, another response would take
place.

Again, on the inside of us we have an equally large realm in
which stimuli can exert their effect. For example, just before dinner the muscles of your stomach begin to contract and expand
rhythmically because of the absence of food. As soon as food is
eaten those contractions cease. By swallowing a small balloon and
attaching it to a recording instrument we can easily register the
response of the stomach to lack of food and note the lack of response when food is present. In the male, at any rate, the pressure
of certain fluids (semen) may lead to sex activity. In the case of
the female possibly the presence of certain chemical bodies can
lead in a similar way to overt sex behavior. The muscles of our
arms and legs and trunk are not only subject to stimuli coming
from the blood; they are also stimulated by their own responses—
that is, the muscle is under constant tension; any increase in that
tension, as when a movement is made, gives rise to a stimulus which
leads to another response in that same muscle or in one in some
distant part of the body; any decrease in that tension, as when the
muscle is relaxed, similarly gives rise to a stimulus.

So we see that the organism is constantly assailed by stimuli—
which come through the eye, the ear, the nose and the mouth—
the so-called objects of our environment; at the same time the inside of our body is likewise assailed at every movement by stimuli
arising from changes in the tissues themselves. Don't get the idea,
please, that the inside of your body is different from or any more
mysterious than the outside of your body.

Through the process of evolution human beings have put on
sense organs—specialized areas where special types of stimuli are

WHAT IS BEHAVIORISM?   13

most effective—such as the eye, the ear, the nose, the tongue, the skin and semi-circular canals.[1] To these must be added the whole muscular system, both the striped muscles (for example, the large red muscles of arms, legs and trunks) and the unstriped muscles (those, for example, which make up the hollow tube-like structures of the stomach and intestines and blood vessels). The muscles are thus not only organs of response—they are sense organs as well. You will see as we proceed that the last two systems play a tremendous rôle in the behavior of the human being. Many of our most intimate and personal reactions are due to stimuli set up by tissue changes in our striped muscles and in our viscera.

HOW TRAINING ENLARGES THE RANGE OF STIMULI

One of the problems of behaviorism is what might be called the ever-increasing range of stimuli to which an individual responds. Indeed so marked is this that you might be tempted at first sight to doubt the formulation we gave above, namely, that response can be predicted. If you will watch the growth and development of behavior in the human being, you will find that while a great many stimuli will produce a response in the new-born, many other stimuli will not. At any rate they do not call out the same response they later call out. For example, you don't get very far by showing a new-born infant a crayon, a piece of paper, or the printed score of a Beethoven symphony. In other words, habit formation has to come in before certain stimuli can become effective. Later we shall take up the procedure by means of which we can get stimuli which do not ordinarily call out responses to call them out. The general term we use to describe this is "conditioning." Conditioned responses will be more fully gone into in chapter II.

It is conditioning from earliest childhood on that makes the problem of the behaviorist in predicting what a given response will

[1] In chapter III we shall see how sense organs are made up and what their general relation is to the rest of the body.

be so difficult. The sight of a horse does not ordinarily produce the fear response, and yet among almost every group of thirty to forty people there is one person who will walk a block to avoid coming near a horse. While the study of behaviorism will never enable its students to look at you and predict that such a state of affairs exists, nevertheless if the behaviorist sees that reaction taking place, it is very easy for him to state approximately what the situation was in the early experience of such a one that brought about this unusual type of adult response. In spite of the difficulty of predicting responses in detail we live in general upon the theory that we can predict what our neighbor will do. There is no other basis upon which we can live with our fellow men.

WHAT THE BEHAVIORIST MEANS BY RESPONSE

We have already brought out the fact that from birth to death the organism is being assailed by stimuli on the outside of the body and by stimuli arising in the body itself. Now the organism does something when it is assailed by stimuli. It responds. It moves. The response may be so slight that it can be observed only by the use of instruments. The response may confine itself merely to a change in respiration, or to an increase or decrease in blood pressure. It may call out merely a movement of the eye. The more commonly observed responses, however, are movements of the whole body, movements of the arm, leg, trunk, or combinations of all the moving parts.

Usually the response that the organism makes to a stimulus brings about an adjustment, though not always. By an adjustment we mean merely that the organism by moving so alters its physiological state that the stimulus no longer arouses reaction. This may sound a bit complicated, but examples will clear it up. If I am hungry, stomach contractions begin to drive me ceaselessly to and fro. If, in these restless seeking movements, I spy apples on a tree, I immediately climb the tree and pluck the apples and begin

WHAT IS BEHAVIORISM? 15

to eat. When surfeited, the stomach contractions cease. Although there are apples still hanging round about me, I no longer pluck and eat them. Again, the cold air stimulates me. I move around about until I am out of the wind. In the open I may even dig a hole. Having escaped the wind, it no longer stimulates me to further action. Under sex excitement the male may go to any length to capture a willing female. Once sex activity has been completed the restless seeking movements disappear. The female no longer stimulates the male to sex activity.

The behaviorist has often been criticized for this emphasis upon response. Some psychologists seem to have the notion that the behaviorist is interested only in the recording of minute muscular responses. Nothing could be further from the truth. Let me emphasize again that the behaviorist is primarily interested in the behavior of the whole man. From morning to night he watches him perform his daily round of duties. If it is brick-laying, he would like to measure the number of bricks he can lay under different conditions, how long he can go without dropping from fatigue, how long it takes him to learn his trade, whether we can improve his efficiency or get him to do the same amount of work in a less period of time. In other words, the response the behaviorist is interested in is the commonsense answer to the question "what is he doing and why is he doing it?" Surely with this as a general statement, no one can distort the behaviorist's platform to such an extent that it can be claimed that the behaviorist is merely a muscle physiologist.

The behaviorist claims that there is a response to every effective stimulus and that the response is immediate. By effective stimulus we mean that it must be strong enough to overcome the normal resistance to the passage of the sensory impulse from sense organs to muscles. Don't get confused at this point by what the psychologist and the psycho-analyst sometimes tell you. If you read their statements, you are likely to believe that the stimulus can be applied today and produce its effect maybe the next day, maybe

within the next few months, or years. The behaviorist doesn't believe in any such mythological conception. It is true that I can give
the verbal stimulus to you "Meet me at the Ritz tomorrow for
lunch at one o'clock." Your immediate response is "All right, I'll
be there." Now what happens after that? We will not cross this
difficult bridge now, but may I point out that we have in our verbal
habits a mechanism by means of which the stimulus is reapplied
from moment to moment until the final reaction occurs, namely
going to the Ritz at one o'clock the next day.

GENERAL CLASSIFICATION OF RESPONSE

The two commonsense classifications of response are "external"
and "internal"—or possibly the terms "overt" (explicit) and "implicit" are better. By external or overt responses we mean the ordinary doings of the human being: he stoops to pick up a tennis
ball, he writes a letter, he enters an automobile and starts driving,
he digs a hole in the ground, he sits down to write a lecture, or
dances, or flirts with a woman, or makes love to his wife. We do
not need instruments to make these observations. On the other
hand, responses may be wholly confined to the muscular and
glandular systems inside the body. A child or hungry adult may be
standing stock still in front of a window filled with pastry. Your
first exclamation may be "He isn't doing anything" or "He is just
looking at the pastry." An instrument would show that his salivary
glands are pouring out secretions, that his stomach is rhythmically
contracting and expanding, and that marked changes in blood
pressure are taking place—that the endocrine glands are pouring
substances into the blood. The internal or implicit responses are
difficult to observe, not because they are inherently different from
the external or overt responses, but merely because they are hidden
from the eye.

Another general classification is that of *learned* and *unlearned*
responses. I brought out the fact above that the range of stimuli

WHAT IS BEHAVIORISM?

to which we react is ever increasing. The behaviorist has found by his study that most of the things we see the adult doing are really learned. We used to think that a lot of them were instinctive, that is, "unlearned." But we are now almost at the point of throwing away the word "instinct." Still there are a lot of things we do that we do not have to learn—to perspire, to breathe, to have our heart beat, to have digestion take place, to have our eyes turn toward a source of light, to have our pupils contract, to show a fear response when a loud sound is given. Let us keep as our second classification then "learned responses," and make it include all of our complicated habits and all of our conditioned responses; and "unlearned" responses, and mean by that all of the things that we do in earliest infancy before the processes of conditioning and habit formation get the upper hand.

Another purely logical way to classify responses is to designate them by the sense organ which initiates them. We could thus have a *visual unlearned response*—for example, the turning of the eye of the youngster at birth toward a source of light. Contrast this with a *visual learned response,* the response, for example, to a printed score of music or a word. Again, we could have a *kinaesthetic* [1] *unlearned response* when the infant reacts by crying to a long-sustained twisted position of the arm. We could have a *kinaesthetic learned response* when we manipulate a delicate object in the dark or, for example, tread a tortuous maze. Again, we can have a *visceral unlearned response* as, for example, when stomach contractions due to the absence of food in the 3 day old infant will produce crying. Contrast this with the learned or visceral *conditioned* response where the sight of pastry in a baker's window will cause the mouth of the hungry schoolboy to water.

This discussion of stimulus and response shows what material we have to work with in behavioristic psychology and why be-

---

[1] By kinaesthetic we mean the muscle sense. Our muscles are supplied with sensory nerve endings. When we move the muscles these sensory nerve endings are stimulated. Thus, the stimulus to the kinaesthetic or muscle sense is a *movement of the muscle itself.*

havioristic psychology has as its goal *to be able, given the stimulus, to predict the response—or, seeing the reaction take place to state what the stimulus is that has called out the reaction.*

IS BEHAVIORISM MERELY A METHODOLOGICAL APPROACH TO THE STUDY OF PSYCHOLOGICAL PROBLEMS, OR IS IT AN ACTUAL SYSTEM OF PSYCHOLOGY?

If psychology can do without the terms "mind" and "consciousness," indeed if it can find no objective evidence for their existence, what is going to become of philosophy and the so-called social sciences which today are built around the concept of mind and consciousness? Almost every day the behaviorist is asked this question, sometimes in a friendly inquiring way, and sometimes not so kindly. While behaviorism was fighting for its existence it was afraid to answer this question. Its contentions were too new; its field too unworked for it to allow itself even to think that some day it might be able to stand up and to tell philosophy and the social sciences that they, too, must scrutinize anew their own premises. Hence the behaviorist's one answer when approached in this way was to say, "I can't let myself worry about such questions now. Behaviorism is at present a satisfactory way of going at the solution of psychological problems—it is really a methodological approach to psychological problems." Today behaviorism is strongly entrenched. It finds its way of going at the study of psychological problems and its formulation of its results growing more and more adequate.

It may never make a pretense of being a *system*. Indeed systems in every scientific field are out of date. We collect our facts from observation. Now and then we select a group of facts and draw certain general conclusions about them. In a few years as new experimental data are gathered by better methods, even these tentative general conclusions have to be modified. Every scientific field, zoölogy, physiology, chemistry and physics, is more or less in a

WHAT IS BEHAVIORISM?

state of flux. Experimental technique, the accumulation of facts by that technique, occasional tentative consolidation of these facts into a theory or an hypothesis describe our procedure in science. Judged upon this basis, behaviorism is a true natural science.

2 The Emerging Science

Defining the Goals, Approaches, and Methods

Lee C. Drickamer and Charles T. Snowdon

Emergence of a New Field of Study

Background material and selections in Part 1 set the stage for the emergence of animal behavior in North America and ethology in Europe as distinct fields of scientific investigation. Around the world, the terms "animal behavior," "comparative animal psychology," and "ethology" are today generally viewed as encompassing the same field of study, but equating these terms is a relatively recent development. We use the term "animal behavior" in this introductory essay. The gulf between apparently separate schools of thought in Europe and North America may not have been as large as we often assume. In Europe, the science of behavior developed from a naturalist tradition, influenced by physiology. The roots of scientific studies of animal behavior in North America involved an interplay between developments in comparative psychology and zoology. The latter was grounded in ecology and also developed from a naturalist tradition. Thus, the various pathways for development of the scientific study of behavior involved concern for the natural lives of the subject species and included both field and laboratory components in the research programs. As several authors, notably Barlow (1989), Dewsbury (1984, 1989), Klopfer (1974), and Thorpe (1979), have outlined, a variety of threads were woven into the fabric of animal behavior from both sides of the Atlantic.

In preparing the list of selections for this volume, we concentrated on materials that were judged most influential in shaping current investigations of behavior. It was not our intention to slight the early twentieth-century developments by excluding papers from this period. Many of the papers from the first half of this century provided in-depth data for a variety of species. Results from these early studies provided baseline information essential for the initiation of synthetic and theoretical work in the mid-1930s. Such pioneering empirical studies included, for example, field investigations of insects (Schneirla 1933; Tinbergen 1935; Wheeler 1923) amphibians (Yerkes 1903), birds (Heinroth 1911; Whitman 1919) and, primates (Carpenter 1934), as well as laboratory work on physiological and behavioral processes (Pavlov 1927; Sherrington 1906; von Holst 1939) and learning (Thorndike 1911; Tolman 1932). Several papers from this period provided contributions that influenced critical thinking in animal behavior for a number of decades, though their influence has waned in more recent years as the concepts have been refined. Among these were Craig's (1918 [see Part 3]) paper on appetitive and consummatory acts as parts of instinctive behavior, proposals concerning approach and withdrawal in relation to stimuli during development (Schneirla 1939), and ideas concerning tropisms and taxes (Loeb 1918). Some of these authors and ideas are touched upon in subsequent sections of this book as we examine the historical foundations for major types of questions asked by ethologists.

The pathways of development for animal behavior in North America and for ethology in Europe ran parallel to one another for much of the first half of this century. There were some contacts, e.g., visits by Niko Tinbergen and others to laboratories in the United States, and cross-citations of a few research publications, but significant integration of these various schools of thought did not occur until the 1950s. Among the papers most responsible for initiating the dialogue was one of those we have included in this volume, Daniel Lehrman's provocative critique of the theory of instinctive be-

havior, published in 1953. This paper reviewed Lorenz's theory of instinct and then examined, in detail, various aspects of that theory in light of existing knowledge concerning the nervous system and development. Lehrman's main premise was that any theory that postulated rigid connections between genetics, specific neural structures, and performance of particular behavior patterns was, in fact, counterproductive with respect to stimulating scientific investigation of behavior development. Lehrman's paper and lively responses from Lorenz were focal points of discussion at international psychological and ethological conferences in the ensuing decade.

The Lehrman paper and others that followed (e.g., Beach 1955; Hinde 1956, 1959) opened the doors to cross-fertilization, extensive visitations between laboratories on both sides of the Atlantic, training of students and postdoctoral fellows in foreign laboratories, and joint research ventures (Dewsbury 1995). The Hinde book of 1956 stands out as a first real attempt to synthesize ethology and comparative psychology, as indicated in his own subtitle. Seen in retrospect, in the perspective of today's global interactions, these events of the 1950s seem very distant. In reconstructing a history of this period with regard to animal behavior, we have had the opportunity to converse with and obtain written views from most of the principals (see Dewsbury 1985). This provides a unique view of scientific interactions as they contributed to the emergence of today's interdisciplinary field of animal behavior.

At least two major developments in the 1930s had great impact on the nascent scientific study of animal behavior. The first of these, an ongoing process, involved individuals who were primarily ecologists, but who became interested in and began to consider behavioral aspects of their various research problems. Examples include W. C. Allee's (1937) studies on life history and behavior of sparrows and an important volume on social behavior in animals (Allee 1938), A. E. Emerson's (1938) report on the evolutionary history of termite nests, and T. C. Schneirla's (1933) comparative investigations on army ants in Panama. These and similar studies played key roles in initiating an ecological perspective on the behavior of animals, which led eventually to the subfield we now call behavioral ecology.

The second event was the appearance of R. A. Fisher's book on genetics and natural selection (1930), still an important treatise for those studying animal behavior. Fisher's work combined knowledge of the theory of evolution by natural selection and an understanding of the Mendelian inheritance with knowledge about population genetics. His theories have served as the starting point for countless studies in animal behavior, receiving a resurgence of attention with the current wave of investigations in behavioral ecology and sociobiology that began in the 1970s (see Part 6).

Establishing Frameworks for the Study of Behavior

During the middle third of the twentieth century, animal behavior began to be identified as a separate discipline, and several investigators offered general frameworks for categorizing and organizing the types of questions to be studied. One such attempt was that by Hess (1962) in which he reviewed the history of ethology as it developed, primarily in Europe, with the concepts of fixed actions patterns (FAPs) and innate releasing mechanisms (IRMs). In particular, and following traditional European views, Hess pointed to the hierarchical organization of behavior, by which he meant that actions were initiated in the form of appetitive behavior that eventually resulted in some form of consummatory act(s). He categorized three approaches to the study of behavior: (a) phylogenetics via use of the comparative method, (b) studies of genetics underlying behavior patterns, and (c) research on neurophysiological mechanisms producing the behavior.

Beach (1960) offered a slightly different scheme, beginning with what he called "species specific" behavior, a departure, in some measure, from American comparative psychology and probably a concession to European views. Species-specific patterns were those that constituted the normal behavioral repertoire, that is, patterns that could be predictably seen in animals of the same species, sex, and age. This general definition could be applied to all types of organisms from invertebrates to mammals. Beach defined the kinds of questions to be asked and tested as determinants of behavior by creating three categories: (a) historical determinants, (b) environmental determinants, and (c) organismic determinants. Historical determinants referred to the development of species-specific patterns and the significance of maturational and experiential processes. Environmental determinants may be direct, as stimuli eliciting particular behavior patterns, or indirect, wherein some cues are viewed as necessary, but not sufficient, for eliciting behavior. Organismic determinants are those that occur inside the individual animal and include both the neural aspects of stimulus and response and the hormonal component, most often studied with respect to sexual and aggressive actions.

A brief example using bird song should help to illustrate these determinants. Young birds of many species, e.g., chaffinches, hear the songs of adults while they are still in the nest and as fledglings, and only later, as reproductively active adults, do they sing the species-specific song. This illustrates historical determinants in Beach's scheme. An environmental determinant in this example would be that a male chaffinch in reproductive condition is stimulated to produce songs by hearing other males in similar condition singing at the start of the breeding season. Organismic determinants would involve the underlying neural processes within the brain and nervous system of the chaffinch that processed incoming stimuli and elicited the singing response, as well as longer-term hormonal changes that prepared the bird for the breeding season and affected its sensitivity to external stimuli such as the songs of conspecifics.

In these two approaches, by Hess and by Beach, it is noteworthy that neither directly addressed questions dealing with the functional significance of the behavior patterns. Beach's scheme appears not to mention in any concrete way the genetic background for what transpires, though a genetic component may be implied in his concept of historical determinants. Hess's scheme, on the other hand, seems to ignore the interactive effects of genetics, maturation, and experience during the course of development. These potential omissions reflect both the contextual period when the papers were written and the differing viewpoints and backgrounds of the authors. There are some similarities, particularly regarding effects of the nervous system in producing observed behavior.

One paper from this era that we have selected stands out in terms of the degree to which it has influenced several genera-

tions of ethologists throughout the world by providing a framework for studying animal behavior. Niko Tinbergen (1963) summarized his views in a paper entitled "On aims and methods of ethology." He organized animal behavior around four questions: (a) causation, by which he meant the underlying physiological explanations, involving, in particular, the nervous and endocrine systems; (b) ontogeny or development, used to include studies of the processes of interaction of the genetic program of the animal with its environment and experiences; (c) survival value or functional significance, built upon the notion that behavior, like morphology and physiology, has been shaped by natural selection; and (d) evolution, incorporating the comparative study of behavior patterns, the phylogeny of similar patterns in related species of animals.

Tinbergen's synthesis of a framework for behavior questions is, perhaps, the most encompassing of those that were put forward during this period. His framework has served as the basis for textbooks in ethology and as the guiding principle for structuring the plenary sessions at several of the international ethology conferences. The first two categories of Tinbergen's scheme are often called proximate explanations of behavior, while the latter two categories are called ultimate explanations. Tinbergen's scheme contains all of the elements of the earlier works by Hess and Beach, though the components are somewhat rearranged and, most important, his scheme goes beyond the earlier versions to include questions that pertain to the evolutionary bases for and consequences of behavior. As a quick examination of the contents for this edited volume will indicate, we have elected to follow Tinbergen's scheme in our arrangement of critical papers in animal behavior. Recently, Dewsbury (1992) has discussed the Tinbergen framework and what has happened to the meanings of the four areas of study in the past several decades. Dewsbury highlights the possibility that an emphasis on how versus why questions (proximate versus ultimate questions) may be leading us a bit astray. He proposes an alternative scheme involving examination of the genesis (genetics, development, evolution), control (hormones, nervous system), and consequences (function, significance) of behavior.

Again, an example, this time involving sexual behavior of rodents, should aid in interpreting the overall framework proposed by Tinbergen. Neural and hormonal processes inside male and female rats mounting, exhibiting lordosis, and copulating constitute the causation. Effects of early experience, including the influence of intrauterine position and prior exposure to members of the opposite sex on successful copulation, constitute developmental questions. The genetic bases for copulation and the manner in which natural selection has influenced the behavior sequence to ensure successful fertilization and progeny production would be evolutionary questions. Finally, the consequences of reproduction in terms of reproductive success and related life-history traits involve functional issues.

As the science of animal behavior grew and became more complex, there were separate theoretical and experimental approaches that became well defined for attacking specific types of problems. These extend, but do not contrast with Tinbergen's. Often, these approaches involved development of new terminology and imaginative experimental test situations; two examples are imprinting and foraging behavior. Early observations by Spalding (1873), Craig (1908), Heinroth (1911), and others (see Part 1) indicated that certain responses of birds in social

situations were dependent upon early rearing conditions. It was necessary to rear birds with members of their own species or they might show attachment only to a foster species or to human caretakers, making breeding difficult or impossible. Lorenz (1935) provided a general framework for studies of imprinting, which has been expanded upon by Hess (1959), Sluckin (1965), and others. Key concepts like critical periods and an understanding of the integration of visual and auditory stimuli affecting behavior have arisen through the extensive studies of imprinting (Gottlieb 1971). Similar phenomena have been reported for insects (Thorpe 1944), fish (Baerends and Baerends-van Roon 1950), sheep (Grabowski 1941), and many other species, as well as for other behavior patterns such as feeding (Burghardt 1967).

A second example is the study of foraging, or how animals find sufficient food to survive and reproduce. Early studies of foraging involved both field work and laboratory studies. Field work focused on what animals consumed and the roles of time and energy in handling and consuming foods (Emlen 1966; Reichmann 1977; Altmann 1991). Laboratory investigators considered various paradigms that involved food as rewards (Herrnstein 1970; Richter 1943). In the past two decades, two major influences have affected our thinking and experimental frameworks regarding foraging behavior and, in fact, many other areas of animal behavior as well: (1) game theory, from which ideas like the evolutionary stable strategy (ESS) are derived (Maynard Smith 1982), and (2) economic metaphors involving terms like cost, benefit, and currency (usually considered to be energy or time in behavioral systems) (Rapport and Turner 1977), and ideas like the marginal value theorem (Schmid-Hempel et al. 1985) and central place foraging (Orians and Pearson 1979). Ideas from game theory and economic theory have proven to be useful in animal behavior because these theories provide a useful and potentially realistic basis for testing hypotheses about foraging behavior, aggressive encounters, finding mates, and other situations involving behavioral choices. Those studying foraging behavior have developed their own concepts and terminology, such as optimal foraging strategy, search image, and risk-sensitive foraging (Schoener 1986; Staddon 1983; Stephens and Krebs 1986). The evolution of thought processes and development of theory used to describe and analyze foraging behavior has benefited other areas of animal behavior, such as investigations of mate choice and habitat selection.

Lastly, an examination of the early general monographs in animal behavior provides a third opportunity to explore the kinds of approaches and frameworks that were promulgated and practiced in this fledgling discipline. Many of these served as textbooks and thus directed the education of future scientists in the field. The earliest books, including *Principles of Animal Psychology* by Maier and Schneirla (1935), *Organization of Behavior* by Hebb (1949), *Learning and Instinct in Animals* by Thorpe (1956), and *The Study of Instinct* by Tinbergen (1951) reflect a range of viewpoints. The first three exhibit an emphasis on learning and, to a lesser extent, development (what we now call proximate mechanisms). Tinbergen's book re-

flects a stronger emphasis on phylogeny and evolutionary functions of behavior. The books by Maier and Schneirla and by Thorpe have a taxonomic arrangement wherein learning and related phenomena are considered first in various groups of invertebrates, then in nonhuman vertebrates, and finally in humans. Subsequent texts, such as *Mechanisms of Animal Behavior* by Marler and Hamilton (1966), *Animal Behavior: A Synthesis of Ethology and Comparative Psychology* by Hinde (1966), *An Introduction to Animal Behavior* by Klopfer and Hailman (1967), and *Ethology, The Biology of Behavior* by Eibl-Eibesfeldt (1970), all reflect the changes brought about by twenty years of additional investigation of animal behavior. Chapters in these four books cover specific topics such as orientation, social behavior, and development, rather than being arranged taxonomically. Material in each chapter summarizes the state of knowledge on that topic, using a broad range of examples from the research literature, including invertebrates and vertebrates. Thus, as the corpus of knowledge broadened, general principles were developed that integrate studies of many taxonomic groups. Today's textbooks continue to reflect to reflect a topical orientation framework. Since the 1970s, there has been a particular emphasis on functional and evolutionary aspects of behavior, a solid counterpoint to the earlier emphasis on proximate mechanisms.

Describing and Quantifying Behavior

As the field of ethology developed, two major related issues came into focus. The first of these was the need to adequately and accurately describe and define behavior. Obtaining agreement on exactly what an animal is doing in a particular

situation is not as easy as it may seem at first glance. Watching male sage grouse (*Centrocerus urophasianus*) performing on a lek while females move through the lek to select mates is a good example. Describing the actions of the male and female grouse in words is difficult enough, but we also need to know that the units that we have selected for behavior observation and description are important with respect to what the birds are actually doing. Practice is required to develop a common language for describing behavior and to make decisions about how large or small the units to be recorded actually are.

The second issue, arising from our descriptive knowledge of behavior, was the realization that animal behavior is a probabilistic phenomenon. That is, there is variation in terms of the performance for virtually any behavior that we want to measure. Examples range from the rate at which insects are brought back to feed fledglings to the likelihood of prey capture in social canid groups of varying sizes. The issue is the extent to which this variation has significant biological meaning. How, for example, do differences in the rate of fledgling feeding relate to the costs and benefits of parental care? Such variation between and among conspecifics can serve as the basis for evolution by natural selection. Because of its probabilistic nature, the study of behavior requires careful attention to the methods of experimental design, sampling, data collection, and statistical analysis.

Galef (Part 1) has provided insights with regard to the manner in which early scientists treated the issue of describing animal actions. Initially, there was a common tendency to use terminology that ascribed human emotions and motivations to the actions and behavior of the subject species. This was followed, during the last years of the nineteenth century and much of the twentieth century, by an almost complete rejection by many investigators of anything that smacked of anthropomorphism. The goal was to attain complete objectivity in describing and interpreting animal behavior. In the last decade we have, perhaps through renewed interest in animal cognition and related phenomena, returned to a measured use of anthropomorphic terminology. Recently, work in the area of cognitive ethology (Griffin 1984; Ristau 1990) has begun to once again explore animal emotions and motivations using methods for the study of behavior that have been developed in the past twenty-five years.

A variety of authors have tackled the thorny issue of describing behavior. As Tinbergen (1963) and Hinde (1970) point out, there are two different levels at which a behavioral action can be described: the first involves a listing of the motor patterns without reference to the consequences of the action, and the second deals with the consequences but not the actual motor patterns. Similar motor patterns, e.g., lowering of the head and antlers by an elk (*Cervus elaphus*), can occur either in the course of male-male combat or when there is a predator threat. Also, a praying mantis (*Mantis religiosa*) seeking food may use a variety of motor patterns, all of which would, from a functional perspective, be labeled as part of the search component of foraging behavior. In practice we usually define behavior patterns by using one of three schemes involving (a) causational linkages between events, (b) functional or adaptive outcomes, and (c) the pathway by which the behavior was acquired, i.e., via evolution or development (Hinde 1970).

Drummond (1981) in his essay on "The nature and description of behavior patterns" covers describing, defining, and naming behavior patterns. Description is a report of specific properties that make up a particular behavior pattern. Defining the behavior involves listing those properties that are needed to identify that the pattern has occurred, including distinguishing it properly from other behavior patterns. Naming involves providing a convenient label for the observed pattern. He provides five domains for the description of behavior: "the location of the animal in relation to some component(s) of its environment; its orientation (the disposition of its structures) with respect to some component(s); the physical topography of the animal; intrinsic properties of its integument; and physical effects induced in the environment" (Drummond 1981, p. 28). He also points quite properly to the role(s) that we play as humans imposing our potentially arbitrary schemes of classification on activities of other animals. There may also be problems that arise because we use several different conceptual schemes to categorize behavior (Purton 1978). Purton notes that the separate classification schemes of form, function, and causation can, unwittingly, be run together, leading to conceptual confusion accompanied by problems of analysis and interpretation. These necessarily subjective decisions regarding which units are most appropriate for recording and analyzing behavior and which schemes offer the most advantages will remain a concern for animal behaviorists well into the future. To aid ethologists, at least two volumes now provide extensive definitions and commentary on a comprehensive range of behavior patterns (McFarland 1981, Immelmann and Beer 1989).

One common theme that occurs in discussions concerning description and analysis of animal actions is that of the most appropriate unit of behavior. Though not universal, one widely used unit has been the fixed action pattern (FAP). As originally conceived by Lorenz (1950) and others, the FAP was a behavior sequence with the following properties: it occurred in an all-or-none fashion, exhibited stationarity in time, independent of environmental control

(once triggered), and performed in exactly the same manner each time by members of a particular species. In a selection reprinted here, Barlow (1968) provides a comprehensive discussion of the FAP and carefully examines whether observed behavior patterns fit the criteria originally established for FAPs. He concludes that a better concept might be the modal action pattern (MAP), this term more properly reflecting the variability among animals when performing a particular action sequence. The MAP can be used as a satisfactory unit of behavior and has been adopted by other authors (e.g., Drummond 1981; Schleidt 1974).

Once a satisfactory solution has been found for describing behavior, the next issue that confronts us in many instances is analysis of sequences of actions. Such sequences may occur within a organism or they may involve interactions of two or more organisms. A selection by Slater (1973), reprinted here, "Describing sequences of behavior," provides a comprehensive review of this subject with regard to both conceptual and practical matters. We chose this selection for several related reasons. It is a good general summary of this issue of analyzing behavioral sequences. It provides solid assessments of the assumptions underlying various methods of sequence analysis and the pitfalls to be encountered. Lastly, it is written by an ethologist using a variety of behavior examples. A key feature of the essay is that Slater stresses causational analyses, clearly noting the difference between these analyses and simple correlational predictions. It is important to emphasize that determining whether two events are correlated (e.g., that lengthening photoperiod in temperate climates results in many invertebrate and vertebrate organisms attaining reproductive condition) is not the same as actually conducting experiments to determine the series of causational links between photoperiod (or something else that is correlated with photoperiod) and the internal hormonal and tissue changes in the organisms leading to successful reproduction. Slater provides a historical perspective for sequence analyses (beginning with the use of flow diagrams), deals with several key types of sequence analyses (Cox and Lewis 1966), and places a heavy emphasis on Markov chain analyses (Cane 1961, Nelson 1964). The behavioral examples are broad, encompassing sequences involved in development of behavior (Fentress 1972), maintenance behavior (Delius 1969), and bird song (Lemon and Chatfield 1971).

In addition to considering definitions and sequences, any attempt to analyze animal activities must address several practical considerations. These include the need for appropriate sampling methods and whether to conduct the study in the field or in the laboratory. Animal behaviorists have developed or adapted a number of sampling methods for use in field and laboratory settings. Such schemes take into account the species being studied and any peculiarities of its behavior, the nature of the question being tested, and the type of statistical treatment that is planned for analyzing the data. Some test questions may involve manipulation of particular independent variables, whereas other tests do not involve any manipulation. Thus, for example, if we are interested in the effect of the number of perching sites for displays of territorial behavior by northern mockingbirds (*Mimus polyglottus*), a non-manipulative study could involve observations on numbers and frequency of use of display sites in connection with measures of territory size and quality. A manipulative test of the same question concerning effects of the number of perch sites could involve either artificially adding or removing potential perch sites or altering the food supply.

Some of these methods issues had been addressed by particular investigators for a number of decades, but no thorough summaries of behavioral sampling techniques used by ethologists occurred until the 1970s (Hinde 1973; Dunbar 1976; Lehner 1979; Martin and Bateson 1986). To exemplify these papers, we have chosen an article by Altmann (1974), "Observational study of behavior: sampling methods." Altmann's pivotal paper provides the most concise and thorough treatment, and is accompanied by numerous references to the use of the sampling methods in the behavior literature. She covers the basic issues involved in selecting an appropriate sampling scheme and scheduling the observations in a manner that fits the question being tested. Historically, the most widely used form of sampling was ad libitum observation. This technique is well suited for obtaining a general picture of what is transpiring in a particular social group or at a particular location, but it has limitations in terms of recording all of what is transpiring and it may be subject to biases toward larger or more readily identified individuals (Drickamer 1974, 1975). A number of techniques are presented, each with features that make it more suitable than ad libitum sampling for particular situations. These include focal animal sampling, all occurrence sampling, sequence sampling, one-zero sampling, and instantaneous or scan sampling.

In addition to questions about sampling method, there are other design considerations that must be addressed prior to conducting a research project. One consideration is whether to observe the same cohorts of subject animals throughout, i.e., a longitudinal design, or whether to watch different sets of subjects at each of a number of specified time periods during the course of the study, i.e., a cross-sectional design (King 1969). Another set of considerations involves what conditions to manipulate (independent variables), for what time periods and at what intensity to manipulate them, and then what variables to measure to assess the effects of the manipulations (dependent variables) and when to measure them (King 1958; Lehner 1979). Also, associated with each experimental design are important considerations of statistical analysis, such as sample size, independence of data points, and pseudoreplication (Kroodsma 1990).

For many questions tested by ethologists, the issue arises of whether to conduct investigations in the field or laboratory. Field settings with subject animals living free and natural lives generally provide the best opportunities for examining functional and evolutionary questions. Laboratory settings provide a degree of control over both the subjects and their surroundings not possible in the field. Controlled conditions may be required to answer some kinds of questions, such as those involving the need to breed particular animals or manipulate aspects of early experience. Such conditions may also be necessary to perform particular manipulations, such as working with implanted electrodes or regular transfusions of blood between organisms. We hasten to note, however, that there are many studies done under field conditions to examine mechanism questions, and numerous laboratory studies have shed light on questions of functional significance. Miller (1977) provided an

excellent summary of the reasoning behind the need for combining naturalistic observations in the field with both field and laboratory testing of particular questions. Thus, there is a trade-off between high internal validity, more characteristic of laboratory research, and high external validity, an ability to generalize the findings, more characteristic of field investigations of behavior (Altmann 1974).

In a selection reprinted here, Rowell (1967) examined the social interactions of captive and wild groups of baboons (*Papio anubis*). We have selected this paper because it exemplifies the need for investigators to be aware of and take into account possible effects of captivity, particularly when they are investigating social behavior. Rowell's contention at the outset was that, for any primate species (and presumably, in the general context, other animal species as well), there is a range of variation recorded for various behavioral phenomena from natural populations and that the caged environment might represent an extension or extreme of that variation. Much higher rates of interaction were recorded for a caged group compared to the free-ranging baboons and, while a linear dominance hierarchy was evident in the captive animals, wild baboons did not exhibit a detectable hierarchy. Similar studies by Bernstein (1967, 1970) and Drickamer (1973) reported on significant variations in rates of behavior patterns, e.g., grooming, aggression, in enclosure groups of pigtailed macaques (*Macaca nemistrina*) and rhesus macaques (*Macaca mulatta*) compared to free-ranging animals.

The Comparative Method

The comparative method has been a tool for the study of ethology for some time. Darwin used it extensively in his book *Expression of the Emotions in Man and Ani-*

mals (1872). European ethologists (e.g., Heinroth 1911; Lorenz 1950; Manning 1971) and animal psychologists and zoologists working in North America (e.g.,

Washburn 1908; Bitterman 1965; Dewsbury 1984; Burghardt 1985) have used it and continue to do so. There appear to be two primary foci for the comparative method. In some studies, careful comparisons are made between the behavior patterns of particular groups of species to ascertain the degrees of similarity and difference. In this approach, a major goal is to study the phylogeny of particular behavior patterns. In other studies, the primary method of comparison is across ecological habitats (including the sorts of challenges and constraints faced by organisms living in such habitats) in order to examine the similarities and differences between the solutions to those constraints that have evolved in various species. Using this approach may involve comparisons among widely divergent species, e.g., vertebrates such as fish, sea lions, and dolphins, that spend all or a portion of their lives in an aquatic habitat. Examination of intraspecific differences could be carried out under either of these foci, to study variation in the performance of a particular pattern of behavior as an indication of evolutionary processes in action, or to test hypotheses about variation in behavioral performance in relation to ecological conditions.

A second use of the comparative approach is more characteristic of some areas of American psychology. This involves making comparisons about similarities that occur across widely different species and including the use of nonhuman animal models for human behavior. This sort of thinking, taken to an extreme, for example by Bitterman (1960, 1965), and also sometimes characteristic of the early books in animal behavior that adopted a taxonomic approach, has been properly criticized by people like Hodos and Campbell (1969). These latter authors decry the type of thinking that re-

sults in a *scala naturae* approach with a hierarchy of organisms proceeding from invertebrates to vertebrates to higher vertebrates, including humans. Most animal behaviorists would now agree that it is not proper to provide the same sorts of tests to a wide range of organisms; the perceptual worlds and abilities of the organisms vary too much, and we may be "asking" some organisms to perform irrelevant or impossible tasks to demonstrate their behavioral capacities (see also Snowdon, Part 3).

In one of the selections reprinted here, Baerends (1958) examines the way in which the comparative method has been used to explore homologies in the behavior of animals. He deals with the problem of selecting comparable units of behavior for making contrasts. A major theme is to compare the criteria used by anatomists in conducting studies of morphological homology with the sorts of criteria that can be established by ethologists for making their behavioral comparisons. Both morphology and behavior can rely on similarity of form and on ontogeny as bases for cross-species assessments of similarities and differences. Baerends notes that behavior patterns are often classified by ethologists by particular category labels, e.g., redirected movements or displacement activities. Such behavior patterns usually involve a number of complex elements, and studying the causal factors that have influenced the evolution of such patterns may prove to be a better avenue of investigation than simply placing these elements into broad categories. That is, because the forces that have shaped behavior may vary widely between species, we can gain new insights into the reasons why we see animals performing certain actions in particular situations through a comparative approach. In his several claims, Baerends can be credited with fostering one of the several key directions, use of the comparative method, that animal behavior has traveled in the several decades since he wrote this paper. Studies of behavior development, neural and hormonal components of behavior, and some of behavioral ecology all have incorporated Baerends's thinking to varying degrees.

Over the course of the last thirty-five years, as use of the comparative method has intensified, two types of problems have arisen. One type of question concerns how broad the spectrum of animals examined needs to be in order to make a study truly comparative. A second type of question deals with the applications of the comparative method, assumptions inherent in its use and design, and statistical issues that accompany attempts to make cross-species comparisons.

Frank Beach (1950) outlined what he perceived to be a serious problem with the failure of animal psychologists of the time to work with a sufficiently varied number of species. His data supported the contention that too much attention was being focused on the laboratory rat (*Rattus norvegicus*) and that, as a result, the generality of the conclusions that could be drawn was becoming more and more limited. At least two other often-cited papers have dealt with this issue of using cross-species contrasts, Hodos and Campbell (1969) and Lockard (1971). We have selected the Hodos and Campbell paper for reprinting here because of the excellent ties that it makes to all of ethology and because it considers, as its underlying theme, the ways in which evolution by natural selection can and cannot function to affect behavior. Thus, Hodos and Campbell have, as their major premise, a need for maintaining vigilance about levels of comparison with regard to whether animal species being compared are from closely related or divergent lineages. A key issue, as they note, is that the number of species needed for a comparative study depends on the question being asked and on the taxonomic distribution of the species being studied. Hodos and Campbell (1969) also anticipated some of the methods issues that have arisen in the past two decades (Gittleman 1989). As more and more data have become available and the machinery is in place to attempt relatively large comparisons through the use of sophisticated statistical procedures, there is great need for care about certain issues. Among the considerations that must be addressed are (Gittleman 1989): (1) Databases employed for cross-species study must be sufficient and appropriate. Sufficient data means that there are no large gaps, necessitating assumptions or analyses with missing data. Appropriate data are those that have been collected by methods that are sufficiently similar to permit valid contrasts. Obtaining data samples that meet this second criterion is difficult because investigators differ in their methods of data collection and in the manner in which they interpret and manipulate the information they have gathered. (2) When a variety of taxonomic subdivisions are to be used for comparisons, care must be taken to conduct analyses at the appropriate taxonomic level. In addition, we should not use multiple data points from one taxon and at the same time only a single datum or a few points from another taxon. The very phylogenetic effects that are being tested by comparisons may be unduly enhanced or hidden by our choices at this stage, leading to weak or incorrect conclusions. (3) Statistical techniques must be examined carefully to avoid problems with underlying assumptions and to ensure that quantitative variables tested have been extracted in accordance with established principles and common sense. An obvious pitfall involves violations of assumptions about the independence of individual data points. (4) Establishing causal links through such analyses can be quite tricky, though with proper techniques for eliminating effects attributable to some variables, or holding some variables constant, some causal sequences can be tested. Allometric effects are an excellent example of a potentially confounding factor when using the comparative method to infer causal links (Gittleman 1989).

The continuing importance and maturation of comparative methods are illustrated nicely by the recent appearance of at least two books and several key articles devoted to this topic (see also Part 6). Ridley (1983) examined use of comparative methods with particular regard to mating behavior in a diversity of animal groups. Harvey and Pagel (1991) provided a synopsis of the use of comparative techniques to study ethology with particular emphasis on the problem of nonindependence of data points. They

also extend the potential utility of comparative methods through careful explanation of some new and potentially important statistical techniques. Martins and Garland (1991) and Garland et al. (1992) have provided additional insights regarding the best way(s) to make phylogenetic comparisons. In particular, these authors recommend a procedure initiated by Felsenstein (1985) for testing the correlated evolution of continuous traits, something that should be highly applicable to behavior.

Thus, the selections we have chosen for Part 2 are designed to provide a foundation for understanding the manner in which animal behaviorists approach their science. The selections also provide some background on the problems that are particular to the study of behavior, due to the probabilistic nature of animal activities. Many of the major frameworks and paradigms used to guide research today are contained within the selections reprinted here.

Literature Cited

Allee, W. C. 1937. Life history of the song sparrow. *Ecology* 18:540–41.

Allee, W. C. 1939. *The Social Life of Animals.* New York: Norton, 293 pp.

Altmann, J. 1974. Observational study of behavior: sampling methods. *Behaviour* 49:277:67.

Altmann, S. 1991. Diets of yearling female primates (*Papio cynocephalus*) predict lifetime reproductive success. *Proceedings of the National Academy of Science* (U.S.) 88:420–23.

Baerends, G. P. 1958. Comparative methods and the concept of homology in the study of behaviour. *Archives neerlandische Zoologie* 13:401–17.

Baerends, G. P., and J. M. Baerends-van Roon. 1950. An introduction to the study of the ethology of Cichlid fishes. *Behaviour Supplement* I:1–243.

Barlow, G. W. 1968. Ethological units of behavior. Pp. 217–232 in D. Ingle, ed., *The Central Nervous System and Fish Behavior.* Chicago: University of Chicago Press.

Barlow, G. W. 1989. Has sociobiology killed ethology or revitalized it? *Perspectives in Ethology* 8:1–45.

Beach, F. A. 1950. The snark was a boojum. *American Psychologist* 5:115–24.

Beach, F. A. 1955. The descent of instinct. *Psychological Review* 62:401–10.

Beach, F. A. 1960. Experimental investigations of species-specific behavior. *American Psychologist* 15:1–18.

Bernstein, I. S. 1967. A field study of the pigtail monkey (*Macaca nemistrina*). *Primates* 8:217–28.

Bernstein, I. S. 1970. Activity patterns in pigtail monkey groups. *Folia primatologica* 12:187–98.

Bitterman, M. E. 1960. Toward a comparative psychology of learning. *American Psychologist* 15–704–12.

Bitterman, M. E. 1965. Phyletic differences in learning. *American Psychologist* 26:396–410.

Burghardt, G. M. 1967. The primacy of the first feeding experience in the snapping turtle. *Psychonomic Science* 7:383–84.

Burghardt, G. M. 1985. The Foundations of Comparative Ethology. New York: Van Nostrand Reinhold.

Cane, V. 1961. Some ways of describing behaviour. Pp. 361–88 in W. H. Thorpe and O. L. Zangwill, *Current Problems in Animal Behaviour.* Cambridge: Cambridge University Press.

Carpenter, C. R. 1934. A field study of the behavior and social relations of howling monkeys. *Comparative Psychology Monographs* 10:1–168.

Cox, D. R., and P. A. W. Lewis. 1966. *The Statistical Analysis of Series of Events.* London: Metheun, 285 pp.

Craig, W. 1908. The voices of pigeons regarded as a means of social control. *American Journal of Sociology* 14:86–100.

Craig, W. 1918. Appetites and aversions as constituents of instincts. *Biological Bulletin* 34:91–107.

Darwin, C. 1872. *The Expression of the Emotions in Man and Animals.* London: Murray, 462 pp.

Delius, J. D. 1969. A stochastic analysis of the maintenance behavior of skylarks. *Behaviour* 33:137–78.

Dewsbury, D. A. 1984. *Comparative Psychology in the Twentieth Century.* Stroudsburg, Pa.: Hutchinson Ross Publishing Company, xii + 413 pp.

Dewsbury, D. A. 1985. *Leaders in the Study of Animal Behavior.* Lewisburg: Bucknell University Press, 512 pp.

Dewsbury, D. A. 1989. A brief history of the study of animal behavior in North America. *Perspectives in Ethology* 8:85–122.

Dewsbury, D. A. 1992. On the problems studied in ethology, comparative psychology, and animal behavior. *Ethology* 92:89–107.

Dewsbury, D. A. 1995. Americans in Europe: the role of travel in the spread of European ethology after World War II. *Animal Behaviour* 49:1649–63.

Drickamer, L. C. 1973. Semi-natural and enclosed groups of *Macaca mulatta:* a behavioral comparison. *American Journal of Physical Anthropology* 39:249–54.

Drickamer, L. C. 1974. Social rank, observability and sexual behavior of male rhesus monkeys (*Macaca mulatta*). *Journal of Reproduction and Fertility* 37:117–20.

Drickamer, L. C. 1975. Quantitative observations of behavior in free-ranging *Macaca mulatta:* methodology and aggression. *Behaviour* 55:209–36.

Drummond, H. 1981. The nature and description of behavior patterns. *Perspectives in Ethology* 4:1–33.

Dunbar, R. I. M. 1976. Some aspects of research design and their implications in the observational study of behaviour. *Behaviour* 58:78–98.

Eibl-Eibesfeldt, I. 1970. *Ethology, The Biology of Behavior.* New York: Holt, Rinehart and Winston, xiv + 625 pp.

Emerson, A. E. 1938. Termite nests—A study of the phylogeny of behavior. *Ecological Monographs* 8:247–84.

Emlen, J. M. 1966. The role of time and energy in food preferences. *American Naturalist* 100:611–17.

Felsenstein, J. 1985. Phylogenies and the comparative method. *American Naturalist* 125:1–25.

Fentress, J. C. 1972. Development and patterning of movement sequences in inbred mice. Pp. 83–132, in J. A. Kriger, ed., *The Biology of Behavior.* Corvallis: Oregon State University Press.

Fisher, R. A. 1930. *The Genetical Theory of Natural Selection.* London: Oxford University Press, 343 pp.

Garland, T. Jr., P. H. Harvey, and A. R. Ives. 1992. Procedures for the analysis of comparative data using phylogenetically independent contrasts. *Systematic Zoology* 41:18–32.

Gittleman, J. L. 1989. The comparative approach in ethology: aims and limitations. *Perspectives in Ethology* 8:55–83.

Gottlieb, G. 1971. *Development of Species Identification in Birds.* Chicago: University of Chicago Press, xi + 176 pp.

Grabowski, U. 1941. Prägung eines Jungschafs auf den Menschen. *Zeitschrift Tierpsychologie* 4:326–29.

Griffin, D. A. 1984. *Animal Thinking.* Cambridge: Harvard University Press, ix + 237 pp.

Harvey, P. H., and M. D. Pagel. 1991. *The Comparative Method in Evolutionary Biology.* Oxford: Oxford University Press, vii + 239 pp.

Hebb, D. O. 1949. *Organization of Behavior.* New York: John Wiley and Sons, xix + 334 pp.

Heinroth, O. 1911. Beträge zur Biologie, namentlich Ethologie und Psychologie der Anatiden. *5th International Ornithological Congress*, 589–702.

Herrnstein, R. J. 1970. On the law of effect. *Journal of Experimental Analysis of Behavior* 13:243–66.

Hess, E. H. 1959. Imprinting. *Science* 130:133–41.

Hess, E. H. 1962. Ethology. Pp. 157–266 in R. Brown, E. Galanter, E. H. Hess, and G. Mandler, eds., *New Directions in Psychology.* New York: Holt, Rinehart and Winston.

Hinde, R. A. 1956. *Animal Behavior, A Synthesis of Ethology and Comparative Psychology.* New York: McGraw-Hill, xvi + 876 pp.

Hinde, R. A. 1959. Some factors influencing sexual and aggressive behaviour in male chaffinches. *Bird Study* 6:112–22.

Hinde, R. A. 1966. *Animal Behavior, A Synthesis of Ethology and Comparative Psychology.* New York: McGraw-Hill, xvi + 876 pp.

Hinde, R. A. 1970. *Animal Behaviour.* New York: McGraw-Hill.

Hinde, R. A. 1973. On the design of check-sheets. *Primates* 14:393–406.

Hodos, W., and C. B. G. Campbell. 1969. *Scala Naturae:* Why there is no theory in comparative psychology. *Psychological Review* 76:337–50.

Immelmann, K., and C. Beer. 1989. *A Dictionary of Ethology.* Cambridge: Harvard University Press, xiii + 336 pp.

King, J. A. 1958. Parameters relevant to determining the effect of early experience upon the adult behavior of animals. *Psychological Bulletin* 55: 46–58.

King, J. A. 1969. A comparison of longitudinal and cross-sectional groups in the development of behavior of deer mice. *Annals of the New York Academy of Science* 159:696–709.

Klopfer, P. H. 1974. *An Introduction to Animal Behavior Ethology's First Century.* Second Edition. Englewood Cliffs: Prentice-Hall, xiv + 332 pp.

Klopfer, P. H., and J. P. Hailman. 1967. *An Introduction to Animal Behavior.* Englewood Cliffs: Prentice-Hall, xiv + 297 pp.

Kroodsma, D. E. 1990. Using appropriate experimental designs for intended hypotheses in 'song' playbacks, with examples for testing effects of song repertoire sizes. *Animal Behaviour* 40:1138–50.

Lehner, P. N. 1979. *Handbook of Ethological Methods.* New York: Garland STPM, xii + 403 pp.

Lehrman, D. S. 1953. A critique of Konrad Lorenz's theory of instinctive behavior. *Quarterly Review of Biology* 28:337–63.

Lemon, R. E., and C. Chatfield. 1971. Organization of song in cardinals. *Animal Behaviour* 19:1–17.

Lockard, R. B. 1971. Reflections on the fall of comparative psychology: Is there a message for us all? *American Psychologist* 26:168–79.

Loeb, J. 1918. *Forced Movements, Tropisms, and Animal Conduct.* Philadelphia: J. B. Lippincott, 209 pp.

Lorenz, K. Z. 1935. Der Kumpan in der Umwelt des Vogels. *Journal für Ornithologie* 83:137–213 and 289–413.

Lorenz, K. Z. 1950. The comparative method in studying innate behaviour patterns. *Symposia of the Society for Experimental Biology* 4:221–68.

Maier, N. R. F., and T. C. Schneirla. 1935. *Principles of Animal Psychology.* New York: McGraw-Hill, 683 pp.

Manning, A. 1971. Evolution of behavior. Pp. 1–52 in J. L. McGaugh, ed., *Psychobiology: Behavior from a Biological Perspective.* New York: Academic Press.

Marler, P. R., and W. J. Hamilton III. 1966. *Mechanisms of Animal Behavior.* New York: John Wiley and Sons, xi + 771 pp.

Martin, P., and P. Bateson. 1986. *Measuring Behaviour.* London: Cambridge University Press, xii + 200 pp.

Martins, E., and T. Garland Jr. 1991. Phylogenetic analyses of the correlated evolution of continuous characters: a simulation study. *Evolution* 45: 534–57.

Maynard Smith, J. 1982. *Evolution and the Theory of Games.* Cambridge: Cambridge University Press, viii + 224 pp.

McFarland, D. 1981. *The Oxford Companion to Animal Behaviour.* Oxford: Oxford University Press, viii + 857 pp.

Miller, D. B. 1977. Roles of naturalistic observation in comparative psychology. *American Psychologist* 32:211–19.

Nelson, K. 1964. The temporal pattern of courtship behaviour in the glandulocaudine fishes (Ostariophysi, Characidae). *Behaviour* 24:90–146.

Orians, G. H., and N. E. Pearson. 1979. On the theory of central place foraging. Pp. 154–77 in D. J. Horn, R. D. Mitchell, and G. R. Stairs, eds., *Analysis of Ecological Systems.* Columbus: Ohio State University Press.

Pavlov, I. P. 1927. *Conditioned Reflexes: An Investigation of the Physiological Activity of Cerebral Cortex.* London: Oxford University Press, 430 pp.

Purton, A. C. 1978. Ethological categories of behaviour and some consequences of their conflation. *Animal Behaviour* 26:653–70.

Rapport, D. J., and J. E. Turner. 1977. Economic models in ecology. *Science* 195:367–73.

Reichmann, O. J. 1977. Optimization of diets through food preferences by heteromyid rodents. *Ecology* 58:454–57.

Richter, C. P. 1943. Total self-regulatory functions in animals and human beings. *Harvey Lectures* 38: 63–103.

Ridley, M. 1983. *The Explanation of Organic Diversity: The Comparative Method and Adaptations for Mating.* Oxford University Press, ix + 287 pp.

Ristau, C. A. 1990. *Cognitive Ethology: The Minds of Other Animals.* Hillsdale: Lawrence Erlbaum Associates, xx + 332 pp.

Rowell, T. E. 1967. A quantitative comparison of the behaviour of a wild and a caged baboon group. *Animal Behaviour* 15:499–509.

Schleidt, W. M. 1974. How fixed is the fixed action pattern? *Zeitschrift für Tierpsychologie* 18:534–60.

Schmid-Hempel, P., A. Kacelnik, and A. I. Houston. 1985. Honeybees maximise efficiency by not filling their crops. *Behavioral Ecology Sociobiology* 17:61–66.

Schneirla, T. C. 1933. Studies on army ants in Panama. *Journal of Comparative and Physiological Psychology* 15:267–301.

Schneirla, T. C. 1939. A theoretical consideration of

the basis for approach-withdrawal adjustments in behavior. *Psychological Bulletin* 37:501–2.

Schoener, T. W. 1986. A brief history of optimal foraging theory. Pp. 5–67 in A. C. Kamil, J. R. Krebs, and H. R. Pullium, eds., *Proceedings of the Second International Foraging Conference.* New York: Plenum Press.

Sherrington, E. S. 1906. *The Integrative Action of the Nervous System.* New York: Cambridge University Press, 411 pp.

Slater, P. J. B. 1973. Describing sequences of behavior. Pp. 131–153 in *Perspectives in Ethology,* vol. 1 (P. P. G. Bateson and P. H. Klopfer, eds.). New York: Plenum Press.

Sluckin, W. 1965. *Imprinting and Early Learning.* Chicago: Aldine Publishing, 147 pp.

Spalding, D. A. 1873. Instinct, with original observations on young animals. *Macmillan's Magazine* 27–282–93. (Reprinted in *British Journal of Animal Behaviour,* 1954, 2:2–11.)

Staddon, J. E. R. 1983. *Adaptive Behavior and Learning.* New York: Cambridge University Press, xiii + 555 pp.

Stephens, D. W., and J. R. Krebs. 1986. *Foraging Theory.* Princeton: Princeton University Press, xiv + 247 pp.

Thorndike, E. L. 1911. *Animal Intelligence.* New York: Macmillan, 297 pp.

Thorpe, W. H. 1944. Some problems of animal learning. *Proceedings of the Linnean Society of London* 156:70–83.

Thorpe, W. H. 1956. *Learning and Instinct in Animals.* Cambridge: Harvard University Press, viii + 493 pp.

Thorpe, W. H. 1979. *The Origins and Rise of Ethology.* New York: Praeger, xi + 174 pp.

Tinbergen, N. 1935. Uber die Orientierung des Bienenwolfes (*Philanthus triangulum* Fabr.). *Zeitschrift für vergleichende Physiologie* 21:699–716.

Tinbergen, N. 1951. *The Study of Instinct.* Oxford: Clarendon Press, xii + 228 pp.

Tinbergen, N. 1963. On aims and methods of ethology. *Zeitschrift für Tierpsychologie* 20:410–33.

Tolman, E. C. 1932. *Purposive Behavior in Animals and Men.* New York: Century.

von Holst, E. 1939. Entwurf eines Systems der Lokomotorischen Periodendildungen bei Fischen. *Zeitschrift für vergleichende Physiologie* 26:481–528.

Washburn, M. F. 1908. *The Animal Mind.* New York: Macmillan, xiii + 431 pp.

Wheeler, W. M. 1923. *Social Life Among the Insects.* New York: Harcourt, Brace, vii + 375 pp.

Whitman, C. O. 1919. The behavior of pigeons. *Carnegie Institution of Washington Publication* 257:1–161.

Yerkes, R. M. 1903. The instincts, habits, and reactions of the frog. *Psychological Review Monographs Supplement* 17:579–638.

A CRITIQUE OF KONRAD LORENZ'S THEORY OF INSTINCTIVE BEHAVIOR

By DANIEL S. LEHRMAN

The American Museum of Natural History and Rutgers University

BEGINNING about 1931, Konrad Lorenz, with his students and collaborators (notably N. Tinbergen), has published numerous behavioral and theoretical papers on problems of instinct and innate behavior which have had a widespread influence on many groups of scientific workers (Lorenz, 1931, 1932, 1935, 1937a; Lorenz and Tinbergen, 1938; Lorenz, 1939; Tinbergen, 1939; Lorenz, 1940, 1941; Tinbergen, 1942, 1948a, 1950; Lorenz, 1950; Tinbergen, 1951). Lorenz's influence is indicated in the founding of the *Zeitschrift für Tierpsychologie* in 1937 and in its subsequent development, and also in the journal *Behaviour*, established in 1948 under the editorship of an international board headed by Tinbergen.

Lorenz's theory of instinctive and innate behavior has attracted the interest of many investigators, partly because of its diagrammatic simplicity, partly because of its extensive use of neurophysiological concepts, and partly because Lorenz deals with behavior patterns drawn from the life cycle of the animals discussed, rather than with the laboratory situations most often found in American comparative psychology. These factors go far toward accounting for the great attention paid to the theory in Europe, where most students of animal behavior are zoologists, physiologists, zoo curators or naturalists, unlike the psychologists who constitute the majority of American students of animal behavior (Schneirla, 1945).

In recent years Lorenz's theories have attracted more and more attention in the United States as well, partly because of a developing interest in animal behavior among American zoologists and ecologists, and partly through the receptive audience provided for Lorenz and his colleague, Tinbergen, by American ornithologists. The ornithologists were interested from the start, especially because a great part of the material on which Lorenz based his system came from studies of bird behavior, but the range of interest in America has widened considerably. Lorenz and his theories were recently the subject of some discussion at a conference in New York at which zoologists and comparative psychologists were both represented (Riess, 1949), and are prominently represented in the recent symposium on animal behavior of the Society of Experimental Biologists (Armstrong, 1950; Baerends, 1950; Hartley, 1950; Koehler, 1950; Lorenz, 1950; Tinbergen, 1950), and extensively used in several chapters of a recent American handbook of experimental psychology which will be a standard sourcebook for some years to come (Beach, 1951a; Miller, 1951; Nissen, 1951).

Because Lorenz's ideas have gained wide attention, and in particular because a critical discussion of these matters should bring usefully into review Lorenz's manner of dealing with basic problems in the comparative study of behavior, a reconsideration of Lorenz's system and school seems very desirable at this time.

LORENZ'S INSTINCT THEORY

We may best represent the general characteristics of the theory under discussion in terms of a case analyzed by Lorenz and Tinbergen (1938). The many subsequent references to this case and the proffered analysis by these authors and their colleagues leave no doubt that the case and its treatment may stand as representative.

Egg-rolling in the Gray Goose

When a gray goose, sitting on its nest, sees an egg that has rolled out of the nest, it reacts in a characteristic fashion. It extends its head toward the egg and then, keeping its head and neck pointed toward the egg and its eyes fixed upon it, stands up and slowly steps forward to stand on the rim of the nest. Next the goose bends its neck downward and forward so that the egg rests against the underside of the bill. It then proceeds to roll the egg back into the nest by shoving it back between its legs, using the underside of the bill. At the same time that this movement of the head and neck is taking place in the sagittal plane, the goose performs side-to-side movements of the head which have the effect of balancing the egg against the under-side of the bill.

The instinctive act

The egg-rolling movement in the sagittal plane may be considered first, without reference to whatever side-to-side movements may occur, since these two types of movement are distinguished very sharply in the theory.

Lorenz and Tinbergen found that the goose's tendency to perform the sagittal movement can be "exhausted" by repeated elicitation, even though observations indicated that the muscles involved evidently are not themselves fatigued. The authors therefore concluded that what is exhausted is a central neural mechanism.

The form of the sagittal movement is always much the same, regardless of variations in the shape of the egg-object or irregularities in the path over which the egg is rolled. Furthermore, when the egg rolls away from its bill, the goose, instead of stopping the sagittal movement and reaching out

toward the egg, frequently continues the sagittal movement to completion much as though an egg were present. The longer one waits after "exhaustion," the easier it is to reelicit the act.

In connection with certain other behavior patterns, Lorenz (1937b) has noted that, after long resting intervals, the animal may perform a complete act without any external stimulus. This performance "in a vacuum" is regarded by Lorenz as the extreme case of the lowering of the threshold of elicitation after long non-exercise of the act. He calls it "Leerlaufreaktion," or "going off in a vacuum" [translated by Tinbergen (1942) as "vacuum activities"].

The sagittal movement thus has the following characteristics: (1) it displays a reaction-specific exhaustibility; (2) although released by stimuli coming from the egg, once released it remains constant in form regardless of variations in stimulation from the environment and even of the presence or absence of the original releasing stimulus; and (3) the threshold for elicitation falls continuously during non-exercise of the act.

The movement in the sagittal plane, distinguished from lateral deviations, is a typical "instinctive act" (Erbkoordination) in Lorenz's system. This "instinctive act," of course, is only a part of the total behavior pattern of egg-rolling. However, Lorenz maintains that every "instinctive" behavior pattern has as its focus such an "instinctive act" or "consummatory act" (Craig, 1918), the performance of which serves as goal for much of the rest of the pattern.

To Lorenz, *the* instinctive act is a rigidly stereotyped innate movement or movement pattern, based on the activity of a specific coordinating center in the central nervous system. In this coordinating center, there is a continuous accumulation of excitation or energy specific for the act. When the animal comes into the appropriate external situation for the performance of the act, stimuli provided by that situation release the energy, the instinctive act is performed, and some or all of the excitation is used up. The center specific for the act thus is able to coordinate the instinctive act completely independently of the receptors, so that once the act is released (i.e., elicited) its performance occurs in complete form, coordinated by impulses from the center and without any chain-reflex character. The function of the stimulus is to release or elicit the act. Once released, the act no longer depends for its form on

ON LORENZ'S THEORY OF INSTINCTIVE BEHAVIOR   339

anything outside the central nervous system. When the animal happens *not* to be in the appropriate stimulus-situation, this reaction-specific energy is presumed to be accumulated, or dammed up. Also, the greater the amount of reaction-specific energy which has accumulated, the more easily may the act be elicited and the more complete will be its form when elicited.

This picture is regarded by Lorenz as a representation of the neurophysiological basis of the above-described functional characteristics of the instinctive act. In particular, accumulation of energy in a neural center capable of determining the form and order of performance of the various movements of the act, independently of the receptors (except for a trigger-like elicitation), is postulated to explain the reaction-specific exhaustibility (using up of the specific energy), the presumed independence of the form of the act from concurrent external stimulation (reaction-specificity of the energy), and the lowering of the threshold during a non-exercise interval (i.e., when an accumulation of reaction-specific excitation is presumed to occur).

The innate releasing mechanism

If energy specific for the instinctive act can accumulate continuously in the neural center specific for that act, why is the act not continuously performed? Tinbergen (1948a) concludes that each coordinating center is normally held under inhibition by another center which functions to block impulses from the coordinating center, save under specific conditions of external stimulation. This postulated inhibiting center is called the "innate releasing mechanism." The effect of an external stimulus which elicits an instinctive act is to release the instinctive center from this inhibition.

For example, the sagittal component of the goose's egg-rolling movement is not performed continuously, even though energy specific for it is being produced continuously in the central nervous system. The movement is only performed in a particular stimulus-situation: i.e., when a smooth-outlined hard object is present near the nest. (In non-experimental situations, such an object in such a place will almost always be an egg that has rolled out of the nest.) This combination of stimuli, which is considered capable of releasing the particular instinctive act from the inhibition under which it is held by the innate releasing mechanism,

is called the "innate releasing pattern." According to Lorenz and Tinbergen (1938), "The innate releasing pattern . . . [is] . . . the innately-determined readiness of an animal to respond to a particular combination of external stimuli with a particular behavior. It thus consists of an innate receptoral correlate of a combination of stimuli which, despite its relative simplicity, characterizes a certain biologically-significant situation sufficiently uniquely so that the animal will not normally perform the appropriate reaction except in that situation."

The view, then, is that the innate releasing mechanism holds the instinctive act under inhibition until there appears a specific innate releasing pattern of stimuli capable of switching off the inhibition and "triggering" an outflow of impulses from the instinctive center to the peripheral effectors.

Also, the higher the level reached by the accumulation of reaction-specific energy in the center, the more difficult is presumed to be any inhibition of the act by the releasing mechanism. Consequently, the less completely does any stimulus-combination need to fit the innate releasing pattern as a whole to elicit the act. This explanation is offered by Lorenz for the fact that the instinctive act is easier to elicit, the more time has elapsed since it was last performed. The Leerlaufreaktion thus is a breaking of reaction-specific energy through the inhibiting barrier, when such energy reaches a very high level.

The taxis

Side-to-side movements of the head, by which the goose keeps the egg balanced against the underside of the bill, unlike the sagittal movement, lack the character of centrally-preformed movement patterns. The side-to-side movements are believed to be elicited independently by contact stimulation of the underside of the bill by the egg being rolled. Whenever the egg rolls off center, a bill movement toward the side of the deviation restores the egg to the path. If the egg happens to roll free so that the bird may continue the sagittal movement without any egg, there are no accompanying side-to-side movements. When the bird is permitted to roll a cylinder, there are usually no side-to-side movements, since a cylinder is unlikely to roll from side to side in its path.

Thus the side-to-side movement is not only *elicited* by external stimuli like the sagittal movement, but is also continuously *oriented* with respect to external stimuli while being performed. In this respect lateral deviations differ fundamentally from the instinctive act, the form of which is determined centrally so that the external stimulus acts as a trigger only.

Movements like the side-to-side movements, which are continuously oriented to stimuli during their performance, are called by Lorenz orienting movements, or *taxes*. A taxis may occur simultaneously with an instinctive act (as in the case of the goose's egg-rolling), or may occur interspersed with instincts in a behavior-chain.

The stimuli releasing the act (innate releasing pattern for the instinctive act), according to Tinbergen (Tinbergen and Kuenen, 1939; Tinbergen, 1942) are not necessarily the same as those guiding it (i.e., the innate receptor pattern for the taxis). In the case of egg-rolling, for example, the instinctive act is released by a combination of visual stimuli and tactual stimuli related to the hardness of the egg, as felt by the tip of the bill in tapping. The (side-to-side) taxis, on the other hand, is released by tactual stimuli on the underside of the bill.

Appetitive behavior

The first part of the goose's reaction to the egg outside the nest is a stretching of the neck forward and downward, toward the egg. This act, according to Lorenz, has a different character from the instinctive act itself. It serves the purpose of getting the animal into the particular situation in which a specific instinctive act can be released. The act thus is truly goal-directed, according to Lorenz, who terms it "appetitive behavior" (Appetenzverhalten). He regards *all* goal-directed behavior as appetitive, in the sense that such acts are directed toward getting the animal into a situation in which some instinctive act can be released. For him, appetitive behavior can be of enormous complexity, involving instincts, taxes, and learned behavior of various kinds. Such behavior normally occurs when the level of excitation in the central nervous system for any instinctive act becomes high enough. This causes the animal to become restless and active. What specific kind of activity may occur depends on the kind of animal, and on which instinct is the source of the appetitive restlessness. For example, a rat set into activity by a high level of energy specific for the instinctive act of eating (i.e., he wants to eat) may turn toward a corner of the cage, walk toward it, pick up a piece of food in its paws, bite at it, and chew it. Now, the whole sequence of behavior in this hypothetical example would be regarded by Lorenz as appetitive to the instinctive act of chewing. For him, turning toward the corner is a taxis, the walking is an instinctive act, picking up the food might be an instinct, turning the head toward the food held in the paws is a taxis, and the chewing an instinct. Which corner the rat turns toward depends on his past experience—the taxis is thus partly learned. The turning of the head toward the food in the paws, however, might be innate. [Tinbergen (1942) points out that some taxes may be learned, others innate. According to Lorenz's system, however, all instinctive acts (as distinct from taxes) are innate.]

For Lorenz, the whole complex of behavior in this example, involving instincts and taxes, learned and innate elements, has at its core the act of chewing, and is motivated by the excitation set up by the neural center for chewing. The appetitive behavior continues until the instinctive act is performed, and the specific energy is thus used up. It is important to note that according to Lorenz the goal of the appetitive behavior is the *performance* of the act, not its biological result. That is to say, in our hypothetical case, the need of which the appetitive restlessness is an expression is reduced not by the introduction of food into the stomach, but by the act of chewing. This is shown more clearly in the case of instinctive acts like courtship displays of birds, which form the goal of appetitive behavior (moving toward the female, orienting to the female, etc.) and which according to Lorenz are subject to Leerlaufreaktionen even though they do not (like chewing and swallowing) result in the satisfaction of an *apparent* peripheral tissue need.

PROBLEMS RAISED BY INSTINCT THEORIES

Even this brief summary brings to light several questions which ought to be critically examined with reference to the theory. These are questions, furthermore, which apply to instinct theories in general. Among them are: (1) the problem of "innateness" and the maturation of behavior; (2) the problem of levels of organization in an organism; (3) the nature of evolutionary levels of behavioral organization, and the use of the comparative method in studying them; and (4) the manner in which physiological concepts may be properly used in behavior analysis. There follows an evaluation of Lorenz's theory in terms of these general problems.

"Innateness" of behavior

The problem

Lorenz and Tinbergen consistently speak of behavior as being "innate" or "inherited" as though these words surely referred to a definable, definite, and delimited category of behavior. It would be impossible to overestimate the heuristic value which they imply for the concepts "innate" and "not-innate." Perhaps the most effective way to throw light on the "instinct" problem is to consider carefully just what it means to say that a mode of behavior is innate, and how much insight this kind of statement gives into the origin and nature of the behavior.

Tinbergen (1942), closely following Lorenz, speaks of instinctive acts as "highly stereotyped, coordinated movements, the neuromotor apparatus of which belongs, in its complete form, to the hereditary constitution of the animal." Lorenz (1939) speaks of characteristics of behavior which are "hereditary, individually fixed, and thus open to evolutionary analysis." Lorenz (1935) also refers to perceptual patterns ("releasers") which are presumed to be innate because they elicit "instinctive" behavior the *first time* they are presented to the animal. He also refers to those motor patterns as innate which occur for the first time when the proper stimuli are presented. Lorenz's student Grohmann (1938), as well as Tinbergen and Kuenen (1939), speak of behavior as being innately determined because it matures instead of developing through learning.

It is thus apparent that Lorenz and Tinbergen, by "innate" behavior, mean behavior which is hereditarily determined, which is part of the original constitution of the animal, which arises quite independently of the animal's experience and environment, and which is distinct from acquired or learned behavior.

It is also apparent, explicitly or implicitly, that Lorenz and Tinbergen regard as the major *criteria* of innateness that: (1) the behavior be stereotyped and constant in form; (2) it be characteristic of the species; (3) it appear in animals which have been raised in isolation from others; and (4) it develop fully-formed in animals which have been prevented from practicing it.

Undoubtedly, there are behavior patterns which meet these criteria. Even so, this does not necessarily imply that Lorenz's *interpretation* of these behavior patterns as "innate" offers genuine aid to a scientific understanding of their origin and of the mechanisms underlying them.

In order to examine the soundness of the concept of "innateness" in the analysis of behavior, it will be instructive to start with a consideration of one or two behavior patterns which have already been analyzed to some extent.

Pecking in the chick

Domestic chicks characteristically begin to peck at objects, including food grains, soon after hatching (Shepard and Breed, 1913; Bird, 1925; Cruze, 1935; and others). The pecking behavior consists of at least three highly stereotyped components: head lunging, bill opening and closing, and swallowing. They are ordinarily coordinated into a single resultant act of lunging at the grain while opening the bill, followed by swallowing when the grain is picked up. This coordination is present to some extent soon after hatching, and improves later (even, to a slight extent, if the chick is prevented from practicing).

This pecking is stereotyped, characteristic of the species, appears in isolated chicks, is present at the time of hatching, and shows some improvement in the absence of specific practice. Obviously, it qualifies as an "innate" behavior, in the sense used by Lorenz and Tinbergen.

Kuo (1932a-d) has studied the embryonic development of the chick in a way which throws considerable light on the origin of this "innate" behavior. As early as three days of embryonic age, the neck is passively bent when the heartbeat causes the head (which rests on the thorax) to rise and fall. The head is stimulated tactually by the yolk sac, which is moved mechanically by amnion contractions synchronized with the heartbeats which cause head movement. Beginning about one day later, the head first bends *actively* in response to tactual stimulation. At about this time, too, the bill begins to open and close when the bird nods— according to Kuo, apparently through nervous excitation furnished by the head movements through irradiation in the still-incomplete nervous system. Bill-opening and closing become independent of head-activity only somewhat later. After about 8 or 9 days, fluid forced into the throat by the bill and head movements causes swallowing. On the twelfth day, bill-opening always follows head-movement.

In the light of Kuo's studies the "innateness" of the chick's pecking takes on a different character from that suggested by the concept of a unitary, innate item of behavior. Kuo's observations strongly suggest several interpretations of the *development* of pecking (which, of course, are subject to further clarification). For example, the head-lunge arises from the passive head-bending which occurs contiguously with tactual stimulation of the head while the nervous control of the muscles is being established. By the time of hatching, head-lunging in response to tactual stimulation is very well established (in fact, it plays a major role in the hatching process).

The genesis of head-lunging to visual stimulation in the chick has not been analyzed. In *Amblystoma*, however, Coghill (1929) has shown that a closely analogous shift from tactual to visual control is a consequence of the establishment of certain anatomical relationships between the optic nerve and the brain region which earlier mediated the lunging response to tactual stimulation, so that visual stimuli come to elicit responses established during a period of purely tactual sensitivity. If a similar situation obtains in the chick, we would be dealing with a case of intersensory equivalence, in which visual stimuli, because of the anatomical relationships between the visual and tactual regions of the brain, became equivalent to tactual stimuli, which in turn became effective through an already analyzed process of development, which involved conditioning at a very early age (Maier and Schneirla, 1935).

The originally diffuse connection between head-lunge and bill-opening appears to be strengthened by the repeated elicitation of lunging and billing by tactual stimulation by the yolk sac. The repeated elicitation of swallowing by the pressure of amniotic fluid following bill-opening probably is important in the establishment of the post-hatching integration of bill-opening and swallowing.

Maternal behavior in the rat

Another example of behavior appearing to fulfil the criteria of "innateness" may be found in the maternal behavior of the rat.

Pregnant female rats build nests by piling up strips of paper or other material. Mother rats will "retrieve" their pups to the nest by picking them up in the mouth and carrying them back to the nest. Nest-building and retrieving both occur in all normal rats; they occur in rats which have been raised in isolation; and they occur with no evidence of previous practice, since both are performed well by primiparous rats (retrieving may take place for the first time only a few minutes after the birth of the first litter of a rat raised in isolation). Both behavior patterns therefore appear to satisfy the criteria of "innateness" (Wiesner and Sheard, 1933).

Riess (pers. com.), however, raised rats in isolation, at the same time preventing them from ever manipulating or carrying any objects. The floor of the living cage was of netting so that feces dropped down out of reach. All food was powdered, so that the rats never carried food pellets. When mature, these rats were placed in regular breeding cages. They bred, but did *not* build normal nests or retrieve their young normally. They scattered nesting material all over the floor of the cage, and similarly moved the young from place to place without collecting them at a nest-place.

Female rats do a great deal of licking of their own genitalia, particularly during pregnancy (Wiesner and Sheard, 1933). This increased licking during pregnancy has several probable bases, the relative importance of which is not yet known. The increased need of the pregnant rat for potassium salts (Heppel and Schmidt, 1938) probably accounts in part for the increased licking of the salty body fluids as does the increased irritability of the genital organs themselves. Birch (pers. com.) has suggested that this genital licking may play an important role in the development of licking and retrieving of the young. He is raising female rats fitted from an early age with collars made of rubber discs, so worn that the rat is effectively prevented from licking its genitalia. Present indications, based on limited data, are that rats so raised eat a high percentage of their young, that the young in the nest may be found under any part of the female instead of concentrated posteriorly as with normal mother rats, and that retrieving does not occur.

These considerations raise some questions concerning nativistic interpretations of nest-building and retrieving in the rat, and concerning the meaning of the criteria of "innateness." To begin with, it is apparent that practice in carrying food pellets is partly equivalent, for the development of nest-building and retrieving, to practice in carrying nesting-material, and in carrying the young. Kinder (1927) has shown that nest-building activity is inversely correlated with environmental temperature, and that it can be stopped by raising the temperature sufficiently. This finding, together with Riess's experiment, suggests that the nest-building activity arises from ordinary food (and other object) manipulation and collection under conditions where the accumulation of certain types of manipulated material leads to immediate satis-faction of one of the animal's needs (warmth). The fact that the rat is generally more active at lower temperatures (Browman, 1943; Morgan, 1947) also contributes to the probability that nest-building activity will develop. In addition, the rat normally tends to stay close to the walls of its cage, and thus to spend much time in corners. This facilitates the collection of nesting material into one corner of the cage, and the later retrieving of the young to that corner. Patrick and Laughlin (1934) have shown that rats raised in an environ-ment without opaque walls do not develop this "universal" tendency of rats to walk close to the wall. Birch's experiment suggests that the rat's experience in licking its own genitalia helps to establish retrieving as a response to the young, as does its experience in carrying food and nesting material.

Maturation-vs.-learning, or development?
The isolation experiment

These studies suggest some second thoughts on the nature of the "isolation experiment." It is obvious that by the criteria used by Lorenz and other instinct theorists, pecking in the chick and nest-building and retrieving in the rat are not "learned" behavior. They fulfil all criteria of "innateness," i.e., of behavior which develops without opportunity for practice or imitation. Yet, in each case, analysis of the developmental process involved shows that the behavior patterns con-cerned are not unitary, autonomously developing things, but rather that they emerge ontogenetically in complex ways from the previously developed organization of the organism in a given setting.

What, then is wrong with the implication of the "isolation experiment," that behavior developed in isolation may be considered "innate" if the animal did not practice it specifically?

Lorenz repeatedly refers to behavior as being innate because it is displayed by animals raised in isolation. The raising of rats in isolation, and their subsequent testing for nesting behavior, is typical of isolation experiments. The development of the

chick inside the egg might be regarded as the ideal isolation experiment.

It must be realized that an animal raised in isolation from fellow-members of his species *is not necessarily isolated from the effect of processes and events which contribute to the development of any particular behavior pattern.* The important question is not "Is the animal isolated?" but *"From what is the animal isolated?"* The isolation experiment, if the conditions are well analyzed, provides at best a negative indication that certain specified environ-mental factors probably are not directly involved in the genesis of a particular behavior. However, the isolation experiment by its very nature does not give a positive indication that behavior is "innate" or indeed any information at all about what the process of development of the behavior really con-sisted of. The example of the nest-building and retrieving by rats which are isolated from other rats but not from their food pellets or from their own genitalia illustrates the danger of assuming "innateness" merely because a *particular* hypothe-sis about learning seems to be disproved. This is what is consistently done by Tinbergen, as, for example, when he says (1942) of certain behavior patterns of the three-spined stickleback: "The re-leasing mechanisms of these reactions are all in-nate. A male that was reared in isolation . . . was tested with models before it had ever seen another stickleback. The . . . [stimuli] . . . had the same re-leaser functions as in the experiments with normal males." Such isolation is by no means a final or complete control on possible effects from experi-ence. For example, is the "isolated" fish unin-fluenced by its own reflection from a water film or glass wall? Is the animal's experience with human handlers, food objects, etc., really irrelevant?

Similarly, Howells and Vine (1940) have re-ported that chicks raised in mixed flocks of two varieties, when tested in a Y-maze, learn to go to chicks of their own variety more readily than to those of the other variety. They concluded that the "learning is accelerated or retarded . . . because of the directive influence of innate factors." In this case, Schneirla (1946) suggests that the effect of the chick's experience with its own chirping during feeding has not been adequately considered as a source of differential learning previous to the ex-periment. This criticism may also be made of a similar study by Schoolland (1942) using chicks and ducklings.

Even more fundamental is the question of what

is meant by "maturation." We may ask whether experiments based on the assumption of an absolute dichotomy between maturation and learning ever really tell us *what* is maturing, or how it is maturing? When the question is examined in terms of *developmental* processes and relationships, rather than in terms of preconceived categories, the maturation-versus-learning formulation of the problem is more or less dissipated. For example, in the rat nest-building probably does not mature autonomously—and it is *not* learned. It is not "nest-building" which is learned. Nest-building develops in certain situations through a developmental process in which at each stage there is an identifiable interaction between the environment and organic processes, and within the organism; this interaction is based on the preceding stage of development and gives rise to the succeeding stage. These interactions are present from the earliest (zygote) stage. Learning may emerge as a factor in the animal's behavior even at early embryonic stages, as pointed out by Carmichael (1936).

Pecking in the chick is also an emergent—an integration of head, bill, and throat components, each of which has its own developmental history. This integration is already partially established by the time of hatching, providing a clear example of "innate" behavior in which the statement "It is innate" adds nothing to an understanding of the developmental process involved. The statement that "pecking" is innate, or that it "matures," leads us *away* from any attempt to analyze its specific origins. The assumption that pecking grows *as* a pecking *pattern* discourages examination of the embryological processes leading to pecking. The elements out of whose interaction pecking emerges are not originally a unitary pattern; they *become* related as a consequence of their positions in the organization of the embryonic chick. The understanding provided by Kuo's observations owes nothing to the "maturation-versus-learning" formulation.

Observations such as these suggest many new problems the relevance of which is not apparent when the patterns are nativistically interpreted. For example, what is the nature of the rat's temperature-sensitivity which enables its nest-building to vary with temperature? How does the animal develop its ability to handle food in specific ways? What are the physiological conditions which promote licking of the genitalia, etc.? We want to

know much more about the course of establishment of the connections between the chick's head-lunge and bill-opening, and between bill-opening and swallowing. This does *not* mean that we expect to establish which of the components is learned and which matured, or "how much" each is learned and how much matured. The effects of learning and of structural factors differ, not only from component to component of the pattern, but also from developmental stage to developmental stage. What is required is a continuation of the careful analysis of the characteristics of each developmental stage, and of the transition from each stage to the next.

Our scepticism regarding the heuristic value of the concept of "maturation" should not be interpreted as ignorance or denial of the fact that the physical growth of varied structures plays an important role in the development of most of the kinds of behavior patterns under discussion in the present paper. Our objection is to the *interpretation* of the role of this growth that is implied in the notion that the *behavior* (or a *specific* physiological substrate for it) is "maturing." For example, the post-hatching improvement in pecking ability of chicks is very probably due in part to an increase in strength of the leg muscles and to an increase in balance and stability of the standing chick, which results partly from this strengthening of the legs and partly from the development of equilibrium responses (Cruze, 1935). Now, isolation or prevention-of-practice experiments would lead to the conclusion that this part of the improvement was due to "maturation." Of course it is partly due to growth processes, *but what is growing is not pecking ability*, just as, when the skin temperature receptors of the rat develop, what is growing is not nest-building activity, *or anything isomorphic with it*. The use of the categories "maturation-vs.-learning" as explanatory aids usually gives a false impression of unity and directedness in the growth of the behavior pattern, when actually the behavior pattern is not primarily unitary, nor does development proceed in a straight line toward the completion of the pattern.

It is apparent that the use of the concept of "maturation" by Lorenz and Tinbergen as well as by many other workers is not, as it at first appears, a reference to a process of development but rather to ignoring the process of development. To say of a behavior that it develops by maturation is tantamount to saying that the obvious forms of learning do not influence it, and that we therefore do not consider it necessary to investigate its ontogeny further.

Heredity-vs.-environment, or development?

Much the same kind of problem arises when we consider the question of what is "inherited." It is characteristic of Lorenz, as of instinct theorists in general, that "instinctive acts" are regarded by him as "inherited." Furthermore, inherited behavior is regarded as sharply distinct from behavior acquired through "experience." Lorenz (1937a) refers to behavior which develops "entirely independent of all experience."

It has become customary, in recent discussions of the "heredity-environment" problem, to state that the "hereditary" and "environmental" contributions are both essential to the development of the organism; that the organism could not develop in the absence of either; and that the dichotomy is more or less artificial. [This formulation, however, frequently serves as an introduction to elaborate attempts to evaluate what part, or even what percentage, of behavior is genetically determined and what part acquired (Howells, 1945; Beach, 1947a; Carmichael, 1947; Stone, 1947).] Lorenz does not make even this much of a concession to the necessity of developmental analysis. He simply states that some behavior patterns are "inherited," others "acquired by individual experience." I do not know of any statement of either Lorenz or Tinbergen which would allow the reader to conclude that they have any doubts about the correctness of referring to behavior as simply "inherited" or "genically controlled."

Now, what exactly is meant by the statement that a behavior pattern is "inherited" or "genically controlled"? Lorenz undoubtedly does not think that the zygote contains the instinctive act in miniature, or that the gene is the equivalent of an entelechy which purposefully and continuously tries to push the organisms's development in a particular direction. Yet one or both of these preformistic assumptions, or their equivalents, must underlie the notion that some behavior patterns are "inherited" as such.

The "instinct" is obviously not present in the zygote. Just as obviously, it is present in the behavior of the animal after the appropriate age. The problem for the investigator who wishes to make a causal analysis of behavior is: How did this behavior come about? The use of "explanatory" categories such as "innate" and "genically fixed" obscures the necessity of investigating developmental *processes* in order to gain insight into the actual mechanisms of behavior and their interrelations. The problem of development is the problem of the development of new structures and activity patterns from the resolution of the interaction of *existing* structures and patterns, within the organism and its internal environment, and between the organism and its outer environment. At any stage of development, the new features emerge from the interactions within the *current* stage and between the *current* stage and the environment. The interaction out of which the organism develops is *not* one, as is so often said, between heredity and environment. It is between *organism* and environment! And the organism is different at each different stage of its development.

Modern physiological and biochemical genetics is fast destroying the conception of a straight-line relationship between gene and somatic characteristic. For example, certain strains of mice contain a mutant gene called "dwarf." Mice homozygous for "dwarf" are smaller than normal mice. It has been shown (Smith and MacDowell, 1930; Keeler, 1931) that the cause of this dwarfism is a deficiency of pituitary growth hormone secretion. Now what are we to regard as "inherited"? Shall we change the name of the mutation from "dwarf" to "pituitary dysfunction" and say that dwarfism is not inherited as such—that what is inherited is a hypoactive pituitary gland? This would merely push the problem back to an earlier stage of development. We now have a better understanding of the origin of the dwarfism than we did when we could only say it is "genically determined." However, the pituitary function developed, in turn, in the context of the mouse as it was when the gland was developing. The problem is: What was that context and how did the gland develop out of it?

What, then, is inherited? From a somewhat similar argument, Jennings (1930) and Chein (1936) concluded that only the zygote is inherited, or that heredity is only a stage of development. There is no point here in involving ourselves in tautological arguments over the definition of heredity. It is clear, however, that to say a behavior pattern is "inherited" throws no light on its *development* except for the purely negative implication that *certain types* of learning are not directly involved. Dwarfism in the mouse, nest-building in the rat, pecking in the chick, and the "zig-zag dance" of the stickleback's courtship (Tinbergen, 1942) are all "inherited" in the sense and by the criteria used by Lorenz. But they are not by any means phenomena of a common type, nor do they arise through the same kinds of developmental processes. To lump them together under the rubric of "inherited" or "innate" characteristics serves to block the investigation of their origin just at the point where it should leap forward in meaningfulness. [Anastasi and Foley (1948), considering data from the field of human differential psychology, have been led to somewhat the same formulation of the "heredity-environment" problem as is presented here.]

Taxonomy and Ontogeny

Lorenz (1939) has very ably pointed out the potential importance of behavior elements as taxonomic characteristics. He has stressed the fact that evolutionary relationships are expressed just as clearly (in many cases more clearly) by similarities and differences in behavior as by the more commonly used physical characteristics. Lorenz himself has made a taxonomic analysis of a family of birds in these terms (Lorenz, 1941), and others have been made by investigators influenced by him (Delacour and Mayr, 1945; Adriaanse, 1947; Baerends and Baerends-van Roon, 1950). This type of analysis derives from earlier work on the taxonomic relations of behavior patterns by Whitman (1898, 1919), Heinroth (1910, 1930), Petrunkevitsch (1926), and others.

Lorenz's brilliant approach to the taxonomic analysis of behavior characteristics has had wide influence since it provides a very stimulating framework in which to study species differences and the specific characteristics of behavior. However, it does not necessarily follow from the fact that behavior patterns are species-specific that they are "innate" as patterns. We may emphasize again that the systematic stability of a characteristic does not indicate anything about its mode of development. The fact that a characteristic is a good taxonomic character does not mean that it developed autonomously. The shape of the skull bones in rodents, which is a good taxonomic character (Romer, 1945), depends in part upon the presence of attached muscles (Washburn, 1947). We cannot conclude that because a behavior pattern is taxonomically stable it must develop in a unitary, independent way.

In addition it would be well to keep in mind

that the species-characteristic nature of many behavior patterns may result partly from the fact that all members of the species grow in the same environment. Smith and Guthrie (1921) call such behavior elements "coenotropes." Further, it is not at all necessary that these common features of the environment be those which seem a priori to be relevant to the behavior pattern under study. Lorenz's frequent assumption (e.g., 1935) that the effectiveness of a given stimulus on first presentation demonstrates an innate sensory mechanism specific for that stimulus is not based on analysis of the origin of the stimulus-effectiveness, but merely on the fact that Lorenz has eliminated the major alternative *he* sees to the nativistic explanation.

Thorpe and Jones (1937) have shown that the apparently innate choice of the larvae of the flour moth by the ichneumon fly *Nemerites* as an object in which to deposit its eggs is actually a consequence of the fact that the fly larva was *fed* on the larvae of the flour moth while it was developing. By raising *Nemerites* larvae upon the larvae of *other* kinds of moth Thorpe and Jones caused them, when adult, to choose preponderantly these other moths on which to lay their eggs. The choice of flour-moth larvae for oviposition is quite characteristic of *Nemerites* in nature. In view of Thorpe and Jones' work, it would obviously be improper to conclude from this fact that the choice is based on innately-determined stimuli. Yet, before their paper was published, the species-specific character of the behavior would have been just as impressive evidence for "innateness" as species-specificity *ever* is.

Taxonomic analysis, while very important, is not a substitute for concrete analysis of the ontogeny of the given behavior, as a source of information about its origin and organization.

Levels of Organization

Levels of "Innateness"

Animals at different evolutionary levels show characteristic differences in the extent and manner of learning. In addition, within the same animal's behavior different activities may be more or less susceptible to the influence of learning, and may be affected in different ways by learning (Schneirla, 1948, 1949a).

Lorenz explains these facts in terms of the richness of the animal's instinctive equipment. As described above, his conception is that instinctive behavior is sharply different from all behavior leading up to the performance of the instinct. This "appetitive" behavior is conceived of as the sole evolutionary source of all learned and intelligent behavior. Thus he says:

"... appetitive behavior, as the sole root of all "variable" behavior, not only is physiologically something fundamentally different from the automatism of instinctive behavior, but ... the two different processes appear as "substitutes" (*vikariierend*) for each other, in that the higher (phylogenetic) development of the one makes the other superfluous and stops its development. The reaching of a higher psychic performance goes hand-in-hand with a reduction of the automatisms that take part in the action, leaving a behavior pattern with the same function as the one originally existing" (Lorenz, 1937a).

Again:

"It is a peculiarity of many behavior patterns of higher animals, that *innate instinctive elements and individually-acquired elements immediately follow each other*, within a functionally unitary chain of acts ... I have characterized this phenomenon as instinct-training interlacement. Similar interlacements occur between instinctive acts and intelligent or insightful behavior. ... The essence of such an interlacement is that, within a chain of innate instinctive acts there is a definite point, which point is innately determined, where a learned act is inserted. This learned act must be acquired by each individual in the course of its ontogenetic development. In such a case, the chain of innate acts has a *gap*, in which, instead of an instinctive act, there is a '*capacity to acquire*' " (Lorenz, 1937a). [All emphases are Lorenz's.]

It is apparent that Lorenz regards differences in the extent to which learning occurs as representing differences in the size of the gaps in the chain of innate behavior. He considers any given "component" of behavior as "innate" or not "innate." This is entirely consistent with his virtual identification of "innate" with "autonomously developing."

However, we have already tried to make it clear that behavior patterns classified as "innate" by *any* criterion do not all fall into the same category with respect to embryonic origin, developmental history, or level of organization. Lorenz notes that more or fewer of the components of behavior may be "innate." But nowhere does he recognize that *one* component may be more or less "innate" or "innate" in one or another *manner*. We may call attention to an important difference between the pecking of the chick and the nest-building of the rat, both behavior patterns which develop without specific practice of the patterns: a major part of the learning which appears to be antecedent to the emergence of pecking in the chick occurs before hatching, while

much of the learning which is antecedent to the emergence of nest-building in the rat occurs after birth.

Shall we call those behavior patterns "innate" which develop before birth and not those which develop after? This would be fruitless in view of the demonstrated existence of prenatal conditioning (Ray, 1932; Gos, 1933; Spelt, 1948; Hunt, 1949), and unsatisfactory in view of the problem of the so-called postnatal "maturation" of various "innate" behavior patterns (Grohmann, 1938). But we must recognize that different behavior patterns may involve learning at different ontogenetic stages to different extents, and in different ways. For example, much less of the behavior of the rat is *directly* a consequence of the specific characteristics of its structure than in the case of the earthworm (Maier and Schneirla, 1935). The involvement of learning in the development of the rat's behavior is different from and occurs at different developmental stages from that of the chick. Further, some responses of the rat (such as licking of a painful spot) are very much less subject to change by learning than others, such as care of young (Sperry, 1945; Uyldert, 1946). These are not differences in the *number* of behavioral elements which are "innate," but rather in the *way* in which the structures are involved in the development of behavior at different evolutionary levels and for different behavior patterns.

Lorenz does not fully utilize the idea of levels of organization of behavior, apparently because his concept of "innateness" is not the result of analysis of the development of behavior; it is in part the result of a preconception that "innate" and "not-innate" are the two categories into which behavior logically falls. Consequently Lorenz and his school have classified behavior as "innate" and "not-innate" on the basis of criteria which when carefully examined appear to be arbitrary. Their category of "innate" therefore includes very different kinds of behavior, which involve learning in many different ways. Lorenz's concept of "innate" behavior represents a lumping-together of many different kinds and levels of behavior on the basis of an essentially phenotypic classification, and the imposition of preconceived categories upon that classification.

Evolutionary Levels

Since Lorenz does not discuss the existence of qualitative differences with respect to modes of development within his category of "innate" behavior it is not surprising that his conception of the evolution of behavior lacks any notion of qualitative change. Lorenz maintains at all levels a sharp distinction between "instinctive acts" and "appetitive behavior" (which includes all oriented, goal-directed, and variable types of behavior at all levels). He says:

"If we consider the unbroken series of forms of corresponding modes of behavior, which extends in a smooth progression from protozoa to man, we must determine that we cannot distinguish between taxis, on the one hand, and, on the other, behavior guided by the simplest intelligence (Einsicht). We cannot here distinguish between taxis and, in the case of our frog, an intelligence which might (anthropomorphically speaking) be limited to the knowledge: 'There sits the fly' " (Lorenz, 1937a).

This is restated in a later paper (Lorenz, 1939): "No sharp line can be drawn between the simplest orienting-reaction and the highest 'insightful' behavior."

It might be pointed out that whether we can distinguish various levels of behavioral organization depends in part on our assiduity in *attempting* to distinguish them. Preconceptions about the number and kind of categories into which behavior ought to fall naturally has an important effect on the kind of examination we make of behavior patterns and the kinds of distinctions we find ourselves able to make among them.

In the quotation above we have translated as "smooth" (progression) Lorenz's word "stufenlose," which might be more literally translated "without steps" or "without levels." This is a gratuitous and very misleading oversimplification on Lorenz's part. The transition from protozoa to man is *not* "step-less." There are *characteristic* structural differences between phyletic levels, and these differences are responsible for *characteristic* differences in the organization of behavior. A protozoan is not like a simpler man. It is a different kind of organism, with behavior which depends in different *ways* on its structure. The analysis of behavior mechanisms at different levels (Schneirla, 1946) shows that it is frequently misleading to speak of *behavior* patterns or elements as homologous when they seem to serve similar (or the "same") functions and have superficially similar characteristics. Analysis of structural organizations out of which the specific behavior patterns emerge shows that similar behaviors at different phyletic levels often are end-products of evolutionary selection leading to the similar behavior, but deriving from different structures so that the underlying processes and mechanisms are not the same.

Lorenz's application of the concept of evolutionary change does not consist of analyzing the different ways in which behavior patterns at different evolutionary levels depend on the structure and life of the organism. It consists rather of abstracting aspects of behavior, reifying them as specific autonomous mechanisms, and then citing them as demonstrations of "evolution" in a purely descriptive taxonomic sense. Taxonomically, this procedure is often extremely valuable, but by its implicit assumption that "elements" of behavior maintain their nature regardless of change in the organization in which they are embedded (more properly, we should say from which they emerge), it hinders rather than helps analysis of the behavior patterns themselves.

Levels of Neural Organization

Lorenz characterizes each instinctive act as depending on a specific center in the central nervous system which continuously produces a type of excitation specific to the act, and which is partly "used up" when the act is performed. He uses the concept "used up" quite literally, even suggesting the existence of act-specific *substances* (Lorenz and Tinbergen, 1938). One of the principal types of evidence used by Lorenz to support this conception is the lowering of the threshold for release of the act as a function of lapse of time since performance of the act. That is, the longer the animal has gone without performing the act, the easier it is to elicit. This is taken by Lorenz as evidence of accumulation of the reaction-specific energy in the central nervous system. Lowering of the intensity of performance upon repeated elicitation is taken as further proof, since it may indicate the using up of the excitation faster than it can be produced.

Lorenz and Tinbergen offer observations along these lines on mammals, birds, fish, and insects. The hunting behavior of the dog (Tinbergen, 1942), food-begging of a young bird (Tinbergen and Kuenen, 1939), fighting in a fish (Tinbergen, 1942), courtship flights of a butterfly (Tinbergen, Meeuse, Boerema, and Varossieau, 1942) are all offered as examples of instinctive acts having this kind of physiological basis.

Lorenz and Tinbergen adduce as physiological evidence for this interpretation a series of studies by von Holst (1935–1937), on the mechanisms of locomotion in fishes. Von Holst observed that almost completely deafferented fishes show some of the coordinations of locomotion. He concluded that the basic movement patterns of locomotion are the result of the accumulation of locomotion-specific energy in the central nervous system independent of peripheral activity, and are the result of a central (non-reflex) coordination.

In his original consideration of von Holst's work, Lorenz (1937b) stated that it would be premature to make positive assertions about the direct relevance of that work to his instinct theory. Lorenz and Tinbergen (1938), at about the same time, stated that the conception of locomotion as an example of instinctive coordination in Lorenz's sense might be a very rough simplification of the facts.

However, over the years since then, the relevance of this kind of evidence in their writings seems to have increased, although some doubt has been thrown on the validity of von Holst's conclusions. One might question the direct relevance of the neural mechanisms of locomotion in fish and amphibians to the explanation of the origins of complex "instinctive" behavior in birds and mammals. Tinbergen is aware of the dangers inherent in the procedure of using physiological evidence from lower evolutionary levels, lower levels of neural organizations, and simpler forms of behavior as analogies for the support of physiological theories of behavior mechanisms at higher and more complex levels. For example, after a description of this aspect of Lorenz's theory, Tinbergen (1948a) says: "These formulations are supported by the entirely independent investigations which have been conducted during the last ten years on the central nervous mechanisms of locomotion. Here, *to be sure on a lower level of integration* [my emphasis— D. S. L.], we are brought to a fundamentally similar position by the researches of von Holst, Weiss, Gray, Lissmann, and others."

In this case, Tinbergen's mention of the fact that the physiological evidence comes from a lower level of integration is actually embedded in an *expanded* use of these data to support theories based on observations at higher levels. This is merely a formal bow to the concept of levels, which appears to strengthen the form of the argument while actually weakening its content.

In point of fact, it is now doubtful whether even so simple a behavior pattern as locomotion, in so simple a vertebrate as a fish, is really organized in the way that Lorenz's instinct theory demands.

Gray and Lissman (1940, 1946a, 1946b; Gray, 1939, 1950; Lissmann, 1946a, 1946b) have studied the effect of deafferentation on locomotion in fish and amphibians. Both regard their evidence as being against the probability of automatic-rhythmic production of coordinations in the central nervous system, even at the fish level, and even for locomotion. Lissmann, in fact, designed his experiments specifically in view of von Holst's observations, and explicitly in view of the use made of the latter by Lorenz. He concluded, on the basis of a *complete* afferent isolation of the central nervous system, that there was no central automatic production of excitation. It should be noted that the experiments of Weiss (1936, 1937a-d, 1941, 1950) support the conclusion that spinal centers in amphibians are so organized that the coordination of locomotor patterns is dependent upon characteristics of the centers. Gray and Lissmann's experiments, however, show that proprioception actually plays a major role in the normal ambulatory rhythms, even of these animals.

Tinbergen (1942) has expanded Lorenz's concept of neural organization to include higher levels of physiological and behavioral organization than the stereotyped "instinctive act" or consummatory act. Tinbergen conceives of instinctive behavior in general as being hierarchically organized in the individual. For example, in the reproductive behavior of the stickleback Tinbergen sees three main levels of organization, hierarchically arranged. The highest level represents the reproductive drive in general. This corresponds to a center at a high level in the nervous system which when activated (by external conditions, hormones, or autonomous cyclicity) sends impulses to a whole group of intermediate centers, making the latter capable of activity. Each of these intermediate centers corresponds to a behavior pattern involved in reproductive activities: fighting, nest-building, courting, parental behavior, etc. Each of these intermediate centers, in turn, activates (or contributes to the disinhibition of) a group of lower centers each of which coordinates a particular act which is released by an innate releasing pattern. For example, the fighting center, when activated by external stimulation (which can only occur when its threshold is lowered by activity in the superordinated center for the reproductive drive in general) puts the animal in "fighting mood" which makes possible the performance under proper stimulus-conditions of each of the acts involved in fighting: biting, chasing, threatening, etc. The latter are the "consummatory acts" or "instinctive acts" of Lorenz. [N. B. The slight differences in terminology sometimes occurring between Tinbergen and Lorenz do not usually represent theoretical or conceptual disagreements.]

Tinbergen cites the work of Hess (1943, 1949) as demonstrating the reality of these autonomous centrally-coordinated centers of activity. By electrically stimulating various points in the hypothalamus, Hess was able to cause cats to perform sleeping, eating, and other behavior which was not related to specific external stimulation, and which ceased upon cessation of the stimulation. Tinbergen regards these observations, with those of von Holst, as demonstrating the reality of Lorenz's picture of centers of automatic-rhythmic production of action-specific excitation.

Hess found, however, that there was considerable variation in the responses to repeated stimulation of a specific spot in the hypothalamus. Stimulation of the same point might elicit quite different responses, depending upon the conditions of afferent inflow. This led Hess to conclude that there was not strict localization of function in the hypothalamus as (he assumed) there is in the cortex. [It might be pointed out that recent discussions of cortical function indicate considerable doubt about the reality of localized functions isomorphic with their behavioral expressions, even in the motor areas of the cortex (Lashley, 1923; Hines, 1947; Clark and Ward, 1948; Clark, 1948).]

Now, a strictly punctate localization of function is not necessary, either in cortex or in hypothalamus, in order that these organs be able to serve organizing and coordinating functions. In the light of Hess's work there is no doubt that the lower-level details and components of many behavior patterns are coordinated and integrated in the hypothalamus. But it is difficult to see how the shifting locus of this integration can be reconciled with the conception of a *center which produces an excitation* specific for the behavior patterns concerned. It is equally difficult to reconcile the fact that the function of a "center" depends partly on the type of afferent inflow present with the notion of the center as a place where excitation is produced for a particular kind of act.

Neither do the researches of Hess, nor those of Gray and Lissmann, support the idea that rhythmicity or cyclicity of behavior is a function of the periodic reaching of a threshold level of energy produced in such centers. As suggested by Gray and Lissmann, rhythmicity of behavior is much more parsimoniously explained in terms of periodic shifts in balance between central and peripheral processes or interaction between different central processes, than in terms of the production of periodic impulses by a single "center" which, in Lorenz's treatment, has the character of a "thing in itself."

Lorenz (1950) describes in some detail a hydraulic model, or analogy, of the instinct mechanism, including a reservoir of excitation and devices for keeping it dammed up (innate releasing mechanism) until appropriate keys unlock the sluices. Hydraulic analogies have reappeared so regularly in Lorenz's papers since 1937 as to justify the impression that they are not really analogies—they are actual representations of Lorenz's conception of the origin and channelling of "instinctive energy." [The basic assumptions—of a special center producing a reservoir of energy specific for each instinct, and of devices for distributing the energy—are very similar to those of MacDougall (1923, 1930).]

There is no neurophysiological evidence for such hydraulics in the brain. Aside from the controversial aspects of the idea of automatic-rhythmic production of excitation, such hydraulic conceptions simply do not conform with what we actually know about the complexities of brain function (cf. Fulton, 1949).

The actual physiological relationships underlying behavior patterns must be *analyzed* for the different behavior patterns concerned. The assumption which underlies Lorenz's approach to the neurophysiological basis of behavior is that the neural events underlying behavior patterns must somehow be isomorphic with the behavior itself. He is thereby led to assume that behavior patterns having similar functional characteristics must be caused by identical neural mechanisms. Lashley (1942) has pointed out the erroneous nature of such reasoning. It is by thus abstracting phenotypic resemblances in behavior at different levels, and by gratuitously transferring physiological explanations from one level to another that Lorenz creates the impression that "instinctive" acts are grounded in a common type of mechanism which is the same at different evolutionary levels.

Levels of Behavioral Function

As already pointed out, a serious question facing all investigators of animal behavior is the extent to which different mechanisms may be assumed to be identical because of the apparent similarities in the behavior patterns they underlie. By this I do not mean to imply that the similarities may be unimportant, but only that functionally similar behavior patterns may be effectuated through very dissimilar causal mechanisms. And if the causal mechanisms hypothesized in the case of one of the behavior patterns are conceptually reified and applied to other patterns or other animals, because of the fact that the similar behavior patterns are subsumed under the same term or included in the same category or concept, the analysis of the mechanisms actually operating in the different cases is seriously hampered. Rather than making a developmental analysis of the processes concretely underlying each behavior pattern, the predominant tendency is to carry out brief studies on a variety of *selected* examples assumed to demonstrate the validity of the a priori "principle" or the reality of the hypothesized structure or "center."

This practice may produce very fallacious results. For example, both the amoeba and the neonate infant will move toward weak stimulation and away from strong stimulation (the amoeba as a whole, the child locally). In both animals, this serves the biological function of bringing the organism into contact with food (and for the child, protection), and away from contact with harmful stimuli. This similar biological utility is a sufficient basis of explanation for the evolutionary development of the similar modes of behavior in the two organisms.

But the *mechanisms* underlying the response in the two animals manifestly must be very different. In the amoeba, the differential response to weak and strong stimuli is caused by the differential effects of the weak and strong stimuli on the sol-gel relationship in the protoplasm of the single cell (Mast, 1926). In the neonate child, the basis is more obscure. Schneirla (1939) has suggested that initially it is the result of differences in arousal-threshold between flexor and extensor muscles of the limbs, so that they respond optimally to different impulse-frequencies in afferent volleys, corresponding to different intensities of stimulation.

These two behavior patterns may seem functionally quite analogous. Can we say that they are homologous? This would obviously be absurd. They represent two totally different kinds of adjustment, both selected (in the evolutionary sense) because they serve the same kind of function. Nor is there a "smooth progression" between the pro-

tozoan response and the human. At each level, the *mechanisms* underlying this characteristic and widely-distributed response (Maier and Schneirla, 1935) are derived from the specific structure of the organism in question. The behavior patterns are not homologous, although they *may* in some cases be based to different extents and in different ways on more or fewer homologous structures. The *analysis* of the behavior at each level must be in terms of its emergence from the structure of organisms at that level, as indicated in the examples of the amoeba and the neonate child, and *not* on superficial comparisons of the behavior with similar behavior patterns at other levels.

Lorenz's concept of "instinct" represents, I think, precisely this kind of undesirable reification of a hypothesized mechanism. Lorenz's use of the term "instinct" does not denote merely a group of behavior patterns characterized by certain common functional characteristics; it denotes a definite class of things—a specific group of homologous structures underlying acts *whose characteristics are isomorphic with those of the structures.* And the nature of the structures is inferred from the behavioral characteristics, supported by physiological evidence the inadequacy of which has already been pointed out.

This reification of the concept of "instinct" leads to a "comparative" psychology which consists of comparing levels in terms of *resemblances* between them, without that careful consideration of *differences* in organization which is essential to an understanding of evolutionary change, and of the historical emergence of new capacities. Thus the lowering of intensity of response as a consequence of repeated elicitation, in the case of certain sexual activities of a butterfly (Tinbergen, Meeuse, Boerema, and Varossieau, 1942) and of a fish (Tinbergen, 1942) is taken in both cases as verifying Lorenz's assumption of the nature of the organizing center for an instinctive act. The fact that some behavior patterns of a butterfly may exhibit functional similarities to some behavior patterns of a fish is interesting as an indication that similar response characteristics may be species-preserving in both cases. But it is not very judicious, and actually is rash, in view of the very different types of organization involved in the structure and the behavior of the two animals concerned, to assume that the mechanisms underlying the two similar response characteristics are in any way identical, homologous, or even similar, or that there is any historical (evolutionary) continuity between them *as such*. Yet this is precisely the basis of Lorenz's whole treatment of "instinct" and evolution.

In addition to distorting comparative study and the study of evolutionary change, this reification of "instinct" has unfortunate effects on the study of ontogenetic development. The development of an "instinctive" act inevitably appears to Lorenz to be 'the self-differentiation of a preformed, autonomous thing. Thus Lorenz sees the developing behavior of the animal as progressing *toward* the full-blown "instinct" rather than as developing *out of* interactions among processes present at that stage. This is a teleology which is inherent in Lorenz's approach, and which cannot be eliminated by his formal attempt to deny teleological and purposive procedures and to exclude the terminology.

For example, Lorenz mentions the development of fighting behavior in ducks. When fighting with another drake, an adult drake will grasp its opponent's neck in its bill and strike at him with a wing. Lorenz noted that ducklings whose wings had not yet feathered would perform the same movements even though the stubby, unfledged wing was not yet long enough to strike the opponent. Lorenz's interpretation (1937a) is that the instinctive act had matured before the full maturation of the structure which was used by it. This interpretation does not explain what the *duckling* is doing; rather, it prevents the investigator from seeing the problem of what it is about the duckling and its situation (and its ontogenetic history) which gives rise to this kind of behavior. This type of theory apparently causes the investigator to look at the process of development in such a way that the problem of the *origin* of this behavior, and its cause and role *in the duckling* are not considered by him at all. In the light of our previous discussion, it would appear that these are the crucial problems, and that a theory which makes them appear as relatively irrelevant to the explanation of the development of fighting behavior must be seriously lacking.

This conceptual merging of very different levels on the basis of superficial similarities permeates the system. For example, the concept of "taxis" as a meaningful class of behavior elements seems to be based on such a procedure. Lorenz defines a taxis as a movement which is continuously oriented with respect to the stimulus (thus distinguishing it sharply from an instinctive act which, once started, is centrally coordinated, independent of the receptors).

Tinbergen (1942) further classifies taxes into several categories, based partly on Kühn's (1919) analysis: (1) tropotaxis, equivalent to Loeb's tropisms, in which the animal turns until the relevant stimulus is equally intense on both sides; (2) telotaxis, a visual orientation based on fixation movements so that either eye can serve as the sole receptor; and (3) menotaxis, like telotaxis except that the orientation, instead of being toward the stimulus, is at a constant angle from it. To these categories of Kühn, Tinbergen added a fourth: pharotaxis, in which the animal is oriented to a part of the visual field defined in terms of its relation to the rest of the field, irrespective of the animal's orientation to the field.

This classification of "taxes" solely in terms of a highly restricted definition of the receptor processes inevitably lumps together many very different processes. For example, our amoeba and newborn infant both show a "turning-to" reaction to mild stimulation. What possible category, based on the characteristics of the turning, could properly include both of these movements as examples of one kind of process? To say that the movement of the child and of the amoeba are both a taxis is to admit that the word "taxis" does not define a group of behavior patterns which have common mechanisms.

Tinbergen (1942) makes this explicit when he says: ". . . in the concept of pharotaxis the part played by mnemic processes is not taken as a criterion, because in tropo-, meno-, and telotaxis the criterion upon which the 'distinction' is based also leaves this topic out of consideration . . . Menotaxis, for instance, can be innate or learned."

What then remains of Tinbergen's classification? Tinbergen himself is aware that the members of any of his taxis-categories probably differ widely in ontogenetic origin and central mechanism. How can the classification be justified? A preliminary classification has heuristic value only if the members of a given class are thought to be representative of similar dynamic processes which can be investigated. That, in fact, is the purpose which Tinbergen assumes for his classifications. But in the case of taxes, the classifications are known to contain different levels of organization and different processes. In this case the classification is based on the analogizing which appears to be basic to the Lorenz approach.

Lorenz (1939) and Tinbergen (1942) have both pointed out that, under the influence of natural selection, widely divergent species may develop similar characteristics which should not be assumed to be homologous. Tinbergen (1948) gives an example of this "convergent evolution": "The most striking example of how far convergencies can go in these phenomena, is given by L. Tinbergen (1939) in his study of the mating behavior of the cuttlefish *Sepia officinalis*. Parallel with the development of eyes in cephalopods (convergent to those of fish) the courtship of the cuttlefish has evolved into a typically visual one closely resembling the courtship of certain sexually dimorphic fish, lizards, and birds.... This state of affairs closely resembles that found by Noble and Bradley (1933) in [the lizard] *Sceloporus*. In both species the male's display is primarily a means of threatening other males...."

It will be noted that Tinbergen specifically notes that the resemblance is caused by convergence, rather than homology. However, his treatment of the behavior patterns involves the implicit assumption that the convergence is one of *mechanisms*. Actually, as far as we know, the convergence is only of *outcomes*. The assumption that the mechanisms underlying these similar outcomes are equally similar is both characteristic and gratuitous.

The Human Level

This analogizing and confusion of levels becomes patently shallow when either Lorenz or Tinbergen discusses human behavior.

For example, Tinbergen (1942) says, "The activation of other drives, too, leads to searching behavior. Classical examples are the searching for a nesting site in birds, for a house in man, etc." It is difficult to see what valid explanatory purpose can be served by such an inappropriate juxtaposition, based on the mere fact that the *outcomes* are similar *from the human point of view*.

Tinbergen (1942), speaking of "instinctive"acts which appear without external stimulation, as the result of extreme lowering of the threshold because of long non-elicitation, says, "Lorenz ... discovered that various activities may occur in cases, where neither proprioceptive stimuli nor hormones could possibly be the driving causes. The simplest instance of this kind of vacuum activity is the hunting behavior of the well-fed dog. As every dog-owner knows, a dog can by no means be prevented from making hunting excursions by supplying it with ample food. Other instances of a similar kind are familiar to us by introspection. Sports,

science and so many other activities certainly have connections with internal factors of this kind."

Here the implication seems to be that, because both are "spontaneous" and neither is mainly caused by proprioceptive stimuli or hormones (itself a gratuitous assumption), therefore the causes of hunting activity in the dog are the same as (or belong to the same class as) those of scientific activity in human beings! It is obvious that this argument is based on the most casual and unanalytical kind of comparison, and a lack of concern with the specific origins of the behavior patterns at issue.

Lorenz (1937b), speaking of the evolutionary relation between instinctive and learned acts, says:

"The presence of an instinctive act also seems to be detrimental to the development of an intelligent process having the same function. At least, it is true of humans. To be convinced of the correctness of this statement, one has only to consider the behavior of highly intelligent men who have otherwise good critical faculties, when they 'fall in love' to carry out the undoubtedly instinctive reaction of mate-selection. The already-mentioned example of the ravens and jackdaws shows that higher psychological development may occur without any reduction of the instinctive, innate members of a behavior chain...."

I include the last sentence to show how very easily Lorenz switches from man to bird without any apparent awareness that he is discussing phenomena which may be very different. The point of Lorenz's statement seems to be that men fall in love irrationally because "falling in love" is an instinctive reaction released by an innately-determined situation. In this case the unreality of the concepts used is apparent to any student of *human* behavior, although it may not be so to one of bird behavior—a fact which itself indicates that the source of the unreality lies partly in Lorenz's merging of different levels on the basis of superficial similarities.

Many other examples of Lorenz's interpretation of human behavior could be cited. For example (Lorenz, 1940), he interprets the relative attractiveness to women of several breeds of dog in terms of the degree to which they fit the innate perceptual pattern releasing instinctive maternal behavior in the human individual! This, again, is entirely derived from a too facile analogy with less complex kinds of animals. Recent work with chimpanzees reared in darkness (Riesen, 1947) and with congenitally blind human beings whose sight had just been restored by surgery (Senden, 1932) indicates that, at least at these phyletic levels, any response to or perception of visual shapes, proportions, sizes, and relationships can only occur as the result of a long and complex learning process. Under these circumstances it is most difficult to assign any meaning whatever to Lorenz's assertion that these responses are based innately on perceptual characteristics of shape and proportion (Hebb, 1949). [Lower mammals apparently require less learning for the establishment of some of their characteristic modes of response to the visual field (Hebb, 1937a, 1937b; Lashley and Russell, 1934).]

The interpretation of human behavior in terms of physiological theory based on lower levels is carried one step further when Lorenz (1940) equates the effects of civilization in human beings with the effects of domestication in animals. He states that a major effect is the involution or degeneration of species-specific behavior patterns and releaser mechanisms because of degenerative mutations, which under conditions of domestication or civilization are not eliminated by natural selection. He presents this as a scientific reason for societies to erect social prohibitions to take the place of the degenerated releaser mechanisms which originally kept races from interbreeding. This is presented by Lorenz in the context of a discussion of the scientific justification for the then existing (1940) German legal restrictions against marriage between Germans and non-Germans.

The directness of Lorenz's application of the concept of innate releasers to human social relations may be gaged by the following quotation (1940): "The face of an Asiatic is enigmatic to us because the physiognomic characteristics to which our innate perceptual patterns respond are not connected with the same behavioral characteristics as in our race. . . . In all likelihood, this function (of recognizing facial characteristics) cannot be substituted for by experience, as has been determined by many people who are acquainted with foreign races."

Social psychologists will all agree that the various degrees of difficulty which different people have in learning to recognize and respond to facial expressions in a culture different from their own is at least partly dependent upon the attitude with which they approach the strange culture to begin with.

The Sources of Motivation

Lorenz states that as the level of action-specific energy in the central nervous system rises the animal is set into activity. He says (1937b), "It is one of the most important and most remarkable features of the instinctive act, that the organism does not wait passively for its release, but *actively seeks these stimuli.*" This active seeking is called "appetitive behavior" (Craig, 1918). It may range from simple turning movements ("taxes") to the most complicated kinds of intelligent behavior. As has already been pointed out, Lorenz regards these as being continuous with each other, both being (at different stages of evolutionary development) means of bringing the animal into a situation containing the stimuli which will release an instinctive act (viz., eating, copulation, etc.).

Lorenz recognizes a few instances of motivation the source of which is peripheral (viz., hunger, defecation, etc.). But he adopts the characteristic procedure of lumping together all "goal-directed" (i.e., adaptive) behavior which does not have an immediately obvious peripheral motivation under the rubric of "appetitive behavior." His conception of "peripheral sources of motivation" is practically limited to the examples just given. I do not recall any reference in any of the writings of either Lorenz or Tinbergen to the autonomic nervous system, or to the possibility of qualitatively different roles of the autonomic nervous system at different phyletic levels, or to the possibility of complex peripheral changes caused by hormones, as sources of motivation (cf. Beach, 1948). One result is that the referring of motivation to the action-specific "centers" in the central nervous system is often like the concept of "innate behavior" itself, simply a substitute for actual analysis of the biology of the specific case. For example, Tinbergen (1951) says that injection of prolactin into a dove has two effects: (1) it causes development of the crop gland, and (2) it causes brooding behavior. It is thus more or less taken for granted that the behavioral effect of the hormone is somehow a specific one; developmental analysis of relationships between broodiness and crop-gland or brood-patch stimulation is excluded by the nature of the instinct theory, and of the consequent theory of motivation. In the case of "dominance" behavior in the chimpanzee Birch and Clark (1950) have shown that behavioral effects of hormones may actually be mediated by peripheral structures in situations where it is not at all apparent a priori that "proprioceptive stimuli" can play a role.

Lorenz regards all purposive (adaptive) behavior as being directed toward the performance of the instinctive acts. For Lorenz, it is the *performance of the instinctive act itself* which serves as the animal's goal. Thus he says (1937a):

"... in a man working with the motive of getting food, the behavior directed toward this goal includes many of the higher psychic performances of which he is capable; the 'motive' (goal)—the instinctive act of 'biting and chewing'—has become drawn back to the end of a long series of acts, without, however, thereby in any way denying its fundamentally instinctive nature."

Thus to Lorenz, the statement "man works in order to be able to have released [by food] the instinctive act of biting and chewing" is the *same* kind of statement as "the frog turns to the right in order to be able to have released [by the sight of a fly in front of him] the instinctive act of flipping out his tongue." He regards these two goal-directed behavior patterns as being (in the evolutionary sense) continuous with each other, and both as having the same kind of relationship to the instinctive act which is the end-member of the behavior chain. However, such a formulation is misleading and of little heuristic value. The actual complexity and variety, and situational relevance, of the sources of human motivation make such statements meaningless, not merely because human motivation is more complicated than that of the frog, but because it is qualitatively different in organization and development.

Tinbergen's equation of the causes of sports and scientific activity with those of hunting in the dog, because both appear to be internal and "self-exciting," is perhaps an extreme example of the result of analogical methods of approach, and of the belief that every behavior must have some center isomorphically corresponding to it in the nervous system.

HYPOTHESIS AND OBSERVATION

The "Innate Releasing Mechanism"

It may be instructive to examine some ways in which Lorenz's theoretical approach is expressed in an investigation of behavior.

Tinbergen and Kuenen (1939) studied the stimulus situations eliciting and directing the gaping (food-begging) movements of young thrushes. The gaping movement consists at first of vertically directed stretching of the neck, and opening of the mouth. The birds are blind at hatching; their eyes do not open until about 9–10 days of age. During this first blind phase gaping can be most easily elicited by tapping or jarring the substrate. When the eyes first open the bird normally lies with its eyes closed, and opens them only when it is gaping. Tinbergen and Kuenen state that the bird will gape in response to a moving visual stimulus as soon as the eyes open, and that the innate releasing pattern for gaping therefore includes visual stimulation. For the first day or so after the eyes open gaping is not *directed* toward a visual stimulus; even though the stimulus will elicit gaping, the gaping is still directed vertically upward. However, after about one day the gaping begins to be directed toward certain defined parts of the visual stimulus (highest, nearest, break in outline, etc.).

Tinbergen and Kuenen's conclusions are that the "centrally-coordinated" instinctive act and the (continuously-directed) taxis mature at different rates, the taxis not maturing until 10 days or so of age while the instinct is fully mature at hatching. In addition, they conclude that the adequate stimulus-situations for releasing and for direction of the act are different, and are both innate.

First, a word about the "maturation of the taxis." It is not clear why the animal's experience during the first day after its eyes open is not an adequate reason for its development of orientation toward the visual stimulus. Tinbergen and Kuenen maintain in their discussion that some of the specific features of the stimulus toward which orientation occurs are not learned by direct experience. However, it is not clear that the *orientation* toward the visual stimulus is not a result of experience. Even their limited discussion of possible learning is based on inferences from incidental observations, indicating that Tinbergen and Kuenen's orientation toward Lorenz's theory led them to discount the serious possibility of learning being involved.

The "innateness" of response to moving visual stimuli is quite ambiguous. It will be recalled that the birds lie with their eyes closed for much of the time just after their eyes have first opened. I quote from Tinbergen and Kuenen's protocols:

"5/26/36. Black Thrush, 9–10 days. Lifting and moving back and forth of the wooden covers evokes no reaction. They gape immediately to a tap on the nest. When the gaping subsides, we move our hand back and forth over the nest, and the birds instantly beg strongly.

"Later on the same day: Tap on the closed box releasing gaping. Subsequent lifting of the cover does not; the eyes are closed. Tap causes gaping. After cessation of the reaction, the animals remain lying with open eyes. We hold over the nest, one after the other: a black disc, a white wooden rod ... and a black wand.... All the objects are reacted to by violent gaping.

"5/10/36. Song Thrush, 10 days. Preliminary lifting

of the cover causes no reaction; light tap does. After the birds begin to calm down, a finger shown over the nest immediately releases gaping."

It is apparent that normally the *first times that young thrushes see visual stimuli they are already in a state of gaping excitement,* since at first their eyes are open only when they gape.

I have verified these findings on young red-winged blackbirds (*Agelaius phoeniceus*), on which I could repeat all of Tinbergen and Kuenen's observations. However, I was able by watching the birds for several consecutive hours to note several occasions on which one or another of the birds was lying quietly with its eyes *open,* when it had not recently gaped. Such birds would *not* gape in response to a moving finger above the head, although they might move their heads to fixate the finger. If I tapped the nest, thus causing gaping, and then moved my finger over the birds when gaping was subsiding, or shortly thereafter, the bird would gape instantly and vigorously. What is meant then by the statement that the birds gape "innately" to visual stimulation? It would be easy to produce ad hoc assumptions about temporary changes in threshold of the innate releasing mechanism as a result of tactual stimulation. But these must be recognized as ad hoc. The possibility should be recognized that any stimulation to which the bird is sensitive will increase the activity of the bird when it is already gaping, and may become associated with gaping, so that the later-apparent specificity of response to visual stimuli may be a consequence, not of innate connections, but of the conditions under which visual sensitivity normally first becomes possible. In addition, these birds must be fed almost every hour, and the possible relevance of association of visual stimulation with food reinforcement should not be overlooked. In this connection we may note the experiments of Padilla (1930), who found that chicks that were kept in the dark and force-fed for the first twelve days of life, so that they had no opportunity to associate pecking behavior with visual stimuli or with food, would when placed in a normal feeding situation starve to death without ever giving any sign of the allegedly "innate" pecking behavior.

It should be noted that the conditional nature of the effectiveness of visual stimuli is indicated by Tinbergen and Kuenen's own protocols, but that evidently these authors have not really considered these facts. This, I think, is because they are

a priori convinced that the developmental process is a maturational one, and that they therefore do not have to analyze its conditions. The Lorenz theoretical approach tends to restrict the recognition of significant details and to obscure possibly relevant features of developmental processes.

Many cases of "innate releasing mechanisms" seem to suffer from a similar approach. It will be recalled that the innate releasing mechanism is regarded as a "preformed neural mechanism" (Lorenz and Tinbergen, 1938) for the release of the instinctive act. Tinbergen refers to the releasing stimuli as "sign stimuli" because they "represent" the biologically appropriate object of the instinctive act. One might ask "Sign of what? Sign to whom?" There is a subtle anthropomorphism about the concept of innate releasing mechanisms which is not at first apparent. For example, Lorenz and Tinbergen (1938), in discussing the egg-rolling of the gray goose, speak of an innate releasing pattern corresponding to the situation "egg outside the nest." Now, "egg outside the nest" is *not* the perceptual situation to the bird—it is the perceptual situation to the human observer. When Lorenz and Tinbergen investigate the effective features of the situation, they are looking for a pattern of stimuli corresponding to a "pattern" which they presume to exist in the central nervous system. Consequently, there is never any analysis of any possible *specific* relationships between effective stimuli and the structure or physiology of the organism concerned. Thus the described stimulus-situations become structured in human terms (bird of prey, vegetation, the parent's head, etc.) instead of in terms indicative of the problems of specific relationships between the structure and function of the animal being investigated. This approach, again, derives from Lorenz's identification of every behavior pattern with a specific hypothetical "center," rather than with the coming into play of specific structural-functional relationships, which may be very different in different kinds of organisms and for different behavior patterns.

For example, Tinbergen (1948b) says: "The escape reactions of many birds from passing birds of prey are a response to a type of movement and to a special characteristic of shape, namely, 'short neck.' " Now, it is certainly true that many birds perform "escape" movements at the sight of a "short-necked" bird flying overhead (Krätzig, 1940). But Tinbergen says "short-necked" rather

ON LORENZ'S THEORY OF INSTINCTIVE BEHAVIOR 357

than, for example, "having a short and long projection at opposite ends, and moving so that the short projection is anterior." His usage is, of course, more convenient, as he makes clear. But in addition it derives from, and in turn reinforces, the Lorenzian notion that the "short-neckedness" is a perceptual "sign" or "sign stimulus" (Tinbergen, 1939) which corresponds innately to a preformed neural "releasing" mechanism. Instead of leading to an analysis of the specific patterns of excitation of sensory elements in the bird's eye which are required for the elicitation of the response, and a further consideration of the effect of such patterning on central nervous activity in the nervous systems of these birds, Tinbergen's terminology requires the identification of the bird's readiness to perform "escape movements" with a preformed "conception" of the short-necked character of hawks. Thus, "the partridge runs for cover from an overhead object with a short neck," and "the goose rolls back to the nest an object lying near the nest which is smooth-contoured and hard-surfaced," become not *definitions* of the problem of how the structure of the various birds makes it possible for them to react to their environment, but rather *solutions* to the question: "What are the characteristics of these two members of the class of innate releasing mechanisms?" The essential assumption of Lorenz's approach is that these two types of behavior are related to environmental stimuli by means of mechanisms that are basically identical except for the perceptual details themselves. When extended (as it is) to the whole animal kingdom, this approach becomes profoundly anti-evolutionary, in spite of Lorenz's concern with "comparative" studies.

Lashley (1949) has noted with some approval Lorenz's studies of releaser patterns. For example he has said:

"A study of complex instincts requires a detailed analysis of the exact stimulus or combination of stimuli which call forth the behavior, combined with descriptions of the behavior elicited. This has been attempted under controlled conditions only for some instinctive behavior of birds (Lorenz, 1935)."

However, a closer examination of Lashley's concepts and those of Lorenz will show that the subsequent development of Lorenz's approach was not at all in the direction anticipated by Lashley. Lashley (1949) says:

"The nesting tern seems to notice no difference when her eggs are dyed . . . but is . . . disturbed if their . . . contour is altered by sticking on a bit of clay or putty . . . smoothness of outline is the essential character of the egg. This is the sort of property that can be most easily interpreted in terms of the inherent tendencies to functional organization in the nervous network.

"I do not mean to imply by this that the geometry of the web of the spider is exactly represented in the spider's brain. . . . The angle of radii may be determined by the angle at which the legs are held (Peters, 1937); the completeness or incompleteness of the orb may depend upon the readiness with which certain postures are assumed in relation to gravity. . . . The simple nest of the rat is piled and pushed about until it satisfies certain sensory requirements of reduced heat loss. The orb of the spider is perhaps a composite of such sensory requirements combined with some specialized geometrical perceptions such as are illustrated by the rat's more ready recognition of a . . . circle than of irregular ink blots."

In contrast to this approach, Lorenz (1935) has used the analogy of a key unlocking a lock, to describe the function of the releasing pattern in releasing an instinctive act. To pursue this analogy, Lashley would regard as the task of the locksmith-investigator to investigate all the characteristics of lockopening devices, including keys, picklocks, and any other means of opening the locks; and to consider these characteristics together with what he knows of the structure of locks, the conditions of their use, their history, etc., in order to gain an understanding of how the functions of the various kinds of locks are related to their structure. To Lorenz on the other hand, all the locks are basically alike, so that investigations of the characteristics of the keys required to open them reveals nothing about internal differences among lock mechanisms, but only about the specific arrangement of tumblers in each lock.

All of this should not be taken to mean that we do not recognize that relatively simple stimuli may sometimes lead to the appearance of quite complex behavior. As a matter of fact, some of the best studies of stimulus-conditions eliciting various types of animal behavior have been carried out by Lorenz and Tinbergen and their associates (e.g., Tinbergen and Perdick, 1950). The point is not to deny the existence of simple stimuli which under some conditions lead to complex behavior. Rather, it is that the assigning of the locus of activity to a hypothetical center in the brain, with characteristics predeterminedly and isomorphically corresponding to those of the stimulus situation, represents an unphysiological way of thinking disguised in physiological terms.

"Vacuum Activities"

The so-called "vacuum activities" or Leer-laufreaktionen are regarded by Lorenz and Tinbergen as evidence of the accumulation of reaction-specific energy in the instinctive center until it "forces" its way through the inhibiting innate releasing mechanism and "goes off" without any detectable external stimulus.

Lees (1949) has cited the example of the cyclical colony activities of the ant *Eciton hamatum* (Schneirla, 1938) as an example of "something akin to 'vacuum activity.'" Colonies of this army ant pass regularly through *statary* and *nomadic* phases, each lasting about 20 days. As Lees points out [based on Schneirla's (1944) description]:

"During the statary phase the bivouac, to which the single queen is confined, remains *in situ* and raiding activities are minimal. During the nomadic phase the position of the bivouac is changed each nightfall and strong raiding parties emerge from the colony. This activity is in no way related to the abundance or scarcity of food in the neighborhood. . . ."

This cyclic behavior thus appears to Lees to have the character of a "vacuum activity," in that it occurs periodically without any noticeable change in the external stimulus-conditions. This is very misleading, for Schneirla's (1938, 1944) analysis of this behavior has shown that the change from statary to nomadic behavior is a consequence of the growth of a great new brood of ants. When the callow workers emerge from their cocoons, their movements stimulate the adult workers to great activity. As the callows mature and cease to be dependent on the adults, their energizing effect is lessened. At this point, the emergence of wriggling larvae from the eggs supplements the diminishing activating effect of the callows on the adults. When the larvae pupate, and become inactive, the adults are no longer subject to trophallactic (Wheeler, 1928) stimulation, and the colony changes to its statary period.

The point that is relevant to our discussion is that Schneirla's analysis leads to a conception that is the *opposite* of that implied by the notion of "vacuum activity." The periodic recurrences are *not* the result of the building up of energy in any animal's nervous system. They are the result of the periodic recurrences of inter-individual stimulating effects. The behavior is not represented "in advance" in *any* of the animals in the colony; it emerges in the course of the ants' relationships with one another and with the environment. There

is no "reaction-specific energy" being built up. The periodicity is a result of the periodicity of the queen's egg-laying, which is *not* a "center" having *any* characteristics corresponding to the behavior. And even this is not a *direct* relationship. If the number of larvae in a colony is experimentally reduced by 50 per cent, thus reducing their total stimulating effect, a normal nomadic phase cannot occur. Recent findings (Schneirla and Brown, 1950) have in fact confirmed the hypothesis that each of the regular large-scale egg-delivering episodes in the queen's function basic to the cycle is a specific outcome of her over-feeding, due to a maximal stimulation of the colony by the brood. This event, occurring inevitably at the end of each nomadic phase, is a "feed-back" type of function, not at all related to the implications of "vacuum activity."

The restrictive nature of such categorical theories as that of Lorenz is very well illustrated by Lees' remarks on *Eciton*. The actual development process leading to the periodic performances of this ant are well understood, and are *known* to have no essential relationship to any "reaction-specific energy" in any nervous system; further they are *known* not to be "innate" as such (Schneirla, 1938). The processes leading to this behavior surely have nothing to do with the processes leading to "vacuum activities" in a fish. Yet the superficial similarity is sufficient to cause Lees to cite the ant's behavior as an example of a type of behavior described for vertebrates. This is a good example of the tendency encouraged by such theories to look for cases fitting the theoretical categories in many types of behavior, rather than analysis of the processes involved in the development of any one behavior pattern.

CONCLUSION

We have summarized the main points of Lorenz's instinct theory, and have subjected it to a critical examination. We find the following serious flaws:

1. It is rigidly canalized by the merging of widely different kinds of organization under inappropriate and gratuitous categories.

2. It involves preconceived and rigid ideas of innateness and the nature of maturation.

3. It habitually depends on the transference of concepts from one level to another, solely on the basis of analogical reasoning.

4. It is limited by preconceptions of isomorphic

ON LORENZ'S THEORY OF INSTINCTIVE BEHAVIOR 359

resemblances between neural and behavioral phenomena.

5. It depends on finalistic, preformationist conceptions of the development of behavior itself.

6. As indicated by its applications to human psychology and sociology, it leads to, or depends on, (or both), a rigid, preformationist, categorical conception of development and organization.

Any instinct theory which regards "instinct" as immanent, preformed, inherited, or based on specific neural structures is bound to divert the investigation of behavior development from fundamental analysis and the study of developmental problems. Any such theory of "instinct" inevitably tends to short-circuit the scientist's investigation of intraorganic and organism-environment developmental relationships which

underlie the development of "instinctive" behavior.

ACKNOWLEDGMENTS

I am greatly indebted to Dr. T. C. Schneirla (who originally suggested the writing of this paper) and to Dr. J. Rosenblatt for many stimulating and helpful discussions of the problems discussed here. Dr. Schneirla in particular has devoted much attention to criticism of the paper at various stages.

The following people also have read the paper, in part and at various stages, and have made many helpful suggestions and comments: Drs. H. G. Birch, K. S. Lashley, D. Hebb, H. Klüver, L. Aronson, J. E. Barmack, L. H. Hyman, L. H. Lanier, and G. Murphy. Since these scientists differ widely in the extent of their agreement or disagreement with various points of my discussion, I must emphasize that none of them is in any way responsible for any errors of omission or commission that may appear.

Present address: Rutgers University, Newark 2, N. J.

LIST OF LITERATURE

ADRIAANSE, M. S. C. 1947. Ammophila campestris Latr. und Ammophila adriaansei Wilcke. Ein Beitrag zur vergleichenden Verhaltensforschung. Behaviour, 1: 1–34.

ANASTASI, A., and J. P. FOLEY, JR. 1948. A proposed reorientation in the heredity-environment controversy. Psychol. Rev., 55: 239–249.

ARMSTRONG, E. A. 1947. Bird Display and Behaviour. Lindsay Drummond & Co., London.

——. 1950. The nature and function of displacement activities. Symp. Soc. exp. Biol., 4: 361–384.

BAERENDS, G. P. 1941. Fortpflanzungsverhalten und Orientierung der Grabwespe Ammophila campestris Jur. Tijdschr. Ent., 84: 68–275.

——. 1950. Specializations in organs and movements with a releasing function. Symp. Soc. exp. Biol., 4: 337–360.

——, and J. M. BAERENDS-VAN ROON. 1950. An introduction to the study of the ethology of cichlid fishes. Behaviour, suppl. 1: 1–242.

BEACH, F. A. 1942. Analysis of factors involved in the arousal, maintenance and manifestation of sexual excitement in male animals. Psychosom. Med., 4: 173–198.

——. 1947a. Evolutionary changes in the physiological control of mating behavior in mammals. Psychol. Rev., 54: 297–315.

——. 1947b. A review of physiological and psychological studies of sexual behavior in mammals. Physiol. Rev., 27: 240–307.

——. 1948. Hormones and Behavior. Hoeber, New York.

——. 1951a. Instinctive behavior: reproductive activities. In Handbook of Experimental Psychology (S. S. Stevens, ed.) pp. 387–434. John Wiley & Sons, New York.

——. 1951b. Body chemistry and perception. In Perception: an Approach to Personality (R. R. Blake and G. V. Ramsey, eds.), pp. 56–94. Ronald Press, New York.

BIRCH, H. G., and G. CLARK. 1946. Hormonal modification of social behavior. II. The effects of sex-hormone administration on the social dominance status of the female-castrate chimpanzee. Psychosom. Med., 8: 320–331.

——, and ——. 1950. Hormonal modification of social behavior. IV. The mechanism of estrogen-induced dominance in chimpanzees. J. comp. physiol. Psychol., 43: 181–193.

BIRD, C. 1925. The relative importance of maturation and habit in the development of an instinct. Pedagog. Semin., 32: 68–91.

——. 1933. Maturation and practise; their effects upon the feeding reaction of chicks. J. comp. Psychol., 16: 343–366.

BREED, F. 1911. The development of certain instincts and habits in chicks. Behav. Monogr., 1: 1–78.

BROWMAN, L. G. 1943. The effect of controlled temperatures upon the spontaneous activity rhythms of the albino rat. J. exp. Zool., 94: 477–489.

CARMICHAEL, L. 1936. A re-evaluation of the concepts of maturation and learning as applied to the early development of behavior. Psychol. Rev., 43: 450–470.

——. 1941. The experimental embryology of mind. Psychol. Bull., 38: 1–28.

——. 1947. The growth of the sensory control of behavior before birth. Psychol. Rev., 54: 316–324.

CHEIN, I. 1936. The problems of heredity and environment. J. Psychol., 2: 229–244.

CLARK, G. 1948. The mode of representation in the motor cortex. *Brain*, 71: 320–331.

——, and J. W. WARD. 1948. Responses elicited from the cortex of monkeys by electrical stimulation through fixed electrodes. *Brain*, 71: 332–342.

COGHILL, G. E. 1929. *Anatomy and the Problem of Behavior*. Cambridge University Press, London.

CRAIG, W. 1918. Appetites and aversions as constituents of instincts. *Biol. Bull.*, *Woods Hole*, 34: 91–107.

CRUZE, W. W. 1935. Maturation and learning in chicks. *J. comp. Psychol.*, 19: 371–409.

DELACOUR, J., and E. MAYR. 1945. The family Anatidae. *Wilson Bull.*, 57: 3–55.

FULTON, J. F. 1949. *Physiology of the Nervous System*, 3d ed. Oxford University Press, New York.

GOS, E. 1933. Les reflexes conditionnels chez l'embryon d'oiseau. *Bull. Soc. Sci. Liége*, No. 4–5: 194–199; No. 6–7: 246–250.

GRAY, J. 1939. Aspects of animal locomotion. *Proc. roy. Soc.*, B, 128: 28–62.

——. 1950. The role of peripheral sense organs during locomotion in the vertebrates. *Symp. Soc. exp. Biol.*, 4: 112–126.

——, and H. W. LISSMANN. 1940. The effect of deafferentation upon the locomotory activity of amphibian limbs. *J. exp. Biol.*, 17: 227–236.

——, and ——. 1946a. Further observations on the effect of deafferentation on the locomotory activity of amphibian limbs. *J. exp. Biol.*, 23: 121–132.

——, and ——. 1946b. The co-ordination of limb movements in the amphibia. *J. exp. Biol.*, 23: 133–142.

GROHMANN, J. 1938. Modifikation oder Funktionsreifung? Ein Beitrag zur Klärung der wechselseitigen Beziehungen zwischen Instinkthandlung und Erfahrung. *Z. Tierpsychol.*, 2: 132–144.

HARTLEY, P. H. T. 1950. An experimental analysis of interspecific recognition. *Symp. Soc. exp. Biol.*, 4: 313–336.

HEBB, D. O. 1937a. The innate organization of visual activity: I. Perception of figures by rats reared in total darkness. *J. genet. Psychol.*, 51: 101–126.

——. 1937b. The innate organization of visual activity: II. Transfer of response in the discrimination of brightness and size by rats reared in total darkness. *J. comp. Psychol.*, 24: 277–299.

——. 1949. *The Organization of Behavior*. J. S. Wiley & Sons, New York.

HEINROTH, O. 1910. Beiträge zur Biologie, namentlich Ethologie und Psychologie der Anatiden. *Int. orn. Congr.*, 5 (Berlin): 589–702.

——. 1930. Ueber bestimmte Bewegungsweisen bei Wirbeltieren. *S. Ges. naturf. Fr., Berl.*, 1929: 333–342.

HEPPEL, L. A., and C. L. A. SCHMIDT. 1938. Studies on the potassium metabolism of the rat during pregnancy, lactation and growth. *Univ. Calif. Publ. Physiol.*, 8: 189–205.

HESS, W. R. 1949. *Das Zwischenhirn. Syndrome Lokalisationen, Funktionen*. Benno Schwabe & Co., Basel.

——, and M. BRÜGGER. 1943. Das subkortikale Zentrum der affektiven Abwehrreaktion. *Helv. physiol. acta*, 1: 33–52.

HINES, M. 1947. The motor areas. *Fed. Proc.*, 6: 441–447.

HOLST, E. VON. 1935. Alles oder Nichts, Block, Alternans, Bigemini und verwandte Phänomene als Eigenschaften des Rückenmarks. *Pflüg. Arch. ges. Physiol.*, 236: 149–159.

——. 1936a. Versuchen zur relativen Koordination. *Pflüg. Arch. ges. Physiol.*, 237: 93–122.

——. 1936b. Vom Dualismus der motorischen und der automatischrhythmischer Funktion im Rückenmark und vom Wesen des automatischen Rhythmus. *Pflüg. Arch. ges. Physiol.*, 237: 356–378.

——. 1936c. Ueber dem "Magnet-Effekt" als koordinierende Prinzip im Rückenmark. *Pflüg. Arch. ges. Physiol.*, 237: 655–682.

——. 1937. Regulationsfähigkeit im Zentralnervensystem. *Naturwissenschaften*, 25: 625–631, 641–647.

HOWELLS, T. H. 1945. The obsolete dogmas of heredity. *Psychol. Rev.*, 52: 23–34.

——, and D. O. VINE. 1940. The innate differential in social learning. *J. abnorm. (soc.) Psychol.*, 35: 537–548.

HUNT, E. L. 1949. Establishment of conditioned responses in chick embryos. *J. comp. physiol. Psychol.*, 42: 107–117.

JENNINGS, H. S. 1930. *The Biological Basis of Human Nature*. Norton & Co., New York.

KEELER, C. 1931. *The Laboratory Mouse*. Harvard University Press, Cambridge.

KINDER, E. F. 1927. A study of the nest-building activity of the albino rat. *J. exp. Zool.*, 47: 117–161.

KOEHLER, O. 1950. Die Analyse der Taxisanteile instinktartigen Verhaltens. *Symp. Soc. exp. Biol.*, 4: 269–303.

KRAMER, G. 1937. Beobachtungen über Paarungsbiologie und soziales Verhalten von Mauereidechsen. *Z. Morph. Ökol. Tiere*, 32: 752–784.

KRÄTZIG, H. 1940. Untersuchungen zur Lebensweise des Moorschneehuhns, *Lagopus l. lagopus*, während der Jugendentwicklung. *J. Orn., Lpz.*, 88: 139–166.

KÜHN, A. 1919. *Die Orientierung der Tiere im Raum*. G. Fischer, Jena.

Kuo, Z. Y. 1932a. Ontogeny of embryonic behavior in Aves. I. The chronology and general nature of the behavior of the chick embryo. *J. exp. zool.*, 61: 395–430.

——. 1932b. Ontogeny of embryonic behavior in Aves. II. The mechanical factors in the various stages leading to hatching. *J. exp. Zool.*, 62: 453–489.

——. 1932c. Ontogeny of embryonic behavior in Aves. III. The structure and environmental factors in embryonic behavior. *J. comp. Psychol.*, 13: 245–272.

——. 1932d. Ontogeny of embryonic behavior in Aves. IV. The influence of embryonic movements upon the behavior after hatching. *J. comp. Psychol.*, 14: 109–122.

Lashley, K. S. 1923. Temporal variation in the function of the gyrus precentralis in primates. *Amer. J. Physiol.*, 65: 585–602.

——. 1938. Experimental analysis of instinctive behavior. *Psychol. Rev.*, 45: 445–471.

——. 1942. The problem of cerebral organization in vision. *Biol. Symp.*, 7: 301–322.

——. 1949. Persistent problems in the evolution of mind. *Quart Rev. Biol.*, 24: 28–42.

——, and J. T. Russell. 1934. The Mechanism of Vision: XI. A preliminary test of innate organization. *J. genet. Psychol.*, 45: 136–144.

Lees, A. D. 1949. Modern concepts of instinctive behaviour (in section on "Entomology"). *Sci. Prog. Twent. Cent.*, 37: 318–321.

Lissmann, H. W. 1946a. The neurological basis of the locomotory rhythm in the spinal dogfish (*Scyllium canicula, Acanthias vulgaris*). 1. Reflex behaviour. *J. exp. Biol.*, 23: 143–161.

——. 1946b. The neurological basis of the locomotory rhythm in the spinal dogfish (*Scyllium canicula, Acanthias vulgaris*). II. The effect of de-afferentation. *J. exp. Biol.*, 23: 162–176.

Lorenz, K. 1931. Beiträge zur Ethologie sozialer Corviden. *J. Orn., Lpz.*, 79: 67–127.

——. 1932. Betrachtungen über das Erkennen der arteigenen Triebhandlungen der Vögel. *J. Orn. Lpz.*, 80: 50–98.

——. 1935. Der Kumpan in der Umwelt des Vogels. *J. Orn., Lpz.*, 83: 137–213, 289–413.

——. 1937a. Ueber den Begriff der Instinkthandlung. *Folia biotheor., Leiden*, 2: 17–50.

——. 1937b. Ueber die Bildung des Instinktbegriffes. *Naturwissenschaften*, 25: 289–300, 307–318, 324–331.

——. 1939. Vergleichende Verhaltensforschung. *Zool. Anz., 12* (Suppl. band): 69–102.

——. 1940. Durch Domestikation verursachte Störungen arteigenen Verhaltens. *Z. angew. Psychol. Charakterkunde* 59: 2–81.

——. 1941. Vergleichende Bewegungsstudien an Anatinen. *J. Orn., Lpz.*, 89 (Sonderheft): 194–294.

——. 1950. The comparative method in studying innate behavior patterns. *Symp. Soc. exp. Biol.*, 4: 221–268.

——, and N. Tinbergen. 1938. Taxis und Instinkthandlung in der Eirollbewegung der Graugans. I. *Z. Tierpsychol.*, 2: 1–29.

MacDougall, William. 1923. *An Outline of Psychology.* Scribner's, New York.

——. 1930. The Hormic Psychology. In *Psychologies of 1930* (Carl Murchison, ed.), pp. 3–36. Clark Univ. Press, Wooster, Mass.

Maier, N. R. F., and T. C. Schneirla. 1935. *Principles of Animal Psychology.* McGraw-Hill Co., New York.

Mast, S. O. 1926. Structure, movement, locomotion and stimulation in amoeba. *J. Morph.*, 41: 347–425.

Miller, N. E. 1951. Learnable drives and rewards. In *Handbook of Experimental Psychology* (S. S. Stevens, ed.), pp. 435–472. John Wiley & Sons, New York.

Morgan, C. T. 1947. The hoarding instinct. *Psychol. Rev.*, 54: 335–341.

Nissen, H. W. 1951. Phylogenetic comparison. In *Handbook of Experimental Psychology* (S. S. Stevens, ed.), pp. 347–386. John Wiley & Sons, New York.

Noble, G. K., and H. T. Bradley. 1933. The mating behavior of lizards. *Ann. N. Y. Acad. Sci.*, 35: 25–100.

Padilla, S. G. 1930. Further studies on the delayed pecking of chicks. *J. comp. Psychol.*, 20: 413–443.

Patrick, J. R., and R. M. Laughlin. 1934. Is the wall-seeking tendency in the white rat an instinct? *J. genet. Psychol.*, 44: 378–389.

Peters, H. 1937. Studien an der Netz der Kreuzspinne (*Aranea diadema* L.). *Z. Morph. Ökol. Tiere*, 1: 126–150.

Petrunkevitsch, A. 1926. The value of instinct as a taxonomic character in spiders. *Biol. Bull., Wood's Hole*, 50: 427–432.

Ray, W. S. 1932. A preliminary study of fetal conditioning. *Child Develpm.*, 3: 173–177.

Riesen, A. H. 1947. The Development of Visual Perception in Man and Chimpanzee. *Science*, 106: 107–108.

Riess, B. F. 1949a. A new approach to instinct. *Sci. & Soc.*, 13: 150–154.

——. 1949b. The isolation of factors of learning and native behavior in field and laboratory studies. *Ann. N. Y. Acad. Sci.*, 51: 1093–1102.

Romer, A. S. 1945. *Vertebrate Paleontology.* University of Chicago Press, Chicago.

362   THE QUARTERLY REVIEW OF BIOLOGY

SCHNEIRLA, T. C. 1938. A theory of army-ant behavior based upon the analysis of activities in a representative species. *J. comp. Psychol.*, 25: 51–90.

———. 1939. A theoretical consideration of the basis for approach-withdrawal adjustments in behavior. *Psychol. Bull.*, 36: 501–502.

———. 1941. Social organization in insects, as related to individual function. *Psychol. Rev.*, 48: 465–486.

———. 1944. The reproductive functions of the army-ant queen as pacemakers of the group behavior pattern. *J.N.Y. ent. Soc.*, 52: 153–192.

———. 1945. Contemporary American animal psychology in perspective. In *Twentieth Century Psychology*. (P. Harriman, ed.) pp. 306–316. Philosophical Library, New York.

———. 1946. Problems in the biopsychology of social organization. *J. abnorm. (soc.) Psychol.*, 41: 385–402.

———. 1948. Psychology, Comparative. Article in *Encyclopedia Brittanica*.

———. 1949a. Levels in the psychological capacities of animals. In *Philosophy for the Future* (R. Sellars and V. J. McGill, eds.). Macmillan & Co., New York.

———. 1949b. Army-ant life and behavior under dry-season conditions. 3. The course of reproduction and colony behavior. *Bull. Amer. Mus. nat. Hist.* 94: 1–81.

———. 1950. The relationship between observation and experiment in the field study of behavior. *Ann. N. Y. Acad. Sci.*, 51: 1022–1044.

———, and R. Z. BROWN. 1950. Army-ant life and behavior under dry-season conditions. 4. Further investigation of cyclic processes in behavioral and reproductive functions. *Bull. Amer. Mus. nat. Hist.*, 95: 263–354.

SCHOOLLAND, J. B. 1942. Are there any innate behavior tendencies? *Genet. Psychol. Monogr.*, 25: 219–287.

SENDEN, M. VON. 1934. Raum-und Gestaltauffassung bei operierten Blindgeborenen vor und nach der Operation. Barth, Leipzig.

SHEPARD, J. F., and F. S. BREED. 1913. Maturation and use in the development of an instinct. *J. Anim. Behav.*, 3: 274–285.

SINNOTT, E. W., L. C. DUNN, and TH. DOBZHANSKY. 1950. *Principles of Genetics*. McGraw-Hill Co., New York.

SMITH, P. E., and E. C. MacDOWELL. 1930. An hereditary anterior-pituitary deficiency in the mouse. *Anat. Rec.*, 46: 249–257.

SMITH, S., and E. R. GUTHRIE. 1921. *General Psychology in Terms of Behavior*. Appleton, New York.

SPELT, D. K. 1948. The conditioning of the human fetus *in utero*. *J. exp. Psychol.*, 38: 338–346.

SPERRY, R. W. 1945. The problem of central nervous reorganization after nerve regeneration and muscle transposition. *Quart. Rev. Biol.*, 20: 311–369.

STONE, C. P. 1947. Methodological resources for the experimental study of innate behavior as related to environmental factors. *Psychol. Rev.*, 54: 342–347.

THORPE, W. H. 1948. The modern concept of instinctive behaviour. *Bull. Anim. Behav.*, 7: 1–12.

———, AND F. G. W. JONES. 1937. Olfactory conditioning in a parasitic insect and its relation to the problem of host selection. *Proc. roy. Soc.*, B, 124: 56–81.

TINBERGEN, L. 1939. Zur Fortpflanzungsethologie von *Sepia officinalis* L. *Arch. néerl. Zool.*, 3: 305–335.

TINBERGEN, N. 1939. On the analysis of social organization among vertebrates, with special reference to birds. *Amer. Midl. Nat.*, 21: 210–234.

———. 1942. An objectivistic study of the innate behaviour of animals. *Bibl. biotheor.*, Leiden, D, 1: 39–98.

———. 1948a. Physiologische Instinktforschung. *Experientia*, 4: 121–133.

———. 1948b. Social releasers and the experimental method required for their study. *Wilson Bull.*, 60: 6–51.

———. 1950. The hierarchical organization of nervous mechanisms underlying instinctive behaviour. *Symp. Soc. exp. Biol.*, 4: 305–312.

———. 1951. *The Study of Instinct*. Oxford University Press, Oxford.

———, and D. J. KUENEN. 1939. Ueber die auslösenden und die richtunggebenden Reizsituationen der Sperrbewegung von jungen Drosseln (*Turdus m. merula* L. und *T. e. ericetorum* Turton). *Z. Tierpsychol.*, 3: 37–60.

———, B. J. D. MEEUSE, L. K. BOEREMA, and W. W. VAROSSIEAU. 1942. Die Balz des Samtfalters, *Eumenis* (=*Satyrus*) *semele* (L.). *Z. Tierpsychol.* 5: 182–226.

———, and A. C. PERDICK. 1950. On the stimulus situation releasing the begging response in the newly hatched Herring Gull chick (*Larus argentatus argentatus* Pont.). *Behaviour*, 3: 1–39.

UYLDERT, I. E. 1946. A conditioned reflex as a factor influencing the lactation of rats. *Acta brev. neerl. Physiol.*, 14: 86–89.

WARDEN, C. J., T. N. JENKINS, and L. H. WARNER. 1936. *Comparative Psychology*, Vol. III. *Vertebrates*. Ronald Press, New York.

WASHBURN, S. L. 1947. The relation of the temporal muscle to the form of the skull. *Anat. Rec.*, 99: 239–248.

WEISS, P. 1936. Selectivity controlling the central-peripheral relations in the nervous system. *Biol. Rev.*, 11: 494–531.

ON LORENZ'S THEORY OF INSTINCTIVE BEHAVIOR 363

——. 1937a. Further experimental investigations on the phenomenon of homologous response in transplanted amphibian limbs. I. Functional observations. *J. comp. Neurol.*, 66: 181–206.

——. 1937b. Further experimental investigations of the phenomenon of homologous response in transplanted amphibian limbs. II. Nerve regeneration and the innervation of transplanted limbs. *J. comp. Neurol.*, 66: 481–535.

——. 1937c. Further experimental investigations of the phenomenon of homologous response in transplanted amphibian limbs. III. Homologous response in the absence of sensory innervation. *J. comp. Neurol.*, 66: 537–548.

——. 1937d. Further experimental investigations of the phenomenon of homologous response in transplanted amphibian limbs. IV. Reverse locomotion after the interchange of right and left limbs. *J. comp. Neurol.*, 67: 269–315.

——. 1941. Self-differentiation of the basic patterns of coordination. *Comp. Psychol. Monogr.* 17: 1–96.

——. 1950. Experimental analysis of coordination by the disarrangement of central-peripheral relations. *Symp. Soc. exp. Biol.*, 4: 92–111.

WHEELER, W. M. 1928. *The Social Insects.* Harcourt, Brace & Co., New York.

WHITMAN, C. O. 1899. Animal behavior. *Biol. Lect. mar. biol. Lab. Wood's Holl*, 1898: 285–338.

——. 1919. The behavior of pigeons. *Publ. Carneg. Inst.*, 257: 1–161.

WIESNER, B. P., AND N. M. SHEARD. 1933. *Maternal Behaviour in the Rat.* Oliver & Boyd, London.

Department of Zoology, University of Oxford

On aims and methods of Ethology

By N. Tinbergen[1])

Received 16 March 1963

Ethology, the term now widely in use in the English speaking world for the branch of science called in Germany „Vergleichende Verhaltensforschung" or "Tierpsychologie" is perhaps defined most easily in historical terms, *viz.* as the type of behaviour study which was given a strong impetus, and was made "respectable", by Konrad Lorenz. Lorenz himself was greatly influenced by Charles Otis Whitman and Oskar Heinroth — in fact, when Lorenz was asked at an international interdisciplinary conference in 1955 how he would define Ethology, he said: "The branch of research started by Oskar Heinroth" (1955, p. 77). Although it is only fair to point out that certain aspects of modern Ethology were already adumbrated in the work of men such as Huxley (1914, 1923) and Verwey (1930), these historical statements are both correct as far as they go. However, they do not tell us much about the nature of Ethology. In this paper I wish to attempt an evaluation of the present scope of our science and, in addition, to try and formulate what exactly it is that makes us consider Lorenz "the father of modern Ethology". Such an attempt seems to me worthwhile for several reasons: there is no consistent "public image" of Ethology among outsiders; and worse: ethologists themselves differ widely in their opinions of what their science is about. I have heard Ethology characterised as the study of releasers, as the science of imprinting, as the science of innate behaviour; some say it is the activities of animal lovers; still others see it as the study of animals in their natural surroundings. It just is a fact that we are still very far from being a unified science, from having a clear conception of the aims of study, of the methods employed and of the relevance of the methods to the aims. Yet for the future development of Ethology it seems to me important to continue our attempts to clarify our thinking, particularly about the nature of the questions we are trying to answer. When in these pages I venture once more to bring this subject up for discussion, I do this in full awareness of the fact that our thinking is still in a state of flux and that many of my close colleagues may disagree with what I am going to say. However, I believe that, if we do not continue to give thought to the problem of our overall aims, our field will be in danger of either splitting up into seemingly unrelated sub-sciences, or of becoming an isolated "-ism". I also believe that I can honour Konrad Lorenz in no better way than by continuing this kind of "soul-searching". I have not hesitated to give personal views even at the risk of being considered rash or provocative.

---

[1]) Dedicated to Professor Konrad Lorenz at the occasion of his 60th birthday.

Ethology a branch of Biology

In the course of thirty years devoted to ethological studies I have become increasingly convinced that the fairest characterisation of Ethology is *"the biological study of behaviour"*. By this I mean that the science is characterised by an observable phenomenon (behaviour, or movement), and by a type of approach, a method of study (the biological method). The first means that the starting point of our work has been and remains inductive, for which description of observable phenomena is required. The biological method is characterised by the general scientific method, and in addition by the kind of questions we ask, which are the same throughout Biology and some of which are peculiar to it. HUXLEY likes to speak of "the three major problems of Biology": that of *causation*, that of *survival* value, and that of *evolution* — to which I should like to add a fourth, that of *ontogeny*. There is, of course, overlap between the fields covered by these questions, yet I believe with HUXLEY that it is useful both to distinguish between them and to insist that a comprehensive, coherent science of Ethology has to give equal attention to each of them and to their integration. My thesis will be that the great contribution KONRAD LORENZ has made to Ethology, and thus to Biology and Psychology, is that he made us realise this close affinity between Ethology and the rest of Biology; that he has made us apply "biological thinking" to a phenomenon to which it had hitherto not been as consistently applied as was desirable. This is, of course, not to belittle LORENZ's concrete, factual contributions, which we all know are massive, but I submit that the significance of all his contributions is best characterised by saying that he made us look at behaviour through the eyes of biologists. I also submit that this is an achievement of tremendous importance and that, if anything deserves the much-abused name of "a major breakthrough", LORENZ's achievement does.

I shall devote the next pages to some remarks on each of these four problems as they apply to behaviour. If these remarks appear to some readers unsophisticated, I should like to remind them of the fact that Ethology is a science in its infancy, where even a little plain common sense can help.

Observation and Description

One thing the early ethologists had in common was the wish to return to an inductive start, to observation and description of the enormous variety of animal behaviour repertoires and to the simple, though admittedly vague and general question: "Why do these animals behave as they do?" Ethologists were so intent on this return to observation and description because, being either field naturalists or zoo-men, they were personally acquainted with an overwhelming variety of puzzling behaviour patterns which were simply not mentioned in behaviour textbooks, let alone analysed or interpreted. They felt quite correctly that they were discovering an entire unexplored world. In a sense this "return to nature" was a reaction against a tendency prevalent at that time in Psychology to concentrate on a few phenomena observed in a handful of species which were kept in impoverished environments, to formulate theories claimed to be general, and to proceed deductively by testing these theories experimentally. It has been said that, in its haste to step into the twentieth century and to become a respectable science, Psychology skipped the preliminary descriptive stage that other natural sciences had gone through, and so was soon losing touch with the natural phenomena.

Ethology was also a reaction against current science in another sense: zoologists with an interest in the living animal, overfed with details of a type of Comparative Anatomy which became increasingly interested in mere homology and lost interest in function, went out to see for themselves what animals did with all the organs portrayed in anatomy handbooks and on blackboards, and seen, discoloured, pickled and "mummified" in standard dissections.

Much of the early ethological work contained a good deal of description and, in these first days of reconnaissance, of taking stock, we tended to think of "ethograms" as hundred-page papers which could contain about all we wanted to know about a species. Even this modest aim, a very sketchy description, was reached for very few species only. We must hope that the descriptive phase is not going to come to a premature ending. Already there are signs that we are moving into an analytical phase, in which the ratio between experimental analysis and description is rapidly increasing. This is a natural outcome of Lorenz's own work, and it is, of course, imperative that work on causation should be intensified and refined. However, we would deceive ourselves if we assumed that there is no longer a need for descriptive work. Misgivings about this wholesale swing towards analysis have been expressed, for instance by Nielsen (1958), who wrote: "In 'modern' Ethology nobody pays the slightest attention to anything but the 'why'. It is a very peculiar situation: we have a science dealing with the causal explanation of observations but the collection of the basic observations is no longer considered a part of the science" (Nielsen 1958, p. 564). While at first glance this is a surprising remark, which very few Ethologists and non-Ethologists will agree with, we cannot brush it aside entirely.

The issue is admittedly not a simple one. Description is never, can never be, random; it is in fact highly selective, and selection is made with reference to the problems, hypotheses and methods the investigator has in mind. In the early days of Ethology these limitations of our descriptions were not always obvious — mainly, I believe, because most of us were not sufficiently conscious of our limited aims, and certainly were not sufficiently aware of the criteria we used for selection.

The variety of behaviours found in the animal kingdom is so vast, and their description is so much more laborious than the description of structure, that selectiveness of description will become increasingly urgent. This will only be possible by a more explicit formulation of the problems we wish to study, and by growing certainty about the nature of the data we need. Yet even with the most economic procedures the amount of description to be done will long remain very large, so large in fact that we shall soon have to resort to a policy of filing descriptive material in libraries or archives (including film libraries) rather than publishing it in the usual journals. Already there are journals which demand a reduction of descriptive material to the absolute minimum required for an understanding of the experiments reported on (or even to less than this minimum); the descriptions (and often the argument behind the work) one has to pick up in personal conversations at conferences.

However, if we overdo this in itself justifiable tendency of making description subject to our analytical aims, we may well fall into the trap some branches of Psychology have fallen into, and fail to describe any behaviour that seems "trivial" to us; we might forget that naive, unsophisticated, or intuitively guided observation may open our eyes to new problems. Contempt for simple observation is a lethal trait in any science, and certainly in a science as young as ours.

On aims and methods of Ethology   413

It seems to me that one of the lessons we can draw from LORENZ's work is that our science will always need naturalists and observers as well as experimenters; we must, by a balanced development of our science, make sure that we attract the greatest possible variety of talent, and certainly not discourage the man with a gift for observation. Instead we should attract such men, for they are rare; we must encourage them to develop their gifts of observation and help them ask relevant questions with respect to what they have seen.

Causation

At an early stage in his work (e.g. 1935, 1937) LORENZ made three statements which I should like to emphasise because I think that modern Ethology derived much of its inspiration from them: (1) animals can be said to "possess" behaviour characteristics just as they "possess" certain structural and physiological characteristics; while LORENZ emphasised this particularly for the relatively stereotyped motor patterns which he called „Instinkthandlungen" we know now that it applies to many other aspects of behaviour; (2) what we call behaviour is, even in its relatively simple forms, something vastly more complex than the types of movements which were then the usual objects of physiological study (and this applied equally to the sensory, the motor and the central nervous processes involved); and (3) the initiation, coordination and cessation of behaviour patterns are controlled by the external world to a lesser extent than reflex-physiologists were at that time prepared to admit. I should like to elaborate these points to some extent.

(1) The first statement was based on an unrivalled store of first-hand experience as well as on much that had already been published, notably in the works of HEINROTH (1911) and of WHITMAN (1919). It led LORENZ to consider behaviour patterns (and by implication the mechanisms underlying them) as *organs,* as attributes with special functions to which they were intricately adapted. This again facilitated causal analysis without interference by either subjectivism or teleology. By subjectivism I mean here the procedure of replying to the question "What causes this behaviour?" by referring to a subjective experience, i.e. a process which per definition can be observed by no one except the subject. It seems to me worth pointing out that Ethology has not yet completely succeeded in freeing itself from subjectivism in this sense. It is true that one rarely meets with it in its crudest form ("the animal attacks because it feels angry"), but in its subtler forms it is still very much with us. Concepts such as "play" and "learning" have not yet been purged completely from their subjectivist, anthropomorphic undertones. Both terms have not yet been satisfactorily defined objectively, and this might well prove impossible; both may well lump phenomena on the one hand, and exclude other phenomena on the other hand (and thus confuse the issue by a false classification) simply because the concepts are directly derived from human experience. In both fields the growing tendency to ignore the term and to return to the phenomena (which are singled out for study because they are suspected of having a different causation than other phenomena), is, I think, an inevitable result of the consistent application of biological thinking to behaviour.

Teleology also can be said to have ceased to be a source of confusion in its cruder forms, in which function was given as a proximate cause, but it may well be a major stumbling block to causal analysis in its less obvious forms. Throughout Biology we tend to classify, and hence to give names, on the basis of criteria of common function. The more complex the behaviour systems we deal with, the more dangerous this can be. For instance, although the more sophisticated ethologist is fully aware of the fact that the term "Innate Releasing Mechanism" refers to a type of function, of achievement, found in many different animals; and that different animal types may well have convergently achieved "mood controlled selective responsiveness" by entirely different mechanisms, the term has given rise to misunderstanding on this point, as LEHRMAN's criticism (1953) showed. And who knows what different mechanisms we are lumping under terms such as learning, displacement activity, drive-reduction?

Another type of difficulty which may have to do with our thinking in terms of function is caused by our habit to coin terms for major functional units such as nest building, fighting or sexual behaviour and treat them as units of mechanism. For instance, while the fact that all fighting acts fluctuate together in the natural situation, as do all components of escape, *does* justify us to use "fighting tendency" and "escape tendency" when we are involved in the first step of analysis of movements caused by the *simultaneous arousal of the two*, as soon as we begin to analyse the causation of each of them separately, it is pre-judging the issue if we take for granted that each is in itself a causally closely-linked complex of components. We may then find that some components have closer links with processes outside this functional system than with other parts of the same functional system (see BEER [1962] on incubation and HINDE [1958, 1959] with respect to nest building). Our habit of giving names to systems characterised by an achievement has made thinking along consistent analytical lines much more difficult than it would have been if we could have applied a more neutral terminology. But rather than to advocate such a dry, non-committal terminology, I would like to accept any frankly functional term, as long as this is done consciously. No physiologist applying the term "eye" to a vertebrate lens eye as well as to a compound Arthropod eye is in danger of assuming that the mechanisms of the two are the same; he just knows that the word "eye" characterises achievement, and no more.

The treatment of behaviour patterns as organs has not merely removed obstacles to analysis, it has also positively facilitated causal analysis, for it led to the realisation that each animal is endowed with a strictly limited, albeit hugely complex, behaviour machinery which (if stripped of variations due to differences in environment during ontogeny, and of immediate effects of a fluctuating environment) is surprisingly constant throughout a species or population. This awareness of the repeatability of behaviour has stimulated causal analysis of an ever-increasing number of properties discovered to be species-specific rather than endlessly variable.

It may not be superfluous to stress that the recognition of the existence of many species-specific behaviour characters does not necessarily imply that all these characters are "innate" in the sense of ontogenetically wholly independent of the environment. It is true that this is often assumed in many ethological publications, and I shall have to return to this when I discuss behaviour ontogeny, but this point is irrelevant to us here. LORENZ's emphasis on the fact that so much in behaviour mechanisms is species-specific remains as fruitful as ever.

(2) LORENZ's emphasis on the complexity of behaviour phenomena (which is only seemingly contradictory to his inclination towards simplifying physiological *explanations*), seems to me still to be of the greatest importance, even though we now take this complexity for granted. Lack of appreciation of this point seems to me to have been one of the most important reasons for

the lack of co-operation between physiologists and ethologists. The magnitude of the gap between the phenomena studied by ethologists and those studied by nenurophysiologistshas been underrated by both parties. The early ethologists underrated the complexity of behaviour mechanisms in various ways, as was evident from our early attempts at "physiologising". One example of this is the lack of any provision for negative feedback in LORENZ's original "psychohydraulics" model (1950); another is provided by my own sketch (1951) of the organisation of the hierarchy in behaviour mechanisms; another again can be found in the original explanation of "displacement activities" (compare, for instance, TINBERGEN 1940, ROWELL 1961 and SEVENSTER 1961). Ethologists are now increasingly avoiding such over-simplifications, without however giving up the application of strictly analytical procedures which were started by LORENZ's work. A corollary of this is the development of concepts suited to the stage of analysis, concepts (and terms) which avoid implying physiological explanations — VON HOLST's (VON HOLST and v. ST. PAUL 1960) *"niveau-adäquate Terminologie"* — without becoming enslaved by such terminology and shutting the door to further analysis. Until recently neurophysiologists, concerned with the analysis of relatively simple processes, were either not considering the more complex phenomena, or were too ready to assume that combinations of the basic phenomena they knew would some day be found to account for behaviour of the intact animal. A striking example of the latter attitude was quoted by VON HOLST and MITTELSTAEDT (1950) in their analysis of the way in which an optomotor response was found not to interfere with "spontaneous" locomotion. Until these authors checked by a simple experiment whether it was true that, under such circumstances, the optomotor response was inhibited (and found that this was not so) this unproven hypothesis seems to have been taken for granted.

The situation is now changing rapidly. The "no-man's land" between Ethology and Neurophysiology is being invaded from both sides. While ethologists are making progress with the "descending" breakdown of complex phenomena, neurophysiologists are "ascending", extending their research to phenomena of greater complexity than was usual 20 years ago. To what extent the latter development has been influenced by Ethology is difficult to say, for who can trace the origin of new fashions in a science? While I am convinced that it would have happened anyway, I am equally convinced that the growth of Ethology has speeded up the process. The *rapprochement* between the two fields has gone so far already that it begins to be difficult, and in some cases even impossible, to say where Ethology stops and Neurophysiology begins. Are DETHIER and ROEDER ethologists or neurophysiologists? And where to put VON HOLST, MITTELSTAEDT and HASSENSTEIN? Several of my colleagues are still inclined to draw a sharp line between the two fields, and to deny some of these workers a place among the physiologists, mainly, I understand, because the mechanisms they describe cannot yet be expressed in physicochemical terms. I believe that this view is a denial of the fact that much which is conventionally called Physiology has not reached that stage either; or was accepted as part of Physiology before physico-chemical explanations were possible or even within sight. What happens throughout this entire field is that *achievements* of complex systems are, after a varying number of analytical steps, described in terms of *achievements* of component systems. If VON HOLST's work on the superposition effect (1937), HASSENSTEIN's work on the interaction of ommatidia responsible for the response to movement (1951, 1957), and VON HOLST's and MITTELSTAEDT's work on reafference (1950) is not Physiology, then why was SHERRINGTON's work Physiology? It

416   N. Tinbergen

is, of course, in itself completely unimportant whether or not one calls a
certain type of work by a special name, as long as one agrees that it has a
place in the progress of science, but the issue has important implications.
I believe that it is doing our science a great deal of harm to impose boundaries
between it and Physiology where there are none, or rather where there is only
a "cline" from behaviour analysis on the one extreme to "Molecular Biology"
on the other. I believe that the only criterion by which these extremes and the
intermediate fields can be distinguished is that of the level of integration of
the phenomena studied. For an understanding of our aims it seems to me much
more important to recognise that fundamental identity of aims and method
unites all these fields. It is the nature of the question asked that matters in
this context, and this is the same throughout. Co-operation between all these
workers is within reach, and the main obstacle seems to be lack of appreciation of the fact that there is a common aim.

(3) LORENZ's third postulate, stressing the part played in the control of
behaviour by internal causal factors, has also had, and is still having, an
effect on analytical studies. Again, the earlier confident statements about the
nature of these internal determinants were, in some respects, premature and
were at best over-simplifications. Thus, we are now far removed from the
simple idea that the effect of hormones on behaviour is no more than a
simple, direct stimulation of target tissues in the c.n.s., for it is clear that
roundabout effects — e.g. those mediated by hormone-induced growth processes in the sensory periphery (BEACH and LEVINSON [1950], LEHRMAN
[1955], and receptor-mediated feedback phenomena (LEHRMAN [1961],
HINDE [1962]) xenter into the causation of hormone-controlled behaviour.
Nor do we believe any more that a complex behaviour system serving one
major function, such as nest building in birds, should necessarily be controlled
by one single, compact "centre" in the c.n.s. Yet it is surprising to see how
interest in "spontaneous" activity of nervous tissue, and in units controlling
entire behaviour patterns has grown, and even how many of LORENZ's suggestions about central control of complex behaviour prove to have more than a
core of truth in them. To take but one example, BLEST's work (1960) on the
interaction between the tendency to fly and the antagonistic tendency to settle
in *Automeris* moths, which by elimination of all the known or suspected alternatives was concluded to be due to direct interaction between parts of the
c.n.s. itself, illustrates a trend which, I am sure, owes much LORENZ's approach.

These briefly mentioned samples do indicate, I believe, how analyses of
behaviour mechanisms which were initiated in the earlier ethological studies
are moving towards a fusion with the fields conventionally covered by Neurophysiology and Physiological Psychology. As far as the study of causation of
behaviour is concerned the boundaries between these fields are disappearing,
and we are moving fast towards one Physiology of Behaviour, ranging from
behaviour of the individual and even of supra-individual societies all the way
down to Molecular Biology. There ought to be one name for this field. This
should not be Ethology, for on the one hand Ethology has a wider scope, since
it is concerned with other problems as well; on the other hand, ethologists
cannot claim the entire field of Behaviour Physiology as their domain, for
they have traditionally worked on the higher levels of integration, in fact almost
entirely on the intact animal. The only acceptable name for this part of the
Biology of Behaviour would be *"Physiology of Behaviour"*, and this name
should be understood to include the study of causation of animal movement
with respect to all levels of integration.

Survival value

Lorenz's thesis that behaviour patterns, i.e. their mechanisms, ought to be considered "organs", and to be studied as such, has also had a beneficial effect on the study of the survival value of behaviour. In the post-Darwinian era, a reaction against uncritical acceptance of the selection theory set in, which reached its climax in the great days of Comparative Anatomy, but which still affects many physiologically inclined biologists. It was a reaction against the habit of making uncritical guesses about the survival value, the function, of life processes and structures. This reaction, of course healthy in itself, did not (as one might expect) result in an attempt to improve methods of studying survival value; rather it deteriorated into lack of interest in the problem — one of the most deplorable things that can happen to a science. Worse, it even developed into an attitude of intolerance: even wondering about survival value was considered unscientific. I still remember how perplexed I was upon being told off firmly by one of my Zoology professors when I brought up the question of survival value after he had asked "Has anyone an idea why so many birds flock more densely when they are attacked by a bird of prey?"

Lorenz was never in danger of conforming to this fashion. He always was much too good a naturalist for this; he further had the good fortune of being taught by Hochstetter, an anatomist with a wide grasp of what Biology is about; finally, he was himself too clear a thinker to confuse teleology with the study of survival value. To him an organ was something which a species had evolved as one of its means for survival, something of which, as a matter of course, both the contribution it made to survival and its causation had to be studied. He has always been equally interested in "What is this good for?" and in "How does it work?". It was partly through this interest in survival value, for instance, that he arrived at his important concept of "Releaser" — an organ adapted to the function of sending out stimuli to which other individuals respond appropriately, i.e. in such a way that survival is promoted. "Releaser" was defined along much the same lines as any other effector, say, a wing, which is an organ adapted to the function of flying, or an endocrine gland, which is a gland adapted to the function of shedding hormones which, by acting on equally adapted "target organs", contribute to the proper co-ordination of functions within the body. It is an illustration of the inability of many biologists to think in terms of survival value that the concept of Releaser as an *organ characterised by a function* is so often misunderstood and confused with "anything which provides stimuli", a confusion due to the failure to see the function of the releaser and to preoccupation with the causation of the behaviour of the reacting animal.

It is through Lorenz's interest in survival value that he appealed so strongly to naturalists, to people who saw the whole animal in action in its natural surroundings, and who could not help seeing that every animal has to cope in numerous ways with a hostile, or at least unco-operative environment. Incidentally, just because Lorenz's work has revived interest in the study of survival value, and because this is an aspect of Ethology which may well fertilise other fields of Biology (where survival value studies are being neglected) I think it is regrettable that his fine new Institute has been named „Institut für Verhaltensphysiologie" — its field of research extends far beyond Physiology.

Being myself both a naturalist and an experimenter at heart, one of my primary interests has always been to find out, if possible by experimentation, how animal behaviour contributes to survival, and I shall therefore enlarge a little more on this theme than on the others.

I have always been amazed, and I must admit annoyed as well, when I met, among fellow-zoologists, with the implied or stated opinion that the study of survival value must necessarily be guesswork, and that exact experimentation on the problem is in principle not possible. I am convinced that this is due to a confusion of the study of natural selection with that of survival value. While I agree that the selection pressures which must be assumed to have moulded a species' past evolution can never be subjected to experimental proof, and must be traced indirectly, I think we have to keep emphasising that the survival value of the attributes of present-day species is just as much open to experimental inquiry as is the causation of behaviour or any other life process.

Our study always starts from an observable aspect of a life process — in the present case, behaviour. The study of causation is the study of preceding events which can be shown to contribute to the occurrence of the behaviour. In this study of cause-effect relationships the observable is the effect and the causes are sought. But life processes also have effects, and the student of survival value tries to find out whether any effect of the observed process contributes to survival if so how survival is promoted and whether it is promoted better by the observed process than by slightly different processes. It is clear that he too studies cause-effect relationships, but in his study the observable is the cause and he tries to trace effects. Both types of worker are therefore investigating cause-effect relationships, and the only difference is that the physiologist looks back in time, whereas the student of survival value, so-to-speak, looks "forward in time"; he follows events after the observable process has occurred. The crux is that both are concerned with a flow of events which can be observed repeatedly, and which thus, unlike the unique events of past evolution, can be subjected to observation and experiment as often as one wishes.

The fact that we tend to distinguish so sharply between the study of causes and the study of effects is due to what one could call an accident of human perception. We happen to observe behaviour more readily than survival, and that is why we start at what really is an arbitrary point in the flow of events. If we would agree to take survival as the starting point of our inquiry, our problem would just be that of causation; we would ask: "How does the animal — an unstable, 'improbable' system — manage to survive?" Both fields would fuse into one: the study of the causation of survival. Indeed, logically, survival should be the starting point of our studies. However, since we cannot ignore the fact that behaviour rather than survival is the thing we observe directly, we have, for practical reasons, to start there. But this being so, we have to study both causation and effects.

The widespread lack of interest in studies of survival value and the opinion that it can never move beyond the level of inspired guesswork are all the more puzzling because the literature contains quite a number of good experimental studies in which the survival value of behaviour has been as well demonstrated as anyone could wish. To mention just a few examples: MOSEBACH-PUKOWSKI (1937) has shown that the habit of crowding of *Vanessa* caterpillars has survival value in affording protection from insectivorous song birds: isolated caterpillars are eaten more readily than those living in a cluster. KRISTENSEN (TINBERGEN 1951) has shown that the "fanning" of male Sticklebacks has survival value by renewing the water round the eggs; if this is prevented the eggs die, and artificial ventilation saves them. BLEST (1957) has shown that sudden display of "eye spots" by certain moths scares away certain predators and thus saves lives. VON FRISCH's studies of the bee dance do

On aims and methods of Ethology 419

not leave doubt about the survival value of this behaviour; it does direct workers to rich food sources unknown to them and thus greatly increases the efficiency of feeding; as LINDAUER (1961) has demonstrated, the dance also plays a part in directing a homeless swarm to a suitable site. Such studies are generally considered both interesting and reliable, and this gives the lie to the argument that survival value cannot be studied experimentally. What then are the reasons for this problem being underworked to such an extent?

First of all, the survival value of many attributes, behaviour and structure alike, is so obvious as to make experimental confirmation ludicrous. One need not starve an animal to death to show that its feeding behaviour has survival value, nor need one cut off a Blackbird's bill to show that this organ is necessary for successful feeding. But one of the reasons why ethologists are so much concerned with survival value is that the "use" of so many behaviour patterns is still completely unknown.

However, the quest for survival value involves, of course, much more than the demonstration that the Blackbird's bill is indispensable to it; one wants to know whether a bill of this size and this shape is best suited to feeding in the environment in which the Blackbird lives; similarly, one needs to understand in detail the suitability of every aspect of its feeding behaviour one sees, and this, of course, is very far from obvious. To think that we understand survival value completely in such cases is to think that, once it is obvious that sex hormones control mating behaviour, we need not inquire into the way they do this, nor into the interaction between various endocrine processes that are involved.

Another important reason for the lack of interest in survival value studies is a practical one. The method to demonstrate survival value of any attribute of an animal is to try whether or not the animal would be worse off if deprived of this attribute. This is easy with structures. For instance BLEST could compare the effect of normal "eye spot"-bearing moths on song birds with that of moths whose eye spots were brushed off. In this test moths without eye spots could safely be regarded as differing from the controls in just this one respect. Similarly, HOOGLAND, MORRIS and TINBERGEN (1957) could show that Sticklebacks without spines were eaten more readily by small Pike than normal Sticklebacks. But how does one make an experimental animal which lacks just one behaviour pattern and is otherwise normal? How, for instance, to make a male bird which does not show aggressive behaviour, or lacks one of its threat postures, while in all other respects behaving normally?

This difficulty can in many cases be overcome, but one has to be aware of many pitfalls. Much of our evidence can come from systematic comparison of the success of animals at times when they do show a certain behaviour and the lack of success when they do not perform it. Thus, if a territory of a male bird is not invaded as long as it fights off intruders, but is invaded when, later in the day, his aggression wanes though he is still there and shows a variety of other behaviour, one has a good indication that it was the aggressive behaviour which kept the territory clear. Or, to take another example, if one can show that a motionless twig caterpillar is not eaten by birds while it is snapped up as soon as it moves (DE RUITER 1952) one can be pretty confident that immobility in this species has survival value.

Yet it is, of course, true that in such "natural experiments" one does not control the feature studied — one never knows which unknown aspect of the animal may have varied with the character studied; the aggressive bird may have made an ultrasonic sound; the moving caterpillar may have given off a scent. Therefore the method of studying the survival value of behaviour is to use dummies and to control their "behaviour". For instance, when a dummy of a male stickleback is either ignored, or merely approached, by a ripe female when moved at random, but elicits following and even the movements of creeping into the nest whenever it is made to move like a "nest-showing" male, even in the absence of any nest, then one has demonstrated the effect of nest-showing and, since eggs not laid in the nest are eaten or abandoned, one has shown that the male's nest-showing contributes to, and is even indispensable for successful reproduction. Of course, an experiment such as this is but the first step, for one needs to know the full story of the cause-effect relationship which makes the female respond the way she does; also, one wants to know not merely whether absence of the behaviour studied has an adverse effect; one also needs to know what kind of deviations from the natural behaviour would reduce the effect — which includes the task of finding out whether the natural movement has the optimal effect and, if not (as is the case in supernormal stimuli) why the behaviour is not "better" — a question which will crop up again in evolutionary studies.

So far, we have only made the barest beginning with this task; there are even many behaviour patterns of which we do not even know the basic answer: has it any function at all? As an illustration let me mention the example of the "rocking" of certain cryptic animals. There are a number of animals which, either as an introduction to the change from motionlessness to movement, or from movement to immobility, perform a series of curious rocking movements. Heinroth (1909) described these for the Nightjar, and mentioned that they are also found in *Phyllium* and in *Dixippus*. Blest (1960) has shown that many Saturniid moths have a similar movement, usually preceding settling. Now these animals are all camouflaged; many of their behaviour characters (immobility by day, background selection, semi-closing of the eyes in the Nightjar, etc.) are obviously adapted to the function of avoiding detection by visually hunting predators. In view of this the habit of rocking seems very strange indeed, for movement in general is a stimulus to which visually hunting predators react, and which these cryptic animals are for the rest at such pains to avoid giving. Therefore the fact that these movements occur in such different animals suggests that they have survival value and are somehow connected with camouflage. I believe that a testable hypothesis can be formulated. De Ruiter (1952) has shown that European Jays ignore twig-shaped caterpillars as long as the latter stay motionless. However, this was only true of Jays which had grown up in a normal environment, and in particular those which had had the opportunity of discovering, by trial and error, that real twigs are inedible. Hand-raised Jays which had not "played about" with twigs took up and tried out twigs and caterpillars alike. We know that many young birds have, at the start, a very "open mind" with regard to food; they respond to an enormous variety of objects, edible and inedible alike, and learn to confine themselves to those they find edible. My suggestion is that we have as yet no more than the faintest idea of the kinds of things such birds learn when young. Heinroth (1909) already suggested that the rocking movements of *Phyllium* might well be harmless because predators might recognise them as passive movements of a leaf slightly moved by the wind. It seems to me quite possible that many young birds actually do learn that certain types of movements are passive, and not indicative of animal prey, and that it might not only be harmless for certain cryptic animals to perform these movements but that it might be definitely beneficial to them because it might ensure that a predator sees them and concludes that they are just vegetable matter. A motionless cryptic animal "hopes to be overlooked"; a rocking cryptic animal makes sure that it is seen and ignored — which means survival.

Critical, scientific zoologists had a way of applying the term "armchair science" to such ideas; yet it is becoming increasingly clear that it was the critics who judged such issues without investigation and even without knowledge of the real events; it is obvious that a well planned study of the ontogeny of the feeding behaviour of certain predators, combined with an accurate study of these rocking movements and the passive movements of inanimate objects could prove the hypothesis to be right or wrong.

Of course, in selecting this example I have applied "shock tactics" by taking a very exceptional type of behaviour. However, I would like to submit, first that we know so little about behaviour that new, equally strange behaviour patterns could well be discovered in large numbers; second, that the problem of survival value applies equally to every detail of behaviour and structure, however self-evident or insignificant it might seem at first glance. For instance, the fact that the Godwit walks differently from the Lapwing would seem too trivial to pay attention to, yet KLOMP (1954) showed that these differences are adaptive: Godwits lift and fold their feet much more than do Lapwings and thus avoid getting their toes caught in the tall grass in which they breed. Lapwings avoid habitats with tall vegetation. A parent Kittiwake does not produce a sound when it is about to feed its young but other gulls do; the Kittiwake is the only species of gull whose chicks stay on the nest and therefore have not got to be called to food (E. CULLEN 1957). HEINROTH (1928) suggested that Starlings and Partridges show remarkably little inhibition from trying to walk or fly straight through the thin metal bars of bird cages because they are adapted to living among grass rather than trees, and grass gives way. The writings of the good naturalists are teeming with such hints, arguments, and occasional demonstrations of the functions of a multitude of aspects of species-specific behaviour. It is also the experience of every good naturalist that the longer one studies a species, the more adaptive aspects of its behaviour one becomes aware of. The phenomena are countless, the field is practically unexplored, and yet without exploring it systematically we cannot hope to understand how behaviour helps animals to survive.

How can we tackle this immense task and catch up with the backlog? Hypotheses can be arrived at in various ways. LORENZ himself has pointed out one which he derives from his particularly fruitful method of raising and keeping animals in freedom, yet in partly artificial surroundings. Under such conditions one observes a number of behaviour patterns which "misfire". The observer's reaction is: "This seems ill-adapted; but it must be good for something" and this makes him try to see the behaviour in its proper context.

The naturalist who studies animals in their natural surroundings must resort to other methods. His main source of inspiration is comparison. Through comparison he notices both similarities between species and differences between them. Either of these can be due to one of two sources. *Similarity* can be due to affinity, to common descent; or it can be due to convergent evolution. It is the convergences which call his attention to functional problems. This method has been applied beautifully by VON HAARTMAN (1957) in his study of adaptations in birds hole nesting. The *differences* between species can be due to lack of affinity, or they can be found in closely related species. The student of survival value concentrates on the latter differences, because they must be due to recent adaptive radiation. An example of this procedure is E. Cullen's elegant study (1957) of the peculiarities of Kittiwakes as compared with other gulls.

Such hypotheses can be made highly probable even without experimentation. E. Cullen's report contains hardly any experimentation, yet her conclusion that, for instance, immobility of the young, their "facing-away" gesture, the tameness of the adult Kittiwakes while on their cliff, their mud-trampling before nest building, and the absence of an acquired attachment to their own young are all adaptive corollaries of cliff-breeding carries conviction because this interpretation is the only one which fits into our general picture.

However, such studies are no more than a beginning; they can be extended and intensified in several ways.

First, as one becomes better acquainted with a species, one notices more and more aspects with a possible survival value. It took me ten years of observation to realise that the removal of the empty eggshell after hatching, which I had known all along the Black-headed Gulls to do, might have a definite function, and that even the length of delay of the response, which varies with the circumstances, and which is on the average longer than that found in the Ringed Plover, may be adaptive (Tinbergen et al. 1962).

Secondly, hunches about survival value must, where possible, be strengthened by experiment. This meets with obstacles of a practical nature, but once the need is obvious, ways can often be found. Thus when one sees that the breeding season of the Black-headed Gulls is much more synchronised than that of Gannets, and one notices at the same time that the late broods of Black-headed Gulls seem to be less successful than the majority; when one further has indications that such late broods perish through heavy predation, one can first systematically compare predation of late and "peak" broods, and ultimately design an experiment to find out whether or not synchronisation has survival value as an anti-predator defence.

Thirdly, the experimental demonstration of survival value involves quite a number of steps. Much of the experimental evidence is not complete, because it has (often of necessity) been done in a situation which differs essentially from the natural context. In order to study the survival value of egg-shell removal in the Black-headed Gull, my co-workers and I demonstrated that gulls' eggs, laid out well scattered over the hunting area of Carrion Crows and Herring Gulls, were found more readily when they hand an empty egg-shell at 4 inches distance than when no egg-shell was added (Tinbergen et al. 1962). However, before we can conclude from this that egg-shell removal reduces predation, we have to consider whether in the natural situation this is its only effect. When a gull removes the egg-shell it leaves its brood unguarded for a few seconds. This, we know, can be critical: neighbouring gulls or Crows sometimes snatch up an egg or a newly-hatched chick in a second or so. It clearly depends on the balance of advantage and disadvantage whether the response is on the whole useful or not. The strict test for this would be a comparison of breeding success of a population of gulls which remove the shell with that of a population which do not, though in all other respects identical to the shell-carrying population. It is just because this is impossible that we have to be content with less good evidence. There are several indications that the advantage outweighs the disadvantage. For instance, whenever a Crow is in sight, the tendency of the gulls to attack it and drive it off dominates that of removing the shell; also, the danger caused by leaving the nest for a few seconds might well be less than that caused by the presence of the tell-tale shells for a long time. It remains true, however, that the ultimate test of survival value is survival itself, survival in the natural environment. This ultimate test has been carried out in very few cases only; a good example, involving a colour adaption rather than a behavioural one, is supplied by the work of KETTLEWELL (1955, 1956) on differential survival of white and black mutants of *Biston betularia* in two different environments: one which favoured the white form, and one which favoured the dark form.

I have argued that survival value has to be studied in its own right, but there are two additional reasons. First, Zoophysiology derived, and again derives much of its inspiration and guidance from knowledge or hunches about survival value. Experiments on the external control of respiration often concentrate on the effect of varying oxygen and CO_2-content of the medium or in the blood; this is because one starts from the knowledge that it is oxygen the animal requires, and CO_2 it must get rid of. The work of VON FRISCH and his school on colour vision in the Honey Bee was set off by VON FRISCH refusing to believe that the colours of flowers had no function; our knowledge of the ability of Arthropods to register the plane of polarisation of light is due to VON FRISCH wondering about the exact function of the bees' dance.

Secondly, the part played by natural selection in evolution cannot be assessed without proper study of survival value. If we assume that differential mortality in a population is due to natural selection discriminating against the less well-equipped (the less "fit") forms, we have to know how to judge fitness, and that can only be done through studies of survival value.

To those, however, who argue that the only function of studies of survival value is to strengthen the theory of natural selection I should like to say: even if the present-day animals were created the way they are now, the fact that they manage to survive would pose the problem of how they do this.

Ontogeny

A newly-hatched Herring Gull pecks selectively at red objects (GOETHE 1937, TINBERGEN & PERDECK 1950) but a human being has to learn to stop when the traffic lights turn red. We have to learn the intricately co-ordinated motor patterns of speech, whereas a Whitethroat raised in isolation produces the complicated normal song of its species (SAUER 1954). It is the contrast between man and animals in the ways they acquire either "knowledge" or "skill" which arouses in most of us an interest in the ontogeny of behaviour. As we all know, the systematic study of behaviour ontogeny has had a slow start, and for a long time was heavily weighted, but differently so in different groups of researchers. While animal psychologists explored the ways in which various types of learning might account for behaviour ontogeny, ethologists emphasised the unlearnt character of many aspects of animal behaviour. Ontogeny was, for a long time, and to a certain extent still is, a field in which there is a real clash of opinion. All concerned agree that a complete understanding of behaviour requires an understanding of its ontogeny, just as morphologists agree that it is not sufficient to understand the adult form, but also the way in which this develops during ontogeny. But there is no agreement about the nature of the problems involved, and while the methods applied by psychologists and ethologists begin to resemble each other so closely as (in some instances) to be indistinguishable, the interpretation of the results gives rise to much discussion.

I believe that this discussion has been and is still being bedevilled by semantics, and that it would be helpful if, instead of discussing the justifica-

424 N. Tinbergen

tion of the use of words such as "innate" and "acquired", of "instinct" (and "instinctive") and "learning" we could return to a statement of the phenomena to be understood and the questions to be asked — indeed I think this is imperative.

I should like to characterise the *phenomenon* as "change of behaviour machinery during development". This is not, of course, the same as a change of behaviour during development; when in spring we see a thrush pick up and smash a snail for the first time in months, this change in feeding behaviour may be due to snails having reappeared for the first time after winter. We can conclude that the thrush itself, i.e. its behaviour machinery, has changed only if the behaviour change occurred while the environment was kept constant. It should be pointed out in passing that systematic descriptions of behaviour ontogeny are still rare and fragmentary.

When we turn from description to causal analysis, and ask in what way the observed change in behaviour machinery has been brought about, the natural first step to take is to try and distinguish between environmental influences and those within the animal. It is about this very first, preliminary step that confusion has arisen.

As in studies of the cyclical behaviour of adult animals, external influences have been studied most, for the simple reason that they are so much more easily manipulated. It is also important to realise that, in ontogeny, the conclusion that a certain change is internally controlled (is "innate") is reached *by elimination*. This is not, of course, a reflection on the validity of a classification — one can perfectly well dichotomise any group of phenomena into one group possessing the character A and the rest not possessing A — but it does reflect on the justification of lumping all examples for which not all environmental influences have been eliminated into a class called "innate", thus suggesting a positive statement where merely a negative statement would be in order. And I submit that most statements about "innateness" of behaviour are based on the elimination of one or some out of several, perhaps many, possible external influences. I am again criticising myself just as much as others, for I am now convinced that I have helped to perpetuate the confusion. If we raise male Sticklebacks in isolation from fellow members of its own species, subject them as adults to tests with dummies, and find (E. Cullen 1961) that they attack red dummies just as selectively as do normal males, we are entitled to say that exposure to red males cannot be responsible for the development of this selectiveness of response. We cannot, however, say anything about the problem whether or not interaction with the environment during "practising" has influenced the form of their fighting movements. When Grohmann (1939) showed that the incomplete flying movements which young pigeons make while growing up (and which might be interpreted as providing "practice") did not influence their flying skill (birds that were prevented from flapping on the nest flew as well as controls on their first attempt), he eliminated a different form of interaction with the environment than Cullen did with her Sticklebacks

It is not helpful and even wrong to apply to both behaviour patterns the term "innate", because in each case only one out of various environmental effects was excluded, and these were different in each case. The conclusion can only be formulated correctly in negative terms, in describing which environmental aspect was shown *not* to be influential.

There is, in addition, another reason for not applying the term "innate" to the fighting behaviour ot the Kaspar Hauser Sticklebacks. Knoll (1953) has shown that the rods of tadpoles raised in darkness do not function pro-

On aims and methods of Ethology   425

perly; exposure to light is required to allow them to become fully functional. We have no information about this problem in Sticklebacks, and this means that, in the absence of evidence with respect to either rods or cones in Sticklebacks, we must allow for the *possibility* that light — an environmental property – is required for the proper "programming" of part of the Stickleback's behaviour machinery. This brings me to another point which I consider important: the term "innate" whether applied to characters, or to differences, or to potentialities, or to developmental processes, is not the opposite of "learnt"; it is the opposite of "environment-induced".

These few considerations seem to me sufficient to conclude that the application of the adjective "innate" to behaviour *characters,* and to do this on the basis of eliminations of different kinds is heuristically harmful.

If I were to elaborate this further I should have to cross swords with my friend KONRAD LORENZ himself — both a pleasure and a serious task requiring the most thorough preparation — but this is not the occasion to indulge in swordplay, and I prefer to continue with my sketch of the procedure which seems to me more fruitful. This seems justified by the fact that the practice of ontogenetic research is not so much dependent on the background of the experimenter as the semantic and theoretical disagreements would lead one to suspect. The difference, for instance, between RIESS's (1950) and EIBL-EIBESFELDT's (1956) work on the ontogeny of nest building in rats is due just as much to a difference in the extent of knowledge of the Rat's normal behaviour as to the theoretical attitudes of the investigators.

A central issue in behaviour development studies seems to me the question raised by the fact that so many behaviour patterns can be said to be at the same time innate and learned, or partly innate and partly learned. EIBL-EIBESFELDT (1955) showed that nut-cracking in Squirrels consists of a series of component acts (manipulating, gnawing, and cracking) each of which develops in naive individuals which have not been able to practice; yet the adaptive integration of the acts into an efficient total pattern has to be learnt. Many similar cases are known, and it was a definite step forward when LORENZ (1937) coined the term „Instinkt-Dressur-Verschränkung" indicating that learning processes were often, so-to-speak, intercalated by non-learnt parts of a behaviour chain. This has given rise to the idea that, if one could only split up behaviour chains in smaller and smaller components, one could always reach a state where some components could be labelled as innate, others as learnt or acquired. I maintain that this may well be unhelpful, since many interactions with the environment which result in increased efficiency are *additive to* some machinery that was already functional. For instance, WELLS (1958) found that a naive young *Sepia* can perform the movements by which it catches *Mysis,* but that both the delay of the "attack" and the selectiveness of the response to stimuli decrease as a result of "having performed". In the sphere of conditioning — to mention a different form of environmental control of ontogeny — something similar is true: the animal is already selectively responsive (in other words it has an "IRM") before it has been conditioned; the conditioning changes a connection that was already present. The story of song development in Chaffinches as revealed by THORPE's work (1962) shows something very similar in the development of motor patterns: there is a definite pattern in the song of naive birds. I cannot see how, in view of such facts, it can be fruitful to look for innate and learned *components,* however small.

It seems to me that, if we return to a description of the phenomenon and the formulation of relatively simple questions, our course is laid out clearly.

The phenomenon (change in behaviour machinery) has to be described; the problem is, how are these changes controlled? As a first step one distinguishes between influences outside the animal and internal influences. External influences are usually detected by manipulating the environment during development, and the way in which such external agents influence the development can be studied along rather conventional lines, although it is obvious that, even in this relatively easier part of our task, we may well be in for some surprises, such as discovering that "rewards" may be of many more different types than known at present. One receives the first indications of *internal* control from demonstrations of the ineffectiveness of certain environmental properties, but the ultimate demonstration of internal control must come from direct interference with internal events. For instance, the development of successful ejaculation in Rats is, as Beach and Levinson (1950) have shown, influenced by sex hormones which promote the growth of sensory papillae on the glans penis. Further insight into the internal control of growth of neural machinery is provided by the fascinating work of Sperry (1959), particularly by his transplantation of peripheral „Anlagen" before their innervation has been completed.

This general procedure, when applied at various levels of integration — to complex patterns, single acts, and even smaller components of the total behaviour machinery — seems to me much more fruitful than either basing a conclusion about innateness on elimination of part of the environmental properties, or proceeding on the assumption that all adaptedness of behaviour is acquired through interaction with the environment. It has been pointed out repeatedly (see Pringle 1951, Lorenz 1961) that there are two methods of "programming" the individual: the evolutionary trial-and-error-interaction with the environment which results in the specialisations of the genetic instructions, and the ontogenetic interaction between the individual and its environment — which, incidentally, takes the form of trial and error only where evolution has not given precise direction to the ontogenetic process.

I believe that such a procedure is in line with that widely applied in experimental embryology, which after all is the science concerned with exactly the same problem with reference to structure. And this takes me back to my starting point: by insisting on a biological approach Lorenz has influenced this aspect of Ethology as much as other aspects, even tough his evaluation of the part played by internal determinants may have been on the optimistic side.

I admitted above that in speaking of "the four problems of Biology" we apply a classification of problems which is pragmatic rather than logical. This is true of ontogeny in two respects at least. First, I have so far been speaking of the causation of ontogeny only, and it is clear that we must apply the question "what for?", the question of survival value to ontogeny as well. That is, we need to ask what the survival value is of the many different types of ontogenetic control that our analysis brings out. As yet we have only a hint of what is in store if we were to apply this question consistently. For instance, it is in some cases easy to see why the control in certain behaviour patterns is largely internal, and why in others interaction with the environment is advantageous. Thus a young Gannet, which has to jump off a high cliff, would be poorly off if he had to acquire the basic pattern of flight the way we acquire a skill such as writing. Similarly, the selective responsiveness to rival males in territorial species might well have to be unconditioned so that it can function at once when a male starts its first breeding cycle. Young song birds on the contrary, which begin by responding to a very wide range of objects when they start feeding independently and gradually learn to take only what has proved to be edible, are by this very "open-mindedness" able to adapt to many different habitats, and learn to select the most abundant food in each, however different this food may be in different habitats.

The study of ontogeny also overlaps, but in another way, with that of causation of cyclical or recurring behaviour in the adult. Some learning processes can occur all through the life of the individual, even though their impact decreases with age. (So do, of course, certain physiological changes, such as the formation of antibodies, or of pigmentation of the skin.) In this respect too ontogeny can be said to continue beyond the period of growth to maturity and the causation of the behaviour of the adult animal therefore grades into that of the phenomena usually classified under ontogeny; the distinction is partly one of the time scale involved. Yet there is sufficient justification to distinguish between the two sets of processes; as is obvious from the fact that one can say that a man is afraid of a flying plane "because he sees it" but also "because he has been bombed out as a child". The main point is to recognise that both statements may be true, that each covers part of the total causal chain involved, and that the question "what made him behave the way he did?" requires a complete answer in which both partial answers are contained.

Evolution

The fact that behaviour is in many respects species-specific, and yet often similar in related species has been recognised by many workers before LORENZ, and the natural conclusion to be drawn from this, namely, that behaviour should be studied comparatively just as structures, with the ultimate aim of elucidating behaviour evolution, had also been drawn. WHITMAN's work (1898, 1919), and that of HEINROTH (1911), HUXLEY (1914, 1923) and VERWEY (1930), preceded LORENZ's contributions. While WHITMAN can be said to have concentrated on questions of homology or common descent, HUXLEY's interest was focussed on the task of testing the theory of natural selection. In a sense, however, all these important studies can be said to have been preliminary, preparatory to a concerted attack on problems of behaviour evolution which has gradually developed since LORENZ (1937) began to emphasise the need for systematic comparative studies.

In this field, too, research began in a rather intuitive way, guided by trends of thought which have been gradually made explicit and which have become increasingly similar to general evolutionary thinking. In this respect, too, Ethology is being incorporated into general Biology.

In some respects the evolutionary study of behaviour suffers from handicaps not met to that extent in that of structure. It needs no repeating that direct documentary evidence in the form of "fossil behaviour" is hardly available. The exploration of behaviour ontogeny as a tool does not seem to be very promising either. This, however, is not a serious handicap in view of the controversial nature of this tool in the study of structural evolution. In other respects ethologists are perhaps better off than students of structure: through their familiarity with the behaviour of many animals in their natural surroundings their attention has been drawn more readily to questions of survival value, and through these to a consideration of the effects of natural selection. When I say that ethologists were "better off" in this respect, I feel I should add in fairness to ourselves that this is due to our own efforts in creating a better opportunity; it is, in principle, easier to experiment on the survival value of structure than on that of behaviour, and the truth is that ethologists, by being in general good naturalists, deserved their good luck.

428   N. Tinbergen

Evolutionary study has, of course, two major aims: the elucidation of the course evolution must be assumed to have taken, and the unravelling of its dynamics.

The first task is being pursued mainly through comparison of groups of closely related species. This limitation to closely allied forms is necessary because it is only here that conclusions about homology (i.e. common descent) can be drawn with any degree of probability. It is due to this restriction that what evidence we have applies to microevolution, particularly to adaptive radiation of relatively recent origin. As I have discussed elsewhere (Tinbergen 1959), the trend here is to apply very much the same methods as those employed by taxonomists: we judge affinity by the criterion of preponderance of shared characters, particularly of those which we consider non-convergent. Once we have hardened the conclusion, often already reached by taxonomists, that a certain group must be monophyletic, we judge the degree of evolutionary divergence by the degree of dissimilarity of those characters that must be considered highly environment-resistant ontogenetically — we try to exclude from our material such differences as are the direct phenotypic consequence of different environments, such as an individually acquired darkening of external coloration under the immediate influence of a moist environment (which, of course, is very different from differences in *ability* to respond to the environment).

The comparative procedure has now been applied to a number of groups and one of the encouraging outcomes is that classifications based on behaviour taxonomy have, on the whole, corresponded very closely to the already existing classifications. Minor differences were found — but these often concerned matters which taxonomists considered not quite settled. I think it is worth emphasising this correlation between the two sets of results because it is again a striking justification of treating behaviour patterns as "organs".

The work on evolution *dynamics* can be said to consist of two major parts. First, the genetic control of species-specific behaviour, about which we know so much less than about that of species-specific structure, is now being studied with all the methods available in genetics; differences between species, subspecies and strains raised in identical environments are registered; the effect of mutations on behaviour are beginning to be explored, and controlled cross-breeding is being done. There seems little doubt as to the general outcome: individuals and populations differ as much in their hereditary behaviour "blueprints" as in their hereditary structural blueprints. The genetic variation on which natural selection can act is there.

The second major task is the study of the influence of selection on behaviour evolution. This task is being tackled in two different ways. One is the study of survival value of species-specific characters, the other is the direct application of a controlled selection pressure and its results over a series of generations.

The study of survival value receives its inspiration from the study of convergencies, and of divergencies within a taxonomic group; hypotheses about survival value are derived from these studies; they can be tested in experiment. The interpretation of the results is worth detailed scrutiny. When one finds that a certain characteristic has survival value — when it has been shown that various deviations from the norm lead to a lower success rate – one can draw one firm conclusion: one can say that one has demonstrated beyond doubt a selection pressure which prevents the species in its present state from deviating. One has really demonstrated the part played by selection in *stabilisation* of the present state. However, the conclusion that this same selection pressure must have been responsible in the past for the *moulding* of the character studied is speculative, however probable it often is. One can support such a conclusion by marshalling supplementary evidence, such as arguing that the environment which exerts the selection pressure must be assumed to have remained constant in this respect for a long time, or showing that the species is even now slightly variable according to area. Thus, E. CULLEN's demonstration of the adaptive nature of many of the Kittiwake's peculiarities (1957) can be used as a pointer to past selection pressures by arguing that the species has probably a fairly long history of cliff-nesting behind it. An example of the second line of argument is BLAIR's demonstration (1955) of the fact that the mating calls of two species of *Microhyla* are more distinct in the areas where they overlap (and where selection against cross-breeding must be assumed to have favoured inter-species distinctness) than in the areas where either species occurs alone.

All I have said above about the study of survival value for its own sake is relevant again here, but I should like to re-emphasize one point. For an assessment of what selection can be assumed to have contributed to the present state of species it is important to realise that selection rewards or penalises isolated bits of animals through rewarding or penalising animals, or breeding pairs, as wholes. Since studies of survival value show us that there are often direct contradictions between different selection pressures, the animal that survives best must be a compromise, and it must be one of our main tasks to try and find all the pressures — favourable and unfavourable alike — that can have affected any character we select for study. In general, we should not only try to pinpoint isolated selection pressures, but study their interaction as well.

In spite of the fact that we shall never be able to prove directly which contributions selection has made in the past, and that therefore any conclusion about the way interaction with the environment has moulded present-day species must remain tentative and, as such, different from conclusions drawn in the field of Physiology, Ontogeny and Survival Value, the ethologist feels that this is no reason to dismiss evolutionary study as just speculation. I believe that this developing branch of Ethology may well have effects on general biological thinking.

The direct application of selection pressures will, with increasing precision of description and measurements, give us an increasing amount of real demonstration of the potentialities of selection. With the growing trend towards experimentation it is important, however, to point out that even the most perfect experiment of this kind does not give us direct proof of what selection has done in the past. The interpretation of such experiments as contributions to evolution theory will always include an extrapolation: while they demonstrate what selection can do, the best they can tell us is that selection can have happened in the way demonstrated, and that the results obtained are not contradictory to what other indirect evidence has led us to suppose. They really deal merely with "possible future evolution", and only indirectly with past evolution. For instance, CROSSLEY's demonstration (1960) that 40 generations of antihybrid selection in partially interbreeding populations of ebony and vestigial mutants of *Drosophila melanogaster* change aspects of mating behaviour of both males and females of both populations does not directly prove that such selection *has* contributed to speciation, but it is in line with ideas developed before these experiments were done. This is, of course, not at all to belittle the relevance of such experiments, but merely to assign them their proper place in the total body of evidence.

Conclusion

I have tried in this paper to give a sketch of what I believe modern Ethology to be about. I have perhaps given Ethology a wider scope than most practising ethologists would do, but if one reviews the various types of investigations carried out by people usually called ethologists, one is forced to conclude that the scope is in fact as wide as I have indicated. This sketch is not meant to be balanced or comprehensive; I have allowed myself to enlarge a little on special issues — on "bees in my bonnet" — such as the relations between Ethology and Physiology; the need to spend more effort on studies of survival value and methods to be employed in such studies; problems and methods of behaviour ontogeny; and the nature of arguments used in the study of evolution — all issues which require further discussion. I have also tried to assess KONRAD LORENZ's contribution to modern Ethology, and have argued that I consider his insistence that behaviour phenomena can, and indeed must, be studied in fundamentally the same way as other biological phenomena to be his major contribution. LORENZ can with justification be said to be the father of modern Ethology — even though he has had his forerunners; there is nothing amazing about every father having had a father.

The central point in LORENZ's life work thus seems to me his clear recognition that behaviour is part and parcel of the adaptive equipment of animals; that, as such, its short-term causation can be studied in fundamentally the same way as that of other life processes; that its survival value can be studied just as systematically as its causation; that the study of its ontogeny is similar to that of the ontogeny of structure; and that the study of its evolution likewise follows the same lines as that of the evolution of form. Moreover, in all these fields LORENZ has done concrete research which demonstrated the great heuristic value of his approach. Yet, although his concrete, factual contributions have been considerable, the impact he has made is due to his sketch of a type of approach, and to new and original hypotheses rather than to the experimental testing of such hypotheses. This is why recent changes of concepts and terminology, revisions of hypotheses, and the reporting of results which are sometimes different from LORENZ's earlier conclusions, have little relevance to the question of the value of his work.

One of the measures of this value which I will mention in passing is the fact that students of human behaviour are showing a growing interest in ethological methods.

Finally, I should like to touch briefly on a matter of terminology. It will be clear that I have used the word "Ethology" for a vast complex of sciences, part of which already have names, such as certain branches of Psychology and Physiology. This, of course, does not mean that I want to claim the name Ethology for this whole science, for this would be falsifying its history; the term really applies to the activities of a small group of biologists. What I have been at pains to develop is the thesis that we are witnessing the fusing of many sciences, all concerned with one or another aspect of behaviour, into one coherent science, for which the only correct name is *"Biology of behaviour"*. Of course this fusion is not the work of one man, nor of the small group called ethologists. It is the outcome of a widespread tendency to apply a more coherent biological approach, which has expressed itself in what may well have been quite independent developments within sciences such as Psychology and Neurophysiology. Among zoologists and naturalists, it is LORENZ who has contributed most to this development, and who has more than any other single person influenced these sister disciplines in this particular way. Finally, the comprehensive view of the aims of the biological study of behaviour has grown more rapidly in Ethology than in any of the other sciences. Yet, in view of the confused "public image" called up by the word Ethology it might well be advisable not to overdo the use of the word. What does seem to me to matter is the growing awareness of the fundamental unity of the Biology of Behaviour, and the realisation that "Ethology" is more than "Physiology of Behaviour", just as "Biology" is more than "Physiology".

Zusammenfassung

Ich habe in diesem Aufsatz kurz anzudeuten versucht, was meiner Ansicht nach das Wesentliche in Fragestellung und Methode der Ethologie ist und weshalb wir in KONRAD LORENZ den Begründer moderner Ethologie erblicken. Hierbei habe ich vielleicht das Arbeitsgebiet der Ethologie weiter gefaßt, als unter Ethologen gebräuchlich ist. Wenn man aber die vielartige Arbeit jener Forscher übersieht, die sich Ethologen nennen, ist man zu dieser weiten Fassung geradezu gezwungen. Ich habe in meiner Darstellung weder Vollständigkeit noch Gleichgewicht angestrebt und, um zur Fortführung des Gesprächs anzuregen, ruhig meine Steckenpferde geritten, vor allem das Verhältnis zwischen Ethologie und Physiologie, die Gefahr der Vernachlässigung der Frage der Arterhaltung, Fragen der Methodik der ontogenetischen Forschung, und Aufgaben und Methoden der Evolutionsforschung.

Bei der Einschätzung des Anteils, den LORENZ an der Entwicklung der Ethologie genommen hat und noch nimmt, habe ich als seinen Hauptbeitrag den bezeichnet, daß er uns gezeigt hat, wie man bewährtes „biologisches Denken" folgerichtig auf Verhalten anwenden kann. Daß er dabei an die Arbeit seiner Vorgänger angeknüpft hat, ist nicht mehr verwunderlich, als daß jeder Vater selbst einen Vater hat.

Insbesondere scheint mir das Wesentliche an LORENZ' Arbeit zu sein, daß er klar gesehen hat, daß Verhaltensweisen Teile von „Organen", von Systemen der Arterhaltung sind; daß ihre Verursachung genau so exakt untersucht werden kann wie die gleich welcher anderer Lebensvorgänge, daß ihr arterhaltender Wert ebenso systematisch und exakt aufweisbar ist wie ihre Verursachung, daß Verhaltensontogenie in grundsätzlich gleicher Weise erforscht werden kann wie die Ontogenie der Form und daß die Erforschung der Verhaltensevolution der Untersuchung der Strukturevolution parallel geht. Und obwohl LORENZ ein riesiges Tatsachenmaterial gesammelt hat, ist die Ethologie doch noch mehr durch seine Fragestellung und durch kühne Hypothesen gefördert als durch eigene Nachprüfung dieser Hypothesen. Ohne den Wert solcher Nachprüfung zu unterschätzen — ohne die es natürlich keine Weiterentwicklung gäbe — möchte ich doch behaupten, daß die durch Nachprüfung notwendig gewordenen Modifikationen neben der Leistung des ursprünglichen Ansatzes vergleichsweise unbedeutend sind.

Nebenbei sei auch daran erinnert, daß eine der vielen heilsamen Nachwirkungen der LORENZschen Arbeit das wachsende Interesse ist, das die Humanpsychologie der Ethologie entgegenbringt — ein erster Ansatz einer Entwicklung, deren Tragweite wir noch kaum übersehen können.

Am Schluß noch eine Bemerkung zur Terminologie. Ich habe hier das Wort „Ethologie" auf einen Riesenkomplex von Wissenschaften angewandt, von denen manche, wie Psychologie und Physiologie, schon längst anerkannte Namen tragen. Das heißt natürlich nicht, daß ich den Namen Ethologie für dieses ganze Gebiet vorschlagen will; das wäre geschichtlich einfach falsch, weil das Wort historisch nur die Arbeit einer kleinen Gruppe von Zoologen kennzeichnet. Der Name ist natürlich gleichgültig; worauf es mir vor allem

432   N. Tinbergen

ankommt, ist darzutun, daß wir das Zusammenwachsen vieler Einzeldiszipli-
nen zu einer vielumfassenden Wissenschaft erleben, für die es nur einen rich-
tigen Namen gibt: *„Verhaltensbiologie"*. Selbstverständlich ist diese synthe-
tische Entwicklung nicht die Arbeit eines Mannes oder gar die der Ethologen.
Sie ist die Folge einer allgemeinen Neigung, Brücken zwischen verwandten
Wissenschaften zu schlagen, einer Neigung, die sich in vielen Disziplinen ent-
wickelt hat. Unter den Zoologen ist es Lorenz, der hierzu am meisten bei-
getragen und zudem manche Nachbardisziplinen stärker beeinflußt hat als
irgendein anderer. Ich bin sogar davon überzeugt, daß diese Einwirkungen auf
Nachbarwissenschaften noch lange anhalten werden und daß die Verhaltensbiologie erst am Anfang ihrer Ontogenie steht.

References

Beach, F. A., and G. Levinson (1950): Effects of androgen on the glans penis and
mating behavior of castrated male rats. J. Exp. Zool. **114**, 159—168 • Beer, C. G. (1962):
Incubation and nest-building behaviour of Blackheaded Gulls II: Incubation behaviour in
the laying period. Behaviour **14**, 283—305 • Blair, W. F. (1955): Mating-call and stage of
speciation in the *Microhyla olivacea-M. carolinensis* complex. Evolution **9**, 469—480 •
Blest, A. D. (1957): The function of eyespot patterns in the Lepidoptera. Behaviour **11**,
209—256 • Blest, A. D. (1960): The evolution, ontogeny and quantitative control of the
settling movements of some New World Saturniid moths, with some comments on distance
communication by Honey-bees. Behaviour **16**, 188—253 • Crossley, S. A. (1960): An experimental study of sexual isolation within the species *Drosophila melanogaster*. Anim.
Behaviour **8**, 232—233 • Crossley, S. A. (Unpublished Doctor's thesis on the effect of
anti-hybrid selection on sexual isolation within the species *Drosophila melanogaster*.
Oxford) • Cullen, E. (1957): Adaptations in the Kittiwake to cliff nesting. Ibis **99**,
275—302 • Cullen, E. (1961): The effect of isolation from the father on the behaviour
of male Three-spined Sticklebacks to models. Tech. (Fin.) Rept. A F 61 (052)—29
U.S.A.F.R.D.C. • Eibl-Eibesfeldt, I. (1955): Über die ontogenetische Entwicklung der
Technik des Nüsseöffnens vom Eichhörnchen *(Sciurus vulgaris L.)* Z. Säugetierk. **21**,
132—134 • Eibl-Eibesfeldt, I. (1956): Angeborenes und Erworbenes im Nestbauverhalten
der Wanderratte. Naturwiss. **42**, 633—634 • Goethe, F. (1937): Beobachtungen und Untersuchungen zur Biologie der Silbermöve auf der Vogelinsel Memmertsand. J. Ornithol. **85**,
1—119 • Grohmann, J. (1939): Modifikation oder Funktionsreifung? Z. Tierpsychol. **2**,
132—144 • Haartman, L. von (1957): Adaptations in hole-nesting birds. Evolution **11**,
339—348 • Hassenstein, B. (1951): Ommatidienraster und afferente Bewegungs-Integra-
tion. Z. vgl. Physiol. **33**, 301—326 • Hassenstein, B. (1957): Über die Wahrnehmung der
Bewegung von Figuren und unregelmäßigen Helligkeitsmustern. Z. vgl. Physiol. **40**,
556—596 • Heinroth, O. (1909): Beobachtungen bei der Zucht des Ziegenmelkers *(Capri-
mulgus europaeus* L.) J. Ornithol. **57**, 56—83 • Heinroth, O. (1911): Beiträge zur Biologie,
namentlich Ethologie und Psychologie der Anatiden. Verh. V. Internation. Ornith. Kongr. Berlin • Heinroth, O. und M. (1928): Die Vögel Mitteleuropas. Berlin • Hinde, R. A. (1958):
The nest-building behaviour of domesticated canaries. Proc. Zool. Soc. Lond. **131**, 1—48 •
Hinde, R. A. (1959): Unitary drives. Anim. Behaviour **7**, 130—141 • Hinde, R. A. (1962):
Temporal relations of broodpatch development in domesticated canaries. Ibis **104**, 90—97 •
Hinde, R. A., and R. P. Warren (1961): Roles of the male and the nest-cup in controlling
the reproduction of female canaries. Anim. Behaviour **9**, 64—67 • Holst, E. von (1937):
Vom Wesen der Ordnung im Zentralnervensystem. Naturwiss. **25**, 625—631, 641—647 •
Holst, E. von, und H. Mittelstaedt (1950): Das Reafferenzprinzip. Naturwiss. **37**,
464—476 • Holst, E. von, und U. von St. Paul (1960): Vom Wirkungsgefüge der Triebe.
Naturwiss. **47**, 409—422 • Hoogland, R., D. Morris and N. Tinbergen (1957): The
spines of sticklebacks *(Gasterosteus* and *Pygosteus)* as means of defence against predators
(Perca and *Esox)*. Behaviour **10**, 205—236 • Huxley, J. S. (1914): The courtship habits
of the Great Crested Grebe *(Podiceps cristatus);* with an addition to the theory of sexual
selection. Proc. Zool. Soc. Lond. 1914, 491—562 • Huxley, J. S. (1923): Courtship
activities in the Red-throated Diver *(Colymbus stellatus* Pontopp); together with a discussion on the evolution of courtship in birds. Jour. Linn. Soc. **35**, 253—291 • Kettlewell, H. B. D. (1955): Selection experiments on industrial melanism in the Lepidoptera.
Heredity **9**, 323—342 • Kettlewell, H. B. D. (1956): Further selection experiments on
industrial melanism in the Lepidoptera. Heredity **10**, 287—301 • Klomp, H. (1954): De
Terreinkeus van de Kievit, *Vanellus vanellus* (L.). Ardea **42**, 1—140 • Knoll, M. (1953):
Über das Tages- und Dämmerungssehen des Grasfrosches nach Aufzucht in veränderten

On aims and methods of Ethology 433

Lichtbedingungen. Z. vgl. Physiol. **35**, 42—67 • Lehrman, D. S. (1953): A critique of Konrad Lorenz's theory of instinctive behavior. Quart. Rev. Biol. **28**, 337—363 • Lehrman, D. S. (1955): The physiological basis of parental feeding behaviour in the Ring Dove *(Streptopelia risoria)*. Behaviour **7**, 241—286 • Lehrman, D. S. (1961): Hormonal regulation of parental behaviour in birds and infrahuman mammals. in: W. C. Young (Ed.): Sex and Internal Secretion, Baltimore • Lindauer, M. (1961): Communication among Social Bees. Cambridge Mass. • Lorenz, K. (1935): Der Kumpan in der Umwelt des Vogels. J. Ornithol. **83**, 137—213, 289—413 • Lorenz, K. (1937): Über die Bildung des Instinktbegriffes. Naturwiss. **25**, 289—300, 307—318, 324—331 • Lorenz, K. (1955): (Discussion remark on p. 77 in: Group Processes, First Josiah Macy Jr Foundation Conference. New York.) • Lorenz, K. (1950): The comparative method in studying innate behaviour patterns. S.E.B. Symposia **4**, 221—269 • Lorenz, K. (1961): Phylogenetische Anpassung und adaptive Modifikation des Verhaltens. Z. Tierpsychol. **18**, 139—187 • Mosebach-Pukowski, E. (1937): Über die Raupengesellschaften von *Vanessa io* und *Vanessa urticae*. Z. Morph. Oekol. Tiere **33**, 358—380 • Nielsen, E. T. (1958): The Method of Ethology. Proc. 10th Internat. Congress of Entomol. **2**, 563—565 • Pringle, J. W. S. (1951): On the parallel between learning and evolution. Behaviour **3**, 174—216 • Riess, B. F. (1950): The isolation of factors of learning and native behavior in field and laboratory studies. Ann. N.Y. Acad. Sci. **51**, 1093—1102 • Rowell, C. H. F. (1961): Displacement grooming in the Chaffinch. Animal Behaviour **9**, 38—64 • Ruiter, L. de (1952): Some experiments on the camouflage of stick caterpillars. Behaviour **4**, 222—232 • Sauer, F. (1954): Die Entwicklung der Lautäußerungen vom Ei ab schalldicht gehaltener Dorngrasmücken *(Sylvia c. communis* Latham) im Vergleich mit später isolierten und wildlebenden Artgenossen. Z. Tierpsychol. **11**, 10—93 • Sevenster, P. (1961): A causal analysis of a displacement activity. Behaviour Suppl. **9** • Sperry, R. W. (1959): The growth of nerve circuits. Sci. American **201**, 68—75 • Thorpe, W. H. (1962): Bird Song. Cambridge • Tinbergen, N. (1940): Die Übersprungbewegung. Z. Tierpsychol. **4**, 1—40 • Tinbergen, N. (1951): The Study of Instinct. Oxford • Tinbergen, N. (1959): Comparative studies of the behaviour of gulls; a progress report. Behaviour **15**, 1—70 • Tinbergen, N., and A. C. Perdeck (1950): On the stimulus situation releasing the begging response in the newly hatched Herring Gull chick *(Larus a. argentatus* Pontopp.). Behaviour **3**, 1—38 • Tinbergen, N., G. J. Broekhuysen, F. Feekes, J. C. W. Houghton, H. Kruuk and E. Szulc (1962): Egg shell removal by the Black-headed Gull, *Larus ridibundus* L.; a behaviour component of camouflage. Behaviour **19**, 74—118 • Verwey, J. (1930): Die Paarungsbiologie des Fischreihers. Zool. Jahrb. Allg. Zool. Physiol. **48**, 1—120 • Wells, M. J. (1958): Factors affecting reactions to *Mysis* by newly hatched *Sepia*. Behaviour **8**, 96—111 • Whitman, Ch. O. (1898): Animal Behavior. Woods Hole • Whitman, Ch. O. (1919): The Behavior of Pigeons. Carnegie Inst. Wash. Publ. **257**, 3, 1—161.

1

Ethological Units of Behavior

George W. Barlow

Department of Zoology and
Museum of Vertebrate Zoology
University of California at Berkeley

Relation of Behavioral Units to Neurophysiology

Behavior may be subdivided into a variety of units. Depending on the objectives of a given study, one might find the appropriate unit at almost any level of integration. Thus for the sociologist, and for some psychologists, the important measure might derive from interindividual responses, or from patterns of behavior of whole groups. Moving down the scale, the behavioral measure could equally well involve some integrated yet arbitrary end point, such as that produced in bar-pressing devices. Such instrumental responses have proved of value in studies of learning, and they have also gained usage in a few ethological studies. An instrumental response, although an arbitrary end point, can still have a great deal of behavior behind it; animals may approach the target differently, and may activate the instrument in a variety of ways. In most cases, however, it will be a useful measure of relatively complex behavior.

The ethologist has by and large resorted to studies of what have been termed "Fixed Action Patterns," the molar unit of ethology. At a still finer level there is the functional unit, as defined by Liem (1967); this is doubtless the smallest unit that could be used as a measure in behavioral studies, consisting of the simplest coaction of a group of muscles, ligaments, bones, and nerves.

Any unit is appropriate if it fits the needs of the study. The more complex and broadly defined units may be adequate for general studies involving relatively broad treatments, such as those which might result from gross manipulation of the endocrine system or extensive ablations of the central nervous system. On the other hand, the finest unit of behavior, for instance the functional unit, obviously has the most direct relationship to neural events. The study of such fine units may overlook the more highly integrative aspects of behavior, however; and these are often of great interest.

217

218   *Barlow*

Somewhere between these extremes lie the instrumental responses and the fixed action patterns. While the instrumental response is excellent for some purposes, it tends to have limited application. Moreover, it is frequently an indirect index of the behavior of interest rather than a direct measure of it.

Many ethological problems have at their center a reliance on the concept of the fixed action pattern. As a unit of behavior it has the advantage that it refers to behavior that is directly observable, and thus no instrumentation is required in order to record it. This also means that the decision about the occurrence of the event has the disadvantage of containing a subjective element, decreasing the precision. Rarely, as in the case of a fish attacking a mirror, the fixed action pattern can be automatically and objectively recorded.

If behavior studies are to have meaning to neurophysiologists, the models that result from the studies must be translatable into neurophysiological language. This means the behavior must be quantified, yielding mathematical models that can be compared with the models produced by the neurophysiologist.

There have been relatively few mathematical models generated by the ethologists. Such models are found largely in the province of the neurophysiologists. I cite, for example, the stimulating work done by Gerstein and his associates (reviewed in Moore, Perkel, and Segundo 1966) involving auto- and cross-correlations of the activity of one or a few neurons. The shape of things to come, however, is more comprehensively communicated in an article by Ashby (1966). These models have been more important in thinking about behavior than the study of behavior has been useful in illuminating neurophysiology. Although stimulating, these models are little more than that; they have not provided us with an understanding of the mechanisms underlying the behavior. The neuron, moreover, is inherently different from the complex fixed action pattern. Models based on the behavior of individual neurons can only be regarded as starting points.

These statistical models contain assumptions that are seldom adhered to in ethological studies. For one, the unit, the event, should be all-or-nothing. Where one is dealing with trains of spikes along a nerve fiber this condition appears to be met, although even here the assumption bears examining.[1] When dealing with fixed action patterns this requirement is seldom fulfilled.

Another limitation of the statistical models is that the phenomenon under study should exhibit "stationarity." This means the statistical parameters should remain unchanged from one sampling point in time to another. As Moore, Perkel, and Segundo (1966) pointed out, this assumption is seldom tested in neurophysiology. The same holds for ethological studies. There have been a few attempts to test for stationarity, however, as in the creeping-through model of Nelson (1965) and in the distribution of attacks to a mirror by a cichlid fish (Skaller 1966).

The Fixed Action Pattern

Lorenz (1950) made a strong case for the importance of units in the development of all branches of science. And he has rightfully pointed to the central role our notion of fixed action patterns has played in the progress of ethology. But just as the atom was once thought to be indivisible and above questioning, much the same attitude prevails toward the fixed action pattern in current thinking in ethology. If the fixed action pattern is to be used to generate models of use to the neurophysiologist, we must examine it closely to see if it qualifies as a basic unit of behavior.

[1] C. H. F. Rowell: personal communication.

As mentioned, there are indeed instances when fixed action patterns exhibit stationarity in time. However, it is also clear that in most instances they do not (Nelson 1965).

It is also obvious that fixed action patterns do not behave as all-or-nothing responses. Marler and Hamilton (1966) reflect this well when they refer to them as "modes of action patterns." Their development of the topic makes it clear that they are referring to events that are statistically definable rather than to absolute events. Surprisingly, there has been no real effort to study the variation in the several parameters of the fixed action pattern, if indeed we can speak of *the* fixed action pattern.

One study, however, has attempted to come to grips with the degree of stereotypy along one dimension. Dane, Walcott, and Drury (1959) filmed the courtship displays of goldeneye ducks, then analyzed them in detail. The only parameter considered was duration. They made no attempt to delimit units other than trying to work with repeatedly recognizable events. They soon discovered they were dealing at times with units nested within units. This was particularly revealing. It led them to identify the sources of variation, as well as the least variable components.

Some students of behavior seem to have drawn from their publication that the displays of the goldeneye duck, and presumably of most anatid ducks, are essentially invariant. The data of Dane, Walcott, and Drury, however, show that the standard deviation is commonly of the order of 10 per cent to 20 per cent of the mean value; occasionally it is less, but sometimes it is considerably more. (The range of values, of course, is usually much greater.) The one nonsignal movement measured, the wing stretch, was extremely variable yet, importantly, readily recognizable as a distinct category of behavior. The significant point is that this is of the proper order of magnitude for variability in most biological measures. One reason the displays seem so fixed is that they happen so quickly.

Theirs is the only systematic quantitative study in ethology, employing films, of the temporal dimensions of what are commonly regarded as fixed action patterns. As is so often the case, it seems necessary to start with an extreme example in order to make a point. The machine-like fixed action patterns of anatid ducks are probably representative of the more sterotyped displays found among animals, as are certain grooming actions of passerine birds (Rowell 1961). Whether or not all the movements we commonly call fixed action patterns exhibit this degree of stereotypy is far from certain.

The Interrelation of the Parts of the Fixed Action Pattern

When an ethologist speaks of a fixed action pattern he usually has in mind a small piece of behavior, but one that is already complex. Russell, Mead, and Hayes (1954) attempted to come to grips with this problem by defining the action pattern in terms of its components, called *acts*. At the risk of oversimplifying, an act is the least divisible unit of behavior making up the fixed action pattern. Several acts together constitute a superact, the fixed action pattern of contemporary usage. These more complex superacts are then arranged into bouts wherein the event is repeated several times, or into sequences wherein they are related to other fixed action patterns. In practice, it often becomes difficult to decide between a very complex fixed action pattern as opposed to a sequence of simple events. Thus it is meaningful to have a definition for the fixed action pattern that will enable one to distinguish it from a sequence of different fixed action patterns. Such a definition is not easy to come by.

It is generally recognized today that the fixed action pattern is indeed a graded response. The word "fixed" refers to the nature of the gradation itself, expressed as "intensity" or "completeness." If a fixed action pattern consists of several components, say *A*, *B*, *C*, and

D, these will occur in some predictable sequence which relates to their thresholds for expression. At a relatively incomplete, or low-intensity, state one would observe only *A*. At increasing levels of completeness, arrangements such as *AB*, *ABC*, and *ABCD* would occur, but never sequences such as *ACD* or *DCA*. We can imagine a fixed action pattern as a film strip containing the complete performance, which must be started at the beginning each time it is run through. Any variation would consist in the variable distance that the film is allowed to travel, but the sequence of pictures would be fixed. The emergence of new components in a pattern as "intensity" increases is entirely predictable, according to this view. Such specificity implies a fixed organization within the central nervous system, as Hinde (1966) has forcefully argued. This argument finds support from the work reviewed by Weiss (1950), who postulated "scores" within the brain according to which complex motor patterns can be orchestrated.

The Relation of Causal Factors to the Components of the Fixed Action Pattern

Hinde (1966) has further argued that, according to this view of a fixed action pattern, "the components all depend on the same causal factors." This statement can be given the status of a definition, and can be used to guide our judgment as to which behavioral chains we should call fixed action patterns. Presumably, a different set of causal factors would be required for a transition to the next fixed action pattern.

Russell, Mead, and Hayes (1954) have attempted to resolve this problem by pointing out how a definition could be achieved. In practice, this might consist in varying all conceivable stimuli that could influence the expression of the particular behavior. However, this would be difficult for two reasons. First, the influence could be only statistically defined, because of the spontaneous underlying variability of the behavior. Second, the number of factors to test is almost infinite; one might be confronted ultimately with the uncomfortable task of attempting to prove a null hypothesis.

The Relation of the Fixed Action Pattern to the Trigger

Another salient characteristic of the fixed action pattern is said to be its independence from environmental control; the trigger that sets it off plays no role in the further expression of the behavior. I quote again from Hinde (1966, p. 17–19): The soliciting posture of the female chaffinch "is remarkably constant in form: although it involves most of the muscles in the body, the relations between their contractions vary little, and they are presumably independent of immediate environmental control." Such variations in the "intensity" of the response may be due to the recent history of the animal, sometimes said to be due simply to its motivational state.

Hinde says further, on page 19, that "each [F.A.P.], although it may consist of a quite complicated spatiotemporal pattern of muscular contractions, cannot be split into successive responses which depend on qualitatively different external stimuli (Lorenz 1935, 1937; Tinbergen 1942). Since such movements depend upon external factors only for their elicitation, they vary in degree of completeness, but not in the relation between their parts." It is not clear from this what constitutes a qualitatively as opposed to a quantitatively different external stimulus. Nonetheless, the reader is left with the impression that the essential feature of a fixed action pattern, with relation to the stimulus that elicits it, is that the stimulus is only important as a trigger, not as a modulator of the behavior that ensues. The distinction here is between peripheral modulators and some organization of the fixed action pattern in the central nervous system. Two separate mechanisms are implied.

Testing the Hypothesis

When the chairman for this session introduced this paper, he felt compelled to respond to some earlier remarks by Professor Arthur Hasler, who had drawn to the attention of the audience a new device for protecting downed pilots from sharks. The apparatus consisted of a float with a sack around the pilot to keep the sharks from detecting chemicals in the water. The chairman, Dr. Arthur Myrberg, described the testing of this apparatus as he had observed it. Shortly before the experiment was to begin, Dr. Myrberg transmitted sound signals into the sea, which resulted in the appearance of many large sharks. Upon seeing so many sharks in the water the experimenters decided to defer testing the apparatus until another day when the sharks were not so abundant.

In many respects we ethologists have been guilty of the same sin. We have tended to pick the examples that will confirm our hypotheses. This is reasonable and understandable in the early stages of development when one needs to stand by ideas of value. But as the field progresses the time comes when hypotheses must be tested in the spirit of attempting to disprove them, rather than the other way around. If the ideas are sound, they will survive the sharks.

In brief, there are three diagnostic properties of a fixed action pattern. (1) It has common causal factors different from those of other fixed action patterns. (2) The stimulation triggering, or releasing, a fixed action pattern ceases to exercise futher control over it. And, (3) the components appear in a predictable sequence in time; while the "intensity" or completeness may vary, the interrelation of the parts may not. (In the literal sense then, a fixed action pattern is not fixed.) The third criterion is relatively easy to deal with. It requires only a careful quantitative analysis of many occurrences of the event. The first and second criteria, however, present formidable difficulties. They require considerable control of the environment and of the animal in a complex interaction.

A closer look at what have been thought to be good fixed action patterns in many animals may well reveal that the fixed action pattern is a special case along a spectrum of behavior leading back to relatively unstructured movements. Even in the instances of those fixed action patterns which seem to fulfil all the requirements so nicely spelled out by Hinde (1966), critical inspection may show that the postulates do not necessarily hold. A good example from my own experience comes from the courtship displays of the cichlid fish *Etroplus maculatus*, the orange chromide.

The central courtship movement of the orange chromide consists of a rapid quivering of the head from side to side, coupled with a pronounced flickering, an opening and closing, of the black pelvic fins. This display can be thought of as having two components, quivering and flickering. Ordinarily the two components occur together and produce a readily recognizable display. The triggering stimulus is the approach of the mate, particularly when the fish is near the presumptive spawning site. In a typical situation the male, or the female, will start quivering and flickering at the same time, the flickering will drop out while the quivering continues, then flickering will reappear shortly before quivering stops. All in all, this looks like a fixed action pattern fitting all the criteria.

Quite by accident I discovered that the coupling of the flickering and the quivering is not fixed. If the relative size of the mate is varied, the quivering and the flickering are influenced differentially. Also, the duration of quivering itself varies. This example indicates that the nature of the stimulus differentially influences the components of this behavior pattern.

As a further example, it sometimes happens when one orange chromide is quivering that the mate will indicate by subtle movements that it intends to groom the animal that is quivering. Without interrupting the quivering the actor changes slightly the amplitude of the movement and prolongs the duration of the quivering, often while deleting the flickering. Thus while the behavior is running its course it is still subject to external modulation by the stimulus that started the behavior in the first place.

It could be argued that quivering and flickering are separate fixed action patterns. Most likely they are. The point is that they nearly always occur together. Most observers would recognize them as a unitary display, ascribing variations between the components to threshold differences.

I purposely dealt with courtship movements because displays such as these are so commonly used as examples in illustrating the nature of fixed action patterns. To the best of my knowledge, there is no single case where all the criteria have been properly tested and fulfilled.

When we direct our attention to other signal movements, particularly those involved in aggressive behavior, it becomes increasingly difficult to apply the criteria for the fixed action pattern. The displays of aggressive behavior are notoriously subject to continuing modulation from the trigger. The hostile behavior of the teleost fish *Badis badis* (Barlow 1962) serves as a good example. A given display preceding a damaging fight consists of many components which may or may not be superimposed one upon the other. As the fight begins, these components tend to occur in a predictable sequence, the *ABCD* situation. But after the fish have been displaying for a while the components become scrambled. As the animals reverberate between the different displays the order may become *ACDB*. Furthermore, under certain circumstances some of the seemingly high threshold components may be seen in the absence of the low threshold components.

It can be argued that all the displays share common causal factors. For instance, Sevenster described how a male stickleback will go through a complete fight in front of an inanimate model that presents a constant feedback.[2] One could argue from this that the elaborate sequence of displays share common causal factors. Certainly the ability to run off a series of displays in the presence of an unchanging stimulus indicates there must be some central patterning responsible for this behavior.

On the other hand, it is equally obvious that the hostile displays are under continuing modulation during their expression. The degree of the spreading of the fins, the angle at which the body is held, the presence or absence and timing of head jerking, are all responsive to the continuing changes in behavior of the opponent.

A similar situation occurs in feral chickens. The degree to which the hen erects the tail feathers and lowers the wing in aggressive display is inversely proportional to the distance from the other chicken. (On the other hand, the display is the same whether it be to another hen, a cock, or an immature fowl.) The nearer the other bird, the more complete the display.[3]

[2] Piet Sevenster: personal communication.
[3] Glen McBride: personal communication.

The display is thus continually modulated, and is expressed as a function of the strength of the stimulation.

The nonsignal movements should also be discussed. There are many examples of stereotyped behavior which have no known signal function. Good examples among fishes are parental fanning of the eggs and larvae, or feeding behavior. The movements vary considerably, both in their amplitude of movement and in the duration of the ongoing event. Sevenster (1961) demonstrated the sensitivity of fanning to immediate environmental stimulation. When *Badis badis* feeds on small worms (*Tubifex*) it employs a predictable sequence of movements for extracting them from the gravel (Barlow 1961). But the precise form of this movement varies in response to the strength of attachment of the worm. If the worm is loosely anchored the fish merely swims forward and upward in a smart movement. If the worm is securely fastened in the bottom, the fish turns on its side as it meets resistance and levers the worm out of the bottom. All intermediates exist depending on the feedback received from the worm. Yet for any degree of attachment by the worm, the nature of the motor pattern is predictable. There is thus a core motor pattern, but one which is continually modulated by the immediate environmental situation.

The essential point I have been attempting to make in this critique is that we need a finer analysis of fixed action patterns. It will doubtless come as a surprise to most nonethologists, that most of our thinking about the fixed action pattern has been done in the absence of hard data. Thus I do not feel embarrassed stressing my point of view in the absence of good data, since this has been the rule. The main issue here is that there are so few data, one way or the other, to support what has been widely accepted as a well-proved and established category of behavior.

I would particularly like to appeal for the use of film analyses in studying the nature of the fixed action pattern. Some surprising results might be obtained. For instance, one might learn more about feeding action patterns than has been previously suspected. It is well known that the action pattern of striking in the praying mantis, once triggered, is no longer under control from the environment (Rilling, Mittelstaedt, and Roeder 1959). When a snake strikes at its prey, one might presume that much the same holds. It is done with blinding speed. One could hardly imagine that a poikilotherm would have the quickness of response necessary to modulate the strike, once triggered. Yet in some films by Frazzetta, which he was kind enought to show me, a python is able to slow the strike in mid-flight when the rat moves, to compensate, to take fresh aim, and to continue in the strike (Frazzetta 1966). This all occurs in just a few frames taken at a high speed, undetected by the eye of the observer. Without the use of film analysis one would have assumed this is a fixed action pattern running its course, immune from further environmental modulation.

Just analyzing behavior on film is not enough. One would have to select a variety of kinds of action patterns, ranging from those seen in courtship displays to aggressive behavior, and on into nonsignal movements. In addition, such an analysis would have to be combined with an experimental approach. A given action pattern would have to be observed in various contexts, and also as it was being influenced by the environment, to determine the degree to which it was or was not independent of continuing environmental cues.

As a note of caution, films are not perfect. Much information is lost and difficult problems of analysis arise.

Of course there are other ways of objectively measuring action patterns. I mention as one example the problem of vocalization. Here the motor pattern may be captured conveniently in three dimensions—duration, frequency, and amplitude. The concept of fixed action pattern would profit from a comparison, say, of vocalizations expressed in the highly stereotyped songs of some birds with the graded responses of primates. There is now evidence to suggest that, even among birds, what appear to be stereotyped vocalizations may be highly graded (Dixon and Stefanski 1965). This problem has not been touched in the vocalization of fishes.

The Separability of the Fixed Action Pattern from Its Orientation

One of the tenets of ethological thinking is that the fixed action pattern has a basic pattern of coordination that is separable from its orientation, usually termed its taxic component. Hinde (1966, p. 20) stated, for instance, that "By definition, the form of a fixed action pattern is independent of the environment."

This concept emanated from the classic experiments on the gray-lag goose conducted by Lorenz and Tinbergen (1939). The goose retrieves a displaced egg by placing her bill just beyond the egg; then with the underside of her bill she prods the egg toward her breast, all the while gently weaving the bill from side to side. If the egg is removed during the process of retrieving, or if a symmetrical cylinder replaces the egg, the head-weaving component disappears and the head comes smoothly back to the breast. The core motor pattern is the bringing of the bill back to the breast. The egg triggers the behavior. The weaving is a response to the erratic path of the eccentric egg. Hence the aiming and the weaving of the head are determined by peripheral input, and are separable from the spatiotemporal coordination of the fixed action pattern.

The question that must be asked is whether all fixed action patterns are so neatly separable into structural and taxic components. In practice a fixed action pattern is usually defined by its form, not necessarily by its coordination. As a consequence it is often difficult to split off the taxic component. For instance, in Hinde's treatment of the problem he cited two examples of the fixed action pattern from many that might have been chosen. One of these is the soliciting posture of the female chaffinch, which seems to fit his criteria rather well. The other example is the response of the young of a mouthbreeding cichlid fish, *Tilapia mossambica*, to the mother when danger threatens. They swim to her, entering her buccal cavity through the mouth or opercular openings. This is a taxic response lacking a distinctive coordination. Not only is the response dependent on the environment, it is unrecognizable in the absence of a mother object. There is no way this simple swimming can be distinguished from any other type of swimming, except by its orientation. In all fairness to Hinde, he does point out several cases where the orientation may be more intimately a part of the fixed action pattern.

Inspection of a wide variety of behaviors should convince one that a number of fixed action patterns are recognizable exclusively by their orientation. The often cited case of the head-standing display of the three-spined stickleback, *Gasterosteus aculeatus*, is a good example. The female Florida flagfish, *Jordanella floridae*, about to spawn responds to the circling male by keeping her tail pointed directly at him, giving rise to a display called T-formation (Mertz and Barlow 1966); there is nothing to distinguish this motor pattern from simple hovering and turning except the orientation itself. When two *Badis badis* engage in hostile displays they characteristically pitch forward at an angle of about 45° while maintaining a head-to-tail orientation; this behavior is recognized through its orientation to the substrate and to the rival (Barlow 1962). As a final example, the orange chromide frequently performs a tailstand in front of its mate, particularly in the early stages of pair formation (personal observation); again, this display is distinguished only by its orientation.

I am convinced that if we study a number of displays, under a variety of situations and in a variety of animals, in many cases the orientation and the form of the fixed action pattern will not be separable. I would further predict that there will be a smooth intergradation from inseparability to separability, as in the case of the gray-lag goose. We must be prepared to visualize the C.N.S. counterpart of a fixed action pattern as more than an undirected core independent of the environment. The taxic component must be regarded in many instances as just as centrally determined as the pattern of coordination. Frequently they will be inseparable. What is determined is not necessarily a precise expression of a movement, but rather a stipulated relation between the organism, its behavior, and key stimuli.

Factors Favoring Stereotypy versus Those Favoring Variability

Two possible reasons for the development of stereotypy, the fixed action pattern, immediately come to mind. One of these is the efficiency of the performance. It is common knowledge from studies of learning that conditioned responses tend to become parsimonious (Adams 1931). (It would be interesting to know if "superstitious behavior" also becomes parsimonious.) Conditioned responses move toward the least expenditure of energy consistent with the problem; in so doing they become relatively stereotyped. Conceivably natural selection may follow a similar path, leading to the simplification of fixed action patterns.

Another factor contributing to the development of stereotypy of behavior is the signal function. Morris (1957) treated this problem lucidly in an article on "typical intensity." He was concerned with the difficulties inherent in the concept of ritualization.

To make his point, Morris offered two simple models. In one of these, the fixed action pattern remains relatively unchanged over a wide range of causal factors (sometimes termed motivational changes). Such a signal has the advantage that it is unambiguous, though relatively poor in information. What has not been sufficiently emphasized is that such a system would be favored when the animals in question mate quickly, or when fights are typically brief and sporadic. Under these circumstances the signals should be unambiguous, facilitating rapid recognition.

The other model put forth by Morris resembles a simple dose-response curve. The "intensity" of the response is directly proportional to the strength of the causal factors. Consequently the intensity or completeness of the behavior reflects the motivation of the sender; the system is said to be rich in information. However, there immediately arises the possibility of ambiguity because of the fine gradations between the various degrees of expression of the behavior. This is just the situation we might expect in pairs of animals that have sustained pair bonds. It might also be expected in animals that engage in complex fights, requiring that each animal continually and accurately assess the state of its opponent. As a corollary, such a system should convey information along many channels and be highly redundant, to minimize the chances of error. And this seems to be the case (Barlow 1962).

The foregoing contained the tacit assumption that the system is deterministic even if the response is graded. What is seldom discussed when considering signal movements is the possibility that there may be selection for a degree of nonpredictability. This might be particularly true in animals that remain paired for long periods preceding copulation or spawning. Under these circumstances the animals could be expected to habituate to one another. Faced with this monotony effect, novelty might be advantageous.

Novelty is commonly associated with arousal and attention (Berlyne 1960). Now there are several ways an animal can introduce novelty into a sequence of courtship movements; it does not require variability in the basic motor pattern. Conceivably, however, the action pattern itself might be variable. On the other hand, the animal could "have its cake and eat it" by keeping certain parameters of the motor pattern constant while varying others. For instance, in the courtship quivering of the orange chromide the fish holds the amplitude and basic form of the movement relatively unchanged while varying such parameters as the duration, the phasing of the pelvic fin movements with the head quivering, the posture (whether horizontal or vertical), and the orientation to the mate.

Another way of reducing monotony is to vary the sequencing of the motor patterns. This is probably more the case than is commonly realized. The numerous flow diagrams given in the literature (e.g., Neil 1964) indicate that the variability in sequencing is much greater than would have been predicted from the original notions of chains of releasers and instincts (e.g., Tinbergen 1951). This is particularly the case among those species where courtship is a relatively continuous event (e.g., Baerends, Brouwer, and Waterbolk 1955; Nelson 1964). (For a similar argument about bird song, see Hartshorne [1958].)

Another hypothesis about fixed action patterns that is often intuitively appreciated but seldom stated is that those movements used in communication are less variable than those which are not. Here again there are few data. If there is some selective advantage in doing movements in the most economical way, therefore in a stereotyped way, then we might expect some of the nonsignal movements to be just as relatively fixed as signal movements. Particularly good examples are to be found among the elaborate grooming movements of mammals (Eisenberg 1963), as well as among activities involved in parental care (Barlow 1964).

Quantitative Models of Behavior

From the introduction one might conclude that I am unduly pessimistic about the application of formal statistical models to behavioral phenomena. Quite the contrary. Those caveats were advanced to create an awareness of the limitations of the models.

The same shortcomings can be found in the application of statistics to almost any real-world problem. It is not so much a question of the identity of the units involved as it is a matter of having an arbitrary criterion so that there is uniformity in the decision-making. In drawing marbles from a bag, for example, it is of little consequence if the marbles are of different colors. The main point is that the differences do not affect the choice.

The application of statistics regularly embraces quite heterogeneous units. In gathering data on accidents involving automobiles and bicycles no great violation of assumptions is done by regarding all automobiles as the same and all bicycles as the same. Useful statistics can be garnered in this way. Indeed, one frequently reads of comparisons between various countries based on such data. On the other hand, an insurance broker would find this gross level of analysis inappropriate to his needs. He would want to segregate the automobiles by age and sex of driver, past history of driver, horsepower of automobile, age of automobile, and so on. As his units become more refined and homogeneous, the power of his predictions increases. Eventually the point of diminishing returns is reached. The same is true with models as they apply to behavior studies. The difficulty again is that if the units are too small and homogeneous, the major phenomenon of interest is lost sight of.

Any quantification of behavior that arrives at some general conclusion is a model. But for fruitful comparisons with neurophysiological investigations the models that seem most promising are those involving the temporal patterning of behavior. This area of study is receiving increasing attention, and we can expect to see growth in this direction in the ethological literature.

The simpler of these models employ only one unit as the measure of the behavior. The number of attacks delivered to an opponent (Heiligenberg 1965*a*) or to a mirror image (Skaller 1966) are examples of particularly convenient units of measure that lend themselves to temporal analysis. A more complex movement is that of parental fanning where several measures may be made of the same event (Barlow 1964; Mertz and Barlow 1966).

More commonly several units are observed jointly. Then the units are analyzed separately to determine their probability of occurrence and distribution in time, or together, largely by means of cross-correlational analysis (Nelson 1964). A particularly popular approach has been the analysis of sequences of behavior without regard to their distribution in time. These are frequently portrayed using flow diagrams, but occasionally they are subjected to multifactorial analysis (Wiepkema 1961; Baerends and van der Cingel 1962).

Verbal Models of Behavior

Much of the recent literature in ethology has been devoted to the elaboration of verbal models of behavior. Most of these arise from a tripartite notion of behavior, building on three major categories of motivation—aggression, sex, and fear. This literature and its conceptual methodology have been well reviewed by Tinbergen (1959), Hinde (1966), and Marler and Hamilton (1966). There is no merit in elaborating further on what has already been said so well.

I bring in verbal models because the need became so obvious during the discussions at this symposium. People working at the neurophysiological level should be aware of some of the thinking of ethologists about problems broader than those involving just the fixed action patterns. I hope my comments will form a bridge between the fixed action pattern and what are commonly referred to as motivational analyses.

In simple terms, the question might be put, Do shared effectors mean shared C.N.S. mechanisms? As an example, the effector system that moves the jaws of a fish is employed in courtship displays, aggressive displays, digging, eating, retrieving larvae, and in many respiratory movements such as yawning and perhaps even coughing. Albrecht (1966) has argued that many of these behaviors in fishes have more in common at higher levels than is commonly suspected. Clearly, biting, digging, and eating are reciprocally influenced in one cichlid fish (Heiligenberg, 1965*b*). And parental retrieving of larvae is remarkably like hunting in *Badis badis* (Barlow 1964). Albrecht has extended this to include courtship displays, arguing that most courtship displays have been derived from aggressive behavior.

It is possible to look at even smaller pieces of behavior and to discover that the same movement is used in many different contexts. Thus the pelvic-fin flickering of the orange chromide may occur in the same way in many different situations, such as during courtship, at the end of a fight, during grooming, and when the parent signals to the young (in the latter case the coordination may be slightly different). Yet all of these seem to have certain causal factors in common; they occur at a time when one could readily hypothesize a strong conflict between fleeing and remaining. In this context, all of these pelvic-fin flickerings might be regarded as displacement grooming.

Much the same could be said of the quivering in the orange chromide. This movement is basically the starting action of swimming. It is highly ritualized into a courtship movement and an invitation to grooming, and it serves as a means of ejecting objects or distasteful sub-stances from the mouth. Starting to swim, courtship, and interindividual grooming all have similar causes, the approach of another orange chromide. On the other hand, it is difficult to relate the head shaking employed when rejecting objects to that seen in courtship and grooming.

It is well to keep in mind that a fish is living in a medium that makes great demands on its locomotory abilities. A fish is highly streamlined and, unlike a bird, is typically relatively devoid of large appendages that can move independently of the body. In a fish, as in most animals, locomotory demands take primacy. Thus the basic coordination upon which most of the display movements are built is fundamentally one that must be a primary part of the locomotory apparatus.

A good example of the primacy of locomotory demands comes from Magnus' (1958) study of the fritillary butterfly. The female flaps her wings at a rate of about 8 beats per sec. The male recognizes the female by the flapping of her colored wings and pursues her. Yet the male will pursue preferentially models that flap at a rate up to tenfold that of the female. One might ask why the female has not evolved a higher rate of wing flapping in order to be more attractive to the males, since such females would have been selected for. The most likely answer is that the locomotory demands have primacy over the signal value.

Other factors must also be considered. Additional primary kinds of behavior are those involved in respiration and in feeding, both of which require the participation of the buccal apparatus. Thus jaw structure will be determined mainly by feeding habits and respiration, usually not by modes of fighting or displays. Some interesting exceptions do nonetheless occur, as in the case of the bizarre jaws of the salmons of the genus *Oncorhynchus;* but in this case the males no longer feed.

The difficulty with such an argument, that common effector means common motiva-tion, is that it reduces all to one. Thus in the end everything is locomotion, respiration, or feeding. Clearly the motivational world of the animal is more structured than this. And yet the arguments are, when more narrowly viewed, quite persuasive. Unfortunately, proof is lacking in most instances. There should be more direct testing of hypotheses as to the degree to which movements expressed by a given effector system share causal factors at some higher level. A good example of such an approach is the recent work done by Heiligenberg (1965*b*) relating eating, digging, and attacking in a cichlid fish.

Such experiments on the causation of behavior should include a detailed analysis of the form and dynamics of the behavior in question and of its temporal patterning. This should be done both with regard to its own expression, and with regard to the interaction with other movements and its equivalence to other movements. Thus if two actions share the same motivation and effector, say biting and eating, then in a study of the bouting of either of these, one should substitute for the other. This is an easy assumption to test by interrelating the probability density functions of the two when both occur.

The other side of the coin must also be examined. That is, Do different effector systems indicate different mechanisms in the central nervous system? It is common knowledge that many animals use different groups of effectors in aggressive behavior, depending on the context. Thus a fish may fight with its mouth, or turn and beat with its tail. These are not controlled the same way, however, biting being much more indicative of attack, and tail beating more related to fleeing.

A better and more provocative example is that of the aggressive responses of a rattle-snake. When two males engage in aggressive behavior, they intertwine their necks as though to test each other's strength; they do not bite one another (Shaw, in Eibl-Eibesfeldt 1961). The aggressive behavior in response to attack by a potential predator is extremely different, however (Bogert 1941). If the putative predator is a large mammal the rattlesnake coils, rattles, strikes at the mammal, and bites it. But if the intruder is a king snake, which feeds on rattlesnakes and is immune to their bite, the rattlesnake shows a remarkably different aggressive behavior. It hides its head under the coil while it raises up one coil and slaps at the king snake with it. The adaptive significance of protecting the head is obvious. The important point is that in all three instances aggressive behavior, as conventionally defined, is obviously involved. Just as obviously, there are differences in the motivational state of the snake.

. At the heart of the matter, of course, is the problem of defining aggressive behavior. The simple criterion of doing harm, or attempting to do harm, lumps many kinds of behavior that are otherwise separable by form and by context. The most suitable arrangement may be a hierarchical organization of behavior that at some point joins all the aggressive behaviors, but further down separates them out according to their causation. Thus one might conclude that different effectors do indeed indicate different motivational sources. But at some point, at a more general level, different effectors may be thought of as being determined by the same motivation.

Conclusions

I am not optimistic that behavioral models will soon lead to the discovery of C.N.S. correlates. At present our quantitative models of behavior are useful in that they lead to more precise, tangible descriptions of behavior, facilitating the search for their counterparts in the nervous system. These models are more important, however, in testing behavioral hypotheses inherent in verbal models.

It is essential to the development of reliable models that more attention be given to the units of behavior. In particular we need to understand better the basic nature of the so-called fixed action pattern. Even the most highly stereotyped of the fixed action patterns will probably always be graded to some degree and, if slow enough, controlled to some extent by the environment.

The ideas behind the fixed action pattern remain useful in thinking about behavior, and in formulating models to describe behavior. But as currently conceived, most motor patterns would be excluded if the criteria outlined in this article were rigidly adhered to. Most likely a gradation exists between the most highly stereotyped behavior patterns and those that are so variable as to defy characterization.

It may be important in the early stages of inquiry to pursue ideas, to select the example to support the hypothesis. But at some juncture good science demands that the effort be devoted to attempting to disprove the hypothesis. By examining a number of behavioral events, a more meaningful conceptualization of behavioral units will emerge.

As a final, and to some a rather minor, point, it might be wise to drop the term fixed from action pattern. In its present usage it is applied uncritically to almost any behavior that has a degree of regularity sufficient to permit one to recognize it. The term appears to have gained popularity as a replacement for innate motor pattern; this term was unfortunate because it contains an explanation that extends beyond the appearance of the behavior. Fixed

230   *Barlow*

action pattern is an improvement in that it was meant to derive solely from the appearance of the behavior. As I have argued, however, the behavior is seldom known to be fixed in the way the criteria demand.

Until there is a better understanding of the mode of expression and control of such behavior it would be preferable to have a less encumbered terminology. As a more general term, I would suggest modal action pattern. This conveys the essential features of the phenomenon: there is a spatiotemporal pattern of coordinated movement, and the pattern clusters about some mode, making the behavior recognizable. (I have purposely avoided the use of the word normative because we know so little about parameters such as central tendency and variance.) Furthermore, the modal action pattern (M.A.P.) is common to the species. The degree of variability and the role of peripheral input remain to be determined in each case. In the meantime, the classification of behavior should be based on the structure of the behavior. With increasing knowledge an articulated classification should evolve. Whether discrete "taxa" or a continuum will emerge remains to be seen.

Acknowledgment

I am grateful to Mildred Eley and to C. Hugh Fraser Rowell for reading the manuscript and offering critical comments. It should be apparent that I am also indebted to Robert Hinde for his timely and lucid restatement (1966) of the problem of the fixed action pattern as well as for his comments.

The writing was done during the tenure of Grant GB-5314 from the National Science Foundation.

References

Adams, D. K. 1931. A restatement of the problem of learning. *Brit. J. Psychol.* 22:150–78.

Albrecht, H. 1966. Zur Stammesgeschichte einige Bewegungsweisen bei Fische, untersucht am Verhalten von *Haplochromis* (Pisces, Cichlidae). *Z. Tierpsychol.* 23:270–302.

Ashby, W. R. 1966. Mathematical models and computer analysis of the function of the central nervous system. *Ann. Rev. Physiol.* 28:89–106.

Baerends, G. P.; Brouwer, R.; and Waterbolk, H. T. 1955. Ethological studies on *Lebistes reticulatus* (Peters): I. An analysis of the male courtship pattern. *Behaviour* 8:249–334.

Baerends, G. P., and van der Cingel, N. A. 1962. On the phylogenetic origin of the snap display in the common heron (*Ardea cinerea* L.). *Symp. Zool. Soc. London* 8:7–24.

Barlow, G. W. 1961. Ethology of the Asian teleost *Badis badis*. I. Locomotion maintenance, aggregation and fright. *Trans. Ill. Acad. Sci.* 54:175–88.

———. 1962. Ethology of the Asian teleost, *Badis badis*. IV. Sexual behavior. *Copeia* 2:346–60.

———. 1964. Ethology of the Asian teleost *Badis badis*. V. Dynamics of fanning and other parental activities, with comments on the behavior of the larvae and postlarvae. *Z. Tierpsychol.* 21:99–123.

Berlyne, D. E. 1960. *Conflict, arousal and curiosity*. New York: McGraw-Hill.

Bogert, C. M. 1941. Sensory cues used by rattlesnakes in their recognition of ophidian enemies. *Ann. N.Y. Acad. Sci.* 41:329–44.

Dane, B.; Walcott, C.; and Drury, W. H. 1959. The form and duration of the display actions of the goldeneye (*Bucephala clangula*). *Behaviour* 14:265–81.

Ethological Units of Behavior   231

Dixon, K. L., and Stefanski, R. A. 1965. An evaluation of the song of the black-capped chickadee. *Amer. Zool.* 5:693 (Abstract).

Eibl-Eibesfeldt, I. 1961. The fighting behavior of animals. *Sci. Amer.* 205(6):112–22.

Eisenberg, J. F. 1963. The behavior of heteromyid rodents. *Univ. Calif. Publ. Zool.* 69:1–100.

Frazzetta, T. H. 1966. Studies on the morphology and function of the skull in the Boidae (Serpentes). Part II. Morphology and function of the jaw apparatus in *Python sebae* and *Python molurus*. *J. Morph.* 118:217–96.

Hartshorne, C. 1958. Some biological principles applicable to song-behavior. *Wilson Bull.* 70:41–56.

Heiligenberg, W. 1965a. The effect of external stimuli on the attack readiness of a cichlid fish. *Z. Vergl. Physiol.* 49:459–64.

———. 1965b. A quantitative analysis of digging movements and their relationship to aggressive behavior in cichlids. *Anim. Behav.* 13:163–70.

Hinde, R. A. 1966. *Animal behavior*. New York: McGraw-Hill.

Liem, K. F. 1967. Functional morphology of the head of the anabantoid teleost fish *Helostoma temmincki*. *J. Morphol.* 121:135–58.

Lorenz, K. Z. 1935. Der Kumpan in der Umwelt des Vogels. *J. Ornith.* 83:137–213, 289–413.

———. 1937. Über die Bildung des Instinktbegriffes. *Naturwissenschaften* 25: 289–300, 307–18, 324–31.

———. 1950. The comparative method in studying innate behaviour patterns. *Symps. Soc. Exp. Biol.* 4:221–68.

Lorenz, K. Z., and Tinbergen, N. 1939. Taxis und Instinkthandlung in der Eirollbewegung der Graugans: I. *Z. Tierpsychol.* 2:1–29.

Magnus, D. 1958. Experimentelle Untersuchungen zur Bionomie und Ethologie des Kaisermantels *Argynnis paphia* L. (Lep. Nymph.): I. Überoptische Auslöser von Anfliegereaktionen und ihre Bedeutung für das Sichfinden der Geschlechter. *Z. Tierpsychol.* 15:397–426.

Marler, P., and Hamilton, W. J., III. 1966. *Mechanisms of animal behavior*. New York: Wiley.

Mertz, J. C., and Barlow, G. W. 1966. On the reproductive behavior of *Jordanella floridae* (Pisces: Cyprinodontidae) with special reference to a quantitative analysis of parental fanning. *Z. Tierpsychol.* 25:537–54.

Moore, G. P.; Perkel, D. H.; and Segundo, J. P. 1966. Statistical analysis and functional interpretation of neuronal spike data. *Ann. Rev. Physiol.* 28:493–522.

Morris, D. 1957. "Typical intensity" and its relation to the problem of ritualisation. *Behaviour* 11:1-12.

Neil, E. H. 1964. An analysis of color changes and social behavior of *Tilapia mossambica*. *Univ. Calif. Publ. Zool.* 75:1–58.

Nelson, K. 1964. The temporal patterning of courtship behavior in the glandulocaudine fishes (Ostariophysi, Characidae) *Behaviour* 24:90–146.

———. 1965. After-effects of courtship in the male three-spined stickleback. *Z. Vergl. Physiol.* 50:569–97.

Rilling, S.; Mittelstaedt, H.; and Roeder, K. D. 1959. Prey recognition in the praying mantis. *Behaviour* 14:164–84.

Rowell, C. H. F. 1961. Displacement grooming in the chaffinch. *Anim. Behav.* 9:38–63.

232 *Barlow*

Russell, W. M. S.; Mead, A. P.; and Hayes, J. S. 1954. A basis for the quantitative study of the structure of behaviour. *Behaviour* 6:153–205.

Sevenster, P. 1961. A causal analysis of displacement activity (fanning in *Gasterosteus aculeatus* L.). *Behaviour Suppl.* 9:1–170.

Skaller, P. G. 1966. *Temporal patterning of the response to a mirror by a cichlid fish.* Master's thesis, University of Illinois, Urbana, pp. 1–41.

Tinbergen, N. 1942. An objectivistic study of the innate behaviour of animals. *Biblioth. Biother.* 1:39–98.

———. 1951. *The study of instinct.* Oxford: Clarendon Press.

———. 1959. Comparative studies of the behaviour of gulls (Laridae): A progress report. *Behaviour* 15:1–70.

Weiss, P. 1950. Experimental analysis of co-ordination by the disarrangement of central-peripheral relations. *Symps. Soc. Exp. Biol.* 4:92–111.

Wiepkema, P. R. 1961. An ethological analysis of the reproductive behaviour of the bitterling. *Arch. Neerl. Zool.* 14:103–99.

Chapter 5

DESCRIBING SEQUENCES OF BEHAVIOR[1]

P. J. B. Slater

Ethology and Neurophysiology Group
School of Biology, University of Sussex
Brighton, England

I. ABSTRACT

A review is given of the methods currently available for analyzing sequences of behavior. Simple flow diagrams based on the frequencies or conditional probabilities of individual transitions are considered to be of restricted usefulness except where the sequence is highly ordered and the different patterns occur at similar frequencies. It is more helpful to compare the data with a random model provided that repetitions of the same behavior, and any transitions which cannot occur, are excluded before the expected number of each type of transition is calculated. Such comparisons are most likely to be helpful if the behavior patterns included are closely related and fall into discrete homogeneous categories. The fact that most behavioral data are unlikely to be stationary is considered to be the main factor limiting this approach. It is suggested that first-order transition analysis and correlative techniques are the best current methods for examining such data. The search for higher-order dependencies is useful only in stationary data, where grouping of acts due to changing causal factors can be assumed to be unimportant. Additional difficulties involved in the analysis of sequences of interaction between individuals are briefly discussed. The major complicating factor here is that the behavior of an individual is likely to be dependent both on that of others and on its own previous behavior.

[1]This research was financed by a grant from the Science Research Council.

131

Some ways of improving current techniques to take account of this are put forward. It is emphasized that sequence analysis provides only a description of the behavior under study and that there are dangers in making causal inferences on the basis of such descriptions alone.

II. INTRODUCTION

The analysis of sequences occupies an important position among the methods available to ethologists, particularly those interested in the causation of behavior. By borrowing and adapting mathematical techniques from other branches of science, they have been able to achieve precise and quantitative descriptions of behavior, and on these to base hypotheses about the causal mechanisms involved.

How can sequence analysis help in the study of causation? It is generally considered by ethologists that the occurrence of two behavior patterns close together in time indicates that they have some causal factors in common. This may not always be the case, for one of the behaviors may bring the animal into a situation appropriate to the other, but in many instances the explanation seems plausible. The causal factors shared may be of two types. First, both behaviors may depend on a particular bodily state, such as the presence of a hormone. Second, both may appear in response to the same or related external stimuli and thus be shown only when these are present. Sequence analysis cannot differentiate between these possibilities, but it can indicate in an objective way the groupings in which behaviors occur and so define the relationships which need to be explained.

Its objectivity is perhaps the greatest asset of the approach. Before such methods were developed, observers depended on their own judgment to determine the affinities between behavior patterns. Because activities which are similar in function, such as courtship and grooming movements, often appear close together in time, this subjective classification led to explanation in terms of specific unitary drives (e.g., courtship or grooming drives). The fact that behaviors of different function sometimes occurred in the midst of these groupings could be accounted for only by assuming their causation to be different, and they were labeled as "displacement activities." These early theories thus depended on the acceptance of a rule and the development of theories to account for exceptions to it. The rule, it should be noted, was based on the doubtful assumption that causation and function are equivalent. The inference that functionally related behaviors occurring close together in time were expressions of a common internal drive also neglected the possibility that their grouping was due to the constraints of the external world: courtship movements may occur only when a female is present and thus appear to be grouped. This point was appreciated by Tinbergen (1951), whose model of causation stressed the role of external stimuli in leading to the grouping of behavior patterns, though he still felt it necessary to argue that each functional group shared a particular drive.

With the appreciation that drive concepts are of limited usefulness in the explanation of causation (Hinde, 1970), it has become necessary to arrive at objective ways of examining associations between acts empirically, so that more realistic models of their causation may be constructed (Slater, in press). Sequence analysis was one of the first methods to be employed in this context, and its continuing attractions make an appraisal of its usefulness long overdue. That such methods are preferable to the rather subjective assessments of the interrelations between behavior patterns which were previously common seems beyond dispute, but one must be aware of their limitations and not be overimpressed by their precision. The mechanisms underlying the sequence of acts shown by an animal are likely to be complex, and to analyze them using methods appropriate to simpler situations, such as the sequence of cars passing along a road, will inevitably involve some rather sweeping assumptions. The less valid the assumptions, the less realistic will be the results. But it is perhaps even more important that the questions asked should be framed with the nature of the behavior, rather than the nature of the mathematical tools available, in mind. It is, unfortunately, only too easy for research using sophisticated mathematical methods to give very little insight into behavior.

The purpose of this chapter is to review the methods which have been employed by ethologists to analyze sequences, to assess their validity and usefulness, and to point to the conclusions which may be drawn from their application. Its main concern will be with the role of these methods in helping to outline the relationships between successive acts shown by the same individual, but, as they have also been used in studies of interactions between animals, additional points raised by this application will be more briefly considered.

III. SEQUENCES WITHIN THE INDIVIDUAL

A. Methods of Analysis

The methods used for analyzing sequences vary according to the complexity of the data and how structured the behavior appears to be at first sight. Basically, they may be split into two groups depending on whether or not they involve comparison with a random model.

1. Analysis of Transition Frequencies and Conditional Probabilities

The simplest type of sequencing of events is a deterministic sequence: in this, the events always follow each other in a fixed order, so that the nature of the preceding act defines precisely the nature of that which will follow. Such sequences are rarely studied in normal behavior, partly because two acts always occurring in a particular order tend to be regarded as a single behavior pattern. An instance of a sequence parts of which are deterministic was, however, described by Isaac and Marler (1963). They found the song of a mistle thrush (*Turdus viscivorus*) to be composed of some 21 possible syllables; within a song, each of these could be followed by only a small range of others and often, deterministically, by only one. Fabricius and Jansson (1963) observed that pushing by courting pigeons (*Columba livia*) was always followed by nest demonstration, but it was useful here to regard the two activities as separate because nest demonstration often occurred without pushing preceding it.

Most behavioral sequences are probabilistic rather than deterministic in form, meaning that while the probability of a given act depends[2] on the sequence of those preceding it, it is not possible to predict at a particular point exactly which behavior will follow. Some such sequences are only marginally less precisely ordered than deterministic ones, and these are usually referred to as "chained responses." In these cases, the probability of a particular event is so markedly altered by the nature of that immediately before it that a flow diagram indicating the frequencies with which different transitions occur provides a good impression of the organization of the behavior. This has proved particularly true of goal-directed sequences, such as courtship, in invertebrates and lower vertebrates (e.g., Brown, 1965; Parker, 1972; Baerends *et al.*, 1955; Hinde, 1955/56). Noirot (1969) has applied a similar technique to the maternal behavior of mice: she looked at the order in which different acts occurred during an observation session, disregarding repetitions of the same act.

Where sequences are not so highly ordered, some transitions may be observed between almost every behavior and every other, so that simple flow diagrams become complex and hard to interpret. A way around this is to include only those transitions which have a high probability of occurrence, a method which was used by Fabricius and Jansson (1963). For each sequence of two acts, they worked out the conditional probability ($p\ B|A$, the probability that B will occur given that A has just occurred), and only where this exceeded 0.1 was an arrow from A to B included in their flow diagram. For each type of act, an impression is thus given of those other behaviors most likely to follow. This method has also been employed to show up differences in sequences between two groups of animals (isolated and socially

[2]The expression "one behavior depends on another" is used throughout this chapter in the statistical sense: the occurrence of the first alters the probability of the second. Where it is intended to imply that the first is in some way causal to the second, this will be referred to as a "sequence effect" rather than a "dependency."

reared guinea pigs, *Cavia porcellus*) by Coulon (1971). It is perhaps best suited to situations like this where comparisons are to be made, for the individual flow diagrams obtained are hard to interpret when behaviors differ strongly from each other in frequency. A large number of transitions, yielding a comparatively complex diagram, may result even from sequences in which the events are independent of one another. This is because the expected conditional probability of a rare act following a common one is low, whereas that of a common act following a rare one is high. In a sequence of 1000 acts, if act A is observed 100 times and B is observed 10 times, and A and B are independent, then one would expect the sequences $A \rightarrow B$ and $B \rightarrow A$ to occur once each (100 × 10/1000), but the expected conditional probabilities are quite different: $p\ A|B = 0.1$ (100/1000); $p\ B|A = 0.01$ (10/1000). In the light of this complication, it is difficult to assess the importance of the different sequences found by Fabricius and Jansson (1963), as they do not give the frequencies of the individual acts.

These simple ways of looking at sequences are thus likely to prove useful only where sequencing is strong and where the different behaviors considered occur at roughly equal frequencies.

2. Comparison with a Random Model

The shortcomings of descriptions based on frequencies and conditional probabilities have led many workers to adopt some of the techniques of Markov chain analysis. A sequence of behavior can be described as a Markov chain if the probabilities of different acts depend only on the immediately preceding act and not on any earlier ones (Cane, 1961). This model is thus appropriate to describe the sequence $A \rightarrow B \rightarrow C$ if the probability that C will follow B is not in any way altered by the nature of A or events prior to A. Such a sequence of events can also, more generally, be referred to as a "first-order" Markov chain, to differentiate this model from ones in which more of the preceding events affect the outcome: in this case, an rth-order Markov chain is one in which the probability of a particular event is significantly affected by the r preceding events (Chatfield, in press). Thus if the probability of C depends on A as well as B, the first-order model is inadequate to account for the structure of the sequence and second-order dependencies can be said to be in operation. If the probability of an event does not depend on any previous events, the events can be said to be independent, and this is sometimes referred to as a "zeroth-order" Markov chain.

Most sequence analysis in behavior is concerned with establishing the existence of, and identifying, first-order dependencies. Here, the matrix of observed transition frequencies is compared with that which would be expected if all acts were independent of one another. For these first-order transitions, the expected values are calculated as for a contingency table (assuming that it is possible for any of the behaviors considered to follow any other), and comparison between the matrices can then be made either in the usual way, using χ^2, or by the use of information theory (Bolles, 1960; Chatfield and Lemon, 1970; Fentress, 1972).[3] If the difference is found to be significant, the hypothesis that the behavior consists of a sequence of independent acts can be rejected. This discovery is not, in itself, surprising; it would be more so if a sequence of independent acts were found, which has not so far been the case. Further analysis is required before any useful description of behavior is obtained, and here two different approaches are possible.

First, the sequence can be analyzed more closely as a whole to decide whether the first-order model is adequate or whether higher-order dependencies affect it. It can be tested, as a start, whether a second-order model is more precise: in the generalized sequence $A \to B \to C$, is the probability of obtaining C after B significantly altered by the nature of A? In a first-order Markov chain, this is not the case, and the frequency of $A \to B \to C$ is that predicted from the frequencies of $A \to B$ and $B \to C$. A difficulty arising here is that N different types of behavior will give N^3 triplet types, and so very extensive data are required before reasonable numbers of each of these are expected. In two cases where triplets have been examined, however, a first-order Markov model has been found to be satisfactory: the frequency of triplets of each type corresponded closely to that predicted from the first-order relationships (Nelson, 1964; Lemon and Chatfield, 1971). As will be seen below, rather special conditions, which are unlikely to arise often in behavioral data, may have to be met before sequences of this type are discovered. The statistical comparison of first- and higher-order models has been thoroughly explored by Chatfield and Lemon (1970; Lemon and Chatfield, 1971; Chatfield, in press).

A second approach to finer analysis is an attempt to detect those first-order transitions which are significantly commoner than their expected values. This is done when the groups in which behaviors occur and the exact relationships between different behavior patterns are the main points of interest. Because such discoveries are behaviorally more interesting than the detection of higher-order dependencies, this approach has greater currency. It carries with it, however, the necessity to examine the discrepancy between the observed and expected values in individual cells of the transition matrix. Because many cells often contain rather low figures and because the observed value in each cell depends on those in others, this is difficult both to carry out and to interpret. The commonest method is a condensation of the whole matrix into a 2×2 table about the cell of interest, followed by a χ^2 test to detect whether that particular transition is more frequent than expected (e.g.,

[3]Strictly speaking, a transition matrix is not a contingency table, as the events included in it are not independent of one another, the second act of each pair being the first act of the next. The use of χ^2 on such data has, however, been validated by Bartlett (1951).

Stokes, 1962; Blurton Jones, 1968). Other techniques have been employed by Andrew (1956), Weidmann and Darley (1971), and Slater and Ollason (1972). In this last case, an attempt was made to overcome the biases involved in massing data from different animals, which is a usual practice where χ^2 is employed in order to boost the figures in individual cells. Perhaps the safest method is a simple inspection of the data, as recommended by Lemon and Chatfield (1971).

Whichever of these tests is used, pairs of behaviors which are commoner than expected can be extracted and a flow diagram constructed along the lines of those used in frequency and conditional probability analyses. This flow diagram has the advantage, however, that the relationships in it are not biased by the frequencies of the individual acts.

A further method may be used if it is of interest to detect asymmetry in particular sequences. If the row and column totals of the transition matrix are the same, then the expected values of the generalized sequences $A \rightarrow B$ and $B \rightarrow A$ will be the same, and tests can be carried out to see if the observed values differ (Blurton Jones, 1968; Slater and Ollason, 1972). While sequence analysis is not the only, or necessarily the best, way of looking for associations between acts, it is certainly the simplest way to detect such asymmetries.

B. Some Comments on the Methods Used

Having given a general outline of the commonest approaches to sequence analysis, it is now appropriate to consider the assumptions underlying them in the wide range of circumstances in which they have been used. Three main points are important here: the validity of using complete matrices, the choice of categories of behavior, and the problem of stationarity.

1. Complete vs. Incomplete Matrices

Comparison with a random model is easy if the matrix considered is a complete one, as shown in Table IA. The expected number of transitions between any pair of behaviors is calculated in the same way as for a contingency table (row total × column total/grand total; Table IB). This then gives the expected number of acts which would succeed each other if all acts were independent of one another. Two considerations are liable to make such a model unreasonable:

a. Repetitions of the Same Act. The complete matrix includes transitions which are repetitions of the same behavior. Figures thus appear on the descending diagonal of the matrix indicating the frequency of transition between A and A, B and B, etc. For each behavior, the observer must develop a criterion with which he can decide when one act ends and the next begins before he can reach a figure for the number of these repetitions. This may be easy for some behaviors: they may almost never be immediately

Table I. Artificial Data Showing the Observed Number of Transitions Between Three Behavior Patterns and the Number Which Would Be Expected If All Acts Were Independent of Each Other

A. Observed Matrix			
	A	*B*	*C*
A	10	20	5	35
B	15	4	6	25
C	10	2	8	20
	35	26	19	80

B. Expected Matrix			
	A	*B*	*C*
A	15.3	11.4	8.3	35
B	10.9	8.1	6.0	25
C	8.8	6.5	4.7	20
	35	26	19	80

repeated, or, if they are, a consistent time interval may separate succeeding events so that they can be simply distinguished. Other behaviors pose more difficult problems: when, for instance, does one act of locomotion end and the next begin? Here, it would be necessary to select some arbitrary time interval during which the animal did not move as indicating a gap between two acts.

In zebra finches (*Taeniopygia guttata*), other behaviors which highlight this problem are preening and singing (unpublished observations). These birds tend to groom in sessions several minutes long, during which preening of the feathers with the bill is interrupted little by other acts, with the exception of scratching of the head. A preening bird lowers the head to a small area of the body, preens one or a few feathers, and raises the head again. The next series of preening movements is most likely to be directed to the same area of the body as the last. There are thus at least three ways of defining an act of preening:

1. The preening of a single feather, several such acts often taking place between each raising of the head.
2. The series of movements between each raising of the head, which may involve several feathers.
3. The series of movements directed to the same area of the body, which may be interrupted by several instances of head raising, depending on how an area of the body is defined.

In the case of song, a series of almost identical phrases is produced, followed by a gap usually lasting for more than 2 sec before the next such series. An act here might be defined as each individual phrase, or as each series of phrases separated by longer than a certain time.

It is clear from these examples that defining an act other than arbitrarily will often be impossible. For some purposes, this may not be a hindrance, as long as the definition used can be consistently applied, but this is certainly not the case with sequence analysis. Here, two observers might apply different criteria, each in themselves perfectly reasonable, to the same data and so reach quite different conclusions. This is illustrated by Table II, where the same data as in Table I are shown except that the number of $A \rightarrow A$ transitions has been boosted from 10 to 100, as might happen if an experimenter with a different criterion for what constituted an act of A had collected the same data. A comparison of the expected values in Table IIB with those in Table IB shows that this difference, in only one cell of the observed matrix, would lead to a totally different set of expecteds.

The results of sequence analysis in relation to a random model using a complete matrix will usually depend on such arbitrary criteria selected by the experimenter. It is interesting to note, for instance, that each of the behaviors studied by Wiepkema (1961) was recorded as being repeated often, whereas this was seldom the case with the data given by Blurton Jones (1968). This tells one more about the criteria chosen by these workers than about the behaviors which they studied. Most of the matrices published in the literature are rather similar to that of Wiepkema in that the act definitions chosen lead to relatively high figures on the descending diagonal (e.g., Bolles, 1960; Baerends et al., 1970; Lemon and Chatfield, 1971). Carrying out sequence analyses on matrices such as these will tend to show just that these acts occur in bouts rather than more interesting relations between different behaviors.

Table II. The Same Data as in Table I But with the Number of $A \rightarrow A$ Transitions Boosted from 10 to 100

A. Observed Matrix			
	A	*B*	*C*
A	100	20	5	125
B	15	4	6	25
C	10	2	8	20
	125	26	19	170

B. Expected Matrix			
	A	*B*	*C*
A	91.9	19.1	14.0	125
B	18.4	3.8	2.8	25
C	14.7	3.1	2.2	20
	125	26	19	170

If arbitrary act definitions are involved or if behavior patterns tend to fall into bouts, the analysis of the complete matrix of transitions will thus be of little help in elucidating the interrelationships between activities. Under these circumstances it is more fruitful to eliminate the entries on the descending diagonal before analysis so that only transitions between different behaviors are considered.

   b. Cells Which Cannot Be Occupied. In some sequences of behavior, there may be transitions which cannot occur, particularly because of environmental constraints. Thus the matrix of Slater and Ollason (1972) could not include transitions from feeding to drinking, as locomotion, another category of behavior included in the study, had to occur between the sources of food and water. Likewise, Baerends *et al.* (1971) recognized in their matrix that a herring gull (*Larus argentatus*) could not show "looking down while not on nest" immediately after "sitting on nest" as well as certain other transitions. These constraints may not occur in many sequences, but they should be recognized where they do so.

   While many workers eliminate the descending diagonal from their transition matrices and some remove other cells which cannot be occupied, few realize that the calculation of expected values must then be altered so that the row and column totals add up correctly without figures appearing in the empty cells. Goodman (1968) gives a method whereby this may be done, and a simpler, though more approximate, method has been developed by Lemon and Chatfield (1971), whose study of song in cardinals (*Richmondena cardinalis*) was the first to use this technique on sequential data. Table IIIA shows the same data as in Tables I and II but without the descending diagonal. The expecteds in Table IIIB are calculated according to the Goodman method; they add up to the correct row and column totals despite the

Table III. The Same Data as in Tables I and II but with Expected Values Calculated Ignoring Transitions Between Each Behavior and Itself

A. Observed Matrix			
	A	*B*	*C*
A	—	20	5	25
B	15	—	6	21
C	10	2	—	12
	25	22	11	58

B. Expected Matrix			
	A	*B*	*C*
A	—	17.8	7.2	25
B	17.2	—	3.8	21
C	7.8	4.2	—	12
	25	22	11	58

presence of empty cells. These expected values give the number of bouts of each type of behavior which would succeed each other type if all bouts were independent of one another. A "bout" is here defined as a sequence of a given type of behavior not punctuated by the occurrence of any of the other types considered.

As these methods are available, there is no hindrance to the analysis of transition matrices in which repetitions of the same act and transitions which are impossible are not considered. While only Lemon and Chatfield (1971) and Slater and Ollason (1972) have so far used them, it is clear that most data on sequences of behavior would benefit from their application.

2. The Choice of Categories of Behavior

An important preliminary to any study, and especially to the analysis of sequences, is the classification of behaviors into discrete categories. Some acts are almost invariant in form and intensity and can easily be defined and recognized; others are more problematic and may require more or less arbitrary definition. However the categories of behavior are defined, it is important that they should be mutually exclusive and that the observer should "be prepared to treat all members of a given category as equivalent" (McFarland, 1971). Several factors may make these conditions hard to comply with. First, several different behaviors may frequently occur simultaneously, making it difficult to cast the data into a transition matrix. Data of this kind are more appropriately analyzed using correlation techniques (e.g., Delius, 1969), and Golani (unpublished manuscript, cited by Golani and Mendelssohn, 1971) has also applied multidimensional scalogram analysis to this problem. Second, the equivalence of behavior patterns within a category may be doubted where they differ greatly in form (as where all grooming movements are grouped under a single heading) or where there is a wide spread of bout lengths within a category. In this instance, long and short bouts may differ in context so that grouping them in one category clouds their different relationships with other activities.

Studies differ a great deal in the diversity of behavior patterns considered and the ways in which these are classified. It is common, for example, for transition matrices to include mainly behaviors from one functional category (comfort movements: Andrew, 1956; Delius, 1969; Fentress, 1972; courtship behavior: Nelson, 1964; Wiepkema, 1961; Coulon, 1971; song syllable types: Isaac and Marler, 1963; Lemon and Chatfield, 1971). Certain other behaviors may be included, but these are mainly taken as markers indicating the beginning and end of sequences of the group under study (e.g., in studies of comfort movements: resting and locomotion by Andrew, 1956; "other behavior" by Delius, 1969). Where such markers are not included, it is often unclear how a sequence is defined, though the careful study of Nelson (1964) is an exception. He established the time interval between two events for the succeeding to be independent of the nature of that before it and took intervals longer than this to be between, rather than within, sequences. Other matrices avoid the problem of sequence definition by including all behaviors shown in a particular situation regardless of function (e.g., at the nest site: Baerends *et al.,* 1970; at a food source: Blurton Jones, 1968; in isolation from conspecifics: Bolles, 1960; Slater and Ollason, 1972). The range of behaviors included can therefore vary from a wide spectrum of very different acts (e.g., sleep, grooming, and feeding) to a variety of very similar acts (e.g., grooming different areas of the body, different song types). All the behaviors included in a matrix of the latter type may be lumped as a single category in matrices of the former type.

For a number of reasons, the application of Markov chain analysis to the whole matrix is likely to prove more fruitful if the behaviors studied are homogeneous within categories, distinct between categories, and yet all closely related to one another:

a. Cane (1959) comments that "If the states of a Markov chain are grouped, the resulting process is not in general a Markov chain." Homogeneity within each category of behavior will make it less likely that such a grouping has been made.

b. If the behaviors considered are all closely related to each other, it may be possible to split them into categories which are, at least approximately, at the same level of organization. If the categories considered are very dissimilar, this may affect the order of the Markov chain found. Birds, for example, tend to wipe their beaks after drinking, presumably stimulated to do so by water left on the outside. If drinking and beak wiping were included in a matrix with various small bodily movements, some of the latter would often be interpolated between them, making it necessary to invoke higher-order relationships to explain the sequencing.

c. It is useful to work out the order of the dependencies governing a sequence only if the rules are the same for the whole matrix. This is more likely to be the case if the different categories can be taken to be distinct from each other and yet closely related. One deterministic sequence in a matrix of acts which are otherwise independent of one another could lead the observed and expected matrices to differ significantly.

d. Markov chain analysis presupposes that the animal is in a steady state (i.e., that the system is stationary), which is much more likely to be the case during sequences of closely related acts which are considered motivationally similar. The problem of stationarity has more widespread repercussions, however, and these will be discussed in the next section.

3. Stationarity

A stationary sequence is one in which the probabilistic structure does not change with time (Cox and Lewis, 1966; Delius, 1969). Nonstationarity can lead to very complex transition matrices: while the rules underlying sequencing may be simple on the short term, if they are not the same throughout the observed data in a transition matrix the matrix will be a complicated sum of the various simple processes involved and extremely hard to interpret.

Checking to see that the data are stationary is therefore a first requirement of rigorous Markov chain analysis as well as of other methods for examining sequences. It is a condition likely to be met seldom in behavioral data, which is one reason why this type of analysis is not often taken far. The use of concepts such as "drive" and "arousal" to explain behavior (Bindra, 1959; Hinde, 1970) and the discovery of daily cycles (e.g., Palmgren, 1950; Aschoff, 1967) and of short-term cycles (Richter, 1927; Wells, 1950) all speak for the long-standing realization among those studying behavior that the data they collect are not stationary. In certain situations, the steady-state assumption may be valid, but this will not often be the case. The shorter the observation period, for example, the more reasonable the assumption, but the more difficult it becomes to obtain adequate data for analysis.

The most likely source of nonstationarity is the existence of a trend within the data, such as might be caused by daily cycles. The effects of such trends may be minimized by testing only for short periods at a consistent time of day, and methods are also available for detecting trends so that the extent to which they affect the data may be assessed. Perkel et al. (1967) suggest breaking the data into segments, analyzing each separately, and then testing to see if they are drawn from the same population. Especially where the types of act considered are heterogeneous, most behavioral data will fail to pass this test: sporadic periods of sleep or long bouts of grooming, which may take place only a few times per day, will tend to occur in some segments but not others.

Moore et al. (1966) point out the complexity of rigorous criteria for establishing stationarity in a simple system such as the firing pattern of a single neuron. Very few workers on sequences of behavior attempt such checks, and the frequent interpretation of results in terms of changing motivation suggests that nonstationarity is accepted. Rough tests for stationarity have been performed in two cases. Nelson (1964), working on courtship in fish, was able to carry out Markov chain analysis on one species studied (*Corynopoma*) after excluding intersequence intervals from the data, but he rejected the possibility of sequence analysis on another (*Pseudocorynopoma*), as the probabilities of different acts showed marked temporal fluctuations. Lemon and Chatfield (1971) proceeded with Markov chain analysis after finding no significant differences in the probabilities of different song types between the first and second halves of the records they analyzed.

The fact that these workers were able to satisfy at least some of the criteria for stationarity with their data and the fact that in both cases the behavior was found to follow a first-order Markov chain may be no coincidence. Moore *et al.* (1966) point out that "the distinction between serial dependence (with an overall stationary sample) and actual nonstationarity is an arbitrary one." Thus data which appear to follow a Markov chain of high order could also be interpreted as nonstationary and *vice versa*. In some behavioral data, triplet sequences of the general form $A \rightarrow B \rightarrow A$ have been found to be commoner than expected (Slater and Ollason, 1972; Fentress, 1972), indicating the inadequacy of a first-order Markov model. It is possible here that a second-order model might be satisfactory, but shortage of data precludes testing for this. Another interpretation would be that periods of time exist during which the two acts involved in the sequence are of high probability, while that of all other behaviors is low, thus leading to alternation between them. This interpretation assumes nonstationarity, i.e., fluctuating probabilities.

C. Alternative Methods of Approach

It will be apparent from the above discussion that there are many reasons why the techniques of Markov chain analysis should be applied only with caution to behavioral data. The most obvious reason for this is that the probabilities of different behavior patterns tend to change with time. Given this difficulty, it is worth considering briefly other methods which, while not primarily concerned with sequencing, may be used to detect associations between behaviors.

Factor analysis is particularly interesting in this context, as it has also been employed by ethologists for the treatment of data in transition matrices (Wiepkema, 1961; Baerends and van der Cingel, 1962; van Hooff, 1970; Baerends *et al.,* 1970). Using this method, a small number of hypothetical variables are extracted, the existence of which as causal factors could account for most of the observed correlations between acts, without the acts themselves being directly related. It is assumed that the measured variables, in this case behavior patterns, do not depend causally on each other but only on the postulated factors (Blalock, 1961). This assumption highlights the sharp contrast between this method and Markov chain analysis. The choice of which of these approaches is used appears to depend mainly on whether the research worker believes "sequence effects" or "motivational states" to be the more important in determining the relationships between behaviors. In the former case, it is thought useful to describe the probability of an act at a particular instant in terms of the sequence of acts which preceded it. The animal is assumed to be in a steady state, which is tantamount to ignoring the possibility that motivational changes occur. On the other hand, if interpretation is in terms of motivational states (underlying variables which are postulated as changing more slowly than the switching in overt behavior), the same data can be treated quite differently using factor analysis. An association between acts is here taken to be indicative of common causal factors underlying the behaviors involved, and the factor on which they have a high loading may be argued to represent a motivational state (e.g., "aggressive," "sexual," "nonreproductive": Wiepkema, 1961; "affinitive," "play," "aggressive," etc.: van Hooff, 1970). If factor analysis is based on a transition matrix, two different models may be derived, depending on whether the frequency of acts following each other or preceding each other is taken as a starting point. The method assumes that sequence effects are unimportant, as these will make the two models differ from one another (Slater and Ollason, 1972). As both motivational changes and sequence effects are likely to occur in most behavior, each of these two descriptions is necessarily imperfect, and cases where either is valid in the strictly mathematical sense may be hard to find.

Factor analysis is open to several other objections, some of which have been discussed by Overall (1964), Andrew (1972), and Slater and Ollason (1972). The use of the method supposes it to be useful to describe behavior in terms of a few underlying variables, and yet it is not altogether clear what these may represent. While Wiepkema (1961) referred to his factors as "tendencies," thus identifying them with a drive type of model, Baerends (1970) gives them weaker status as "areas of higher density within the causal network," thus recognizing that their existence may depend on external as well as internal variables. It is doubtful whether the extraction of factors which are themselves of complex causation advances understanding. At the descriptive stage of analysis, it appears preferable to examine the relationships between individual acts, as these can be directly assessed without theoretical implications and do not place constraints on the type of model of organization which may emerge from subsequent experimental work.

Analysis based on transitions between behaviors is not the only way of detecting associations between individual acts: an alternative approach which has much to recommend it is the use of correlation techniques. This involves measuring the frequencies of different behavior patterns within a series of time units and then correlating between each pair of acts to determine whether they are positively or negatively associated. Stationarity remains a problem here, though perhaps a less acute one, while some of the other objections to sequence analysis are not important: two behaviors which occur concurrently are no problem, nor does it bias the results if acts at different levels of organization are included in the analysis. The main difficulty is that the results may depend on the choice of time unit. While two acts may be significantly correlated over 1-hr periods, this is not necessarily true when their correlation is assessed over 10-sec units. To allow for this, it is therefore preferable to analyze the data separately for a number of different time intervals, as has been done by Baerends *et al.* (1970). With this approach, it becomes possible to detect much looser associations than would be achievable with sequence analysis.

Ordinary correlation techniques give no indication of whether acts tend to follow each other in a particular order. If one behavior leads to another, correlations of this type can only indicate that they are in some way associated. A modification of the method can, however, extract such information; this is the cross-correlation procedure which is discussed in detail by Delius (1969). Here the frequencies of different acts are correlated, not just for the same time unit, but with progressive lags introduced between them. Thus if behavior B tends to occur 5 sec after behavior A and the time unit used is 1 sec, a positive correlation will be found when the data for A are moved forward 5 sec in relation to B. Heiligenberg (1973) has carried out the most elegant study to date employing this procedure and, using 1-sec time units, has been able to demonstrate both sequence effects and looser associations between seven behavior patterns in a cichlid fish (*Haplochromis burtoni*).

Correlative techniques can thus give useful information on a number of facets of behavioral organization, including sequence effects. They are particularly suitable where the behaviors under study are diverse and where their associations are liable to be due to common causation rather than sequencing. A full analysis of this sort requires a considerable amount of data (Heiligenberg's results were based on 360 hr of observation), and, unlike the simpler forms of sequence analysis, the amount of calculation involved makes the use of a computer essential. Where an approximate guide to associations is required, or where sequencing is strong, analysis of first-order transitions between acts is clearly more practicable.

IV. SEQUENCES OF INTERACTION BETWEEN INDIVIDUALS

The discussion so far has centered around sequences of behavior within the individual in contexts where the external world is presumed to be relatively unchanging. Some of the examples used were concerned with animals in a social situation (e.g., the studies of fish courtship by Wiepkema, 1961, and Nelson, 1964); in these cases, the assumption was made that the presence of other animals, while perhaps stimulating the behavior under study, had

Describing Sequences of Behavior   147

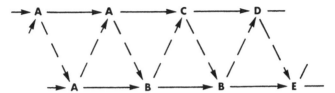

Fig. 1. A simple model of an interactive sequence involving two
individuals.

little effect on its sequence. Nelson (1964), for instance, stated that the part
played by the female in glandulocaudine courtship is slight, and therefore felt
justified in examining sequences within the male as if he were in a constant
environment.

Most studies of social sequences have a different primary aim: to
demonstrate that the behavior of one animal *is* affected by that of others,
and so obtain evidence for the role of different behavior patterns in communication. At first sight, this is an easier task than the analysis of within-
individual sequences, for data of this sort can easily be cast into a complete
matrix such as Table I, the acts listed down the side being those shown by
one individual and those along the top being the subsequent behaviors of
another. The objections to including figures on the descending diagonal do
not apply: these entries refer to those cases where the performance of a
behavior by one animal is followed by the same behavior from the other.

Other difficulties do, however, arise, including those concerned with the
choice of categories and with stationarity which have already been discussed.
Stationarity may be a particularly serious problem, for as MacKay (1972)
points out, many of the most interesting behaviors in communication are
those which change in probability as the interaction proceeds. A further
difficulty, which is often neglected, is that the behavior of an animal in a
social situation is likely to depend partly on the sequence of acts that it has
already shown as well as on the behavior of others. Figure 1 shows a simple
model of an interaction between two animals designed to take account of
both these influences.[4] Studies of within-individual sequences, such as those
of Wiepkema (1961) and Nelson (1964), suppose that the effects portrayed by
dotted lines are of little importance, whereas some studies of interactions
between animals pay only passing attention to the possible influence of the
effects illustrated by complete lines (e.g., Hazlett and Bossert, 1965). The
most thorough analysis of interactive sequences, that by Altmann (1965)

[4]The model is, of course, too unsophisticated for many purposes. It assumes, first, that each
behavior depends only on those shown by the individual and by his companion immediately
previously and, second, that the interaction is like a game of chess, the animals making
alternate moves.

on social communication in the rhesus monkey (*Macaca mulatta*), does not attempt to differentiate between these two possible influences. His transition matrices are based on the order of events within a group of monkeys, making no distinction between animals; thus two consecutive behaviors may be by the same or by different individuals. Using data of this sort, Altmann has taken Markov chain analysis further than any other worker, his examination extending to quadruplets of acts (third-order relationships). The confounding of these two effects was perhaps inevitable in a field study where several animals might be in view and each could show any one of 120 different behavior patterns, but it means that his results cannot be interpreted easily. In common with a number of other studies, some of sequences within the individual (e.g., Chatfield and Lemon, 1970; Fentress, 1972) and some of interactive sequences (e.g., Hazlett and Bossert, 1965), Altmann's analysis used the techniques of information theory. Such methods can also be used to assess the relative importance of within- and between-individual influences. This can be done by deriving measures of information transfer separately for the transitions shown in Fig. 1 as dotted lines and those shown as complete lines, and comparing the two values so obtained. The study by Dingle (1969) includes such calculations: his conclusion that the two influences were of equal importance in interactions between mantis shrimps (*Gonodactylus bredini*) warns against ignoring the effects of either of them. It should be noted in passing that measures of information transmission, while a useful way of expressing the correlation between events, do not necessarily imply a causal relationship between them (MacKay, 1972).

Consideration of a whole matrix of interindividual transitions can indicate only that the behavior of one animal is in some way linked to that of others. Ethologists usually want to ask more specific questions: e.g., does one animal tend to perform behavior *B* more often than expected after another has shown behavior *A*? If within-individual effects are of no importance, an unlikely finding, this could reasonably be tested by comparing the number of times an event of *B* follows an event of *A* with that expected on a random model. But if, as is quite possible, *A* and *B* tend to fall into bouts, the number of such occurrences may be high purely by chance. If a bout of *B* happens to start just after a bout of *A*, many transitions between the two behaviors will be scored even though the animals may be acting independently of one another. Thus if successive events of *B* are not independent of each other, it is clearly invalid to suggest that each of them was separately stimulated by *A*. It is more valid to ask whether more *bouts* of *B* started during or just after each event of *A* than expected. A bout of *B* must be defined so that the first event in one bout is statistically independent of previous occurrences of that behavior in the same animal. Approximate methods for determining the interval between events necessary for this to be the case have been provided by Duncan *et al.* (1970) and Wiepkema (1968), but have not yet been used in sequence analysis. The closest attempt has been in the study by Wortis (1969) of interactions between ring dove (*Streptopelia risoria*) parents and their chicks. She accepted that only the start of a particular behavior should be scored, but took a gap of only 2 sec as indicating a gap between bouts. This criterion was apparently chosen arbitrarily, and it seems unlikely that 2 sec would be adequate for the second event to be independent of the first. A different way of getting around this problem has been used by Heckenlively (1970), who examined encounters between pairs of crayfish (*Orconectes virilis*). For each encounter, he scored only a single transition, that between the first act of one animal and the response to it of the other, and all transitions could thus be considered truly independent of each other.

The analysis of interactive sequences is clearly a more difficult task than that of sequences within the individual because of these additional complications. Furthermore, in this case other methods do not afford the prospect of achieving an improvement. Sequence analysis seems an obvious first step in the study of communication, for here the suggestion is being made that the behavior of one animal is in some way causal to the behavior of another, while this is not necessarily true of two successive behavior patterns shown by the same individual.

V. DISCUSSION

In this chapter, I have mentioned various ways in which sequences of behavior can be analyzed and some of the assumptions which underlie each of them. It is now pertinent to outline the circumstances in which they are likely to prove helpful and to discuss the conclusions which may be drawn as a result of their application.

The simplest form of sequence analysis, that in terms of frequencies or conditional probabilities, can usefully be applied only in circumstances where sequence effects are strong, the acts following each other in an almost fixed order. If the behavior is less structured than this and, in particular, when the different acts considered occur at markedly different frequencies, a better approach is to compare the observed first-order transition frequencies with those which would be expected on a random model, a method equivalent to the initial stage of Markov chain analysis. This is preferable because it takes into account the frequencies of the individual acts and thus extracts those transitions which are commoner than expected, rather than those which are common in absolute terms. In the past, ethologists have not applied this method as carefully as they might have done, particularly those who have based their analyses on complete matrices of transitions. But, while this methodological problem can be overcome, other difficulties are harder to cope with.

There are several reasons why two behaviors may tend to succeed each other more often than expected. Of these, the two most likely are that they both share causal factors and so occur only when these are present, or that the first act stimulates the second in some way. Data in which behaviors are grouped due to changing causal factors fail to fulfill the statistical requirement of stationarity and will thus be unsuitable for rigorous Markov chain analysis. Despite this, the examination of first-order relationships will provide a rough guide to the groupings into which different activities fall. While first-order transition analysis may therefore be of some usefulness in nonstationary data, the extraction of higher-order dependencies is here pointless, for the statistical dependencies between successive acts may be just a byproduct of their shared causation rather than indicating that rules govern the sequence as such. Correlative techniques using various different time units offer a more effective way of assessing the associations between acts when these are thought to result from changing causal factors.

In situations where internal and external causal factors are thought not to change with time, the data are more likely to pass the test of stationarity, and here it is probable that the sequential dependencies found result from causal dependencies between the successive acts, or between the states of which they are the overt indicators. In these circumstances, the analysis of first-order dependencies is a first step toward discovering the rules governing sequencing. These may then be fully elucidated by a search for higher-order dependencies.

Many behavioral sequences will not fall neatly into one of these categories or the other; perhaps the most profitable approach here is to combine first-order transition analysis to look for associations liable to result from sequence effects, with correlation techniques, to find the looser associations which might result from shared causal factors.

It is important to stress that none of these methods provides more than a description of the behavior under study. That two acts tend to occur in sequence need not imply a similarity of causation, nor that one generates the other in some way: they could be associated by exclusion, or because of particular characteristics of the experimental situation. The choice of descriptive method may bias the researcher toward a particular type of explanation. For example, if $A \rightarrow B$ more than $B \rightarrow A$ in a transition analysis, it suggests that A generates a state appropriate to B, or increases the probability of B. This result can, however, also be explained on a motivational model in which the acts differ in threshold (Bastock and Manning, 1955; Hinde and Stevenson, 1969). Conversely, a symmetrical relationship between two behaviors such that both $A \rightarrow B$ and $B \rightarrow A$ are common could be taken as indicating common causation or that each of the two acts stimulated occurrence of the other. The description of behavior provides hypotheses but not explanations; a frequent mistake made by those using factor analysis is the attempt to provide a causal explanation based on purely descriptive data. On the other hand, the detailed knowledge of the associations between behaviors made possible by the other methods discussed here may prove an invaluable source of hypotheses; the way in which these associations change in response to experimental intervention can then be used to build models of the causation of behavior.

Similar arguments apply to the description of interactive sequences. Even after valid ways have been found of demonstrating that behavior B by one animal tends to occur after behavior A by another more often than expected, this discovery is not evidence that A causes B, though this is one possible explanation. Other possibilities are that a behavior occurring close to A in time stimulates B, or a behavior associated with B, or even that the two behaviors tend to occur synchronously due to some event in the more distant past, such as social facilitation of other activities or the onset of daylight (Andrew, 1972). Experiments are essential if communication is to be demonstrated. Nevertheless, the analysis of sequences of behavior in normal interactions is a useful initial stage before such experiments are carried out to indicate the acts which are likely to be important in communication. The fact that most studies on this topic have failed to distinguish clearly between intra- and interindividual effects does not argue against the usefulness of the general approach.

VI. ACKNOWLEDGMENTS

I am grateful to Professor R. J. Andrew, Dr. P. P. G. Bateson, Dr. C. Chatfield, Mrs. J. C. Ollason, and Dr. L. M. Rosenson, all of whose comments have helped to improve this article.

VII. REFERENCES

Altmann, S. A. (1965). Sociobiology of rhesus monkeys. II. Stochastics of social communication. *J. Theoret. Biol.* **8**:490–522.
Andrew, R. J. (1956). Normal and irrelevant toilet behaviour in *Emberiza* spp. *Brit. J. Anim. Behav.* **4**:85–91.
Andrew, R. J. (1972). The information potentially available in mammalian displays. In Hinde, R. A. (ed.), *Non-Verbal Communication*, Cambridge University Press, Cambridge, England, pp. 179–204.
Aschoff, J. (1967). Circadian rhythms in birds. *Proc. 14th Internat. Ornithol. Congr.*, pp. 81–105.

152   P. J. B. Slater

Baerends, G. P. (1970). A model of the functional organization of incubation behaviour. *Behaviour Suppl.* **17**:263–312.

Baerends, G. P., and van der Cingel, N. A. (1962). On the phylogenetic origin of the snap display in the common heron (*Ardea cinerea* L.). *Symp. Zool. Soc. Lond.* **8**:7–24.

Baerends, G. P., Brouwer, R., and Waterbolk, H. T. (1955). Ethological studies on *Lebistes reticulatus* (Peters): I. An analysis of the male courtship pattern. *Behaviour* **8**:249–334.

Baerends, G. P., Drent, R. H., Glas, P., and Groenewold, H. (1970). An ethological analysis of incubation behaviour in the herring gull. *Behaviour Suppl.* **17**:135–235.

Bartlett, M. S. (1951). The frequency goodness-of-fit test for probability chains. *Proc. Cambridge Philos. Soc.* **47**:89–95.

Bastock, M., and Manning, A. (1955). The courtship of *Drosophila melanogaster*. *Behaviour* **8**:85–111.

Bindra, D. (1959). *Motivation: A Systematic Reinterpretation*, Ronald Press, New York, 361 pp.

Blalock, H. M. (1961). *Causal Inferences in Nonexperimental Research*, University of North Carolina Press, Chapel Hill, 200 pp.

Blurton Jones, N. G. (1968). Observations and experiments on causation of threat displays in the great tit (*Parus major*). *Anim. Behav. Monogr.* **1**:75–158.

Bolles, R. C. (1960). Grooming behavior in the rat. *J. Comp. Physiol. Psychol.* **53**:306–310.

Brown, R. G. B. (1965). Courtship behaviour in the *Drosophila obscura* group. Part II. Comparative studies. *Behaviour* **25**:281–323.

Cane, V. (1959). Behaviour sequences as semi-Markov chains. *J. Roy. Stat. Soc. Ser. B* **21**:36–58.

Cane, V. (1961). Some ways of describing behaviour. In Thorpe, W. H., and Zangwill, O. L. (eds.), *Current Problems in Animal Behaviour*, Cambridge University Press, Cambridge, England, pp. 361–388.

Chatfield, C. Statistical inference regarding Markov chain models. *Appl. Stat.* (in press).

Chatfield, C., and Lemon, R. E. (1970). Analysing sequences of behavioural events. *J. Theoret. Biol.* **29**:427–445.

Coulon, J. (1971). Influence de l'isolement social sur le comportement du cobaye. *Behaviour* **38**:93–120.

Cox, D. R., and Lewis, P. A. W. (1966). *The Statistical Analysis of Series of Events*, Methuen, London, 285 pp.

Delius, J. D. (1969). A stochastic analysis of the maintenance behaviour of skylarks. *Behaviour* **33**:137–178.

Dingle, H. (1969). A statistical and information analysis of aggressive communication in the mantis shrimp *Gonodactylus bredini* Manning. *Anim. Behav.* **17**:561–575.

Duncan, I. J. H., Horne, A. R., Hughes, B. O., and Wood-Gush, D. G. M. (1970). The pattern of food intake in female brown leghorn fowls as recorded in a Skinner box. *Anim. Behav.* **18**:245–255.

Fabricius, E., and Jansson, A-M. (1963). Laboratory observations on the reproductive behaviour of the pigeon (*Columba livia*) during the pre-incubation phase of the breeding cycle. *Anim. Behav.* **11**:534–547.

Fentress, J. C. (1972). Development and patterning of movement sequences in inbred mice. In Kiger, J. A. (ed.), *The Biology of Behavior*, Oregon State University Press, Corvallis.

Golani, I., and Mendelssohn, H. (1971). Sequences of pre-copulatory behaviour of the jackal (*Canis aureus* L.). *Behaviour* **38**:169–192.

Goodman, L. A. (1968). The analysis of cross-classified data: Independence, quasi-independence and interactions in contingency tables with or without missing entries. *J. Am. Stat. Ass.* **63**:1091–1131.

Hazlett, B. A., and Bossert, W. H. (1965). A statistical analysis of the aggressive communications systems of some hermit crabs. *Anim. Behav.* **13**:357–373.

Heckenlively, D. B. (1970). Intensity of aggression in the crayfish, *Orconectes virilis* (Hagen). *Nature* **225**:180–181.

Heiligenberg, W. (1973). Random processes describing the occurrence of behavioral patterns in a cichlid fish. *Anim. Behav.* **21**:169–182.

Describing Sequences of Behavior **153**

Hinde, R. A. (1955/1956). A comparative study of the courtship of certain finches (Fringillidae). *Ibis* **97**:706–745; **98**:1–23.

Hinde, R. A. (1970). *Animal Behaviour: A Synthesis of Ethology and Comparative Psychology,* McGraw-Hill, New York, 876 pp.

Hinde, R. A., and Stevenson, J. G. (1969). Integration of response sequences. In Lehrman, D. S., Hinde, R. A., and Shaw, E. (eds.), *Advances in the Study of Behavior,* Vol. 2, Academic Press, New York, pp. 267–296.

Isaac, D., and Marler, P. (1963). Ordering of sequences of singing behaviour of mistle thrushes in relation to timing. *Anim. Behav.* **11**:178–188.

Lemon, R..E., and Chatfield, C. (1971). Organization of song in cardinals. *Anim. Behav.* **19**:1–17.

MacKay, D. M. (1972). Formal analysis of communicative processes. In Hinde, R. A. (ed.), *Non-Verbal Communication,* Cambridge University Press, Cambridge, England, pp. 3–25.

McFarland, D. J. (1971). *Feedback Mechanisms in Animal Behaviour,* Academic Press, London, 279 pp.

Moore, G. P., Perkel, D. H., and Segundo, J. P. (1966). Statistical analysis and functional interpretation of neuronal spike data. *Ann. Rev. Physiol.* **28**:493–522.

Nelson, K. (1964). The temporal pattern of courtship behaviour in the glandulocaudine fishes (Ostariophysi, Characidae). *Behaviour* **24**:90–146.

Noirot, E. (1969). Serial order of maternal responses in mice. *Anim. Behav.* **17**:547–550.

Overall, J. E. (1964). Note on the scientific status of factors. *Psychol. Bull.* **61**:270–276.

Palmgren, P. (1950). On the diurnal rhythm of activity and rest in birds. *Ibis* **91**:561–576.

Parker, G. A. (1972). Reproductive behaviour of *Sepsis cynipsea* (L.) (Diptera: Sepsidae). I. A preliminary analysis of the reproductive strategy and its associated behaviour patterns. *Behaviour* **41**:172–206.

Perkel, D. H., Gerstein, G. L., and Moore, G. P. (1967). Neuronal spike trains and stochastic point processes. I. The single spike train. *Biophys. J.* **7**:391–418.

Richter, C. P. (1927). Animal behavior and internal drives. *Quart. Rev. Biol.* **2**:307–342.

Slater, P. J. B. A reassessment of ethology. In Broughton, W. B. (ed.), *Biology of Brains,* Academic Press, London, in press.

Slater, P. J. B., and Ollason, J. C. (1972). The temporal pattern of behaviour in isolated male zebra finches: Transition analysis. *Behaviour* **42**:248–269.

Stokes, A. W. (1962). Agonistic behaviour among blue tits at a winter feeding station. *Behaviour* **19**:118–138.

Tinbergen, N. (1951). *The Study of Instinct,* Clarendon Press, Oxford, 228 pp.

van Hooff, J. A. R. A. M. (1970). A component analysis of the structure of the social behaviour of a semi-captive chimpanzee group. *Experientia* **26**:549–550.

Weidmann, U., and Darley, J. (1971). The role of the female in the social display of mallards. *Anim. Behav.* **19**:287–298.

Wells, G. P. (1950). Spontaneous activity cycles in polychaete worms. *Symp. Soc. Exptl. Biol.* **4**:127–142.

Wiepkema, P. R. (1961). An ethological analysis of the reproductive behaviour of the bitterling. *Arch. Néerl. Zool.* **14**:103–199.

Wiepkema, P. R. (1968). Behaviour changes in CBA mice as a result of one goldthioglucose injection. *Behaviour* **32**:179–210.

Wortis, R. P. (1969). The transition from dependent to independent feeding in the young ring dove. *Anim. Behav. Monogr.* **2**:1–54.

OBSERVATIONAL STUDY OF BEHAVIOR: SAMPLING METHODS

by

JEANNE ALTMANN [1]

(Allee Laboratory of Animal Behavior, University of Chicago, Chicago, Illinois, U.S.A.)

(Rec. 15-III-1973)

CONTENTS

	Page
I. Introduction	227
II. Sampling variables	231
III. *Ad libitum* sampling	235
IV. Sociometric matrix completion	240
V. Focal-animal sampling	242
VI. Sampling all occurrences of some behaviors	247
VII. Sequence sampling	248
VIII. One-zero sampling	251
IX. Instantaneous and scan sampling	258
Summary	261
Bibliography	262
Zusammenfassung	265

I. INTRODUCTION

This is an observer's guide: in it I will present sampling methods for use in direct observation of spontaneous social behavior in groups of men or other animals. All observational sampling methods known to me will be described, and their uses and limitations indicated.

A. SCOPE

In what follows I shall assume that the observer has a group of spontaneously interacting individuals to watch, that he has formulated one or more questions about social behavior, that he knows what behaviors he wishes to study, and that he has found suitable methods for recording such behaviors.

1) During the preparation of this paper the author was supported by research grants GB-27170, from the National Science Foundation, and MH-19,617, from the Public Health Service. The encouragement and criticisms of my husband, Stuart ALTMANN, were of utmost value at all stages of the research and writing. The manuscript benefitted from critical reading of an earlier version by Joel COHEN, Glenn HAUSFATER, James LOY, Donald SADE, Montgomery SLATKIN, Thomas STRUHSAKER and Stephen WAGNER.

Thus, I will not be concerned here with the logistics of such research nor with the problems of choosing research questions and defining behavior categories. No attempt will be made to cover statistical aspects of experimental design, such as adequacy of sample size, factorial design, and so forth. Instead, the focus will be on a question that arises at an earlier stage in research planning, namely, how does the choice of sampling method restrict the kinds of behavior processes that can be studied? Or, conversely, given a particular behavioral research problem what sampling methods are suitable for it?

I shall assume that the research question has been sufficiently well-formulated that the investigator can identify the relevant sample space, *i.e.* the set of events that must be sampled in order to answer the question. Let me illustrate. Suppose that we are interested in studying aggression in a group of monkeys. We might begin by formulating a question such as this: Are males more aggressive than females? At this stage, the question is too ambiguous for a sampling method to be chosen. For example, we need to specify which behaviors will be classed as aggressive, and which will not. We need to make clear whether the question refers to all age classes or, say, just to adults. Even then, there are numerous reasonable interpretations of the question, such as:

(i)   On the average, do males spend more of the day involved in aggressive behavior than do females?
(ii) Do males initiate aggressive bouts more often than do females?
(iii) Are the aggressive acts of males more serious, more intense, more potentially destructive?
(iv) Do the behavioral acts of males include a higher proportion of aggressive acts than do those of females?
(v)   Is the response to an aggressive act more likely to be an aggressive act if the recipient is a male?

The choice of one or more of such formulations depends on an evaluation of their relevance to the original question. That evaluation depends, in turn, on numerous questions about the behavioral or biological significance of sex differences in aggression. Such evaluation lies outside the scope of this paper. But an unambiguous formulation of the research question is a prerequisite to the kinds of sampling decisions that will be discussed here: different formulations will usually involve sampling from different sample spaces.

The observer needs to know how to gather data that will answer such specific research questions. Should he repeatedly scan the group, recording each individual's behavior at the time that it is seen? Should he watch each individual in turn, each for a fixed amount of time? If an individual continues in one observation period the behavior that he began in a previous period, should the behavior be recorded again? Should the observer record every occurrence of a behavior, or only whether it occurred at least once in the observation period? In this paper, I shall examine such sampling alternatives and provide guidelines for choosing among them in observational studies.

Sampling decisions are made whenever the student of social behavior cannot continuously observe and record all of the behavior of all of the members of a social group, and must therefore settle for a partial record. However, even in the most systematic of observational behavior studies, only partial descriptions of the sampling procedure are provided. Seldom has an author provided justification for his choice of sampling method.

We suspect that the investigator often chooses a sampling procedure without being aware that he is making a choice. Of course, he does not thereby escape the consequences of the choice: the data that result from any sampling method can only answer certain classes of questions. From the standpoint of the behavioral questions, a given question can be answered only by data obtained through certain kinds of sampling methods.

B. MANIPULATIVE *VS* NON-MANIPULATIVE RESEARCH

The method of direct observation plays a curious and unique role in the behavioral sciences. It is at once the necessary link between laboratory research and "real-world" behavior, and the bane of our aspirations for more accurate, more objective information about behavior.

From time to time, one hears the claim that accurate studies of behavior can be made only in the laboratory, and that quantitative research on behavior is not practicable in the context of ongoing, real-life situations. Such a restriction on research would mean that the behavioral sciences would forever forsake any hope of knowing whether their most powerful theories have any relevance to the world of behavior outside the laboratory. Unless we develop methods for field research that are comparable in sensitivity to those of the laboratory, the behavioral sciences will become progressively more isolated from the very behavior that their theories are supposed to explain.

A primary function of research design is to maximize the validity of the conclusions (WEBB *et al.*, 1966), *i.e.*, to minimize the number of plausible alternative hypotheses that are consistent with the data. It is useful to distinguish between internal validity, which deals with statements about the sample, and external validity, which deals with interpretations and generalizations from the sample to other situations or populations. For example, housing experimental animals under identical conditions would represent an attempt to increase internal validity, whereas the process by which these animals were chosen in the first place or the similarity of the housing to their natural habitat would affect the external validity.

Internal validity is an essential component of external validity: to the extent that we have not eliminated alternative explanations for the results within our sample, we cannot rule them out of any generalization or interpretation derived from the sample. However, some conclusions depend more heavily on the generality (external validity) of the results than do others: internal validity should not be purchased through complete loss of external validity.

Laboratory research on behavior has usually emphasized internal validity, but in such research we ignore at our peril the question of whether high internal validity has been gained by an inordinate sacrifice of generality and relevance. The choice of animals and the artificial world in which they are placed may distort the results, or the experimental task that is presented to the animals may be largely irrelevant to an understanding of how these animals solve their own problems.

In contrast, observational field studies of behavior tend to show the converse imbalance: low internal validity but, ostensibly, high external validity. A major source of these imbalances is this: external validity has been largely ignored in laboratory studies, and internal validity, in field research. The assumption apparently has been that internal validity requires manipulation, by the scientist, of subjects and behavior and that external validity depends on a naturalistic setting and the absence of such manipulations.

Attempts to correct this imbalance have recently been made by utilizing information from field studies to design laboratory experiments (*e.g.*, MASON & EPPLE, 1969) and by bringing some of the manipulations of a laboratory experiment into the natural field situation (*e.g.*, HALL, 1965; KUMMER, 1971; MENZEL, 1969). This approach is one way, but not the only way, of increasing the internal validity of field research; it will not be discussed in what follows.

The primary function of experimental controls is to reduce or eliminate alternate hypotheses, and it is this general function of methodology to which the field worker should look in his attempt to increase the internal validity of non-experimental field studies. Observational research may require the development of research tactics that are particularly suited to its needs. Of these needs, one of the most important is for sensitive, non-destructive methods of studying social processes (BARKER, 1963).

As SCHNEIRLA (1950) has pointed out, controls are not absent in field situations; instead, they are usually "observation-selective" rather than "manipulative". What, then, are the non-manipulative controls that are available? Can we use them in such a way that we increase the internal validity of observational studies of behavior, without losing external validity?

Sampling decisions offer the student of behavior a prime opportunity to increase internal validity through means which are non-manipulative and are therefore less likely to alter or destroy the social system that is being studied. Use of such controls — in particular, sampling decisions — in observational studies of social groups can increase the validity of comparisons both within and between studies, whether observational or experimental, field or laboratory.

II. SAMPLING VARIABLES

Before turning to specific sampling techniques, let us consider the major variables that distinguish existing sampling methods and that are most crucial for sampling decisions. As noted previously it will be assumed that the observer has a well-formulated research question, and that he has at least a preliminary catalog of behaviors that are under study. It will also be assumed that certain preliminary sampling decisions have been made: the study locale and population have been selected.

A. BEHAVIOR RECORD

E v e n t s *vs* s t a t e s .

Behaviors may be regarded either as events or as states. Events are instantaneous; states have appreciable durations. Of course, in reality, the performance of any kind of behavior takes some amount of time, however brief. But if we consider behaviors at the moment of their onset, or at any other single defining instant, then we are recording events. We can, for example, record that an animal assumes a sitting posture, an act that occurs at an instant (and is therefore an event), or that the animal is seated (a state).

Our choice between regarding behaviors as states or as events depends upon the questions about behavior that we are attempting to answer. In particular, questions about frequencies of behavior, such as questions ii, iv, and v on p. 228, entail considering the behavior as events [2]. Once the

---

2) The term 'frequency' will be used in this paper to mean *number of occurrences,* in accord with convention in the statistical literature. It has a different meaning in some other contexts. Thus, 'gene frequency' is used by population geneticists to refer to a relative frequency or *proportion.* In the physical sciences, 'frequency' commonly refers

investigator has decided on a defining event, such as onset, for a particular behavior, that behavior is not scored in a sample session unless the defining event occurs during the session, even though the behavior is otherwise "in progress" during the session. On the other hand, any question involving the duration of a behavior, or the percent of time spent in some activity (*e.g.* question iii, p. 228) is a question about states. To answer questions involving duration one could time each occurrence directly, perhaps using a standard stopwatch. Alternatively, if an exclusive and exhaustive classification of states has been made, one could record transition times (*i.e.* onsets and terminations), thus preserving frequency and sequence information as well as that of durations and time spent in various activities. For information only on percent of time spent in a particular state, one could merely cumulate durations of the state of interest (*c.g.* by means of a cumulative stopwatch) and record the total sample time. The extent to which various sampling techniques are suited to answering each of these two basic types of questions will be discussed in the sections on individual techniques.

Completeness of frequency record.

If events rather than states are scored, the sampling procedures may be divided into three classes with respect to the completeness with which frequencies of behaviors are recorded. In one class of procedures (see, *e.g.*, Sections V and VI), a complete or total frequency record is kept. By this it is meant that during a sample, all occurrences of the behavior of interest for some subset of group members are recorded. In a second type of sampling, partial frequencies are obtained; such records usually consist of an unknown percentage of the total occurrences, which is variable from sample to sample, and from individual to individual. *Ad Lib. Sampling* (Section III) usually results in such records. Finally, the observer may record, during each sample period, the fact that the behavior occurred at least once (scored as *one*) or did not occur (scored as *zero*). Thus, a score of "one" may represent one occurrence or a multitude of occurrences. Such sampling is discussed under the heading "One-zero Sampling" (Section VIII).

Frequency and rate.

In the behavioral literature, many comparisons that are presented in terms of frequency differences are actually statements about rates. For example, we cannot justifiably claim that the dominant male copulated more often

---

to number of occurrences per second, and thus to a *rate*. In some common English expressions (*e.g.* "I frequently sunbathe") it seems that the intended meaning is sometimes *percent of time,* and other times *rate, i.e.* number of occurrences per unit time.

than did the subordinate, until we know the amount of time that was devoted to sampling the behavior of each, or, at least, that the two amounts were equal. The question that is being answered is whether the first male copulated more frequently *per unit time* than did the other, *i.e.* we are comparing rates.

If one knew that observation times were equal for all individuals, then the frequencies themselves could be compared directly, ignoring the time base if rates were not of interest. However, particularly in field situations, individuals are seldom observed for exactly the same amount of time, and often for very different amounts of time. This may be the result of circumstances beyond the observer's control, or a direct result of the sampling procedure. Under these circumstances, rate comparisons are the obvious solution. Thus, implicitly (if observation times are equal) or explicitly, rates are often being used in behavior studies. Time is often an important variable.

There are, however, other statements or questions about behavior that are based on frequency *per se,* and not on rates. Most such statements are essentially statements about conditional probabilities, in which the condition is the occurrence of acts of a particular kind. For example, the statement, "Males hit more often than do females," is a statement about rates, whereas, "The aggressive acts of males more often involve hitting than do those of females," is not: the latter is equivalent to, "The probability of hitting in aggressive acts by males is higher than it is in aggressive acts by females." Here is another example: the statement, "Adult males threaten juveniles more often than they threaten infants," is based on a comparison of rates, whereas, "Infants run away, when threatened by adult males, more often than do juveniles," is not [3]).

Since some of the sampling methods that will be described in this paper can be used to estimate rates of interaction and others cannot, it is important that the investigator know in each case which type of question he is posing.

Content of record.

Behavioral records may contain records only of the occurrences of behaviors of interest, or they may in addition, include a variety of other data. A record is usually made of the date and time of each sample session onset and termination. These times determine the length of intervals between samples, the duration of the samples and the seasonal and diurnal distribution of the samples, all of which are important. Particularly common in studies of social behavior are records of: (1) the actors, (2) the receivers or object

---

3) Questions that involve true frequency comparisons may still require consideration of time base, in order to insure an unbiased sample of the conditional events.

234   JEANNE ALTMANN

individuals, (3) the sequence of events or states, without times of occurrence, (4) the time of occurrence, thus also including sequential information, (5) duration of behavioral state (see p. 231-232), without recording the actual time of onset and termination, or (6) onsets and terminations for some or all behaviors of interest. In addition, the record may include contextual data, such as habitat, weather, predominant group activity, distances to or identities of neighbors, or the size and membership of the subgroup in which behavior occurred. As we shall see, the choice among such characteristics of the data will, in turn, narrow the choice of appropriate sampling techniques.

B. SCHEDULING SAMPLE SESSIONS

Scheduling of session onsets.

A sample session may be scheduled to begin at a predetermined time. One possibility is that the sample onsets are chosen as a stratified random sample (*e.g.* with a fixed number of samples per hour, beginning at times chosen randomly within each hour); another is that they are scheduled at a regular time (say, once an hour, on the hour, for all daylight hours), or after a fixed time has elapsed since the termination of the previous sample. Alternatively, the sample sessions may be scheduled to begin, not at a particular time, but whenever a particular behavior occurs (*e.g.* whenever a particular pair of individuals interact, or whenever the animals enter a particular habitat). Finally, there may be no scheduling rule: observations may be made on an "*ad lib.*" basis. It is highly unlikely that the scheduling of such *ad lib.* samples will ever be independent of the behavior (see Section III). Furthermore, the nature and extent of the dependence is less likely to be known than in the behavior-dependent case just mentioned.

Scheduling session terminations.

A sample session may be scheduled to terminate after a fixed time period, after occurrence of a fixed number of behaviors, after a particular class of behaviors or interactions has terminated, or it may continue so long as the animals are in view. Alternatively, there may be no termination rule. Once again, such *ad lib.* termination produces a sample with unknown and perhaps variable dependence on the behavior being sampled. For example, ALTMANN & ALTMANN (1970) suggested that a mid-day peak in their data on baboon social behavior might be due to the fact that the observers were more likely to take a mid-day break if nothing of special interest was occurring. As a result, those mid-day periods during which observations were made probably were a biased sample of mid-day periods, and behavior that was recorded then probably was a biased sample of mid-day behavior.

Number of individuals per session.

If all occurrences of behaviors of interest are recorded for a particular individual during an entire sample period, that individual will be referred to as a *focal individual* for that sample period. Or, there may be a focal sub-group, which can range in size from one individual to the entire group. Alternatively, there may be no focal individuals, either because only partial frequencies are being recorded or because attention is focused on first one individual, then another, the choice usually determined on an *ad lib.* basis throughout the sample session. Any form of *Ad Lib.* Sampling introduces the problem of unknown and probably variable biases mentioned previously (See also Section III).

Selecting focal individuals.

The choices among potential focal individuals or among focal sub-groups can be randomized utilizing a table of random numbers *(e.g.* with individuals picked at random from all individuals in the group), it may be a stratified random sample *(e.g.* with a predetermined number of focal individuals randomly chosen within each of a fixed number of age-sex classes), it may be regular *(e.g.* rotating according to a fixed schedule through all of the individuals or all the individuals of a class), or irregular, with the individual chosen on the basis of some behavioral criterion, for example, the first pair to interact (see *e.g.* Section VII, Sequence Sampling) or the closest readily-visible individual. Such behavior-determined selection of individuals may decrease the amount of sample time in which no behaviors of interest occur, but it will also introduce dependence between the samples of behavior and of participants. The choice of selection criteria can best be determined by the demands of the particular research question.

III. *AD LIBITUM* SAMPLING

In field studies of behavior, perhaps the most common form of behavior record consists of what I shall call "typical field notes", or "*Ad Lib.* Sampling." Of course, the same type of record can be obtained in the laboratory by means of non-systematic sampling or informal observations. Such records are the result of unconscious sampling decisions, often with the observer recording "as much as he can" or whatever is most readily observed of the social behavior of a group in which behaviors, individuals and often the times for behavior sessions are chosen on an *ad libitum* basis. Presumably this is the technique used in most observational studies in which no mention is made of the sampling method.

Two kinds of assumptions often seem to be implicit in attempts to utilize data obtained from *Ad Lib.* Sampling for the purposes of quantitative anylysis of behavior. The frequency of two classes of behavior may be compared, as in: "Female A grooms more often that she threatens," or "Rhesus monkeys groom more often than they fight." The assumption made here is that the chance that a behavior would be recorded does not depend upon the class of behavior — that grooming is no more likely to attract attention than fighting, or at least that the magnitude of such a difference is negligible, relative to the actual differences in the frequencies with which the behaviors occur. Second, comparisons may be made across age-sex classes, as in: "Adult females groom more frequently than adult males." Here the assumption is that, for this class of behavior, the likelihood that a behavior would be observed and recorded does not depend on the age-sex class of the individual involved. Of course, a statement may involve simultaneously a comparison across behaviors and across age-sex classes, as in "Males fight more than females groom." Such conclusions involve simultaneously making both of the above assumptions about lack of bias, neither of which can be justified with this sampling technique.

These uses of *ad lib.* data are not, of course, the only ones. They have been pursued here as examples because they are common, particularly in first stages of quantification in field studies. The same line of argument can be used in making explicit the assumptions underlying any use of data. If data are to be used to answer a particular question, the observer should ask, What are the assumptions underlying such use and for which sampling techniques are such assumptions reasonable? Asking whether the assumptions are reasonable is just another way of looking at the question of whether alternative hypotheses have been ruled out. If, for example, *Ad Lib.* Sampling of behavior yielded a higher frequency of aggression by males than females, the results might be explained by the greater conspicuousness of male aggression, unless it is reasonable to assume that the sampling was independent of the sex of the actors.

With *Ad Lib.* Sampling, it is rarely possible to determine which differences in data are due to true differences between individuals, age-sex classes, or behaviors, and which due merely to biases in sampling. When comparing the results of one such study with those of another, we cannot tell which differences were due to differences in what could be seen, which to differences in what was selected for recording, and which to actual differences in the populations.

In any field study some data probably will consist of such records, which may be of considerable use as illustrative material and because of their heuristic value in searching for ideas and in planning systematic sampling of behavior. Often, too, rare but significant events are recorded during such nonsystematic sampling periods. But studies which consist only, or even primarily, of such records leave open too many alternative hypotheses that might account for the data. Without some form of systematic sampling procedure, there appears to be no way to avoid the bias that results when the observer's attention is attracted by certain types of behavior or certain classes of individuals.

If we could assume that the biases in *Ad Lib.* Sampling were of constant direction and magnitude over time, within and between studies, we might be able to compare such data despite these biases. Unfortunately, such an assumption will seldom be justifiable. This becomes particularly significant when one examines, as we shall in what follows, the unsuccessful attempts to correct for bias in such sampling in order to utilize the results.

Several authors (*e.g.*, ALDRICH-BLAKE, 1970; CHALMERS, 1968a; SADE, 1966) have suggested that a major bias in *Ad Lib.* Sampling of individuals results from the fact that some members of a group are more readily observed than others, and that this bias results directly from differences in the proportion of time that each individual is visible, rather than from individual differences in, say, size or activity level. CHALMERS (1968a) and SADE (1966) have attempted to provide a measure of such differential visibility.

In CHALMERS' (1968b) study of mangabeys in Uganda, he, too, attempted to measure differential visibility for various age-sex classes. He writes:

"Censuses were taken at half hourly intervals on the monkeys visible. These noted, among other things, the number of monkeys of each age/sex class in sight."

CHALMERS compared frequency of appearance in such "censuses" (probably Instantaneous or Scan Samples, see Section IX) with that expected on the basis of group composition (as determined by an independent sample) and found significant deviation. CHALMERS next calculated the frequencies of various social behaviors that would be expected if the "census" frequencies adequately represented the differential observability, if differential observability were the only source of bias in sampling individuals and if animals of any one age-sex class were as likely as those of any other to engage in such behavior. He compared the expected with the observed frequencies and suggested that deviation from the "census" distribution provides evidence that members of the age-sex classes differed in the frequency with which they engaged in such behavior.

In his studies of the rhesus monkeys of Cayo Santiago, SADE (1966) measured differential observability as follows:

238   JEANNE ALTMANN

"During each hour of observation a two minute period was chosen at random during which I noted down each individual member of Group F that I could find and identify."

These samples were taken by the observer as he walked through the group. SADE refers to such records as "time samples" (see footnote p. 252). He considers the results of such "time samples" to be a measure of the likelihood that a record will be obtained if an animal did some behavior of interest during observations of behavior, and that these "time sample" results can therefore be used to correct for individual sampling bias.

In the only published attempt to utilize such "time sample" data, HAUS-FATER (1971, 1972 and personal communication) proceeded as follows. For each animal, he took the number of "time samples" in which it was seen and divided that number into the smallest non-zero number of this kind in the population. This gives a putative correction factor: when the number of encounters in which the individual was observed to participate is multiplied by this factor, a new number is obtained that is presumed to be the individual's corrected number of encounters relative to other individuals (except those individuals that were never observed during time samples).

Both of these methods are attempts to deal with and correct biased sampling of individuals. These authors have singled out one possible component of such bias: the fact that some individuals are available for observation more of the time than are others.

The implicit assumption is that there exists a positive number, K (an "observability constant"), such that for each animal, i, K times $p(O_i)$, the probability that i will appear during an observability sample, is equal to the probability that a record will be made, during a behavior sample of i's participation in a behavior under study. It then follows that if n_i/N, the proportion of observability samples in which i appeared, is used as an estimate of $p(O_i)$ for each individual that appeared in at least one observability sample; then multiplying N/Kn_i times x_i, the observed number of participations by i during behavior sampling will yield an estimate of the true number of participations by i. (Multiplying all of these values by any non-zero constant will yield *relative* rather than true participations. HAUSFATER (1971, 1972) did this by multiplying them by $n_m K/N$ where n_m is the frequency of appearance in observability samples for the animal that appeared least often, but at least once.)

But what are the grounds for this basic assumption? In order to justify the use of observability samples, one would require either direct testing of this basic assumption or testing of other assumptions from which it could be derived. No such direct or indirect test has been published. Consider the following line of argument, which may represent the rationale that was used:

(i) The observability sample is assumed to provide an accurate measure of relative observability in behavior samples with which it is used, *i.e.,* the proportion of time an animal is visible during behavior sampling is assumed to be adequately estimated by the proportion of Observability Samples in which he appears. Procedural differences in these two types of sampling can result in a difference between the relative amount of time that an individual is visible in each and, therefore, in failure of this assumption. For example, such differences in observability will be present to the extent that the observer scans the group differently in the two kinds of samples.

(ii) Even if the observability samples provide an accurate measure of the proportion of time that each individual was visible during behavior samples, we still need to assume that there is a relation between such observability and the probability that an individual's behavior would be recorded. Observability samples provide an accurate correction only to the extent that the probability of a behavior being recorded if any individual performs that behavior is directly proportional to the percent of time that the individual is visible.

There are at least three sources of failure to obtain such a consistently proportional relationship in *Ad Lib.* Sampling. (a) Individual or class-specific differences in observability may vary greatly from one kind of behavior to another. For example, a subordinate male but not a dominant one might tend to mount in concealment, and thus would be obscured at such times, not just from other members of his group, but from the observer as well, despite the fact that both males may be visible to the observer the same percent of the day. (b) Some forms of behavior may affect the observability of any individual that participates in them. For example, extensive chases may bring both participants into view, thus making it irrelevant whether one of the participants is seen in, say, twice as many observability samples as is the other. (c) Any time that more is visible than can be recorded, sampling decisions remain and, in the absence of systematic sampling, the observer's preferences will come into play. Thus, if some individuals or behaviors (or a combination of the two) are rarely seen, the observer may unconsciously "compensate" by paying more attention and recording more when they are seen. This might, for example, lead to an overestimate of the frequency, or relative frequency, of a rarely seen form of behavior, particularly if "correction" factors were applied to the data.

Under any of these circumstances, application of a correction factor may

not merely fail to provide a complete correction: it may increase the existing
bias. The result would be a poorer estimate of what actually occurred than
the original data would have provided.

In summary, then, the application of observability correction involves a
number of assumptions. Justifying the use of these or similar corrections
would require evidence that these assumptions were reasonable under the
sampling and behavioral conditions, or at least, that the use of the observability
corrections would result in a reduction in bias over the amount in the *Ad
Lib.* Sampling alone. One obvious source of such evidence for the validity
of these corrections would be a comparison of their results with the results
of a sampling method that is unbiased with respect to individuals. But then,
the corrections will usually be superfluous. The more productive approach
is to look for sampling techniques that are unbiased with respect to the main
variables of interest. In what follows I shall consider the extent to which
other existing techniques enable the observer to avoid various biases.

IV. SOCIOMETRIC MATRIX COMPLETION

In some studies, *Ad Lib.* Sampling has been supplemented by making
additional observations on particular pairs of individuals. This has been
accomplished by spending more time with these individuals, or, *e.g.,* by
experimentally provoking a fight by means of competition in pairs of individuals for whom the original sample size was considered inadequate. The
results of such sampling are usually published in the form of a "sociometric
matrix," that is, a contingency table in which actors (*e.g.,* aggressors, or
winners) are represented by the rows, and recipients (*e.g.,* losers of fights)
by the columns, and in which the cell entries indicate the frequencies of the
corresponding (dyadic) interactions. In these studies the object has been
to establish, for each pair, the direction and degree of one-sidedness of some
relationship, such as groomer-groomee or winner-loser of fights (see *e.g.,*
ALEXANDER & BOWERS, 1969; SADE, 1966; MISSAKIAN, 1972, in which this
technique apparently was used). In such studies, then, a sample usually is
considered to be inadequate if the data in a pair of cells are small in number,
or large in number but nearly equal.

When data for such a sociometric matrix are obtained by this sampling
procedure, the result is not a contingency table in the usual sense, but simply
a compact way of tabulating data. No biological interpretation can be given
to the row or column totals. Each cell frequency reflects both the effects of
the animals' choices among partners in dyadic interactions and the effects of
attempts by the observer to boost the frequencies of certain cells. Consequently, one cannot directly compare each cell with every other cell: they

do not repersent the results of unbiased sampling of dyadic interactions. Another consequence is that the row totals are probably a biased sample of the distribution of acts by the members of the group; similarly, the column totals are probably a biased sample of acts received by the members of the group. Any row or column will contain such biases if it includes cells some of whose data result from the supplementary observations.

Certain kinds of questions cannot in general be answered by such data, *e.g.* for any two individuals A and B, "Does individual A do more grooming than individual B?," or even, "If A grooms, is he more likely to groom B than C?" The former would have to be answered through unbiased samples of grooming bouts (or, at least, of groomers) in the group as a whole, the latter, by at least an unbiased sample of A's grooming bouts.

However, if, for each pair of individuals, the observer can assume that the data represent an unbiased sample of their relations in their paired encounters (*e.g.* grooming sessions between A & B), that the outcome (A grooms B, or B grooms A) of any trial (grooming event) is independent of the outcome of any other, and that the probability of each outcome remains constant from trial to trial, then each cell of the matrix can be compared with the corresponding cell of the transposed matrix, treating the two cells as the components of a binomial distribution. For example, the frequency with which any individual, A, groomed any other individual, B, can be compared with the frequency with which B groomed A. We could also ask whether A is more likely to be the groomer in his grooming interactions with B than in those with C. We could look at the "linearity" (transivity) of grooming; *e.g.* if, for any three members of the group, A, B, and C, among whom A usually grooms B (rather than vice versa), and B usually grooms C, is it true that A usually grooms C more often than C grooms A? As a note of caution, we observe that the latter two questions will usually involve comparisons of binomial probability estimates with different sample sizes and hence different confidence intervals.

It would be preferable, then, to present such data in the form of a table that brings together the data for each pair of individuals, thereby facilitating binomial testing or other comparisons and avoiding the temptation to treat the data as if they constituted entries in a conventional contingency table or matrix.

In summary, the technique of Matrix Completion of *Ad Lib.* Samples is particularly suited to studies in which the basic problem of interest is the direction and degree of one-sidedness in the relations of each pair of individuals, but is ill-suited to answering many other types of questions about behavior. If the observer's main interest is in such asymmetry problems, if he feels that the binomial assumptions are satisfied, *and* if he feels he can obtain much larger sample sizes with this technique, then it might be the technique of choice. One alternative, Focal Animal Sampling (Section V), would provide relatively unbiased data both on degree of asymmetry and on many other aspects of behavior as well.

Perhaps the best solution would be to begin with Focal Animal Sampling (rather than *Ad Lib.* Sampling) and then to supplement these data as needed to insure an adequate sample for each pair. This supplementary sampling might consist of additional Focal Sampling of particular individuals. If doing so required that unacceptably large amounts of time be devoted to those individuals, then the observer might work on other aspects of the study until individuals of a pair in question moved near each other, at which time sampling on them would begin. The data from the Focal Animal Sampling would then be available for other kinds of analysis for which its relative lack of bias would be advantageous.

V. FOCAL-ANIMAL SAMPLING

I use the term Focal-Animal Sampling to refer to any sampling method in which (i) all occurrences of specified (inter)actions of an individual, or specified group of individuals, are recorded during each sample period, and (ii) a record is made of the length of each sample period and, for each focal individual, the amount of time during the sample that it is actually in view. Once chosen, a focal individual is followed to whatever extent possible during each of his sample periods.

This kind of sampling has been used in field situations, by BEER (1961-63) in studies of gulls, by WOOTTON (1972) in studies of stickleback fish, by STRUHSAKER in studies of elk (1967) and vervet monkeys (1971), by FISLER while observing Cayo Santiago rhesus (1967), by RICHARD (1970) in a study of howler and spider monkeys, by SAAYMAN (1972) in studies of chacma baboons, and by ALTMANN & ALTMANN (in preparation) in studies of social behavior of Amboseli baboons. The work of DOYLE *et al.* (1969) on lesser bushbabies, of PLOOG and his colleagues (PLOOG *et al.*, 1963; PLOOG, 1967; HOPF, 1972) on squirrel monkeys, of ROSENBLUM & KAUFMAN (1967) on pigtail and bonnet macaques and of SCRUTTON & HERBERT (1970) on talapoin monkeys, provide examples of its use in studies of caged colonies. Examples from research on human behavior include the work of SWAN (1938), WASHBURN (1932), Mapheus SMITH (1931, 1933), CHAPPLE and his colleagues (1963), BLURTON-JONES (1967, 1972), and McGREW (1972).

In a Focal-Animal Sample, the sampling of non-social behavior is relatively straightforward. I shall discuss below some of the problems of sampling social behavior. Most such behavior is directed ("addressed"); I shall distinguish between the actor or sender, and the object or receiver of each social act.

Under some conditions and at least for some behaviors, one may reasonably assume that a complete record is obtained not only of the focal animal's actions, but also of behaviors directed to him by others. Then a Focal-Animal Sample on animal i provides a record of all acts in which i is either the actor *or receiver*. This means that both during animals i's focal samples and animal j's focal samples we are recording all interactions between i and j: either sample or both together would provide the necessary data for estimating their rate of interaction (see Frequencies and Rates, p. 232, 233).

Under other circumstances, it may be possible to record all acts by the focal individual, but not all those directed toward him by others (*e.g.* silent threats). For those behaviors for which records are incomplete, it may still be reasonable to assume that the sample distribution of, say, senders is unbiased. However, to estimate the rate of interaction between the two animals, i and j, we would then need to use i's sample for those acts directed from i to j, and j's sample for the rate at which acts are received by i from j (see Frequencies and Rates, p. 232, 233).

Focal subgroups.

Although Focal-Animal Sampling as defined above does not exclude the possible use of a focal (sub)group of several animals, such sampling will usually be practicable only when it is possible to keep every member of the focal subgroup under continuous observation during the sample period. The reason for this is that the sample space in a focal sample consists of those dyadic interactions in which at least one participant is a member of the focal group. If only one of two focal animals were visible for, say, 5 minutes, that period can still be used in estimating the interaction rate between the two of them but if both were out of sight during some time, no interaction between the two would have been available for observation during that time. Consequently, focal group sampling requires that the concurrent observation time be known for every pair of focal individuals. (Similarly, such time records would be necessary for every triple of focal individuals if triadic interactions were also under study.) Under most circumstances, the only condition under which such a record can be obtained is that in which all the individuals in the sample group are continuously visible throughout the sample period.

Beyond the problem of time records, there is a further reason for having only one focal individual per sample period. One of the great strengths of Focal-Animal Sampling is that in order to stay with a focal individual, the observer follows him and obtains observations on him in situations in which he would not ordinarily be under observation. This advantage would be lost by having multiple focal animals: if one of three focal individuals moved out of sight, pursuing that one would usually mean losing the others.

Thus, if one is working with observational conditions that are less than perfect, Focal-Animal Sampling should be done on just one focal individual at a time, or at most a pair (*e.g.* mother and young infant). The following discussion of the Focal-Animal Sampling method will assume that there is only one focal individual in each sample; however, all of the principles that are described also apply *mutatis mutandis* to Focal-Animal Sampling on subgroups and on whole groups.

Time records.

For some research problems, time may not be an important variable (*e.g.* in a study of the response to a particular behavior, or of the order of behaviors in a sequence of interactions). For such a study one might want to use Focal-Animal Sampling in which no record is kept of the length of the sample period or of the time that the focal individual is in view. However, many questions about behavior are known to involve a time base (See Frequencies and Rates, p. 232, 233). For others, we may not know ahead of time whether time is a variable that can be ignored. Thus, in an exploratory study, an observer will do well to record time information, even if internal behavior-conditional aspects of the activities are of primary interest, or the session onset and termination rules are behavioral ones (see p. 234).

The simplest method for obtaining time records in Focal-Animal Sampling is to sample for a predetermined amount of time, keeping records of the amount of time during each sample session that the focal individual is visible and being sampled ("time in"), or else of the amount of time that it is out of sight ("time out"). A cumulative stopwatch is useful for this purpose. Alternatively, the end of the sample period can be determined by other stopping rules, such as after a predetermined amount of "time in". Or sampling might be terminated according to some behavior-dependent stopping rule, such as after the behavior under study has occurred a predetermined number of times. Such a technique could be used to guarantee an adequate sample size (see p. 234). However, the observer should bear in mind the need for a stopping rule that is independent of the behavior parameters that will be investigated (p. 234). Independence might be obtainable with a behavior-dependent stopping rule if few and explicit research questions have been formulated ahead of time and none will be added later. Otherwise, sampling for a predetermined amount of time will usually be the method of choice.

The choice of sample session length will depend upon several considerations. An upper limit on the length of the sample sessions will be set by observer fatigue. If sample sessions are too long, it becomes increasingly difficult to keep one's eyes and attention fixed on a single individual, and the accuracy of the records is affected. Of course, fatigue depends in part on one's familiarity with the species and its repertoire. Much greater mental effort is required to encode unfamiliar behavior. Fatigue will be affected by the number of behavior categories to be recorded, the rapidity and subtlety of these behaviors, and the amount of contextual and sequential information to be gathered.

If the durations of behavior are of interest, then the sessions should be long enough to obtain an adequate estimate of the distribution of durations. The differences between many common statistical distributions is revealed in the tails of the distributions (COCHRAN, 1954). Similarly, if sequential constraints are under investigation, the sample period should be long enough to include an adequate sample of the longest sequences that are of interest. If only frequencies are of interest, then the length of each session is theoretically immaterial. Of course, the total "time in" for all samples which are to be pooled must be long enough to provide adequate estimates for the least frequently occurring behavior under study.

Scheduling focal individuals.

Depending on the nature of the research problem, there are various possibilities for the focal individuals that will be covered in a study: all members of a group, all members of certain age-sex classes or some other subset (say, females with neonates), some members of particular subsets, and so on. Assumptions about variability within and between individuals and classes will affect scheduling decisions — whether, for example, one samples five males for twenty samples each, twenty males for five samples each, or even one hundred males for one sample each. Likewise, if the observer does not wish to assume that diurnal variability in behavior is negligible, focal individuals might be sampled at the same time of day or each one at several periods (say, once an hour) during the day. Otherwise, daily sample periods could be assigned to individuals without regard to this variable. Thus, the assignment of individuals to sample periods and the scheduling of sample periods will depend on both the questions being asked and the assumptions that one is willing to make (see also p. 235).

In field research, it is not always possible to recognize individuals. Under

246   JEANNE ALTMANN

those conditions, it is not feasible to make an unbiased selection of focal individual, to take individual variability into account, or to study such variability. Nevertheless, a kind of Focal-Animal Sampling can be carried out, choosing at random among visible individuals, utilizing a table of random numbers, then continuing to sample that individual so long as it is possible to keep track of him. Such a procedure might be preferable to resorting to other, even more biased sampling methods.

Behavior record.

When using Focal-Animal Sampling in studies of the social behavior of primates or other highly social animals, so much data may be obtained that one pushes the limit of the observer's ability to process information. I have already discussed the problem of observer fatigue. In addition, we have found, with the baboons and macaques that we have observed, that it is not possible simultaneously to record all social behaviors, their durations, the sender and receiver, the distance relations of the participants, their neighbor relationships, and the temporal pattern of the behaviors even for one focal individual per sample. Data on temporal patterning is particularly difficult to obtain: in one study (ALTMANN & ALTMANN, in prep.) we found that even with two observers, one 15-minute sample per hour was near the upper limit of our capacity when obtaining an accurate record, with some 5 dozen social behavior categories, of who did what to whom and in what order, as well as keeping track of most nonsocial behavior, durations, and time-out periods. In a more recent study in which we utilized 40-minute Focal-Animal Samples, we were able to collect extensive data on neighbor relationships and on social interactions (including the behavior, and the age-sex class of the social partner) of wild baboons, but without obtaining complete sequential records or much information about the durations of behaviors. In that study we took a ten-minute rest after such a sample and then reversed the roles of primary and secondary observers before taking a second 40-minute sample.

Montgomery SLATKIN (personal communication) has been able to obtain data on the durations of behavioral states in baboons, as part of his field study of baboon time budgets. He utilized the classification of the activity states of the individual into 5 exclusive and exhaustive categories. Then, during Focal-Animal Sampling, he recorded all transition times and the behavior state following each transition. Such records required two observers. The primary observer kept his eyes on the focal individual, punched a stopwatch at every transition, and dictated the behavior. The assistant drove the vehicle and recorded the transition times and activities. (A stopwatch with an extra "marker" hand was invaluable for this study.)

In another study, Thomas STRUHSAKER was able to obtain data on the duration of behavioral states in focal mother-neonate pairs of vervet monkeys (STRUHSAKER, 1971).

In summary, with appropriate choice of focal individuals, sample periods and behavior records, Focal-Animal Sampling will usually be the technique of choice. It can provide relatively unbiased data relevant to a wide variety of questions about spontaneous social behavior in groups. Since observation is usually made on one animal per session, to the exclusion at those times of detailed information about other (inter)actions in the group, this technique is least suited to answering questions about behavioral synchrony. For most such questions, the methods discussed in Sections VI and IX would be more appropriate, but for studying behavioral synchrony among neighbors, Focal-Animal Sampling might be the method of choice.

VI. SAMPLING ALL OCCURRENCES OF SOME BEHAVIORS

Under some conditions, it is possible to record all occurrences of certain classes of behaviors in all members of the group during each observation period. Such samples have been obtained by ROWELL (1967, 1968) in her studies of a caged social group of baboons, by CRAIG et al. (1969) in studies of agonistic behavior among birds in a field situation, and by LINDBERG (1971), who obtained data on the frequency of all agonistic vocalizations for 20 hours, in a field study of rhesus monkeys. Such records are generally possible only if (i) observational conditions are excellent, (ii) the behaviors are sufficiently "attention-attracting" that all cases will be observed, and (iii) the behavioral events never occur too frequently to record. For example, in our studies of baboons in Kenya, (ALTMANN & ALTMANN, in prep.) we were able to keep such frequency records for agonistic encounters that involved a vocalization and for sexual mounting between adults. Even for these two categories, we did not always obtain complete records of the individuals that were involved and certainly not of the complete sequences of behavior, since our notes on many of the occurrences began with the actual agonistic vocalization or when one individual was seen mounted on another.

For behaviors that can be sampled in this way, what kind of information can such sampling provide?

(i) With a wise choice of sample periods, it can provide accurate information about the rate of occurrence (and temporal changes in the rate) of such behavior in the group as a whole. If all participants can be identified at each occurrence of the behavior under study, this sampling technique is equivalent to Focal-Animal Sampling (Section V) on the whole group with respect to this particular behavior, and provides data of the kind discussed in that section. When not all identifications are possible, the data that do include identification will be an unbiased sample of the distributions of those behaviors among individuals (or classes) if there were no differential identifiability, *i.e.* if identifiability were random with respect to individuals (or independent of their age-sex classes). By the same token, they will be an unbiased sample of the outcomes of dyadic interactions, and could therefore be used to answer questions, of the sort described in Section IV, about the degree of one-sidedness of relations.

(ii) This is not the technique of choice for many kinds of sequential analysis. However, if the sequential information that is desired is the sequelae of some behavior that can be sampled in this way, *e.g.* the response to vocal threats, one could start each sample with an observation on such behavior, then record what happens next.

(iii) This sampling technique is appropriate for studies of behavioral synchrony if the observational and recording conditions are such that occurrences of the behavior can be recorded even if they are simultaneous. Actual time of occurrence, rather than just the frequency within an interval or the number of simultaneous occurrences, would, of course, provide more fine-grained information as to the temporal distribution of the behavior. This information may be of interest in itself, or it might be needed in order to test certain assumptions that are made when using other sampling techniques.

VII. SEQUENCE SAMPLING

In Sequence Sampling, the focus of observation is an interaction sequence, rather than any particular individual(s). A sample period begins when an interaction begins. During the sample, all behaviors under study are recorded, in order of occurrence. The sample continues until the interaction sequence terminates or is interrupted, and the next sample begins with the onset of another sequence of interactions.

Sequence Sampling has been used in studies of social behavior in crabs (Hazlett & Bossert, 1965) and monkeys (S. Altmann, 1965). In both studies, sequential dependencies in communicative interactions were of primary interest. Hazlett & Bossert (1965: 359) write:

> "To carry out the observations reported here, groups of 25 to 100 individuals were placed in an observation aquarium, and after 15 minutes, observations were started. Whenever the movements of one or two of the crabs were such that it appeared they would subsequently come

into social contact (HAZLETT, in press), recording was started. When one crab deviated from its path before the animals came into social contact, *i.e.* there was no observable interaction between the two crabs, the recording was stopped and discarded in later calculations. The movements and displays of the two interacting individuals were recorded until one or both crabs moved away from one another, either by some form of retreat, climbing over the other crab or by moving past one another. If the interacting pair was interrupted by a third individual, recording was stopped and the results discarded in later calculations."

Presumably, the first dyadic interaction to occur after observations began was the one that was recorded first, the next interaction that began after the first one terminated was recorded second, and so forth. Note that the Sequence Sampling of Hazlett and Bossert is essentially Focal-Animal Sampling (in which both members of the pair may be considered focal individuals) that differs from conventional Focal-Animal Sampling in that behavior-contingent rules were used for starting and stopping a sample. However, because of the effects of these rules on the records, the results will not be equivalent to Focal-Animal Sampling with, say time-contingent starting and stopping rules.

ALTMANN (1965) "tried to sample at random from among the monkeys," although "no systematic randomizing technique was used." If a selected individual was not interacting, ALTMANN chose another individual, continuing until an interacting individual was located. The sequence of interactions was then recorded, continuing "until the interaction process terminated, or until it was no longer possible for me to see everything that was going on." In that study, an interaction process was not judged to have terminated solely because the initial individual (or any other individual) left the interaction group, and thus this is not Focal-Animal Sampling, as we use that term (p. 242). For example, a play group might persist for some time, despite the fact that various individuals entered or left the play group during that period. The sequence of interactions within such an interacting group was recorded so long as it was, in some sense, unbroken.

Initially it seemed to me that this method was biased toward events that occur in sequences which, as a class, take up proportionately more time. Consequently, it would be biased toward acts, individuals, or sequential constraints that are different in such sequences. Thus, if two individuals spend the same amount of time interacting, but one is involved only in long interaction sequences and the other only in ones whose durations are half as long (so that there are twice as many of the latter) it appeared that the observer would be just as likely to choose an interaction of the one as of the other, but he would then spend twice as long with (and record more acts of) the one involved in longer sequences. Similarly, if some behavior is more (or less) likely to occur in those longer sequences than in the short ones, it seemed that this sampling method would not provide good estimates of relative frequencies of behavior patterns. In this example, such sampling would therefore be biased toward the characteristics of events in long sequences.

Crucial in this line of argument are the probabilities of choosing sequences of various lengths. If the choice of the first (or any other) sequence to be sampled is made at random from among ongoing sequences, then the probabilities will in fact be equal to the relative amount of time taken up by the set of all sequences of that length (duration), as indicated in the example above. If, however, the observer always begins sampling at the onset of a sequence and chooses the next sequence to sample at random among sequence onsets or in any other way that samples sequences of each length in proportion to their *frequency of occurrence,* the resulting data will be unbiased with respect to sequence length: the total time spent with sequences of, say, duration d_i, will be proportional to d_i times f_i, where f_i is the frequency of sequences of length d_i. Then the time spent with sequences of different lengths, not the probability of choosing such sequences, will be in proportion to the total time taken up by sequences of that length.

Done this way, Sequence Sampling would still present several problems. It requires a method for choosing sequences that satisfies the above-mentioned criterion and a way of identifying the beginning and end of each sequence. If the observer always chooses the next available sequence onset, or one that occurs in the vicinity of the last one, the sampled sequences may not be independent of each other. Yet, it is not obvious how one might pick sequences at random. In addition, getting a record of a sequence from its beginning places heavy reliance on the ability of the observer to anticipate those circumstances under which interactions are likely to occur.

ALTMANN's method in particular illustrates both advantages and disadvantages of Sequence Sampling. The sampling procedure that was used enables the observer to stay with and record social interactions, the persistence of which does not require the continued participation of any one individual. For example, monkey A aggresses against monkey B, who redirects the aggression to C, who then enlists the collaboration of D. The method thereby provides information about the sequential structure of social interactions that is not provided by Focal-Animal Sampling or any other sampling procedure described here. Another advantage is that large samples of social interactions can be obtained by this method: because the observer takes the next available interaction in the group, his time is seldom spent without available data.

However, specifying criteria for identifying the beginning and end of a sequence may be difficult. The sequence definition that S. ALTMANN used depends on the fact that the behavior of one individual may influence the behavior of another, which may, in turn, affect the next reaction, and so forth. The resulting chains of influence are Altmann's interaction sequences, and it is these chains that he followed in Sequence Sampling. Regarded in this way, interactions may have two properties that present sampling problems, branching and converging. If a sequence branches (*e.g.* if a play-group divides in two without a break in the interaction), which branch should the observer follow? Or, if two sequences should join into one (*e.g.* if an individual goes from one interacting group to another without a break) how can the conjoint influences be sampled? (HAZLETT and BOSSERT avoided both these problems by restricting their sampling to sequences involving interactions between just two individuals.)

In summary, the primary advantage of Sequence Sampling is that it enables the observer to obtain large samples of social behavior and to sample sequences of interaction that may persist regardless of the continued participation of any one individual. The primary disadvantages stem from problems in selecting sequences and identifying their beginning and end.

VIII. ONE-ZERO SAMPLING

A. BACKGROUND

During the 1920's a systematic sampling method was developed for studying spontaneous behavior in children, (OLSON, 1929), and was referred to as "time sampling." GOODENOUGH, one of the earliest advocates of this sampling method, defined "time sampling" as:

> "... the observation of everyday behavior of an individual or a group of individuals for definite short periods of time and the recording of the occurrence or non-occurrence of certain specified and objectively defined forms of behavior during each of these periods." (GOODENOUGH, 1928, p. 23.)

The common features of the technique are the following. (i) In each sample period, occurrence or non-occurrence (rather than frequency) is scored. (ii) Interactions of just a single individual or pair of individuals are recorded in each sample period. (iii) *Occurrence,* for most users of this technique, has meant "in process" at any time during the sample period, *i.e.* a sampling of states rather than events (see p. 231). (iv). The sample periods are usually short (*e.g.* 15 secs.), with about 20 sample periods in succession. Such batches of samples may then be repeated, perhaps twice a day over the period of the study [4]).

Observational studies of children in social groups were relatively common for several years, particularly during the 1930's (WRIGHT, 1960). In such research, "time sampling" (*i.e.* One-Zero Sampling) predominated. However, only one of these early studies (OLSON, 1929) utilized One-Zero Sampling for animal investigations.

During this early period of development, a number of workers investigated methodological questions. Those who criticized "time sampling," as well as proponents of the technique, focused on secondary questions, such as the appropriate length for the sample periods, adequate sample sizes, changes in the state of the system over time, observer agreement, choice of behavior categories, and so forth (cf. ARRINGTON, 1943; M. SMITH, 1931; OLSON & CUNNINGHAM, 1934). At no point, however, was the basic rationale for One-Zero Scores questioned.

In later years, observational studies became relatively less common in child behavior research (WRIGHT, 1960). Among studies that continued to utilize observational techniques, increased emphasis was placed on rating scales and on controlled, one- or two-person settings. Observers tended to use interpretive behavior categories, such as *seeks attention*, rather than relatively non-interpretive motor patterns, such as *hits*. A few workers turned to other observational sampling techniques (see e.g. Sections VII and IX), while others (*e.g.* BISHOP, 1951) continued to use One-Zero Sampling.

Observational studies, of both human and non-human behavior, have become increasingly popular in recent years. In the study of animal behavior, One-Zero Sampling has been rediscovered and widely used, *e.g.* in a field study by KUMMER (1965, 1968), in a study of caged cats by COLE & SHAFER

---

4) The terms "time samples" and "time sampling" have been variously used by different writers. SADE (1966) refers to his observability samples as *time samples*. In studies of human behavior, *time sampling* has been used by some to refer only to One-Zero Scoring, as described by GOODENOUGH, and by others to mean almost any sampling in which a fixed time unit of observation is maintained (see *e.g.* OLSEN & CUNNINGHAM, 1934; CONNOLLY & SMITH, 1972). In this paper I use the terms "time sample" and "time sampling" only when, for clarity, it seems advantageous to use the same term as a particular author in discussing that author's work — in which case these terms are used in quotation marks. I refer to SADE's samples simply as SADE's *observability samples*, to those techniques defined by GOODENOUGH as *One-Zero Sampling*, and I have labeled other sampling techniques utilizing a time base according to their distinctive features (*e.g.* Focal-Animal Sampling, Section V).

(1966), in studies of caged groups of monkeys by Lindburg (1969), Menzel (1963), Bernstein & Draper (1964), Bernstein (1968), and Rhine & Kronenwetter (1972), by Hinde and his students (*e.g.* Hinde, 1964, 1967) and by Hansen (1966) and others at the University of Wisconsin (*e.g.* Mitchell, 1968a, 1968b; Seay, 1966; Suomi, *et al.*, 1971). This last group of investigators refer to the method as a "Hansen system". Recent examples in child behavior research include the work of Hutt (1966), and Richards & Bernal (1972).

It should be noted here that some workers (*e.g.* Arrington, 1943; Kummer, 1968) indicate that at some times (or for some behaviors) they actually recorded all occurrences; but their data was tabulated, compared, and presented as One-Zero Scores. The discussion that follows refers to One-Zero Scores, regardless of whether they result from the method of recording or of tabulating. I consider the technique in detail because of its widespread use in observational studies of humans and of caged animals, because of indications that it is now beginning to be used by a number of field primatologists, and because of my serious reservations about its value in most situations.

B. INTERPRETATION OF SCORES

Authors that use One-Zero Scores usually state, for each individual or specified class of individuals, the number of sample intervals and the number (or percentage) of intervals that included at least one occurrence of the behavior(s) in question. These scores may be combined or averaged over several sample sessions or for several individuals. In some cases, the differences between scores are tested by means of non-parametric tests, such as the Mann Whitney U Test (see *e.g.* Richards & Bernal, 1972), or scores for different behaviors subjected to correlational analysis (see *e.g.* Mitchell, 1968b).

Frequency of communicative acts and time spent in various states are two common variables measured in behavioral research, and are assumed to be important to the animals. Most users of One-Zero Scoring, implicitly or explicitly, seem to have assumed that these scores provide good measures of one or both of these variables. Examples of such use of these scores can be found in Bishop (1951), Hinde & Spencer-Booth (1964, 1967), in which the scores are treated as behavior frequencies, and in Hinde & Spencer-Booth (1964, 1967), and Kummer (1968), in which they are treated as representing percent of time spent in an activity. It is too easy for both author and reader to forget that a One-Zero Score is not the frequency of *behavior* but is the frequency of *intervals* that included any amount of time spent in that behavior. Such lapses occur in SMITH and CONNOLLY's recent review of "time sampling" as well as in the papers just cited. Nor is the percentage of intervals the same as the percentage of time spent in an activity. In what follows, we shall examine the relationship between these scores and the frequency, duration, and proportion of time spent in specified activities.

Of the authors cited in this section, all who are explicit about their scoring method indicate that they scored states. From what I can infer, most others did so, too. However, it is conceivable that some observers would score events; in seminars, when I have discussed One-Zero Scoring of states, several people have suggested One-Zero event recording as a way of removing the defects of this scoring system. For that reason, I also discuss here the case in which the recording is of events.

State scores.

As I noted previously, most of the studies that have utilized One-Zero Scores have scored presence or absence of states, rather than of events. That is, an act that began in one interval, continued through a second, and terminated in a third would result in three scores, one for each of these intervals. But three occurrences, each with onset and termination in the same interval, would yield only one score for all three of them. Thus, there is in general no direct relationship between such scores and the true frequencies. However, SMITH & CONNOLLY (1972) suggest that under special conditions, One-Zero Sampling will provide data on frequency and durations of behaviors. They write:

> "If the time sample period is much less than the behaviour duration (bout length) then the distinction between frequency and all-or-none recording vanishes. Use of sequential samples gives information on both number of occurrences and durations."

What is assumed in such use is a one-to-one correspondence between any onset of behavior and the corresponding record for two consecutive intervals, in which the first, taken just before the onset, contains no score (zero) and the second contains a score (one). Likewise, it is assumed that a score (one) in one interval followed by the absence of one (zero) in the next interval bears a one-to-one correspondence to the termination of behavior. Another way of looking at it is that one assumes that only one onset or termination (but not both) of a behavior being scored can occur in one interval. The probability of more than one onset in an interval must be negligible, as must the probability of both an onset and a termination (either of the same occurrence of a behavior, or the termination of one occurrence and the onset of the next). This not only requires that the sample intervals be much shorter than the "usual" behavior duration, as SMITH and CONNOLLY suggest, but much shorter than the "usual" intervals between behaviors as well. How much shorter for any level of probability to be considered "negligible" will depend on the distributions of behavior onsets, of durations, and of the length of intervals between behaviors. Obtaining adequate information on these distributions would require extensive sampling by an unbiased method (such as those discussed in Sections V and VI). If adequate samples were then available, the One-Zero Samples would usually be superfluous.

Do these One-Zero Scores provide a good measure of percent of time spent in a given behavior? The precent of intervals containing a One-Zero Score is used by some researchers as a measure of the percent of time spent in a behavior. This would be correct only if the behavior in question took up all of the time in each interval in which it was scored, and none of the time in the others.

The percent of intervals including a score will be an upper bound on the percent of time spent. How close the true value is to the upper bound will depend on how much of a "scored" interval is in fact taken up on the average by the behavior. Clearly, the shorter the intervals, relative to the behavior durations, the closer this upper bound will be to the time spent. However, durations are likely to vary between individuals, over time for the same individuals, and from one behavior to another — the very classes which are usually being used for comparisons.

Thus differences in two such scores cannot be attributed to differences in the proportion of time spent in an activity (or to differences in frequencies of occurrence) unless it is known that the scores provide consistent measures of such. In the absence of that knowledge, providing an interpretation for the scores remains a problem.

A few workers recognize that these scores do not represent either frequencies or time spent. MITCHELL (1968b) addresses himself to this point:

> "It is emphasized here that a Hansen frequency is not a true frequency of occurrence. When it is stated that the Hansen system was utilized to measure visual orients of a mother toward her infant it is not meant that each and every glance at the infant was recorded. Only one visual orient was recorded whether the mother looked at her infant once during a fifteen second interval or several times during that interval. Since there are 60 fifteen second intervals in a fifteen minute test session, an upper limit has been imposed on the number of times a behavior can

occur. This procedure allows the experimenters to observe several behaviors at a time without sacrificing observer reliability, but the numbers which result reflect a little of both the duration and the frequency of a behavioral act, not just frequency alone."

Thus, the resultant scores may be greater than, equal to, or less than, the true frequencies. As MITCHELL has observed above, the numbers reflect a little of both the duration and the frequency of behavior. However, that seems to be a weakness of such scores, rather than an advantage: they do not provide accurate information about either.

Event scores.

Would the interpretation become any more sound if events rather than states were scored? First, could we determine or estimate the percent of time spent in a behavioral state? Even total frequency records during the sample intervals would not enable us to do so unless we also had information about the distribution of durations of the behavioral events in question. Therefore, One-Zero Scores could be used to estimate time spent only if such information about durations was used in combination with a technique for estimating the true frequencies from the One-Zero Scores.

Consider, then, how One-Zero Scores are related to frequency. If events rather than states were sampled — that is, if an occurrence was recorded only if the defining event for that behavior occurred in that interval — then the "score", which is the number of sample periods with such a record of occurrence, tells us that at least that many events occurred in that session. That is, we would have a lower limit on the number of occurrences. But surely, unless we know that this lower limit is close to (or bears a known fixed relationship to) the true frequency of occurrence, such a lower limit tells us little. The relationship between this lower limit and the true frequency may vary from individual to individual or over time for one individual.

Is there any other way that we can utilize these scores to get at frequencies? ALTMANN & WAGNER (1970) suggest that we look at the problem as one of estimation of the rate of occurrence of the behaviors. The assumption again is that *only events* have been scored. They write as follows:

"Suppose that the temporal distribution of the behavior can be described by a Poisson process; we will return later to what this implies. If so, then the probability p_0 of no occurrence of the behavior in an interval of length T is given by

$$p_0 = \frac{(\lambda t)^0}{0!}\, e^{-\lambda t} = e^{-\lambda t}, \tag{1}$$

where λ is the mean rate of occurrence of events, and e is the base of natural logarithms. From eq. (1) we have $\log_e p_0 = -\lambda t$, and thus $\lambda = (-\log_e p_0)/t$. The maximum likelihood estimate of p_0 is obtained from the number n of intervals in which the behavior did not occur divided by N, the total number of intervals. Thus, λ can be estimated as follows:

$$\lambda = -\frac{\log_e (n/N)}{t} \qquad (2)"$$

ALTMANN and WAGNER point out that:

"Use of the Poisson distribution implies that the behavior occurs randomly at a constant rate, that the chance of two or more simultaneous occurrences of the behavior is negligible, and that the chance that a particular behavior will occur during an interval is independent of the time that has elapsed since the last occurrence of that behavior."

However, even if one feels confident that the data could be approximated by a Poisson distribution, the estimates obtained from such scores would usually not be as good as those obtained from the true frequency distribution: too many data have been discarded (see FIENBERG, 1972).

I have recently learned that this approach is used in estimation of density within bacterial samples and insect populations in which the spatial frequency distribution is assumed to be Poisson. Its use in that context apparently was first suggested by FISHER (1935).

If it is not reasonable to assume that the behavior has a Poisson distribution, and if the actual frequency distribution is not known and one cannot reasonably guess at those properties that could be used to relate the One-Zero Scores to the true frequencies, then there is no basis for a frequency interpretation of a One-Zero Score of events, and *a fortiori* no basis for using them to estimate percent of time spent in an activity.

C. EASE OF SCORING AND OBSERVER AGREEMENT

One-Zero Scoring has been advocated on two other grounds: that such scoring is easier to do, and that greater observer agreement results. As for the first, the observer should ask himself whether the effort saved is worth the information lost. Even under field conditions, or in studies of human social groups, it may be possible to obtain complete frequency and duration records, instead of One-Zero Scores. CHAPPLE *et al.* (1963) did so in studies of patients on a psychiatric ward, and STRUHSAKER (1971) was able to do so for a study of infant vervet monkeys in their natural habitat in Africa, while JENSEN and his colleagues (*e.g.* JENSEN *et al.* 1967) did so for caged mother-infant pairs of monkeys, even though several laboratory studies of monkey infant behavior and mother-infant interactions have resorted to One-Zero Scores. If ease of use is the deciding factor, other techniques often would be preferable, such as Instantaneous and Scan Sampling (Section IX) which is easier to do than One-Zero Sampling.

As for the second claim, it is true that two independent observers, watching the same behavior, may get One-Zero Scores that closely agree; but as I have indicated, such scores will not, in general, be an accurate indication of either the frequency or the proportion of time spent on a behavior. In fact, it seems the poorer that One-Zero Scores are as measures of these variables, the greater will be the observer agreement. Thus, if three maternal glances occurred in an interval, two observers will agree on a check for that interval as long as both saw at least one glance. But if exact frequencies were being recorded, they would agree only if both saw all three glances. By the same token, if a behavior took up, say, half an interval, the One-Zero recorders would agree as long as each thought it took up some part of the interval. Greater agreement does not guarantee more information.

In short, neither ease of use nor observer agreement *per se* provide an adequate justification for the use of this technique. Despite this, and despite the absence in such scores of reliable information about frequency and time spent, an observer might maintain that such scores are good predictors of other phenomena; this would have to be demonstrated in each case. For those who consider frequency and duration of behaviors and percent of time spent in various states as variables of interest, alternative sampling methods should be considered (see sections V and IX).

IX. INSTANTANEOUS AND SCAN SAMPLING

Instantaneous Sampling is a technique in which the observer records an individual's current activity at preselected moments in time (*e.g.* every minute on the minute throughout the day). It is a sample of states, not events.

Such sampling has been used to study the percent of time spent in various activities by caged golden hamsters (M. P. M. RICHARDS, 1966), squirrels (C. C. SMITH, 1968), adult male baboons (SLATKIN, unpublished), and humans (BINDRA & BLOND, 1958; SMITH & CONNOLLY, 1972, and presumably CONNOLLY & SMITH, 1972).

Instantaneous Sampling can be used to obtain data from a large number of group members, by observing each in turn. Moreover, if the behavior of all visible group (or subgroup) members are sampled within a very short time period the record approaches a simultaneous sample on all individuals.

We shall refer to such Instantaneous Sampling on groups as Scan Sampling. If such sampling is done frequently, data are obtained on the time distribution of behavioral states in the whole social group. In particular, data are obtained on behavioral synchrony in the group. Such data are almost impossible to obtain by most other sampling techniques (cf. Sections V and VI).

Such Scan Sampling has been used by COHEN (1971b) in his studies on subgroups of children, and by COHEN (1971a) and S. ALTMANN (unpublished) in studies of subgroups in yellow baboons and gelada monkeys, respectively, by CHALMERS (1968a, b) in his attempt to estimate differential visibility of monkeys (see Section III of this monograph), by CHALMERS (1968a) and by S. & J. ALTMANN (1970) to sample diurnal variation in activities and synchrony of activities in mangabey and baboon groups, respectively.

Censuses are essentially Scan Samples in which one obtains data on population parameters (such as age-sex distribution) as well as total group size. In censuses one is not usually concerned with approximating an instantaneous sample because the change from one age-state, for example, to another is quite long relative to the time necessary to complete the census.

A. PRACTICAL CONSIDERATIONS

In an ideal Instantaneous Sample, each individual's state would be instantly noted. If, in addition, the state of the entire group is of interest (as in studies of subgroups or of synchrony), then ideally the state of every individual in the group would be noted at the same moment in time: the scan should be instantaneous. In practice, however, the observation, classification, and recording of a state takes time, and so does scanning from one individual to the next. The observer should try to scan each individual for the same brief period of time, for otherwise, a scan sample is equivalent to a series of short Focal-Animal samples of variable and unknown durations. In order to keep sampling time brief, the categories that are recorded should be easily and quickly distinguished. For this reason, it is in general more suited to studies of non-social behavior (with all social activities lumped into one state, as SLATKIN did in his study), or to situations in which social behaviors can be lumped into a few easily distinguished categories.

Of the reports cited above in which Scan Sampling was utilized, none indicate the time spent per individual, or whether an attempt was made to keep the times brief and even. Several indicate the amount of time per complete scan. COHEN (personal communication) took 5-10 seconds for his scans of subgroups in nursery school children, and about 45 seconds for a group of savannah baboons. S. ALTMANN (personal communication) took three to seven minutes to scan and record subgroups involving up to 331 individuals in herds of gelada monkeys that ranged up to 425 individuals. S. & J. ALTMANN (unpublished) took 45-60 seconds for scans of a group of baboons (about 40 individuals) in which we noted behavior, rather than subgroup sizes; 60 seconds were required when we obtained both kinds of data (behavior and subgroup affiliations) for a group of about 38 baboons with about 30 individuals visible for each sample.

B. ESTIMATING PERCENT OF TIME

A primary use of Instantaneous Sampling is in studies of the amount or percent of time that individuals devote to various activities. The percent of time is estimated from the percent of samples in which a given activity (state) was recorded. As I noted in the summary of One-Zero Sampling (p. 258), Instantaneous Sampling is at least as easy as One-Zero Sampling and, unlike One-Zero Sampling, readily provides data appropriate to estimating percent of time spent in various activities. In most of the studies cited above, Instantaneous Sampling was used to obtain such estimates. SMITH & CONNOLLY (1972, pp. 70-71) explicitly chose Instantaneous Sampling over One-Zero Sampling for this purpose.

SLATKIN was interested in baboon time budgets and considered it crucial for his purposes that he stay with an individual throughout a day. While Focal-Animal Sampling would have been the ideal method, he found it impossible to record behavioral and activity transition time data throughout an entire day. With the aid of an assistant, he did such Focal-Animal Sampling for selected half-hour periods (see p. 228); during the rest of the day he did Instantaneous Sampling at one-minute intervals.

C. ESTIMATING RATES AND RELATIVE FREQUENCIES

Instantaneous Samples are discrete samples of states, *i.e.*, of ongoing behaviors. They are not samples of events, or transition times between states. It is true that under some sets of *ad hoc* assumptions about the distributions of the transition times, or of the durations of the states, it is possible to use Instantaneous Sample data to estimate transition rates, but without such assumptions such data by themselves provide no information whatever about rates of events or transitions — except of course that the number of consecutive samples exhibiting differing states does give a crude lower bound for the number of transitions. In the special case where the interval between Instantaneous Samples is short enough that no more than one transition can occur between consecutive samples, the resulting data are essentially equivalent to that of Focal-Animal Sampling for rate and relative frequency estimates, but have a greater margin of error for duration estimates. (See the corresponding discussion for One-Zero Sampling, p. 254). However, utilizing such sufficiently short intervals will usually be no easier than Focal-Animal Sampling, while providing less information. If events, rates or relative frequencies are of primary importance in a study, the sampling method of choice would be one in which transition times, or other events, are recorded (see, *e.g.*, Focal-Animal Sampling, Section V). The necessary data are then provided directly.

SUMMARY

Seven major types of sampling for observational studies of social behavior have been found in the literature. These methods differ considerably in their suitability for providing unbiased data of various kinds. Below is a summary of the major recommended uses of each technique:

Sampling Method	State or Event Sampling	Recommended Uses
1. *Ad Libitum* (p. 235)	either	Primarily of heuristic value; suggestive; records of rare but significant events.
2. Sociometric Matrix Completion (p. 240)	event	Asymmetry within dyads.
3. Focal-Animal (p. 242)	either	Sequential constraints; percent of time; rates; durations; nearest neighbor relationships.
4. All Occurrences of Some Behaviors (p. 247)	usually event	Synchrony; rates.
5. Sequence (p. 248)	either	Sequential constraints.
6. One-Zero (p. 251)	usually state	None.
7. Instantaneous and Scan (p. 258)	state	Percent of time; synchrony; subgroups.

In this paper, I have tried to point out the major strengths and weaknesses of each sampling method.

Some methods are intrinsically biased with respect to many variables, others to fewer. In choosing a sampling method the main question is whether the procedure results in a biased sample of the variables under study. A method can produce a biased sample directly, as a result of intrinsic bias with respect to a study variable, or secondarily due to some degree of dependence (correlation) between the study variable and a directly-biased variable.

In order to choose a sampling technique, the observer needs to consider carefully the characteristics of behavior and social interactions that are relevant to the study population and the research questions at hand. In most studies one will not have adequate empirical knowledge of the dependencies between relevant variables. Under the circumstances,

262   JEANNE ALTMANN

the observer should avoid intrinsic biases to whatever extent possible, in particular those that direcly affect the variables under study.

Finally, it will often be possible to use more than one sampling method in a study. Such samples can be taken successively or, under favorable conditions, even concurrently. For example, we have found it possible to take Instantaneous Samples of the identities and distances of nearest neighbors of a focal individual at five or ten minute intervals during Focal-Animal (behavior) Samples on that individual. Often during Focal-Animal Sampling one can also record All Occurrences of Some Behaviors, for the whole social group, for categories of conspicuous behavior, such as predation, intergroup contact, drinking, and so on. The extent to which concurrent multiple sampling is feasible will depend very much on the behavior categories and rate of occurrence, the observational conditions, *etc*. Where feasible, such multiple sampling can greatly aid in the efficient use of research time.

BIBLIOGRAPHY

ALDRICH-BLAKE, F. P. G. (1970). Problems of social structure in forest monkeys. — In: J. H. CROOK (ed.), Social Behaviour in Birds and Mammals. New York, Academic Press.

ALEXANDER, B. K. & BOWERS, J. M. (1969). Social organization of a troop of Japanese monkeys in a two-acre enclosure. — Folia primat. 10, p. 230-242.

ALTMANN, S. A. (1965). Sociobiology of rhesus monkeys. II: Stochastics of social communication. — J. theoret. Biol. 8, p. 490-522.

—— & ALTMANN, J. (1970). Baboon Ecology. African Field Research. — Bibl. Primatol., No. 12. Basel, S. Karger and Chicago, Univ. Chicago Press.

—— & —— (In prep.) Baboon social behavior.

—— & WAGNER, S. S. (1970). Estimating rates of behavior from Hansen frequencies. — Primates 11, p. 181-183.

ARRINGTON, R. E. (1943). Time sampling in studies of social behavior: a critical review of techniques and results with research suggestions. — Psychol. Bull. 40, p. 81-124.

BARKER, R. G. (1963). The stream of behavior as an empirical problem. — In: R. G. BARKER (ed.), The Stream of Behavior. New York, Appleton-Century-Crofts, p. 1-22.

BERNSTEIN, I. S. (1968). The lutong of Kuala Selangor. — Behaviour 32, p. 1-16.

—— & DRAPER, W. A. (1964). The behaviour of juvenile rhesus monkeys in groups. — Anim. Behav. 12, p. 84-91.

BEER, C. G. (1961). Incubation and nest building behaviour of black-headed gulls. I: Incubation behaviour in the incubation period. — Behaviour 18, p. 62-106.

—— (1962). Incubation and nest-building behaviour of black-headed gulls. II: Incubation behaviour in the laying period. — Behaviour 19, p. 283-304.

—— (1963). Incubation and nest-building behaviour of black-headed gulls. III: The pre-laying period. — Behaviour 21, p. 13-77.

—— (1963). Incubation and nest-building behaviour of black-headed gulls. IV: Nest-building in the laying and incubation periods. — Behaviour 21, p. 155-176.

BINDRA, D. & BLOND, J. (1958). A time-sample method for measuring general activity and its components. — Can. J. Psychol. 12, p. 74-76.

BISHOP, B. M. (1951). Mother-child interaction and the social behavior of children. — Psychol. Monogr.: General and Applied, H. S. CONRAD (ed.), 65, p. 1-34.

BLURTON-JONES, N. G. (1967). An ethological study of some aspects of social behaviour of children in nursery school. — In: D. MORRIS (ed.), Primate Ethology. Chicago, Aldine, p. 347-368.

—— (1972). Categories of child-child interaction. — In: BLURTON-JONES, N. (ed.), Ethological Studies of Child Behaviour. Cambridge, Cambridge University Press, p. 97-127.

CHALMERS, N. R. (1968a). Group composition, ecology and daily activities of free living mangabeys in Uganda. — Folia primat. 8, p. 247-262.

—— (1968b). The social behaviour of free living mangabeys in Uganda. — Folia primat. 8, p. 263-281.

CHAPPLE, E. G., CHAMBERLAIN, A., ESSER, A. & HANDKLINE, N. S. (1963). The measurement of activity patterns of schizophrenic patients. — J. nerv. ment. Dis. 137, p. 258-267.

COCHRAN, W. G. (1954). Some methods for strengthening the common χ^2 tests. Biometrics 10, p. 417-451.

COHEN, J. E. (1971a). Social grouping and troop size in yellow baboons. — Proc. 3rd Int. Congr. Primat. Vol. 3. Basel, S. Karger, p. 58-64.

—— (1971b). Casual Groups of Monkeys and Men; Stochastic Models of Elemental Social Systems. Cambridge, Harvard Univ. Press.

COLE, D. D. & SHAFER, J. N. (1966). A study of social dominance in cats. — Behaviour 28, p. 39-53.

CONNOLLY, K. & SMITH, P. K. (1972). Reactions of pre-school children to a strange observer. — In: BLURTON-JONES, N. (ed.), Ethological Studies of Child Behaviour. London, Cambridge Univ. Press, Ch. 6.

CRAIG, J. V., BISWAS, D. K. & GUHL, A. M. (1969). Agonistic behavior influenced by strangeness, crowding and heredity in female domestic fowl (*Gallus gallus*). — Anim. Behav. 17, p. 498-506.

DOYLE, G. A., ANDERSSON, A. & BEARDER, S. K. (1969). Maternal behaviour in the lesser bushbaby (*Galago senegalensis moholi*) under semi-natural conditions. — Folia primat. 11, p. 215-238.

FIENBERG, S. E. (1972). On the use of Hansen frequencies for estimating rates of behavior. — Primates, 13, p. 323-326.

FISHER, R. A. (1935). The Design of Experiments. — London, Oliver & Boyd.

FISLER. G. F. (1967). Nonbreeding activities of three adult males in a band of free-ranging rhesus monkeys. — J. Mammal. 48, p. 70-78.

GOODENOUGH, F. L. (1928). Measuring behavior traits by means of repeated short samples. — J. juv. Res. 12. p. 230-235.

HALL, K. R. L. (1965). Experiment and quantification in the study of baboon behavior in its natural habitat. — In: VAGTBORG, H. (ed.), The Baboon in Medical Research, p. 29-42.

HANSEN, E. W. (1966). The development of maternal and infant behavior in the rhesus monkey. — Behaviour, 27, p. 107-149.

HAUSFATER, G. (1971). Seasonal changes in participation in the intergroup encounters of free-ranging rhesus monkeys (*Macaca mulatta*). — Paper delivered at 40th annual meeting of the American Association of Physical Anthropologists, Boston.

—— (1972). Intergroup behaviour of free-ranging rhesus monkeys (*Macaca mulatta*). — Folia primat. 18, p. 78-107.

HAZLETT, B. A. & BOSSERT, W. H. (1965). A statistical analysis of the aggressive communications systems of some hermit crabs. — Anim. Behav., 13, p. 357-373.

HINDE, R. A., ROWELL, T. E. & SPENCER-BOOTH, Y. (1964). Behaviour of socially living rhesus monkeys in their first six months. — J. Zool. 143, p. 609-649.

—— & SPENCER-BOOTH, Y. (1968). The study of mother-infant interaction in captive group-living rhesus monkeys. — Proc. Roy. Soc., London, Ser. B, 169, p. 177-201.

HOPF, S. (1972). Study of spontaneous behavior in squirrel monkey groups: Observation techniques, recording devices, numerical evaluation and reliability tests. — Folia primat. 17, p. 363-388.

HUTT, C. (1966). Exploration and play in children. — Symp. Zool. Soc. London 18, p. 61-81.

264   JEANNE ALTMANN

JENSEN, G. D., BOBBITT, R. A. & GORDON, B. N. (1967). The development of mutual
    independence in mother-infant pigtailed monkeys, *Macaca nemestrina*. — In: S. A.
    ALTMANN (ed.), Social Communication Among Primates, p. 43-53.
KUMMER, H. (1968). Social Organization of Hamadryas Baboons. A field study. Chicago,
    University of Chicago Press and Basel, S. Karger, Bibl. Primatol. No. 6.
—— & KURT, F. (1965). A comparison of social behavior in captive and wild
    hamadryas baboons. — In: VAGTBORG (ed.), The Baboon in Medical Research.
    Austin, Univ. Texas Press, p. 65-80.
—— (1971). Primate Societies. Group Techniques of Ecological Adaptation. Chicago,
    Aldine-Atherton.
LINDBURG, D. G. (1969). Behavior of infant rhesus monkeys with thalidomide-induced
    malformations: A pilot study. — Psychon. Sci. 15, p. 55-56.
—— (1971). The rhesus monkey in North India: an ecological and behavioral study. —
    In: ROSENBLUM, L. A. (ed.), Primate Behavior. Developments in Field and
    Laboratory Research, 2, p. 1-106.
MASON, W. A. & EPPLE, G. (1969). Social organization in experimental groups of
    Saimiri and *Callicebus*. — In: C. R. CARPENTER, (ed.), Proc. 2nd Int. Congr.
    Primat., 1 (Behavior), Basel, S. Karger, p. 59-65.
McGREW, W. C. (1972). Aspects of social development in nursery school children with
    emphasis on introduction to the group. — In: N. BLURTON-JONES (ed.), Ethological
    Studies of Child Behaviour. London, Cambridge Univ. Press, Ch. 5.
MENZEL, E. W., Jr. (1963). The effects of cumulative experience on responses to novel
    objects in young isolation-reared chimpanzees. — Behaviour 21, p. 1-12.
—— (1969). Naturalistic and experimental approaches to primate behavior. — In:
    E. P. WILLEMS & H. RAUSH (eds.), Naturalistic Viewpoints in Psychological
    Research. New York, Holt, Rinehart & Winston, p. 78-121.
MISSAKIAN, E. A. (1972). Geneological and cross-genealogical dominance relations in a
    large group of free-ranging rhesus monkeys (*Macaca mulatta*) on Cayo Santiago.
    — Primates 13, p. 169-180.
MITCHELL, G. D. (1968a). Attachment differences in male and female infant monkeys. —
    Child Develop. 39, p. 611-620.
—— (1968b). Intercorrelations of maternal and infant behaviors in *Macaca mulatta*. —
    Primates 9, p. 85-92.
OLSON, W. C. (1929). The Measurement of Nervous Habits in Normal Children. —
    Minneapolis, Univ. Minnesota Press.
—— & CUNNINGHAM, E. M. (1934). Time-sampling techniques. — Child Dev. 5,
    p. 41-58.
PLOOG, D. W. (1967). The behavior of squirrel monkeys (*Saimiri sciureus*) as revealed
    by sociometry, bioacoustics, and brain stimulation. — In: S. A. ALTMANN (ed.),
    Social Communication Among Primates. Chicago, Univ. Chicago Press, p. 149-184.
——, BLITZ, J. & PLOOG, F. (1963). Studies on social and sexual behavior of the
    squirrel monkey (*Saimiri sciureus*). — Folia primat. 1, p. 29-66.
RHINE, R. J. & KRONENWETTER, C. (1972). Interaction patterns of two newly formed
    groups of stumptail macaques (*Macaca arctoides*). — Primates, 13, p. 19-33.
RICHARD, A. (1970). A comparative study of the activity patterns and behavior of
    Alouatta villosa and *Ateles geoffroyi*. — Folia primat. 12, p. 241-263.
RICHARDS, M. P. M. (1966). Activity measured by running wheels and observation
    during the oestrous cycle, pregnancy and pseudo-pregnancy in the golden hamster. —
    Anim. Behav. 14, p. 450-458.
—— & BERNAL, J. F. (1972). An observational study of mother-infant interaction. —
    In: N. BLURTON-JONES (ed.), Ethological Studies of Child Behaviour. London,
    Cambridge Univ. Press. Ch. 7.
ROSENBLUM, L. A. & KAUFMAN, I. C. (1967). Laboratory observations of early

OBSERVATIONAL SAMPLING   265

mother-infant relations in pigtail and bonnet macaques. — In: S. A. ALTMANN (ed.), Social Communication Among Primates. Chicago, Univ. Chicago Press.

ROWELL, T. E. (1967). A quantitative comparison of the behaviour of a wild and a caged baboon group. — Anim. Behav. 15, p. 499-509.

—— (1968). Grooming by adult baboons in relation to reproductive cycles. — Anim. Behav. 16, p. 585-588.

SAAYMAN, G. S. (1972). Effects of ovarian hormones upon the sexual skin and mounting behaviour in the free-ranging chacma baboon (*Papio ursinus*). — Folia primat. 17, p. 297-303.

SADE, D. S. (1966). Ontogeny of social relations in a group of free-ranging rhesus monkeys (*Macaca mulatta* Zimmerman). — Dissertation, Univ. Calif., Berkeley.

SCHNEIRLA, T. C. (1950). The relationship between observation and experimentation in the field study of behavior. — Ann. N. Y. Acad. Sci. 51, p. 1022-1044.

SCRUTON, D. M. & HERBERT, J. (1970). The menstrual cycle and its effect on behaviour in the talapoin monkey (*Miopithecus talapoin*). — J. Zool. 162, p. 419-436.

SEAY, B. (1966). Maternal behavior in primiparous and multiparous rhesus monkeys. — Folia primat. 4, p. 146-168.

SINGH, S. D. & MANOCHA, S. N. (1966). Reactions of the rhesus monkey and the langur in novel situations. — Primates 7, p. 259-262.

SLATKIN, M. (in preparation). Activity patterns in two baboon species (*Theropithecus gelada* and *Papio cynocephalus*).

SMITH, C. C. (1968). The adaptive nature of social organization in the genus of tree squirrels *Tamiasciurus*. — Ecol. Monogrs 38, p. 31-63.

SMITH, Mapheus. (1931). A study of the unsupervised behavior of a group of institutional children. — Nashville, Marshall & Bruce Co.

—— (1933). A method of analyzing the interaction of children. — J. juv. Res., 17, p. 78-88.

SMITH, P. K. & CONNOLLY, K. (1972). Patterns of play and social interaction in preschool children. — In: N. BLURTON-JONES (ed.), Ethological Studies of Child Behaviour. London, Cambridge Univ. Press. Ch. 3.

STRUHSAKER, T. T. (1967). Behavior of elk (*Cervus canadensis*) during the rut. — Zs. f. Tierpsychol. 24, p. 80-114.

—— (1971). Social behaviour of mother and infant vervet monkeys (*Cercopithecus aethiops*). — Anim. Behav. 19, p. 233-250.

SUOMI, S. J., HARLOW, H. F. & KIMBALL, S. D. (1971). Behavioral effects of prolonged partial social isolation in the rhesus monkey. — Psychol. Rep. 29, p. 1171-1177.

SWAN, C. (1938). Individual differences in the facial expressive behavior of pre-school children: a study by the time-sampling method. — Genet. Psychol. Monogr. 20, p. 557-650.

WASHBURN, R. W. (1932). A scheme for grading the reactions of children in a new social situation. — J. genet. Psychol. 40, p. 84-99.

WEBB, E. J., CAMPBELL, D. T., SCHWARTZ, R. D. and SECHREST, L. (1966). Unobtrusive Measures: Nonreactive Research in the Social Sciences. — Chicago, Rand McNally.

WOOTTON, R. J. (1972). The behaviour of the male three-spined stickleback in a natural situation: A quantitative description. — Behaviour 41, p. 232-241.

WRIGHT, H. F. (1960). Observational child study. — In: P. H. MUSSEN (ed.), Handbook of Research Methods in Child Development. New York, J. Wiley & Sons, p. 71-139.

ZUSAMMENFASSUNG

In der Literatur finden sich hauptsächlich sieben Methoden, um Beobachtungen vom Sozialverhalten zu erheben. Sie leisten objektiv Unterschiedliches und sind daher nicht für jeden Untersuchungsweck gleich geeignet. Die Tabelle faßt diese Unterschiede zusammen:

266   JEANNE ALTMANN

Erhebungsmethode	Aufzeichnung: Verlaufsdauer oder -wechsel	Anwendungsbereich
1. Je nach Gelegenheit (S. 235)	beides	Hauptwert liegt in erster Bestandsaufnahme und im Auffinden von Hinweisen für die Anlage einer mehr systematischen Untersuchung; Entdeckung sehr seltener, aber theoretisch wichtiger Verhaltensweisen.
2. Soziometrische Matrixausfüllung (S. 240)	-wechsel	Asymmetrie der Wechselbeziehungen von Zweiergruppen
3. Konzentration der Beobachtung auf jeweils ein bestimmtes Tier (Fokus-Tier) (S. 242)	beides	Verlaufsregeln, % Zeit je Verhaltensweise; Dauer des Einzelablaufs; Häufigkeit; Verhalten zum nächsten Gruppenmitglied
4. Aufzeichnung jedes Auftretens einer bestimmten Verhaltensweise in der Gruppe (S. 247)	meist: -wechsel	Synchronisierung; Häufigkeit
5. Aufeinanderfolge von Verhaltensweisen (S. 248)	beides	Feststellung von Verlaufsordnungen und -regeln
6. Auftreten oder Nichtauftreten von Verhaltensweisen in festgelegten Beobachtungsintervallen (S. 251)	meist: -dauer	nutzlos
7. Gesamtfeststellung aller gleichzeitig ablaufenden Aktivitäten (S. 258)	-dauer	Synchronisierung; Zeitaufwand je Verhaltensweise; Untergruppen

In dieser Zusammenstellung habe ich versucht, die hauptsächlichen Stärken und Schwächen jeder Erhebungsmethode darzulegen. Ob sich in einer gegebenen Forschungssituation eine Fehlerquelle erheblich oder nur unwesentlich auswirkt, hängt von dieser Situation und nicht von der gewählten Methode ab.

Allerdings sind manche Methoden hinsichtlich einer größeren Anzahl von Variablen mehr fehlerbelastet als andere. Wenn man eine bestimmte Erhebungsmethode auswählt, ist daher die erste Frage, ob die Methode gegenüber jenen Variablen zu fehlerhaften Ergebnissen führen könne, die man zu untersuchen wünscht, oder ob ein solcher Fehlereinfluß indirekt eintreten kann, weil eine der untersuchten Variablen in Korrelation mit einer anderen steht, die ihrerseits von der Methode nur unsicher erfaßt wird.

Bei der Auswahl der Erhebungsmethode muß sich der Untersucher sorgfältig überlegen, welche besonderen Verhaltenseigentümlichkeiten und gegenseitigen Beziehungen der Gruppenmitglieder sowohl für die untersuchte Population wie die besondere Fragestellung von Bedeutung sind. In den meisten Fällen wird man empirisch nicht genügend über die wechselseitige Abhängigkeit der zu untersuchenden Variablen wissen. Gerade dann soll sich der Untersucher bemühen, immanente Fehlerquellen möglichst auszuschließen, ganz besonders solche, welche die zu untersuchenden Variablen unmittelbar betreffen.

OBSERVATIONAL SAMPLING   267

Schließlich läßt sich in einer Untersuchung oft mehr als nur eine Methode anwenden, nacheinander wie auch — unter günstigen Bedingungen — gleichzeitig. Z.B. konnten wir gleichzeitig (Methode 7) in Abständen von 5-10 Minuten Identität und Individualdistanzen der sich in der Nähe aufhaltenden Gruppenmitglieder feststellen, während wir ständig ein bestimmtes Tier im Auge hielten (Methode 3). Ebenso kann man oft, während man sich auf ein Tier konzentriert (Methode 3), zugleich jedes Auftreten bestimmter, besonders auffallender Verhaltensweisen (Beutefang, Trinken, Verhalten zu fremden Gruppen usw.) innerhalb der ganzen Gruppe mitaufnehmen (Mehode 4). Das Ausmaß, bis zu dem man mehrere Methoden nebeneinander verwenden kann, hängt natürlich von den zu beobachtenden Verhaltensweisen, ihrer Häufigkeit, den allgemeinen Beobachtungsbedingungen und ähnlichem ab. Die gleichzeitige Anwendung verschiedener Methoden kann — soweit dies möglich ist — dazu beitragen, die für die Untersuchung verfügbare Zeit wesentlich besser als sonst auszunutzen.

Anim. Behav., 15, 1967, 499–509

A QUANTITATIVE COMPARISON OF THE BEHAVIOUR OF A WILD AND A CAGED BABOON GROUP

By T. E. ROWELL

Zoology Department, Makerere University College, Kampala, Uganda

It is much easier to study the social behaviour of monkeys in captive groups than of those in the wild, but we need a measure of the similarity of behaviour in cages and in the wild to assess the general validity of cage studies. As more field studies become available, the possibility arises that the social organization of a species may vary in different habitats in its natural range (Crook & Gartlan, 1966; Gartlan & Brain, 1967; Hall, 1963; Rowell, 1967). Compare also Hall & DeVore (1965) with Rowell (1966c), and Jay (1965) with Sugiyama (1964). Cage environments can be regarded as an extreme of the range of environments in which a species can survive and breed. In this sense the problem of the differences in behaviour caused by captivity and that of variability of behaviour in relation to ecology are only aspects of the same problem, which is essentially the question of meaning, if there is one at all, of the word 'normal' when applied to primate social behaviour.

Hierarchical organization, one particular aspect of social behaviour, has been accepted as 'normal' in studies on caged primates in general, and has been given varying amounts of attention in field studies. In baboon species hierarchies have been described in caged groups by Kummer (1956) and Rowell (1966b). In field studies DeVore (1965) and Hall & DeVore (1965) emphasized hierarchies among adult males, though Hall (1962) did not stress this aspect of male behaviour to the same extent. None of these authors nor Kummer & Kurt (1963) pay much attention to possible hierarchies among other sections of the populations studied in the field. As the cage hierarchies included all adults and even older juveniles, an interesting difference between wild adults and wild and caged groups is suggested which it is relatively easy to observe and measure. Discussion of the following observations is therefore centred on the occurrence of hierarchies in the two populations. Direct comparison of inter-male behaviour was not possible, but particular attention has been paid to the behaviour of wild adult males since they have previously been especially considered in this context.

The same units of social behaviour were observed in both wild and caged populations, and no behaviour patterns were seen in one and not the other. Comparison must therefore be in terms of relative frequencies of behaviour patterns and their distribution among age and sex classes of interacting baboons.

Method

Animals

The wild population lived in and near the Ishasha riverine forest in Uganda. Their ecology has been described elsewhere (Rowell, 1966a). Most of the social behaviour discussed here was seen in 'S' and 'V' troops whose composition at the end of the study is shown in Table I. Social behaviour was recorded during about 300 hr of 'good' observation—that is with the observer near enough and the vegetation open enough for a satisfactory record of social interactions to be made (out of about 700 hr of observations during 3 years). It was rare that a whole troop could be seen at once; an estimated average of ten or twelve animals was usually visible. It is assumed that sightings were of a random selection of individuals from the population, so that any one age/sex class was available for interaction in the proportion in which they were present in the population. Method, timing, and conditions under which the field observations were made are described in the ecological study (Rowell, 1966a).

The caged group included ten animals past weaning age, as well as some babies (Table I). Interactions with an eleventh animal, an aged male in the group for the first third of the study, have been ignored here.). Three hundred hours of observation were made on this group over 3 years, excluding the time in which introduction of new animals was taking place. The social organization of this group has been described elsewhere (Rowell, 1966b). It was characterized by an obvious and clear-cut hierarchical structure. The same data are used here as in the previous paper where details of recording methods are given.

500   A N I M A L   B E H A V I O U R ,   1 5 ,   4

Table I. Composition of the Wild and Caged Populations at the End of the Studies

	'S' troop	'V' troop	Total wild	Cage
Adult males	5	17	22	1
Adult females	5	16	21	5
Juveniles:				
male	11	2		2
female	4	2		2
unknown	1	8		0
Total juveniles	16	12	28	4
Total	26	45	71	10
+ infants <1 year old (excluded from study)	6	13	19	4

It will be seen that the data from the two populations are roughly comparable in terms of hours of baboon-observation (about 3000 in each case). They differ of course in innumerable ways, the most important of which are probably size and composition of the groups, and differences in approach to observation necessitated by the differences between the two environments. To attempt an evaluation of all the differences would not be useful with this limited material, although such an assessment might be made when more quantitative comparative studies become available. Theoretical aspects of this problem have been discussed elsewhere (Rowell, 1967).

Behaviour Patterns Selected

The majority of interactions involved only two animals. Where there were more than two, however, as when A and B chase C, the interaction was broken up into two-animal units, and counted as 'A chases C' and 'B chases C'. Interactions were classified into two types: 'approach-retreat', which included the basic pattern of one animal approaching and the other avoiding, and 'friendly', where one animal did not attempt to avoid the other, as in grooming, presenting, or most mating behaviour. This distinction is considered fundamental for reasons discussed elsewhere (Rowell, 1966b). An interaction may of course include a variable number of behaviour patterns on the part of each animal. Seven behaviour patterns which are frequent components of the first type of inter-

action and thirteen of the second type, discussed here are listed in Tables V and VI. This is not a complete list of observed behaviour. All noises have been excluded because their recording was not considered reliable enough. Uncommon patterns (less than ninety records from the caged group) have been lumped (as in the composite 'making aggressive contact') or are not discussed.

Limitations of Comparison

(a) Interactions with nursing babies (up to 12 or 15 months) are not included because they are nearly always complicated by simultaneous interaction with the mother.

(b) Adult females are not differentiated with regard to reproductive status; some changes in behaviour with their reproductive cycles have already been discussed by the author (Rowell, 1966c). In both populations females gave birth at intervals of just over a year; so all females were usually pregnant or in the acyclic lactation interval.

(c) 'Juveniles' are not differentiated by age or sex, first because this was not always known in the field observations and second in order to keep the whole comparison reasonably simple. Where a juvenile interaction was typical of one class of juveniles, this is noted in the text.

Thus three classes of animals were recognized, males, females and juveniles, and hence there were six classes of interaction. Where a distinction is drawn between actor and reactor there were nine classes.

Results

Total Interactions

In roughly the same amount of time, nearly four times as many interactions were seen in the caged group as in the wild. This difference *might* be partly attributed to better conditions of observation in the cage, but this was certainly not the only source of difference. Foraging (or exploring) occupies more time in the wild, and so does moving from place to place. To one used to caged monkeys, wild baboons do seem to spend a great deal of time sitting by themselves and doing nothing. This is of course limiting the sense of the word interaction to mean overt communication or contact. In a wider sense, group cohesion in an open space implies continual interaction between all members of the group.

Distribution of Interactions Between Classes

The wild population was of course much larger than the caged, but the main difference in composition was in the proportion of adult males (Table I). In the wild there were more or less equal numbers of males, females and juveniles, but in the cage there was only a single adult male. Thus 8 per cent of interactions seen in the wild were between adult males, for which there was no cage equivalent. Correspondingly, the proportion of interactions which included the adult male (with a female or juvenile, was also much smaller—28 per cent in the cage, 46 per cent in the wild. The difference in distribution of interactions between classes can be seen in Table II. The difference is unlikely to have arisen by chance ($P<0.01$).

It we assume, for the moment, that the behaviour of the wild animals represents a 'norm' for baboon social behaviour, how did the caged group resolve the problem of its distorted demographic structure? As can be seen from Table III, the differences in each class are not predictable from each other nor from the differences in composition. Thus the roles of the

Table II. Total Interactions Seen, and Their Distribution Between Classes of Baboons

Population	Total	Male—male	Female—female	Male—female	Male—juvenile	Female—juvenile	Juvenile—juvenile
Wild	1,901 (100%)	158 (8%)	180 (10%)	615 (32%)	265 (14%)	439 (23%)	244 (13%)
Cage	7,344 (100%)	—	2,188 (30%)	1,442 (20%)	569 (8%)	2,567 (35%)	578 (8%)

The probability of the proportions of interactions in each class seen in the two populations being two observations from the same distribution is less than 0·01 (χ^2)

Table III. Interactions of the Three Classes of Animals, and Their Distribution Among Classes of Partners in the Two Populations

	Adult males*		Adult females†		Juveniles‡	
	Wild	caged	wild	caged	wild	caged
Adult males	158 (15%)	—	615 (50%)	1,442 (23%)	265 (28%)	569 (15%)
Adult females	615 (59%)	1,442 (72%)	180 (15%)	2,188 (35%)	439 (46%)	2,567 (69%)
Juveniles	265 (26%)	569 (28%)	439 (36%)	2,567 (42%)	244 (26%)	578 (16%)
Totals	1,038 (100%)	2,011 (100%)	1,234 (100%)	6,197 (100%)	948 (100%)	3,714 (100%)

*There is no significant difference between wild and caged males in the proportions of interactions with females and juveniles.
†Wild and caged females distributed their interactions amongst classes differently (χ^2 gives $P<0.01$).
‡Wild and caged juveniles distributed their interactions differently (χ^2 gives $P<0.01$).

demographic classes have a degree of flexibility.

The interaction of adult males with females and juveniles were in similar proportions in the two populations. The caged male did not apparently use the large juvenile male as an adult male substitute; in fact he increased his interaction with females slightly more than with juveniles. It would seem that adult male interactions are in some degree not replaceable by interaction with other classes.

Females and juveniles in the cages interacted less with the adult male than their wild counterparts. They made up most of the difference by increasing their interactions with females; the high proportion of female-female interactions in the cage group is the most striking difference between the populations.

Relatively few juvenile-juvenile interactions were recorded in the cage. This is probably an artefact of the small group size and the way the data were selected. In the wild juveniles tended to play with others of about the same size; in the cage four animals were too few to make a satisfactory play school and the small babies were usually also involved; as has been mentioned, interactions including babies were excluded in this analysis.

The Proportion of Approach-retreat Interactions

In the wild population a quarter of all interactions were of the approach-retreat pattern. In the cage the proportion was one-third (Table IV) and was significantly greater.

This difference was not distributed evenly through all classes of interaction. The greatest difference is again in the female to female class, caged females being much more aggressive towards each other. The males and juveniles were also more aggressive towards females in the cage, and juveniles towards each other. On the other hand, both males and females were slightly *less* aggressive towards juveniles in the cage, while the proportion of juvenile-to-male interactions which were of the approach-retreat pattern happened to be identical in the two populations.

It will be noted that only in interactions between wild adult males and between males and juveniles of both populations was the approach-retreat pattern more common than 'friendly' interactions.

Behaviour Patterns

We have seen that the frequency of interaction was much greater in the cage than in the wild, that interactions were distributed differently among interaction classes, and that, overall, the proportion of approach-retreat interactions was higher in the cage. We also saw earlier that the same range of communicative behaviour patterns was used in both environments. The classification into 'friendly' and 'approach-retreat', although basic, is of course extremely crude, and the quality of social organization in the two populations might differ greatly because of differences in the proportion of different behaviour patterns used in the two groups of interactions: if, for example, approach-retreat interactions were mainly non-agonistic avoiding responses in one population, but mainly aggressive contact in the other, or if 'friendly' behaviour were mainly appeasement gestures in one population, but mainly play in the other.

Many behaviour patterns occurred mainly in one interaction class (e.g. mounting in the male-to-female class, embracing in the female-to-female class) so that differences already noticed in relative frequency of interactions in the classes would obviously affect the total frequency with which each behaviour pattern was seen. Each interaction class was therefore considered separately.

Comparison is made between the proportion of interactions which included a behaviour pattern in the two populations. These proportions are of course independent of each other since each interaction may contain any number of behaviour patterns. Approach-retreat and friendly interactions are considered separately. In the 'paired' classes (e.g. male-to-female and female-to-male) the proportion used is of the total interactions between the two types of baboon in both directions. For example, the forty-six approach-retreat interactions between males and females in the wild included three instances in which a female chased a male; prop. $= \frac{3}{46} = 0.07$. A test of difference in proportions was used, entering z tables, and differences whose probability of arising by chance were less than 0.01 were assumed to be real.

Behaviour patterns in approach-retreat interactions (Table V). The most striking feature is the similarity in the structure of approach-retreat interactions in the two populations as measured by the frequency of their component behaviour patterns: out of fifty-six comparisons, only fifteen differed. The same behaviour patterns were used in the study of the social structure of the caged group (Rowell, 1966b), and

Table IV. The Proportion of Approach-retreat to Friendly Type Interactions in the Two Populations

	Total		♂♂		♀♀		♂/♀		♀/♂		♂/J		J/♂		♀/J		J/♀		J/J	
	W	C	W	C	W	C	W	C	W	C	W	C	W	C	W	C	W	C	W	C
N	1,901	7,344	158		180	2,188	299	850	317	592	109	311	156	258	163	901	276	1,666	244	578
prop.	0·25	0·33	0·63		0·23	0·46	0·14	0·33	0·01	0·04	0·74	0·64	0·13	0·13	0·3	0·27	0·1	0·27	0·48	0·35
W. prop. −C. prop	−0·07*		—		−0·23*		−0·19*		−0·03		0·1		0		0·03		−0·17*		0·13*	

Where the difference in the proportions of approach-retreat interactions in the two populations (W. prop. − C. prop.) is unlikely to have occurred by chance ($P < 0.01$), it is marked with an asterisk (*)

W=Wild; C=Caged; N=Number of interactions of each class; prop.=Proportion of approach-retreat type interaction.

Table V. Behaviour Patterns in Approach-retreat Interactions of Each Class in the Two Populations

	♂♂	♀♀		♂/♀		♀/♂		♂/J		J/♂		♀/J		J/♀		JJ		Total instances		W. prop. −C. prop $P<0.01$
	W	W	C	W	C	W	C	W	C	W	C	W	C	W	C	W	C	W	C	
Grin	0·18	0·12	0·03	0·02	0·01	0·04	0·09	0·01	0·004	0·1	*0·1	0·01	0·1	0·03	0·03	0·03	*0·33	45	245	2
Avoid	0·46	0·37	0·44	0·1	0·01	0·69	*0·37	0·02	0·004	0·21	*0·35	0·21	0·2	0·04	0·06	0·07	0·05	149	819	2
Supplant	0·22	0·32	0·21	0·11	*0·28	0·02	0·01	0·09	0·17	0·1	0·004	0·1	0·03	0·06	0·07	0·14	0·06	80	420	1
Chase	0·34	0·17	0·2	0·11	0·17	0·07	0·01	0·42	*0·14	0·28	*0·06	0·27	0·17	0·1	*0·28	0·73	*0·63	224	747	4
Aggressive contact	0·05	0	0·05	0·11	0·06	0·15	*0·01	0·03	0·07	0·13	0·03	0·13	0·11	0·1	*0·28	0·19	*0·41	62	489	3
Punish	0	0·07	0·03	0·1	0·09	0	0·003	0·08	0·06	0·04	0	0·04	0·03	0·01	0·01	0·03	0·02	23	103	0
Facial threat	0·2	0·2	0·21	0·02	0·01	0	0·02	0·15	*0·02	0·3	0·06	0·3	*0·09	0·09	0·05	0·27	*0·02	109	340	3
N in class	100	41	1,009	All ♂/♀ a-r interactions W=46 C=302		All ♀/♂ a-r interactions W=302		All ♂/J a-r interactions W=101 C=231				All ♀/J a-r interactions W=78 C=687				118	201			

N=number of approach-retreat interactions in each class; W=wild; C=cage population. The proportion is given for each behaviour pattern in each interaction class, = the proportion of N which included this behaviour. An asterisk(*) indicates W. prop. − C. prop., unlikely to have arisen by chance ($P < 0.01$). Since the behaviour patterns are not mutually exclusive in an interaction the sum of the proportions is greater than 1 in any column. The first two behaviour patterns are of retreating, the rest of approaching partners.

it is interesting that behaviour patterns which correlated less well with the apparent ranking of that group showed more frequent differences in the proportions seen in the two populations (Table V, last column).

Some of the differences can be partly explained on common-sense grounds in comparing the properties of the two environments, although this does not necessarily mean that surroundings can be disregarded in their effect on social organization. For example, chasing was a more frequent component in the wild in general, although caged juveniles chased females more than wild ones. Even though 'chasing' in the cage included jumping at the other baboon which made only token movements of fleeing, it could be argued that chasing was inhibited in the cage by sheer lack of space—in the wild a chase might continue for over a 100 yards and then be reversed with equal vigour. Similarly, 'supplanting' tended to be more frequent in the cage (although only significantly so in the male-to-female class), perhaps because the choice of desirable places was much more limited.

The definition of the approach-retreat interaction has the advantage of being easy to apply, but on the other hand it certainly includes behaviour where the subjective impression given is of play rather than 'serious' behaviour. Chasing and making aggressive contact (mouthing, pulling fur, poking) were frequent components of such interactions. Most juvenile-to-adult female approach-retreat behaviour was in fact 'teasing', typically by juvenile males; but the females would respond by avoiding or even grinning and geckering fear responses—or by hitting or threatening the teaser, depending on the individual. Any response seemed to delight the tormentor equally. Such behaviour was much more frequent in the cage; in the wild, young males usually left the females when the group was stationary, and occupied their abounding energies in such activities as shinning up the 80-ft smooth pole trunks of forest trees with others of the same age group. Again, aggressive contact between caged juveniles was mainly the males teasing the young females.

The seven approach-retreat behaviour patterns are not mutually exclusive within an interaction, but most 'avoid' occurred without any apparent communication from the approaching animal, which was usually on some other errand. A high proportion of approach-retreat interactions between males and both juveniles and females were of this type; in the wild population two thirds of male-female approach-retreat was merely females avoiding males (much of it in a sexual context, involving swollen females), but in the cage the proportion was significantly lower.

Facial threat was a more frequent component of approach-retreat interactions in the wild than in the cage, mainly because of threats by all classes of juveniles. This may again reflect a rather lower level of more active aggression. But in fact more facial threats directed to the observer were recorded in the wild than between baboons; they were not frequent.

Approach-retreat behaviour was sufficiently similar in the two populations for a descriptive summary of the interactions of each class to apply to both:

(1) Between females the non-agonistic patterns of avoiding and supplanting were most common, with frequent reinforcement by facial gestures of threat and grinning.

(2) Between males and females again the non-agonistic pattern, females avoiding males, was the most frequent pattern.

(3) Between males and juveniles chasing and threatening by males and reciprocal chasing by juveniles, but avoiding of males by juveniles was also common (in the cage the adult male would incite the juveniles to chase him).

(4) Again females often simply avoided juveniles, but chasing and making aggressive contact were more common, especially in the cage, and facial gestures from the females, threat in the wild and grinning in the cage were common.

(5) Between juveniles, chasing was by far the most common pattern.

Behaviour patterns in friendly interaction (Table VI). Friendly behaviour in the two populations was rather more different than approach-retreat behaviour, as measured by the difference in proportions of various patterns: out 104 comparisons, 43 reached the $P<0.01$ criterion of difference.

Again the behaviour patterns involving changes in relative positions were probably affected by the properties of the environment—both associating and following were recorded relatively more frequently in the wild. In the cage a baboon could not be more than 50 ft from any other. Following in particular, a behaviour of males with swollen females, was hardly necessary in these circumstances. However, females associated with juveniles *more*

Table VI. Behaviour Patterns in Friendly Interactions of Each Class in the Two Populations

	♂—♂	♀—♀		♂—♀		♀—♂		♂—J		J—♂		♀—J		J—♀		J—J		Total instances		W. prop. −C. prop. P<0.01
	W	W	C	W	C	W	C	W	C	W	C	W	C	W	C	W	C	W	C	
Touch	0·03	0·01	*0·93	0·01	0·02	0	0·004	0·02	0·05	0·01	*0·05	0·01	0·02	0·01	0·02	0·01	0·02	18	252	2
Nose to genitals	0	0	*0·05	0·02	*0·07	0·002	0·01	0·02	0·03	0·01	0·02	0·003	0·005	0·01	*0·03	0	0·01	21	237	3
Hand to genitals	0·12	0·01	*0·05	0·03	*0·09	0·01	*0·04	0·04	0·06	0·01*	0·04	0·01	0·02	0·01*	0·05	0·01	0·01	43	368	5
Intention mount	0·09	0	0·01	0·08	0·09	0	0	0·04	0·05	0	0·01	0·003	0·001	0·02	*0·06	0·02	0·02	71	254	1
Mount	0·15	0	0·01	0·1	*0·25	0	0	0·04	0·04	0	0·01	0	0	0·22	*0·16	0·23	*0·01	177	613	3
Embrace	0	0·04	*0·17	0·002	0·001	0	0·001	0·01	0·02	0·01	0·02	0·01*	0·04	0·02	*0·04	0·06	0·02	28	372	3
Present	0·43	0·28	0·26	0·04	0·02	0·4	*0·34	0·01	0·02	0·47	*0·32	0·09	0·07	0·11	0·07	0·1	0·06	479	1133	2
Soliciting grooming	0·03	0·06	0·02	0·05	*0·01	0·02	0·03	0·006	0·003	0	0·003	0·008	0·004	0·06	*0·01	0·01	0·003	75	93	2
Lipsmack	0·19	0·06	*0·22	0·02	*0·2	0·01	0·01	0·02	*0·2	0·01	*0·05	0·02	*0·04	0·01	0·02	0·01	0·01	50	707	5
Groom	0	0·53	0·62	0·24	*0·19	0·26	*0·18	0·07	0·06	0·31	*0·06	0·35	*0·24	0·27*	0·15	0·37	0·28	688	2028	5
Follow	0	0·04	0·08	0·21	*0·07	0·03	0·02	0·006	0·002	0·07	*0·02	0·003	0·01	0·13*	0·07	0	0·003	205	370	3
Play-wrestle	0	0	0·001	0	0·001	0·004	0	0	*0·34	0·01	*0·45	0·003	*0·03	0	*0·07	0·46	0·29	62	584	5
Associate	0·28	0·37	*0·09	0·07	*0·04	0·08	*0·03	0·01	0·003	0·06	0·02	0·02	*0·04	0·08	0·06	0·04	0·03	206	402	4
N in class	58	139	1179	W = 569 (All ♂/♀ friendly interactions)		C = 1140		W = 164 (All ♂/J friendly interactions)		C = 338		W = 361 (All friendly ♀/J interactions)		C = 1880		126	377		43

Symbols as in Table V.

frequently in the cage than in the wild. Again this explanation of the difference does not imply that it was irrelevant to the overall social organization differences.

Grooming was a more frequent component in the wild than in the cage; and its associated behaviour patterns, soliciting grooming, and presenting, which is often followed by grooming and seems to have soliciting grooming as one of its functions, also tended to be relatively more frequent in the wild. The wild environment was more itchy than the cage, on the observer's personal assessment; ticks, grass seed and various biting-flies were frequent sources of irritation so that grooming and being groomed might be more highly motivated there (only social grooming is included here, not attention to own skin). In group-living birds (Sparks, 1964, on the red avadavat), however, no increase in the amount of allopreening was observed when the feathers were dirty, and lacking direct evidence on primates this explanation is discounted. On the other hand, Rowell & Hinde (1963) showed an inverse correlation between grooming and stress in a caged rhesus population.

Lip-smacking was more frequent in the cage population, significantly so in five out of eight classes. This is an apeasement gesture used to reassure other animals of higher or lower rank of the friendly intentions of the lip-smacker. In the same experiment, Rowell & Hinde found a positive correlation between stress and the amount of lip-smacking recorded.

Friendly forms of physical contact, except grooming, were consistently more frequent in the cage, though only significantly so in thirteen out of thirty-two comparisons. This form of behaviour included both contacting the genital area with nose or hand, which was most usually directed at swollen females and might be regarded as having sexual significance, and also gently touching other parts of the body with the hand and embracing, which typically occurs between females. The latter two look like gestures of reassurance, and touching the genital area may perhaps also be regarded as such.

Finally, two behaviour patterns were not consistent in the direction of difference between the populations; mounting of females by males was a more frequent component of interactions in the cage. But mounting by juveniles was relatively more frequent in the wild, while intention mounting by juveniles was more frequent in the cage.

Play by adults with juveniles was more common in the cage than in the wild; play between juveniles was recorded less frequently in the cage. As far as the caged juveniles are concerned this difference reflects the inadequate size of their play school, the poverty of physical environment for non-social play, and the recording bias discussed earlier. Play initiated by caged adults, especially the adult male, was probably the result of boredom, and in the male's case a lack of peers.

Interactions between adult males. Since the notorious aggressiveness of males towards each other is the reason that the cage group was constructed with only one male, the problem is pertinent to this discussion even though no wild-cage comparison of inter-male behaviour could be made. In fact when an aged male was added to the group the resident adult male showed no aggressive behaviour, although the newcomer made frequent gestures of fear or submission while being investigated with much lip-smacking, touching and mounting intention movements by the male; he was rapidly accepted as a group member. In contrast two adult females added to the group after the end of the study were severely mauled before being accepted, in spite of all animals being heavily dosed with tranquillizers.

No hierarchy was noticed among the adult male groups in the wild population during the study. In the best known 'S' troop thirty-seven approach-retreat interactions between pairs of adult males were observed, twenty-five of these between known partners, a further four where one partner only was recognized, and on eight occasions neither protagonist was known. For every pair of males that were observed in approach-retreat interaction more than once, both were seen to retreat at different times. The small numbers alone would suggest that this type of behaviour was probably not an important component of social behaviour of the whole troop. However, the observations are extremely unlikely to have been collected at random from a group of males actually organized in a rigid hierarchy: perhaps one male could be ranked higher than the rest having approached nine times and retreated only four. It could of course be argued that a hierarchy existed but was so well established that no overt behaviour was required to demonstrate it, but to accept this would be purely an act of faith.

DeVore (1965) stressed the co-operation of

males in agonistic behaviour and discussed hierarchy mainly in terms of multi-male interactions. Such interactions were rare in this population, and when they occurred were characterized by the frequency with which the males 'changed side'. Such interactions also involved sub-adult males, and were frequently broken off by one male seizing and cuddling a passing infant. There was much movement of adult males between troops. The extent of this was not realized until late in the study when frequent and fairly complete tallies of the troops could be made, partly because the movements were not accompanied by social disturbance or changes in overall troop numbers, and partly because it happened that they did not at first occur between the two best known troops, but between these and their less known neighbours on either side. In retrospect it seems probable that the rather high proportion of unidentified males was due to these frequent unrecognized exchanges. For example, S troop nearly always contained five males, occasionally four or six. Only $\male$B is known to have been present during the whole study. $\male$F was present almost 90 per cent of the time, $\male$L for the middle part of the study, perhaps 60 per cent. $\male$P, G and E were present only for two or three 2-week observation periods each; two other recognizable males were also recorded in the troop but were certainly absent most of the time—one of them was recognized in two troops on 2 successive days. In V troop, six males that had been steady members during the whole study had moved on a later census visit; one of these was recognized in another troop. Males that moved were not distinguished in behaviour—they included frequent and infrequent copulators, for example. Such a degree of mobility in itself practically precludes the formation of rigid hierarchies of males. In contrast, no movements of females between troops were detected.

What of the pattern of interaction between males? Adult male primates have often been regarded as having a peculiar relationship but the psychological effect on the observer of their larger size and often more dramatic appearance must be taken into account. The interactions between males in the wild population were compared with those between females. Approach retreat interactions were relatively and absolutely more common between males than between females (Tables II and IV). The interactions in the two sexes were similar, however, in that for no behaviour pattern the difference in pro-

portion of interactions containing it attained high statistical significance ($P<0.01$).

Friendly interactions were more difficult to compare, because males did not follow or embrace or groom each other, while females did not mount or intention-mount. Males touched each others' genitals more frequently than did females; the most striking difference, however, was in lip-smacking, which was much more frequent between males—almost as frequent as lip-smacking between *caged* females. Touching, soliciting grooming and associating did not differ, while neither males nor females played, or nosed genitals.

'Interaction', in the limited sense used here, did not reflect the immense amount of co-operation between males in a group. This took the form of paying attention to each others' interactions with the environment; if juveniles registered alarm or excitement the nearest male would investigate the cause and the others would act on his assessment of its importance. Thus the males provided a continuous intensive policing of the environment for the troop. Adult males were not, however, seen to defend the troop; in fact being larger they usually outdistanced the rest in a flight from danger, the last animals being females carrying the larger babies. Nor did they lead troop movements more often than adult females, though one might wait for stragglers to catch up, and thus form a rear-guard of sorts. This is in contrast to the observation of Washburn & DeVore, 1961.

Discussion

The social organization of the caged group had a straight-line hierarchy as an obvious component (Rowell, 1966b); in the wild population no suggestion of hierarchical organization was detected. In the preceding sections differences in behaviour between the two populations, which were quantitative rather than qualitative, have been described. The wild population apparently differed also from previously described wild baboons, at least in inter-male behaviour, although data for full comparison are not available. Discussion of the relative importance of the behaviour differences in producing different social organizations must remain speculative until more detailed comparisons of populations are available, but some points seem worthy of further attention.

The first striking difference is the much greater frequency of interactions of all kinds in the cage. Associated with this was a greater relative

508   ANIMAL BEHAVIOUR, 15, 4

frequency of approach-retreat interactions. Female-female interactions showed the biggest difference in both these features, and caged females had amongst themselves the most markedly rigid hierarchy (Rowell, 1966b). The hypothesis is put forward that overall frequency of social interactions in primate societies is positively correlated with the proportion of approach-retreat interactions, and that this may be a basic determinant of the type of social organization developed.

The make-up of approach-retreat interactions seen in the two populations was rather similar, and those behaviour patterns which differed most in frequency were those which had been shown to be least well correlated with rank order in the caged group. In both populations non-agonistic approach-retreat patterns were the more numerous. It would seem that the overall frequency of this type of interaction, rather than the form they take, is the more important variable.

Friendly behaviour differed more in structure in the two populations and most of the differences were in a direction which suggest a higher level of stress in the caged group. The latter made more appeasement or conciliation gestures while the wild group did relatively more grooming, which is an activity typical of unstressed monkeys (Rowell & Hinde, 1963).

Hierarchy has been described as a stress-reducing device (e.g. Collias, 1953) and if this is accepted it implies a reduction of stress in the caged group from a hypothetical, much higher level; the actual finding is of a positive correlation between stress and hierarchy in this single comparison. The view that an apparently non-hierarchically organized group has in fact a covert hierarchy too well accepted to require expression is not regarded as helpful.

Wild adult males seemed to have an organization among themselves quite different from that reported by Hall & DeVore (1965) in the population in Nairobi National Park, with no apparent ranking. It is possible that the environment of the Ishasha baboons, with a rich food supply, abundant cover and no visitors supplying competition-engendering titbits, is less stressful than that at Nairobi and that we have here another correlation between stress and hierarchy formation. Approach-retreat interactions were more common than friendly ones between the Ishasha males. Compared with females in the same group, inter-male approach-retreat was more common than inter-female, but

the structure of the interactions did not differ significantly. A comparison of friendly behaviour shows more differences, and many of these lie in the same direction as the comparison between caged and wild females: conciliation gestures like lip-smacking were more common in males than females, and males did not groom each other, unlike females. This suggests that the males were in fact rather more 'tense' among themselves than the females, although the overall impression they gave was primarily one of co-operation.

Summary

(1) A wild and a caged baboon population were compared on the basis of roughly 300 hr of observation on each population. The effect of different population compositions is discussed.

(2) Comparison was of the number and type of interactions seen between adult males, adult females and weaned juveniles and of the incidence of twenty common behaviour patterns in these interactions.

(3) Interactions were nearly four times more frequent in the cage than in the wild.

(4) Approach-retreat (as opposed to friendly) interactions were more common in the cage. This increase was not even throughout all eight classes of interactions for which comparison was available, but was especially marked in female-female interactions.

(5) The structure of approach-retreat interactions in the two populations did not differ much within each interaction class. Most differences occurred in those patterns which are also included in play behaviour and were shown elsewhere to be little correlated with rank order in the caged group.

(6) Friendly behaviour showed more differences, many of which were consistent with a higher level of stress in the caged group.

(7) Wild adult males had no detectable hierarchy and frequently moved between troops. Their approach-retreat behaviour was similar to that of wild adult females amongst themselves; their friendly behaviour differed in a way that suggested a higher degree of tension than between females. The role of the males in the group in forming a highly co-operative policing system is described.

(8) Results are discussed in terms of the relation between hierarchy formation and environmental stress.

Acknowledgments

The author was research assistant to the late Professor C. F. A. Pantin, F.R.S., financed by the S.R.C. Much help was given by the staff of Uganda National Parks and by the Makerere Zoology Department in which author and caged baboons were guests.

REFERENCES

Collias, N. E. (1953). Social behaviour in animals. *Ecology*, 34, 810–811.

Crook, J. H. & Gartlan, J. S. (1966). Evolution of primate societies. *Nature, Lond.*, 210, 201–203.

DeVore, I. (1965). Male dominance and mating behaviour in baboons. In *Sex and Behaviour* (ed. by F. A. Beach). New York: John Wiley & Sons.

Gartlan, J. S. & Brain, C. K. (1967). Ecology and social variability in *Cercopithecus aethiops* and *C. mitis*. *Wenner Grenn Symposium* (in press).

Hall, K. R. L. (1962). The sexual, agonistic and derived social behaviour patterns of the wild Chacma baboon. *Proc. zool. Soc. Lond.*, 139, 283–327.

Hall, K. R. L. (1963). Variations in the ecology of the Chacma baboon. *Symp. zool. Soc. Lond.*, 10, 1–28.

Hall, K. R. L. & DeVore, I. (1965). Baboon social behaviour. In *Primate Behaviour* (ed. by I. DeVore), pp. 53–110. New York: Holt Rinehart & Winston.

Jay, P. (1965). The common langur of North India. In *Primate Behaviour* (ed. by I. DeVore), pp. 197–249. New York: Holt Rinehart & Winston.

Kummer, H. (1956). Rang-Kriterien bei Mantelpavianen. *Rev. Suisse Zool.*, 63, 288–297.

Kummer, H. & Kurt, F. (1963). Social units of a free-living population of Hamadryas baboons. *Folia Primat.*, 1, 4–19.

Rowell, T. E. (1966a). Forest living baboons in Uganda. *J. Zool. Lond.*, 149, 344–364.

Rowell, T. E. (1966b). Hierarchy in the organisation of a captive baboon group. *Anim. Behav.*, 14, 430–443.

Rowell, T. E. (1966c). Behaviour and reproductive cycles of baboons and macaques. In *Behaviour and Social Communication of Primates* (ed. by Altman). University of Chicago Press.

Rowell, T. E. (1967). Social organisation of primates. In *Primate Ethology* (ed. by Morris). London: Weidenfeld & Nicholson.

Rowell, T. E. & Hinde, R. A. (1963). Responses of rhesus monkeys to mildly stressful situations. *Anim. Behav.*, 11, 235–243.

Sparkes, J. H. (1964). Flock structures of the red Avadavat with particular reference to clumping and allopreening. *Anim. Behav.*, 12, 125–136.

Sugiyama, Y. (1964). Behavioural development and social structures in two troops of Hanuman langurs. *Primates*, 6, 214–247.

Washburn, S. L. & DeVore, I. (1961). The social life of baboons. *Sci. Am.*, 204, 62–71.

(*Received 8 November 1966; revised 4 February 1967; Ms. number: 704*)

COMPARATIVE METHODS AND THE CONCEPT OF HOMOLOGY IN THE STUDY OF BEHAVIOUR

by

G. P. BAERENDS

(Zoological Laboratory, State University, Groningen)

Biological phenomena can be compared for three different purposes (see VAN DER KLAAUW, 1947). Firstly, to obtain classifications of various kinds, the criteria for which depend on the special aims of the different investigators. Secondly, to detect distinct objective types or principles leading to similarity in structure or machinery. Thirdly, to investigate the phylogenetic origin of the phenomena. Although each purpose has a value of its own, I feel that the working-out of the problem deepens in the order in which these purposes are given here; this sequence also corresponds with the historical development of comparative studies. The first kind of comparison is the most primitive, it provides a necessary and useful, but not very firm basis for further advancement. In morphology the second one has led to typology, while it is practised in comparative physiology to help us understand the basic functional principles according to which the physiological systems are working. However, I see as the ultimate object of the comparative method the acquisition of insight into the evolution of the phenomena studied, whether taxonomic groups, organs, parts of organs or behaviour elements.

Comparative studies of behaviour starting from the second purpose have produced ample evidence that behavioural mechanisms of the same functional type occur ever and again in various, often very different, systematic groups. This is for instance true for the releasing mechanism, the instinctive activity and for the hierarchical system of instincts of different order. It should be emphasized here that correspondence in function does not necessarily involve similarity in structure or common descent. It does not imply more than the statement that different taxonomic groups have similar physiological systems, such as respiratory systems in which separate mechanisms for the uptake, the transport and the delivery of gases can be distinguished.

In this paper I will restrict my considerations to one group of behaviour mechanisms: the motor systems underlying behaviour. Our knowledge of these systems, as derived from non-comparative causal and analytical behaviour studies and as far as relevant in this paper, I would briefly summarise as follows:

Behaviour consists of the activities of muscles and glands. To activate a muscle different motorneurons have to co-operate; for movements of a part of the body these activities have to be co-ordinated by motor centres superimposed on the motorneurons (WEISS, 1950). Such co-ordinations can again be intregrated in units of greater complexity, varying from simple mono-synaptic reflexes through locomotory automatisms and instinctive activities to instincts of different order.

As LORENZ (1937 a, b) has emphasized, the level of the instinctive activity is the most obvious to the student of behaviour, because of the stereotyped form of these activities. LORENZ, and TINBERGEN (1938) showed that these elements can be separated into two components: the fixed pattern ("Erbkoordination", "Bewegungsnorm") and the directive component ("Taxis", "Reaktionsnorm") which differ mainly in this sense that the latter, unlike the former, is continuously susceptible to external stimuli. It can be proved that several instinctive activities of an animal share common causal factors and for that reason can be grouped in superordinate systems ("instincts"). Again, instincts of lower order can be grouped under instincts of higher order. Thus, a hierarchical system underlies all behaviour, from the motor units up to the most complicated instincts (TINBERGEN, 1942a, 1950, 1951; BAERENDS, 1941). The same element of lower order may be subordinated to different superimposed mechanisms. The elements of different order are not sharply distinguished from each other by qualitative differences; I am inclined to consider them more as quantitatively different levels of integration (BAERENDS, 1956).

The activation of all behavioural elements is brought about by external and internal stimuli in co-operation. The internal factors are thought to be of different kind, for instance hormones, autonomous impulses and impulses from already activated superimposed systems (instincts). Moreover, experience in combination with conditioning processes plays a role in determining the internal readiness to act.

Behaviour elements of the same level often interact with each other. As has been shown by TINBERGEN (1952a, 1952b, 1954) and his pupils VAN IERSEL, MOYNIHAN (1955) and MORRIS (1956), this interaction, and particularly inhibition, can have the following effects on the ensuing behaviour:

1. The occurrence of incomplete movements (intention movements, DAANJE, 1950). Such movements may also appear when the total amount of information from internal and external stimuli is low. When the information increases the activity unfolds itself in such a way that new muscle co-ordinations are added in a stereotyped order (BAERENDS, BROUWER and WATERBOLK, 1955). This makes it possible to distinguish different intensity stages of an activity.

2. Different intention movements can occur in combination, either simultaneously of successively (ambivalence; compromise, ANDREW, 1956b).

3. Movements that, at first view, seem "irrelevant" in the context of behaviour can be brought about (displacement movements). Originally, the occurrence of displacement activities was tentatively explained by assuming that the "blocked energy" of the mutually inhibiting instincts would find an outlet through a behaviour pattern of another instinct and thus "allochthonously" activate this latter pattern (KORTLANDT, 1940; TINBERGEN, 1940). Recently, however, another hypothesis is gaining ground, in which it is suggested that —because the two instincts are in conflict— their inhibiting effects upon other instincts become reduced and can even diappear (VAN IERSEL and BOL, 1958). Which of the other instincts comes out, depends, on the degree of desinhibition, on its momentaneous state of activation and on facilitating influences of the external situation and of the posture of the animal.

4. The behaviour can appear to its full extent, but instead of being directed towards the releasing situation it is performed towards another object (re-direction) (BASTOCK, MORRIS and MOYNIHAN, 1953).

All four phenomena can be present together. For instance BAERENDS and BAERENDS-VAN ROON (1950) have described how during the beginning of courting a *Cichlasoma meeki* male—by the presence of the female simultaneously stimulated to attack and to lead—was performing inhibited attack movements (ad. 1 and 2) directed, not towards the female, but towards plants (ad. 4) and mixed with a quivering movement which was considered to be displacement digging or displacement tearing off and pushing away plants (ad. 3).

The principles of this functional machinery have been found to determine the final form of the behaviour patterns in all animals studied so far in sufficient detail. I will show that their knowledge will be a great help to understand how different behaviour elements can develop from a common origin. With RAY-LANKESTER (1870) I will use here the term homologous to indicate when structures or behaviour patterns are considered to be genetically related, in so far as they have a single

404   G. P. BAERENDS

representative in a common ancestor. As is usual in ethology, I shall, therefore, not use the term homology in its typological sense. Our use, moreover, implies that we cannot distinguish, as Von Bertalanffy (1934) did, different kinds of homology in accordance with the methodological criteria used. We have to apply criteria that speak for derivation from the same thing; in the attempt to find such arguments for common descent we have to be careful not to incur into a vicious circle.

When searching for criteria it will be wise to look at the much older disciplin of comparative anatomy, because there, in contrast to ethology, the criteria can often be checked against palaeontological evidence. It will be interesting to see how far the criteria used in morphology can also be applied to comparative behaviour studies. Moreover, it will be interesting to find out whether there are criteria exclusively applicable to the latter. It seems appropriate to make such an attempt in the present volume, published in honour of Prof. C. J. van der Klaauw, who introduced me into the fields of comparative anatomy and theoretical biology and who encouraged in his laboratory at Leiden the development of ethology as a new separate branch of zoology.

The difference between a behaviour element and an anatomical structure is not as great as it seems at first view. Behaviour is determined by nervous connections and differences in thresholds of synapses, thus anatomical and physiological characteristics, while in their turn anatomical structures are brought about by physiological processes. Many behaviour elements are hereditary according to the same laws as morphological structures (Hinde, 1956; Von Hörmann-Heck, 1957), and have proved their value for taxonomic use, e.g. in Anatids: Heinroth (1911) and Lorenz (1941), in Finches: Andrew (1956), in Drosophila: Weidmann (1951), in Ammophila: Adriaanse (1947), and in grasshoppers: Jacobs (1953). This particularly applies to the instinctive activities which are also the most promising elements for comparative studies. Open for comparison are:

1. Behaviour elements of individuals of the same species.
2. Behaviour of individuals of different genotypes.
3. Behaviour elements of different related species.
4. Behaviour elements of one individual, either at different ontogenetical phases or on different occasions in the same phase.

Comparisons of type 1 give information on the extent of individual differences; whether these are genotypical or phenotypical can be learned from comparisons of type 2. In both types there will never be doubt that the elements compared have a common origin. However, for comparisons of type 3 and 4 we must always ask in how far similar looking activities are homologous or analogous. In the latter case it might be of interest to distinguish further between a parallel and a convergent development, which terms, according to NOVIKOFF (1935), express whether the correspondence of the elements is mainly thought to be caused respectively by developmental or by selective factors. This problem lies outside the scope of the present paper; we will only remark that this already rather theoretical distinction is made still more difficult by the fact that the developmental factors are not free from the influence of selective agencies.

The criteria used for distinguishing homology have been discussed by VAN DER KLAAUW (1947) and REMANE (1956). I will follow their design in spotting homology between similarities in behaviour, as derived from comparisons of the categories two, three or four.

It should be pointed out here that in this comparison only ethological data may be used. Although ethologists have sometimes made speculations about the physiological or even structural mechanisms underlying behaviour, apart from the weakness of such data at the present state of our knowledge, their use mixed with behavioural data for making conclusions about homology would be illegitimate as they belong to another class.

A. SIMILARITY IN STRUCTURE

a. *Form*. By its particular pattern of muscular activities the "Erbkkoordination" is at least as rigidly fixed as morphological structures. Each of the composing elements has its own threshold; when the activity unfolds under increasing drive these different parts are added in a stereotyped sequence. Consequently, behaviour should be described, by preference quantitatively, in terms of muscle patterns co-ordinated in space and time. Comparison is made easier when the "Erbkoordination" can be stripped of its steering components. In principle this can be done, experimentally by controlling the external situation, and deductively after one has obtained insight into the function of the taxis elements (BAERENDS and BAERENDS-VAN ROON, 1950). Moreover, when comparing, one has to realize that alterations of the original pattern may have occurred in correspondence with a secondarily obtained signal function.

Quantitative details of the activities may be a great help for their identification. For instance, the male of *Lebistes reticulatus* performs in the two phases of courtship differently looking sigmoid displays: leading and checking. It could be shown by quantitative observations that the sequence of the constituent parts of the sigmoid of leading was the same as that of checking. Moreover, the relative frequency at which these parts appeared in the pattern was proved to be the same for both types of sigmoid. These arguments in favour of homology were supported when finally the differences between the displays could be understood as adaptations to different signal functions. Some changes were shown to correspond to new mechanical demands, others served the required conspicuousness (BAERENDS, BROUWER and WATERBOLK, 1955). As it is very unlikely that behaviour patterns with qualitatively and quantitatively exactly the same fixed "core" will have descended through parallel or convergent development from different origins, "form" must be a very reliable criterion for homology.

b. *Internal causal factors.* In the introduction we have mentioned that activities are often performed under the influence of (motivational) impulses from one or more superposed instincts. If similar activities can be shown to occur under influence of the same instincts this can be used as an indication of homology. For instance, several Cichlid species have a similar defensive frontal display which appears during boundary fights under the combined influences of aggressive and fleeing motivations. We also find in the same species, when they are leading young, another more or less similar activity, "calling young", with an identical motivation.

However, one has to be very careful in using this criterion. Firstly, conformity of motivating instincts may lead to a convergent development of similar taxis components. Particularly if the latter dominate a relatively simple "Erbkoordination", the resulting acts may look very similar, without having a common descent. Secondly, as has been proved by KRUIJT (pers. comm.) for chickens of the red jungle fowl, in the course of the development the same behaviour pattern may come successively under the control of different instincts. Thirdly, as has often been suggested (LORENZ, 1951; TINBERGEN, 1952; BAERENDS, 1956), the process of ritualization may be connected with a change in the superposing factors. Therefore, a difference in internal causation can hardly be used as an argument against homology whereas a correspondence is a weak argument, useful only when supported by other evidence.

c. *External causal factors.* Similarity in the releasing situations is of value as an additional argument; difference, however, can never be used to reject the possibility of homology. I am inclined to consider a correspondence of sensitive periods, in which the releasing mechanism has to be modified by conditioning, as a criterion for homology, but unfortunately the investigations have not yet advanced far enough to enable me to illustrate this idea with well analysed examples. Moreover, this criterion leads us to a discussion of homology of afferent mechanisms which we have planned to leave outside the scope of this paper.

B. SIMILARITY IN ONTOGENY

The fact that only little attention has been paid to the ontogeny of instinctive activities makes a discussion on the applicability of this criterion difficult and rather premature.

A first necessity using ontogeny is to distinguish real ontogenetical stages from cases of submaximal intensity due to insufficient drive. Because so often the drive increases with age, for instance when hormones are involved, this distinction is far from easy and will often require careful observations and experimental tests. KUO's (1932) work on the embryonic movements of embryo's, SAUER's (1953) study of the development of song in the Common Whitethroat and BLAUVELT's (1954) analysis of mother–child relations in goats come nearest to these wishes, but are not yet in a stage to permit interspecific comparison. Also the observations by KORTLANDT (1940a) on the early appearance of normal adult activities in the Cormorant can only be considered as a beginning. His descriptions are not yet detailed enough for our special purpose, because he was primarily interested in the time and sequence of the appearance of the different activities (i.e. the development of a super act) and not so much in the developmental stages of one single act. The ethologists have often called the instinctive activities "inborn". This was to indicate that they need not be learned by imitation of the behaviour of more experienced animals; it was not meant to imply (like some critics seem to have understood, e.g. LEHRMAN, 1953) that such activities would not develop, like organs and their parts, under the constant interaction of genetic and environmental factors (BEACH, 1954). It is, however, true that the ontogeny of behaviour has been neglected and the critisicm that this was promoted by the rather careless use of the term "inborn" is probably true. Ontogeny might prove to be a very important method in the study of homology. Particularly, a picture of the kind of environmental stimuli, and of when and how they act, might give important clues.

It might show that developmental stages of instinctive activities are rarely observable in animals. I have the impression that most activities are already complete as soon as they appear in a recognizable state, even when the executing organ is not yet fullgrown and the behaviour still without function. However, as KORTLANDT (1940a) has shown for the cormorant, complexes of elementary instinctive activities (super acts like nestbuilding, feeding, fighting or courting) develop gradually and a comparative study of the process of integration of the different constituent parts into such a complex is most promising. KRUIJT (pers. comm.) has found in fowl that during this period of integration motor patterns are linked with motivational factors which in later stages can be substituted for other superposed mechanisms. Conditioning—in mammals often through experience during "play"—must be of great importance in this context. KRUIJT found that there are special periods in which the animal is most sensitive to incorporate such experience.

C. SIMILARITY IN TOPOGRAPHY

In terms of behaviour this means correspondence of the places where the activities occur in the behaviour repertoire of the animal. This criterion is particularly important if this appearance of the activity is not directly released by external factors. To give some examples: *Tilapia mossambica* and *T. nilotica* perform comparable but rather different looking "tail wagging" movements which, however, occur at exactly the same places in the ceremonies by which the males of these fishes lead the females towards the nest. The same is true for quivering in *Hemichromis bimaculatus* and digging in *Geophagus brasiliensis* (which both occur in the early courtship of these Cichlid fishes), for the zigzag dance in different stickleback species, and for wing-flapping in pheasants. Showing the nest entrance in the three-spined stickleback looks very different from fanning. It may, however, be considered homologous with the latter, because at the corresponding place in its behaviour repertoire the ten-spined stickleback performs a nest-showing activity which is hardly different from the fanning movement (TINBERGEN, 1952b, 1954).

D. LINKAGE BY INTERMEDIATE STAGES

In cases where the complete form of an activity in species A looks rather different from that of a comparable activity in a related species B, intermediate forms between A and B can often be found in other species of the same taxonomic group. Moreover, linkages can frequently be established by extending the comparison also to different intensity stages of the activities. In agonistic encounters of low intensity the Cichlid fish *Astronotus ocellatus*, while standing beside its opponent, displays with a vertical quivering of the tail fin, which is visually supported by a lateral eye-spot, situated dorsally on each side of the fin basis. This movement occurs, though far less conspicuously, in a great many related Cichlids, where it is often combined with a mumbling movement of the jaws and slight nodding of the head *(Aequidens portalegrensis)*. In *Cichlasoma severum* this movement regularly precedes the mutual gripping of the jaws at the start of a fight and seems to be an incomplete intention stage of it. Only through these intermediate connections the vertical quivering of *Astronotus* can be understood as a ritualized incipient attack.

The possibility of comparing different stages of intensity renders to ethology a criterion for the establishment of homology which is lacking in morphology. This method resembles somewhat the ontogenetical approach, but is certainly not identical with it for, as we have explained above, intensity stages and ontogenetic stages belong to different classes.

The comparison of sufficient intermediate stages will prevent one to homologize unjustly two similar looking activities in species A and Z, when the activity considered in A and that in Z—through intermediate stages in a series of related species—can be identified with two separate activities in a species M, existing there beside each other ("Vorbei-gleiten"; REMANE, 1956). As far as present evidence goes this seems to be the case in some aerial displays in the terns. In several tern species high flight and fish flight are interwoven; in some of these the high flight and in others the fish flight has developed more prominently. It is, however, unlikely that high flight and fish flight have been derived from the same origin, as in the black tern (BAGGERMAN et al., 1955) the high flight and the fish flight occur both independently in what seem to be rather primitive forms.

Apart from these main criteria REMANE (1956) mentions a number of additional ones which also apply to behaviour studies, viz.:

1. The probability of homology of similar structures increases with the degree of universality of their distribution among related species, but decreases the more frequent they are in unrelated forms.

2. The probability of homology of similar structures in two related species increases with the presence of other independent conformities. In my opinion this criterion is not entirely different from the first part of the former one, for the independent conformities will usually already serve as arguments for the conclusion that the species are closely related.

Animals with homologous structures may live under quite different conditions. For convergencies, however, similar environmental circumstances are necessary. A knowledge of the selective factors exerting pressure on the animal will, therefore, help to judge how far in a certain case the possibility of convergency has to be taken into consideration. Moreover, strong similarities are more likely to have originated from homologous elements than from analogous patterns that developed convergently.

Our considerations lead to the conclusion that in comparative ethology it is most essential for homology that the patterns of muscle contractions should be largely identical. If these show important differences, which cannot be explained as secondary adaptations to specialised functions, homology is less likely, even when the topography and function of the elements are identical.

410   G. P. BAERENDS

In morphology organs may be considered homologous (i.e. of common descent) even if the composing elements have been substituted, provided the structures compared are identical as morphological and often also as functional units and if the substitution takes place at a relatively low level of integration. Examples are stomachs built up of different kinds of tissues and orbits comprising different sets of skull bones. When comparing behaviour elements we often meet differences that remind us of this phenomenon of replacement. On page 406 we have discussed differences in causation, particularly in the superposed motivational factors, and argued that they do not compel us to reject the possibility of homology. If we consider the pattern as most essential, this implies that the effector organs need not necessarily be the same. One could think of different related animals, one of which at the firing of a definite neuron would show a decoloration due to contraction of superficial capillaries, whereas in another species which has developed chromatophores in this region, activation of the same neuron would cause these chromatophores to act.

Another change in the effector that does not exclude homology occurs when the same nervous pattern activates in the same way muscles which, in the species compared, are derived from different segments, as for instance in corresponding fins of different fish. Just as we call such fins homologous we must consider identical movements of such fins as homologies.

Thinking of fins, which are often built up from segments of similar morphological structure, we come to the question whether we may speak, in behaviour studies, of serial homology. This can certainly be done for similar movement patterns of serially homologous organs of an animal, like the rays of a fin, the appendages of a lobster or the legs of a millipede.

Just as we can find similar structures in different positions on an animal we can also find similar behaviour patterns at different places of its behaviour repertoire, e.g. displacement activities. In case of structures we may speak of serial homology. The behaviour patterns, however, if they fulfil the criteria given above, must be considered homologous in the phylogenetical sense, for in both cases the same nervous pattern and the same effectors perform the activity, even if the superposing motivation might have changed through substitution.

Homoiologous structures are similar, but not homologous, characters of homologous organs. We could expect homoiologous behaviour patterns to develop superimposed on homologous activities, for instance on displacement activities derived from the same origin, when both have secondarily and independently acquired superficially similar signal functions.

In morphology organs having a very different appearance (like the endostyl and thyroid) can often be proved to be homologous. In behaviour, where the comparison of final and complete activities is of such paramount importance, it is much less likely that very dissimilar forms can be recognised as homologous. Therefore, as was already pointed out by TINBERGEN (1942b), in ethology homologizing will remain restricted mainly to similar activities in related groups.

DE BEER (1938) has pointed out that in morphology the precursor of homologous organs in related species may be phenotypically absent in the common ancestor, for instance in the horns of the Titanotheria. This implies that homology in morphology ultimately depends on the identity of the still largely unknown system of developmental potentialities that through the interaction of genetic and environmental influences leads to a special form. The same must be true for homology of behaviour patterns. The same developmental principles are a causal link in ontogeny as well as in phylogeny: new species are formed under the influence of selection when these principles are acting on a genetic basis, slightly changed by mutation. This explains why some ontogenetic facts can be used to study phylogeny and why earlier phylogenetical stages can be detected in ontogeny. But, ontogeny is not recapitulated phylogeny (HAECKEL), it is the other way round: phylogeny is modified ontogeny (DE BEER, 1940), new forms develop through variations and extentions on already existing causal systems.

Comparative studies of behaviour have already been carried out in a number of cases. I will mention here the studies by HEINROTH (1911) and LORENZ (1941) on the Anatids, by HINDE (1955) on finches, by TINBERGEN (1959) and this pupils on gulls and terns, by SEITZ (1940, 1942) and BAERENDS and BAERENDS-VAN ROON (1950) on Cichlid fishes and by JACOBS (1953) on grasshoppers.

All this work has shown that the study of homology of behaviour elements is possible at different levels of integration. This applies in the first place to comparison between different species. In many of such cases the criteria discussed above, have given strong evidence that related species possess behaviour elements that must be of common descent. A difficult problem, however, is to deduce from the comparison of recent species which of a series of gradually changing homologous activities is primitive and which is specialized. Sometimes knowledge of intensity stages of the activity or of causal factors may be helpful in this respect.

Each activity has an origin in evolution that deserves to be studied. However, until now, attention has almost exclusively been directed towards activities with a signal function, because the problem was more conspicuous and fascinating there. To study the derivation of signal activities, comparison of similar behaviour elements in the same animal is the obvious method, although interspecific comparison is sometimes very helpful when in a group of related animals homologous behaviour elements have diverged differently far from the common origin (nest showing and fanning in the three-spined stickleback). Our knowledge about the causation of behaviour, derived from analytical experimental studies and summarised in the first section of the paper, can be of great help here as the forces causing behaviour in the individual must also have played their part in evolution. Their role is similar and additional to that of the developmental factors. These kind of studies have already given us a picture of how the evolution of signal behaviour must take place.

The simplest type of signal activities are those activities which, apart from their direct function, have at the same time signal value. Examples are feeding movements that stimulate members of the same species to concentrate on the spot to search for food (as KEENLEYSIDE, 1955, has shown for the stickleback; see also MORRIS, 1956), and attacking or fleeing activities that, apart from direct damaging effects, stimulate or inhibit certain types of behaviour in the adversary.

We have seen that in cases of low motivation or inhibition a behaviour element may show itself in an incomplete form. In evolution such intention movements have often been fixed at a certain stage, they lost the direct function of the complete acts and became specialized as signals to release specific reactions in other animals, usually conspecific partners. MORRIS (1957) has pointed to the fact that in such cases the activities became frozen at a "typical intensity".

As mentioned before, sometimes the same external situation activates more than one instinct or instinctive activities and in such cases ambivalent behaviour results. This too has been a basis for building up signal activities. Well known examples are: of successive ambivalence the zigzag dance of the sticklebacks, and of simultaneous ambivalence the upright (threatening) attitude of the gulls (TINBERGEN, 1952a, b). A second possibility for the development of signal activities when conflicting tendencies are simultaneously activated is the occurrence of displacement, through which quite different behaviour patterns, borrowed from elsewhere, can be brought into the picture.

In the ethological literature derived signal activities are usually labelled according to one of the categories: intention movements, ambivalent movements, displacement movements, redirected movements, reversed movements, etc. We are now inclined to think that one signal activity hardly ever fits into a single one of these categories, but that the causal factors, typical for *several* of these, have played a role in building up the signal movement as it is performed. In an encounter between animals, tendencies to approach and to retreat will always be involved simultaneously. The ultimate posture of the animal then often results from a compromise between the taxes corresponding with both conflicting tendencies (see also ANDREW, 1956b). Moreover, at the same time, mutual inhibition of both instincts will lead to incompleteness and to displacement, and elements of both may be superimposed on each other. Which activity is brought in through displacement, depends on the internal factors involved in the conflict, on the external stimuli present, and on postural facilitation (TINBERGEN 1952b; MORRIS, 1956; ANDREW, 1956 b, c). The redirection effect may accentuate definite external stimuli in the total situation. For instance, in a boundary fight of the herring gull ambivalent intention activities can be detected in the posture of the birds, fear for each other causes redirection which leads to attacking grass tufts and this external stimulus facilitates nestbuilding through the mechanisms of displacement. GROEN (1955) and MORRIS (1956) have pointed to the fact that, in close connection with these centrally innervated movements, also autonomously innervated activities occur (changes of respiration, perspiration, flushing, vomiting and pilomotoric reactions) which can also become incorporated in the final activity. Finally I want to mention here a very peculiar effect discovered by TINBERGEN and MOYNIHAN (1952) in the head-flagging of the black-headed gull. In this movement, which serves the appeasement between partners, the bill (a fight-weapon) is suddenly pointed away from the other. This "reversal" might be explained by assuming that an intention movement of flight became ritualized at a very early stage.

Like the pituitary, which is a unit, in the course of ontogeny built up out of heterogeneous elements, most signal activities are also of heterogeneous origin. Whether the different causal factors which in the course of the evolution must have led to the present shape of the signal activity are still at work when it appears in its ritualized state, is unknown. There is some evidence (KRUIJT, pers. comm.) that the external factors remain of great influence and some indication that the motivational situation may change.

In the example just given I have suggested that the boundary fight in herring gulls contains nestbuilding as a displaced element, and that the introduction of nestbuilding into this context was facilitated by redirection. One could ask now what is the origin of nestbuilding. I am inclined to think that the redirected aggression towards grass tufts, caused by the appearance of intruders in or near the area, may have led to the development of nesting behaviour as soon as it proved to

414   G. P. BAERENDS

have some selective advantages. This behaviour once established as a fixed pattern could then again be introduced through displacement into the fighting ceremony.

SUMMARY

This paper gives a discussion of methods which can be used to study the phylogeny of behaviour elements, in particular motor activities. The criteria practised in morphology and ethology to establish homology (in the sense of common descent) are compared. Similarity in form and topography are equally useful as criteria in both disciplins; similarity in ontogeny, already of great value in morphology is expected to increase in importance in ethology as the study of the development of behaviour grows. The possibility to compare motor patterns at different stages of intensity (to be sharply distinguished from stages of ontogeny) offers a special opportunity in behaviour studies for the use of "linkage by intermediate stages" as a criterion for homology. Knowledge of the causation of a behaviour element may help the study of phylogeny in several ways: (a) it facilitates the choice of really comparable units, (b) it shows principles for derivation of activities, (c) conformity of causation can be used with care to distinguish between homology and analogy. The most essential element for comparison is the pattern, a possible replacement of motivational factors or effectors need not necessarily exclude homology.

It is demonstrated that comparative and causal analytical ethological studies have already led to a picture of how behaviour elements, particularly those acting as social releasers, develop in the course of evolution. Now that we recognize the complex origin and character of such behaviour patterns it is advisable no longer to classify them into definite categories like displacement-, intention-, redirection-, or reversed activities, etc.; it will be more correct to suffice with a description of what agents have contributed to the form of the activity.

The thoughts given here arose in discussions with colleagues and collaborators which I wish to thank here collectively. I realise that my ideas are in many respects premature but I hope that they will help to improve our thinking about these problems.

REFERENCES

ADRIAANSE, A., 1947: *Ammophila campestris* Latr. und *Ammophila adriaansei* Wilcke. Behaviour, **1**, 1–34.

ANDREW, R. J., 1956a: Intention movements of flight in certain passerines, and their use in systematics. Behaviour, **10**, 179–204.

ANDREW, R. J., 1956b: Some remarks on behaviour in conflict situations, with special reference to *Emberiza* spp. Br. J. anim. Behav., **4**, 41–45.

ANDREW, R. J., 1956c: Normal and toilet behaviour in *Emberiza* spp. Br. J. anim. Behav., **4,** 85–91.

BAERENDS, G. P., 1941: Fortpflanzungsverhalten und Orientierung der Grabwespe *Ammophila campestris* Jur. Tijdschr. v. Entomol., **84,** 68–275.

BAERENDS, G. P., 1956: Aufbau des tierischen Verhaltens. Kükenthal's Handb. Zoologie, **10,** 3, 1–32.

BAERENDS, G. P. and J. M. BAERENDS-VAN ROON, 1950: An introduction to the ethology of Cichlid fishes. Behaviour, suppl. **1.**

BAERENDS, G. P., R. BROUWER and H. Tj. WATERBOLK, 1955: Ethological studies on *Lebistes reticulatus*. I. An analysis of the male courtship pattern. Behaviour, **8,** 249–335.

BAGGERMAN, B., G. P. BAERENDS, H. S. HEIKENS and J. H. MOOK, 1956: Observations on the behaviour of the black tern (*Chlidonias n. niger* L.) in the breeding area. Ardea, **44,** 1–70.

BASTOCK, M., D. MORRIS and M. MOYNIHAN, 1953: Some comments on conflict and thwarting in animals. Behaviour, **6,** 66–84.

BEACH, F., 1954: Ontogeny and living systems. Group processes, 9–74.

BEER, G. R. DE, 1938: Embryology and Evolution. Evolution, essays presented to E. S. GOODRICH, Oxford 1938, 57–78.

BEER, G. R. DE, 1940: Embryos and ancestors. Clarendon press.

BERTALANFFY, B. VON, 1934: Wesen und Geschichte des Homologiebegriffes. Unsere Welt, **28.**

BLAUVELT, H., 1954: Dynamics of the mother-newborn relationship in goats. Group processes, 221–258.

DAANJE, A., 1950: On locomotory movements in birds and the intention movements derived from them. Behaviour, **3,** 48–99.

GROEN, J., 1955: De mens als psycho-somatische totaliteit van reactiepatronen. Ned. Tijds. Psychol., **10,** 187–223.

HEINROTH, O., 1911: Beiträge zur Biologie der Anatiden. Verh. 5 Int. Ornithol. Kongr. Berlin.

HINDE, R. A., 1955: A comparative study of the courtship of certain finches (Fringillidae). Ibis, **97,** 706–745.

HINDE, R. A., 1956: The behaviour of certain cardueline F_1 inter-species Hybrids. Behaviour, **9,** 202–213.

HÖRMANN-HECK, S. VON, 1957: Untersuchungen über den Erbgang einiger Verhaltensweisen bei Grillenbastarden. Z. Tierpsychol., **14,** 137–183.

IERSEL, J. J. A. VAN and A. C. A. BOL, 1958: Preening of two tern species. A study on displacement activities. Behaviour, **13,** 1–88.

JACOBS, W., 1953: Verhaltensbiologische Studien an Feldheuschrecken. Z. Tierpsychol., suppl. **1.**

KEENLEYSIDE, M. H. A., 1955: Some aspects of the schooling behaviour of fish. Behaviour, **8,** 183–249.

KLAAUW, C. J. VAN DER, 1947: Inleiding. Leerboek der Vergelijkende Ontleedkunde van de Vertebraten, ed. 3, 1, 1–14. N.V. Oosthoek, Utrecht.

KORTLANDT, A., 1940 a: Eine Übersicht der angeborenen Verhaltungsweisen des Mitteleuropäischen Kormorans, ihre Funktion, ontogenetische Entwicklung und phylogenetische Herkunft. Arch. Néerl. Zool., **4,** 401–442.

KORTLANDT, A., 1940 b: Wechselwirkung zwischen Instinkten. Arch. Néerl. Zool., **4,** 443–520.

KUO, Z. Y., 1932: Ontogeny of embryonic behaviour in Aves. I and II: J. exp. Zool., **61,** 395–430 and **62,** 453–489. III and IV: J. comp. Psychol., **13,** 245–272 and **14,** 109–122.

416   G. P. BAERENDS

LEHRMAN, D. S., 1953: A critique of Konrad Lorenz's theory of instinctive behaviour.
    Quart. Rev. Biol., **28**, 337–363.
LORENZ, K., 1935: Der Kumpan in der Umwelt des Vogels. J. f. Ornithol., **83**,
    137–213 and 289–413.
LORENZ, K., 1937 a: Über die Bildung des Instinktbegriffes. Naturwissenschaften,
    25, 289–300; 307–318 and 324–331.
LORENZ, K., 1937 b: Über den Begriff der Instinkthandlung. Folia bioth., **2**, 17–30.
LORENZ, K., 1941: Vergleichende Bewegungsstudien an Anatiden. J. f. Ornithol.,
    89, Sonderheft, 194–294.
LORENZ, K., 1950: The comparative method in studying innate behaviour patterns.
    Symp. Soc. Exp. Biol., **4**, 221–269.
LORENZ, K., 1951: Über die Entstehung auslösender "Zeremonien". Vogelwarte,
    16, 9–13.
LORENZ, K. and N. TINBERGEN, 1938: Taxis und Instinkthandlung in der Eiroll-
    bewegung der Graugans. I. Z. f. Tierpsychol., **2**, 1–29.
MORRIS, D., 1956: The function and causation of courtship ceremonies. L'instinct
    dans le comportement des animaux et de l'homme. Masson et Cie, Paris,
    261–286.
MORRIS, D., 1956: The feather posture of birds and the problem of the origin of
    social signals. Behaviour, **9**, 75–113.
MORRIS, D., 1957: "Typical intensity" and its relation to the problem of ritualisation.
    Behaviour, **9**, 1–12.
MOYNIHAN, M., 1955: Some aspects of reproductive behavior in the black-headed
    gull and related species. Behaviour, suppl., **4**, 1–201.
NOVIKOFF, M., 1935: Homomorphie, Homologie und Analogie. Anat. Anz., **80.**
RAY-LANKASTER, E., 1870: On the use of the term homology. Ann. Mag. Nat. Hist.,
    6, Ser. 4.
REMANE, A., 1956: Die Grundlagen des natürlichen Systems der vergleichenden
    Anatomie und der Phylogenetik. Leipzig, Akad. Verlagsges.
SAUER, F., 1953: Die Entwicklung der Lautäusserungen vom Ei ab schalldicht
    gehaltener Dorngrasmücken im Vergleich mit später isolierten und mit wild-
    lebenden Artgenossen. Z. Tierpsychol., **11**, 10–93.
SEITZ, A., 1940: Die Paarbildung bei einigen Cichliden. I. Die Paarbildung bei
    Astatotilapia strigigena Pfeffer. Z. Tierpsychol., **4**, 40–84.
SEITZ, A., 1942: Die Paarbildung bei einigen Cichliden. II. Die Paarbildung bei
    Hemichromis bimaculatus Gill. Z. Tierpsychol., **5**, 74–101.
TINBERGEN, N., 1940: Die Übersprungbewegung. Z. Tierpsychol., **4**, 1–40.
TINBERGEN, N., 1942 a: An objectivistic study of the innate behaviour of animals.
    Bibl. Biother., **1**, 39–98.
TINBERGEN, N., 1942 b: Signaalbewegingen bij dieren. Vakbl. Biol., **23**, 7, 1–6.
TINBERGEN, N., 1950: The hierarchical organization of nervous mechanisms under-
    lying instinctive behaviour. Symp. Soc. Exp. Biol., **4**, 305–312.
TINBERGEN, N., 1951: The study of instinct. Oxford, Clarendon press.
TINBERGEN, N., 1952 a: A note on the origin and evolution of threat display. Ibis,
    94, 160–162.
TINBERGEN, N., 1952 b: "Derived" activities; their causation, biological significance,
    origin and emancipation during evolution. Quart. Rev. Biol., **27**, 1–31.
TINBERGEN, N., 1954a: Psychology and ethology as supplementary parts of a science
    of behavior. Group processes, 75–167.
TINBERGEN, N., 1954b: The origin and evolution of courtship and threat display.
    In: J. HUXLEY, Evolution as a process. Allen and Unwin Ltd., London, 233–250.
TINBERGEN, N., 1959: Behaviour, in preparation.

TINBERGEN, N. and M. MOYNIHAN, 1952: "Head-flagging" in the black-headed gull: its origin and evolution. Brit. Birds, **46,** 19–22.

WEIDMANN, U., 1951: Über den systematischen Wert von Balzhandlungen bei *Drosophila*. Rev. Suisse. Zool., **54,** 502–511.

WEISS, P., 1950: Experimental analysis of co-ordination by the disarrangement of central-peripheral relations. Symp. Soc. Exp. Biol., **4,** 92–111.

PSYCHOLOGICAL REVIEW

SCALA NATURAE:
WHY THERE IS NO THEORY IN COMPARATIVE PSYCHOLOGY [1]

WILLIAM HODOS [2] AND C. B. G. CAMPBELL

Walter Reed Army Institute of Research *Center for Neural Sciences, Indiana University*
Washington, D. C.

The concept that all living animals can be arranged along a continuous "phylogenetic scale" with man at the top is inconsistent with contemporary views of animal evolution. Nevertheless, this arbitrary hierarchy continues to influence researchers in the field of animal behavior who seek to make inferences about the evolutionary development of a particular type of behavior. Comparative psychologists have failed to distinguish between data obtained from living representatives of a common evolutionary lineage and data from animals which represent divergent lineages. Only the former can provide a foundation for inferences about the phylogenetic development of behavior patterns. The latter can provide information only about general mechanisms of adaptation and survival, which are not necessarily relevant to any specific evolutionary lineage. The widespread failure of comparative psychologists to take into account the zoological model of animal evolution when selecting animals for study and when interpreting behavioral similarities and differences has greatly hampered the development of generalizations with any predictive value.

Nearly two decades have passed since Beach (1950) presented his classic paper "The Snark was a Boojum" in which he deplored the decline of comparative psychology as a result of "excessive concentration upon a single species," namely, the albino rat. His paper appears to have stimulated a renewed interest in an animal psychology which is more broadly comparative than the rat learning studies which were prevalent in the 1940s and 1950s. Rhesus monkeys and White Carneaux pigeons have now been

added to the animal psychologist's standard menagerie. Occasional studies of behavior in teleost fish, reptiles, and carnivores have also appeared in psychological journals and some attempts at comparison across species have been made. In addition, several textbooks and collections of readings in comparative psychology recently have been published. However, much of the current research in comparative psychology seems to be based on comparisons between animals that have been selected for study according to rather arbitrary considerations and appears to be without any goal other than the comparison of animals for the sake of comparison. This rather tenuous approach to research has apparently been brought about by the absence of any broad theoretical foundation for the field. Such a theoretical foundation, though partly or even totally incorrect, would at least have the virtue of en-

[1] The authors wish to express their gratitude to their colleagues and students for their helpful comments on this paper and to J. Z. Young and the Oxford University Press for their permission to reproduce the phylogenetic trees shown in Figures 1-4.

[2] Requests for reprints should be sent to William Hodos, Department of Experimental Psychology, Walter Reed Army Institute of Research, Walter Reed Army Medical Center, Washington, D. C. 20012.

337

couraging a systematic study of animal behavior rather than the current haphazard manner of operation.

The purpose of this paper is to point out some of the factors which have hindered theoretical development in comparative psychology and to suggest some ways of remedying the situation. Many of the concepts that will be discussed are not novel; indeed, they would be regarded as rather elementary by students of such fields as systematic biology, paleontology, physical anthropology, etc. However, even a casual examination of the recent literature in the behavioral and neural sciences leads one to the conclusion that many experimenters are greatly misinformed about these fundamental concepts. As a result, a number of unjustified conclusions about behavioral and neural evolution have been drawn from the data of comparative studies.

The Scala Naturae

In attempting to find order in an apparently chaotic universe, Aristotle (1910, 1912a, 1912b; Ross, 1949) attempted various organizational schemes for classifying animals. These classifications were based on such characteristics as number of legs, whether or not the organism appeared to possess blood, whether they were oviparous or viviparous in their reproductive mechanisms, etc. Aristotle also proposed that the various categories of animals might be arranged on a graded scale of complexity or perfection, with man at the top. Although Aristotle did not advocate any such ranking of the animals within each category, later scholars expanded his suggestions so that there came to be general acceptance of the concept that all animals could be ranked on a unitary, graded, continuous dimension known as the *scala naturae* or Great Chain of Being (Lovejoy, 1936; Wightman, 1950). The lowest position on the *scala naturae* was occupied by sponges and other creatures considered to be essentially formless. At the intermediate levels were insects, fish, amphibians, reptiles, birds, and various mammals. At the top of the scale was man. Furthermore, as Lovejoy (1936) points out

in his extensive treatment of the history of this subject, the *scala naturae* eventually became involved in theological formulations which considered that God was perfect and all other creatures were merely progressively less perfect copies. Thus, angels were somewhat imperfect copies, man more imperfect, apes still more imperfect, and so on, down the scale to the "formless" sponges.

The attractiveness of the notion of a *scala naturae* is attested to by its persistence throughout the centuries and its influence on contemporary scientific thought. The recent literature in the life sciences dealing with comparisons between different groups of animals abounds with references to a hierarchy called the "phylogenetic scale," which appears to be the modern counterpart of the *scala naturae*. According to their relative positions on this scale, animals are designated as "subprimate" or "submammalian" or "higher animals" or "lower animals." However, there seems to be no compilation of the phylogenetic scale to which one could refer to answer such questions as "Is a porpoise a higher animal than a cat?" Nevertheless, in a recent textbook, Waters (1960) characterizes comparative psychology as "the study of behavior wherever exhibited along the phylogenetic scale . . . [p. 14]." Likewise, Ratner and Denny (1964), in a discussion of the evolution of behavior, state that "As one climbs the scale from fish to primates the principle seems best stated as follows: The higher the phyletic level, the greater the multiple determination of behavior [p. 680]." The meaning of such terms as "phyletic level" is usually not specified but the implication seems to be that an organism's phyletic level is determined by its proximity to man on the *scala naturae*. Apparently, these writers and numerous others have confused the *scala naturae* with another organizational arrangement of animals, one based on probable lines of evolutionary descent: the phylogenetic tree. However, as Simpson (1958a) has pointed out, the phylogenetic tree is a genealogy. It is based on the data of paleontology and comparative morphology and represents the current state of knowledge about the course of evolution of the various animal species.[3] Like any other historical survey, it is subject to change with the acquisition of new data and by itself gives no indication of the relative status of the individuals listed with respect to any gradational arrangement. On the other hand, the *scala naturae* or phylogenetic scale is a hierarchical classification. While such a hierarchy can provide interesting information about relative performance and relative degrees of structural differentiation, it tells us nothing about evolutionary development since it is unrelated to specific evolutionary lineages. Thus, to say that amphibians represent a higher degree of evolutionary development than teleost fish is practically without meaning since they have each followed independent courses of evolution. Moreover, one can find characteristics in which teleosts exceed amphibians as well as vice versa. For example, the central nervous system of teleost fishes in many ways exhibits a greater degree of differentiation and specialization than does that of amphibians (Ariëns Kappers, Huber, & Crosby, 1960). Indeed the general absence of amphibians from recent comparative studies of learning or intelligence such as those of Bitterman (1965a, 1965b) suggests that their behavior patterns may be relatively inflexible. The difficulties encountered in the training of these organisms are discussed by McGill (1960) and van Bergeijk (1967).

An important feature of the *scale naturae* is the concept of a smooth continuity between living animal forms rather than the discontinuities implicit in the theory of evolution as a result of the divergence of evolutionary lines and the extinction of many intermediate forms. This continuity, which Lovejoy (1936) calls the "principle of unilinear gradation," has also had a profound influence on contemporary research design and theorization in the comparative life sciences. For example, in a paper on the evolution of learning, Harlow (1958) speculates that

simple as well as complex learning problems might be arranged into an orderly classification in terms of difficulty, and that the capabilities of animals on these tasks would correspond roughly to their positions on the phylogenetic scale [p. 283].

A survey of the literature reviews and research reports of the past 10 years might lead one to the conclusion that meaningful statements could be made about the evolution of some morphological, physiological, or behavioral characteristic by comparing goldfish, frog, pigeon, cat, and man. As Simpson (1958a) puts it,

some such sequence as dogfish-frog-cat-man is frequently taught as "evolutionary," *i.e.*, historical. In fact the anatomical differences among these organisms are in large part ecologically and behaviorally determined, are divergent and not sequential, and do not in any useful sense form a historical series [p. 11].

Another characteristic of the *scala naturae,* which seems to be implicit in discussions of a phylogenetic scale, is the notion that man is the inevitable goal of the evolutionary process and that once he has evolved, the phylogenetic process ends. These ideas have been succinctly expressed in the following lines by Emerson:

> Striving to be man, the worm
> Mounts through all the spires of form.

While this view of the animal kingdom may have a certain amount of face validity, it too runs contrary to the currently accepted data on the course of evolutionary history which indicate that primates represent only one of the many lines of vertebrates which have evolved and survive today.

The Phylogenetic Tree

Figure 1 presents a phylogenetic tree showing the approximate times of origin and probable lineages of the various classes of living vertebrates and some related groups of animals (Young, 1962). The animals represented across the top form an approximation of the *scala naturae* or phylogenetic scale. Note that the sequence of animals from left to right is completely arbitrary. A very different sequence would result from merely having some evolutionary lines branch to the left instead of the right. This

---

[3] The term "phylogenetic tree" is used here in a generic sense since such trees, constructed by various evolutionary theorists, would differ in some details.

340   W. Hodos and C. B. G. Campbell

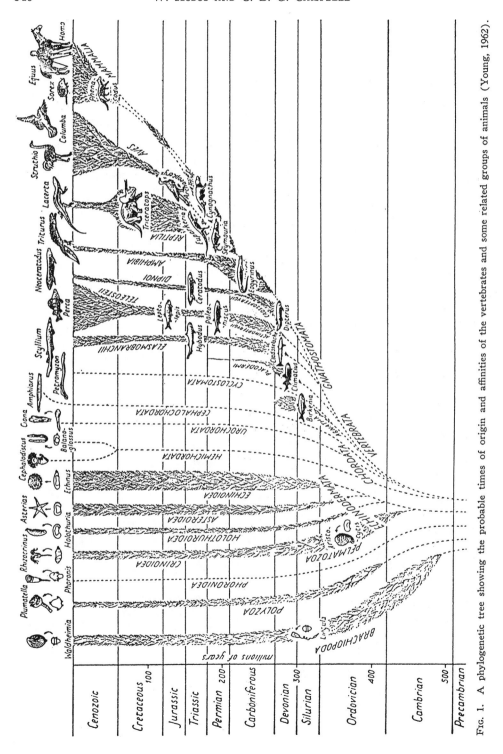

Fig. 1. A phylogenetic tree showing the probable times of origin and affinities of the vertebrates and some related groups of animals (Young, 1962).

would in no way alter the schematic representation of evolutionary lineages.

Several additional features of this phylogenetic tree should be noted. First, the evolutionary line of vertebrates leading to mammals, which begins in the Cambrian period, passes only through lobe-finned fishes (crossopterygians), amphibians, and reptiles. Second, the line of fishes which gave rise to amphibians evolved fairly early in fish evolution and followed a course of development quite separate from that of other fishes. The teleost fishes, so often used as a basis for evolutionary comparison with "higher vertebrates," are descendents of a line of development which is collateral to that giving rise to tetrapods. Therefore, no teleost fish ever was an ancestor of any amphibian, reptile, bird, or mammal. Likewise, birds represent another line of specialization from the reptiles and cannot be regarded as ancestral to mammals.

A similar situation exists in the phylogenesis of mammals. As may be seen in Figure 2, which represents a phylogenetic tree of mammals (Young, 1962), primates evolved as a specialized branch of the insec-

tivore line (i.e., shrews, moles, hedgehogs, etc.). Carnivores and rodents, which are frequently compared with primates, have followed independent courses of development from the primate line and from each other since the late Cretaceous or early Paleocene periods. Rats were never ancestral to cats nor were cats ancestral to primates; rather, each represents a different evolutionary lineage. Therefore, from the point of view of the phylogenesis of primate characteristics, the rat-cat-monkey comparison is meaningless.

Figure 3 represents a phylogenetic tree of primates (Young, 1962). The earliest primates, the prosimians, appear to have developed as a specialization of the line of the insectivores. The living prosimians (tarsiers, lorises, and lemurs) retain some insectivore characteristics (LeGros Clark, 1959). A comparison could therefore be made between living insectivores, prosimians, cercopithecid (Old World) monkeys, pongids (great apes), and hominids (men) which would give some clue to patterns of evolution in the human lineage. Such a comparison was recently made by Diamond

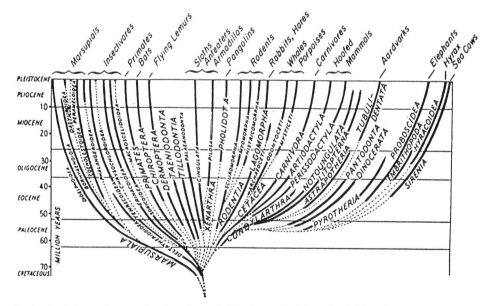

Fig. 2. A phylogenetic tree showing the probable times of origin and affinities of the orders of mammals (Young, 1962). (Common names have been added at the top.)

342   W. Hodos and C. B. G. Campbell

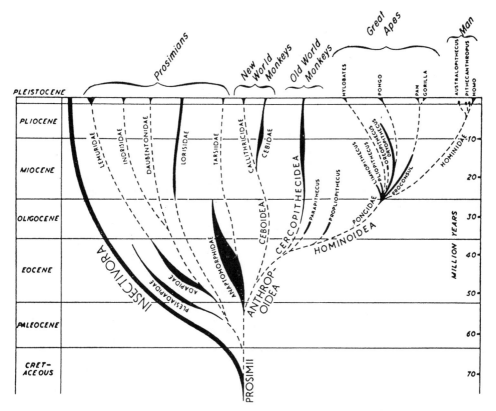

Fig. 3. A phylogenetic tree showing the probable times of origin and affinities of the primates (Young, 1962). (Common names have been added at the top.)

(1967) as an attempt to infer the course of evolution of neocortex in man.

Although primates are more closely related to each other than to other mammalian orders, there is still considerable variation and specialization within the primate order. Scott (1967) has recently warned that

Subhuman primates are not small human beings with fur coats and (sometimes) long tails. Rather, they are a group which has diversified in many ways, so that they are as different from each other as are bears, dogs and racoons in the order Carnivora. The fact that an animal is a primate therefore does not automatically mean that its behavior has special relevance to human behavior [p. 72].[4]

---

[4] The term "subhuman" connotes relative position on the *scala naturae* or phylogenetic scale. A term more in keeping with the conceptual framework of the phylogenetic tree would be "nonhuman." Similar implications are carried in the

EVOLUTIONARY INFERENCES FROM THE STUDY OF LIVING ANIMALS

Even though one may select animals for study that are descendants of a common evolutionary lineage, the question remains as to what can be learned about evolution, behavioral or otherwise, from the study of living animals. The rationale for such a study is based on the assumption that the living animals that have been selected are sufficiently similar to the ancestral forms that some inferences about the ancestral forms can be drawn. Our knowledge of ancestral vertebrates is based on fossilized skeletons. Soft tissues, including the central nervous system, do not usually fossilize; nor does

---

term "subprimate" which should be replaced by "nonprimate" and "submammalian" which should be replaced by "nonmammalian."

behavior. Therefore, a living animal species which appears to be skeletally very similar to fossil ancestors, may in fact have undergone some change in its soft tissues or behavior. Thus, conclusions about evolution, even though they are based on the study of living animals selected because they are descendents of a continuous evolutionary lineage, do retain some degree of uncertainty. However, for the study of behavioral evolution, there are no alternatives. The behavior of ancestral organisms can only be inferred from the behavior of living organisms which appear to be structurally similar and which inhabit similar environments. A method based on these assumptions has worked with reasonably good success in the study of morphological evolution and there seems to be good reason to believe that it will serve psychologists as well as it has served morphologists (Mayr, 1958; Simpson, 1958a).

Two general approaches can be used to collect data which are meaningful for the study of behavioral evolution. The first is the historical or "phylogenetic" approach similar to that used in comparative anatomy. In this instance, an attempt is made to infer changes in behavioral characteristics or patterns through time by comparing living animals which form a quasi-evolutionary series within a common lineage. An important consideration in such comparisons is the fact that the animals which have been chosen to represent ancestral groups are not the actual ancestors, but merely descendents of them. They are suitable for phylogenetic comparisons because they possess many characteristics which are primitive; that is, unchanged from ancestral forms (Simpson, 1961). The primitive behavior pattern of a group can often be inferred by comparing the behaviors of the living members of the group and looking for elements of the behavior pattern common to all. However, the greater the degree of diversity and specialization within a group, the greater the need for studying more variants. Furthermore, these elements should be sought in related groups and in living representatives of ancestral groups.

In using the "phylogenetic" method, the student of behavior has the same problems which face morphologists in the study of evolution. Some attempt must be made to determine which characteristics of the behavior are primitive, that is, derived from the behavior of the ancestors of the species, and which are specializations of the species studied. However, Simpson (1961) has cautioned that the concepts of primitive and specialized are meaningful only when related to a particular taxon, lineage, or phylogeny. Therefore, a behavioral characteristic which is found as a specialization in one animal group may be a primitive trait in another.[5]

In the case of living animals, the assumption is usually made that the more primitive characteristics a given species has, the more likely it is to resemble the ancestral form. However, the behaviorist has no behavioral fossil record to corroborate his decisions. He must decide on the basis of the properties of the behavior itself. Moreover, Klopfer and Hailman (1967), in their discussion of behavioral phylogenesis, conclude that

there is no good reason to assume that the most complex form has been evolved from the simpler one, since there are many cases of secondary simplification of morphological structures among animals. Also, one cannot infer direction of evolution from the relative abundance of variants; one cannot know a priori, whether the variation represents the end of an evolutionary process of adaptation (the commonest variant being the most advanced one) or merely the beginning (the commonest variant being the most primitive) [p. 187].

A case in point is the study by Doty, Jones, and Doty (1967). These investigators compared learning set performance in four species of carnivores. They found that domestic cats (which rank with domestic dogs as the "standard" carnivores) acquired the learning set more slowly than did weasels, skunks, or ferrets. While these data are of considerable interest, there is, unfortunately, no way of concluding from the behavior itself whether the inferior learning set performance of the domestic cats represents a primitive state, comparable to that which presum-

[5] See LeGros Clark (1959) and Simpson (1961) for a fuller discussion of the concepts of specialized and primitive.

344   W. Hodos and C. B. G. Campbell

ably existed in ancestral carnivores or whether it represents a more recent secondary simplification of an older, more complex form of behavior.

One possible means of resolving this problem would seem to be to fall back on the assertion that behavior is subject to the same evolutionary principles as any other characteristic of the organism. As Nissen (1958) has argued

it seems just as logical and possible that an adaptive behavior should give selective value to a related structural character as that an adaptive anatomic feature should lend selective advantage to behavior which incorporates or exploits that structure [p. 186].

We might further assume that an animal which is morphologically primitive will be behaviorally primitive and one which has developed morphological specializations will have also developed appropriate behavioral specializations. This assumption seems justified in view of the strong emphasis which paleontologists have placed on the relations between form and function in survival (Colbert, 1958; Simpson, 1958b).

Attempts to set up quasi-evolutionary series of living animals for the purpose of determining trends in behavioral or nervous system evolution have usually been unsatisfactory because the animals chosen for the series were highly specialized species of widely divergent lineages. For example, von Békésy (1960) traced the course of evolution of the organ of Corti from bird to alligator to duckbill to man. Unfortunately, specific conclusions about the evolution of the organ of Corti are unjustified from this group of animals since birds are not ancestral to alligators or to mammals. Moreover, there are virtually no paleontological data on the monotremes, and their relationship to the other mammalian orders is obscure (Romer, 1966). Similarly, Bishop (1958), in a discussion of the evolution of the cerebral cortex, chose the brain of a snake to represent the reptile group from which mammals evolved. Snakes are among the most specialized of the living reptiles (Goin & Goin, 1962) and first appear in the fossil record in the Cretaceous

period, 100 million years after the origin of the reptilian group which gave rise to the mammals (Romer, 1966).

Quasi-evolutionary series also often suffer from being too restrictive; that is, they attempt to represent all of vertebrate evolution with four or five species and sometimes even fewer. Bishop (1958), for example, based his evolutionary conclusions on three brains, one of which was the unlikely composite of a cat and a monkey.

Aside from the uncertainties inherent in using living animals to approximate actual evolutionary sequences and the necessity of choosing species which retain the most primitive traits, another problem arises. In some instances no living representatives of truly ancestral groups exist. Students of evolutionary patterns within the placental mammals, including the primates, however, are fortunate in that relatively unchanged direct descendants of the ancestral insectivores that gave rise to other placental mammals are living today. All of the living insectivores have diverged from ancestral insectivores to some degree, but the Erinaceidae (hedgehogs) are the most primitive and retain a relatively primitive eutherian status (Anderson & Jones, 1967). Study of this group of animals in some detail will probably be very profitable. Marsupials and placentals presumably had a common ancestry (Romer, 1966) and comparative studies of the primitive members of these two groups might allow some inferences to be made about the earliest mammals. Although monotremes were probably never ancestral to marsupials or placentals (Romer, 1966), they do retain many more reptilian characteristics than do the latter groups of mammals. The study of monotremes may therefore provide some insight into the behavioral characteristics of the therapsid reptiles from which all mammals evolved.

Comparative studies of behavioral or neural evolution of fishes have a particular advantage in that there are over 25,000 species of living fishes (Herald, 1961), a number of which closely resemble ancestral forms. Furthermore, fishes have undergone a considerable degree of evolution and occupy a great variety of ecological niches. In contrast, the living amphibians are relatively poor in number of species and diversity of terrain occupied and are generally regarded as a decadent class containing many relict groups (Darlington, 1957; Goin & Goin, 1962).

The crocodilians are living descendents of the archosaurian reptiles which also gave rise to birds (Romer, 1966) and appear to have remained relatively unchanged. Although the avian and crocodilian lineages are divergent, comparisons of the behavior and central nervous systems of these groups might be useful in view of their common archosaurian heritage.

There are unfortunately no surviving representatives of the reptile groups which gave rise to the mammals. These therapsid reptiles disappeared from the fossil record in the Jurassic period and since their lineage branched off from the stem reptiles very early, none of the living reptiles can be considered to be very closely related to mammals. Although turtles seem to be relatively unchanged since the Triassic period, they are highly specialized. *Sphenodon*, a surviving rhynchocephalian, also seems to be little changed since the Mesozoic, and the crocodilians mentioned previously are only slightly modified following their origin from thecodonts in the Triassic period (Romer, 1966). Comparative studies on the more primitive reptiles might allow some inferences about the behavioral repertoire of the ancestral reptiles from which the mammalian lineage is also derived. However, this would be a "weak inference" in the terminology of Klopfer and Hailman (1967) since it would rest on the assumption that the most frequently observed characteristics are primitive.

In his discussion of the evolution of intelligence, Bitterman (1965a, 1965b) compares the performance of teleost fish, turtles, pigeons, rats, and monkeys on probability and discrimination reversal learning tasks and characterizes their performance as "fishlike" or "ratlike." However, these comparisons do not permit generalizations to be made about the evolution of intelligence or any other characteristic of these organisms since they are not representative of a common evolutionary lineage. Figure 4 is a diagrammatic representation of two vertebrate phylogenetic trees. The left tree shows the approximate evolutionary relationships among the various animals studied by Bitterman in the visual reversal problem and the right tree shows the evolutionary relationships among the animals in the visual probability problem. The squares indicate the animals whose behavior was characterized by Bitterman as "ratlike" and the triangles indicate the animals whose behavior was characterized as "fishlike." Assuming for the moment that these animals are sufficiently similar to ancestral forms that some specific evolutionary generalizations could be made from them, one might conclude that the "ratlike" pattern in visual probability learning is limited to mammals. However, since no rat was ever an ancestor of any monkey, it is not clear whether the rat and monkey independently evolved the "ratlike" pattern or whether it was inherited from a common reptilian ancestor. The absence of this pattern in turtles would not support either possibility since turtles are so far from the line (or lines) of reptiles that were ancestral to mammals. In the case of the visual reversal learning data, the "ratlike" performance pattern is present in pigeons, rats, and monkeys, but absent in turtles. Again, no firm conclusions can be drawn. On the one hand, the assumption might be made that the "ratlike" pattern evolved independently in birds and mammals, since turtles, which share a remote common ancestry with birds and mammals in the stem reptiles (Romer, 1966), possess the presumably primitive "fishlike" pattern. On the other hand, the absence of the "ratlike" pattern in turtles may represent a secondary simplification of behavior. Still a third possibility must be considered in the analysis of Bitterman's data; that is, whether the "fishlike" pattern is a primitive characteristic at all. In the absence of data on the distribution of the "fishlike" pattern in amphibians and nonteleost fish, it is not possible to determine whether this pattern was

346   W. Hodos and C. B. G. Campbell

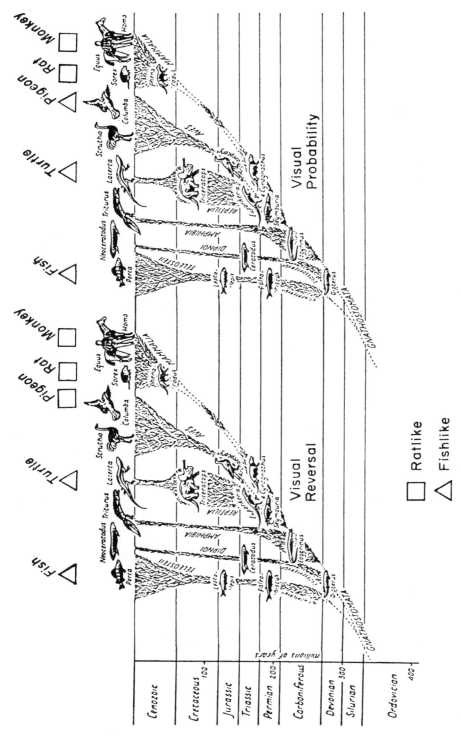

Fig. 4. Data from the table presented by Bitterman (1965a) plotted on segments of the phylogenetic tree shown in Figure 1 to indicate their relationship to approximate lineages. (The characterization of the behavior as "ratlike" or "fishlike" is Bitterman's notation.)

independently acquired by teleosts and turtles or inherited from a common ancestor.

Similar problems attend Harlow's (1959) interpretation of evolutionary trends in learning set performance in primates. Harlow compares data from New World monkeys, Old World monkeys, chimpanzees, and humans. Old World monkeys and chimpanzees, although specialized in their own ways, are reasonable enough representatives of human ancestors (Romer, 1966) that meaningful conclusions about the phylogenesis of man's behavior may be drawn. However, the New World monkeys appear to have evolved from New World prosimians which are not closely related to the Old World prosimians from which are derived Old World monkeys, apes, and men. Consequently, New World monkeys should not be used as representatives of man's ancestors (Campbell, 1966; Romer, 1966). A more appropriate representative would have been one or more species of the Old World prosimians.

The comparative studies of Bitterman, Harlow, and others discussed in this paper have been selected to illustrate some of the complexities involved in the interpretation of evolutionary relationships, not as a critique of their research. Unless comparative psychologists are prepared to deal with the details and intricacies of the evolutionary history of vertebrates, meaningful descriptive statements or theoretical formulations of specific behavioral phylogenies will not be forthcoming.

Analysis of Adaptation

A second and equally important approach to the study of behavioral evolution is through the analysis of adaptation. This method is based on the study of living animals, selected because they possess differing degrees of specialization (adaptation) with respect to some particular characteristic such as development of sense organs or central nervous system, the amount of postnatal care given to offspring, complexity of courtship patterns, etc. Such animals need not be descendants of a common evolutionary lineage and consequently any conclusions drawn

from such comparisons will be applicable only to general principles of adaptation and survival. They will give no direct clues to specific sequential patterns of evolutionary development in specific lineages. For example, Weiskrantz's (1961) description of the differential effect of lesions in the visual system in frogs, birds, rats, cats, dogs, monkeys, and man does not provide us with a picture of the sequential evolutionary history of the visual system and its function in the line of vertebrates leading to man; however, it is very useful in understanding the general relationship between the development of certain structural adaptations and their function in behavior. The same may be said of Woolsey's (1958) comparative studies of the sensory-motor cortex in rat, rabbit, cat, and monkey. They are extremely useful in relating structure and function, but are of no value in describing the sequence of evolutionary development in this system since each of these animals reached its respective degree of cortical development quite independently of any of the others.

Commenting on problems of the analysis of morphological evolution, Davis (1954) observed that most studies of comparative anatomy were concerned with determining the particular course of evolution rather than the reasons that one course was followed and not another. In his view, comparative studies should also be aimed at the mechanisms of adaptation. Similarly in comparative psychology, the study of analogous behavior in animals of divergent groups may be very useful in formulating generalizations about behavioral adaptation to specific problems of survival. Such generalizations might have broad applicability to a number of lineages of the phylogenetic tree and would greatly aid in the interpretation of data obtained through the phylogenetic approach.

Thus, the phylogenetic and adaptation approaches are not mutually exclusive. Data obtained through one approach can be used to augment conclusions based on the other. However, the experimenter must be clear as to which method of comparison he is using if he is to avoid the interpretive pitfalls described above.

W. HODOS AND C. B. G. CAMPBELL

CRITIQUE OF THE PHYLOGENETIC MODEL

King and Nichols (1960) object to the use of a phylogenetic model as a basis for the study of comparative psychology for three reasons: First, one apparently cannot predict behavior from one phylogenetic level to another. Second, the experimenter is required to make inferences about the behavior of extinct forms. Third, some taxonomic groups are differentiated only by characteristics which appear to be relatively unrelated to behavior. They suggest instead that comparative psychology develop its own system of classification based on behavior. However, such a classification would simply be a behavioral hierarchy and since it would not necessarily be related to evolutionary lineages, it would be nearly useless in understanding the evolution of behavior. Unless comparisons are made between organisms of a common evolutionary lineage, the relationship between the evolution of structure and the evolution of behavior will never become discernible. The apparent failure of the zoological model of evolution to lead to meaningful predictions of relative behavioral performance has been due to the fact that many comparative psychologists, neuroanatomists, physiologists, etc., have applied the inappropriate phylogenetic scale model rather than the appropriate phylogenetic tree model. Furthermore, they have generally failed to recognize the important taxonomic principle that there are diverse sources of similarity (both morphological and behavioral) among various organisms. Only one of these sources is inheritance from common ancestors and is usually referred to as "homology." On the other hand, similarities may also be due to the independent evolution of similar characteristics by more or less closely related forms. This similarity, not due to inheritance from a common ancestor, is referred to as "homoplasy." Homoplasy is a generic term which includes such forms of similarity as parallelism, convergence, analogy, mimicry, etc.[6] When nonhomologous characters serve similar *functions,* whether or not they are similar in appearance, they

are referred to as "analogous." Thus, the hands of a racoon and a man are homologous as anterior pentadactyl appendages and homoplastic in their particular appearance as hands. They are not homologous as hands since they were evolved independently of each other. Finally, they are analogous in their functions as prehensile organs.

There would then seem to be little point in looking for parallels between a hierarchy of behavioral complexity and the phylogenetic tree unless the behaviors were homologous. Behaviors that are convergent or analogous may provide insights into general mechanisms of survival and adaptation, but will not shed any light on the specific behavioral history of any particular group of organisms.

Another deterrent to theoretical development in comparative psychology has been the typological approach to behavior. As Mayr (1968) has described it, "When the learning psychologist speaks of The Rat or The Monkey, or the racist speaks of The Negro, this is typological thinking [p. 597]." The typological approach carries with it the implication that the particular species being investigated is a generalized representative of the entire order or class when in fact that species may be highly specialized and not at all representative. Simpson (1958a, 1958b) has also discussed this point. Bitterman (1965a, 1965b) has regrettably used typological designations of "ratlike" and "fishlike" as a shorthand characterization of the behavior of the teleosts, reptiles, birds, and mammals which he has studied in probability learning and reversal learning problems. Unfortunately, this typology implies that the "fishlike" behavior of the particular fishes which Bitterman studied is representative of most fishes or possibly all fishes. Considering the diversity of specialization in teleost fishes, one would not be surprised to find several or even many species which had evolved the capacity for behavior analogous to that which Bitterman has called "ratlike."

CONCLUSIONS

Beach (1950) has suggested that any experiments in which a nonhuman species

[6] For a more extensive discussion of homoplasy and homology, see Simpson (1961).

is compared with another nonhuman species or with humans should be considered as being within the realm of comparative psychology. His main point is that the term comparative psychology should be reserved for experiments in which organisms of different species are *compared*. We would like to expand this definition by further suggesting that any experiments carried out with nonhuman subjects, in which no attempt is made to compare these subjects to other species of animals or humans, be regarded as simply animal psychology. In other words, the term comparative psychology should be reserved for experiments in which interspecies comparisons are made.

Schneirla (1949) has stated that

> The general problem of the animal psychologist is the nature of behavioral capacities on all levels of accomplishment. He must contrive to understand how each animal type functions as a whole in meeting its surrounding conditions: what its capacities are like and how they are organized . . . [p. 245].

This would seem to be a reasonable goal for animal and comparative psychologists alike. The difference between the two would be the emphasis of the latter on the similarities and differences between various taxonomic groups of organisms. However, Ratner and Denny (1964) warn that "capricious comparisons" will add little to our understanding of systematic differences and similarities among species. This is not to say that questions such as "Do pigeons acquire learning sets faster than rats?" are without meaning if there is a specific reason for comparing those particular organisms and the suggestion is not made that the outcomes of the comparisons necessarily imply anything about the evolution of these behaviors. However, our opinion is that if comparative psychology is ever to develop a theoretical model capable of predicting behavior, experimenters will have to specify their independent variable more precisely than "species difference." Similarly, an experimenter who reports differences in visual acuity in several species that have varying degrees of differentiation of the visual cortex might be criticized on the grounds that these animals also vary in the near point of accommodation

which could be a confounding variable. This example illustrates the point that comparative psychologists have the same problems as other experimenters in determining that their chosen independent variable is indeed the only relevant variable operating.

What then can we hope for as attainable goals for comparative psychology? First, as Schneirla (1949) has suggested, would be a description of the behavioral capacities of organisms throughout the animal kingdom. A second would be the search for systematic trends in behavior which hopefully would vary reliably with other taxonomic indexes, since form and function in nature are inextricably interrelated. A third would be an attempt to reconstruct the historical development of behavior as best as can be done from the data of paleontology and the study of living organisms which resemble, as closely as possible, ancestral forms of particular evolutionary lines. A fourth goal would be the analysis of the general mechanisms of adaptation and survival. These goals can best be attained by ridding ourselves of concepts like the phylogenetic scale, higher and lower animals, nonrepresentative behavioral typologies, and other notions which have had the effect of oversimplifying an extremely complex field of research.

REFERENCES

Anderson, S., & Jones, J. K. *Recent mammals of the world*. New York: Ronald, 1967.

Ariëns Kappers, C. U., Huber, G. C., & Crosby, E. C. *The comparative anatomy of the nervous system of vertebrates, including man*. New York: Hafner, 1936. (Republished: 1960.)

Aristotle. *Historia animalium*. (Trans. by D. W. Thompson) Oxford: Clarendon, 1910.

Aristotle. *De partibus animalium*. (Trans. by W. Ogle) Oxford: Clarendon, 1912. (a)

Aristotle. *De generatione animalium*. (Trans. by A. Platt) Oxford: Clarendon, 1912. (b)

Beach, F. A. The snark was a boojum. *American Psychologist*, 1950, 5, 115–124.

Bishop, G. H. The place of cortex in a reticular system. In H. H. Jasper et al. (Eds.), *Reticular formation of the brain*. Boston: Little, Brown, 1958.

Bitterman, M. E. The evolution of intelligence. *Scientific American*, 1965, 212, 92–100. (a)

Bitterman, M. E. Phyletic differences in learning. *American Psychologist*, 1965, 20, 396–410. (b)

350   W. Hodos and C. B. G. Campbell

Campbell, B. G. *Human evolution.* Chicago: Aldine, 1966.

Colbert, E. H. Morphology and behavior. In A. Roe & G. G. Simpson (Eds.), *Behavior and evolution.* New Haven: Yale University Press, 1958.

Darlington, P. J. *Zoogeography: The geographical distribution of animals.* New York: Wiley, 1957.

Davis, D. D. Primate evolution from the viewpoint of comparative anatomy. *Human Biology,* 1954, **26**, 211–219.

Diamond, I. T. The sensory neocortex. In W. D. Neff (Ed.), *Contributions to sensory physiology.* Vol. 2. New York: Academic Press, 1967.

Doty, B. A., Jones, C. N., & Doty, L. A. Learning set formation by mink, ferrets, skunks and cats. *Science,* 1967, **155**, 1579–1580.

Goin, C. J., & Goin, O. B. *Introduction to herpetology.* San Francisco: Freeman, 1962.

Harlow, H. F. The evolution of learning. In A. Roe & G. G. Simpson (Eds.), *Behavior and evolution.* New Haven: Yale University Press, 1958.

Harlow, H. F. Learning set and error factor theory. In S. Koch (Ed.), *Psychology: A study of a science.* Vol. 2. New Haven: Yale University Press, 1959.

Herald, E. S. *Living fishes of the world.* New York: Doubleday, 1961.

King, J. H., & Nichols, J. W. Problems of classification. In R. H. Waters, D. A. Rethlingshafter, & W. E. Caldwell (Eds.), *Principles of comparative psychology.* New York: McGraw-Hill, 1960.

Klopfer, P. H., & Hailman, J. P. *An introduction to animal behavior: Ethology's first century.* Englewood Cliffs, N. J.: Prentice-Hall, 1967

LeGros Clark, W. E. *The antecedents of man.* Edinburgh: Edinburgh University Press, 1959.

Lovejoy, A. O. *The great chain of being.* Cambridge: Harvard University Press, 1936.

Mayr, E. Behavior and systematics. In A. Roe & G. G. Simpson (Eds.), *Behavior and evolution.* New Haven: Yale University Press, 1958.

Mayr, E. The role of systematics in biology. *Science,* 1968, **159**, 595–599.

McGill, T. E. Response of the leopard frog to electric shock in an escape-learning situation. *Journal of Comparative and Physiological Psychology,* 1960, **53**, 443–445.

Nissen, H. W. Axes of behavioral comparison. In A. Roe & G. G. Simpson (Eds.), *Behavior and evolution.* New Haven: Yale University Press, 1958.

Ratner, S. C., & Denny, M. R. *Comparative psychology.* Homewood, Ill.: Dorsey, 1964.

Roe, A., & Simpson, G. G. (Eds.). *Behavior and evolution.* New Haven: Yale University Press, 1958.

Romer, A. S. *Vertebrate paleontology.* Chicago: University of Chicago Press, 1966.

Ross, W. D. *Aristotle.* London: Methuen, 1949.

Schneirla, T. C. Levels in the psychological capacities of animals. In R. W. Sellars, V. H. McGill, & M. Farber (Eds.), *Philosophy for the future.* New York: Macmillan, 1949.

Scott, J. P. Comparative psychology and ethology. *Annual Review of Psychology,* 1967, **18**, 65–86.

Simpson, G. G. The study of evolution: Methods and present status of theory. In A. Roe & G. G. Simpson (Eds.), *Behavior and evolution.* New Haven: Yale University Press, 1958. (a)

Simpson, G. G. Behavior and evolution. In A. Roe & G. G. Simpson (Eds.), *Behavior and evolution.* New Haven: Yale University Press, 1958. (b)

Simpson, G. G. *Principles of animal taxonomy.* New York: Columbia University Press, 1961.

van Bergeijk, W. Anticipatory feeding behavior in the bullfrog (*Rana catesbeiana*). *Animal Behaviour,* 1967, **15**, 231–238.

von Békésy, G. Experimental models of the cochlea with and without nerve supply. In G. L. Rasmussen & W. F. Windle (Eds.), *Neural mechanisms of the auditory and vestibular systems.* Springfield, Ill.: Charles C Thomas, 1960.

Waters, R. H. The nature of comparative psychology. In R. H. Waters, D. A. Rethlingshafter, & W. E. Caldwell (Eds.), *Principles of comparative psychology.* New York: McGraw-Hill, 1960.

Weiskrantz, L. Encephalization and the scotoma. In W. H. Thorpe & O. L. Zangwill (Eds.), *Current problems in animal behaviour.* Cambridge: Cambridge University Press, 1961.

Wightman, W. P. D. *The growth of scientific ideas.* Edinburgh: Oliver & Boyd, 1950.

Woolsey, C. N. Organization of somatic sensory and motor areas of the cerebral cortex. In H. F. Harlow & C. N. Woolsey (Eds.), *Biological and biochemical bases of behavior.* Madison: University of Wisconsin Press, 1958.

Young, J. Z. *The life of vertebrates.* Oxford: Oxford University Press, 1962.

(Received July 12, 1968)

3 Development and Learning

Genetic Influences, Learning, and Instinct

Charles T. Snowdon

Historical Background

The study of ontogeny has occupied a major place in the history of animal behavior. Is behavior determined primarily by genetic mechanisms or are environmental influences most important in behavioral development? The dispute between nature and nurture has led to many fractious and antagonistic debates (see Lehrman 1953, discussed above in Part 2). The strongest proponents of the nature position most frequently have been identified as those trained in the tradition of European ethology; the strongest proponents of the nurture position typically were those trained in the tradition of American learning psychology.

In reality there is no simple dichotomy between nature and nurture or between European and American ethologists. The attribution of one tradition as resulting from European ethology and the other from American psychology is a carica-

ture that does not reflect the historical record. In Britain, Spalding (1875) demonstrated what we now call imprinting. This form of learning occurs during a limited period of development and may be restricted to particular classes of stimuli, a clear interaction of nature and nurture. Lloyd Morgan (1896) introduced ideas of trial-and-error learning along with the idea that successful responses were reinforced and unsuccessful responses were inhibited.

In North America, a generation of psychologists was influenced by the English animal behaviorists. E. L. Thorndike began his studies of trial-and-error learning at Harvard shortly after a visit by Lloyd Morgan. J. B. Watson, the quintessential spokesman for controlled laboratory studies of behavior, actually began his career as a field biologist studying the nesting and mating behavior of terns. In response to psychologists like McDougall, who proposed a theory that attempted to describe all complex behavior as a system of instincts, Watson and other psychologists focused their research on rigorously controlled laboratory experiments evaluating behavioral plasticity. Often the behavior studied bore little resemblance to that expressed in more natural conditions. Other North Americans continued to study behavior in natural contexts, integrating concepts of both genetic determination and the importance of experience.

One of the most influential American zoologists was Charles Whitman (1898), whose study of instinctive behavior in insects and pigeons led him to conclude that notions of regularity and fixity of instincts were great exaggerations. Whitman's writings subsequently influenced the thinking of Konrad Lorenz. Hailman's (1967, 1969) demonstration of the interaction between instincts and plasticity showed clearly the interaction of perceptual and motor predispositions with learning during the ontogeny of the pecking response of gull chicks.

Wallace Craig, one of Whitman's students, was frustrated with instinct theory and focused instead on how external factors could affect behavior. Craig was one of the first to recognize that the behavior of an individual can be directed or influenced by behavior from social companions. The paper reprinted here (Craig 1918) was a precursor to many of the ideas that have directed thinking and research in animal behavior in the twentieth century. Craig distinguished between appetitive and consummatory behavior. Appetitive behavior is an active sequence of behavior that begins as trial-and-error activities directed toward a particular goal. The trial-and-error behavior may eventually be organized into smooth sequence of learned behavior as to appear like a series of reflexes. Consummatory responses were thought to be reflexive acts. Thus, a hungry animal might engage in a variety of nonstereotyped activities to locate food, and these would be shaped by reinforcement into a series of predictable actions. The actual ingestion of food is a consummatory response governed by reflexes such as salivation, chewing, and swallowing. Snowdon (1969) supported Craig's distinction of appetitive and consummatory behavior by showing that rats trained to feed themselves through an intragastric tube continued to make licking, chewing, and swallowing movements during self-injection even though no food was passing through the mouth or esophagus. An animal actively searches for a certain stimulus condition (food, receptive mate) through trial-and-error behavior that then permits the consummatory response.

A second important feature of the Craig paper is its statement of a motivational theory. Craig's discussion of appetites and aversions presages Schneirla's (1959) approach-withdrawal theory of motivation. Craig's discussion of a struggle between conflicting appetites preceded discussions of conflict in the ethological literature. He also includes one of the first descriptions of vacuum activity. Craig constructed a cyclic concept of motivation; one based on the view that appetitive behavior has a physiological basis. The absence of a stimulus thus leads to an increased physiological responsiveness that leads to restlessness and search behavior. When the appropriate stimulus is found, consummatory behavior is elicited, and the physiological state is altered. The organism then has a surfeit of the stimulus and develops an aversion to the stimulus. The avoidance of the stimulus leads to a state of physiological quiescence. This cyclic model of motivation has a clear relationship to Lorenz's subsequent hydraulic model of motivation (Lorenz 1950). By invoking an aversive process that arises during consumption, however, Craig differed sharply from Lorenz and provided the precursor of the opponent-process model of motivation of Solomon and Corbit (1974). The opponent-process model has received little attention in recent animal behavior research, but this model has been important in explaining a variety of human motivational issues, such as ad-

dictions. Craig was many decades ahead of his time, both in describing the necessary interrelatedness of learned and instinctive behaviors and in formulating motivational issues in ways that modern theorists have had to rediscover. Craig's focus on the physiological basis of appetitive behavior has resulted in an extensive literature on the physiology of motivation (e.g., Satinoff and Teitelbaum 1983).

Methods of Studying Development

We can acknowledge Craig's focus on the interaction of learned and instinctive activities on motivated behavior, but it is still important to consider these activities separately. Which aspects of behavior are determined genetically, which by experience, and which result from the interaction of genes and environment? Several methods have been developed to disentangle the relative contributions of genes and environment. The gamut of research in behavioral development includes such examples as surgically rotating the eyes of embryonic frogs and salamanders or transplanting skin patches from dorsum to ventrum. These experiments showed that, after a certain embryological stage, no further modification of behavior is possible (Sperry 1968, Weiss 1947, Jacobson 1970). Others conducting developmental studies demonstrated that population-specific dialects of bird song were learned early in life (Marler and Tamura 1964), that salmon imprinted on the odors of their home streams (Hasler 1966), and that cultural transmission of tool use in chimpanzees occurred through active teaching (Boesch 1993; McGrew 1992).

Lorenz (1965) proposed the isolation experiment as the ideal method. If an organism reared in isolation and deprived of the ability to obtain experience dem-

onstrates appropriate behavior, then that behavior has to be innate. Lorenz conveniently argued, however, that failure to demonstrate a behavior after isolation does not prove that the behavior must be acquired through experience, only that the isolation conditions were inappropriate. For example, some research studies showed deficits in high-level processing of visual information by animals deprived of visual input. These deficits subsequently were shown to be the result of retinal degeneration in the isolated animal (Riesen 1961). If the deprivation experience prevents input of appropriate releasing stimuli, then the failure to perform the appropriate behavior does not imply that learning is necessary. It is easy to find real or hypothetical reasons to account for a subject in a deprivation study failing to perform adequately. The isolation experiment can be used only to prove that a behavior is innate and can never be used to prove that experience is necessary. Thus, isolation experiments are rarely useful.

Cross-fostering methods have greater power. If an animal can be cross-fostered to a different species or subspecies and still display a species-specific trait, a genetic basis can be inferred. If an animal acquires the trait of a foster parent, then early experience must play an important role in behavioral development. Cross-fostering studies are often difficult to do. Not all parents will accept conspecific offspring that are not their own, let alone heterospecific offspring. The retention of species-typical behavior even after cross-fostering still may not indicate innateness: prenatal or pre-hatching experience can be of critical importance (see discussion of Gottlieb [1968] below). Also, when two species can be cross-fostered easily, the target behaviors of interest may not differ sufficiently between the species for cross-fostering to provide conclusive data on the role of experience (see, for example, Masataka and Fujita [1989] versus Owren et al. [1992] on vocal development in rhesus and Japanese macaques). Thus, while cross-fostering methods improve upon isolation experiments, cross-fostering still may not disentangle the contributions of nature and nurture.

Genetic Control of Behavior

Despite the problems with deprivation and cross-fostering studies, there are several methods of direct genetic manipulation—making use of spontaneous mutations, creating hybrids between species or populations, and selective breeding techniques—that can be used to determine the genetic control of behavior. There is a long history of research on selective breeding for specific behavioral traits in domestic animals (see Price 1984). Several laboratory studies have used directional selection for traits such as maze learning in rats (Tyron 1940), and phototactic behavior (Hirsch 1962) and mating speed (Manning 1961, 1963) in fruit flies (*Drosophila*). Other studies have examined breed differences in emotionality in dogs (Scott and Fuller 1965) or used specific mutations to explain the cellular processes controlling behavior in *Drosophila* (see Benzer 1973). The papers reprinted here go beyond genetic studies to relate the results to either evolutionary processes or physiological processes.

Margaret Bastock's dissertation paper (Bastock 1956) is an elegant example of experimental research where all variables can be carefully controlled and where many alternative hypotheses were considered and tested. While many papers have documented mutations that affect behavior patterns, few have taken Bastock's additional step of asking what effects the mutation might have for subsequent evolutionary change. She focused on the yellow mutant of the fruit fly (*Drosophila melanogaster*), rare in the wild but occurring frequently in captive stocks. Bastock first demonstrated that yellow males take twice as long to initiate courtship as wild-type males and that these mutants have a much lower percentage of courtship success than wild-type males. As both sexes must cooperate for sperm transfer to occur, the reduced performance of yellow males may be due to some deficit in male courtship, to lowered responsiveness of females to yellow males, or to an interaction of both variables.

Bastock described courtship patterns in wild-type males to provide a baseline for evaluating the performance of yellow males, and she conducted experiments to rule out the effects of odor and color differences and evaluated female responsiveness to different males. The results suggested that females do not respond differently to yellow males because of odor or color differences, and that the reduced mating success must be due to reduced courtship vigor by the yellow males. Bastock concluded that yellow males have reduced sexual motivation though she was not able to specify the physiological basis for this reduced motivation.

The small change in courtship behavior of yellow mutants is similar to differences between species and subspecies, suggesting that such changes might be evolutionarily significant. Yellow-type females were much more responsive to courtship of both wild and yellow males. Although a mutant such as the yellow mutant in fruit flies would not survive within a heterogeneous population, such mutations could survive in conjunction with concurrent changes in female responsiveness, if these mutant populations were to be isolated, or if there were other selection pressures favoring the mutants. Bastock argued that this was the first demonstration of a mutation that affected the frequency of elements in a behavior pattern, rather than affecting the entire pattern, and that such mutations might be quite common in behavioral evolution.

In a different kind of genetic study, Bentley and Hoy (1972) examined the stridulation patterns (or songs) of male crickets. Their goals were to determine the mechanisms of genetic transmission of the patterns and also to develop inferences about the nature of the neuronal oscillators that generate the stridulation patterns. From the time of Sherrington's (1906) work on spinal reflexes onward, insights concerning physiological mechanisms have frequently resulted from careful behavioral description. Behavioral analysis is an important tool in understanding physiology. Bentley and Hoy provided very detailed parametric descriptions of cricket songs, defining chirp-type, pulses, trills, and phrases and measuring the rate of pulses and inter-pulse intervals. This detailed description is necessary both for the genetic analysis and in order to develop inferences about the neuronal mechanisms producing the song.

Cricket stridulation is highly stereotypic among conspecifics and also shows a high degree of species specificity. Males reared in isolation from other conspecific crickets produce normal, species-specific songs. Each species can be readily identified on the basis of stridulation pattern alone. Hybrids between two species produce intermediate patterns that retain aspects of each parent. The nature of the hybrid pattern, however, suggested a polygenic rather than single-gene involvement for each song parameter. Furthermore, song inheritance is sex-linked. The songs of hybrids more closely resembled the songs of males of their mother's species than they did the songs of their father's species. A similar pattern of inheritance has been described for certain hybrid phenotypes of squirrel monkeys (Newman and Symmes 1983).

The second part of the Bentley and Hoy paper considered hybrids of two species in which the chirp pulses are grouped into units of two or three, each unit being separated by silent periods. In a third species, pulse rate is continuous. The most parsimonious explanation for a pulsed chirp song had been that two oscillators with different rates are interacting to produce a chirp pattern. Only a single oscillator is presumed to be functioning in the non-chirping species. By examining the patterns of hybrids of chirping and nonchirping species, especially by measuring the interpulse intervals, Bentley and Hoy concluded that the two chirping species are governed by different neuronal mechanisms. The song of one species can be explained by a dual oscillator theory but the song of the second species is best explained by an alteration in a single oscillator. This study shows how a combination of genetic manipulation with precisely described behavior can lead to strong inferences about underlying physiological controls.

Interactions of Genes and Environment

Some behavior appears to be under strict genetic control. More commonly, behavior is a product of epigenesis—the interaction of genetic and environmental mechanisms. Several phenomena attest to this interaction. Tyron (1940) used selective breeding to produce two strains of rats, one that was "maze bright" and one that was "maze-dull," as measured by performance on a standard series of mazes. These strains are "bright" or "dull," however, only with respect to the specific mazes that were used in these experiments. Whatever behavior was selected in these studies, it was not general learning ability or intelligence. Furthermore, maze performance abilities can be modified by environmental variables. Rosenzweig, Bennett, and Diamond (1972) and Greenough (1975) have demonstrated that rats reared in an "enriched" environment, one with many objects and other rats present, had a greater number of brain cells and a thicker cortex than rats housed in isolation in traditional metal and wire-mesh cages. The enriched rats showed greater performance in learning tasks. Cooper and Zubek (1958) showed that "maze dull" rats reared in an enriched environment performed just as well as "maze bright" rats reared in a traditional environment.

These results are now joined by many others demonstrating that genetic differences in behavior can be modulated by early experience. For example, Galef (1971) compared aggressive behavior of wild and domestic rats and their hybrids and found that hybrids had levels of aggression intermediate to both parents. Wild rats cross-fostered to domestic rat mothers still displayed high levels of aggressive behavior. Both of these results suggested a genetic determination of aggressive behavior in wild rats. However, when Galef repeatedly presented wild rats with a specific stimulus used in his aggression tests, the wild rats displayed minimal levels of aggression to that stimulus. Thus, rats exposed each day to a novel rat intruder did not show the normal high levels of aggression toward intruders, while rats exposed to mice each day failed to attack and kill mice as other wild rats did. A genetic predisposition to aggressive behavior, therefore, could be modified by specific experience in early development.

The demonstration that a genetic predisposition can be modified by nongenetic effects illustrates the concept of epigenesis. A second area where epigenesis is clearly demonstrated is the phenomenon of imprinting. Imprinting is a phenomenon where learning is essential, but only for a very restricted period in development. The learning that occurs during the sensitive period is permanent. No further learning or reinforcement is required. Furthermore, not all stimuli are equally effective for imprinting. Many precocial birds develop an attachment to a specific object during a critical period (12–16 hours after hatching in chickens). Birds learn about stimulus characteristics of an attachment object and persist in following that object afterwards. Chickens, ducks, and geese will develop an attachment to a human being, to a different species or even to a moving inanimate object. Stimulation by novel objects before and after this sensitive period are ineffective. Imprinting is stronger if the stimulus is moving, produces sounds, and if the chick is able to follow the stimulus (Hess 1973).

Another well-studied example of epigenesis is learning of song in birds. Thorpe (1958) studied song learning in chaffinches and found that learning occurred during a critical period, songs from other species of finches were rejected, and performance of song was dissociated in time by several months from the learning of song. Thorpe also found that four young birds housed together without tutoring did somehow work out a species-typical song.

We have reprinted the paper by Marler and Tamura (1964), probably the most famous and influential paper on song learning. Young white-crowned sparrows reared in isolation have highly abnormal songs. Birds that were exposed to song stimuli only before 20 days of age or after 100 days of age failed to develop normal song. In contrast, birds exposed to only a few minutes of tape recorded song each day for a 2–3 week period between 20 and 100 days of age acquired normal adult song. Birds that received tutoring of songs of different but related species failed to develop that song, although birds from one dialect region could easily acquire the songs of white-crowned sparrows from a different dialect region.

The nature of this phenomenon is remarkable on several accounts. First, male song birds do not sing until the spring of their second year of life, so some memory trace must represent the tutored song heard early in life. Second, there must be species-specific perceptual mechanisms that filter through appropriate song models while eliminating inappropriate song models. Third, there must be temporal or maturational constraints on the developing brain to limit song learning to a very short period during development.

The significance of the Marler and Tamura paper results first from their choice of a species that has distinct geographical dialects. Birds from each of three distinct but nearby populations have dialects that are equally learnable by birds whose parents are from a different dialect group. Hence, this study suggests a possible mechanism for the cultural transmission of behavior. Second, Marler (1970) has used these results to develop bird song ontogeny as a model system for understanding human language development. He noted several similarities: the species specificity of song and language, a sensitive period for exposure to song or speech, the need to learn from tutors, the need to be able to hear both the tutor and oneself while practicing song or speech, and the importance of practice phases (subsong and plastic song for birds, babbling for humans) in acquiring vocal competence. Marler's theoretical ideas closely paralleled those of Lenneberg (1967) and Chomsky (1957) on the biological (as opposed to purely social or cultural) foundations of human language. Thus, considerable synergy has resulted between animal behaviorists and developmental psycholinguists who collectively have produced an enormous body of literature on both bird song and speech development.

Subsequent research has modified the original model of song acquisition. Studies on other populations of white-crowned sparrows by Petrinovich and Baptista (1987) have shown that the use of live tutors can extend the critical period for song learning beyond the period when tape-recorded tutoring song is effective, and at least a few birds have been able to learn songs of alien species when their live tutors are of that species. Cowbirds, which are nest parasites and are not exposed to conspecific adult tutors early in life, nevertheless respond to subtle social cues from females to shape their song development (West and King 1988). Other studies on calls (not song) illustrate that some birds can modify vocalizations in response to flock formation (Nowicki 1989) or to pairing with a new mate (Mundinger 1970). Despite the elaborations and modifications made in the ensuing years, the Marler and Tamura model of song acquisition in sparrows has had a major impact on how both zoologists and psychologists think about vocal development.

Two major proponents of an epigenetic view of behavioral development have been Gilbert Gottlieb and Zing-Yang Kuo (see Kuo 1967). Gottlieb (1968) summarized a series of exciting studies on the prenatal behavior of birds. He demonstrated that prenatal sensation and prenatal action is often critical to normal behavioral development. We often assume that behavioral development begins after birth or after hatching, so demonstration of prenatal sensation and prenatal behavior is extremely important. Behavior that may seem at birth to be fixed or innate may, in fact, have its origin in prenatal experience.

Kuo is credited with developing a highly creative technique that allows visualization of chick embryos though a window in the shell. Kuo (1932) showed that rhythmic movements of the fetal heart produced movements of the head that, he argued, prepared the chick for pecking at food after it hatched. By using this window method in conjunction with careful stimulation from different modalities, the ontogeny of sensitivity within an embryo can be determined. Gottlieb and colleagues showed that duck and chick embryos are sensitive to stimulation at a very early embryonic age. Moreover, there was a clear progression of the responsiveness of different sensory systems, progressing from non-visual photic sensitivity to tactile to vestibular to proprioception to audition to vision. Both auditory and tactile stimuli can be used to produce conditioned responses in late embryos.

The significance of documenting prenatal sensation and movement was enhanced by interpreting these developmental changes in terms of their subsequent functional effects. Vince (1964, 1966) showed that clicking sounds made by quail embryos were important in coordinating hatching. Gottlieb (1968) showed that prenatal auditory perception in ducks played a critical role in hatching. On the day before hatching, duck embryos show an increased rate of oral activity and vocal responsiveness to maternal duck calls but not to calls of siblings or of other species. This pre-hatching response occurred only in communally incubated embryos, however, and not in embryos incubated individually. This differential responsiveness suggests that the perception of sounds from sibling eggs also contributes to conditioning an individual's response to the maternal call.

Gottlieb's demonstration of the significance of pre-hatching communication between mothers and embryos required a new, more ecological, interpretation of laboratory studies of imprinting. In the natural environment, duck eggs hatch at different times during the day, with hatching being somewhat synchronized by an interaction of vocal exchanges between the mother and embryos. The mother does not leave the nest until the last egg has hatched, which means there is a wide range of post-hatching ages within a brood of ducklings when they first are able to follow a moving stimulus. For some individuals, the first opportunity to follow the mother will occur during the time when Hess (1973) hypothesized that fear of novel stimuli prevents imprinting from occurring. Although the first opportunity to follow the mother may not occur within the laboratory-determined critical period, the ducklings have heard maternal calls for two or three days prior to hatching. Thus, in nature, the real stimuli for imprinting are presented over a broader time-span than the effective critical period that was obtained from laboratory studies. Thus, prenatal sensory and motor processes permit prenatal learning that can be critical to an individual's ability to function normally and survive in its natural environment.

Stimulated by the important findings of Kuo and Gottlieb, the study of prenatal sensory and motor capacities has been extended to mammals. Smotherman and Robinson (1988) developed an elegant technique for studying the sensorimotor development of embryonic rats, and other scientists have shown that the human fetus can learn to recognize its mother's voice.

Naturalistic Studies of Learning

Although the previous section demonstrates the importance of considering the interaction of heredity and learning in the context of an organism's natural environment, the preponderance of research on learning has been conducted in controlled laboratory environments where the phenomena often appeared to have little relevance to real-world behavior (see Schwartz 1984 for a review).

However, many zoologists and some psychologists retained an interest in how learning occurred in an animal's natural environment and contributed to that animal's survival and reproductive success. These researchers also created laboratory studies that mimicked the learning that occurs naturally. In an exciting study that was among the first to demonstrate complex learning in natural environments, Tinbergen (1938) showed that a solitary sand wasp could learn the local features of the site where she built a nest. The female, upon emerging from the nest, makes a single brief flight around the nest before flying off to find provisions for her egg. Tinbergen manipulated the local cues and found that, if he moved the features while the wasp was away, she would return to the precise spot where her nest would have been if the array had not been moved. Thus, the wasp learns about the local cues near her nest and forms a spatial map during the brief flight around the nest as she emerges.

Wasps showed rudimentary concept learning as well. If a circular array of pine cones around a nest opening during the female's emergence was replaced with a circular array of flat chips and a triangular array of pine cones, the wasp would return to the circular array of the novel objects, chips, and not to the familiar pine cones placed in a new array.

There are many other examples of complex learning and memory in natural environments. Homing behavior, long thought to be innate, was shown to involve learning when adult birds that were displaced managed to return to their normal migratory grounds while displaced yearling birds did not (Perdeck 1958). Salmon that leave their natal stream and spend two-to-five years feeding in the ocean, can use memories of odor cues to return precisely to their natal stream (Hasler 1966). Many species of birds cache seeds and must remember enough of the cache locations to feed themselves through winter and spring (e.g., Balda and Kamil 1989).

In nonhuman primates, evidence of complex learning in natural situations is abundant. Seyfarth and Cheney (1986) have suggested that observational learning and social reinforcement are the mechanisms whereby young vervet monkeys learn the proper use of, and proper response to, the predator-specific alarm calls of vervet monkeys. Mineka and Cook (1993) have shown that fear of snakes is not innate in monkeys but can be rapidly learned by observing another monkey respond fearfully to a snake. Itani (1958) described the acquisition and spread of food-washing behavior in groups of Japanese macaques. Galef (1990) has argued that operant conditioning practices inadvertently used by human caretakers shaped food-washing behavior in monkeys. A subsequent study by Huffman (1984), however, described a similar spread of stone-rubbing behavior in Japanese macaques where feeding reinforcement could not be involved. McGrew (1992) documented population-specific patterns of tool use in wild chimpanzees and argued that cultural differences rather than ecological variables must be used to explain differences in tool use patterns.

We have selected two studies that are excellent examples of using controlled laboratory techniques to study learning phenomena that have natural relevance. Jane van Zandt Brower published a series of papers on mimicry derived from her dissertation research. Brower (1958) used monarch and viceroy butterflies to document the mechanisms of Batesian mimicry using Florida scrub jays as subjects. Batesian mimicry is the phenomenon where a non-toxic prey has similar visual, auditory, or olfactory features to a toxic prey. The adaptive significance of this imitation of a toxic animal by a non-toxic species results from predators being able to learn and remember from aversive encounters they have had with toxic prey items. Through processes of generalization, the predator subsequently avoids the non-toxic mimic as well as the toxic model.

Brower was the first to provide an experimental demonstration of how mimicry can work. In the present study alternate trials presented butterflies known to be palatable to experimental jays with toxic models being given on the other trials. After 50 presentations of the toxic monarch butterfly, the palatable mimic was presented on occasional trials. Although the experimental birds varied in the degree of responsiveness to both model and mimic, no bird ever ate either a model or a palatable mimic, and three of the four birds ceased responding to both model and mimic. In contrast, the four control birds eventually did not differentiate between palatable viceroy butterflies that were not native to Florida from other local non-mimetic palatable butterflies. Thus, prior experience with a toxic model does protect a palatable mimic from being preyed upon. The lack of response to both monarch and viceroy butterflies by the experimental jays is not due either to an inability to handle the butterflies or to a lack of hunger, as each of the non-mimetic palatable butterflies was always eaten. Thus, basic principles of learning, generalization and memory are necessarily involved in the evolution of mimicry as an adaptive predator defense.

Our second example of laboratory experimentation that revealed learning paradigms and their natural relevance is Garcia and Koelling (1966). This paper had a long history of rejection by psychologists, primarily because the results contradicted several then-current principles of animal learning. One of these principles was that any stimulus could easily be associated with any positive or negative reinforcer. Garcia and Koelling's work showed that this concept of "stimulus equality" was based on a profound ignorance of which stimuli were important cues to an animal in its natural environment. Another long-held principle disputed by this paper was that, if reinforcement was to be effective, the reinforcer had to be administered immediately after the stimulus. Again, the effectiveness of a delayed reinforcer made sense in the context of the animal's natural world. Garcia and Koelling noticed that ionizing radiation (such as used in cancer chemotherapy) had aversive effects on animals, leading to the avoidance of flavors and tastes closely associated with the x-irradiation. By pairing x-irradiation, lithium chloride, and electric shock with "bright-noisy" or "tasty" water, Garcia and Koelling found clear differences between stimulus type and reinforcer type. Rats exposed to "tasty" water paired with nausea induction showed a subsequent reduction of fluid intake as did rats that were shocked while drinking "bright-noisy" water. Rats with "bright-noisy" water paired with nausea induction and rats with "tasty" water paired with shock, showed no subsequent avoidance of the water. Thus, stimuli were not equally associated with different reinforcers. An ecological explanation can account for the results. Taste and smell cues should be associated with gastrointestinal disturbance while peripheral cues like visual and auditory stimuli should be associated with distal reinforcers such as pain stimuli.

A second finding is that delayed shock and the delayed effects of lithium chloride ingestion or x-irradiation did not slow the aversive learning process. Until the Garcia and Koelling study, there was no evidence that long-delayed reinforcement could be used effectively in conditioning studies.

The Garcia and Koelling paper spawned a vast number of empirical studies and theoretical explanations leading to a renewed interest in the biological and evolutionary constraints on learning (see Seligman and Harger 1972; Hinde and Stevenson-Hinde 1973). Results from these studies provided explanations for several puzzling phenomena. Rozin and Kalat (1971) explained hungers for very specific nutrients, such as thiamine, on the basis of taste-aversion learning principles. In particular, they explained why animals can respond to a variety of vitamin and mineral depletions by seeking foods that replenish the deficiency, even though these animals appear to have specific detection mechanisms only for sodium. Rozin and Kalat suggested that animals experiencing a nutritional deficiency learn to avoid the foods that lead to the deficiency and to seek out novel foods. Whenever a novel food source produced symptom reduction, the animals continued to ingest that food. Thus animals do not need specific sensory mechanisms to detect each essential nutrient; instead a more general learning paradigm will suffice.

The power of the Garcia and Koelling paradigm is that it can be applied to a variety of species, including humans. Snowdon (1977) applied this paradigm to understand the maladaptive behavior of lead ingestion in children. His experiments were based on the knowledge that calcium deficient rats and monkeys ingested lead at higher levels than animals with normal nutrition. Normal rats learned to avoid a flavored solution that was associated with an intubation of lead solution, while calcium deficient rats failed to learn an aversion to lead. Lead solutions presumably alleviate the symptoms of calcium deficiency on a short-term basis, providing a positive reinforcement that can promote continued lead ingestion even to toxic levels. Thus, the high incidence of lead ingestion in some human children might be explained as an initially adaptive response to calcium deficiency. Bernstein (1978) also used principles derived from the Garcia and Koelling study to develop ways to help cancer chemotherapy patients overcome the aversive conditioning to foods that chemotherapy often produces.

Social and Emotional Development

Harry Harlow has made many important contributions to animal behavior. He developed creative ways to test the learning and cognitive behavior of nonhuman monkeys using the Wisconsin General Test Apparatus. He also was the first to demonstrate the phenomenon of learning to learn, that monkeys could learn rules that would help them solve future problems without trial-and-error learning (Harlow 1949). Much of Harlow's work challenged traditional notions of motivation. The classic behaviorists related all actions to the solving of a few basic biological needs: feeding and drinking, fleeing predators, fighting, and reproducing. Infant attachment to a mother was thought to be a secondary drive based on the mother's providing food to her infant. Harlow and Zimmermann (1959) summarize several years of research with rhesus monkeys and surrogate mothers, demonstrating that characteristics of mothers other than food resources are important for attachment. When presented with a wire surrogate where milk could be obtained and with a terry cloth mother without milk, infants spent most of their time with the terry cloth surrogate. When the infants were placed in a novel environment or threatened with a strange object, these monkeys appeared to derive comfort from the cloth surrogate. Harlow and Zimmermann concluded that this effect could not be explained by secondary reinforcement from feeding and that other biological drives must make contact comfort reinforcing.

These studies have spawned a host of additional research using nonhuman primates as models of human attachment processes as well as social and emotional development. The surrogate-reared monkeys proved to be deficient in adult social skills, such as mating and infant care, so many subsequent studies have examined the roles that mothers, fathers, and infant peers play in normal social development. Sociosexual and infant care skills appeared to be normal in rhesus monkeys within a captive context if an infant was reared with its mother alone. This contrasts with marmosets and tamarins where young animals must have direct experience in helping an adult care for a new infant in order to develop competent parental skills (Epple 1978). In a study of the development of resiliency in infants (Mineka, Gunnar, and Champoux 1986), groups of young monkeys were reared in environments where they could push levers, pull chains, or perform other actions to get small food rewards. These monkeys were much more resilient to brief separations from their peers and showed less fear in strange situations than monkeys that received the same food incentives but did not have to push levers or pull chains. The experience of acting on the environment and producing results (what might be called work) leads to more competent and resilient primates.

Other researchers have examined brief separations from attachment objects (either peers or mothers) to discover that a great variety of ecological variables are important. Infants whose mothers leave while they remain in a group are much less affected by separation than infants who leave while mothers stay in a group (Hinde and Spencer-Booth 1971). Infants who are separated from their mothers but can see and hear them are more active and vocalize more, but have lower stress-hormone levels, than infants who have no contact with mothers during separation (Levine and Wiener 1988). Bonnet macaques (*Macaca radiata*), where infants interact with many other group members, show less behavioral stress to separation and weaning than do pigtail macaques (*Macaca nemestrina*), a species in which mothers prevent infants from interacting with other group members (Rosenblum and Kaufman 1968). Surprisingly, brief separations can affect the immune system for as long as several months, and a variety of social and environmental variables can accentuate or ameliorate the immune response (Coe 1993).

The work reviewed in Harlow and Zimmermann (1959), as well as subsequent research, has led to a better understanding of how to rear primates successfully in captivity and how to improve the well-being of captive primates. The

research also has direct relevance to human child-rearing in that separations and reunions parallel leaving children with baby-sitters or at day care centers and that providing many different social companions reduces the distress at separation. As monkeys that gain predictability and control over their environment by working for small food rewards are more resilient to potentially stressful situations, having children work for extra rewards might be valuable. Moreover, the demonstration of the importance of tactile stimulation to well-being in monkeys has led to recommendations for child rearing that have changed drastically since the 1940s. No longer are children left alone with minimal contact, as was common practice 40–50 years ago. High degrees of tactile contact are now encouraged between infants and caretakers. What may seem to be common sense today was not common practice a few decades ago. The work described in Harlow and Zimmermann has had a major impact on human child-rearing practices.

Literature Cited

Balda, R. P., and A. C. Kamil. 1989. A comparative study of cache recovery by three corvid species. *Animal Behaviour* 38:486–95.

Bastock, M. 1956. A gene mutation which changes a behavior pattern. *Evolution* 10:421–39.

Bentley, D. R., and R. Hoy. 1972. Genetic control of the neuronal network generating cricket song patterns. *Animal Behaviour* 20:478–92.

Benzer, S. 1973. Genetic dissection of behavior. *Scientific American* 229:24–37.

Bernstein, I. L. 1978. Learned taste aversions in children receiving chemotherapy. *Science* 200:1302–3.

Boesch, C. 1993. Aspects of transmission of tool-use in wild chimpanzees. Pp. 171–83 in K. R. Gibson and T. Ingold, eds., *Tools, Language and Cognition in Human Evolution*. Cambridge: Cambridge University Press.

Chomsky, N. 1957. *Syntactic Structures*. The Hague: Mouton.

Coe, C. L. 1993. Psychosocial factors and immunity in nonhuman primates: a review. *Psychosomatic Medicine*, in press.

Cooper, R., and J. Zubek. 1958. Effects of enriched and restricted early environments on the learning ability of bright and dull rats. *Canadian Journal of Psychology* 12:159–64.

Craig, W. 1918. Appetites and aversions as constituents of instincts. *Biological Bulletin* 34:91–107.

Epple, G. 1978. Reproductive and social behavior of captive marmosets with special reference to captive breeding. *Primates in Medicine* 10:50–62.

Galef, B. G. 1971. Social factors in the poison avoidance and feeding behavior of wild and domesticated rat pups. *Journal of Comparative and Physiological Psychology* 75:341–57.

Galef, B. G. 1990. Tradition in animals: Field observations and laboratory analyses. Pp. 74–95 in M. Bekoff and D. Jamieson, eds., *Interpretations and Explanations in the Study of Behavior: Comparative Perspectives.* Boulder: Westview Press.

Garcia, J., and R. A. Koelling. 1966. Relation of cue to consequence in avoidance learning. *Psychonomic Science.* 4:123–24.

Gottlieb, G. 1968. Prenatal behavior in birds. *Quarterly Review of Biology* 43:148–174.

Greenough, W. T. 1975. Experiential modification of the developing brain. *American Scientist* 63:37–46.

Hailman, J. P. 1967. The ontogeny of an instinct: the pecking response in chicks of the laughing gull (*Larus atricilla L.*) and related species. *Behaviour,* Supplement 15, 1–159.

Hailman, J. 1969. How an instinct is learned. *Scientific American* 221:98–106.

Harlow, H. F. 1949. The formation of learning sets. *Psychological Review* 56:51–96.

Harlow, H. F., and R. R. Zimmermann. 1959. Affectional responses in the infant monkey. *Science* 130:421–32.

Hasler, A. D. 1966. *Underwater Guideposts: Homing of Salmon.* Madison: University of Wisconsin Press.

Hess, E. H. 1973. *Imprinting.* New York: D. Van Nostrand.

Hinde, R. A., and Y. Spencer-Booth. 1971. Effect of brief separation from mother on rhesus monkeys. *Science* 173:111–18.

Hinde, R. A., and J. Stevenson-Hinde. 1973. *Constraints on Learning: Limitations and Predispositions.* New York: Academic Press.

Hirsch, J. 1962. Behavior genetics and individuality understood. *Science* 142:1436–42.

Huffman, M. A. 1984. Stone play of *Macaca fuscata* in Arashiyama B troop: Transmission of a nonadaptive behavior. *Journal of Human Evolution* 13:725–35.

Itani, J. 1958. On the acquisition and propagation of a new food habit in a troop of Japanese monkeys at Takasakiyama. *Primates* 1:131–48.

Jacobson, M. 1970. *Developmental Neurobiology.* New York: Holt, Rinehart, and Winston.

Kuo, Z.-Y. 1932. Ontogeny of embryonic behavior in aves: IV. The influence of embryonic movements upon behavior after hatching. *Journal of Comparative Psychology* 14:109–22.

Kuo, Z.-Y. 1967. *The Dynamics of Behavior Development: An Epigenetic View.* New York: Random House.

Lenneberg, E. H. 1967. *Biological Foundations of Language.* New York: Wiley.

Levine, S., and S. G. Wiener. 1988. Psychoendocrine aspects of mother-infant relationships in nonhuman primates. *Psychoneuroendocrinology* 13:143–54.

Lorenz, K. Z. 1950. The comparative method in studying innate behavior patterns. Pp. 221–68 in *Society of Experimental Biology Symposium on Physiological Mechanisms in Animal Behavior.* New York: Academic Press.

Lorenz, K. 1965. *Evolution and Modification of Behavior.* Chicago: University of Chicago Press.

Manning, A. 1961. The effects of artificial selection for mating speed in *Drosophila melanogaster. Animal Behaviour* 9:82–92.

Manning, A. 1963. Selection for mating speed in *Drosophila melanogaster,* based on the behavior of one sex. *Animal Behaviour* 11:116–120.

Marler, P. 1970. Birdsong and human speech: Can there be parallels? *American Scientist* 58:669–74.

Marler, P., and M. Tamura. 1964. Culturally transmitted patterns of vocal behavior in sparrows. *Science* 146:1483–86.

Masataka, N., and K. Fujita. 1989. Vocal learning of Japanese and rhesus monkeys. *Behaviour* 109:191–99.

McGrew, W. C. 1992. *Chimpanzee Material Culture.* Cambridge: Cambridge University Press.

Mineka, S., and M. Cook. 1993. Mechanisms involved in the observational conditioning of fear. *Journal of Experimental Psychology: General* 122:23–38.

Mineka, S., M. Gunnar, and M. Champoux. 1986. The effects of control on the early social and emotional development of rhesus monkeys. *Child Development* 57:1241–56.

Morgan, L. 1896. *Habit and Instinct.* London: Edward Arnold.

Mundinger, P. C. 1970. Vocal imitation and individual recognition of finch calls. *Science,* 168:480–82.

Newman, J. L., and D. Symmes. 1982. Inheritance and experience in the acquisition of primate acoustic behavior. Pp. 259–78 in C. T. Snowdon, C. H. Brown and M. R. Petersen, eds., *Primate Communication.* New York: Cambridge University Press.

Nowicki, S. 1989. Vocal plasticity in captive blackcapped chickadees: The acoustic basis of call convergence. *Animal Behaviour* 37:64–73.

Owren, M. J., J. A. Dieter, R. M. Seyfarth, and D. L. Cheney. 1992. Evidence of limited modification in the vocalizations of cross-fostered rhesus (*Macaca mulatta*) and Japanese (*M. fuscata*) macaques. Pp. 257–70 in T. Nishida, W. C. McGrew, P. Marler, M. Pickford and F. B. M. de

Waal, eds., *Topics in Primatology: Volume 1: Human Origins*. Tokyo: University of Tokyo Press.

Perdeck, A. C. 1958. Two types of orientation in migrating starlings *Sturnus vulgarus* L. and chaffinches *Fringella coelebs* L. as revealed by displacement experiments. *Ardea* 46:12–37.

Petrinovich, L., and L. F. Baptista. 1987. Song development in the white-crowned sparrow: modification of learned song. *Animal Behaviour* 35:961–74.

Price, E. O. 1984. Behavioral aspects of animal domestication. *Quarterly Review of Biology* 59:1–32.

Riesen, A. H. 1961. Stimulation as a requirement for growth and function in behavioral development. Pp. 57–80 in D. W. Fiske and S. R. Maddi, eds., *Functions of Varied Experience*. Homewood: Dorsey Press.

Rosenblum, L. A., and I. C. Kaufman. 1968. Variations in infant development and response to maternal loss in monkeys. *American Journal of Orthopsychiatry* 38:418–26.

Rosenzweig, M. R., E. L. Bennett, and M. C. Diamond. 1972. Brain changes in response to experience. *Scientific American* 226:22–29.

Rozin, P., and J. W. Kalat. 1971. Specific hungers and poison avoidance as adaptive specializations in learning. *Psychological Review* 78:459–86.

Satinoff, E., and P. Teitelbaum. 1983. *Handbook of Behavioral Neurobiology, Vol. 6: Motivation*. New York: Plenum Publishing.

Schneirla, T. C. 1959. An evolutionary and developmental theory of the biphasic processes underlying approach and withdrawal. *Nebraska Symposium on Motivation* 7:1–42.

Schwartz, B. 1984. *Psychology of Learning and Behavior*. New York: W. W. Norton.

Scott, J. P., and J. L. Fuller. 1965. *Genetics and the Social Behavior of the Dog*. Chicago: University of Chicago Press.

Seligman, M. E. P., and J. L. Harger. 1972. *Biological Boundaries of Learning*. New York: Appleton-Century Crofts.

Seyfarth, R. M., and D. L. Cheney. 1986. Vocal development in vervet monkeys. *Animal Behaviour* 34:1640–58.

Sherrington, C. S. 1906. *The Integrative Action of the Nervous System*. New Haven: Yale University Press.

Smotherman, W. P., and S. R. Robinson. 1988. *Behavior of the Fetus*. Caldwell: Telford Press.

Snowdon, C. T. 1969. Motivation, regulation and the control of meal parameters with oral and intragastric feeding. *Journal of Comparative Physiology and Psychology* 69:91–100.

Snowdon, C. T. 1977. A nutritional basis for lead pica. *Physiology and Behavior* 18:885–93.

Solomon, R. L., and J. D. Corbit. 1974. An opponent process theory of motivation. I. Temporal dynamics of affect. *Psychological Review* 81:119–45.

Spalding, D. A. 1875. Instinct and acquisition. *Nature* 12:507–8.

Sperry, R. W. 1968. Plasticity of neural maturation. *Developmental Biology Supplement* 2:306–27.

Thorpe, W. H. 1958. The learning of song patterns by birds with especial reference to the song of the chaffinch, *Fringella coelebs*. *Ibis* 100:535–70.

Tinbergen, N. 1938. On the orientation of the digger wasp *Philanthus triangulum* Fabr. III. Selective learning of landmarks. Reprinted in N. Tinbergen, *The Animal in Its World: Field Studies*. 1972. Pp. 146–96. London: Allen and Unwin.

Tyron, R. C. 1940. Studies in individual differences in maze ability: VIII. The specific components of maze ability and a general theory of psychological components. *Journal of Comparative Physiology and Psychology* 30:283–335.

van Zandt Brower, J. 1958. Experimental studies of mimicry in some North American Butterflies, Part 1: The monarch *Danaus plexippus* and viceroy *Limenitis archippus archippus*. *Evolution* 12:32–47.

Vince, M. A. 1964. Social facilitation of hatching in bobwhite quail. *Animal Behaviour* 12:531–34.

Vince, M. A. 1966. Artificial acceleration of hatching in quail embryos. *Animal Behaviour* 14:389–94.

Weiss, P. 1947. The problem of specificity in growth and development. *Yale Journal of Biology and Medicine* 19:235–78.

West, M. E., and A. P. King. 1988. Visual displays by female cowbirds affect the development of songs in males. *Nature* 334:244–46.

Whitman, C. O. 1898. Animal Behavior. *Biology Lectures, Marine Biology Laboratory (Woods Hole)*: 285–338.

APPETITES AND AVERSIONS AS CONSTITUENTS OF INSTINCTS.

WALLACE CRAIG,

UNIVERSITY OF MAINE.

GENERAL ACCOUNT OF APPETITE AND AVERSION.

The overt behavior of adult animals occurs largely in rather definite chains and cycles, and it has been held that these are merely chain reflexes. Many years of study of the behavior of animals—studies especially of the blond ring-dove (*Turtur risorius*) and other pigeons—have convinced me that instinctive behavior does not consist of mere chain reflexes; it involves other factors which it is the purpose of this article to describe. I do not deny that innate chain reflexes constitute a considerable part of the instinctive equipment of doves. Indeed, I think it probable that some of the dove's instincts include an element which is even a tropism as described by Loeb. But with few if any exceptions among the instincts of doves, this reflex action constitutes only a part of each instinct in which it is present. Each instinct involves an element of appetite, or aversion, or both.

An appetite (or appetence, if this term may be used with purely behavioristic meaning), so far as externally observable, is a state of agitation which continues so long as a certain stimulus, which may be called the appeted stimulus, is absent. When the appeted stimulus is at length received it stimulates a consummatory reaction, after which the appetitive behavior ceases and is succeeced by a state of relative rest.

An aversion (example 7, p. 100) is a state of agitation which continues so long as a certain stimulus, referred to as the disturbing stimulus, is present; but which ceases, being replaced by a state of relative rest, when that stimulus has ceased to act on the sense-organs.

The state of agitation, in either appetite or aversion, is exhibited externally by increased muscular tension; by static and phasic contractions of many skeletal and dermal muscles, giving rise to bodily attitudes and gestures which are easily recognized signs or "expressions" of appetite or of aversion; by restlessness; by activity, in extreme cases violent activity; and by "varied effort" (Lloyd Morgan, '96, 7, 122, 154; Stout, '07, 261, 267).

In the theoretically simplest case, which I think we may observe in doves to some extent, these states bring about the appeted situation in a simple mechanical manner. The organism is disturbed, actively moving, in one situation, but quiet and inactive in another; hence it tends to move out of the first situation and to remain in the second, obeying essentially the same law as is seen in the physical laboratory when sand or lycopodium powder on a sounding body leaves the antinodes and comes to rest in the nodes.

But pigeons seldom are guided in so simple a manner. Their behavior involves other factors which must be described in connection with appetite and aversion.

An appetite is accompanied by a certain *readiness to act*. When most fully predetermined, this has the form of a chain reflex. But in the case of most supposedly innate chain reflexes, the reactions of the beginning or middle part of the series are not innate, or not completely innate, but must be learned by trial. The end action of the series, the consummatory action, is always innate. One evidence of this is the fact that in the first[1] manifes-

[1] To see the appetitive nature of an instinct, it is necessary in some cases to observe an individual animal carefully during its first performance of the act in question. But the performance may be so quick that the observer is quite unable to analyze it. Analysis may be aided by preventing the animal from attaining the consummatory situation for a time, so that the appetitive phase is prolonged, as it were magnified. My cripple dove (example 5, p. 99) afforded just this aid to analysis. The literature is full of reports of instinctive behavior which might well be further analyzed. Consider for example the case of the young moorhen cited by Lloyd Morgan ('96, 63) which had never previously dived, but on being suddenly frightened by a puppy, dived like a flash. That act was too quick for us to analyze it. But if we could successfully impede the diving of a young moorhen so as to prolong the phases of the act, I think it probable that we should find an appetite for the consummatory situation (that of being under water) and a restless striving until it is attained; and that some details in the series of actions, details which in a normal dive are very sure to be hit upon by accident, are not innately predetermined. When one sees the first performance of an instinctive act take place very quickly and with apparent perfection, this does not prove that there is an innate chain reflex determining every detail of the act.

tation (also, in some cases, in later performances) of many instincts, the animal begins with an *incipient consummatory action*, although the appeted stimulus, which is the adequate stimulus of that consummatory action, has not yet been received. I speak of an incipient "action" rather than "reaction," because it seems clearly wrong to speak of a "reaction" to a stimulus which has not yet been received. The stimulus in question is obtained only after a course of appetitive, trial-and-error behavior. When at last this stimulus is obtained, the consummatory reaction takes place completely, no longer incipiently. Then the appetitive behavior ceases; in common speech we say the animal is "satisfied."

One may observe all gradations between a true reflex and a mere readiness to act, mere facilitation. Thus, in the dove, a stimulus from food in the crop may cause the parent to vomit the food or to feed it to young: there are all gradations from an immediate crop-reflex, in which the food is vomited upon the ground, through intermediate cases in which the parent is much disturbed by the food in his crop, but appetitively seeks the young and induces them to take the food; to other cases is which the parent is only ready to feed the young if importuned by them; and finally to cases in which the stimulus from the crop does not even cause facilitation, and the parent does not disgorge the food at all, even if importuned by the young.

While an appetite is accompanied by readiness for certain actions, it may be accompanied by a distinct *unreadiness* for certain other actions, and this is an important factor in some forms of behavior. It is altogether probable that this unreadiness is due in some cases to the fact that the activity of certain neurones *inhibits* the activity of certain other neurones. It is now well-known, too, that unreadiness may be due to the condition of the internal secretions. And the mutual exclusion of certain forms of instinctive behavior is inevitable, due to the incompatibility (Washburn, '16) of their motor components.

Unreadiness may be accompanied by aversion, and vice versa; but either of these may occur without the evident presence of the other. An aversion is sometimes accompanied by an innately determined reaction adapted to getting rid of the disturbing stimulus, or—this point is of special interest—by two alternative reactions which are tried and interchanged repeatedly until the disturbing stimulus is got rid of (see example 7, page 100).

The escape from a disturbing situation or the attainment of an appeted one is accomplished, in case of some instincts, far more surely and more rapidly after one or more experiences. In the first performance of an appetitive action, the bird makes a first trial; if this fails to bring the appeted stimulus he remains agitated and active, and makes a second trial, which differs more or less from the first; if this fails to bring the appeted stimulus he remains still active and makes a third trial; and so on until at last the appeted stimulus is received, the consummatory reaction follows, and then the bird comes to rest. In later experience with the same situation, the modes of behavior which were followed immediately by the appeted stimulus and consummatory reaction are repeated; those which were not so followed tend to drop out.

If a young bird be kept experimentally where it cannot obtain the normal stimulus of a certain consummatory reaction, it may vent that reaction upon an abnormal or inadequate stimulus, and show some satisfaction in doing so; but if the bird be allowed at first, or even later, to obtain the normal stimulus, it will be thereafter very unwilling to accept the abnormal stimulus. That this is true of the sex instinct has been shown in a former article (Craig, '14). It is true also of the appetite for a nest. Thus a female dove which has never had a nest, nor material to build one, lays eggs readily on the floor; but a dove that has had long experience with nests will withhold her egg if no nest is obtainable. The male dove similarly, if he has never had a nest, goes through the brooding behavior on the floor; but an experienced male is unwilling to do so, and shows extreme anxiety to find a nest. These examples illustrate the fact that the bird must *learn* to obtain the adequate stimulus for a complete consummatory reaction, and thus to satisfy its own appetites.

There is often a struggle between two appetites, as when a bird hesitates, and it may hesitate for a long time, between going on the nest to incubate and going away to join the flock, eat, etc. By watching the bird one can predict which line of behavior it will follow, for each appetite is distinguished by its own expressive signs (consisting partly of the incipient consummatory action), and one can see which appetite is gaining control of the organism.

These outward expressions of appetite are signs of physiological states which are but little known. Since my own observations have been on external behavior only, I say little about the internal states. They are probably exceedingly complex and numerous and similar to the physiological states which in the human organism are concomitants of appetites,[1] emotions, desires. They doubtless include stimulations from interoceptors and proprioceptors; perhaps automatic action of nerve centers; perhaps readiness or unreadiness of neurones to conduct. It is known that some of the periodic appetites are coincident with profound physiological changes. Thus Gerhartz ('14) found that during the incubation period in the domestic fowl the metabolism of the body as a whole is at a low ebb. In some cases a stimulus from the environment is the immediate excitant of an appetite; especially stimulation of a distance-receptor may arouse appetite for a contact stimulus, as when the sight of food arouses appetite for the taste of it. But probably in every case appetite is dependent upon physiological factors. And in many cases the rise of appetite is due to internal causes which are highly independent of environmental conditions, and even extremely resistant to environmental interference.

Appetitive behavior in vertebrates is evidently a higher development of what Jennings ('06, p. 309) calls the positive reaction in lower organisms; aversive behavior in vertebrates corresponds to what Jennings (p. 301) calls negative reactions.

The attempt to distinguish between instinct and appetite, as in Baldwin's Dictionary ('01), is not justified by the facts of behavior. Baldwin says : "Appetite is distinguished from instinct in that it shows itself at first in connection with the life of the organism itself, and does not wait for an external stimulus, but appears and craves satisfaction." These characteristics,

---

[1] Hunger furnishes a typical case of appetitive behavior (Carlson, '16; Ellis '10, 198–199). Carlson makes a distinction between hunger and appetite. The distinction he finds is certainly real, but the use of words is unfortunate, for hunger is clearly one kind of appetite.

here ascribed to appetite, are the very ones which I have observed in the instinctive behavior of pigeons. The instincts of pigeons satisfy Baldwin's further description of appetite in that each appears first as a "state of vague unrest" involving especially "the organs by which the gratification is to be secured"; and "a complex state of tension of all the motor . . . elements whenever the appetite is aroused either (a) by the direct organic condition of need, or (b) indirectly through the presence or memory of the object." This last point is illustrated, e. g., by doves learning to drink (example 1, page 97), in whom the sight of the water-dish at a distance aroused the drinking actions by associative memory. I have observed appetitive behavior as Baldwin describes it in nearly all the instinctive activities of doves, and I think that sufficient observation will reveal it in all their instincts.

The most thorough attempt to distinguish instincts from appetites and to show the logical consequences of such distinction, in all the literature to which I have access, is in an old article by Professor Bowen ('46). This article is still worth study, to suggest the conclusions to which one is logically led if he denies that instincts contain any element of appetite. These conclusions, taken almost literally from Bowen, may be summarized as follows: (1) (P. 95) "If the name of instinct be denied to these original and simple preferences [appetites] and aversions, there will appear good reason to doubt whether man is ever governed by instinct, whether all his actions are not reducible to passion, appetite, and reason." (P. 115) The "passions" of man can not be concomitants of instinct. (2) (P. 117) "Instinct is not a free and conscious power of the animal itself. It is, if we may so speak, a foreign agency, which enters not into the individuality of the brute." (P. 118) Instinct "has no effect on the rest of their conduct, which is governed by their own individuality." (3) Bowen contends with logical consistency that if instinct contains no appetitive factor, the ends toward which instincts work, as seen by an observer, are not ends for the agent; that therefore the agent has no power to make the instinctive behavior more effective. In short, instinctive behavior is not susceptible of improvement by intelligence. (4)

Bowen concludes that the intellect and the "passions" of man are not products of evolution. (5) It may be added that even Bowen, strive as he did to separate appetite from instinct, was compelled to admit that the attempt at such separation leads one into difficulties and disputed cases. In contravention of Bowen's conclusions I contend: (1) That much of human behavior is instinctive. (2) That Bowen's description of instinct as "a foreign agency, which enters not into the individuality" is true of reflex action, such as coughing or sneezing, but is not true of instinctive behavior, which is extremely different from such mere reflexes. (For a fuller statement on this point, see below page 1c6. See also Hobhouse, '15, 98–99.) (3) That, of the useful results toward which instincts tend, some, not all, are ends for the agent. For they are the objects of appetites, and the animal strives and learns to attain them. (4) That human conative behavior evolved from the instinctive appetitive behavior of lower animals.

In another article I hope to publish soon a further discussion of the literature.

EXAMPLES.

1. The case of doves learning to drink, as described in detail in a former article (Craig, '12), illustrates appetite. The observed appetitive behavior was aroused by stimulation of distance-receptors, such as the sight of the water-dish being brought to the cage, and of the man bringing it; these acted as appetizers. Each dove, as soon as it had learned to associate such stimuli with the drinking situation, responded to these stimuli by making drinking movements (incipient consummatory action) at once without going to the water dish. The first drinking movements failing to bring water, the dove repeated these movements again and again, sometimes walking a few steps, sometimes turning round, until after many trials and many errors it did get its bill into the water, received the stimulus from water in the mouth (appeted stimulus), whereupon the drinking movements (consummatory reaction) were made not incipiently but completely, the water being swallowed, after which the bird rested and appeared satisfied.

2. A good example of appetitive behavior is seen in the way in

which a young male dove locates a nesting site for the first time. The first thing the observer sees is that the dove, while standing on his perch, spontaneously assumes the nest-calling attitude, his body tilted forward, head down, as if his neck and breast were already touching the hollow of a nest (incipient consummatory action), and in this attitude he sounds the nest-call. But he shows dissatisfaction, as if the bare perch were not a comfortable situation for this nest-dedicating attitude. He shifts about until he finds a corner which more or less fits his body while in the tilted posture; he is seldom satisfied with his first corner, but tries another and another. If now an appropriate nest-box or a ready-made nest is put into his cage, this inexperienced dove does not recognize it as a nest, but sooner or later he tries it, as he has tried all other places, for nest-calling, and in such trial the nest evidently gives him a strong and satisfying stimulation (the appeted stimulus) which no other situation has given him. In the nest his attitude becomes extreme; he abandons himself to an orgy of nest-calling (complete consummatory action), turning now this way and now that in the hollow, palpating the straws with his feet, wings, breast, neck, and beak, and rioting in the wealth of new, luxurious stimuli. He no longer wanders restlessly in search of new nesting situations, but remains satisfied with his present highly stimulating nest.

3. Fetching straws to the nest is apparently due to an appetite for building them into the nest. The dove has an innate tendency to pick up straws, and an innate tendency to build them into the nest (consummatory reaction); but it has apparently no innate tendency to carry a straw to the nest, no innate "chain" of reflexes. When an experienced bird finds a straw he seizes it repeatedly and toys with it, sometimes making movements resembling those by which he would build the straw into the nest. He seems thus to get up an appetite for building the straw in, and when this appetite is sufficiently aroused he flies to the nest, guided by associative memory, and performs the consummatory reaction completely. A young female, no. 70, which I observed picking up a straw for the first (?) time, on her 54th day, showed the lack of a "chain reflex." For she continued toying with the straw an excessively long time, not carrying it at all,

though she happened to be very near the nest. This was the more remarkable as she had a well-formed habit of going to the nest on all occasions. At length she did go to the nest with her straw, and made well-ordered movements to build it in.

4. The male and the female dove take regular turns in sitting on the eggs. The male is seized by the appetite for brooding about 8 or 9 A. M., and the female about 5 P. M., the state evidently being brought on in each case by physiological causes which are part of the daily physiological rhythm. When either one, *e. g.*, the female, comes to the side of the nest prepared to enter and sit, she already has somewhat the attitude of the sitting bird, the body sunk down on the legs and the feathers fluffed out (incipient consummatory action). If her sitting appetite be thwarted, as by her mate refusing to budge from his position, she shows restlessness and makes intelligent efforts to obtain possession of the nest. When at last her mate yields his place, she steps exultingly into the center of the nest and settles herself on the eggs with many movements indicative of satisfying emotion (complete consummatory reaction).

A broody hen of course illustrates the same principle.

5. It is an interesting fact, exhibited in a variety of instincts, that a young bird may make feints of performing actions which it has never yet performed. Thus the young dove makes feints of flying before it has ever flown. This was illustrated in a peculiarly instructive manner by one of my young doves, no. 46, which developed cripple wings and was unable to fly. When placed in a box with sides $3\frac{1}{2}$ inches high it was just able to jump on the edge. Nevertheless, when its roosting instinct developed, it endeavored strenuously every evening to fly to the perch which was some inches above its head. It looked at the perch and aimed at it with perfect definiteness, opening its wings and making feints of flying. In the evolution of birds, there can be no doubt, flying developed gradually from jumping. The new movements of flying were gradually intercalated into the interval between the initial action, leaping from the ground, and the final action, landing again upon the feet. The young dove to this day shows *first* the incipient end action, aiming at the perch to be alighted on, and only after it has launched itself

toward this end situation does the "chain" of flight reactions take place.

6. In the pigeons the order of activities culminating in the sexual act is, first display, second billing, third copulation, with numerous details each finding a place in the succession. Yet the sexual tendency is mainfestly present from the beginning of the "chain," and the preliminary steps are directed, with much guidance by experience, toward securing the stimulation required for discharging the sexual reflex. In absence of the normal stimulus to the consummatory reaction, the instinct manifests itself in marked appetitive behavior, and, especially in inexperienced birds (Craig, '14), in those imperfect consummatory reactions known as perversions and auto-erotic phenomena. The behavior of the sexual appetite is now so well known that it may be cited as the type of appetitive behavior; and to readers who are familiar with modern analyses of the sex instinct I may make my whole article clearest by saying that all the appetitive mechanisms I have mentioned, and ₁ believe all the instincts of the dove, behave in the same manner as that of sex, in regard to appetitive manifestations and anticipation of the consummatory reaction.

7. I shall take space to describe only one example of aversion—the so-called jealousy of the male dove, which is manifested especially in the early days of the brood cycle before the eggs are laid. At this time the male has an aversion to seeing his mate in proximity to any other dove. The sight of another dove near his mate is an "original annoyer" (Thorndike, '13, Chap. IX.). If the male sees another dove near his mate, he follows *either of two* courses of action; namely, (*a*) attacking the intruder, with real pugnacity; (*b*) driving his mate. gently, not pugnaciously, away from the intruder. When he has succeeded either in conquering the stranger and getting rid of him, or in driving his mate away from the stranger, so that he has got rid of the disturbing sight of another dove in presence of his mate, his agitation ceases. If we prevent him from being successful with either of these methods, as, by confining the pair of doves in one cage and the third dove in plain sight in a contiguous cage, then he will continue indefinitely to try both methods. If we leave all three

doves free in one pen, the mated male will try the mettle of the intruder and conquer him if he can; if he fails, he will turn all his energies into an effort to drive his mate away from the intruder. Or if in former experiences he has learned to gage this individual intruder, if he conquered him before he will promptly attack him now, but if defeated by him before he will now choose the alternative of driving his mate away. In sum, the instinctive aversion impels the dove to thoroughly intelligent efforts to get rid of the disturbing situation.

8. In some cases the seeking of a certain situation involves both appetences and aversions in considerable number. Thus, when the day draws to a close, each dove seeks as its roosting-place a perch that is high up, with free space both below it and above it, with no enemies near, with friendly companions by its side, but these companions not too close, not touching (except in certain cases of mate, nest-mate, or parent). The endeavor to achieve this complex situation, to secure the appeted stimuli and to avoid the disturbing ones, keeps the birds busy every evening, often for an hour or more.

CYCLES.

Instinctive activity runs in cycles. The type cycle, as it were a composite photograph representing all such cycles, would show four phases as follows.

Phase I.—Absence of a certain stimulus. Physiological state of appetite for that stimulus. Restlessness, varied movements, effort, search. Incipient consummatory action.

Phase II.—Reception of the appeted stimulus. Consummatory reaction in response to that stimulus. State of satisfaction. No restlessness nor search.

Phase III.—Surfeit of the said stimulus, which has now become a disturbing stimulus. State of aversion. Restlessness, trial, effort, directed toward getting rid of the stimulus.

Phase IV.—Freedom from the said stimulus. Physiological state of rest. Inactivity of the tendencies which were active in Phases I., II., III.

Some forms of behavior show all four phases clearly. The following are examples.

Sex.—(Phase I.) The dove, either the male or the female, shows sexual appetite and invites the mate to sexual activity. Gradually they lead up to (Phase II.) the consummatory sexual act. (Phase III.) After the sexual act, in some cases one bird shows marked aversion, *e. g.*, by striking at the mate. Either the male or the female may show aversion. In some species, signs of aversion after the sexual act seem to be a normal and regular occurrence. In other species they are shown only by a bird whose mate, having failed of satisfaction, invites to further sexual activity. (Phase IV.) The pair usually become sexually indifferent for a considerable time after each copulation.

Brooding.—(Phase I.) The dove shows the brooding appetence, goes to the nest, and, if need be, struggles to obtain possession of it. (Phase II.) It sits throughout its customary perood, during which it often resists efforts of the mate to relieve it. (Phase III.) At the end of this period, in contrast, it comes off at a slight sign from the mate, runs about, flaps its wings, and thus shows its joy in being off. This may be interpreted as a sort of mild aversion for the nest. (Phase IV.) It goes away and becomes temporarily indifferent to the nest.

In other cases, one or other of the phases is not clearly present, so that there are various sorts of incomplete cycles, such as the following.

(*a*) When the bird shows appetitive behavior but fails to obtain the appeted stimulus, the appetite sometimes disappears, due to fatigue or to drainage of energy into other channels; in which case, Phase II. is not attained.

But many instinctive appetites are so persistent that if they do not attain the normal appeted stimulus they make connection with some abnormal stimulus (see page 94); to this the consummatory reaction takes place, the tension of the appetite is relieved, its energy discharged, and the organism shows satisfaction. This is of course *compensation*, in the sense in which that word is used in psychiatry. But the abnormal stimulus is usually inadequate or incomplete, the relief or discharge is imperfect, the satisfaction is marred by the fact that some of the constituent elements of the appetite, failing to receive their appeted stimuli, are still in Phase I. and abnormally active, while at the same time other elements have already reached Phase III., aversion.

(b) Some forms of behavior consist of appetite and satisfaction which are not, in ordinary cases, followed by any distinct aversion. For example, the drinking cycle shows clearly: (Phase I.) appetite for water; (Phase II.) the drinking reaction, with expression of satisfaction; (Phase IV.) indifference. The dove when it finishes drinking shows no distinct sign of aversion (Phase III.) except withdrawing the bill from the water. But if the observer takes this dove then gently in the hand and re-submerges its bill in the water, it shows marked aversion, struggling to withdraw the bill and to shake the water out of it.

(c) On the other hand it may seem that there are some forms of behavior, e. g., fear, in which Phases I. and II. are lacking; that there is no appetite for the fear stimuli and no satisfaction in them; that when the slightest of these stimuli is received it at once arouses (Phase III.) aversive behavior. Yet it is an interesting fact that even in these cases a slight degree of appetite and satisfaction may be present. Children seek and enjoy a little fear. A dove, when it hears the alarm cry from other doves, at once endeavors to see the alarming object. Even pain is (in man) to some degree, sought and enjoyed.

In actual life the cycles and phases of cycles are multiplied and overlapped in very complex ways.

For example, when a certain satisfaction has been attained, this, instead of leading at once to a state of surfeit and aversion, may lead to further appetite, which leads to a second satisfaction, and so on. Thus Phase I. and Phase II. continue to alternate, constituting a "circular reaction" (Baldwin). I have seen a pair of house sparrows copulate thirteen times in immediate succession, and know by the sound of their voices that I did not see the beginning of the series. In many cases such circular reaction serves to rouse the organism to a high state of appetite and readiness for action.

Smaller cycles are superposed upon larger ones. For example, when a female bird is building a nest, so long as she is in the nest she is in a certain nest-building attitude, a high state of satisfaction, which constitutes the consummatory reaction (Phase II.) of a large cycle. But each time she reaches for a straw, seizes it, and tucks it into the nest, she exhibits thus a little cycle containing a little appetence followed by its own satisfaction.

The time occupied by a cycle varies extremely, from cycles measured in seconds to those that occupy a year or even longer. The relative duration of the phases also is extremely variable. In some cases the appeted situation is attained without delay, and Phase I. thus passes so rapidly as to be overlooked by the observer. In other cases the bird strives hard to overcome great obstacles which stand in the way of the attainment of the appeted stimulus, consequently Phase I. is of long duration. Phase II. may last, in the case of drinking, about one second; in the case of incubation, about three weeks.

It should be stated, too, that the phases are not sharply separated; each passes more or less gradually into the next. Thus, from Phase IV. of one cycle in a series to Phase I. of the succeeding cycle, there is often a gradual rise of appetite; active search for satisfaction does not commence until a certain intensity of appetite is attained. This is what is known in pedagogical literature as "warming up." This gradual rise of the energy of appetite is followed (Phase II.–III., or II.–IV.) by its sudden or gradual discharge. This rise and discharge are named by Ellis ('03), in the case of the sex instinct, "tumescence" and "detumescence." They are important phases in the psychology of art, in which sphere they are named by Hirn ('00) "enhancement" and "relief." The discharge (Phase II.) is also exemplified in "catharsis" in art and in psychiatry.

The cycles in the behavior of birds are fundamentally the same phenomenon as the cycles in human behavior. Human cycles are enriched by an intelligence far surpassing that of doves, but this is a difference of degree only. If the dove's cycles are determined largely by instinct, habit, physiological conditions, and not intelligence, so are some human cycles, as those of sleeping, eating, drinking, sex. F. H. Herrick ('10, 83) emphasizes the fact that a bird may scamp one cycle in order to begin another. Thus, birds may abandon young which are not yet weaned, because their appetence for a new brood has set in. But the same principle works, though not quite so crudely, in human life; as in the case of a mother who grows indifferent or even somewhat hostile toward her older children each time a new child is born. Herrick emphasizes also the fact that when anything disturbs the bird in the progress of a cycle, she very often gives up that cycle and begins a new one. Thus, a cedarbird who has just completed her nest one day finds a man examining it; she forthwith abandons that nest and begins to build another. But, again, the same phenomenon appears in human behavior. A man begins to build a house; when he has progressed far with the building he meets some horrible experience in it which "turns him against" it, and nothing will induce him to proceed with that house; he abandons it and begins to build elsewhere. The cedarbird has had a, to her, horrible experience which has turned her against her nest; that nest has lost its *value* for her; the sight of it now, instead of arousing her appetence, arouses aversion.

C. J. Herrick ('15, p. 61) says that many of these cyclical activities of birds are "simply complex chain reflexes." The reason he gives for this statement is that "each step in the cycle is a necessary antecedent to the next, and if the series is interrupted it is often necessary for the birds to go back to the beginning of the cycle. They cannot make an intelligent adjustment midway of the series." But all this, in some degree, is true of the behavior of human beings toward their mates, their nests, and their young. This has been illustrated in the preceding paragraph, and a few illustrations are here added. As to mates: When the cordial relation between a husband and wife is, by some mischance, broken, the pair may make an "intelligent adjustment" if the difficulty is not too great. But birds also make such adjustments constantly, when the difficulty is not too great. And with human beings, as with birds, the difficulty may be insurmountable; in which case, the husband and wife separate for a week, a month, or a year, after which period of rest (Phase IV.), they can commence a new cycle with Phase I., courtship. As to their nests: The fact of homesickness proves that the behavior of a human being toward his or her home runs in a series which conforms to Herrick's statements. As to behavior toward the young: The inability of human parents to make "an intelligent adjustment midway of the series" is shown by the fact that they cannot arouse the fullest degree of parental behavior toward an adopted child unless they adopt the child in its infancy. These facts do not prove that the human behavior in question consists of mere chain reflexes. Neither do the similar facts as to avian cycles prove that the avian behavior consists of mere chain reflexes.

The birds in their cycles exhibit attention (using this and all the following terms in a strictly behavioristic sense), intelligence, memory, intensely emotional behavior, conflict of tendencies, hesitation, deliberation (of course an elementary sort of deliberation), rise, maintenance, and decline of appetences, behavior conformable to certain laws of valuation. All these forms of behavior function in bringing about the consummatory situations of the cycles. Thus the instinctive behavior of birds, so far from consisting of mere chain-reflexes, and having no relation to "individuality" (Bowen, vide ut supra, p. 97), is in reality very highly integrated, and is the very core of the bird's individuality.

All human behavior runs in cycles which are of the same fundamental character as the cycles of avian behavior. These appear in consciousness as cycles of attention, of feeling, and of valuation.

This description is true not only of our behavior toward objects specifically sought by instinct, such as food, mate, and young, but also of our behavior toward the objects of our highest and most sophisticated impulses. Consider, for example, the course of a music-lover's feelings and attention in the case of a symphony concert. Before the concert, if his internal state is favorable (Phase I.), he is all eagerness, desire, interest. He goes to the concert-hall, chooses a good seat for hearing, and in every way shows appetitive behavior. (Phase II.) The music begins, he pays close attention, and feels satisfaction. (Phase III.) If the concert continues too long, he is surfeited, his pleasure diminishes, he even feels some unpleasantness, and his attention turns away, which is of course a form of aversion. (Phase IV.) When the music at length ceases he feels restfulness, relief, and his attention goes elsewhere. This cycle of the whole concert is overlaid by a complex system of epicycles, each extending through one symphony, one movement, or a smaller division, down to the measure and the beat. This is only one illustration of the fact that the entire behavior of the human being is, like that of the bird, a vast system of cycles and epicycles, the longest cycle extending through life, the shortest ones being measured in seconds. This view helps us to understand the laws of attention; for example, the law that attention cannot be held continuously upon a faint, simple stimulus. For as soon as such a stimulus is brought to maximum clearness, which constitutes the consummatory situation, the appetite for it is quickly discharged and its cycle comes to an end. This familiar fact shows that we, like the birds, are but little able to alter the course of our behavior cycles.

BIBLIOGRAPHY.

Baldwin, J. M.
  '01 Dictionary of Philosophy and Psychology. New York. Art. " Appetite."
Bowen, Francis.
  '46 Instinct and Intellect. North Amer. Review, 63, 91–118.
Carlson, A. J.
  '16 The Control of Hunger in Health and Disease. Chicago.
Craig, W.
  '12 Observations on Doves Learning to Drink. Jour. Animal Behav., 2, 273–279.
  '14 Male Doves Reared in Isolation. Jour. Animal Behav., 4, 121–133.
Ellis, Havelock.
  '03 Studies in the Psychology of Sex. Vol. 3, Analysis of the Sexual Impulse, etc. Philadelphia.
  '10 Studies in the Psychology of Sex. Vol. 6, Sex in Relation to Society. Philadelphia.
Gerhartz, H.
  '14 Ueber die zum Aufbau der Eizelle notwendige Energie (Transformationsenergie). Pfluger's Arch., Bd. 156, 1–224.
Herrick, C. J.
  '15 An Introduction to Neurology. Philadelphia.
Herrick, F. H.
  '10 Instinct and Intelligence in Birds. Popular Science Monthly, 76, 532–556, 77, 82–97, 122–141.
Hirn, Y.
  '00 The Origins of Art. London.
Hobhouse, L. T.
  '15 Mind in Evolution. Second edition. London.
Jennings, H. S.
  '06 Behavior of the Lower Organisms. New York.
Morgan, C. Lloyd.
  '96 Habit and Instinct. London.
Stout, G. F.
  '07 A Manual of Psychology. Second edition. London.
Thorndike, E. L.
  '13 The Original Nature of Man. (Educational Psychology, Vol. 1.) New York.
Washburn, M. F.
  '16 Movement and Mental Imagery. Boston.

as its input. The input fed to the computer was picked up from a bipolar pair of electrodes on the scalp, one mounted over the inion and the other about 5 cm anterior. The output of the computer, after it averaged the activity which followed 50 saccadic movements, was fed to an X-Y plotter. The ongoing brain activity, the raw eye-movement record, and the computer trigger signal were constantly monitored on an ink writer.

Continuously illuminated fixation targets were the only stimuli presented to the three observers: maximum retinal illuminance of the targets was 2×10^5 trolands. The illuminance could be varied in steps of 1.0 log unit over a range of 4.0 log units with neutral filters. The dimmest stimulus used was estimated by psychophysical means to be between 1.0 and 2.0 log units above absolute threshold. For one observer (J.K.) a single circular fixation spot 20 minutes in diameter produced reliable responses. The electroencephalogram records of the other two observers (K.G. and V.G.) had considerable "alpha" activity which made it difficult to evaluate the responses in the averaged records with this target. With the last two subjects a larger, more complicated target elicited useful records. This target had an overall diameter of 5 degrees and an internal grid structure consisting of 53 circular spots 20 minutes in diameter.

The experimental procedure was as follows. After alignment of the optical systems, the observer fixated the target at its lowest illuminance. The computer was activated, and the average of 50 samples of electrical brain signals was written on the plotter. The stimulus was then set at the next higher illuminance level and the process was repeated. An experimental session consisted of one ascending and one descending series of illuminances covering the whole 4.0 log unit range available. Each observer served in ten sessions.

A typical set of averaged records is given in Fig. 1. Each record showed a response consisting of a negative wave with an implicit time (or peak latency) of the order of 100 msec, followed by a positive wave which reached its peak by 250 msec. The waveform was sharply defined at the higher illuminances and more rounded and slower at the lower illuminances.

Averaged potentials recorded from the scalp exhibit complex and ephemeral waveforms. Since the purpose of our study was to determine whether any stimulus-determined response followed saccades, quantitative analysis was limited to the prominent negative-positive complex found in all records. Two measures were taken: (i) the implicit time or peak latency to the lowest negative point, and (ii) the amplitude from the lowest negative point to the highest positive point within the first 250 msec following the saccade. These measures are presented in Fig. 2. The implicit time of all observers was a monotonic decreasing function of fixation-target illuminance. In the case of J.K., the amplitude was a monotonic increasing function of illuminance. The amplitude curves for the other two observers were not simple functions of illuminance.

Because the observer is unaware of saccadic eye movements or of changes in the stimulus occurring during fixation, it is hard to attribute the responses to startle or variation in attention. The results indicate that a true evoked response is produced by the retinal image displacement accompanying saccadic eye movement, lending support to the idea that saccadic eye movements help to maintain vision. The results suggest that the nervous-system discharge accompanying vision may not be solely continuous, but instead may be characterized by more or less discontinuous bursts related in time to saccades. This contrasts with the observer's reports that the fixation target appeared steady, a result which may have implications for theories of perception.

KENNETH GAARDER*
Chestnut Lodge Research Institute,
Rockville, Maryland

JOHN KRAUSKOPF
VIRGIL GRAF
University of Maryland,
College Park

WALTER KROPPL
JOHN C. ARMINGTON
Walter Reed Army Institute
of Research, Washington, D.C.

References and Notes

1. L. A. Riggs, F. Ratliff, J. C. Cornsweet, T. N. Cornsweet, J. Opt. Soc. Amer. 43, 495 (1953); R. W. Ditchburn and B. L. Ginsborg, Nature 170, 36 (1952).
2. J. Krauskopf, J. Opt. Soc. Amer. 47, 740 (1957).
3. L. A. Riggs, J. C. Armington, F. Ratliff, ibid. 44, 315 (1959). For more detail, see J. Krauskopf, T. N. Cornsweet, L. A. Riggs, ibid. 50, 572 (1960).
4. Mnemotron Computer of Average Transients.
5. W. J. Rietveld, Acta Physiol. Pharmacol. Neerl. 12, 373 (1963).
6. Supported in part by PHS grant MH 06554 to the Chestnut Lodge Research Institute (K.G.) and contract DA-49-193-MD-2327 between the Office of the Surgeon General, U.S. Army, and the University of Maryland (J.K. and V.G.).
* Present address: Clinical Neuropharmacology Research Center, St. Elizabeths Hospital, Washington, D.C. 20032.

† October 1964

Culturally Transmitted Patterns of Vocal Behavior in Sparrows

Abstract. Male white-crowned sparrows have song "dialects," acquired in about the first 100 days of life by learning from older males. In the laboratory an alien white-crowned sparrow dialect can be taught. Once the song is established further acoustical experience does not change the pattern. White-crowned sparrows do not copy recorded songs of other sparrow species presented under similar conditions.

The white-crowned sparrow, Zonotrichia leucophrys, is a small song bird with an extensive breeding distribution in all but the southern and eastern parts of North America (1). Ornithologists have long remarked upon the geographical variability of its song. Physical analysis of field recordings of the several vocalizations of the Pacific Coast subspecies Z. l. nuttalli reveals that while most of the seven or so sounds which make up the adult repertoire vary little from one population to another, the song patterns of the male show striking variation (see 2).

Each adult male has a single basic song pattern which, with minor variations of omission or repetition, is repeated throughout the season. Within a population small differences separate the songs of individual males but they all share certain salient characteristics of the song. In each discrete population there is one predominant pattern which differs in certain consistent respects from the patterns found in neighboring populations (Fig. 1). The term "dialect" seems appropriate for the properties of the song patterns that characterize each separate population of breeding birds. The detailed structure of syllables in the second part of the song is the most reliable indicator. Such dialects are known in other song birds (3).

The white-crowned sparrow is remarkable for the homogeneity of song patterns in one area. As a result the differences in song patterns between populations are ideal subjects for study of the developmental basis of behavior. If young male birds are taken from a given area, an accurate prediction can be made about several properties of the songs that would have developed if they had been left in their natural environment. Thus there is a firm frame of reference with which to com-

pare vocal patterns developing under experimental conditions. Since 1959 we have raised some 88 white-crowned sparrows in various types of acoustical environments and observed the effects upon their vocal behavior. Here we report on the adult song patterns of 35 such experimental male birds. The several types of acoustical chamber in which they were raised will be described elsewhere.

In nature a young male white-crown hears abundant singing from its father and neighbors from 20 to about 100 days after fledging. Then the adults stop singing during the summer molt and during the fall. Singing is resumed again in late winter and early spring, when the young males of the previous year begin to participate. Young males captured between the ages of 30 and 100 days, and raised in pairs in divided acoustical chambers, developed song

patterns in the following spring which matched the dialect of their home area closely. If males were taken as nestlings or fledglings when 3 to 14 days of age and kept as a group in a large soundproof room, the process of song development was very different. Figure 2 shows sound spectrograms of the songs of nine males taken from three different areas and raised as a group. The patterns lack the characteristics of the home dialect. Moreover, some birds from different areas have strikingly similar patterns (A3, B2, and C4 in Fig. 2).

Males taken at the same age and individually isolated also developed songs which lacked the dialect characteristics (Fig. 3). Although the dialect properties are absent in such birds isolated in groups or individually, the songs do have some of the species-specific characterisitcs. The sustained tone in the introduction is generally, though not always, followed by a repetitive series of shorter sounds, with or without a sustained tone at the end. An ornithologist would identify such songs as utterances of a *Zonotrichia* species.

Males of different ages were exposed to recorded sounds played into the acoustical chambers through loudspeakers. One male given an alien dialect (8 minutes of singing per day) from the 3rd to 8th day after hatching, and individually isolated, showed no effects of the training. Thus the early experience as a nestling probably has little specific effect. One of the group-raised isolates was removed at about 1 year of age and given 10 weeks of daily training with an alien dialect in an open cage in the laboratory. His song pattern was unaffected. In general, acoustical experience seems to have no effect on the song pattern after males reach adulthood. Birds taken as fledglings aged from 30 to 100 days were given an alien dialect for a 3-week period, some at about 100 days of age, some at 200, and some at 300 days of age. Only the training at the age of 100 days had a slight effect upon the adult song. The other groups developed accurate versions of the home dialect. Attention is thus focused on the effects of training between the ages of about 10 and 100 days. Two males were placed in individual isolation at 5 and 10 days of age, respectively, and were exposed alternately to the songs of a normal white-crowned sparrow and a bird of a different species. One male was exposed at 6 to 28 days, the other

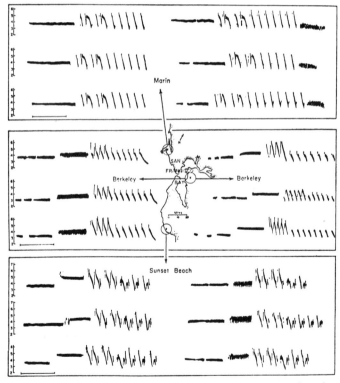

Fig. 1. Sound spectrograms of songs of 18 male white-crowned sparrows from three localities in the San Francisco Bay area. The detailed syllabic structure of the second part of the song varies little within an area but is consistently different between populations. The introductory or terminal whistles and vibrati show more individual variability. The time marker indicates 0.5 second and the vertical scale is marked in kilocycles per second.

at 35 to 56 days. Both developed fair copies of the training song which was the home dialect for one and an alien dialect for the other. Although the rendering of the training song is not perfect, it establishes that the dialect patterns of the male song develop through learning from older birds in the first month or two of life. Experiments are in progress to determine whether longer training periods are necessary for perfect copying of the training pattern.

The training song of the white-crowned sparrow was alternated in one case with the song of a song sparrow, *Melospiza melodia*, a common bird in the areas where the white-crowns were taken, and in the other case with a song of a Harris's sparrow, *Zonotrichia querula*. Neither song seemed to have any effect on the adult patterns of the experimental birds. To pursue this issue further, three males were individually isolated at 5 days of age and trained with song-sparrow song alone from about the 9th to 30th days. The adult songs of these birds bore no resemblance to the training patterns and resembled those of naive birds (Fig. 3). There is thus a predisposition to learn white-crowned sparrow songs in preference to those of other species.

The songs of white-crowned sparrows raised in isolation have some normal characteristics. Recent work by Konishi (4) has shown that a young male must be able to hear his own voice if these properties are to appear. Deafening in youth by removal of the cochlea causes development of quite different songs, with a variable broken pattern and a sibilant tone, lacking the pure whistles of the intact, isolated birds. Furthermore, there is a resemblance between the songs of male white-crowned sparrows deafened in youth and those of another species, *Junco oreganus*, subjected to similar treatment. The songs of intact juncos and white-crowns are quite different. Konishi also finds that males which have been exposed to the dialect of their birthplace during the sensitive period need to hear themselves before the memory trace can be translated into motor activity. Males deafened after exposure to their home dialects during the sensitive period, but before they start to sing themselves, develop songs like those of a deafened naive bird. However, once the adult pattern of singing has become established then deafening has little or no effect upon it. Konishi infers that in the course of

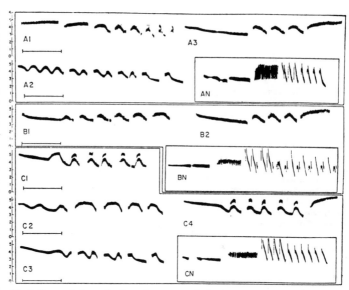

Fig. 2. Songs of nine males from three areas raised together in group isolation. *A1* to *A3*, Songs of individuals born at Inspiration Point, 3 km northeast of Berkeley. *B1* and *B2*, Songs of individuals born at Sunset Beach. *C1* to *C4*, Songs of individuals born in Berkeley. The inserts (*AN*, *BN*, and *CN*) show the home dialect of each group.

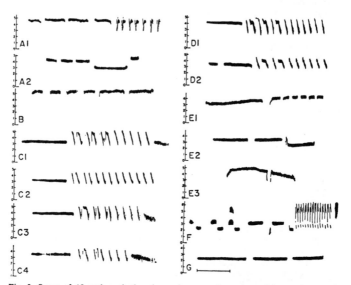

Fig. 3. Songs of 12 males raised under various experimental conditions. *A1* and *A2*, Birds raised in individual isolation. *B*, Male from Sunset Beach trained with Marin song (see Fig. 1) from the 3rd to the 8th day of age. *C1* to *C4*, Marin birds brought into the laboratory at the age of 30 to 100 days. *C1*, Untrained. *C2* to *C4*, Trained with Sunset Beach songs; *C2* at about 100 days of age, *C3* at 200 days, *C4* at 300 days. *D1*, Bird from Sunset Beach trained with Marin white-crowned sparrow song and a Harris's sparrow song (see *G*) from the age of 35 to 56 days. *D2*, Marin bird trained with Marin white-crowned sparrow song and a song-sparrow song (see *F*) from the age of 6 to 28 days. *E1* to *E3*, Two birds from Sunset Beach and one from Berkeley trained with song-sparrow song from the age of 7 to 28 days. *F*, A song-sparrow training song for *D2* and *E1* to *E3*. *G*, A Harris's sparrow training song for *D1*.

crystallization of the motor pattern some control mechanism other than auditory feedback takes over and becomes adequate to maintain its organization. There are thus several pathways impinging upon the development of song patterns in the white-crowned sparrow, including acoustical influences from the external environment, acoustical feedback from the bird's own vocalizations, and perhaps nonauditory feedback as well.

Cultural transmission is known to play a role in the development of several types of animal behavior (5). However, most examples consist of the reorientation through experience of motor patterns, the basic organization of which remains little changed. In the development of vocal behavior in the white-crowned sparrow and certain other species of song birds, we find a rare case of drastic reorganization of whole patterns of motor activity through cultural influence (6). The process of acquisition in the white-crowned sparrow is interesting in that, unlike that of some birds (7), it requires no social bond between the young bird and the emitter of the copied sound, such as is postulated as a prerequisite for speech learning in human children (8). The reinforcement process underlying the acquisition of sound patterns transmitted through a loudspeaker is obscure.

PETER MARLER
MIWAKO TAMURA
Department of Zoology,
University of California, Berkeley

References and Notes

1. R. C. Banks, *Univ. Calif. Berkeley Publ. Zool.* **70**, 1 (1964).
2. P. Marler and M. Tamura, *Condor* **64**, 368 (1962).
3. E. A. Armstrong, *A Study of Bird Song* (Oxford Univ. Press, London, 1963).
4. M. Konishi, in preparation.
5. W. Etkin, *Social Behavior and Organization Among Vertebrates* (Univ. of Chicago Press, Chicago, 1964).
6. W. Lanyon, in *Animal Sounds and Communication, AIBS Publ. No. 7*, W. Lanyon and W. Tavolga, Eds. (American Institute of Biological Sciences, Washington, D.C., 1960), p. 321; W. H. Thorpe, *Bird Song. The Biology of Vocal Communication and Expression in Birds* (Cambridge Univ. Press, London, 1961); G. Thielcke, *J. Ornithol.* **102**, 285 (1961); P. Marler, in *Acoustic Behaviour of Animals*, R. G. Busnel, Ed. (Elsevier, Amsterdam, 1964), p. 228.
7. J. Nicolai, *Z. Tierpsychol.* **100**, 93 (1959).
8. O. H. Mowrer, *J. Speech Hearing Disorders* **23**, 143 (1958).
9. M. Konishi, M. Kreith, and J. Mulligan cooperated generously in locating and raising the birds and conducting the experiments. W. Fish and J. Hartshorne gave invaluable aid in design and construction of soundproof boxes. We thank Dr. M. Konishi and Dr. Alden H. Miller for reading and criticizing this manuscript. The work was supported by a grant from the National Science Foundation.

14 September 1964

Mangabey x and b Wave Electroretinogram Components: Their Dark-Adapted Luminosity Functions

Abstract. The temporal separation of x and b components in the electroretinogram of the dark-adapted eye of the sooty mangabey, Cercocebus torquatus atys, permits an uncomplicated calculation of luminosity functions. Flicker electroretinogram studies indicate enhancement of the photopic blue sensitivity.

The electroretinogram (ERG) comprises several individual components which have been treated in great detail by several authors (1–3). Our paper concerns the two major corneal positive components, the x and b waves. Motokawa and Mita (4) were the first to discover an early positive wave in the response to brief flashes of red light, which they called the x wave. In subsequent investigations Adrian (5) showed that the x wave had a considerably shorter latency than the b wave and was photopic in nature. Armington (6) related the x wave to red sensitivity, demonstrating that in man the spectral sensitivity function of the x wave fits the CIE (Comm. Intern. de l'Eclairage, ICI) photopic function only in the deep red. The ERG from the dark-adapted primate eye is usually dominated by the b component, which has been related to rod function and rhodopsin. The smaller x wave can usually be seen in response to flashes of light of long wavelength, but appears, if it can be detected at all, only as a notch in the leading edge of the rising b wave at wavelengths shorter than about 590 mμ. However, the ERG of the dark-adapted eye of the sooty mangabey, *Cercocebus torquatus atys*, reveals a distinct separation of the x and b components of the response at all wavelengths; this permits direct determination of the spectral sensitivity of both components.

The four adult mangabeys used in this study were lightly anesthetized and placed in a stereotaxic instrument. A silver–silver chloride ring electrode was placed around the limbus of each eye. The left eye was covered and the right pupil was dilated with a 1 percent solution of cyclopentolate hydrochloride; the lids were retracted. The animal was placed in a light-tight, electrically shielded box and aligned in the optical system in a manner that placed the final lens focus on the node of the eye, presenting the animal with

a Maxwellian view subtending about 20 degrees of visual angle.

The light source was a 6-volt, 18-amp tungsten ribbon filament bulb. The filament image was condensed and focused on a point aperture. The diverging beam from the point source was collimated, passed through appropriate filters, and then refocused on the node of the animal's eye. An electrically operated flag-shutter interrupted the beam at the point aperture. Flickering light with a light-to-dark ratio of 1:1 was provided by inserting at the aperture an episcotister driven by a constant-speed motor. Various combinations of disks and motors provided any desired flicker rate from 2 to 35 cy/sec. Single-flash studies were conducted, with a stimulus duration of 200 msec. Interstimulus intervals were 1 minute or greater, depending on the intensity series. Light composed of narrow-band wavelengths was provided by interference filters or Wratten color filters. Flash intensity was controlled with neutral-density filters. All filters were calibrated with a spectrophotometer and equated for equal energy transmission in the construction of luminosity curves. The filters used gave test flashes at eight spectral points with peak band pass at 452, 490, 505, 538, 576, 606, 633, and 646 mμ.

The ERG was displayed on four

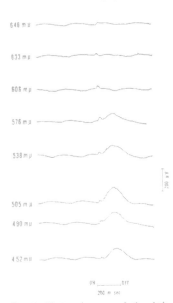

Fig. 1. Electroretinogram of the dark-adapted mangabey eye to a series of equal energy stimuli; animal No. 2.

PRENATAL BEHAVIOR OF BIRDS

By Gilbert Gottlieb

*Psychology Laboratory, Division of Research,
North Carolina Department of Mental Health,
Raleigh, North Carolina, 27602*

ABSTRACT

An analysis of behavioral, neurological, and biochemical research indicates that the sensory systems of the avian embryo become functional in the following order: (1) non-visual photic sensitivity; (2) tactile sensitivity; (3) vestibular sensitivity; (4) proprioception; (5) audition; (6) vision. There is no firm information about the onset of gustation and olfaction in the avian embryo. With the possible exception of vision, there would appear to be adequate stimulation during the normal course of incubation to excite all the sensory systems. Whether such stimulation actually plays a role in regulating species-typical sensory and perceptual development is not known. While it seems doubtful that depriving the embryo of relevant stimulation, or prematurely exposing it to such stimulation, would alter the absolute sequence in which the sensory systems develop, such a manipulation might exert some effect on the relative time of onset of the particular sensory system involved, and a major effect on the development of adaptive perceptual discrimination abilities within the modality thus affected.

An experiment involving sub-total auditory deprivation demonstrated in duck embryos that normally occurring auditory stimulation is essential to the development of species-specific auditory perception. Although partially deprived embryos were not capable of making the usual perceptual discriminative response, they were capable of hearing. Similarly deprived embryos were able to make the species-specific auditory discrimination after hatching, indicating that the sub-total deprivation induces a lag in the development of an embryo's discriminative ability. These findings are consonant with the view that exposure to relevant stimulation plays an important role in regulating the development of adaptive perceptual discrimination abilities in the avian embryo.

Dedicated to Zing-Yang Kuo on the occasion of his 70th birthday

WHILE there are a number of interesting aspects to the problem of the genesis of behavior, the role, if any, played by sensory stimulation in embryonic behavior has become an overriding issue in recent years. The questions are: does sensory stimulation influence the movements of the embryo, or are certain motor movements independent of such stimulation (i.e., spontaneous)?; does sensory stimulation encountered during embryonic development influence behavior after hatching?

In recent years the spontaneous aspect of motor movement in avian embryos has received careful theoretical analysis (Hamburger, 1963) and convincing experimental support (Hamburger, Wenger, and Oppenheim, 1966). These experiments have shown that the temporal periodicity of embryonic movement (alternating periods of activity–inactivity) is regulated by some feature of the embryo's motor system which is essentially independent of sensory instigation. That is, if the sensory ganglia which feed stimulation to certain motor neurons are removed, the cyclic activity–inactivity of the de-afferented portion of the embryo continues unabated. Because the cyclic activity occurs in the absence of sensory stimulation, this feature of behavior can be said to be spontaneous.

148

Whether the spatial pattern of embryonic movement (flexion, extension, coordination) is the same with and without the sensory ganglia is not known. It is also not yet known whether, in the *intact* embryo, sensory stimulation influences any other aspect of the otherwise spontaneous activity–inactivity cycles. These problems as well as other aspects of embryonic motor and neurological development have recently been extensively reviewed by Corner and Bot (1967), Gottlieb (1968), Jacobson (1967), and Sperry (1965).

In order to round out the picture of prenatal behavior in birds, the present review deals with problems of sensory and perceptual development. Specifically, the purpose of the review is (1) to synthesize the behavioral, neuroanatomical, biochemical, and electrophysiological information bearing on the sequential order of sensory development in the avian embryo; (2) to indicate certain hypotheses and problems concerning the ontogeny of sensory and perceptual development which require further study; and (3) to report the first direct evidence that the temporal development of species-specific perception is regulated by sensory stimulation encountered in prenatal ontogeny.

METHODOLOGICAL CONSIDERATIONS

Most of the research on birds has been conducted with domestic chick embryos (*Gallus gallus*); there have been a few studies of domestic duck embryos (*Anas platyrhynchos*) and one study of pigeon embryos (*Columbiana* sp.). As pointed out by Hooker (1952), the chick egg, like other bird eggs, affords deceptively easy access to the embryo. Since the difficulties lie not in exposing the embryo but in maintaining it in a near normal state during an extended observation period, it is necessary to describe some basic methodological problems before presenting the review of experimental findings. In most of the studies considered here, the presentation of procedural details was very incomplete. While some investigators seem to have been naive with respect to the procedures they employed, others seem to have been intent on not revealing details of their procedures. In any event, the incomplete presentation of critical procedural details makes replication difficult and the reasons for failure of replication

uncertain. In preparing this review I have tended to ignore minor inconsistencies and to state plainly the existence of major ones.

The most fundamental methodological considerations concern subject populations, chronological staging, incubation practices, and observation and stimulation procedures.

Subjects

In the poultry science literature, it is well known that hatchability, fertility, normal growth and development, and embryonic mortality depend on season of year, stability of incubator temperature and humidity, and a number of other factors (Abbott and Craig, 1963; Kosin, 1964). Although there are normative stage series available for domestic chick (Hamburger and Hamilton, 1951) and domestic duck embryos (Koecke, 1958), these criteria are not unequivocal, especially in the late stages of incubation. In actual practice it is only by observing the morphology and general condition of many embryos at various stages of development that allows one to formulate standards of normalcy for a given population (i.e., eggs from a single supplier incubated under highly standardized conditions). The results from experiments conducted under seasonal or incubative conditions when fertility is low or embryonic mortality is high are likely to be misleading.

Chronological Staging

In order to know the age of an embryo accurately, it is at the very least necessary that the eggs be incubated in the laboratory from the start. Individual variability is reduced if the eggs are submitted to an empirically determined pre-incubation refrigeration procedure (Gottlieb, 1963). There are three ways of determining the age of embryos, and considerable confusion exists in the literature because it is usually not clear which method has been used. Some investigators apparently consider the embryo to be one day old during the first 24 hours of incubation, while others regard the embryo as one day old (plus some hours) during the second 24 hours of incubation. Other investigators use the morphological tables, so they may report age in days based on morphology, irrespective of length of incubation; this latter practice does not seem wise because an embryo incubated for 10 days which has the appearance of an 8-day or 9-day embryo may be atypical in other ways.

In the present review the age of the embryos will be cited according to the following example: 26 hours of incubation=Day 1, 2 hours. Under this system, domestic chicks hatch on Day 20 (the 21st day of incubation), domestic ducks hatch on Day 27 (the 28th day of incubation), and carrier pigeons hatch on Day 15 (the 16th day of incubation).

Incubation

Most investigators do not report any details of incubation. Since temperature, humidity, position of eggs, turning of eggs, and type of incubator (Abbott and Craig, 1963) influence development, it seems certain that some differences in results among investigators, especially time of first appearance of a given activity or response, can be attributed to incubation differences. In a forced-draft incubator, optimal temperature is 37.5°C. at a relative humidity between 60 and 70 per cent. Low temperature (35.8°C.) is associated with the development of cardiac defects and other malformations, while high temperature (39.4°C.) is related to structural defects of the eye and nervous system (de la Cruz, Campillo-Sainz, and Muños-Armas, 1966). Since the egg is removed from the incubator for observation, it is necessary that the observation set-up include proper temperature and humidity control. Clark and Clark (1914) and Tuge (1937) report severe changes in embryonic activity in relation to temperature fluctuations between 35°C. and 41°C.

Observation

There is no ideal way to observe embryonic behavior, since direct observation of the embryo entails some interference with its normal physiological environment. Based on minimal disturbance to the embryo and its surrounding membranes, however, some observation procedures are better than others. A few investigators have resorted to the unnecessarily heavy-handed procedure of cracking the egg and dumping the embryo into a pre-warmed saline bath. This procedure may be appropriate for sacrificing the embryo for strictly anatomical research purposes, but it is the most damaging one in terms of behavioral study.

The embryo's integrity is intimately connected to the state of its yolk sac, amnion sac (surrounds embryo), and chorioallantois (vascularized membrane adhering to shell membranes). Thus, in making observation windows in the shell, most investigators have realized that particular care must be taken to interfere with these structures as little as possible.

There are two main techniques for making windows in the egg shell: Kuo's liquid vaseline technique and the lateral window technique (Fig. 1). Kuo's method (1932) seems to interfere least with ongoing embryonic activity. It involves removal of the shell and outer shell membrane over the air-space at the large end of the egg. This procedure leaves the inner shell and chorioallantoic membranes intact. When the opaque inner shell membrane is painted with warm liquid vaseline it becomes transparent and the embryo can be seen. Becker (1942) has criticized this technique on the grounds that anoxemic conditions result from interference with gaseous exchange in the region of the air-space. It is difficult, however, to evaluate the validity of Becker's criticism for two reasons: first, Becker used a high incubation temperature (39.5°C); second, he did not put a shell cap over the painted window when he returned the eggs to the incubator. Therefore, the membranes were subject to excessive and rapid drying, and embryonic mortality was higher than usual. In recent years an improvement in the Kuo technique negates objections to the use of liquid vaseline for rendering the inner shell membrane transparent. Specifically, by the careful use of forceps it is possible to peel the inner shell membrane away from the chorioallantoic membrane. Since the chorioallantoic membrane itself is transparent, it is not necessary to apply liquid vaseline to observe the embryo.

The other main technique for making an observation window is to cut away the shell and shell membranes on the side of the egg (Fig. 1). To avoid hemorrhaging the chorioallantois (CA), some investigators first draw the CA away from the shell membranes (e.g., Gottlieb and Kuo, 1965). Other investigators do not separate

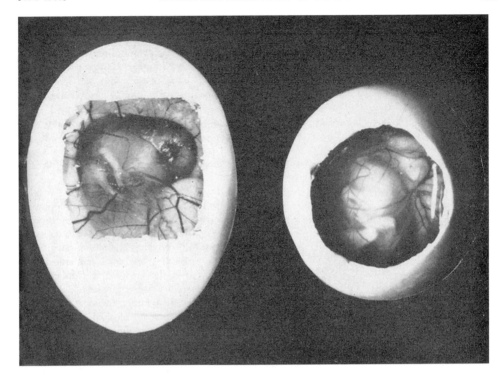

Fig. 1. Two Types of Observation Windows

Lateral observation window in shell of ten-day-old Peking duck embryo (left) and observation window at large end of egg in a 14-day-old Peking duck embryo (right). If proper precautions are taken and a glass cover slip is sealed in place over the hole in the shell, the behavior of the same embryo can be observed throughout its 27-day incubation period.

the CA from the shell membranes before cutting the shell; the CA, which is part of the embryo's circulatory and respiratory system, is thus unnecessarily wounded. As mentioned above, the CA membrane is transparent so it is not necessary to apply liquid vaseline to it to render the embryo visible.

Stimulation

While stimulating embryos with sound or light is a relatively simple non-interruptive procedure, stimulating them with electricity, light touch, or mechanical pressure involves partial destruction of the chorioallantois and amnion in order to gain direct access to the embryo's body surface. Likewise, recording brain waves (EEG), retinal activity (ERG), muscular activity (EMG), or cochlear microphonics usually involves a rather gross interference with the embryo's integrity. At late

stages (2–3 days before hatching), however, these electrophysiological procedures seem less damaging in the sense that many embryos survive if proper precautions are taken.

SENSORY DEVELOPMENT

Information on the avian embryo's response to cutaneous, vestibular, proprioceptive, gustatory, auditory, and visual stimulation is reviewed in this section. New results on prenatal-postnatal interrelationships are presented in the final section.

Cutaneous Sensitivity

This modality is probably the first one to become functional with respect to sensorimotor reflex activity. In Day $6\frac{1}{2}$ or Day 7 chicks, stimulation of the oral region with a single loop of baby hair usually elicits gross and local movements of a withdrawal character (Orr and Windle, 1934; Visintini and Levi-Montalcini, 1939; Hamburger and Balaban, 1963). In pigeons, Tuge (1937) observed this response beginning about Day 5, and I have observed a similar response in the Peking duck beginning about Day 9. In all cases the embryo does not respond to each application of the stimulus and it does not always respond in exactly the same way. It would be interesting if the failure of the embryo to react to each application of the stimulus reflects the existence of periodic fluctuations in tactile sensitivity as has been reported in amphibian embryos (Coghill and Watkins, 1943). If such a periodicity is actually present in birds, it could prove to be the sensory counterpart of the temporal periodicity observed in embryonic motor movement.

In the development of each of the three above-mentioned avian species the chronological stage at which the embryos become more or less consistently reactive to cutaneous stimulation is the same — approximately one-third of the way through incubation (Day $6\frac{1}{2}$ in chicks, Day 5 in pigeons, and Day 9 in Peking ducks). This fact is particularly interesting, since each species has a different incubation period and pigeons are an altricial species while chicks and ducklings are precocial upon hatching. From histological studies (Windle and Orr, 1934; Visintini and Levi-Montalcini, 1939), the appearance of cutaneous reflex sensitivity coincides with the establishment of multisynaptic sensorimotor reflex arcs around Day $6\frac{1}{2}$ in the chick; the neuroanatomy of the other two species has not been studied, but presumably the same neuroanatomical correlates would be in evidence for the mediation of this reflex.

From the comparative point of view, it is of interest that in the human fetus the first trigeminal reflex mechanism is completed at about 15 per cent of the gestation period (5.5 weeks), which corresponds in time with the appearance of the first overt response to cutaneous stimulation of the oral region (Humphrey, 1964). At this time the growing nerve tips are well below the surface epithelium. Since the oral or snout region is the first one to become sensitive in all vertebrates thus far examined (fish through man), and the developmental spread of cutaneous sensitivity follows

the same cephalo-caudal pattern in all vertebrates, this problem deserves further study in birds. The neuroanatomical work of Tello (1923), for example, shows the trigeminal ganglion to be in a rather advanced stage at only 70–72 hours in the chick embryo. At this time the trigeminal nerve is already divided into its three branches (maxillary, mandibular, and opthalmic), and it contains many bipolar neuroblasts supplied with a perfectly impregnated neurofibrillar network. The peripheral portion of each branch has reached its point of distribution and a descending pathway has been established between the bulb and the second neuromere of the hindbrain. Since this rather advanced neuroanatomical development of the trigeminal nerve at only three days suggests the possibility of function before Day $6\frac{1}{2}$ or 7 in the chick, the onset of (pre-reflexive) cutaneous sensitivity mediated via the trigeminal nerve deserves a more thorough analysis than it has heretofore received.

For the sake of clarity, it should be mentioned that the functional status of a sensory system can be demonstrated in the absence of an immediate motor response to stimulation — i.e., when later behavior can be shown to be a consequence of earlier sensory stimulation. This approach has been overlooked because the traditional procedure of behavioral embryology has been to ascribe functional status to a sensory system exclusively on the basis of the overt manifestation of sensorimotor reflexes. Because of primary interest thus far in the development of reflexes, no research has been addressed to the question of whether a sensory system can process "information" in the absence of an immediate reflex response (i.e., before the establishment of effective connections with the motor system). The pre-reflexive functional capacity of a sensory system can be demonstrated in three ways: when sensory enrichment accelerates development of the system, when sensory deprivation slows it down, or when a perceptual preference can be shown to be the consequence of earlier exposure to a certain pattern of sensory stimulation. (While it is also feasible to attempt to record evoked potentials from a sensory system during the pre-reflexive stage, in order for positive results to have functional significance some correlation with behavior must ultimately be made.)

A further analysis of the onset of cutaneous sensitivity in the avian embryo is required to substantiate several aspects of the hypothesis that embryonic self-stimulation plays an important role in prenatal as well as postnatal behavior (Gottlieb and Kuo, 1965; Holt, 1931; Kuo, 1967; Lehrman, 1953; Schneirla, 1965 and 1966). This hypothesis rests heavily on the assumption that at the early stages of development cutaneous and pressure stimulation is effective in influencing either the current or later behavior of the embryo or in influencing both. To the extent that these systems are completely non-functional in early stages, no influence can be validly claimed for them (e.g., before Day $6\frac{1}{2}$ in the chick). If the sensory side is adequately differentiated before this time, however, it is possible that later behavior (sensory and perceptual) can be traced to early stages. For current motor behavior to be influenced, a complete sensorimotor arc must be in evidence, as suggested by Hamburger (1963). However, Hamburger's claim that no aspect of the chick embryo's behavior can be affected by cutaneous stimulation before Day $6\frac{1}{2}$ is no less speculative than statements which suggest the contrary. There is no a priori reason for assuming that a sensory system has to be connected to a motor system in order for (later) sensory and perceptual function to be affected by stimulation which occurred prior to the period of effective motor connections. This aspect of the problem requires further study before it can be known whether cutaneous stimulation can influence the chick embryo before Day $6\frac{1}{2}$, the pigeon embryo before Day 5, and the duck embryo before Day 9.

Tactile conditioning has been studied by Gos (1935) in chick embryos beginning on Day 17. Her procedure involved habituating the embryos' response to cutaneous stimulation (feather or cotton ball) and then re-establishing the response to cutaneous pressure by using electric shock as reinforcement. In the habituation phase, the tactile stimulus was applied until the embryo failed to respond for 100 consecutive applications. The embryos required 10–70 applications on Day 17, 7–32 applications on Day 18, 25–43 applications on Day 19, and 38–44 applications on Day 20. The 101st application was followed by electric shock and the criterion for re-establishment of the re-

sponse was five consecutive unreinforced responses (i.e., five consecutive responses to the cutaneous stimulus alone). The number of CS-UCS pairings required to reach the criterion at each age is shown in Fig. 2; the CS (conditioned stimulus) was a feather or cotton ball and the UCS (unconditioned stimulus) was electric shock. As can be seen there was a progressive and orderly decline in the required mean number of CS-UCS pairs as a function of age (24, 20, 16, 12 pairings on Day 17, 18, 19 and 20 respectively). Gos states that a typical response to the electric shock could be distinguished from the embryos' spontaneous movements only by Day 17; otherwise it may have been possible to demonstrate tactile conditioning as early as Day 15 in the chick.

Although conditioning methods are useful for revealing the embryo's capacity for association learning or development of the central nervous system, it does not seem likely that such methods reveal anything about the actual nature of prenatal behavioral development. For example, there is no evidence to suggest that the avian embryo is exposed to strong noxious stimulation (like electric shock) during development, although the embryo certainly is consistently exposed to cutaneous stimula-

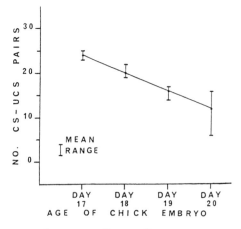

Fig. 2. Progress of Tactile Conditioning in the Chick Embryo during the Final Four Days of Incubation

A smaller number of CS (conditioned stimulus)-UCS (unconditioned stimulus) pairs indicates a more rapid rate of conditioning. (Data from Gos, 1935.)

tion from the very earliest stages (Gottlieb and Kuo, 1965; Holt, 1931; Kuo, 1967).

Vestibular Sensitivity

Vestibular excitability is first observed in the chick embryo around Day 8. Visintini and Levi-Montalcini (1939) placed opened eggs on a rotating disc for two minutes and observed post-rotational nystagmus of the embryo's head (head nystagmus, not optic nystagmus). During rotation all embryonic movement usually ceased. The post-rotational movements consisted of a series of rapid lateral excursions of the head (head-shaking) which were more restricted in extent than the head-turning of the embryo prior to rotation. The nystagmus began about 20–30 seconds after rotation and lasted for only a few seconds. Upon cessation of nystagmus the embryo resumed its pre-rotational pattern of motility with a temporary increase in the frequency of head-turning.

The onset of vestibular sensitivity coincides with the arrival of the large diameter fibers (Cajal) of the vestibular nerve in the ampullate crest, where spoon-shaped connections are formed. At the same time the efferent fibers penetrate the medulla, where they synapse with the nucleus (Deiters) of the acoustic area (Visintini and Levi-Montalcini, 1939).

During the normal course of incubation in nature, the hen's settling movements on the nest have the effect of turning the eggs and could thereby excite the embryo's vestibular system. In the laboratory or hatchery, it is conventional to turn the eggs regularly by manual or mechanical means during incubation. Since egg-turning does temporarily displace the embryo, it seems possible that the threshold of the vestibular apparatus might be reached under normal conditions of embryonic development. However, this particular point has not yet been experimentally demonstrated by using linear acceleration velocities which approximate those pertaining to the usual circumstances of incubation.

The other possibility for vestibular stimulation comes from the embryo's own head and body movements (self-stimulation). In this connection it is interesting to note that in the final quarter of the incubation period chick embryos which have been deprived of their vestibular ganglia (1) show a significantly elevated level of activity and (2) fail to perform the rotatory and head movements which are essential to hatching (John Decker, 1968, pers. commun.). Thus, if these results are not due to secondary, post-operative degeneration in the motor system itself, an intact vestibular system would appear to be necessary (1) for inhibiting abnormally high levels of activity and (2) for guiding the coordination of movements relevant to hatching.

Proprioception

Visintini and Levi-Montalcini (1939) observed the first local proprioceptive muscle reflexes in the chick embryo around Day 10. The appearance of these reflexes coincides with the formation of the first known simple reflex arc (monosynaptic), the direct collaterals of the reflex–motor type having arrived at the anterior horn cells.

Proprioception is a basic form of self-stimulation and it occurs during normal development as a natural consequence of the embryo's movements. The necessity of such movements for normal skeletal growth has been demonstrated by various experiments in which the embryo's musculature has been paralyzed in one way or another (Drachman and Sokoloff, 1966). These experiments, however, have not delineated the extent to which intact muscle stretch or muscle spindle reflexes are necessary for the promotion of normal skeletal growth. Other evidence (Hamburger, Wenger, and Oppenheim, 1966) indicates that autonomous (non-reflexive) discharges of the motor system are adequate for the sculpturing of bony joints.

Taste

Based on the presence of excretory substances in the egg, both chicks (Kuo, 1967) and ducks (Gottlieb and Kuo, 1965) begin digestive processes rather early in development (about Day 10 in chicks and Day 13 in Peking ducks). It seems likely that amniotic fluid is orally ingested, though this has not actually been demonstrated. Whether taste receptors are functional at this early stage is unknown. According to Sedláček's work (1962) with the chick, by Day 16 it is possible to establish a rudimentary conditioned swallowing reflex using a 3000 cps tone as the conditional stimulus and saccharin solution as the unconditioned stimulus. Since other solutions were not used, it is not possible to say definitely whether taste was involved or not. (Introduction of the saccharin solution into the embryo's beak produced tactile or mechanical stimulation which could have been an effective unconditioned stimulus by itself.)

*Audition: Neuroanatomy and
Cochlear Microphonics*

There is not yet any behavioral information on the earliest age at which the embryo's immediate or subsequent behavior can be influenced by auditory stimulation. Several lines of non-behavioral evidence suggest that the chick embryo's peripheral and associated acoustic centers may be sufficiently differentiated by Day 12 or 13 to respond to auditory stimulation. Beginning about Day 13 Vanzulli and Garcia-Austt (1963) were able to record microphonic potentials from the cochlea in response to low frequency sounds (between 100 and 250 cps). During the course of development the upper range of the frequency response increased daily, so that by the time of hatching microphonics were recorded for tones around 4000 cps. Both the lower and upper limits of the frequency response in these experiments must be regarded as conservative since the research was performed under non-optimal temperature conditions (20–22° C.). This methodological deficiency may account for the fact that Sedláček (1964a) has successfully used a 3000 cps tone for conditioning chick embryos as early as Day 16, while Vanzulli and Garcia-Austt were unable to record cochlear microphonics in response to 3000 cps until somewhat later in development.

In a neuroanatomical experiment (otocyst extirpation) with chick embryos, Levi-Montalcini (1949) deprived the secondary acoustic centers (nucleii angularis, magnocellularis, and laminaris) of afferent root fibers from the cochlear nucleus and observed their growth and differentiation over the course of embryonic development. The differentiation of the three secondary acoustic centers was unaffected up

to 11 days. After that point, the nucleus angularis, the only one of the three acoustic nucleii which receives all its afferent synaptic connections from the cochlear root fibers, began to degenerate. This result indicates that the afferent root fibers from the cochlear nucleus begin to play a major role in the completion of differentiation and maintenance of nerve cells of the angularis around Day 11. Whether the dependence of the angularis on afferent fibers from the cochlear nucleus is trophic or excitatory cannot be ascertained at the present time. The remarkably advanced differentiation of the peripheral and associated acoustic centers by Day 12 and 13 (Levi-Montalcini, 1967, pers. commun.), plus the dependence of the nucleus angularis on afferent fiber connections from the cochlear nucleus at this stage, suggest, however, that the auditory system of the chick may be capable of receiving sensory stimulation about Day 12 or 13. Further attesting to this possibility, the work of Rebollo and Casas de Roncagliolo (cited by Vanzulli and Garcia-Austt, 1963) shows that the tectorial membrane is fully developed by Day 11, and the scala tympani and basilar membrane are fully developed by Day 12–13. Also, on Day 12 the basilar membrane is freed from the underlying mesenchyma so it is in a position to vibrate. Finally, between Day 12 and 14 the sensory cells complete their differentiation and assume the characteristics of adult cells. In a more recent study Knowlton (1967) has made a detailed descriptive analysis of the very early growth and differentiation of the chick embryo's auditory system, and her findings lend further substance to the above picture.

Since it may be relevant to the onset of audition, the development of electrical activity in the chick embryo's brain should be mentioned here. The first electrocerebral potentials are recorded around Day 12 or 13 (Garcia-Austt, 1954; Peters, Vonderahe, and Huseman, 1960). At this time the electrical activity is sporadic and of low amplitude. By Day 15 the EEG pattern changes considerably and it is possible to pick up potentials from the optic lobes and cerebellum as well as the cerebrum (Peters, Vonderahe, and Huseman, 1960). Garcia-Austt (1954) distinguishes an "early phase" beginning on Day 12–13 (low voltage) and a "late phase" beginning around Day 15 (high voltage), and this division seems to agree with the recordings published by Peters, Vanderahe, and Huseman (1960).

Audition: Behavioral Aspects

The first behavioral evidence of hearing in chick embryos was published by Gos (1935). She employed the same procedure here as with her work on tactile conditioning. It is of interest that she believed that habituation (100 rings of a bell without a response) occurred as early as Day 10 in the chick and that association processes began as early as Day 15. She presents conditioning data from Day 17–20 only, however, and her results are shown in Fig. 3. The main differences in number of CS-UCS pairings to criterion (five consecutive unreinforced responses to CS alone) are between the Day 17–18 embryos and the Day 19–20 embryos. It is of interest that the same disjunction may have been present during the habituation trials. It took up to 71 applications of the bell to habituate the response in both the Day 17 and 18 embryos, while it took only 35 (Day 19) and 45 applications (Day 20) in the older embryos. This sort of difference is more definitively explicated in the auditory conditioning research of Sedláček reviewed below.

Hunt (1949) briefly extended Gos' work. While he found no evidence of conditioning prior to the fourteenth day, 17 of 19 chicks treated after Day 14 appeared to be conditioned to the bell. (Electric shock was used as the UCS by both Gos and Hunt; Hunt did not first habituate the response to the bell as did Gos.)

The most extensive prenatal auditory conditioning work on the chick has been performed by Sedláček (1964a). He used a 3000 cps tone as the CS and local beak or head movements as the conditioned response (CR) in exploring various parameters of conditioning in the chick embryo beginning on Day 16. [Since the embryo is quite active, it is necessary to ask how the conditioned response is isolated from movements which may not be attributable to the conditioned stimulus. In answer to this question, Sedláček (1965, pers. commun.) stated that he used four criteria:

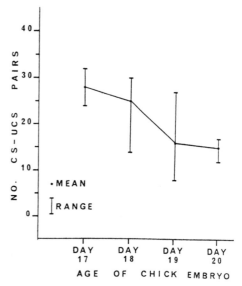

FIG. 3. AUDITORY CONDITIONING IN THE CHICK EMBRYO

Smaller number of CS-UCS pairs indicates a more rapid rate of conditioning. (Data from Gos, 1935.)

"(a) The single isolated auditory stimulus doesn't evoke the local motor reaction; (b) the latent period between the start of conditioned stimulus and conditioned motor reaction must be shorter than the beginning of unconditioned stimulus; (c) the statistical coincidence between the conditioned stimulus and conditioned response; (d) the conditioned response can be extinguished (inhibited) during the applications of conditioned stimulus without the reinforcement."]

Because of the importance of the intensity of the UCS (electric shock) in demonstrating conditioning at various ages, in his first experiment Sedláček examined the effectiveness of three levels of shock from Day 16–20. He found that a five-volt electric shock (USC) was more effective than 20 or 40 volts from Day 17–20. At Day 16, however, 40 volts was significantly more effective than 20 volts or five volts. The effectiveness of the 40-volt UCS in promoting conditioned responses progressively decreased on each day (30 per cent CR on Day 16 to 14 per cent CR on Day 20). The 20-volt UCS progressively increased in effectiveness on each day (24 per cent CR on Day 16 to 58 per cent CR on Day 20), while the five-volt UCS showed an even greater increase in effectiveness on Day 16 (28 per cent CR) to Day 20 (73 per cent CR).

In his next experiment, Sedláček varied the intensity of the 3000 cps tone using a 20-volt electrical shock as reinforcement. As can be seen in Fig. 4, on all days the highest sound pressure level (68 db) produced the greatest proportion of conditioned responses. There was a substantial lowering of the auditory threshold on the day of hatching (Day 20), but there was still a statistically reliable difference between 68 db and 55 db at that time.

In another experiment, Sedláček examined the effect of duration of the 3000 cps tone on the formation of the conditioned response. The tone and shock were presented sequentially. As shown in Fig. 5, the shortest duration of the CS (3 sec) was most effective on Day 16 and 17 and least effective on Days 18, 19, and 20. The five-second duration was more effective than the seven-second duration on

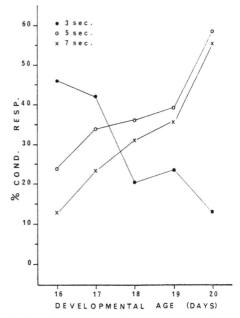

FIG. 5. EFFECT OF THE DURATION OF THE CONDITIONED STIMULUS ON THE PROPORTION OF CONDITIONED RESPONSES IN CHICK EMBRYOS DURING THE LAST FIVE DAYS OF INCUBATION

Conditioned stimulus (3000 cps tone) and unconditioned stimulus (electric shock) were presented sequentially. (Data from Sedláček, 1964a.)

Days 16 and 17, and there were no differences between them on Days 18 through 20.

In still another experiment, Sedláček examined the influence of the time of coincidence of the CS and UCS on the percentage of conditioned reactions. The CS was a 3000 cps tone at 68 db and the UCS was a 20-volt electrical shock. As shown in Fig. 6, the five-second CS-UCS overlap was significantly more effective than one second on Days 17, 18, and 19. There was no difference on Day 16, and on Day 20 the five-second overlap was significantly less effective than that of one second.

From the results of these and other experiments, Sedláček (1964a) concluded that there are three stages in the development of the "temporary connection" during the final five days of incubation in the chick. The first stage (Day 16) involves the summation reflex, the second stage (Day 17–19) involves the dominant reflex, and the final stage (Day 20) involves a genuine conditioned reflex. The terms *sum-*

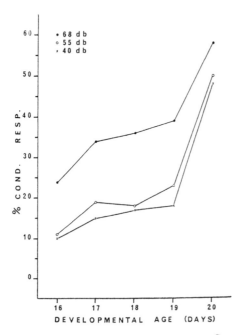

FIG. 4. INFLUENCE OF THE INTENSITY OF THE CONDITIONED STIMULUS (3000 CPS TONE) ON THE PROPORTION OF CONDITIONED RESPONSES IN CHICK EMBRYOS DURING THE FINAL QUARTER OF THE INCUBATION PERIOD (Data from Sedláček, 1964a.)

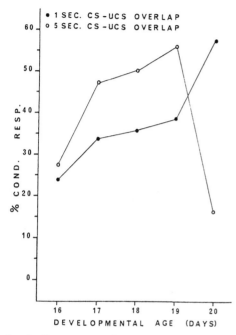

Fig. 6. Effect of the Degree of Temporal Overlap of the Conditioned Stimulus (3000 cps tone) and the Unconditioned Stimulus (Electric Shock) on the Formation of Conditioned Responses in Chick Embryos during the Last Five Days of Incubation

(Data from Sedláček, 1964a.)

mation reflex and *dominant reflex* refer to the supposed neurological correlates of the conditioning process. The summation reflex is regarded as the most primitive form of temporary connection and refers to the relative inability of the nervous system to preserve traces of previous excitation. In such instances, acquisition of the conditioned response proceeds slowly, never reaches a high level, and is very dependent upon high intensity stimuli. In the succeeding or next highest form of temporary connection, the nervous system is said to be able to form a dominant focus of excitation and therefore conditioning proceeds somewhat more rapidly and successfully, although the "dominant center" is still characterized by a relatively high degree of inertia. Sedláček's work thus appears to indicate that the effects of auditory stimulation encountered during these periods may differ markedly, (1) in terms of the fetus' immediate response, and (2) in

setting up possible trace effects on postnatal behavior. The effects of prenatal conditioning on postnatal behavior have not yet been explored — as suggested by the term "temporary connection," these effects may be quite transient, especially in view of the rapidly changing response tendencies of the embryo. Again, it seems likely that aversive conditioning is not directly relevant to the actual process of prenatal behavior development, but this technique does reveal interesting and potentially valuable information on the functional state of the nervous system.

Auditory research which has direct relevance to the actual development of behavior has been published recently by Gottlieb (1965a), Grier, Counter, and Shearer (1967), and Vince (1966a, b, 1968). My results demonstrated that both chick and duck embryos show an increase in activity (oral movements, vocalization, and heart rate) when exposed to the maternal call of their species on the day before hatching. As will be seen in the final section of the present paper, the behavioral changes represent a discriminative perceptual response and not simply a response to sound per se. On the other hand, heart rate changes do represent a non-discriminative reaction to auditory stimulation of any kind.

Grier, Counter, and Shearer (1967) exposed chick eggs to an intermittent 200 cps tone from Day 13–18; this stimulation increased the postnatal responsiveness of the embryos to 200 cps vs. 2000 cps. This finding may be relevant to the actual development of audition because (1) the cochlear microphonics research and the histological evidence show the organic possibility of the embryo's reacting to low frequency sounds; and (2) embryos are routinely exposed to such sounds during incubation. Low frequency noise is produced by motors in incubators, by the knocking of eggs together in the nest, and from diverse other sources in nature. Scraping and other frictional sounds produced both inside and outside the egg may be another source of activation of the embryo's auditory system. Since all the embryos in the study by Grier, Counter, and Shearer were exposed to the 200 cps tone for the entire five-day period, it is not possible to say whether exposure as early as Day 13 was effective or not. These investigators also directly observed the embryo's response to tonal stimulation on various days of the incubation period, but they used only one embryo at each age and their procedure was too crude to allow any firm statement on the overt onset of auditory sensitivity in chick embryos.

The research of Vince (1966a, b, 1968) deals with the last few days of incubation in quail eggs. Her results show that the "clicking" sound produced in conjunction with respiration plays a role in the synchronization of hatching in multi-egg clutches. It is not yet known whether this is solely an auditory phenomenon or whether vibration is also involved under normal conditions. Experimentally, it was possible to show the effects of either sound or vibration at the rate of 3 stimuli per second. Other avian species (chicks, ducks) besides quail produce clicking sounds, but the synchronization effect is apparently restricted to quail. Vince's most recent studies (1966b, 1968) provide evidence that such stimulation accelerates hatching when it occurs within the range of 1.5 to 60 per second, while acceleration declines with rates above and below this range. Thus, it is not stimulation per se which accelerates hatching; rate of stimulation is an important factor.

To summarize the basic work on the development of audition in the chick embryo, the cochlear microphonics research and neuro-anatomical evidence suggest that the chick may be sensitive to low frequency sounds as early as Day 12–13. (This would correspond to Day 16–17 in duck embryos.) There is as yet no firm behavioral evidence for auditory sensitivity at these early ages. Chicks exposed to low frequency sounds from Day 13–18, however, show an increased responsiveness to those sounds after hatching. Chicks show some capability for rudimentary auditory conditioning by Day 16 and their auditory threshold drops considerably on the day of hatching (Day 20). The increased capacity for auditory conditioning around Day 19–20 in the chick is correlated with the establishment of pulmonary respiration. That respiration may be causally involved is suggested by the finding that guinea pig fetuses cannot be conditioned while umbilical circulation is intact but only after the umbilicus is ligated (Sedláček, Halváčková, and

Švehlova, 1964), thereby forcing the fetus to breathe. Also, vagal deafferentation in guinea pigs and chick embryos (Sedláček, 1964b) severely reduces the possibility of conditioning. According to Sedláček, the respiratory apparatus triggers the vagal nucleii in the medulla which, in turn, send afferent impulses into the central system, thereby contributing to an increase in CNS "arousal."

Non-Visual Light Sensitivity

By means of some as yet obscure mechanism, the chick embryo reacts non-visually to light shortly after the onset of motility (i.e., before its eyes are functional). Beginning late on Day 3, Bursian (1964) exposed embryos to 10–20 sec of light from a 500-watt photo lamp at intervals of one to two minutes depending on the frequency of the embryo's movements. To assure that the embryo was responding to light and not to heat, Bursian used a heat-insulating filter. The light served to increase the frequency and duration of embryonic motility from late on Day 3 to some time on Day 6. Thereafter, from late on Day 6 through Day 8, the light increasingly served to inhibit or arrest the embryo's movements. The effect of the light may have been mediated through photoreceptors in the skin, muscle, brain, or spinal cord. Such photosensitive processes have been demonstrated in the embryonic muscle of clam larvae (Lábos, 1966), in the skeletal muscle of frogs (Rosenblum, 1960), and in the brain of immature ducks (Benoit and Ott, 1944–45). In view of the histological evidence which indicates the closure of the multi-synaptic cutaneous reflex arc around Day $6\frac{1}{2}$ in the chick (Windle and Orr, 1934; Visintini and Levi-Montalcini, 1939), it is most interesting that this age corresponds to the time when the influence of light begins to become inhibitory rather than excitatory. Whether this is a coincidence, or whether it reflects the cutaneous mediation of light after Day $6\frac{1}{2}$, needs to be investigated.

In further studies involving different wavelengths of light, Bursian (1965) found that green light (540 mμ) significantly increased the duration of movement in Day 4 chick embryos, while red (630 mμ) and blue light (420 mμ) had no effect at this early age. On Day 6 the effects of the green light became inhibitory and remained inhibitory even at the highest intensity of illumination. It was only at the highest intensity that red light became effective; it exerted an excitatory effect on Day 6. Blue light remained ineffective at both high and low intensities on Day 6. Thus, the chick embryo appears to be capable of making a non-visual discriminative response to different wavelengths of light as early as Day 6 of development.

To digress momentarily, the results of two other widely disparate studies suggest that around Day 6 some extremely critical or delicately balanced changes take place in the chick embryo and that this stage may represent a particularly significant transitional period in development. First, the pattern of the chick's metabolic rhythm around Day 6 is apparently peculiar to that particular stage of development (Johnson, 1966). Second, exposure of the chick embryo to X-irradiation on Day 6 exerts a deleterious effect on the chick's following-response after hatching. The same exposure before or after Day 6 does not affect the hatchling's ability to follow (Strobel, Clark, and Macdonald, 1968). While it is possible that the findings of Windle and Orr, Visintini and Levi-Montalcini, Bursian, Johnson, and Strobel are unrelated or stem from different sources, the fact that they all occur on or about Day 6 suggest that this period may be particularly important in the chick's development.

Another experiment which indicates non-visual sensitivity to photic stimulation concerns shortening of the incubation period as a function of exposing eggs to incandescent light. Lauber and Shutze (1964) found that chick eggs exposed to incandescent light during the first week of incubation hatched somewhat sooner than those exposed during the second or third week of incubation. The eggs exposed to light during the first week hatched about eight hours earlier than non-exposed eggs. Eggs exposed to light during all three weeks of the incubation period began hatching a full 24 hours (Day 19) before the non-exposed eggs and 16 hours before the eggs which were exposed during the final two weeks of incubation. In comparing the effects of incandescent and fluorescent light, Lauber and Shutze found

that both kinds of light accelerate embryonic development, but incandescent light is most effective. (While the emission spectra of the lamps were not reported, the fluorescents were three to four times brighter than the incandescents in these experiments.) Finally, exposure to fluorescent light during only the second week of incubation retarded hatching.

The findings of Lauber and Shutze clearly show the differential sensitivity of the embryo to light stimulation in terms of (1) type of light, (2) age at exposure, and (3) length of exposure. It is not known whether exposure to photic stimulation actually caused the visual system to begin functioning sooner than usual; however, exposure did foster anatomical development of the eyeball. On six measures of growth the right eyeball was larger than the left one in the stimulated group, while this difference was not consistent in the non-stimulated embryos. Because of the embryo's position in the egg during the last-half of incubation, it is the right eye which can receive the most photic stimulation. As far as it goes, this finding is consistent with the bi-directional version of the structure–function relationship (Gottlieb, 1968) which holds that appropriate stimulation fosters prenatal physio-anatomical growth and differentiation. "Appropriate stimulation" is meant to take cognizance of the fact that stimulative effects need not always be beneficial. Whether such effects are beneficial or deleterious clearly depends both on the kind of stimulation and the developmental status of the embryo. This point is demonstrated by the difference in the direction of the effect of fluorescent light, depending on whether the embryo was exposed during the second week of incubation or for the entire period of incubation. [If chick *hatchlings* are exposed to continuous incandescent light for 10 weeks after hatching, a glaucoma-like condition appears in the eyes along with an actual reduction of thickness in the nerve fiber layer and the layer of rods and cones in the retina (Lauber, Schutze, and Mc Ginnis, 1961).]

Other evidence (Källén and Rüdeberg, 1964) suggests that the incidence of abnormal mitoses is increased when the chick embryo is exposed to a 60-watt incandescent light during the first 24 hours of incubation. In the control embryos

42 abnormal anaphases (chromosomal breakages) were observed out of 409 studied, while 34 abnormal anaphases out of 186 were found in malformed embryos from the treated group. Källén and Rüdeberg found a marked decrease in the occurrence of malformations which followed exposure of eggs to light after the first 24 hours of incubation. By the end of 48 hours of incubation, no deaths or malformations were induced by exposure to the light. The possibility that an uneven distribution of heat (rather than light) may have caused the malformations cannot be discounted, however, especially in view of a study by Tamimie and Fox (1967), which demonstrated that continuous exposure of eggs to a single 100-watt incandescent light during incubation caused a high precentage of mortality, delay in hatching, and malformations. Lauber and Shutze employed three 40-watt bulbs to disperse the light in their incubator, and continuously recorded the incubator temperature during the experiment. It seems possible, therefore, that the uneven incubator temperature generated by a single source of light may have been the primary factor in producing abnormalities in the Källén-Rüdeberg and Tamimie-Fox studies, in both of which temperature was neither adequately monitored nor controlled. (Another factor which may have contributed to uneven heat or inadequate heat circulation in the Tamimie-Fox study was the use of metal dividers in the incubator to prevent light from penetrating to certain egg trays.) In view of the conflicting results, further studies of the effects of light vs. heat on embryonic development would seem in order. In this connection, it may also be important to examine the effects of different types of fluorescent light, or at least to be alert to this aspect of the problem for methodological reasons. Although the findings of Ott (1964) seem almost too astounding, his preliminary observations suggest the possibility of important differences in the effect of white, pink, and black fluorescent light on embryonic development.

A recent report by Menaker (1968) supports both the validity and importance of non-visual light sensitivity in birds. Menaker blinded adult house sparrows and found that their activity rhythms could subsequently be entrained to a 24-hour visible light–dark cycle.

Most important for control purposes, the sparrows did not entrain to the cycles of temperature, noise, infrared radiation, or vibration that were associated with the light cycle. Therefore, the sparrow must possess an extra-retinal photoreceptor which is coupled to its biological "clock." Thus, in view of the findings with embryos reviewed above, the possibility is open that circadian activity rhythms may be entrained via non-visual light sensitivity during embryonic development.

Visual Sensitivity: Behavioral Aspects

There has been only one carefully controlled study of a visually mediated behavioral response to light in chick and duck embryos. Twelve to 40 hours before hatching, Oppenheim (1968a) removed the shell over the right eye of the embryos and exposed them to an intense incandescent light (approx 1500 foot candles) for one minute. During this period both species of embryos showed a reliable increase in their baseline rate of oral activity (beak- or bill-clapping), and this activity returned to the pre-stimulation baseline in the five-minute post-stimulation period. An effective heat-insulating filter between the light and the embryo assured that the embryos were responding to the light and not to an increase in heat.

Oppenheim's results are in agreement with the observations of Lindeman (1947) and Kuo (1932), who noted pupillary constrictions and eyelid movement in chick embryos in response to light one to two days before hatching. (These latter studies did not preclude heat as an effective factor, nor did they make use of an objective means of recording.) Now that the recording of oral activity has proven to be a satisfactory technique for a behavioral investigation of vision as well as audition (Gottlieb, 1965a), a further objective examination of the embryo's visually mediated behavioral response to light can be undertaken. The lower age limit of overt visual responsiveness needs to be determined and differential sensitivity to intensity and wavelength would certainly repay examination, especially in tracing out the prenatal origins of the pecking preferences which both chicks (Hess, 1956) and ducklings (Kear, 1964) exhibit after hatching.

Oppenheim (1968b) has also made a beginning in this direction. He determined the color preferences of visually naive ducklings which had been incubated, hatched, and reared in total darkness, and attempted to modify their naive preference by exposing them to lights of another color on the day before hatching. Based on exposure the day before hatching, the Peking duckling's postnatal pecking preference for green was resistant to modification. Oppenheim's results suggest that the duck embryo's color preferences may already be firmly established on the day before hatching. If so, any attempt to trace out the development of that preference, or to modify it, must begin at an earlier stage.

One of the most basic unresolved differences in conceptions of prenatal behavior (Gottlieb, 1968) concerns the temporal aspect of the development of adaptive behavior which is evinced after hatching. One group of theories (predetermined epigenesis) holds that such behavior develops abruptly, while other theories (probabilistic epigenesis) predict the gradual development of such behavior.

The term *predetermined epigenesis* was adopted from Needham (1959, footnote to p. 213), who used it in a historical context to characterize the modern conception of morphogenesis. While some readers may decry the introduction of new terms to characterize theories of development, this practice seems necessary insofar as it is agreed that all modern conceptions of development are epigenetic (i.e., no embryologist today openly avows a belief in preformation). In keeping with this trend, Waddington (1952) has used the term epigenetics to designate all aspects of the causal analysis of morphological development, and it seems appropriate to extend this usage to include all causal analyses of behavioral development.

In the case of the development of color perception, predeterministic theory would predict that neonatally adaptive preferences should be present at or shortly after the original manifestation of visual responsiveness in the duck embryo. Probabilistic theory would predict that such preferences are at first relatively undifferentiated and become sharpened gradually over the course of time. In the instance of a green color preference in ducklings, it is doubtful whether this reflects a perceptual preference specifically adapted to a particular class of food objects; it

seems more likely that it reflects a fundamental feature of the visual sensitivity of the duckling. While all young Anatidae thus far examined show a pecking preference for green objects when tested in the laboratory (Kear, 1964), all members of this family do not subsist on a diet of green food under natural conditions.

Study of the prenatal ontogeny of color perception is one way in which the major assumptions of our very primitive theoretical conceptions can be tested; from the results of such deductively designed experiments we may eventually be able to arrive at a more adequate and unified conception of behavioral development. The nature-nurture controversy is "dead" only insofar as it is ignored or its major assumptions remain untestable. Some progress has been made in stating the key assumptions of these conceptions in testable terms (Gottlieb, 1968); as these assumptions are tested it should be possible to forge a new view of the development of prenatal behavior and thereby break down the purely conceptual dichotomy (nature-nurture) which many workers find inadequate as well as vexing. (The experiments presented in the final section of this paper bear directly on this issue.)

Visual Sensitivity: Electrophysiological Aspects

A great deal of electrophysiological work has been done on the development of visual function in the chick embryo and, to a lesser extent, on the duck embryo. Although this work has been carefully executed with respect to (gross) recording techniques, it is apparent that all of the embryos were exposed to an indeterminate amount of light prior to the experiments (i.e., during incubation). Since it is known that exposure to light plays a role in accelerating electrophysiological changes in the visual system in both chick embryos (Peters, Vanderahe, and Powers, 1956) and duck embryos (Paulson, 1965), the failure to control this variable must be taken into account when evaluating the results of studies of electrophysiological aspects of visual sensitivity — i.e., there is probably more flexibility than indicated with respect to the age of onset of the various visual phenomena. The following summary describes in a general way the research findings of Garcia-Austt, 1954; Garcia-Austt and Patetta-Queirolo, 1961a; Peters, Vanderahe, and Powers, 1958; and Peters, Vanderahe, and Huseman, 1960. [Recently, Sedláček (1967) has replicated the work on evoked potentials from the optic lobe of chick embryos, using a more precise aging procedure than has heretofore been employed in electrophysiological research with chicks.]

While sporadic electrical activity in the cerebrum begins about Day 12 in the chick, it is not until Day 14–15 that the optic lobes begin to show irregular signs of low amplitude electrical activity. Potentials from the retina do not appear until Day 17–18. At the same time, the chick shows reflexive ocular movements to light stimulation and these can be distinguished from spontaneous eye movement. When the afferent optic pathways are severed, retinal activity and spontaneous eye movements persist, while ocular movements elicited by light disappear.

The earlier appearance of electrical activity in the optic lobes than in the retina is a consistent finding in chick embryos. By Day 18 the patterns of electrical activity in the optic lobe are fundamentally similar to that of a newly hatched chick. Although electrical activity in the retina begins later than optic lobe activity, retinal functioning apparently matures at a faster rate than optic lobe function.

> Gross recording techniques were used in all the studies cited. Microelectrode studies might reveal an earlier onset of activity in both the tectum and retina, and such studies could also change the current picture that shows retinal activity beginning later than optic lobe activity.

By Day 19 the pattern of retinal activity in response to light is similar to that of a hatchling. Thus, the postnatal type of EEG appears about Day 18 in the chick embryo, while the postnatal type of ERG appears about Day 19. Work on the duck embryo has not been extensive and the earliest onset of EEG and ERG is uncertain (Paulson, 1965). Paulson's results, however, clearly indicate the shortening of latency and duration of evoked electrical activity in the retina and optic lobe as a consequence of prior exposure to visual stimulation, and this finding is predictable on the basis of a probabilistic view of behavioral epigenesis (Gottlieb, 1968). Specifically, these results with duck embryos, as well as similar findings by Peters, Vanderahe,

and Powers (1958, p. 465) with chick embryos, suggest that the timing of critical periods in behavioral development is not solely attributable to organic growth and differentiation but is also influenced by the stimulative history of the embryo. The existence of critical periods is ascribable to organic factors; whether a critical or sensitive period is shorter or longer, or occurs earlier or later, is ascribable to stimulative factors. It is by virtue of the uncertain interplay between organic and stimulative factors that sensitive periods in behavioral development become probabilistic occurrences rather than absolutely predictable events. Although not widely appreciated, this point has been repeatedly demonstrated in experiments on imprinting where the duration or timing of *the* critical period has been shown to shift according to changes in the stimulative environment in which the hatchling is kept prior to testing (e.g., Gottlieb and Klopfer, 1962; Haywood, 1965; Moltz and Stettner, 1961).

From the standpoint of postnatal behavior — specifically, visual perception — the most relevant electrophysiological research on embryonic retinal development has been performed by Garcia-Austt and Patetta-Queirolo, 1961b. Their work shows that the physiological basis of certain visual perceptual preferences of newly hatched chicks is evident one to two days before hatching (Day 18-19). On both Days 18 and 19, the maximum amplitude of the electroretinogram is evoked by a visual flicker rate of 3 per sec. No response was elicited by a flicker rate less than 1.5 per sec and increasing the rate beyond 3 per sec decreased the amplitude of the ERG. When the approach-response of chicks is tested to different visual flicker rates after hatching (Simner, 1966), it is found that at eight hours they respond maximally to rates between 2 and 4 per sec; they become more responsive to lower and higher rates only at later ages (26–76 hours).

The concordance of prenatal retinal activity and early postnatal behavior in response to a visual flicker rate of approximately 3 per sec warrants further examination. There are three questions which require an answer before the nature of the psychophysiological relationship can be fully understood: (1) Does a flicker rate of 3 per sec maximally activate the *behavior* of the embryo? (2) Is there a shift in the ERG between 8 and 26 hours after hatching, corresponding to the initiation of responsiveness to higher and lower flicker frequencies? Garcia-Austt and Patetta-Queirolo (1961b) report that no optimal flicker frequency is found in the "newborn" chick, but the exact age of the chicks at testing was not specified. (3) Since an *auditory* flicker rate of around 3 per sec is preferred in postnatal behavioral tests in both chicks and ducklings (Collias and Collias, 1956; Tolman, 1967), does this rate actually represent or tap a general systemic (central neural) preference beyond the peripheral sense organs?

Such a general phenomenon (trans-sensory flicker rate preference) would support the cardiac self-stimulation hypothesis (Simner, 1966) which predicts that rate preferences in various sensory modalities are determined by the effect of the embryonic heart beat on the various sensory systems during embryonic differentiation. While Simner's cardiogenic hypothesis is quite speculative and is very possibly based on correlations which have no causal connection, it does indicate a relationship which, if verified, would have some significance for comparative as well as ontogenetic studies of behavior: namely, a lawful relationship may exist between heart rate and flicker rate preference, such that species with a relatively high heart rate prefer high rates of flicker and species with a relatively low heart rate prefer low rates of flicker. Such a correlational statement need not imply a causal mechanism to be useful for predictive purposes. Whether the embryonic heart beat induces flicker rate preferences or whether the embryo's heart rate itself simply reflects a general systemic rhythmicity remains to be determined. The heart rate of both chick and duck embryos slowly rises during the early stages of incubation until it reaches 160–260 beats per minute, the typical range for both species during the final three-quarters of the incubation period (Romanoff, 1960; Gottlieb and Kuo, 1965; Gottlieb, unpubl. — based on 59 duck embryos and 108 chick embryos). Thus, if the cardiogenic hypothesis of flicker rate preference is true, two predictions require support: (1) Individual differences in embryonic heart rate should be directly correlated with individual differences in flicker rate preferences, and (2) exogenous manipulations of heart rate during the embryonic period should alter postnatal flicker rate preferences in the direction of the prenatal manipulation. If heart rate simply reflects a general systemic rhythm, however, the first prediction would be true while the second one would be false. [A recent technologically advanced study of heart rate in chick embryos, using a ballistocardiograph (Cain, Abbott, and Rogallo, 1967), indicates much less variability than heretofore reported, so a test of (1) may not be practicable.]

The final experiments on the electrophysiological aspects of vision in the embryo concern the effect of varying wavelengths of light on the embryonic ERG. According to the results of Garcia-Austt and Patetta-Queirolo (1961b), the ERG of the Day 17–18 chick embryo is unresponsive to wavelengths between 390 and 510 mμ (blue to green), while it is responsive to white light, green to orange light (480–610 mμ), and dark red light (600–700 mμ). On the next day of embryonic development, when the postnatal-type of ERG appears, the chick embryo's retina begins to show a low level of responsiveness to blue light. In terms of flicker fusion, on the day before hatching the highest fusion frequency was obtained for dark red followed in decreasing order by green-orange, white, and blue-green light. Since the requisite controls for brightness differences were apparently not made in these studies, the results cannot be taken as unequivocally demonstrating differential color sensitivity in the embryo. To further analyze the ontogeny of color perception, it would also be important to determine whether the embryo is capable of making a differential behavioral response to various wavelengths prior to hatching. Specifically, it would be of value to determine whether there are different changes in rate of oral activity in response to the different colors, since this behavioral indicator has already proven useful in demonstrating auditory discrimination in highly developed duck embryos (data presented in final section of this review).

In terms of postnatal color preferences in chicks, the published studies have been poorly controlled either in terms of prior exposure of the egg or hatchlings to light, in the failure to control for brightness, or the lack of control of positional biases in multiple choice apparatuses. Although one might tentatively predict from the prenatal ERG results that red and blue might differ markedly in their effectiveness in the early postnatal color perception of the chick, the results on this point are equivocal: some investigations show red to be significantly more effective than other colors while other studies do not (cf. Gray, 1961; Schaeffer and Hess, 1959). Carefully controlled behavioral studies of both the embryo and the hatchling are required before the relationship of the prenatal ERG results to the ontogeny of color perception can be specified.

Visual System: Neuroanatomical and Biochemical Aspects

These aspects have been rather thoroughly studied in the chick embryo. Shifferli's work (1948) indicates that the fibers of the optic nerve are myelinated by Day 16, and the retina and optic lobe have almost completed differentiation by Day 18 (Coulombre, 1955; Cowan and Wenger, cited by Oppenheim, 1967a). However, the electrophysiological evidence from gross recordings (Garcia-Austt and Patetta-Queirolo, 1961a; Peters Vonderahe, and Powers, 1958) suggests that the optic lobes become active somewhat earlier (Day 14–15) than the retina (Day 17–18). Nissl bodies first appear in the ganglion cells of the retina on Day 16, in the amacrine cells on Day 17, and in the bipolar cells on Day 18 (Rebollo, 1955). In terms of biochemical development, the retina shows a marked increase in cholinesterase, acetylcholine, alkaline phosphatase, and glutamotransferase beginning around Days 16 and 17 (Lindeman, 1947; Rebollo, 1955; Rudnick and Waelsch, 1955). The optic lobe shows a significant increase in cholinesterase activity beginning as early as Day 12, with another spurt between Days 18 and 20 (Vernadakis and Burkhalter, 1967). In view of the anatomical and biochemical findings, it seems possible that the visual system of the avian embryo is rather highly developed several days before hatching, so an extensive behavioral analysis of the embryo's perceptual capabilities might prove fruitful in answering questions about the ontogeny of color discrimination, flicker preferences, and form discrimination.

TABLE 1

Sequence of sensory development in chick embryos: behavioral evidence

SEQUENCE	DAY	STAGE OF INCUBATION
1. Non-visual photic sensitivity	3	15%
2. Tactile sensitivity	6½	33%
3. Vestibular sensitivity	8	40%
4. Proprioception	10	50%
5. Audition	12	60%
6. Vision	19	95%

Note: While the sequence of development is fairly well established, time of onset is tentative.

SENSORY DEVELOPMENT: CONCLUSIONS

It is now possible, with some certainty, to arrange the sequence of sensory development in the chick embryo as follows: (1) non-visual photic sensitivity, (2) tactile sensitivity, (3) vestibular sensitivity, (4) proprioception, (5) audition, (6) vision. The tentative time of functional onset of each modality, as indicated by the presently available behavioral evidence is shown in Table 1.

Since the overwhelming majority of research has been conducted on the domestic chick embryo, work on other avian forms would be very desirable (lest the domestic chick become to behavioral embryology what the domestic rat has become to experimental psychology). The meagre amount of research on duck and pigeon embryos supports the sequential picture of sensory development gained from the chick embryo, at least with respect to the early development of cutaneous sensitivity and the late development of audition and vision. It is not known when gustation or olfaction begin to function in the avian embryo; although not yet definitely established, it seems likely that taste develops shortly after auditory sensitivity.

EVOLUTIONARY AND ONTOGENETIC IMPLICATIONS OF SENSORY DEVELOPMENT

From the comparative point of view, it would be of particular interest to determine if the time of onset of the various sensory functions is similar (relative to length of incubation) in both altricial and precocial species of birds. In light of the available evidence, it is tentatively assumed that the sequential or serial order of sensory development is the same in both types of birds.

From the evolutionary point of view, it would be of special significance to determine the extent to which the sequence of sensory development is the same for all vertebrates, despite the diverse ecological and social conditions which confront different species during later stages of life. The main evolutionary changes in the embryology of behavior may be largely restricted to (1) the time of functional onset of the different sensory systems (relative to length of incubation or gestation) and (2) the development of different perceptual capabilities or preferences within each sense modality. From my own acquaintance with the comparative studies, it seems that the sequential order of sensory development has the possibility of the widest generality (similarity) across the range of vertebrate species, with temporal onset relative to length of incubation or gestation period next, and perceptual discrimination ability showing the widest dissimilarity among species. If this proposal is true, the serial order of sensory development would be the most conservative feature of the evolution of embryonic behavior in vertebrates and the development of perceptual discrimination abilities would be the most labile feature in such a system.

Translated into ontogenetic terms, the above proposal implies that if the embryo is prematurely exposed to relevant stimulation during development or indefinitely deprived of such stimulation, there would be little or no effect on the absolute sequence in which the sensory systems develop. However, such a manipulation might have some effect on the time of onset of the particular sensory system involved, and there would be major effect on the development of perceptual discrimination abilities within the sensory modality thus manipulated. For example, withholding relevant auditory stimulation from the avian embryo should affect the development of its usual auditory discrimination or perception abilities, but should not alter the time of onset of auditory sensitivity to such an extent that audition would develop later than vision. The experiments described

in the next section provide some support for this point of view, while also demonstrating that normally occurring embryonic stimulation is necessary to the development of adaptive auditory perception in the duck embryo.

ONTOGENY OF AUDITORY PERCEPTION
IN THE DUCK EMBRYO

If Peking ducklings hatched and reared in the laboratory are presented with two maternal replicas, one emitting the maternal attraction call of their species and the other emitting the maternal attraction call of some other species, the ducklings unerringly follow the replica which is emitting the maternal call of their own species (Gottlieb, 1965b). Further, in tests with a single maternal replica, both ducklings and chicks are more likely to follow the replica when it is emitting the species' maternal call in contrast to the maternal call of some other species (Gottlieb, 1965b). Since this auditory perceptual discrimination ability manifests itself (1) in the absence of previous experience with the maternal call, (2) is present during the appropriate period of development, and (3) is biologically adaptive, it fits the main criteria defining instinctive behavior. Although Holt (1931), Kuo (1967), Lehrman (1953), Schneirla (1966), and others have repeatedly discussed the possibility that instinctive or species-specific behavior might have its roots in prenatal ontogeny, this possibility has never been experimentally demonstrated. Consequently, I decided to trace the embryonic background of the Peking duckling's ability to selectively respond to the maternal call of its own species. This pursuit involved two steps: first, a determination of whether communally incubated duck embryos are capable of making a discriminative behavioral response to the appropriate maternal call prior to hatching and, if so, to determine whether exposure to naturally occurring auditory stimulation (embryonic "peeping") is requisite to the establishment of the discriminative response. (Bird embryos can vocalize shortly before or after their bill or beak penetrates the air space at the large end of the egg. The Peking duck embryos used here usually begin vocalizing three to four days before hatching.)

Communally incubated duck embryos were exposed individually to maternal attraction calls of various species on the day before hatching and their behavioral response (bill-clapping and vocalization) and physiological reactivity (heart beat) objectively recorded according to procedures already described in detail (Gottlieb, (1965a). In brief, the procedure involved opening the egg at the large end and fixing the recording electrodes to the embryo, followed by a 15–30 minute acclimation period during which the adequacy of the behavioral and physiological recordings was checked (Fig. 7). After the acclimation period a baseline rate of activity (oral, vocal, heart) was obtained for a 10-minute pre-stimulation period. The baseline activity period was immediately followed by a 30-second period during which the embryo was exposed to one of the maternal calls (at a sound pressure level of 68–74 db in all cases) and any change in its rate of activity recorded. Recording continued for a five- or a ten-minute period after stimulation to determine if activity returned to the baseline rate in the event of a change during the stimulation period. The statistical reliability of any changes during the 30-second stimulation period was evaluated by the Wilcoxon matched-pairs, signed-ranks test. In view of the great degree of individual variability in embryonic behavior, the Wilcoxon test is particularly appropriate as it takes into account the direction and magnitude of change shown by each individual during the stimulation period. If the direction of individual changes during the stimulation period is not consistent, the results will not achieve statistical reliability even though the magnitude of group change (+ or −) may be large in one direction or the other. For convenience of presentation, the experimental findings are shown in terms of group means and standard deviations in Tables 2 and 3. Eighty embryos were used in the first experiment, twenty embryos each being exposed individually to only one of the four calls.

As shown in Table 2, 12–24 hours before hatching Peking duck embryos (a highly domesticated form of Mallard, *Anas platyrhynchos*), show an increase in rate of oral activity and vocalization which is specific to the Mallard maternal call. Heart rate shows a non-specific increase to all maternal calls, and the brooding-like call of a sibling of the species evokes an increase in vocalization and heart rate but not

TABLE 2

Evidence of auditory discrimination in duck embryos (Day 26-27): rate per min of oral activity, vocalization, and heart beat before, during and after 30 sec exposure to maternal calls of various species

TYPE OF ACTIVITY	MALLARD CALL [N = 20]			SIBLING CALL [N = 20]			WOOD DUCK CALL [N = 20]			CHICKEN CALL [N = 20]		
	Before	During	After	Before	During	After	Before	During	After	Before	During	After
Oral activity	32.8	53.3***	32.0	34.8	45.9	30.8	39.2	39.6	31.9	19.4	20.6	19.3
(SD)	(18.4)	(24.7)	(25.2)	(44.7)	(73.6)	(31.8)	(30.4)	(41.1)	(22.0)	(12.9)	(17.8)	(13.1)
Vocalization	2.6	24.0***	4.0	4.0	7.6*	3.5	2.5	4.8	2.3	3.0	9.4	2.1
(SD)	(2.2)	(25.1)	(5.5)	(4.7)	(7.4)	(4.1)	(2.8)	(7.1)	(1.7)	(4.7)	(13.3)	(2.8)
Heart beat	217.8	243.8***	227.8	212.0	227.9***	216.1	236.9	245.0*	237.6	215.9	238.3***	211.1
(SD)	(24.3)	(34.0)	(29.1)	(33.4)	(38.4)	(35.8)	(34.2)	(30.7)	(33.3)	(32.3)	(37.6)	(27.7)

Designation of statistical reliability of changes from baseline: *P<0.05; **P<0.01; ***P<0.005
SD = standard deviation
[Where no asterisks appear, the change from the baseline rate during the stimulation period was not statistically significant (P>0.05).]

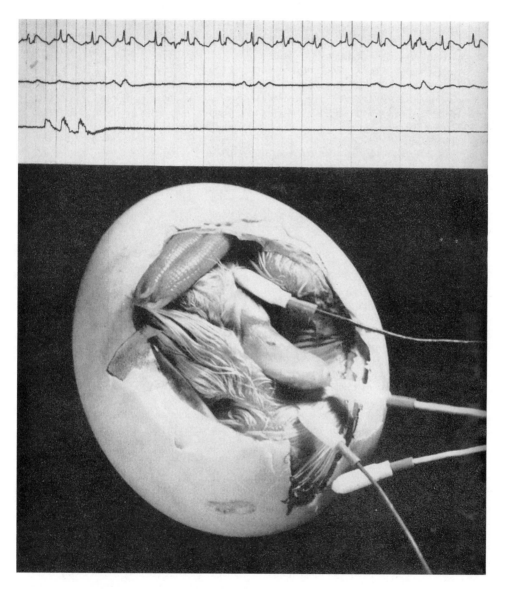

FIG. 7. DUCK EMBRYO WITH RECORDING ELECTRODES IN PLACE ON THE DAY BEFORE HATCHING
(Day 26)

Heart beat is shown on top line, the second line depicts oral activity (bill-clapping), and occurrence of embryo's vocalizations is shown on third line. Note that bill-clapping and vocalization are independent measures since the embryo can vocalize without making overt bill movements.

an increase in oral activity (bill-clapping). Since the embryo, as well as the hatchling, can vocalize without making overt bill movement, vocalization and oral activity are independent behavioral measures. None of the extra-specific maternal calls provoked an increase in either vocalization or oral activity. Thus, the communally incubated duck embryo is capable of making a discriminative behavioral response to the maternal call of its species on the day before hatching, as indicated by a selective increase in oral activity.

We next endeavored to determine if prior exposure to normally occurring auditory stimulation ("peeping" of self and sibs) is essential if the embryo is to make a discriminative response to the appropriate maternal call, as indicated by an increase in oral activity. For this aspect of the study we placed 22 Peking duck eggs in auditory isolation in individual compartments of a specially constructed sound-proof incubator on Day 23, 0 hours, and tested their response to the Mallard maternal call on Day 26–27. Before doing this experiment, it had been determined that Peking duck embryos are capable of vocalizing as early as Day 23; that is, even before their bill penetrates the air space at the large end of the egg. By placing the eggs in individual isolation, the embryos were prevented from hearing the vocalizations of sibs, but they could hear their own self-produced vocalizations. Since highly developed duck embryos typically vocalize in response to alterations in the sensory environment during the normal course of communal incubation, and isolation severely reduces such alterations, the isolated embryos probably vocalized less than those incubated communally. However, we did not objectively measure the amount of self-produced vocalizations in the isolated embryos.

As shown in Table 3, when the isolated duck embryos were tested individually on Day 26–27 for their response to the Mallard maternal call, they showed an increase in vocalization and heart beat, but no increase in bill-clapping. This is the same pattern of response which communally incubated embryos showed upon exposure to the call of a sibling (see Table 2). Since an increase in bill-clapping has previously

TABLE 3

Absence of increase in oral activity in response to species' maternal call in duck embryos which were isolated on Day 23 and tested on Day 26–27: subtotal auditory deprivation

TYPE OF ACTIVITY	MALLARD CALL [N = 22]		
	Before	During	After
Oral activity	52.1	64.7	60.1
(SD)	(35.2)	(55.1)	(36.5)
Vocalization	5.2	15.9*	4.5
(SD)	(8.3)	(21.0)	(5.2)
Heart beat	265.1	287.0**	263.8
(SD)	(59.0)	(43.0)	(48.2)

Reliability of changes from baseline: *$P<0.05$; **$P<0.01$. The change in oral activity was not reliable ($P=0.20$).
SD = standard deviation.

been shown to be the discriminative aspect of the response to the maternal call of the species, it can be concluded that the isolated duck embryos were not capable of the usual auditory discrimination ability manifested by communally incubated embryos. Therefore, these results indicate that normally occurring embryonic auditory stimulation plays a role in regulating the embryo's ability to respond selectively to the maternal call of its species.

In comparing the baseline rates of activity in the isolated and communally incubated embryos (Tables 2 and 3), it can be seen that the isolates showed higher rates of activity during the pre-stimulation period on each of the three measures. From other findings it can be ascertained that the high baseline rate of oral activity in the isolates did not represent a ceiling which precluded the possibility of a significant increase during the 30-second stimulation period. Specifically, a group of 20 communally incubated embryos tested on Day 25 had a very high baseline rate of oral activity (80.3 per min) and showed a statistically reliable increase to 107.9 per min ($P=0.05$) when exposed to the Mallard maternal call for 30 seconds. Thus, it may be concluded that the failure of the isolated embryos to show an increase in oral activity was not due to their relatively high baseline rate (52 per min), but to their sub-total auditory isolation from Day 23 to Day 26–27.

While the embryos in the above study were isolated from other eggs, they were able to hear their own self-produced vocalizations prior to the time they were tested for their response to the species' maternal call. Thus, the auditory deprivation was partial and involved only a reduced amount of exposure to relevant stimulation during a presumed formative period of species-specific auditory perceptual development. Assuming the results of the isolation experiment were due to the sub-total auditory deprivation and not to some extraneous feature of the isolation situation, similarly deprived embryos should be able to respond discriminatively to the appropriate maternal call some time after hatching. This prediction stems from the hypothesis presented in the previous section which stressed the importance of exposure to relevant stimulation during ontogeny in regulating the development of perceptual discrimination abilities. Admittedly, the extent of the lag or retardation in auditory perceptual development cannot be precisely predicted in the present case, although the hypothesis does imply that the effects of sub-total deprivation should not be as severe as the effects of total deprivation. Thus, to test for the effects of deprivation with the procedures presently at our disposal, a further experiment was undertaken to determine if sub-total auditory deprivation only induces a lag in species-specific auditory perceptual development and, if so, the extent of the retardation. To examine this question, 39 Peking duck embryos were kept in individual isolation from Day 23 to 14–30 hours after hatching (Day 27–28), at which time they were placed individually in a simultaneous choice situation involving their approach-response to the Mallard maternal call or the Wood duck maternal call emanating from speakers which were not visible to the duckling. In the five-minute choice test, 21 of the ducklings responded and all of them approached the Mallard call exclusively. Thus, with reference to the initial isolation experiment, the present result indicates that the sub-total auditory deprivation from Day 23 to the time of testing only induced a more or less predictable temporal lag in the development

of the embryo's ability to respond discriminatively to the appropriate maternal call.

Although the results of the initial isolation experiment offer the first direct evidence that at least one form of species-specific embryonic behavior has its background in sensory stimulation encountered in prenatal ontogeny, whether the effect of the deprivation is simply to raise the threshold of a latent discrimination or whether exposure to peeping involves an actual transformation of the duckling's auditory system is presently unknown. Further experiments are required to determine the exact nature of the effect produced by the normal amount of exposure to embryonic peeping. Experiments with hatchlings (Gottlieb, 1966) have shown that increased exposure to the peeping of siblings enhances the duckling's response to the maternal call, which at the very least suggests a lowering of excitatory threshold through such prior experience. Since other studies (Gottlieb, 1965b) have shown that it is not simply prior exposure to any sound at all which enhances the duckling's responsiveness to the maternal call, the possibility must be held open that exposure to peeping or to sounds which resemble peeping actually transforms the embryo's auditory system in a way that goes beyond a mere alteration of excitatory threshold. Thus, the question remains open whether exposure to peeping simply serves to sensitize the embryo to the maternal call or whether such exposure plays a role in actually structuring the embryo's perception of its maternal call. This question can only be answered by depriving the isolated embryo of hearing its own vocalizations; in that case it would be possible to specify precisely the role of self-stimulation in preparing the hatchling to respond discriminately to the maternal call.

Whatever the specific mechanism turns out to be, the results of the auditory deprivation experiments are in accord with one of the proposals raised in the previous section of this paper — i.e., that depriving the embryo of relevant stimulation can exert an effect on the development of the usual perceptual discrimination abilities within the sensory modality thus affected. The fact that the sub-totally deprived embryos showed an increase in heart rate and vocal activity when exposed to the maternal call indicates that audition itself was functional; the lack of activation of bill-clapping indicates that some aspect of perceptual discrimination was altered by the partial auditory deprivation. The fact that similarly deprived embryos were able to discriminate the maternal call of their own species after hatching indicates that the sub-total depreviation merely induced a predictable temporal lag in the development of the embryos' discriminative ability. These findings are consonant with the view that exposure to relevant stimulation plays an important role in regulating the embryo's ability to manifest species-specific perception.

Questions on the nature of the effects of total sensory deprivation on species-specific perception can be answered by further research into the neonatal abilities of embryos which have been so deprived.

ACKNOWLEDGMENTS

The experiments reported in the last section were supported in part by United States Public Health Service Research Grant HD-00878 from the National Institute of Child Health and Human Development. The careful technical assistance of Carole S. Ripley, Patricia Willis, Evelyn Strickland, and Jo Ann Winfree is gratefully acknowledged.

LIST OF LITERATURE

ABBOTT, U. K., and R. M. CRAIG. 1963. The laboratory preparation of normal avian embryos. *Poultry Sci.*, 42: 429–437.

BECKER, R. F. 1942. Experimental analysis of the vaseline technique of Kuo for studying behavioral development in chick embryos. *J. Genet. Psychol.*, 60: 153–165.

BENOIT, J., and L. OTT. 1944–45. External and internal factors in sexual activity. *Yale J. Biol. Med.*, 17: 27–46.

BURSIAN, A. V. 1964. [The influence of light on the spontaneous movements of chick embryos.] *Bull. Exper. Biol. Med.*, 7: 7–11. [In Russian.]

———. 1965. Primitive forms of photosensitivity at early stages of embryogenesis in the chick. *J. Evol. Biochem. Physiol.*, 1: 435–441. [Translation.]

CAIN, J. R., U. K. ABBOTT, and V. L. ROGALLO. 1967. Heart rate of the developing chick embryo. *Proc. Soc. Exper. Biol. Med.*, 126: 507–510.

CLARK, E. L., and E. R. CLARK. 1914. On the early pulsations of the posterior lymph hearts in chick embryos: Their relation to the body movements. *J. Exper. Zool.*, 17: 373–394.

COGHILL, G. E., and R. W. WATKINS. 1943. Periodicity in the development of the threshold of tactile stimulation in Amblystoma. *J. Comp. Neurol.*, 78: 91–111.

COLLIAS, N. E., and E. C. COLLIAS. 1956. Some mechanisms of family integration in ducks. *Auk*, 73: 378–400.

CORNER, M. A., and A. P. C. BOT. 1967. Normal developmental patterns in the central nervous system of birds. III. Somatic motility during the embryonic period and its relation to behavior

after hatching. In C. G. Bernhard and J. P. Schadé (eds.), *Progress in Brain Research: Developmental Neurology*, vol. 26, p. 214–236. Elsevier, Amsterdam.

COULOMBRE, A. S. 1955. Correlations of structural and biochemical changes in the developing retina of the chick. *Amer. J. Anat.*, 96: 153–190.

CRUZ, MARÍA DE LA, C. CAMPILLO-SAINZ, and S. MUÑOZ-ARMAS. 1966. Congenital heart defects in chick embryos subjected to temperature variations. *Circul. Res.*, 18: 257–262.

DRACHMAN, D. B., and L. SOKOLOFF. 1966. The role of movement in embryonic joint development. *Develop. Biol.*, 14: 401–420.

GARCIA-AUSTT, E. 1954. Development of electrical activity in cerebral hemispheres of the chick embryo. *Proc. Soc. Exper. Biol.*, 86: 348–352.

GARCIA-AUSTT, E., and M. A. PATETTA-QUEIROLO. 1961a. Electroretinogram of the chick embryo. I. Onset and development. *Acta Neurol. Latinoamer.*, 7: 179–189.

———, and ———. 1961b. Electroretinogram of the chick embryo. II. Influence of adaptation, flicker frequency and wavelength. *Acta Neurol. Latinoamer.*, 7: 269–288.

GOS, M. 1935. Les réflexes conditionnels chez l'embryon d'oiseau. *Bull. Soc. Roy. Sci. Liége*, 4: 194–199; 246–250.

GOTTLIEB, G. 1963. Refrigerating eggs prior to incubation as a way of reducing error in calculating developmental age in imprinting experiments. *Anim. Behav.*, 11: 290–292.

———. 1965a. Prenatal auditory sensitivity in chickens and ducks. *Science*, 147: 1596–1598.

———. 1965b. Imprinting in relation to parental

and species identification by avian neonates. *J. Comp. Physiol. Psychol.* 59: 345–356.

——. 1966. Species identification by avian neonates: Contributory effect of perinatal auditory stimulation. *Anim. Behav.* 14: 282–290.

——. 1968. Conceptions of prenatal behavior. In L. R. Aronson, D. S. Lehrman, J. S. Rosenblatt, and E. Tobach (eds.), *Development and Evolution of Behavior*, vol. 1. W. H. Freeman, San Francisco. [In press.]

GOTTLIEB, G., and P. H. KLOPFER. 1962. The relation of developmental age to auditory and visual imprinting. *J. Comp. Physiol. Psychol.*, 55: 821–826.

GOTTLIEB, G., and Z.-Y. KUO.. 1965. Development of behavior in the duck embryo. *J. Comp. Physiol. Psychol.*, 59: 183–188.

GRAY, P. H. 1961. The releasers of imprinting: Differential reactions to color as a function of maturation. *J. Comp. Physiol. Psychol.*, 54: 596–601.

GRIER, J. B., S. A. COUNTER, and W. M. SHEARER. 1967. Prenatal auditory imprinting in chickens. *Science*, 155: 1692–1693.

HAMBURGER, V. 1963. Some aspects of the embryology of behavior. *Quart. Rev. Biol.*, 38: 342–365.

HAMBURGER, V., and M. BALABAN. 1963. Observations and experiments on spontaneous rhythmical behavior in the chick embryo. *Develop. Biol.*, 7: 533–545.

HAMBURGER, V., and H. L. HAMILTON. 1951. A series of normal stages in the development of the chick embryo. *J. Morphol.*, 88: 49–92.

HAMBURGER, V., E. WENGER, and R. OPPENHEIM. 1966. Motility in the chick embryo in the absence of sensory input. *J. Exper. Zool.*, 162: 133–160.

HAYWOOD, H. C. 1965. Discrimination and following behavior in chicks as a function of early environmental complexity. *Percept. Mot. Skills*, 21: 299–304.

HESS, E. H. 1956. Natural preferences of chicks and ducklings for objects of different colors. *Psychol. Rep.*, 4: 477–483.

HOLT, E. B. 1931. *Animal Drive and the Learning Process*, vol. 1. 307 p. Henry Holt Co., New York.

HOOKER, D. 1952. *The Prenatal Origin of Behavior*. 143 p. Univ. of Kansas Press, Lawrence.

HUMPHREY, T. 1964. Some correlations between the appearance of human fetal reflexes and the development of the nervous system. In D. P. Purpura and J. P. Schadé (eds.), *Progress in Brain Research*, vol. 4, p. 93–133. Elsevier, Amsterdam.

HUNT, E. L. 1949. Establishment of conditioned responses in chick embryos. *J. Comp. Psychol.*, 42: 107–117.

JACOBSON, M. 1967. Starting points for research in the ontogeny of behavior. In M. Locke (ed.), *Major Problems in Developmental Biology* (25th Growth Symposium), p. 341–383. Academic Press, New York.

JOHNSON, L. G. 1966. Diurnal patterns of metabolic variations in chick embryos. *Biol. Bull.*, 131: 307–322.

KÄLLÉN, B., and S.-I. RÜDEBERG. 1964. Teratogenic effects of electric light on early chick embryos. *Acta Morphol. Neerl-Scand.* 6: 95–99.

KEAR, J. 1964. Colour preference in young Anatidae. *Ibis*, 106: 361–369.

KNOWLTON, V. Y. 1967. Correlation of the development of membranous and bony labyrinths, acoustic ganglia, nerves and brain centers of the chick embryo. *J. Morphol.*, 121: 179–208.

KOECKE, H. V. 1958. Normalstadien der Embryonalentwicklung bei der Hausente (*Anas boschas domestica*). *Embryologia*, 4: 55–78.

KOSIN, I. L. 1964. Recent research trends in hatchability-related problems of the domestic fowl. *World's Poultry Sci. J.*, 20: 254–268.

KUO, Z.-Y. 1932. Ontogeny of embryonic behavior in Aves: I. Chronology and general nature of behavior of chick embryo. *J. Exper. Zool.* 61: 395–430.

——. 1967. *The Dynamics of Behavior Development*. 240 p. Random House, New York.

LÁBOS, E. 1966. Energetic aspects of the photosensitization of embryonic muscle by xanthene dyes. *Comp. Biochem. Physiol.*, 17: 353–362.

LAUBER, J. K., and J. V. SHUTZE. 1964. Accelerated growth of embryo chicks under the influence of light. *Growth*, 28: 179–190.

LAUBER, J. K., J. V. SHUTZE, and J. McGINNIS. 1961. Effects of exposure to continuous light on the eye of the growing chick. *Proc. Soc. Exp. Biol. Med.*, 106: 871–872.

LEHRMAN, D. S. 1953. A critique of Konrad Lorenz's theory of instinctive behavior. *Quart. Rev. Biol.*, 28: 337–363.

LEVI-MONTALCINI, R. 1949. The development of the acoustico-vestibular centers in the chick embryo in the absence of the afferent root fibers and of descending fiber tracts. *J. Comp. Neurol.*, 91: 209–241.

LINDEMAN, V. F. 1947. The cholinesterase and acetylcholine content of the chick retina, with special reference to functional activity as indicated by the pupillary constrictor reflex. *Amer. J. Physiol.*, 148: 40–44.

MENAKER, M. 1968. Extraretinal light perception in the sparrow. I. Entrainment of the biological clock. *Proc. Nat. Acad. Sci. U. S.*, 59: 414–421.

Moltz, H., and L. J. Stettner. 1961. The influence of patterned-light deprivation on the critical period for imprinting. *J. Comp. Physiol. Psychol.,* 54: 279–283.

Needham, J. 1959. *A History of Embryology.* 304 p. Abelard-Schuman, New York.

Oppenheim, R. W. 1968a. Light sensitivity in chick and duck embryos. *Anim. Behav.,* in press.

——. 1968b. Color preferences in the pecking response of newly hatched ducklings (*Anas platyrhynchos*). *J. Comp. Physiol. Psychol.,* in press.

Orr, D. W., and W. F. Windle. 1934. The development of behavior in chick embryos: The appearance of somatic movements. *J. Comp. Neurol.,* 60: 271–285.

Ott, J. N. 1964. Some responses of plants and animals to variations in wavelengths of light energy. *Ann. N. Y. Acad. Sci.,* 117: 624–635.

Paulson, G. W. 1965. Maturation of evoked responses in the duckling. *Exper. Neurol.,* 11: 324–333.

Peters, J. J., A. R. Vonderahe, and T. H. Powers. 1956. The functional chronology in developing chick nervous system. *J. Exper. Zool.,* 133: 505–518.

——, ——, and ——. 1958. Electrical studies of functional development of the eye and optic lobes in the chick embryo. *J. Exper. Zool.,* 139: 459–468.

Peters, J. J., A. R. Vonderahe, and A. A. Huesman. 1960. Chronological development of electrical activity in the optic lobes, cerebellum, and cerebrum of the chick embryo. *Physiol. Zool.,* 33: 225–231.

Rebollo, M. A. 1955. Some aspects of the histogenesis of retina. *Acta Neurol. Latinoamer.,* 1: 142–147.

Romanoff, A. L. 1960. *The Avian Embryo.* 1305 p. Macmillan, New York.

Rosenblum, W. I. 1960. The stimulation of frog muscle by light and dye. *J. Cell. Comp. Physiol.,* 55: 73–79.

Rudnick, D. and H. Waelsch. 1955. Development of glutamotransferase and glutamine synthetase in the nervous system of the chick. *J. Exper. Zool.,* 129: 309–326.

Schaefer, H. H. and E. H. Hess. 1959. Color preferences in imprinting objects. *Z. Tierpsychol.,* 16: 161–172.

Schneirla, T. C. 1965. Aspects of stimulation and organization in approach/withdrawal processes underlying vertebrate behavioral development. In D. S. Lehrman, R. A. Hinde, and E. Shaw (eds.), *Advances in the Study of Behavior,* vol. 1., p. 1–74. Academic Press, New York.

——. 1966. Behavioral development and comparative psychology. *Quart. Rev. Biol.,* 41: 283–302.

Sedláček, J. 1962. Functional characteristics of the centre of the unconditioned reflex in elaboration of a temporary connection in chick embryos. *Physiol. Bohemoslov.,* 11: 313–318.

——. 1964a. Further findings on the conditions of formation of the temporary connection in chick embryos. *Physiol. Bohemoslov.,* 13: 411–420.

——. 1964b. Temporary connections in chick embryos and vagal deafferentation. *Physiol. Bohemoslov.* 13: 421–424.

——. 1967. Development of optic evoked potentials in chick embryos. *Physiol. Bohemoslov.,* 16: 531–537.

Sedláček, J., V. Halváčková, and M. Švehlová. 1964. New findings on the formation of the temporary connection in the prenatal and perinatal period in the guinea-pig. *Physiol. Bohemoslov.* 13: 268–273.

Shifferli, A. 1948. Ueber Marksheidenbildung im Gehirn von Huhn und Star. *Rev. Suisse Zool.,* 55: 117–212.

Simner, M. L. 1966. Cardiac self-stimulation hypothesis and the response to visual flicker in newly hatched chicks: preliminary findings. *1966 Proc. Amer. Psychol. Assoc.,* 1: 141–142.

Sperry, R. W. 1965. Embryogenesis of behavioral nerve nets. In R. L. DeHaan and H. Ursprung (eds.), *Organogenesis,* p. 161–186. Holt, Rinehart, and Winston, New York.

Strobel, M. G., G. M. Clark, and G. E. MacDonald, 1967. Ontogeny of the following response: A radiosensitive period during embryological development of the domestic chick. *J. Comp. Physiol. Psychol.,* 65:314–319.

Tamimie, H. S., and M. W. Fox. 1967. Effect of continuous and intermittent light exposure on the embryonic development of chicken eggs. *Comp. Biochem. Physiol.,* 20: 793–799.

Tello, J. F. 1923. Les différenciations neuronales dans l'embryon du poulet, pendant les premiers jours de l'incubation. *Trab. Lab. Invest. Biol.* (Madrid, 21: 1–90.

Tolman, C. W. 1967. The effects of tapping sounds on feeding behavior of domestic chicks. *Anim. Behav.,* 15: 145–148.

Tuge, H. 1937. The development of behavior in avian embryos. *J. Comp. Neurol.,* 66. 157–179.

Vanzulli, A. and E. Garcia-Austt. 1963. Development of cochlear microphonic potentials in the chick embryo. *Acta Neurol. Latinoamer.,* 9: 19–23.

Vernakakis, A., and A. Burkhalter. 1967. Acetylcholinesterase activity in the optic lobes of chicks at hatching. *Nature,* 214: 594–595.

VINCE, M. A. 1966a. Potential stimulation produced by avian embryos. *Anim. Behav.*, 14: 34–40.

——. 1966b. Artificial acceleration of hatching in quail embryos. *Anim. Behav.*, 14: 389–394.

——. 1968. Effect of rate of stimulation on hatching time in Japanese quail. *British Poultry Sci.*, 9: 87–91.

VISINTINI, F. and R. LEVI-MONTALCINI. 1939. Relazione tra differenciazone strutturale e funzionale dei centri e delle vie nervose nell'embrione di pollo. *Schweiz. Archiv. Neurol. Psychiat.*, 43: 1–45.

WADDINGTON, C. H. 1952. *The Epigenetics of Birds.* 272 p. Univ. Press, Cambridge.

WINDLE, W. F. and D. W. ORR. 1934. The development of behavior in chick embryos: Spinal cord structure correlated with early somatic motility. *J. Comp. Neurol.*, 60: 287–307.

A GENE MUTATION WHICH CHANGES A BEHAVIOR PATTERN [1]

Margaret Bastock

Dept. of Zoology and Comparative Anatomy, Oxford

Received May 1, 1956

Introduction

It is tempting to assume that behavior patterns have evolved in the same manner as other features of animal populations, by the selection of inheritable differences. This idea is supported by those comparisons which have been made between the behavior patterns of closely related animals, one of the best examples of which is the review by Spieth (1952) of the courtship behavior of 101 species and subspecies of the genus *Drosophila*. Spieth compares the degree of morphological with that of behavioral divergence and he finds that the two roughly agree so that mating behavior confirms the validity of the existing classification as presented by Patterson and Wheeler (1949).

Yet it is impossible to be certain, however strong the implication about the evolution of behavior, unless one can demonstrate the existence of 'genetic' variations in behavior upon which selection could act. It is true that some gene mutations are known which affect behavior in some way. Thus one mutation in *Panaxia dominula* has influenced its mating 'preferences' (Sheppard, 1952), the ebony mutant in *Drosophila melanogaster* mates more successfully in the dark than in the light (Rendel, 1951), and there is a mutation which affects tameness in rats (Keeler and King, 1942). But the differences in the courtship behavior of the *Drosophila* species studied by Spieth and also by Milani (1951a, b) are of a more specific nature. Courtship is a complex behavior pattern and the variations often

occur only in the relative frequency of the elements concerned, some elements occurring more often in one species than another. Sometimes elements are omitted or new ones are added and sometimes they may change in form, but the most common difference between the most closely related species concerns this difference in frequency. The investigations reviewed by Moynihan (1953, 1955) on the differences between closely related gulls and the descriptions given by Clark, Aronson and Gordon (1954) on the differences between the platyfish (*Xiphophorus* (*Platypoecilus*) *maculatus*), and the swordtail (*Xiphophorus helleri*), suggest changes of the same nature, and yet no single gene mutation is known which alters a behavior pattern in this way. However since considerable detailed investigation is necessary to reveal differences in frequency or patterning, this is not so very surprising. Indeed some of the mutations listed above might well have such effects; they simply have not been investigated. It seemed worth while therefore to examine more closely one example of a gene mutation affecting behavior and to ask two questions, (1) how does it bring about its effect? (this is of considerable genetical interest), (2) what part might it play in evolution? (this is of considerable evolutionary interest).

Material Used

The mutant which I chose to investigate was the yellow mutant of *Drosophila melanogaster*. This is a recessive sex-linked gene located at one end of the X chromosome, the adults being clearly distinguished from wild type because the body is yellow, instead of the normal gray,

[1] This work is part of that submitted for the degree of Doctor of Philosophy, in the University of Oxford. March 1955.

Evolution 10: 421–439. December, 1956. 421

and because the transverse bands on the body are dark brown instead of black. The larvae are also recognizable because their setae and mouth parts are yellow-brown instead of dark brown; in fact yellow insects seem to be generally deficient in melanin in spite of the fact that they contain more of the enzyme tyrosinase (responsible for the melanin reaction) than do wild type insects (Graubard, 1933). The mutant arises frequently, but although yellow insects are very successful as laboratory stocks they are not frequent in nature, probably because of their lack of mating success. This was first noted in 1915 by Sturtevant, who showed that yellow males are usually unsuccessful in competition with wild males for females. It seemed possible that this could be because their behavior had changed in some way so as to make them less stimulating.

I demonstrated the nature of the effect by comparing the success of single pair matings of wild and yellow males with wild females. My two stocks (wild and yellow) had previously been made genetically similar by intercrossing for seven generations, and in this, as in all other experiments and observations quoted in this paper, the flies were four or five days old, males and females having been kept for this time after emergence in separate vials. Figure 1 is a cumulative graph in which the percentage of successful males of each type is plotted against time. Clearly although the yellow males do eventually mate they take much longer to do so. Moreover although yellow males take slightly longer before they begin to court (an average of 9.6 minutes as compared with 4.9 minutes), it is also true that they have to court for longer before they can mate. Yellow males take on the average 10.5 minutes from the beginning of courtship to copulation, wild males only 6.0 minutes. These figures were all obtained from one series of experiments because each pair was watched and the times of introduction. courtship commencement,

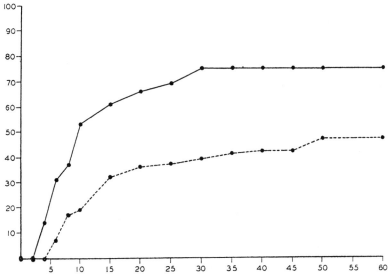

Fig. 1. Mating success compared for wild and yellow males with wild females in paired matings. The wild and yellow stocks had been previously crossed for 7 generations. Average time from introduction to copulation: wild male × wild female: 9.1 minutes; yellow male × wild female: 18.9 minutes. Abscissa: time in minutes from introduction. Ordinate: % successful males. Dashed line: yellow male × wild female. Solid line: wild male × wild female.

copulation, were noted for each. All the differences quoted above are significant at the 1% level.

It is clear, therefore, that yellow males must be less stimulating in some way than wild ones. This could be because they look or smell different or it could be because their courtship has changed. Because the courtship is so prominent a feature of *Drosophila* pre-copulatory behavior it is likely that it plays a dominant role in stimulating the females and it therefore seemed worth while investigating this first. One point is clear; yellow males fall very little short in the persistency of their courtship. I watched 19 wild males and 19 yellow males separately courting wild females and recorded their activity every 1½ seconds for a period of 2½ minutes each. The wild males courted for an average of 92% of the time, the yellow ones for 83%. This difference is significant at the 5% level but it seems too small to account entirely for the reduced success of the yellows. I therefore looked for a possible reduction in the stimulating quality of the courtship itself.

DESCRIPTION OF COURTSHIP

The courtship of *Drosophila melanogaster* is too well known to need a detailed description. When a male is put into an observation cell with a virgin female (preferably after both have been isolated for about 3–5 days from eclosion), he will very soon approach her and tap her with his fore-legs. After tapping, he may not immediately begin to court; he may clean or walk away. If so, then he will sooner or later tap again and eventually he will follow this by courtship. After starting courtship, however, he rarely taps a second time except perhaps after a long break.

At the beginning of courtship the male stands facing the female, usually in the position from which he has approached, whether it be at the side of, in front of or behind her. (Spieth says that he always circles to the rear immediately but I do not agree with this, and Sturtevant also observed that he would stand in any position relative to the female.) If the female moves off, the male will follow, and this of course brings him behind her, but often the female stands still and then he does not at first change his position. I have called this part of the courtship *orientation*, irrespective of whether the male is following or standing.

The wing display usually follows after a few seconds of orientation. One wing is brought out until it is at right angles to the male's body, then it is vibrated rapidly up and down for a few seconds before being returned to rest. This may be repeated several times at short intervals during which time the male may be following the female or standing facing her. Eventually if he is not already behind her, he will move into this position, although he may first circle right round her head, and sometimes, even from an original position behind her, he may move first to one side and then to the other before finally staying behind. It is noticeable, during these maneuvers, that he always vibrates the wing nearest to the head of the female, usually changing wings as he passes her head or her tail.

Finally, once the male is behind the female, he licks her genitalia with his proboscis and attempts to copulate. This involves curling his abdomen under his body and jerking it towards the vaginal plates of the female meanwhile grasping her abdomen with his forelegs. He will not succeed however in inserting his phallus unless the female cooperates and spreads her vaginal plates.

The wing display I have called *vibration*, and licking and attempted copulation are collectively termed *licking*. I have not separated these two because it appears that licking is always followed by attempted copulation unless the female is able to prevent it (by kicking or twisting her abdomen or running away).

During vibration the male is always orientating, and during licking he is both orientating and vibrating; these three ele-

424   MARGARET BASTOCK

ments are therefore not distinct but are superimposed upon one another. Figure 2 gives a representation of the patterning of these elements in time. It shows records of the courtships of four males, the activity of each being noted every 1½ seconds. It will be seen that a given male changes rather regularly from one activity to another, but that males vary in the proportions and in the bout lengths of the different elements.

The Stimulating Elements of the Courtship

To return to the problem of the yellow males, it seemed of interest to discover the relative stimulating value of these elements, and since vibration is the most prominent and the most peculiar feature of the courtship, I investigated this first. It is easy to eliminate vibration alone, for males whose wings have been removed (about 2 days previously under ether) will court just as persistently as normal ones. Detailed observations on ten wingless males showed that they courted for 89% of the time as compared with 86%

for the winged controls. Moreover they court quite normally. One can see them 'vibrating' their wing stumps, and wingless males licked an average of 5.8 times in a 2½ minute observation period as compared with the 6.1 times of winged males. These differences are not statistically significant.

Figure 3 gives the results of some mass mating experiments which I performed with 5 males (all of one type) and 10 females (all of one type) together in vials for two hours, the females being dissected afterwards to discover which had been fertilized. It can be seen that wingless males are very much less successful than normal ones (32% success in 2 hours as compared with 70% : P < 0.01). They are equally unsuccessful if the experiments are done in the dark; indeed light does not appear to make very much difference to the success of either type of male, the figures in the dark being 31% for wingless and 75% for winged males. This agrees with the findings of Spieth and Hsu (1950) that visual stimuli are not important in the courtship of *Dro-*

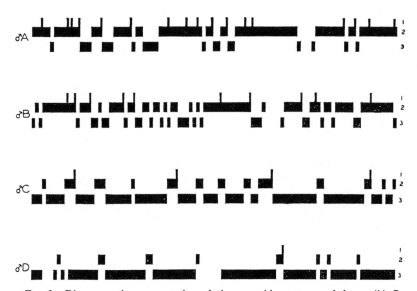

Fig. 2. Diagrammatic representation of the courtship patterns of four wild *D. melanogaster* males, each with a virgin wild type female. 1. licking; 2. vibration; 3. orientation.

MUTATION AND BEHAVIOR PATTERN   425

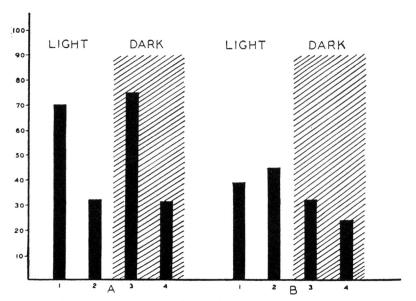

Fig. 3. Results of experiments performed simultaneously with winged and wingless males, normal and antennaless females. 5 males and 10 females per vial. 9 vials in each category. Abscissa: A. normal ♀♀; 1. winged ♂♂, 2. wingless ♂♂, 3. winged ♂♂, 4. wingless ♂♂. B. antennaless ♀♀. 1, 2, 3, and 4, as for A. Ordinate: % females fertilized.

sophila melanogaster. It also shows that the stimuli provided by wing vibration are not visual, for both the undiminished success of winged insects in the dark and my own observations confirm that court-ship is quite normal in the dark.

The other experiments shown in figure 3 were run simultaneously with those just quoted, but antennaless rather than normal females were used (like the wings, the antennae were removed from anes-thetized insects 2 days before the experi-ments and to counteract possible effects of ether, all normal flies used in these ex-periments were also anesthetized at the time of the operations). It is clear that antennaless females do not discriminate between winged and wingless flies, either in the light or in the dark (such differ-ences as appear in the figures are not sig-nificant). Moreover it seems that the success of both types of male with these females is reduced by the same amount as that of wingless males with normal fe-males. In other words it is likely that

the same stimulus is lacking in both cases, i.e. the stimulus produced as a result of vibration, and that this is perceived by the antennae of the female. These findings explain the role of the antennae in the mating behavior of females of this species demonstrated by Mayr (1950) and Begg and Packman (1951). The fact that the two latter authors demonstrated that re-moval of the male's antennae also reduced mating success suggests that olfactory stimuli are important also to them.

The degree of the drop in success is evidence that vibration is a very im-portant stimulating element in the court-ship and that orientation and licking (which are present in wingless males) are relatively less important. Licking probably is of some stimulus value but these experiments suggest that it plays a relatively minor role, thus agreeing with Spieth's suggestion (1952) that licking is more important in the species of the sub-genus *Drosophila,* where it tends to be more prolonged, than in the sub-genus

Sophophora of which *melanogaster* is a member.

Thus it is apparent from this investigation that any courtship deficient in vibration is likely to be considerably less stimulating than one containing a high proportion of this element. Now an examination of the courtship records shown in Figure 2 shows that individual males do differ both in the amount, and in the bout length of this element, males A and B giving higher values of each than C and D. The males pictured here were all wild males, but males like C and D are relatively rare, while A and B may be described as typical. It seemed worth while to see if this was also true of yellow males.

THE DIFFERENCE BETWEEN THE COURTSHIPS OF YELLOW AND WILD MALES

In these experiments I wished to avoid as far as possible the effect of genes other

TABLE 1. *Analysis of courtship of wild and yellow males with wild females*
Experiments with 'brother' males

All insects *D. melanogaster*

100 records made for each pair at intervals of $1\frac{1}{2}$ seconds

	Wild males × wild females						Yellow males × wild females					
	T.R.T.	O%	V%	L%	Av. V+L	Av. O	T.R.T.	O%	V%	L%	Av. V+L	Av. O
Day 1	100	78	19	3	2.7	6.5	95	76	20	4	2.6	6.7
	98	73	17	10	3.8	4.4	98	94	5	1	2.5	25.8
	68	62	37	1	4.3	4.2	95	82	12	6	2.8	5.8
	81	75	19	6	3.5	5.6	86	90	7	3	2.6	11.0
	88	80	18	2	3.1	8.5	83	77	18	5	3.3	4.8
	83	72	27	1	3.4	5.1	70	71	28	1	2.7	6.6
Day 2	96	69	24	7	4.6	5.7	30	73	14	13	3.3	5.8
	99	81	15	4	3.0	6.3	97	78	16	6	3.3	5.6
	82	82	17	1	3.0	10.8	93	80	17	3	2.9	7.3
	100	60	32	8	4.4	3.1	100	65	25	10	3.3	3.2
Day 3	86	87	12	1	2.8	7.6	85	86	13	1	2.8	13.8
	97	72	22	6	3.6	5.2	51	67	27	6	2.8	4.0
Day 4	100	74	20	6	3.6	5.6	85	78	16	6	2.8	6.2
	88	73	24	3	3.4	6.7	89	72	22	6	2.9	4.1
Day 5	100	51	44	15	6.2	2.4	98	73	24	3	3.1	4.5
	94	74	17	9	3.6	4.5	92	74	16	10	4.1	4.8
	100	61	24	15	6.9	2.9	92	75	14	11	2.7	4.1
Day 6	95	68	23	9	3.3	4.7	75	79	16	5	2.9	4.9
	85	74	20	6	4.9	5.0	68	69	24	7	3.2	3.0
Av. for 19 pairs	92	72	22	6	3.9	5.5	83	77	18	6	2.9	6.9

T.R.T. = no. of records out of 100 in which courtship activities were shown.
O% = percentage orientation in the courtship records.
V% = percentage vibration in the courtship records.
L% = percentage licking in the courtship records.
Av. V+L = average bout length of vibration plus licking expressed in units of $1\frac{1}{2}$ secs.
Av. O = average bout length of orientation (calculated in the same way as Av. V+L).

Results of analysis of variance: Significance of differences between male types

T.R.T., P < 0.05; O%, P < 0.05; V%, P < 0.05; L% not significant; Av. V+L, P < 0.01; Av. O, not significant.

than yellow. I therefore crossed my two stocks together (yellow females × wild males) and then mated the daughter heterozygous females to their brother yellow males. The sons of these females were wild and yellow in equal numbers, and it may be assumed that any other genes for which their parents were heterozygous (i.e. 'stock' differences) should assort independently of the wild and yellow genes, unless very closely linked. I did not in any case expect there to be many such differences as it was only a few generations since I had crossed my two stocks together for 7 generations.

Table 1 shows the results of comparisons of the courtships of wild and yellow males. The insects were recorded for the first 2½ minutes of their courtship with wild females, each couple being alone in an observation cell and the male's behavior being noted every 1½ seconds. Wild and yellow males were recorded in turn and all those recorded on a given day were of the same age, i.e. either four or five days from emergence. From the records I could count the total number of courtship activities (disregarding periods when the male was not courting) and I could then calculate the percentage of that courtship which consisted of each of the three elements, orientation, vibration and licking. By this means I could assess the stimulating quality of the courtship itself, irrespective of the effect of breaks. It can be seen that yellow males show a slightly lower percentage of vibration (the difference is significant at the 5% level) and this is compensated, not by licking, which is probably also stimulating, but by orientation (the difference is again significant at 5%). An even greater difference is found if one calculates the average bout length of vibration. Reference to Fig. 2 will show that this really means the average bout length of vibration plus licking, since licking is momentarily superimposed during a vibration bout. The bouts shown by wild males are considerably longer than those shown by yellow ones and this difference is significant at 1% level. The

orientation bouts also appear to be longer for yellow males although this is not significant on these figures. However I have presented evidence elsewhere (Bastock and Manning 1955) that vibration bout length is negatively correlated with orientation bout length, so that it appears that these mutants tend to vibrate for shorter periods and at longer intervals than do wild males. This could be very important if perhaps a greater concentration of the stimulating elements is significant to females. Bursts of scent or sound might need to be of a certain duration before they are of any stimulus value, and their effect may tend to die down between bursts.

THE NATURE OF THE EFFECT OF THE MUTATION

Having established a difference in behavior which could well account for the reduced success of yellow males, it is not legitimate to cease the investigation at this point. In the first place, it is possible that changed behavior, changed color, and perhaps changed scent, are all separate effects of this mutation and perhaps not only one, but several or all of them may diminish mating success. But even more important is the possibility that the behavioral change is not a primary effect of the mutation at all. A male's courtship behavior is never an entirely automatic process; it is determined, at least in part, by the stimuli received from the courted object. It is therefore quite possible that a female, reacting against the changed appearance or scent of a yellow male, may either fail to give attractive stimuli or give instead repelling stimuli. Thus the male's deficient courtship behavior may be explained simply as a different reaction to different stimuli, and there may be no fundamental difference between the two males in this respect at all.

This second possibility can be investigated in various ways.

(1) In the first place it is possible to consider the behavior of the females dur-

MARGARET BASTOCK

TABLE 2. *Number of records of repelling movements made by melanogaster and simulans females
with wild and yellow melanogaster males*

Out of a total of 100 records per pair

Experiments with 'brother' males

A.

Wild male × wild *melanogaster* female						Yellow male × wild *melanogaster* female					
Pair	J	K	FL	TW	EX	Total	Pair	J	K	FL	TW	EX	Total
(1)	1	3	1	1	0	6	(2)	0	10	0	0	0	10
(3)	3	0	7	5	0	15	(4)	2	3	5	0	5	15
(5)	6	0	0	3	4	13	(6)	6	1	4	3	0	14
(7)	3	4	4	0	0	11	(8)	1	0	6	1	0	8
(9)	2	5	2	3	0	12	(10)	2	1	0	1	1	5
Total	15	12	14	12	4	57	Total	14	15	15	5	6	52

B.

Wild male × *simulans* female						Yellow male × *simulans* female					
Pair	J	K	FL	TW	EX	Total	Pair	J	K	FL	TW	EX	Total
(11)	0	8	10	20	0	38	(12)	2	7	0	7	0	16
(13)	2	7	1	19	0	29	(14)	1	13	0	5	1	20
(15)	1	9	0	6	1	17	(16)	0	13	3	0	0	16
(17)	3	17	5	1	0	26	(18)	0	18	0	1	0	19
(19)	2	9	1	3	0	15	(20)	0	15	6	6	1	28
Total	8	50	17	49	1	125	Total	3	66	9	19	2	99

J = jumping.
K = kicking.
FL = flicking wings.
TW = twisting abdomen.
EX = extruding genitalia.

ing the courtship. Although females give no obvious acceptance or 'attractive' stimuli, an unwilling female gives clearly recognizable repelling movements. She may kick at the male, she may twist her abdomen so that he cannot copulate, she may flick her wings (a movement similar to that made by males courted by other males), she may extrude her genitalia (a movement made only rarely by virgins but frequently by fertilized females) and finally she may jump or fly suddenly away. It seems legitimate to suppose that if females are for some reason reacting against yellow males, then they will show more of these movements when courted by them. Table 2A shows that this is not so. Here the repelling movements made during 2½ minutes of courtship

are tabulated for 10 females, five of them courted by wild males and five by yellow ones. In fact slightly fewer of these movements are made to yellow males (although the difference is not significant), and it is also clear that there is no great difference in the proportion of the different types of movement made to the two males.

(2) It is also possible to measure how far a female tends to stand still or run around the cell while being courted by the two types of male. This is important because I have shown that significantly more vibration and less orientation are directed to a standing than to a moving female (see fig. 4). (The figures are 49% orientation and 41% vibration given to a standing female; 68% orientation and 24% vibration given to a moving one;

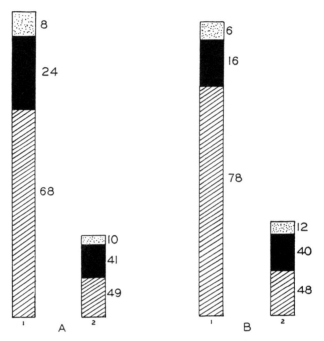

Fig. 4. Diagram illustrating the proportion of time spent running or standing by females courted by wild or yellow males (indicated by the height of columns). The composition of the courtship under these two conditions of females is shown by shading of the columns. The figures are calculated from 2½ minute records of 23 pairs of each type. A. wild type males; 1. ♀ running for 79% of the courtship, 2. ♀ standing for 21% of courtship. B. yellow males; 1. ♀ running for 76% of the courtship, 2. ♀ standing for 24% of the courtship. Key: hatchured areas: % orientation; solid areas: % vibration; stippled areas: % licking.

differences significant at the 1% level in each case.) This may be because the male cannot keep so close to the female and is therefore less stimulated. Whatever the reason, if it could be shown that females courted by yellow males tend to run for a greater proportion of the time, this might well explain the difference in courtship pattern. However, once again, this is not the case. 23 females courted by yellow males, each for 2½ minutes ran for 76% of the time, while a similar number of females courted by wild males ran for 79% of the time. There is no significant difference between these figures but in any case its direction is that which would favor yellow males giving more, not less, vibration in their courtship.

(3) Another line of approach to this problem is to consider the courtship given by these two types of male to females which reject them both (the assumption being that they would reject both equally). Ideally one should compare their courtship to a standard stimulus object, such as a model, but I have not succeeded in inducing *melanogaster* males to court models. (Milani (1951b) induced such behavior only in those species of the *obscura* group whose mating success decreased considerably in the dark, i.e. only in those which gave great 'weight' to visual stimuli. Since *melanogaster* males respond mainly to olfactory and tactile stimuli, my failure is not surprising.) *Melanogaster* males will court dead fe-

430   MARGARET BASTOCK

males, but their behavior to these is highly variable, probably due to the degree of decomposition, which is difficult to control. They do however court the females of the sibling species *Drosophila simulans* very persistently and apparently normally, although they are rarely accepted by them. So I compared the behavior of the two males to these females, in experiments performed exactly as for *melanogaster* females, using 'brother' males as before. Table 3 shows the results and gives an exactly similar picture of the distinction between the courtship of the two male types, except for the fact that, because of higher variability, in this case only the difference in average vibration bout length is statistically significant. This however is significant at the 1% level. Confirmation that *simulans* females do tend to reject both males equally comes from counts of inhibitory stimuli, detailed in Table 2B. Once again slightly more were given to wild than to yellow males.

(4) Finally it seemed worth while comparing the relative success of wild and yellow males in the dark and with antennaless females. If the females react against yellow males because of changed scent or changed appearance then one would expect the elimination of these stimuli to reduce the difference in success between the two males. Figures 5 and 6 show the results of pair mating experiments designed to test this. In neither case is the difference reduced, although in the case of antennaless females the suc-

cess of both types of male is considerably lowered. Clearly then neither changed appearance nor changed scent is a primary cause of the yellow male's failure. It cannot be argued that these factors play no role, only that in the absence of either type of stimulus, differences between the males still remain. It is probable that these are behavioral differences.

The above arguments therefore strongly suggest that a less stimulating courtship is a characteristic of the yellow male and does not seem to be attributable to a difference in the behavior of the courted object to the two types of male. It is necessary at this point to say something about the situation when yellow females are used instead of wild ones. These females accept yellow males almost as readily as wild ones, and what difference does exist can be attributed mainly to the fact that yellow males do not begin courting quite so soon. There is very little difference in the length of actual courtship required before copulation, and this is undoubtedly puzzling if we believe that yellow males are less stimulating. It could be explained if the yellow females were much more highly receptive than the wild ones, i.e. they would not then require so much stimulation. But although yellow females taken from yellow stock bottles often *are* more receptive than wild type ones, I have evidence which I give in detail later (see page 436) that this is due to genes other than yellow, and must be selected for in yellow stocks. I have not yet solved this problem as I have not yet sufficient de-

TABLE 3. *Analysis of the courtship of wild and yellow* melanogaster *males with* simulans *females.*
Experiments as in table 1 using 33 males of each type. Only the average results are shown.
Experiments with 'brother males'

	Wild males × *simulans* females							Yellow males × *simulans* females					
	T.R.T.	O%	V%	L%	Av. V+L	Av. O		T.R.T.	O%	V%	L%	Av. V+L	Av. O
Av. for 33 males	86	61	29	10	5.2	4.1	Av. for 33 males	90	64	27	9	4.2	4.3

Results of analysis of variance: Significance of differences between male types
Av. V+L, P < 0.01. All others not significant.

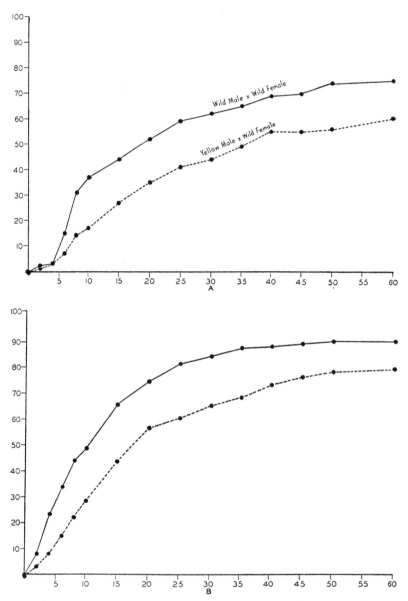

Fig. 5. A. Mating success compared for wild and yellow males with wild females in DARK in pair matings. Average time from introduction to copulation in DARK: wild male × wild female: 16.0 minutes; yellow male × wild female: 20.9 minutes. B. Controls tested simultaneously in the LIGHT. Average time from introduction to copulation in LIGHT: wild male × wild female: 11.4 minutes; yellow male × wild female: 17.6 minutes. The wild and yellow stocks had been previously intercrossed for 7 generations. Abscissa: time in minutes from introduction. Ordinate: % of successful males. Dashed line: yellow male × wild female. Solid line: wild male × wild female.

432   MARGARET BASTOCK

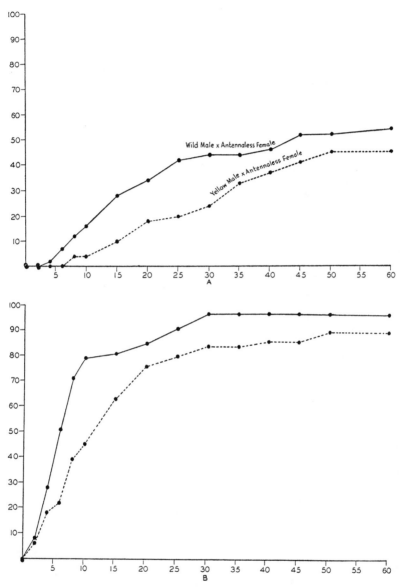

Fig. 6. A. Mating success compared for wild and yellow males with wild ANTEN-
NALESS females in pair matings. Average time from introduction to copulation: wild
male × antennaless female: 19.3 minutes; yellow male × antennaless female: 27.3 minutes. Dashed line: yellow male and antennaless female. Solid line: wild male and
antennaless female. B. Controls tested simultaneously with NORMAL females. The
wild and yellow stocks had been previously intercrossed for 7 generations. Average
time from introduction to copulation: wild male × normal female: 8.4 minutes; yellow male and normal female: 12.5 minutes. Dashed line: yellow male and normal female. Solid line: wild male and normal female. Abscissa: time in minutes from introduction. Ordinate: % successful males.

tailed analyses of courtships directed to these females. Those I have, however, suggest that yellow females may stand still more when courted, and since there is less difference between the two courtships when directed to a standing female (see fig. 4), this could reduce the distinction between the two types of male. Thus this behavioral difference is to some extent dependent upon the behavior of the courted object and it will be greatest when that object is very active. But there is still a fundamental difference between wild and yellow males, for they behave differently to the same object, but more differently to some objects than to others.

The way in which this gene mutation brings about this difference is of course of great interest. There are certain clues in the results which I have already quoted. Thus besides changing their pattern of courtship, yellow males also take slightly longer to begin courting and court slightly less persistently (see p. 423). Also the results just quoted showing that their success is still reduced even with antennaless females means that besides being deficient in the major stimulus of wing vibration (perceived by the female's antennae), they are also deficient in other minor stimuli which serve to rouse the female when vibration is absent. All this suggests that the behavior of the yellow male is generally less vigorous than that of the wild one, and that the reduction of vibration is only one aspect of this.

This idea is amply confirmed by other observational evidence which is difficult to quantify. For instance, when following, yellow males rarely keep up so well with the females, and in wing vibration too, the whole activity seems less efficient, often the wing is not extended to its full 90° angle with the body, and it looks as if the amplitude of the vibrations may be reduced. The licking too seems a weaker movement and often, if the female is running fast, or if she kicks, a yellow male will fail to bring his proboscis in contact with her genitalia. These observations are in line with those of Sturtevant

(1915) for *melanogaster* and Rendel (1945) for *subobscura* who both suggested that yellows were more lethargic in their courtship.

The changed pattern of courtship is therefore clearly associated with reduced general courtship vigor. Is it possible to ascribe both to the same cause? In another paper (Bastock and Manning, 1955), we have investigated the organization of the behavior pattern of courtship. I have described earlier that the three elements, orientation, vibration and licking, are superimposed upon one another, and we examined the possibility that each element appears at a different threshold of a common motivation. Figure 7 is a diagrammatic representation of our hypothesis. Three assumptions are made: (1) that the motivation fluctuates regularly, (2) that the elements occur at different thresholds of this motivation, and (3) that during courtship the motivation remains for most of the time within the limits set by these thresholds. If the first two assumptions are accepted, the third is justified by the fact that courtship is mainly continuous with few breaks.

Predictions can be made from this hypothesis which can be tested. We made, and verified three such predictions. (1) Vibration bout length will be inversely correlated with orientation bout length, i.e. if the motivation is for long periods above the threshold for vibration, it will tend to be below it for short periods and v.v. We obtained an inverse correlation from records of a large number of males ($r = -0.61$, $P < 0.01$). (2) There will be a positive correlation between vibration bout length and licking, i.e. if the motivation is for a longer period above the threshold for vibration then it will rise for longer above that for licking. Here $r = 0.64$, $P < 0.01$. (3) It will be possible to correlate high vibration bout length, taken as a measure of high average motivation, with some independent assessment of this. For an independent measure we took high 'total response time' over a long period, i.e. the amount of time

434   MARGARET BASTOCK

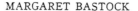

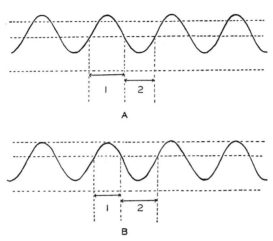

Fig. 7. Diagram of the suggested way in which a change in average sexual motivation affects the courtship pattern. Motivation is assumed to fluctuate regularly in relation to the three thresholds at which the elements appear. A. High average motivation produces courtship typical of wild type males. B. Low average motivation produces courtship pattern typical of yellow males. 1. Bout length of vibration and licking. 2. Bout length of orientation.

which a male would spend courting a non-responsive object (e.g. a *simulans* female) before he became 'exhausted.' The proposition is that those males which would court for the longest time would show the 'high intensity' pattern of courtship, with long vibration bouts. In fact we obtained a linear regression of vibration bout length upon total response time which is significant at the 1% level.

The hypothesis therefore seems to be supported, and it is clear that, if it is true, we can regard the pattern of wild males as that typical of insects with high average motivation, that of yellows typical of low motivation. And this agrees with the impression obtained from the observations quoted above of low sexual vigor in yellow males.

The conclusion therefore seems to be that this gene mutation has in some way lowered the sexual motivation of yellow males. Just what this implies in physiological terms must be a subject for further investigation. Although our hypothesis was concerned with specifically sexual motivation we do not necessarily infer from this that the gene mutation has affected either some sexual hormone or some nervous threshold. It could of course have some peripheral effect upon an organ or tissue which indirectly affects sexual motivation. For instance it might diminish the amount of sperm in the gonads and thus reduce a stimulus (perhaps nervous) received from these organs. On the other hand it might affect the efficiency of the sense organs which receive stimuli from the female and thus reduce this source of motivation.

One could however imagine a much more general effect of the mutation. Perhaps it simply lowers physiological or mechanical efficiency, so that yellow males simply cannot keep up so well with rapidly moving females and thus are less stimulated by them. Hence again, their sexual motivation would be lowered. A further theory along these lines however could deny my hypothesis and suggest that sexual motivation is not involved at all. For it could be imagined that yellow males simply can not vibrate their wings for such long periods as wild ones; perhaps they become fatigued after a few seconds and must then pause before they can vibrate again. Thus although impulses might be sent to the effector organs (i.e. the motivation might be high) the muscles might be unable to respond.

Certain points can be made about all these suggestions, although I have as yet made no experiments. The last idea can, I think, be denied at once by reference to the figures of table 3 where both males are courting *simulans* females. Both types of male give much longer vibration bouts to these females than to *melanogaster* ones. I have given reasons elsewhere (Bastock and Manning, 1955) why this should be so, but it is clear at once that yellow males are physically capable of giving vibration bouts of longer duration than those which they direct to *melanogaster* females. It could be suggested that yellow males are incapable only of the more difficult feat of vibrating for long periods while running and that perhaps *simulans* females run less often or less quickly, but I think this also unlikely. It may be true that *simulans* females run less often, but yellow males give long bouts of vibration to them even while running and it is at least not obvious that they run more slowly.

The idea that the gonads are affected is also made unlikely by the fact that yellow males seem able to fertilize successfully as many females in succession as do wild ones. Also Maynard Smith of University College, London, has found that mutant males of *D. subobscura* which lack a testis are just as successful in mating as wild ones (personal communication). This suggests that the gonads do not make a significant contribution to sexual motivation at least in this species.

The other suggestions remain as possibilities. The idea that yellow males are stimulated less because they fail to keep up so well with very active females is tempting in view of the fact that it is mainly while the female is running that

the yellow males fail. However it is necessary to be cautious for it is clearly difficult to distinguish cause and effect in such a case. Careful investigation is necessary before it can be decided whether an insect is primarily inactive and hence poorly stimulated or whether it is primarily poorly motivated and hence less vigorous in pursuit. Many wild type males fail to keep up with females at the beginning of courtship yet quickly develop this capacity as they become 'warmed up.' In this case at least, the cause is most likely to be poor motivation.

The effect of this mutation upon the courtship behavior pattern may therefore be called 'direct' in the sense that it does not act by inducing the courted object to behave differently. However it is possible that it is 'indirect' in the sense that it may cause no change in the nervous system. The motivation may be altered because of some peripheral or even general physiological change. It would be interesting to know how many changes in behavior patterns could be and have been brought about by such indirect means, i.e. by changes in motivation attributable to a physiological or even morphological effect upon something other than the nervous system. Even the form of movements might be changed in this way. It has been shown by Tinbergen (1952) and others that many display activities have dual motivation. It is easy to imagine that were the strength of one of the sources of motivation to change, the form of the movement concerned might also change.

Evolutionary Implications

Finally the evolutionary implications of these findings merit some discussion. This gene mutation certainly brings about a change in the courtship behavior pattern of males which is very similar to differences found to exist between species and subspecies. However, it also has the effect of reducing the stimulating qualities of the males, at least as far as normal females are concerned. Clearly such a mutant, even although advantageous, would

436   MARGARET BASTOCK

have difficulty in establishing itself in a population. The most likely way in which a pure yellow population could arise, then, would be by the colonization of an isolated habitat, by, perhaps, a single yellow fertilized female. Alternatively, an already existing population, in a similar isolated niche, might be reduced by severe selection to a few yellow insects, supposing these to have some advantage in this situation.

The hypothesis that yellow (or any other mutant which has a similar change in its courtship) might, in some situations, be advantageous, is one which has never been seriously considered before. Yet positive evidence comes from Kalmus (1941), who shows that yellow insects are better able to withstand starvation in saturated air than are wild ones.

Thus it is by no means impossible that a pure yellow population could arise, differing in behavior from its parent population because the yellow gene (selected for other reasons) incidentally caused this. The fact that there are many observations of differences in mating vigor between species and subspecies of *Drosophila,* supports the idea that this sort of thing may well have happened. For example Mayr (1946) says that *D. persimilis* males are much more sluggish in their courtship than are those of *D. pseudoobscura.* Spieth (1952) gives many more examples in his review, and he points out that one cannot explain all these cases by assuming that laboratory conditions are more optimal for some species than others. For he points out that many of the examples show a reverse relationship between male courtship and female receptivity, and this suggests a 'gearing' effect rather than bad conditions.

This 'gearing' idea leads on to the proposition that an initial change in courtship behavior such as this might well stimulate the selection of other, perhaps more far-reaching changes in the behavior concerned. In my yellow *Drosophila* stocks changes in receptivity certainly seem to occur. When I first observed

tained these stocks it was clear that yellow females were much more highly receptive than wild ones. The percentage success in one hour which I obtained from pair matings was as follows:

wild male × wild female	62%
yellow male × wild female	34%
wild male × yellow female	87%
yellow male × yellow female	78%

Here the difference between the females is highly significant and on an analysis of variance (P < 0.01).

Later, however, I crossed my wild stock to my yellow stock for 7 generations, in such a way that the wild stock became genetically similar to the yellow except for yellow and closely linked genes.

The results of similar experiments were then as follows:

wild male × wild female	75%
yellow male × wild female	47%
wild male × yellow female	81%
yellow male × yellow female	59%

The female difference is no longer significant and it will be seen that by crossing to yellow stocks, I have raised the receptivity of wild females until it almost equals that of yellow females. High female receptivity is not therefore a characteristic of the yellow gene, and it is possible that it had been selected for in the yellow stock to counteract the effect of the deficient courtship.

However other changes might also or alternatively take place, for it might be dangerous in nature for females to become too highly receptive. Not only does this increase their chances of mating with males of other species but it prevents their operating the system of sexual selection (suggested by Richards (1927), Merrell (1949, 1953) and others) against males too inferior morphologically or physiologically to court adequately. If a male ceases to vibrate adequately it is possible that selection might favor females giving more weight to other stimuli (e.g. visual) received perhaps during orientation. In turn this could favor the incorporation of more 'visual' elements into the courtship of the males. So the courtships of the 'new' population might progressively diverge from that of the old until the two might become sexually isolated.

Much of this is mere speculation, but there are indications to be found in comparisons of many closely related species which suggest that this may have occurred. For instance, *D. simulans* has a very similar courtship pattern to that of its sibling species *D. melanogaster*. The main difference is that *simulans* males are less active in their courtship than are *melanogaster*, and they tend to stand for long periods beside the female before they suddenly vibrate and copulate. However *simulans* females probably pay more attention to visual stimuli given in courtship, for mating success is considerably reduced in the dark in this species, unlike *melanogaster* (Spieth and Hsu, 1950). It is also interesting to find that *simulans* has added a new visual element to its courtship, which does not occur in *melanogaster*. To quote Spieth (1952) 'when the (*simulans*) male is in the act of circling the female, he often stops either directly in front of her or in front of one of her eyes and rapidly scissors both wings, gradually increasing the amplitude of the wings until they are extended 70° to 90° and then holds them briefly in this position, after which he drops the wings to the normal position and continues with his circling.'

All these differences could have evolved in the way which I have suggested above, stimulated by an original, perhaps incidental difference in courtship vigor. Admittedly, this is not the only possible explanation, but neither is this the only case of its kind. Spieth's review suggests many cases not only of differences in courtship vigor between species but also of differences in the sense modality given the greatest 'weight.' Moreover he finds that the most closely related species and subspecies tend to have very similar courtships but differ at the sensory level and

he comes to the conclusion that sensory divergence nearly always precedes gross divergence in behavior pattern in *Drosophila*. This supports my hypothesis, if it can be shown that minor divergences in frequency or vigor usually accompany the sensory divergence. My contention is that the courtships described by him as 'similar' may well differ in a manner similar to that in which the courtship of the yellow *melanogaster* differs from that of the wild type.

The great interest behind this hypothesis lies in the fact that it suggests a way in which sexual isolation could evolve between two populations which remain in isolation from one another, and which, perhaps, have no differences which could cause their hybrids to be at a disadvantage and thus select for isolation on meeting. This possibility was first favored by Müller (1942, 1948) and later stressed by Mayr (1948). It contrasts with Dobzhansky's view (1951) that only vestiges of sexual isolation are likely to arise as automatic by-products of divergence, selection always being necessary to complete it. The evidence supporting Dobzhansky's view is that sexual isolation between *Drosophila* species is strongest where they overlap and becomes weaker as the distance between them increases (King (1947), Dobzhansky and Koller (1938)), but other workers, e.g. Wharton (1942), Mainland (1942), Spieth (1951), find apparently no such geographical correlation in other cases. Moreover the chief theoretical argument against the evolution of such mechanisms in isolation, is that they demand too many parallel and correlated changes, to have arisen without the stimulus of selection. If however one imagines that one small change could stimulate the selection of others, this difficulty is removed.

My conclusion, therefore, although speculative, is that this gene mutation in *Drosophila melanogaster does* change the pattern of courtship behavior in a manner which may well have been used in evolution. Such an effect of a .single gene

438   MARGARET BASTOCK

mutation upon the frequency of elements in a behavior pattern has, I think, not previously been demonstrated. Although its effect is deleterious to mating success nevertheless it could be incidentally incorporated in an isolated population if the mutant concerned were useful in any other sense. It might then stimulate the selection of other divergences on behavior such as might ultimately bring about the sexual isolation of the population concerned, or it might alternatively form a basis upon which sexual isolation could be built by selection.

Summary

Yellow mutant males of *Drosophila melanogaster* have reduced success in fertilizing normal females.

There is a change in the pattern of courtship which could explain this. The wing display (vibration), which is shown to provide important stimuli perceived by the females' antennae, is given in shorter bouts and at longer intervals.

The idea that this behavioral change is indirect is dismissed. It is possible that females might react against changed color or scent, and, by repelling yellow males, thus influence their courtship. This is not supported by

(1) analyses of female behavior;
(2) comparisons of male behavior to *D. simulans* females which refuse both normal and yellow;
(3) comparisons of male success with females lacking visual or olfactory stimuli.

Consideration of the possible causal mechanism underlying the courtship pattern suggests that each of the three elements (orientation, vibration and licking) might occur at different thresholds of a fluctuating sexual excitation. A typical 'yellow' pattern with fewer and shorter bouts of 'higher' elements could be explained in terms of low average sexual excitation. This mutation thus changes a behavior pattern probably by affecting sexual motivation. How it does this is not yet known.

The nature of the change is of interest because it parallels behavioral differences found between closely related species of animals. Such a change may sometimes be incorporated incidentally into a population if the mutants concerned are selected for other reasons. A courtship change, such as this, might stimulate further evolution which could bring about the sexual isolation of the mutant population. For the reduction of an important stimulus from males might lead females to give weight to other stimuli (perhaps of another sense modality), which in turn would put a selective premium on other elements of the courtship.

Acknowledgments

This work was done under the supervision of Dr. N. Tinbergen, to whom I am grateful for much encouragement and advice.

I am grateful also to Prof. A. C. Hardy, F.R.S., for facilities in this department, and to the Nuffield Foundation, for from their grant to Dr. Tinbergen, most of my equipment and technical assistance were provided.

My thanks are especially due to Aubrey Manning and Pieter Sevenster for cooperation in experiments and to Dr. A. J. Cain, Dr. R. Milani, Mr. J. Maynard Smith, Dr. W. M. S. Russell, Dr. P. M. Sheppard, Dr. H. Spurway and Dr. U. Weidmann for many discussions and advice.

I am also grateful to Dr. Ernst Mayr for demonstrating to me certain techniques, and to Dr. M. R. Sampford and Mr. J. M. Hammersley for assistance in statistical calculations.

Literature Cited

Bastock, M., and A. Manning. 1955. The courtship of *Drosophila melanogaster*. Behavior, **8**: 85–111.

Begg, M., and E. Packman. 1951. Antennae and mating behavior in *Drosophila melanogaster*. Nature, **168**: 193.

CLARK, E., L. ARONSON AND M. GORDON. 1954. Mating behavior patterns in two sympatric species of *Xiphophorin* fishes: their inheritance and significance in sexual isolation. Bull. of the Am. Mus. of Nat. Hist., 103: 141–225.

DOBZHANSKY, TH. 1951. Genetics and the Origin of Species. New York. 3rd Ed. Revised.

——, AND P. C. KOLLER. 1938. An experimental study of sexual isolation in *Drosophila*. Biol. Zentral., **58**: 589–607.

GRAUBARD, M. A. 1933. Tyrosinase in mutants of *Drosophila melanogaster*. J. Genetics, 27: 199–218.

KALMUS, H. 1941. The resistance to desiccation of *Drosophila* mutants affecting body colour. Proc. Roy. Soc., 130: 185–201.

KEELER, C. E., AND H. D. KING. 1942. Multiple effects of coat color in the Norway rat, with special reference to temperament and domestication. J. Comp. Psychol., 34: 241–250.

KING, J. C. 1947. Interspecific relationship within the *guarani* group of *Drosophila*. EVOLUTION, 1: 143–151.

MAINLAND, G. B. 1942. Genetic relationships in the *Drosophila funebris* group. The Univ. of Texas Publ. 4228: 74–112.

MAYR, E. 1946. Experiments on sexual isolation in *Drosophila*. VII. The nature of the isolating mechanisms between *Drosophila pseudoobscura* and *Drosophila persimilis*. Proc. Nat. Acad. Sci., 32: 128–137.

——. 1948. The bearing of the new systematics on genetical problems. The nature of the species. Advances in Genetics, 2: 205–237.

——. 1950. The role of the antennae in the mating behaviour of female *Drosophila*. EVOLUTION, 4: 149–154.

MERRELL, D. J. 1949. Selective mating in *Drosophila melanogaster*. Genetics, 34: 370–389.

——. 1953. Selective mating as a cause of gene frequency changes in laboratory populations of *Drosophila melanogaster*. EVOLUTION, 7: 287–296.

MILANI, R. 1951a. Osservazioni sul corteggiamento de *Drosophila subobscura*. Collin. Inst. Lombardo di Sci. e Lett., 84: (1–12) (page numbers as in reprint).

——. 1951b. Osservazioni comparativo ed esperimenti sulle modalita del corteggiamento nelle cinque species europes del gruppo "*obscura.*" Ibid., 84: (1–12) (page numbers as in reprint).

MOYNIHAN, M. 1953. Some aspects of the reproductive behaviour of the black-headed gull and related species. Dissertation for D.Phil., University of Oxford.

——. 1955. (As above.) Behaviour Supplement. IV.

MÜLLER, H. J. 1942. Isolating mechanisms, evolution and temperature. Biol. Symp., 6: 71–122.

——. 1948. Evidence for the precision of genetic adaptation. The Harvey Lectures, 43: 165–229.

PATTERSON, J. T. AND M. R. WHEELER. 1949. Catalogue of described species belonging to the genus *Drosophila*, with observations on their geographical distribution. The Univ. of Texas Publ. 4920: 207–233.

RENDEL, J. M. 1945. Genetics and cytology of *Drosophila subobscura*. II. Normal and selective mating in *D. subobscura*. J. Genetics, 46: 287–303.

——. 1951. Mating of ebony vestigial and wild type *Drosophila melanogaster* in light and dark. EVOLUTION, 5: 226–230.

RICHARDS, O. W. 1927. Sexual selection and allied problems in the insects. Biol. Rev., 2: 298–364.

SHEPPARD, P. M. 1952. A note on non-random mating in the moth *Panaxia dominula* (L). Heredity, 6: 239–241.

SPIETH, H. T. 1951. Mating behavior and sexual isolation in the *Drosophila virilis* species group. Behaviour, 3: 105–145.

——. 1952. Mating behavior within the genus *Drosophila* (*Diptera*). Bull. of the Am. Mus. of Nat. Hist., 99: 401–474.

——, AND T. C. HSU. 1950. The influence of light on seven species of the *Drosophila melanogaster* species group. EVOLUTION, 4: 316–325.

STURTEVANT, A. H. 1915. Experiments on sex recognition and the problem of sexual selection in *Drosophila*. J. of Animal Behaviour, 5: 351–366.

——. 1921. The North American species of *Drosophila*. Carn. Inst. of Wash. Publ. 301: 1–150.

TINBERGEN, N. 1952. Derived activities, their causation, biological significance, origin and emancipation during evolution. Quart. Rev. Biol., 27: 1–32.

WHARTON, L. 1942. Analysis of the *repleta* group of *Drosophila*. The Univ. of Texas Publ. 4228: 23–52.

GENETIC CONTROL OF THE NEURONAL NETWORK GENERATING CRICKET (*TELEOGRYLLUS GRYLLUS*) SONG PATTERNS

By DAVID R. BENTLEY & RONALD R. HOY*

Department of Zoology, University of California at Berkeley

Abstract. Hybrid field crickets between *Teleogryllus* species and *Gryllus* species were produced to examine: (a) the genetic system controlling the neuronal network underlying song production; and (b) the neuronal mechanism responsible for the superimposed rhythms of chirping songs. Pulse numbers, intervals, and progressions were subjected to statistical and graphic analyses. *Teleogryllus* songs are controlled by a complex, polygenic, multichromosomal system, even at the level of 'unitary', acoustical parameters. Genes regulating certain pattern characteristics are on the X-chromosome'. The superficially similar chirping songs of two *Gryllus* species appear due to different neuronal mechanisms. One song appears to reflect a single rhythm generator with some relaxation oscillator properties, while the rhythms of the other song seem to be caused by the mixed effect of two resonant type oscillators.

The behaviour of an animal is one of its most sophisticated and critical attributes. A variety of studies (Ewing & Manning 1967) have shown that behaviour is under differing degrees of genetic control. The number of cells, the time intervals, and the complex interaction with the environment involved in the generation of behaviour result in its being remote from the direct effects of genes. Consequently, little is known about the mechanisms by which genes exert their effects on behaviour.

Since much of this control must be mediated by the nervous system, considerable attention is being focussed on the effects of genes on neurones. Most of these studies have been based upon well known genetic systems, but are hindered by the technical difficulties of working with large numbers of neurones (Sidman 1968) or very small neurones (Hotta & Benzer 1969; Ikeda & Kaplan 1970). We are exploring an alternative approach of using a relatively well-known nervous system involving small numbers of large, individually identifiable neurones, and are using the system underlying cricket song patterns.

The calling song, used by males to attract females, offers a number of unique advantages for genetic study. It is remarkably stereotyped throughout the lifetime of the individual and among different individuals; as a communication device, its efficacy in attracting females depends upon its uniformity. The song pattern is produced by a group of neurones whose activity is typical of the central programming or fixed

*Present address: State University of New York at Stony Brook.

pattern generating systems recently described for many invertebrate behaviours (Wilson 1968; Dorsett, Willows & Hoyle 1969). Such neurones are organized in a network lying within the central nervous system and, upon reception of an appropriate triggering input, can generate a particular pattern of motor output without any phasic sensory information. In fact, a cricket central nervous system isolated from all phasic input can produce a calling song pattern indistinguishable from that of the intact animal (Bentley 1969a). Therefore, the output of these networks is very well buffered from the influence of the environment and provides a correspondingly precise monitor of the genotype. Furthermore, the sound pulses of the songs can be quickly and easily recorded from large numbers of animals, and are an accurate indication of the activity of identified motor neurones.

A considerable amount of information is already available on most aspects of the system, including: the behaviour (Alexander 1961), the role of the higher central nervous system (Huber 1960; Otto 1968), the operation of the motor apparatus (Bentley & Kutsch 1966; Kutsch 1969), intracellular activity in the pattern generating network (Bentley 1969a, 1969b), post-embryonic development of patterned activity (Bentley & Hoy 1970), the reception of sound patterns (Zaretsky 1971; Stout 1970) and the neuro-anatomy of an homologous system (Bentley 1970). Several previous studies have been conducted upon the genetics of cricket song, and these have recently been thoroughly reviewed by Alexander (1968).

In the current study, we are concerned with two main questions: (a) can hybridization be used to reveal the basic features of the genetic system controlling song pattern?; (b) can manipulation of the genotype be used to test hypotheses about the structure of the neuronal pattern generating network?

Methods

We investigated the inheritance of calling song patterns in two cricket genera, *Teleogryllus*, which has complex songs, and *Gryllus*, which has simple songs. Stock colonies of five species raised in large numbers in the laboratory were obtained from the following sources: *T. commodus* and *T. oceanicus* (Dr T. Hogan, Australia), *G. rubens* (Dr T. Walker, Florida), *G. campestris* (Dr F. Huber, Germany), *G. armatus* (captured in California). From the stock colonies, we set the following successful crosses:

(1) *T. commodus* ♀ × *T. oceanicus* ♂ ⎫ recipro-
(2) *T. oceanicus* ♀ × *T. commodus* ♂ ⎭ cal cross
(3) *G. armatus* ♀ × *G. rubens* ♂
(4) *G. campestris* ♀ × *G. rubens* ♂

For each cross, twelve animals from each species were placed in a container for 3 months. Eggs were collected in peat moss filled dishes, and nymphs were raised in separate containers. Several hundred offspring were obtained from each *Teleogryllus* cross. About fifty nymphs from cross number three reached maturity, while cross number four produced only a single adult male.

For examination of acoustical behaviour, males from the parental and hybrid stocks were isolated soon after the final moult. Usually, they began to produce the calling song within 2

weeks. From five of each parental type and ten of each hybrid type (excepting the single *G. campestris* hybrid), several minutes of steady calling, involving several hundred sound pulses, were recorded on tape (*Teleogryllus* songs were recorded at 24·5 ± 1°C; *Gryllus* at 24·5 ± 2°C). The tapes were replayed while the sound pulses were viewed on an oscilloscope screen, and a portion of uninterrupted singing was filmed on oscillograph paper at a speed of 4 ms per mm. On a section of film containing 400 to 600 sound pulses, the duration of each inter-pulse interval (defined as extending from the beginning of one pulse to the beginning of the next) was measured. In the case of the *Teleogryllus* songs, for example, this provided about 15,000 data points,

The data were displayed in two basic forms: first, in inter-pulse interval histograms which revealed the number, duration, variability, and relative abundance of intervals in each song pattern (Figs 4 and 7). Secondly, the durations of successive intervals were plotted to display the patterned arrangement of intervals, and in particular, progressive changes of intervals within sub-elements of the song (Fig. 5). Most of our conclusions about the structure of the parental songs and the inheritance of song characters are drawn from these two types of plots, although additional measurements were made directly from the film in some cases, especially those involving phrase structure (Fig. 1).

The statistical procedure for each character analysed was as follows: the mean and variance of data from each individual were obtained. Within each genotype, the variances from different individuals were compared by Bartlett's test (and found not to be significantly different).

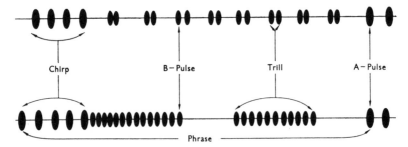

Fig. 1. Diagram of structural components and terminology of *Teleogryllus* songs: upper line=*T. oceanicus*; lower line=*T. commodus*. Inter-chirp interval=interval between onset of A-pulses. Intra-trill interval=interval between onset of B-pulses. Inter-trill=interval between onset of last B-pulse in one trill and the first pulse of the next trill.

ANIMAL BEHAVIOUR, 20, 3   480

The data from individuals within the same geno-type were then pooled, and a pool mean and variance obtained (for selected characters, the validity of pooling was also checked by com-paring different genotypes using pairs of in-dividuals with Wilcoxon's signed rank test). The pool means and variances from different genotypes were compared by pairs: first, the variances were compared with a two-variance F test. If they were not significantly different, the means were compared with a t-test; if the variances were significantly different, the means were compared with the Fisher-Behrens test.

In order to determine: (a) the precision with which the sound pulses reveal the activity of identified motor neurones; and (b) whether the relationship between motor neurone activity and sound pulses is the same in *Teleogryllus* species as in the intensively studied *Gryllus* species, muscle action potentials were recorded from identified muscles during calling. Since each muscle action potential is caused by a single motor neurone action potential, and since there are only one to three motor units in the muscles involved, this procedure can be employed to determine the firing patterns of identified motor neurones (Bentley & Kutsch 1966). Recordings were made from 25 μm insulated wires inserted through the cuticle into the muscles, and action potentials were displayed and filmed with conventional electrophysiological apparatus (Fig. 2).

Results

Teleogryllus Hybrids

Three lines of evidence indicate that the song pattern is under rigid genetic control. First, the songs of individuals with the same genotype (members of the same species do not, of course, have identical genotypes, and will normally carry different alleles at many loci; they do have a common karotype) are very similar. Data in Table I demonstrate that such song patterns are highly stereotyped. This homogeneity is main-tained despite a variety of environmental con-ditions, including changes in diet, crowding, light cycles, temperature, and time of year. Therefore, it appears that the similarity of patterns is due to the common genotype. Secondly, songs of individuals with different genotypes are marked different (Table I; Figs 3 to 5). This is most clearly illustrated by Fig. 5, which displays the arrangement of successive inter-pulse intervals; parental and hybrid pat-terns are strikingly different. The heterogeneity is maintained even when animals are raised in very similar environmental conditions, and again seem due to the genotype. Thirdly, individuals produced the calling song pattern characteristic of their genotype without ever having heard it before. For example, the first hybrids to mature produced a song characteristic of all subsequent hybrids, and different from either parental type. Therefore, the song pattern appears to be well isolated from environmental influences, and is a correspondingly fine monitor of the genotype. This suggests that other neuronal pattern gen-erators now known in invertebrate behaviour may be under similar genetic control.

The relationship between the discharge of identified motor units and the occurrence of the sound pulse is illustrated in Fig. 2. Well before the sound pulse, the wings are opened by the group of muscles which also depress the wings in flight; the subalar muscle is representative of these. Subsequently, the flight elevator

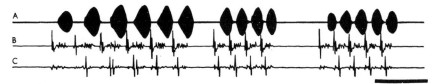

Fig. 2. Electromyograms recorded from *T. oceanicus* ♀ × *T. commodus* ♂ hybrid during the generation of the calling song. A = sound pulses; B = muscle action potentials recorded from the first (lower threshold) unit of the subalar muscle, a wing opener. C = muscle action potentials recorded from the second of the three units of the promotor muscle. This muscle closes the wings to produce the sound pulses. Each action potential reflects a single impulse in the motor neurone innervating the unit. Note that: (a) firing of the second promotor unit is omitted when weak sound pulses occur at the beinning of the chirp or trill, (b) the second promotor unit fires twice to produce the particularly loud sound pulses in the centre of the chirp. The position of the sound pulse can be used as an accurate monitor of the activity of this population of neurones. Time cal: 100 ms.

group, represented by the promotor muscle, fires to close the wings and produce the sound pulse. The units innervating each muscle are arranged in a recruitment hierarchy. The particular subalar unit monitored has the lower threshold of two units innervating the muscle, and fires with each wing opening. The promotor unit monitored is the second of three units; failure of the unit to fire is followed by weak sound pulses, such as the initial pulse(s) of chirps or trills. Single discharges of the unit are followed by moderately intense pulses, and double discharges precede the most intense sound (Fig. 2). This demonstrates that the activity of these units is homologous with the large body of information on *Gryllus* stridulation and flight systems, and on the locust flight system (Wilson 1968). In addition, it indicates the precision with which the sound pulses can be used to monitor the activity of the nervous system.

Since a major problem of behavioural genetics has been the difficulty of obtaining precise, reproducible measurements, it is worth emphasizing the unusual stereotypy of the song patterns. Some parameters, particularly in reiterated portions of the song, are somewhat variable; for example, in the case of the number of trills in the *T. oceanicus* song, or the number of pulses in *T. commodus* trills (Table I). However, parameters which are probably important for song recognition, such as the intervals between pulses, are extremely regular. Such intervals characteristically vary only a few milliseconds, and some of the variation is due to progressive changes during a phrase (Fig. 5) and to the error factor introduced by estimating the time of neuronal discharge from the sound pulse. Therefore, the sound patterns are uniquely quantifiable and reproducible forms of behaviour.

Structure of the *Teleogryllus* Calling Song

In order to describe the inheritance of features of the songs it is necessary to discuss the pattern in more detail. *Teleogryllus* songs are among the most complex produced by crickets. While this makes description difficult it provides a large number of parameters to observe. Basically, the song consists of two types of sound pulses, grouped into chirps and trills which are in turn arranged in a repeating phrase (Fig. 1).

The most elementary unit of the song is a sound pulse. Each pulse is produced by closing the wings once, and reflects an impulse or burst

Table I. *Teleogryllus* **Hybrid Song Pattern Parameters**

Parameter	Genotype	Mean	Standard deviation	N
Intra-chirp interval	O	66·8	6·7	290
	O/C	57·5	6·5	712
	C/O	60·4	6·1	807
	C	52·1	6·8	489
Intra-trill interval	O	41·0	1·3	740
	O/C	33·0	3·8	2833
	C/O	36·7	5·0	2450
	C	31·7	3·5	1840
Inter-trill interval	O	122·8**	14·1**	813
	O/C	136·9	23·6	837
	C/O	154·0	38·8	693
	C	160·9	60·6	119
No. of A-pulses per chirp	O	4·8	0·9	100
	O/C	4·2	0·6	100
	C/O	4·9	0·7	100
	C	5·9	0·9	100
No. of B-pulses per trill	O	2·0*	0·1*	250
	O/C	4·5	1·5	194
	C/O	4·9	2·0	200
	C	10·7	5·3	147
No. of B-pulses per phrase	O	20·9	3·0*	38
	O/C	18·4	4·9	28
	O/C	19·4	4·7	37
	C	21·5	10·4	34
No. of trills per phrase	O	9·4*	2·3	100
	O/C	4·2	1·0	100
	C/O	4·8	1·2	100
	C	2·3	1·2	100
Phrase repetition rate	O	28·5	2·4**	77
	O/C	37·5	5·3	105
	C/O	33·0	7·0	70
	C	41·7	18·1	102

O = *T. oceanicus* (5 individuals); O/C = *T. oceanicus* ♀ × *T. commodus* ♂ (10 individuals); C/O = *T. commodus* ♀ × *T. oceanicus* ♂ (10 individuals); C = *T. commodus* (5 individuals). Intervals in milliseconds, repetition rate in phrases per minute.

*Hybrids significantly different from both parental types.
**Hybrids significantly different from each other and from both parental types.

of impulses in the wing closer motor neurones. There are two types of sound pulses: A-pulses are longer and more intense. They are grouped into chirps, each chirp containing four to seven pulses. B-pulses are shorter and softer; they are grouped into trills consisting of from two to twenty pulses. Chirps and trills are arranged in the largest repeating unit of the song, a phrase. In *T. oceanicus* each phrase consists of a chirp followed by about ten trills, each composed of

ANIMAL BEHAVIOUR, 20, 3 482

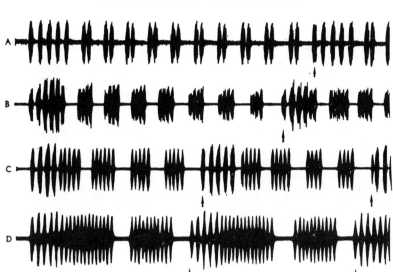

Fig. 3. Sound pulse patterns in the calling song of *Teleogryllus* species and F₁ hybrids. A=*T. oceanicus*; B=*T. oceanicus* ♀ × *T. commodus* ♂ F₁ hybrid; C=*T. commodus* ♀ × *T. oceanicus* ♂ F₁ hybrid. D= *T. commodus*. The hybrid songs are intermediate in most of the features in which the parental species differ: note (a) number of pulses per trill; (b) number of trills per phrase; (c) phrase repetition rate (each trace starts at the beginning of a phrase, and arrows indicate the beginnings of subsequent phrases). Some of the parameters of the song are sex-linked rather than intermediate (see text). Time cal: 0·5 s.

two pulses. The *T. commodus* phrase consists of a chirp followed by one to four long trills (Fig. 3).

A number of more subtle features of the patterns are revealed by the intervals between pulses. There are three basic intervals, the intra-chirp interval between A-pulses, the intra-trill interval between B-pulses, and the inter-trill interval between trills or between trills and chirps. Each interval has a characteristic value for each species (Table I). In addition, there are controlled progressions, or gradual changes in successive intervals, within the phrase (Fig. 5). For example, the inter-trill intervals gradually lengthen during the phrase; the intra-chirp intervals, in *T. commodus*, lengthen and then shorten in a parabolic fashion; the intra-trill intervals remain very constant. These progressions are, of course, responsible for a considerable amount of the variation noted in Table I, so the song pattern is even more highly controlled than these statistics indicate. It seems likely that the progressions reflect changes in key neurones, such as fatigue, which are responsible for termination or resetting of chirps

and phrases, and therefore play a critical role in pattern determination.

The transitional intervals between chirps and trills in the two *Teleogryllus* species differ in a way which makes an obvious acoustical impression but is not so apparent when examining the data visually. The *T. oceanicus* song sounds continuous because the interval between the last trill in a phrase and the next chirp is about equal to the normal inter-trill intervals; on the other hand, the *T. commodus* song often sounds like a series of discrete phrases because the corresponding interval is distinctively longer than the inter-trill intervals. The interval following the chirp is also basically different, since in *T. commodus* the chirp switches directly into the trill with no intervening long interval (Fig. 3).

A final critical difference in the two song types is the degree of variability; the *T. oceanicus* pattern is much more regular than that of *T. commodus*. In fact, *T. commodus* has a secondary calling song consisting solely of chirps delivered at much longer intervals, and this can grade toward the song we studied by shortening the intervals and inserting trills. The degree of variability is genetically controlled and has turned out to be one of the more interesting characters.

Inheritance Patterns

We examined the inheritance of all of the characters of the calling song which primarily reflected neurone firing patterns (characters such as the pitch or carrier frequency of the song which are partially determined by the structure of the exoskeleton were not treated). This involved eighteen parameters of the numbers of pulses, arrangements of pulses, interval durations, and interval progressions. The statistics on eight representative parameters are presented in Table I. The inheritance of these parameters fell into three categories: (a) parameters not significantly different among any of the four genotypes; (b) hybrid characters different from both parental types, but not from each other; and (c) hybrids significantly different from both parental types, and also from each other. The implications of these situations will be treated separately.

Cases in which the parameters were not significantly different in any of the genotypes are trivial in the sense that no information about the nature of the genetic system can be derived from them. However, some are quite interesting from an evolutionary point of view. It was immediately apparent that the number of A-pulses/phrase was essentially the same in both parental species and their hybrids. It was more surprising that the number of B-pulses/phrase was also the same. This means that the total number of motor neurone bursts/phrase is the same in both species, and suggests that the striking dichotomy of pattern has evolved simply by an adjustment of intervals rather than by the deletion or insertion of pulses.

Distinct Hybrid and Parental Patterns

In general, when parental species characteristics differed significantly, the hybrid characters were about intermediate between the parental types (Fig. 3). This situation is well illustrated by the number of pulses/trill, or the number of trills/phrase (Table I). We found no cases where the hybrid parameter was indistinguishable from one of the parental types. Therefore, no parameters are controlled by monofactorial inheritance involving a simple dominant-recessive situation. This would not eliminate the possibility that some parameters are inherited by monofactorial systems involving genes with equal penetrance or expressivity. However, the pattern of inheritance is characteristic of, and strongly suggests, a polygenic system (on-going analysis of backcrosses confirms this hypothesis). Therefore, it appears that each song parameter is affected by a number of genes, resulting in a stable genetic arrangement well buffered against mutation.

Sex-Linked Characters

The third type of inheritance, in which the reciprocal hybrids were significantly different from each other as well as from the parental species, was found in the following parameters: the duration of the inter-trill interval, the variability of the inter-trill interval, and the variability of the phrase repetition rate (Table I; Fig. 4). In each of these cases, the hybrid song is more similar to that of the male of the maternal species than to the paternal species (only male crickets sing). No cases were found where hybrid parameters more closely resembled those of the paternal species. Crickets lack a Y-chromosome, and have XX, XO sex determination (Alexander 1968). Therefore, when reciprocal crosses are made between two species (using each species first as the maternal parent and then as the paternal parent), the two types of hybrid male offspring have identical genotypes except that each has the X-chromosome from the maternal species. Consequently, differences in acoustical phenotype between the two types of hybrids must be due to genes located on the X-chromosome. Since the *Teleogryllus* species examined have twenty-nine chromosomes (Leroy 1967), delimitation of the genes controlling these three characters to a single chromosome represents a considerable degree of localization.

The existence of sex-linked characters also has a number of broad implications for the nature of the genetic system underlying song production. First, since the control of some characters is sex-linked while that of others is not, more than one chromosome must be involved in the system; it is multichromosomal. Second, the presence of sex-linked and non-sex-linked parameters demonstrates that different sets of genes are involved in the control of different operations of the neural network; even this highly delimited behaviour pattern is not controlled by a single set of linked genes. Finally, the sex-linked characters which we have observed are not completely sex-linked.

ANIMAL BEHAVIOUR, 20, 3   484

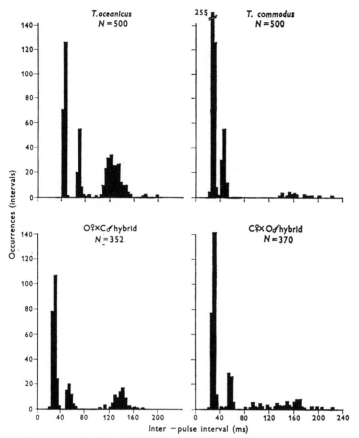

Fig. 4. Inter-pulse interval histograms of calling songs of *Teleogryllus* species and F₁ hybrids. The histograms contain all of the intervals in a period of steady singing, and indicate: (1) the intervals are divided into three modes (intra-trill; intra-chirp, inter-trill), (2) the precision with which each mode is maintained, (3) the relative abundance of each mode, (4) the absolute durations of intervals. Note the sex-linked control of the inter-trill interval (third mode). In *T. oceanicus*, the interval is abundant, has a sharp peak (little variation), and a mean at about 120 ms, while in *T. commodus*, the interval is rare, has a broad distribution, and a mean at about 160 ms; in all three features, the hybrid interval resembles that of its maternal parent. Such histograms were constructed for many individuals, and these appeared to be representative of their types.

For example, the inter-trill interval distribution of the hybrid is clearly distinguishable from that of the maternal parent (Fig. 4; Table I). Therefore, although this single character is influenced by genes on the X-chromosome, it is also influenced by genes on other chromosomes. Consequently, even the more highly restricted set of genes regulating a single song parameter is distributed on more than one chromosome.

In summary, the genetic system which controls the construction of the neuronal network underlying stridulation does not appear to be simple in any of the aspects observed in this study. There is no simple dominance or monofactorial inheritance; rather, the neuronal network is regulated by a widely distributed, polygenic system.

Gryllus **Hybrids**

The second problem we wished to explore was whether manipulation of the genetic material could be used to investigate the nature of the neuronal pattern generating systems. A recurrent problem in neurophysiology has been that of the superimposed rhythms prominent in many behaviours. A typical example is chirping in crickets: chirps are produced at one rate, and each chirp is composed of sound pulses delivered at a second, faster rate. For example, in *G. campestris*, motor neurones must generate impulses at the chirp rate of about three cycles/s, which are modulated at the pulse rate of about thirty cycles/s to form the bursts underlying sound pulses. The problem is whether: (a) the two rhythms are generated by separate neuronal mechanisms and simply mixed to drive the motor neurones; or (b) there is only a single rhythm generating mechanism, and the pause between chirps is due to some inherent property of this mechanism such as accumulating refractoriness.

The question was approached by hybridizing each of two chirping species with a non-chirping species which generates uninterrupted trains of pulses. Previous data demonstrated that when parental parameters differ, an intermediate expression is reached in the hybrid. Consequently, the particular features of the non-chirping song with which an accommodation is reached in the hybrid could reveal: (a) whether the two types of chirping song are based upon the same neurological mechanism; and (b) whether some aspect of the more primitive non-chirping song has been modified to produce chirps, or a totally new rhythmic element has been added.

Crickets produce an enormous variety of songs, but the patterns fall into a series of distinct categories (Alexander 1962). From different categories, two chirping species were selected whose songs contain similar numbers of pulses per chirp but are separated on the basis of other features. *Gryllus campestris* is a 'B$_1$-type'. chirper whose calling song consists of an uninterrupted sequence of short, widely spaced chirps (Fig. 8). Virtually all of the neurophysiological work on crickets has been done with songs of this 'type'. The other chirping species, *G. armatus*, produces a 'C-type' song consisting of short chirps with a brief interchirp interval which are delivered in groups or trains (Figs 6 and 8). The non-chirping species, *G. rubens*, produces long, uninterrupted trains of pulses. The trains are traditionally called trills, although they differ from chirps only by having more sound pulses, not different kinds of pulses as in the *Teleogryllus* species. The trills of *G. rubens* and chirp trains of *G. armatus* vary greatly but are of comparable length and contain roughly the same number of pulses (Fig. 8).

G. campestris × *G. rubens* **hybrid.** This hybrid was marked by very low viability; of several dozen eggs laid, only about a dozen nymphs hatched. Most of these died during the course of development, until only a single adult male was obtained. This animal did produce a calling song, but the sound pulses were quite irregular because the wings were deformed. As a result, accurate measurements could only be made of the number of pulses and of chirp rate, but not of the inter-pulse intervals.

Hybrid chirps were intermediate in length

Table II. *Gryllus* **Hybrid Song Pattern /Parameters**

Parameter	Genotype	Mean	Standard deviation	N
Inter-pulse interval	R	19·8	3·6	439
	A/R	17·4	2·0	837
	A	14·6	1·7	785
Inter-chirp interval	A	52·8	7·8	352
	A/R	39·2*	4·8	474
Chirp frequency	C	3·35	0·3	204
	C/R	3·18†	0·9	260

A = *G. armatus*; A/R = *G. armatus* × *G. rubens* hybrid; R = *G. rubens*; C = *G. campestris*; C/R = *G. campestris* × *G. rubens* hybrid. Data from five animals of each type excepting the single C/R. Intervals in milliseconds, frequencies in cycles per second.

*Mean of A inter-chirp interval and R inter-pulse interval = 33·7.

†Hybrid has two interval modes, one determining chirp frequency, and a second longer interval between groups of chirps (mean = 857; SD = 130; N = 73).

ANIMAL BEHAVIOUR, 20, 3   486

Table III. Chirp Length of *Gryllus* Calling Songs

	No. of pulses/chirp					
	2	3	4–9	10–50	51–120	201–500
A	2·9	96·0	<1·0			
A/R	28·3	62·9	8·8‡			
R				9·3	75·4	15·3
C/R			59·0	41·0*		
C		<1·0	99·0†			

A=*G. armatus* (4 animals; 1016 chirps); A/R=*G. armatus* × *G. rubens* hybrid (6 animals; 2275 chirps); R=*G. rubens* (5 animals; 1027 chirps); C/R= *G. campestris* × *G. rubens* hybrid (1 animal; 206 chirps); C=*G. campestris* (4 animals; 1007 chirps). Data are percentage of chirps to each class.
*Mean, 9·8; SD 3·8.   †Mean, 3·8; SD 0·1.   ‡Mean, 2·8; SD 0·7.

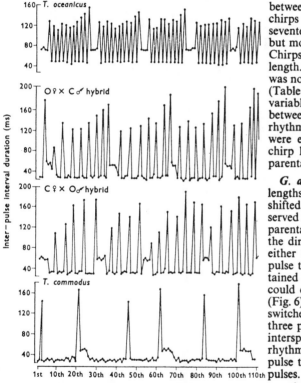

Position of interval in song

Fig. 5. Successive interval display of the calling songs of *Teleogryllus* species and their F₁ hybrids, starting at a random position in the song. This illustrates the pattern in which the various intervals occur, and particularly the progressive changes in pattern during the course of the phrase. Note (1) the arrangement of short (intra-trill), medium (intra-chirp), and long (inter-trill) intervals in the different songs; (2) the uniformity of the intra-trill interval, (3) the progressive lengthening of the inter-trill interval during the phrase (particularly clear in the hybrids); (4) the progressive shortening of the intra-chirp interval in *T. commodus* (compare with *T. oceanicus*).

between the pulse trains of *G. rubens* and the chirps of *G. campestris*. With a range of five to seventeen pulses, the hybrid chirps were variable but most contained about ten pulses (Table III). Chirps in the hybrid occurred in trains of irregular length. Within the trains, the chirp rhythm rate was not significantly different from *G. campestris* (Table II). The interval between trains was variable but was comparable to the interval between *G. rubens* trills (Fig. 8). Therefore, rhythmic components of both parental songs were expressed in the hybrid, and the hybrid chirp length was intermediate between that of parental pulse trains.

G. armatus × G. ruberis hybrids. The chirp lengths of hybrids from the C-type chirper shifted in the opposite direction from that observed in the B₁-type hybrids. Changes from the parental chirp length were predominantly in the direction of fewer pulses than are found in either the *G. armatus* chirps or the *G. rubens* pulse trains (Table III). Most individuals maintained a clear chirping rhythm, but the chirps could contain either two pulses or three pulses (Fig. 6). Some hybrids were consistent, but others switched between sequences of two pulse and three pulse chirps with occasional longer chirps interspersed. A few individuals lost the chirp rhythm altogether and generated irregular pulse trains varying in length from one to nine pulses.

A second major contrast with the B₁-type hybrids was that the chirping rate of the parent was not found in any of the C-type hybrids; all hybrids chirped faster, or not at all. As expected, the inter-pulse interval of the hybrids was intermediate between the corresponding parental intervals (Fig. 7). However, the inter-chirp interval of the hybrids was intermediate between the *G. armatus* inter-chirp interval and the *G. rubens* inter-pulse interval. In most hybrids, there was a clear distinction between inter-

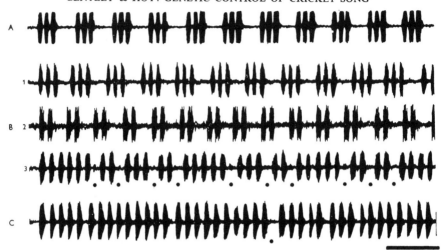

Fig. 6. Calling songs of *Gryllus* species and their F₁ hybrids illustrating sound pulse patterns. A=*G. armatus*. The song is organized into three pulse chirps. B=*G. armatus* × *G. rubens* hybrids. (1) Individual with rhythmically occurring three-pulse chirps. (2) Individual with rhythmically occurring two-pulse chirps. (3) Individual lacking rhythmical, uniform chirps, but with irregularly inserted inter-chirp intervals (dots). C=*G. rubens*. Pulses are not organized into chirps, but occasional long intervals do occur (dot). No hybrids retained the *G. armatus* chirp frequency, and the mean hybrid pulse group was shorter than in either parent. Time cal: 100 ms

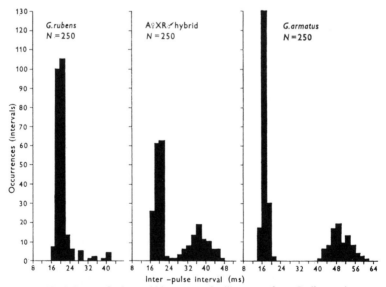

Fig. 7. Inter-pulse interval histograms of calling songs from *Gryllus* species and their F₁ hybrids. The histograms contain the durations of intervals during a period of steady singing. The *G. rubens* song contains one mode, the inter-pulse interval, with occasional longer intervals: the *G. armatus* song contains both inter-pulse and inter-chirp intervals. The hybrid song has an inter-chirp interval which is intermediate in length between the inter-chirp interval of *G. armatus* and the inter-pulse interval of *G. rubens*. The hybrid inter-pulse interval is intermediate in length between the inter-pulse intervals of the parent species.

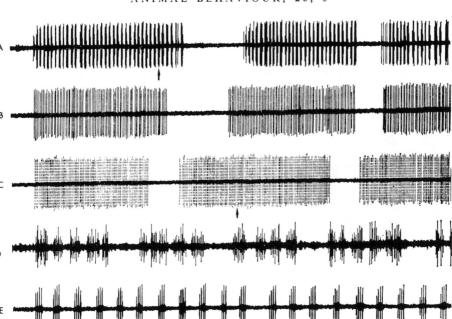

Fig. 8. Calling songs of *Gryllus* species and F₁ hybrids illustrating the pulse train structure. A=*G. armatus*. Arrow indicates a single chirp containing three sound pulses; B=*G. armatus* × *G. rubens* hybrid. Most chirps contain two or three pulses. C=*G. rubens*. Arrow indicates one of the irregularly occurring pauses in the pulse train. *G. armatus* chirp trains and *G. rubens* trills contain similar numbers of pulses and are of comparable length (as are the intervals between them); D=*G. rubens* × *G. campestris* hybrid. Note that (1) chirps are longer than in *G. campestris*, (2) chirp frequency is about the same as *G. campestris*, (3) long intervals occur which are about the same length as the intervals between *G. rubens* pulse trains. E=*G. campestris* chirps are not organized into trains. Time cal: 1 s.

pulse and inter-chirp intervals, but the more irregular the hybrid song, the more overlap there was between these two intervals. No consistent relationship was found between the length of a chirp and the duration of the preceding or the following interval. In summary, the C-type hybrids lost the chirp rate of their chirping parents, and tended to produce shorter pulse trains than either parent. Neurophysiological implications of these data will be considered in the discussion.

Discussion

Genetic System

The most thorough previous studies of cricket song genetics are those of Fulton (1933) on *Nemobius* species, Bigelow (1960) on *Gryllus* species, and Leroy (1966) on *Teleogryllus*

species. The general conclusion of these studies was that cricket songs are controlled by a complex, polygenic system, and our study further supports this. Similar conclusions were drawn from investigations of grasshopper (Perdeck 1958) and *Drosophila* (Ewing 1969) acoustical systems. Some of the more specific conclusions of these studies require additional comment.

Bigelow's (1960) experiments have been cited as evidence for sex-linked control of chirping (Ewing & Manning 1967; Alexander 1968). This impression appears to have been partially produced by the observation that when females of a chirping species (*G. veletis*, *G. assimilis*) are hybridized with males of a non-chirping species (*G. rubens*), the hybrids chirp. This is also true of our hybrids. However, it is not evidence that the genes controlling the character are on the X-chromosome since they could be on any maternal chromosome. The question of whether the particular maternal chromosome involved is the X-chromosome can only be resolved by reciprocal crosses. Bigelow (1960) did perform one reciprocal cross, and reported that both types of hybrid songs included intermediate length trills. Although no comparison of hybrids based on actual measurements was made, he had the impression from listening that they were more similar to their maternal parents, and drew an appropriately tentative conclusion: 'It is possible that some chirp-rate genes are carried on the X-chromosome . . .' (p. 513).

In her study of *Teleogryllus* hybrids, Leroy (1966) reported many of the features which we observed. She found no evidence for sex-linkage, but this would not be expected on the basis of the measurements she made. Leroy (1966), Bigelow (1960), and Cousin (1961) also measured a number of morphological characters associated with the acoustical system, such as the number of teeth in the stridulatory file. They found that these were also under polygenic control.

Hörmann-Heck's (1957) study of *G. campestris* hybrids contains the only report in the literature of single gene effects on cricket acoustical behaviour. It is not clear exactly what aspect of courtship behaviour she was investigating (Alexander 1968), but it was not one of the three well-defined song types of the species. There was no evidence for monofactorial control of calling song in our *G. campestris* hybrids, and it seems worth emphasizing that the normal cricket songs all appear to be under polygenic regulation.

Neuronal Mechanisms

The superimposed rhythms seen in cricket chirping can be explained by two widely divergent hypotheses, or by a variety of compromises between them. One hypothesis states that the two rhythms, underlying pulses and chirps, are produced by two separate resonant-type neuronal oscillators (see Wilson & Waldron 1968, for discussion of oscillators). Each of these would have a strong inherent rhythm, and although both would excite the motor neurones (directly or indirectly), neither would be adequate to drive them alone. The motor neurones would only discharge when inputs from both oscillators arrived simultaneously, and consequently would fire in bursts at the pulse rate which occurs in groups at the chirp rate. The alternative hypothesis states that there is only a single neuronal

oscillator, active at the pulse rhythms. However, the oscillator would be subject to some factor, such as accumulating refractoriness, which would result in termination of the burst sequence; following a silent period, the oscillator would become active again. Therefore, the neuronal mechanism as a whole would behave as a relaxation oscillator. The accumulating 'factor' could reflect a variety of physiological phenomena acting at different sites within the nervous system, and the silent period could involve an active inhibition which, if triggered, might produce an interval of fixed length. However the system worked, it would not involve a second inherent oscillator generating the chirp rhythm. Therefore, both hypotheses agree on the presence of the pulse rhythm oscillator, and the key distinction is the existence of a second, chirp rhythm, oscillator.

Hybrids were produced between a non-chirping species and each of two types of chirping species. In the hybrid song, an intermediate state should be reached between each of the two rhythms of the chirping parent and some parameters of the non-chirping parents' song. The particular parameters selected for this accommodation might reveal whether the two chirping songs are homologous, the type of neuronal mechanisms involved, and which of the above hypotheses is more likely to be correct.

Several lines of evidence indicate that the B_1-type chirp of *G. campestris* does have a second rhythm generating mechanism underlying the chirp, and is more closely allied with the two oscillator hypothesis. First, the chirp rate of *G. campestris* is retained in the hybrid despite the much larger number of pulses/chirp and correspondingly shorter silent period (Fig. 8). This indicates that the chirp rhythm is not due to factors such as accumulating fatigue and a subsequent refractory period in the pulse rhythm generator, but to the effects of a second inherent oscillator. The actual chirp length appears to be an intermediate state between the short *G. campestris* chirps and the long *G. rubens* pulse trains. Although firm conclusions cannot be drawn from a single hybrid individual, additional support for this view is available from other sources.

In the *G. campestris* song, gaps sometimes occur whose duration is equivalent to an inter-chirp interval plus one chirp duration plus a second inter-chirp interval, and the next chirp occurs precisely where it would have been had the song not been interrupted (Bentley

ANIMAL BEHAVIOUR, 20, 3   490

1969a; Kutsch 1969). Again, this suggests that a second inherent oscillator is present for timing chirps, and that for some reason it has failed to trigger one chirp (gaps equivalent to two missing chirps also occur). Such evidence is difficult to explain except on a two oscillator hypothesis. Finally, intracellular recordings in the appropriate ganglion during generation of the calling song pattern have revealed two sets of reciprocally active neurones. One of these fires during the inter-chirp interval, and is inhibited during the chirp, while the other fires during the chirp (but not in phase with the pulses) and is inhibited during the inter-chirp interval (Bentley 1969a). Therefore, neurones whose activity is appropriate for a chirp rhythm oscillator are present in the neuropile. Altogether, there is strong evidence that two neuronal oscillators are involved in the generation of the B_1-type chirping song of G. campestris.

The G. armatus hybrids differ from the G. campestris hybrids in two fundamental ways. First, the G. armatus hybrid chirps do not reach an intermediate length with the G. rubens trains; instead, they tend to be shorter than the pulse trains of either parent. Second, the G. armatus chirp rate is not retained in the hybrid, and in some hybrid individuals the chirp rhythm is lost altogether. This indicates basic differences in the neuronal mechanisms underlying the two types of chirps, and a number of considerations suggest that the crux of this difference is the lack of a second, chirp rhythm, oscillator in C-type chirpers. In particular: (a) The tendency of the hybrids to produce shorter chirps than either parent suggests that the neuronal mechanism underlying chirp production is reaching an intermediate state with respect to the pulse generating mechanism of G. rubens, instead of with the chirp or pulse train controlling mechanism. (b) This is supported by the fact that the inter-chirp interval of the hybrid is almost exactly intermediate between the inter-chirp interval of G. armatus and the inter-pulse interval of G. rubens. (c) Considered in the light of the single oscillator hypothesis, the various hybrid songs can be arranged in a continuous series. In G. armatus, the 'factor' causing the pulse rhythm oscillator to stop periodically would accumulate over three pulses; in G. rubens, it accumulates on a pulse by pulse basis (actually, the presence of occasional pauses within the trills suggests that such a phenomenon may be marginally operative even in G. rubens; Figs 7 and 8). The three-pulse hybrids would be

most similar to G. armatus while the two-pulse hybrids would be closer to G. rubens. In the irregular hybrids, the situation would be at the 'noise level' between the hybrids where accumulation occurs over two pulses, and G. rubens where accumulation occurs only over single pulses; thus, long intervals would be randomly inserted whenever accumulation crossed the threshold for whatever stops the oscillator, producing irregular length chirps (irregularly occurring intervals are particularly difficult to account for with a two oscillator hypothesis). (d) The insertion of standard length inter-chirp intervals in an all-or-none fashion in the irregular songs indicates a triggering effect. Such an effect would account for the unusual degree of hybrid variability, since threshold phenomena are often quite sensitive to slight variations. Also it would explain the lack of correspondence between the number of pulses/chirp and the duration of adjacent intervals. (e) If the G. armatus chirp train is based upon a single pulse rhythm oscillator, it suggests that the C-type chirp evolved from a trilling ancestor by simply interrupting the train, and that the G. rubens pulse trains and G. armatus chirp trains are, as they appear, homologous. The success with which the single oscillator hypothesis accounts for the characteristics of the hybrid song, indicates that it is a much closer approximation of the neuronal mechanism generating the C-type chirps than is the two oscillator model.

In summary, the Gryllus hybridization data demonstrate that the superficially similar chirping patterns of the C-type and B_1-type species are based upon quite different neuronal systems, and provide considerable evidence concerning the nature of the mechanisms involved. A similar procedure could be undertaken with other parameters, and by careful selection of species for hybridization, they could be shifted in different directions. Since there are over 2000 species of crickets, this type of manipulation of the genetic material could be developed into a useful tool for investigating the structure of pattern generating neuronal networks.

Prospects

It now seems clear that a genetic analysis of cricket stridulation will be a serious undertaking requiring a large expenditure of effort and time. The most obvious next steps are attempts to resolve the effects of single genes, either through inducement of single gene mutations, or through selection for single gene replacements. The analysis would be facilitated by the large number of offspring produced per female (to at least 1500 eggs in *T. commodus*; Dr W. Loher, personal communication), by the ease of raising crickets, and by the simplicity of monitoring and measuring the performance of the nervous system. However, the relatively lengthy generation time, with a minimum of about 6 weeks, would be a major obstacle. Nevertheless, there are several reasons for thinking that the undertaking would be worth the effort.

Crickets provide an unusual opportunity for exploiting the small systems approach, which has recently become so successful in neurobiology in an attempt to relate the genetics, development, neurophysiology, and behaviour of a single animal. The crux of the problem lies in finding a workable compromise between features which are advantageous for neurobiology and for genetics, respectively. The preparations whose small numbers of large, individually identifiable neurones have made them so productive for neurophysiology, such as the crayfish, and the molluscs, *Aplysia* and *Tritonia*, are not suitable for genetics because of prohibitive generation times and the difficulty of raising them. On the other hand, the best analysed genetic preparations, such as the mouse (Sidman 1968) and *Drosophila* (Ikeda & Kaplan 1970) are technically difficult for neurophysiology because of the small size or large number of neurones, or both.

The size restriction will become critical if genes manifest themselves by shifting subcellular parameters along a continuum, rather than by gross lesions such as the loss of classes of neurones, single neurones, or classes of synapses. The latter effects may be true of genes operative early in development which are concerned with the gross organization of the nervous system (Sidman 1968). However, it seems unlikely that the specific intercellular connections and set-points of intracellular variables which control motor output patterns and sensory information processing will be determined by single genes. For example, the *Teleogryllus* hybridization studies establish a graded series of patterns between the two parental songs, and therefore, the neurones generating homologous pattern units can be viewed as opposite ends of a continuum. Since the sub-units of the song are continuously variable, it seems likely that the neuronal properties underlying them are also graded, such as rates of adaptation, resting potential,

or synaptic efficacy. The discovery that the units are under polygenic control supports this view, and suggests that pattern diversity has evolved through small, sub-cellular changes rather than gross lesions. Additional evidence is provided by Hotta & Benzer's (1969) work with visual system mutants of *Drosophila*. Evidently, the most probable malfunctions of their two mutants are higher and lower than normal settings of the membrane potential, rather than complete loss of some cellular property or structure. If such sub-cellular variables are a common site of gene action, then preparations with large neurones will be highly advantageous.

Since crickets seem to be in the unusual position of being acceptable from both a genetic and a neurophysiological point of view, we feel that they offer a very promising preparation for analysing the genetic basis of behaviour.

Acknowledgments

This paper is dedicated to Donald M. Wilson, an inspiring individual, scientist and teacher, who was killed in a rafting accident in June 1970 at the age of thirty-six.

We thank Dr F. Huber, Dr T. Walker, Dr T. Hogan and Dr R. D. Alexander for providing crickets. Mrs Alma Raymond supplied indispensable technical aid. Support was provided by Public Health Service research grant NS09074 to D.R.B., and National Institutes of Health Post-doctoral fellowship No. 1-F2-GM-21, 592-01 to R.R.H.

REFERENCES

Alexander, R. D. (1961). Aggressiveness, territoriality, and sexual behavior in field crickets (Orthoptera: Gryllidae). *Behaviour,* **17,** 130–223.

Alexander, R. D. (1962). Evolutionary change in cricket acoustical communication. *Evolution,* **16,** 443–467.

Alexander, R. D. (1968). Arthropods. In: *Animal Communication* (Ed. by T. A. Sebeok). Bloomington: Indiana University Press.

Bentley, D. R. (1969a). Intracellular activity in cricket neurons during generation of song patterns. *Z. vergl. Physiol.,* **62,** 267–283.

Bentley, D. R. (1969b). Intracellular activity in cricket neurons during the generation of behavior patterns *J. Insect Physiol.,* **15,** 677–700.

Bentley, D. R. (1970). A topological map of the locust flight system motor neurons. *J. Insect Physiol.,* **16,** 905–918.

Bentley, D. R. & Hoy, R. R. (1970). Post-embryonic development of adult motor patterns in crickets: a neural analysis. *Science, N.Y.,* **170,** 1409–1411.

Bentley, D. R. & Kutsch, W. (1966). The neuromuscular mechanism of stridulation in crickets (Orthoptera: Gryllidae). *J. exp. Biol.,* **45,** 151–164.

ANIMAL BEHAVIOUR, 20, 3 492

Bigelow, R. S. (1960). Interspecific hybrids and speciation in the genus *Acheta* (Orthoptera: Gryllidae). *Can. J. Zool.*, **38**, 509–524.

Cousin, G. (1961). Analyse des équilibres morphogenetiques des types structuraux spécifiques et hybrides chez quelques Gryllidas. *Société Zool. de France*, **86**, 500–521.

Dorsett, D. A., Willows, A. O. D. & Hoyle, G. (1969). Centrally generated nerve impulse sequences determining swimming behaviour in *Tritonia*. *Nature, Lond.*, **224**, 711.

Ewing, A. W. (1969). The genetic basis of sound production in *Drosophila pseudoobscura* and *D. persimilis*. *Anim. Behav.*, **17**, 555–560.

Ewing, A. W. & Manning, A. (1967). The evolution and genetics of insect behaviour. *Ann. Rev. Entomol.*, **12**, 471–494.

Fulton, B. B. (1933). Inheritance of song in hybrids of two subspecies of *Nemobius fasciatus* (Orthoptera). *Ann. Entomol. Soc. Am.*, **26**, 368–376.

Hörmann-Heck, S. von (1957). Untersuchungen über den Erbgang einiger Verhaltensweisen bei Grillenbastarden. *Z. Tierpsychol.*, **14**, 137–183.

Hotta, Y. & Benzer, S. (1969). Abnormal electroretinograms in visual mutants of *Drosophila*. *Nature, Lond.*, **222**, 354–356.

Huber, F. (1960). Untersuchungen über die Funktion des Zentralnervensystems und insbesondere des Gehirnes bei der Fortbewegung und der Lauterzeugung der Grillen. *Z. verg. Physiol.*, **44**, 60–132.

Ikeda, K. & Kaplan, W. D. (1970). Unilaterally patterned neural activity of gynandromorphs, mosaic for a neurological mutant of *Drosophila melanogaster*. *Proc. nat. Acad. Sci. U.S.A.*, **67**, 1480–1487.

Kutsch, W. (1969). Neuromuskuläre Aktivität bei verschiedenen Verhaltensweisen von drei Grillenarten. *Z. vergl. Physiol.*, **63**, 335–378.

Leroy, Y. (1966). Signaux acoustiques, comportement et systématique de quelque espèces de Gryllides (Orthopteres, Ensiferes). *Bull. Biol. Fr. Belg.*, **100**, 63–134.

Leroy, Y. (1967). Garnitures chromosomiques et systémique des Gryllinae (Insectes Orthopteres). *C.r. hebd Seanc Acad. Sci., Paris*, **264**, 2307–2310.

Otto, D. (1968). Untersuchungen zur nervösen Kontrolle des Grillengesanges. *Zool. Anz. Suppl.*, **31**, 585–592.

Perdeck, A. C. (1958). The isolating value of specific song patterns in two sibling species of grasshoppers (*Chorthippus brunneus* Thumb. and *C. biguttulus* L.). *Behaviour*, **12**, 1–75.

Sidman, R. (1968). Development of interneuronal connections in brains of mutant mice. In: *Physiological and Biochemical Aspects of Nervous Integration* (Ed. by F. D. Carlson). Englewood Cliffs, N.J.: Prentice-Hall Inc.

Stout, J. F. (1970). Response of interneurons of female crickets (*Gryllus campestris*) to the male's calling song. *Am. Zool.*, **10**, 502.

Wilson, D. M. (1968). The nervous control of flight and related behavior. *Adv. Insect Physiol.*, **5**, 289–338.

Wilson, D. M. & Waldron, I. (1968). Models for the generation of the motor output pattern in flying locusts. *Proc. I.E.E.E.*, **56**, 1058–1064.

Zaretsky, M. D. (1971). Patterned response to song in a single central auditory neuron of a cricket. *Nature, Lond.*, **229**, 195–196.

(*Received* 15 *April* 1971; *revised* 29 *July* 1971; *MS. number:* A1153.)

EXPERIMENTAL STUDIES OF MIMICRY IN SOME NORTH AMERICAN BUTTERFLIES

PART I. THE MONARCH, *DANAUS PLEXIPPUS*, AND VICEROY, *LIMENITIS ARCHIPPUS ARCHIPPUS* [1]

JANE VAN ZANDT BROWER

Department of Zoology, Yale University

Received May 25, 1957

INTRODUCTION

When the vast literature on mimicry is considered, the striking lack of experimental evidence for its existence is remarkable. Morgan (1896, 1900), while studying learning in chicks, designed mimicry experiments which showed that they can associate an unpleasant feeding experience with a particular color and pattern. Thereafter, the chicks would also reject modifications ("mimics") of the same color pattern. More recently, Mühlmann (1934) used dyed and treated mealworms as "models," and partially dyed ones as "mimics" in experiments with several species of passerine birds in an attempt to show how similar a model and mimic must be to deceive the birds. The work of Mostler (1935) on wasp-mimicry showed that birds could learn to reject wasps, and then also would mistake like-colored insects for wasps and reject them. Working on the problem of beetle mimicry, Darlington (1938) carried out experiments with captive *Anolis* lizards, which showed that they rejected the model beetle, and also mimic beetles. The laborious efforts of Finn (1895, 1896, 1897a, 1897b), Marshall (1902), Pocock (1911), Swynnerton (1919), Carpenter (1921, 1942), and Jones (1932) are paramount among the many attempts to determine the relative edibility of mimetic butterflies to various vertebrate predators. Both Marshall and

Swynnerton reported some evidence for the existence of mimicry in the complex African butterfly fauna, on the basis of feeding experiments with captive Baboons and captive Rollers (*Coracias garrulus* Linné) respectively, but the major parts of both works were devoted to the more general problem of cryptic *versus* warning coloration in butterflies.

As Poulton (1909) pointed out, in North America there are three relatively simple examples of supposed mimicry in butterflies. The purpose of the present study is to describe an experimental investigation of mimicry in these North American butterflies, and it will be presented as three papers which deal separately with each of these three mimicry complexes. The same methods were employed for each complex, and the three parts were carried out as a continuous investigation over a period of 60 days. The first series of experiments was carried out with *Danaus plexippus* (Linné) and *Limenitis archippus archippus* (Cramer), presumed model and mimic, respectively. *D. Plexippus* is known respectively as the Monarch and *L. a. archippus* is known as the Viceroy; they will hereafter be referred to by their common names. The second series of experiments was a study of the model, *Battus philenor* Linné, and its mimics, *Papilio troilus* Linné, *P. polyxenes* (Fabricius), and the black female form of *P. glaucus* Linné. The third series was a study of the model, *Danaus gilippus berenice* (Cramer), and its mimic, *Limenitis archippus floridensis* (Strecker). The

[1] Submitted to the Department of Zoology, Yale University, in partial fulfillment of the requirements for the degree of Doctor of Philosophy, June 1957.

experiments were designed to investigate the following relationships between models and mimics and a selected bird species, to be used as a caged predator:

1—the reaction to models by individually caged experimental birds;

2—the reaction to mimics by the same experimental birds, after they had had initial laboratory experience with the models;

3—the reaction to mimics by individually caged control birds, which had had no prior laboratory experience with the models.

An experimental study demonstrating these three points could be expected to show the effectiveness of mimicry, within the limitations of the experimental conditions.

MATERIALS AND METHODS

The technical difficulties inherent in these experiments were minimized by the choice of an especially favorable situation in which to undertake the investigation. The Archbold Biological Station, near Lake Placid, Florida, provided excellent laboratory facilities, without which the work would have been impossible. In addition the required butterflies are relatively abundant in this area of Florida.

The Florida Scrub Jay, *Cyanocitta coerulescens coerulescens* (Bosc), was selected as the predator in the mimicry experiments. It is non-migratory and is found in areas in which Sand Pine (*Pinus clausa* (Engelm.) Vasey), scrubby oaks (*Quercus myrtifolia* Willd., *Q. geminata* Small, and *Q. catesbaei* Michx.), Saw Palmetto (*Serenoa repens* (Bartr.) Small), and Dwarf Wax Myrtle (*Myrica pumila* Michx.) predominate (Bent, 1946, and Amadon, 1944). The diet of the Scrub Jay is said to be about 60% animal matter, on the basis of analyses of stomach contents of sixteen birds (Bent, 1946), and includes beetles, butterflies, moths, caterpillars, and grasshoppers. Sprunt (1954) stated that the animal matter is 50% and did not note

the presence of Lepidoptera in the stomach contents. Amadon (1944) observed Scrub Jays feeding in nature at the Archbold Biological Station, primarily on acorns and insects. The male and female Scrub Jays have similar plumage and size, though the males may be slightly brighter in color and larger than the females. Amadon also observed that in his experience a hiccup sound is peculiar to the female Scrub Jay. These birds are particularly easy to trap, and readily adapt to life in a cage (Amadon, 1944, and Bent, 1946).

Eight Florida Scrub Jays were trapped on the Archbold Biological Station property between 14 April and 17 April 1956. Since the young are known to hatch in early to mid April, it may be assumed that these Scrub Jays were all at least one year old. As noted above, the sex of these jays is difficult to determine from plumage, and even from size. However, in an attempt to ascertain the sex of the nine individuals used in the experiments, all but one were weighed prior to release. The following weights were recorded on 17 June 1956:

No. of bird	Weight in gms.	Tentative sex determination
C-1	84.0	male
C-2	66.5	female
C-3	83.0	male
C-4	77.0	female (hiccup)
E-1	88.0	male
E-2	74.0	female (hiccup)
E-3	71.0	female
E-4	—	mate of E-3
E-4A	77.5	female

From a generalization based on weight alone, C–2, E–3, and E–4A (a replacement for E–4, captured 1 June 1956) seemed to be females. E–4, the probable wild mate of E–3, was not weighed prior to release. On the basis of the hiccup sound mentioned by Amadon (1944), C–4 and E–2 could be considered females. C–1, C–3, and E–1 could have been males.

Each bird was confined in a cubic cage thirty inches on a side, with sides and top covered with aluminum-painted wire of

34   JANE VAN ZANDT BROWER

one-half inch square mesh. The cages were arranged in the laboratory in two racks of four each, within two adjacent enclosures 6' wide × 13' 5" deep × 7' 8" high. Each enclosure had a door (3' 6" wide × 7' high) at one end, and both the door and the top of each enclosure were made of one-half inch square mesh. The eight bird cages were identical in every possible respect, as were the two enclosures. The individual bird cages were separated from one another by opaque cardboard, so that at no time could one bird see another. In each enclosure there was a pair of white porcelanized reflectors, each with a 100 watt incandescent bulb, fastened on the wall opposite the rack bearing the cages. These lights were on during all the experiments, and off when experiments were not in progress. Daylight entered the laboratory through a large skylight of northern exposure. The aluminum-painted steel tray floor of each of the eight bird cages was covered with sifted white sand obtained from the area where the birds were caught. Two perches at heights of 8" and 16" were provided in each cage, and each had a water bottle accessible at all times.

The eight birds were given a regular feeding at 5:30 P.M. daily. The standard laboratory diet consisted of approximately: 5 cc scratch (cracked corn and wheat), 1 cc pebbles, 2 cc Purina chow (for hens), 2 pecan meats, 1 peanut, 2 sunflower seeds, and 4 Scarabaeid beetles (*Phyllophaga prununculina* Burmeister). Occasionally, one or more additional items were given simultaneously to all birds. These included hard boiled egg, egg shell, lettuce, chopped meat, bread, and insects of various orders.

Experimental Insects

Insects of the orders Coleoptera, Hemiptera, Orthoptera, and Lepidoptera were attracted by a 15 watt G-E black light stationed on an outside platform of the laboratory. These were collected

each night and stored in a cold room (about 3° C.) for use in the feeding experiments with the Scrub Jays. It is of interest that wild birds, including the Florida Scrub Jay, fed regularly at dawn on the remaining insects that had been attracted to the light the night before. The butterflies used in the experiments were obtained by collecting in the field, or by rearing the adults from eggs laid by confined female butterflies, or from larvae found in the field and reared in the laboratory. Supplies of butterflies were also stored in glassine envelopes in the cold room for use as needed. The particular source of each species will be noted as the experiments are described. No effort was devoted at this time to a study of the proportions of models and mimics in given localities.

Preliminary Experiments

To determine whether or not any erratic behavior existed among the eight Scrub Jays, all of the birds were given identical preliminary tests with several orders of insects. All birds responded similarly to representative Coleoptera (e.g., *Phyllophaga prununculina* and *Dyscinetus morator* Fabricius, Scarabaeidae, *Epicauta tenuis* (Leconte), Meloidae); Orthoptera (e.g., Acrididae, and *Gryllotalpa hexadactyla* Perty, Gryllidae); Hemiptera (e.g., *Lethocerus uhleri* (Montandon), Belostomatidae); Lepidoptera (e.g., *Pholus fasciatus* (Sulzer) and *Xylophanes tersa* (Linné), Sphingidae; *Phoebis sennae eubule* (Linné), Pieridae; *Papilio glaucus* Linné (yellow female and male), *P. palamedes* (Drury), and *P. marcellus* (Cramer), Papilionidae). In each trial for each bird, a stopwatch was used to record the time in seconds for the bird to seize an insect, and if the bird then ate the insect, this time was also recorded. Similarity among all the birds in their reaction times in seizing the insects, and also in their responses to the insects after seizure, indicated that they were a reasonably uniform group with which to conduct experiments on mimicry. The set of mean reaction times, based on twelve trials for each of the eight birds, gave an observed standard deviation of 1.34 seconds, with a mean value of 3.31 seconds and a range of from 1.5 to 5.0. A table in David, Hartley, and Pearson (1954) showed that for eight observations the ratio of range to standard deviation must be 3.399 at the 5% significance level; the ratio of range to standard deviation here is 2.61 so that there is no significant difference among the eight observations, i.e., the birds' mean reaction times in seizing insects.

During the preliminary trials, two readily available butterflies, *Papilio glaucus* (yellow female and male) and *P. palamedes* which are not known to be involved in mimicry either as models or as mimics, were eaten in every case by all eight birds. It was therefore decided that these two species, obtained as living adults by local field collecting, would be used as the edible control butterflies throughout the course of the experiments. Hereafter they will be referred to as "non-mimetic butterflies." In all of the experiments the butterflies were immobilized by pinching the thorax before they were placed with a pair of forceps on the floor of a bird cage. For uniformity and stability all butterflies were presented lying on their sides, with wings together dorsally. Therefore in the experiments with models and mimics only mimicry in the characters of the underside of the wings was being tested. The immobilization of the butterflies and their sideways placement in the cages were arbitrarily decided upon to eliminate the following variables: (1) a mobile butterfly might be more or less difficult for a bird to catch depending on its amount of activity and its location in a cage; (2) mimicry might be more or less effective depending on whether a bird saw the upper or lower wing surfaces of a temporarily resting butterfly, or both surfaces of a flapping butterfly. The role of behavior of models

and mimics in increasing or diminishing the effectiveness of mimicry would be interesting in itself (e.g., it is known that two forms of the African butterfly, *Hypolimnas dubius* de Beauvais, which mimick two closely allied *Amauris* models, differ from one another in behavior as well as color and pattern, as discussed by Ford, 1953), but the present experiments do not attempt to investigate this aspect of the problem. Every bird was allowed two minutes to react to each butterfly presented. At two minutes an uneaten butterfly was removed. Whenever possible, living butterflies were used, but the data indicated that the reaction to any given species of butterfly was the same whether live or dead specimens were presented to the birds. In a few trials, dead and rather dried specimens, of non-mimetic butterflies only, had to be used, and rarely a slightly moldy non-mimetic butterfly was used in the absence of fresh material. Even these butterflies were quite acceptable as food to the jays. If a living model or mimic remained untouched throughout a trial, it was used again in successive trials. If a model or mimic was in any way torn or injured by a bird, however, it was not used again.

THE MIMICRY EXPERIMENTS

Of the eight Scrub Jays, four were randomly selected as experimental birds; the other four were control birds. The mimicry experiments were designed so that each trial consisted of giving a bird a pair of butterflies in succession, one non-mimetic butterfly and one model or one mimic. In each trial, the order of presentation of the non-mimetic butterfly and the model or mimic was determined by a random number table, with the use of consecutive digits in a different vertical column for each bird. If a digit read 0–4, a non-mimetic butterfly was given first, followed by a model (or mimic), but if a digit read 5–9, a model (or mimic) was given first, followed by a non-mimetic butterfly. By this method

36   JANE VAN ZANDT BROWER

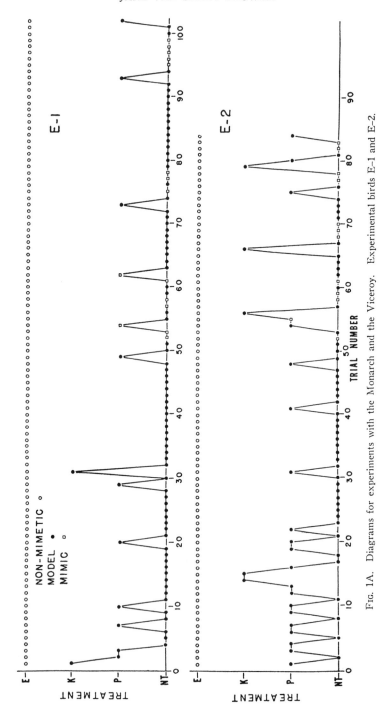

FIG. 1A.   Diagrams for experiments with the Monarch and the Viceroy.   Experimental birds E-1 and E-2.

EXPERIMENTAL STUDIES OF MIMICRY   37

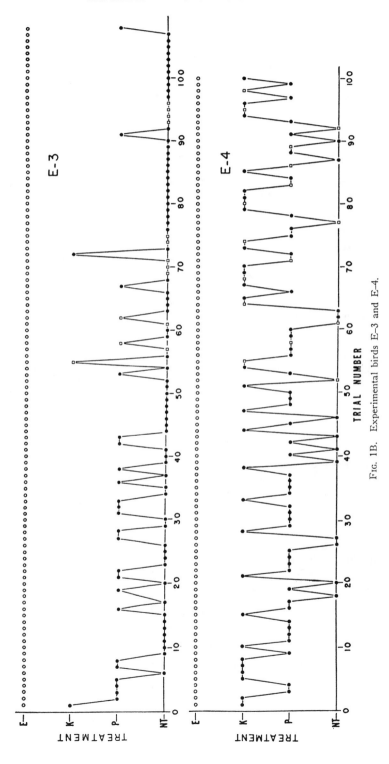

Fig. 1B. Experimental birds E-3 and E-4.

38   JANE VAN ZANDT BROWER

runs of more than two of each kind of butterfly were eliminated, and the following consecutive trials were possible (model or mimic $= M$; non-mimetic butterfly $= N$): (1a) M,N; (1b) M,N. (2a) N,M; (2b) N,M. (3a) M,N; (3b) N,M. (4a) N,M; (4b) M,N. Use of these randomized pairs avoided subjective repetition of any sequence in the presentation of the butterflies to the birds, and prevented the birds from reacting to the butterflies on the basis of an expected order of sequences. After each bird was given the first butterfly in a trial, it did not receive the second until each of the other birds had received its first butterfly. Thus for each bird the time interval between trial 1a and 1b was about the same as it was between trial 1b and 2a. The design of the experiments did not eliminate the possibility that the birds might predict the second butterfly given in each trial on the basis of the first one given, but the data suggested that this was not the case.

The experimental birds were given models and non-mimetic butterflies, randomly in pairs as described, until the birds had developed a relatively reliable pattern of behavior toward the model species. At this time, the mimic was substituted for the model in one trial. Subsequent substitutions of the mimic for the model depended on the particular pattern of behavior of the individual bird, and will be discussed when the data are considered.

The control birds were given mimics and non-mimetic butterflies randomly in pairs, by the same procedure used for the experimental birds. The difference between the two sets of birds was that only the experimental birds were given experience with models.

Experiments with the Monarch and Viceroy

The classic example of mimicry among North American butterflies is the Monarch-Viceroy situation, cited by Walsh and Riley (1869). In spite of a small amount of evidence to the contrary, the Monarch generally has been considered relatively unpalatable to vertebrate predators. Poulton (1909) commented upon the relative edibility of the Viceroy. He was of the opinion that the Viceroy is a Müllerian mimic, unpalatable like its model, basing his views on the supposed origin of the Viceroy from a warningly colored ancestor, like *Limenitis arthemis* (Drury), and on the assumption that warningly colored butterflies are unpalatable to predators. Walsh and Riley (1869) assumed that the Viceroy is a Batesian mimic, edible to vertebrate predators. It seemed fitting to begin the mimicry experiments with this long-debated problem of the Monarch and the Viceroy.

Geographic Distribution and Sources of Butterflies

The range of the Monarch covers North America, northeastward to central Ontario (Klots, 1951). The Viceroy is more limited in its distribution, ranging from central Canada south to South Carolina, Georgia, and Louisiana. The species possibly forms a north-south cline with *Limenitis archippus floridensis*, the mimic of the southern danaid, *Danaus gilippus berenice*. The latter mimic and model both occur in Florida. The Viceroy (*L. archippus archippus*) does not reach central Florida, which makes it certain that the Florida Scrub Jays used in these experiments had never had experience with the Viceroy in nature. However, the birds may have attempted to eat Monarchs, which are present at the Station in winter and spring.

The source of Viceroys for the feeding experiments was over-wintering hibernacula collected in Litchfield Co., Conn., in February 1956. The larvae from these were reared to the adult stage at the Archbold Biological Station. The Monarchs were obtained locally in Florida by collecting living adults and by rearing adults from eggs obtained from females in outdoor insect cages, and also from larvae collected in the field. Both male and female butterflies were used in the experiments, but because female Viceroys resembled Monarchs more closely in size, females of this species were used in most trials. It should be noted that the relative size of models and mimics in nature is probably of little significance compared to their similarities of color and pattern, and that the effect of size on mimicry could assume a disproportionate importance under experimental conditions. The sex of a butterfly *per se* did not affect the reaction of any bird. A total of approximately 1,000 butterflies was used.

RESULTS

The basic data for the feeding experiments with Monarchs and Viceroys are presented in figures 1A. and 1B., diagrams E–1, E–2, E–3, and E–4, and in figure 2, diagrams C–1, C–2, C–3, and C–4. The four categories of reaction to the butterflies by the birds were selected on the basis of observed behavior of the birds. "NT" means that the butterfly was not touched by the bird during the two minute test period. The NT category sometimes was accompanied by characteristic behavior on the part of a bird which will be discussed below. The "P" category means that the bird pecked the butterfly but not to the point of any injury which in nature would impair the insect's ability to reproduce. Pecking could imply tasting, but the contact of the beak of the bird with the external surface of the butterfly is what is meant here. The "K" (killed) category is used when a bird had torn the body of the butterfly so that in the wild it would have died, or could not have reproduced, but the bird left the torn specimen uneaten. "E" refers to the fact that the bird had eaten the butterfly, leaving only the wings and legs, which were picked off and discarded.

From an examination of the diagrams of the experiments (see figs. 1A and 1B;

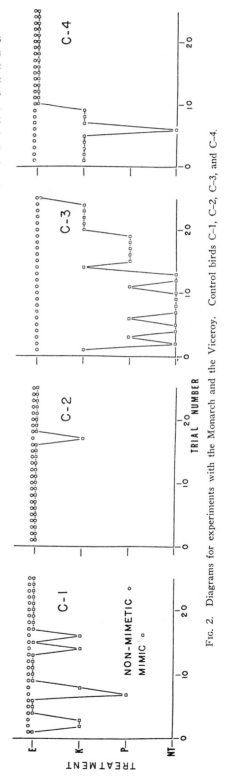

FIG. 2. Diagrams for experiments with the Monarch and the Viceroy. Control birds C–1, C–2, C–3, and C–4.

40   JANE VAN ZANDT BROWER

fig. 2), it will be evident that in all trials for both experimental and control birds, the non-mimetic butterfly (white circle) of each pair of butterflies in any given trial was eaten. In only four trials in this entire investigation, a bird was reluctant to eat a non-mimetic butterfly. When such hesitation occurred, the trials were stopped for a few minutes, and were not continued again until the bird readily accepted and ate the non-mimetic butterfly. Therefore every recorded trial indicates that a non-mimetic butterfly was eaten by a bird. The model, the Monarch, (black circle) was given in couplet with a non-mimetic butterfly to the four experimental birds for at least 51 trials prior to the substitution of the mimic, Viceroy, (white square) for the Monarch. The Monarch was not eaten by the four experimental birds in any trial. The diagrams for E–1, E–2, and E–3 show that during the initial trials, the Monarchs were pecked, killed, and not touched, and that the "not touched" category became more frequent as the birds learned that the Monarch was not palatable. All four birds reacted in their initial trial with the Monarch by killing or pecking the butterfly. This suggests either that the four experimental birds had not had experience with Monarchs in the field prior to capture, or, if they had experienced Monarchs in the wild, they had not remembered the unpalatability of Monarchs on sight alone under the cage conditions. That the lesson of unpalatability was never finally learned in the course of the experiments is seen in the repeated lapses of E–1, E–2, and E–3 into pecking or killing a Monarch after several trials during which the Monarch was not touched. On the basis of sight alone, E–4 was outstanding in its failure to remember for more than two successive trials that the Monarch was unpalatable.

The question of when to substitute Viceroys for Monarchs was complicated by the fact that the birds showed these periodic lapses when a model was again

pecked or killed. In order to minimize the chance that a Viceroy would fall on the observed "error" pattern, i.e., the killing or pecking of the model, the presentation of each mimic was determined by the individual error pattern of each bird. A Viceroy was substituted for a Monarch soon after a bird had made an error on a Monarch. This meant that a bird had just been exposed to the unpleasant qualities of a Monarch before being offered a Viceroy. Because experience with each bird was the only guide, the technique of substitution of the mimic necessarily improved with time. For E–1, from trial 63 on, for E–2, from trial 57 on, and for E–3, from trial 63 on, no Viceroy was given unless it followed (1) a trial in which a Monarch was pecked or killed, and (2) a trial in which a Monarch was not touched, in that order. With the design of the experiments adjusted so that any pattern of natural pecking and/or killing lapses would be likely to fall on the model, as can be seen on the diagrams, the mimic was not touched.

Although E–4 did not learn to refuse the Monarch on sight, its treatment of the Viceroy shows no discrimination of the latter from the Monarch. The Viceroy was not eaten by any of the four experimental birds, and was not even killed by E–1 and E–2.

The four control birds, C–1, C–2, C–3, and C–4, were given trials with the Viceroy (mimic) and non-mimetic butterflies. Because the number of Viceroys available was limited, each control bird could be offered only 25. After these 25 Viceroys had been used, each control bird was given only the non-mimetic butterfly in each trial for the total number of trials shown in the "no. seize" column for each control bird on table 2. Each of the control birds ate every non-mimetic butterfly offered, although only the 25 which correspond to the Viceroy trials are shown in fig. 2, diagrams C–1, C–2, C–3, and C–4. The reaction to the Viceroy by C–1 indicates that it was palatable to that

EXPERIMENTAL STUDIES OF MIMICRY 41

TABLE 1. *Reactions of the experimental and control birds to Viceroys*

A. Comparison of no. "not touched" *vs.* no. "pecked—killed—eaten"

	Control birds				Experimental birds			
	C-1	C-2	C-3	C-4	E-1	E-2	E-3	E-4
NT	0	0	9	1	14	12	12	4
P—K—E	25	25	16	24	2	1	3	12

B. Comparison of no. "not touched—pecked—killed" *vs.* no. "eaten"

	Control birds				Experimental birds			
	C-1	C-2	C-3	C-4	E-1	E-2	E-3	E-4
NT—P—K	6	1	24	9	16	13	15	16
E	19	24	1	16	0	0	0	0

bird in most of the trials (diagram for C-1). C-2 ate the Viceroy in all but one trial as the diagram shows. In the case of C-3, most of the Viceroys were not touched, or were pecked or killed, and in only one trial, the last, was a Viceroy eaten. In the first nine trials, C-4 killed Viceroys, and once did not touch a Viceroy, but from trial 10 on, it ate all Viceroys.

STATISTICAL ANALYSIS OF THE DATA

Table 1,A shows a comparison of the treatment of the Viceroy for the category "NT" compared with the lumped categories "P-K-E" for all eight birds. On the hypothesis that the experimental birds and the control birds reacted to the Viceroy in the same way, a chi-squared test gave a value of 80.81, d.f. $= 7$, with the Yates correction factor, which was used in all chi-squared tests (table 1,A). The probability of obtaining these results, or worse ones, given this hypothesis, is less than .001. It may therefore be said that the two groups of birds did not react to the Viceroy in the same way, and that the control birds pecked, killed, and ate Viceroys significantly more than the experimental birds did. Similarly, when the data were analyzed to compare the categories "NT-P-K" with "E," the chi-squared value was 116.81, d.f. $=7$, P less than .001 (table 1,B). The control birds ate significantly more Viceroys than the experimental birds did; in fact, no Viceroys were eaten by the latter group.

In addition to an analysis of the response to the Viceroys by the experimental and control bird groups, con-

TABLE 2. *Reaction times of jays in seizing butterflies (in seconds to the nearest whole second)*

	Monarch					Viceroy					Non-mimetic butterflies				
Birds	Mean time seize	Min. time	Max. time	No. seize	No. NT	Mean time seize	Min. time	Max. time	No. seize	No. NT	Mean time seize	Min. time	Max. time	No. seize	No. NT
C-1	—	—	—	—	—	8	1	100	25	0	3	1	14	92	0
C-2	—	—	—	—	—	5	2	11	25	0	5	2	25	95	0
C-3	—	—	—	—	—	10	1	36	16	9	1	1	3	101	0
C-4	—	—	—	—	—	11	2	46	24	1	6	1	76	94	0
E-1	9	1	31	12	76	14	6	22	2	14	1	1	3	102	0
E-2	27	1	102	25	48	20	20	20	1	12	1	1	2	84	0
E-3	36	1	119	24	69	22	4	41	3	12	2	1	16	108	0
E-4	12	1	88	72	12	30	2	112	12	4	3	1	10	100	0

sideration was also given to the reaction times of the birds in seizing the butterflies. The measure of time which has particular application for mimicry is the interval between the time a bird first saw a butterfly placed in its cage to the time when the bird first pecked or seized the butterfly. The mean time in seconds that each bird took to seize the Monarch, the Viceroy, and the non-mimetic butterflies used in the present experiments was calculated, and is shown on table 2. To indicate the range of seconds for each bird in seizing the butterflies of each group, the minimum and maximum times in seconds are also given. Since the experimental birds did not touch many Monarchs and Viceroys, the number of butterflies seized (from which the means were calculated) and also the number that were not touched are given in separate columns.

As table 2 shows, the time which the experimental birds took to seize Monarchs and Viceroys was considerably greater than that for non-mimetic butterflies, which suggests that the birds did not discriminate the Viceroy from the Monarch. If a similar hesitation in seizing a Monarch and its mimic were shown by an experienced bird in nature, escape of the butterfly from predation would be likely because of the time factor alone.

These data also suggest a correlation with the relative palatability of the various butterflies. The mean-times-seize of all the birds for the palatable non-mimetic butterflies are less than those of the control birds for the Viceroy.

BEHAVIOR OF THE SCRUB JAYS

In the description of the categories of reaction to the butterflies by the birds, it was noted that the "NT" category was sometimes accompanied by characteristic behavior on the part of a bird. This consisted of the bird ruffling its feathers and shaking its body, and/or by what I termed a "hop routine." The hop routine was a repetition of a series of hop move-

ments by a bird, from perch, to side of cage, to floor, to perch, which often persisted for the full two minutes of a test, if an unpalatable model or its mimic had been presented.

A detailed examination of the occurrence of the feather-ruffling reaction in the individual birds shows a definite correlation between the reaction and the unacceptability of a given butterfly species to a bird. For the experimental birds, the feather-ruffling often occurred when either a Monarch or a Viceroy was presented. Thus E–1 gave this reaction in 77 out of 88 trials (88%) with the Monarch, and in 13 out of 16 trials (81%) with the Viceroy. The reaction was given by E–2 in 43 out of 73 trials (59%) with the Monarch, and in 9 out of 13 trials (69%) with the Viceroy. E–3 gave the reaction in 42 out of 94 trials (45%) with the Monarch, and in 5 out of 15 trials (33%) with the Viceroy. E–4 showed this reaction in 6 out of 84 trials (7%) with the Monarch, and in 2 out of 16 trials (13%) with the Viceroy. The reaction was never given by any of the birds when a non-mimetic butterfly (palatable) was presented.

The control birds also gave this same reaction when the Viceroy was presented, although on the whole less frequently than the experimental birds did. C–1 gave the reaction for 3 out of 25 Viceroys (12%); C–2 for 0 out of 25 Viceroys (0%); C–3 for 19 out of 25 Viceroys (76%); C–4 for 3 out of 25 Viceroys (12%). As might have been expected on the basis of the experimental birds, C–3, the bird that reacted to the Viceroy as an unpalatable butterfly during all but the last trial, also gave a high proportion of feather-ruffling reactions to the Viceroy.

LEARNING AND MEMORY IN THE SCRUB JAYS

Although a study of learning in the Florida Scrub Jays was not the primary purpose of this inquiry, a few words should be devoted to this aspect of the experiments. De Ruiter (1952) described, in his work with Jays (*Garrulus glandarius* Linné) and Chaffinches (*Fringilla coelebs* Linné), how the birds became used to the inedibility of small sticks, and refused to peck at them. After experience with the twigs, the birds also refused to peck at caterpillars bearing close similarity to the twigs. Only if a caterpillar were pecked by accident would the birds, thus rewarded, continue to hunt for caterpillars. If they found only sticks again, they soon ceased pecking. This behavior has been termed an example of habituation by Thorpe (1956). In so far as the Scrub Jays found the Monarch butterfly inedible in a series of 50 trials, one might have expected habituation to develop in the course of the experiments. Indeed, the theory of mimicry suggested by Müller (1879, 1881) assumes that young birds learn by experience which butterflies are inedible and remember what they have learned. However, the repeated lapses of the Scrub Jays into pecking the Monarch show a persistent trial-and-error learning pattern. Swynnerton (1915) in referring to the mistaken attacks of his caged birds on unpalatable insects noted: ". . . no bird could be too old or too experienced to make continual mistakes of this kind." Similar "forgetfulness" is reported by Sadovnikova (1923). She found that in maze learning in passerine birds, individual birds would occasionally show periods of "forgetting" when they entered blind alleys which they had previously learned to avoid. The pattern of apparent forgetfulness shown by the Scrub Jays thus may be of a general nature, or may be associated with cage experiments. The experimental behavior can not be extended to imply that such short-term forgetfulness (day to day) exists among Scrub Jays in the wild; this is not known.

In spite of the short-term lapses of memory seen in the Scrub Jays, the data did show that the birds could remember to reject a Monarch and a Viceroy on sight alone after a period of about two weeks. E–1 rejected a Monarch and a Viceroy after 16 days of 82 trials with unrelated butterflies of the *Battus* mimicry complex; E–2, after 19 days of 84 such interim trials; E–3, after 15 days of 82 such interim trials. Tests of memory in the Scrub Jays after longer periods had elapsed were not possible in these experiments.

Discussion

The first point to be considered is that of the reaction to the model, the Monarch, by the four experimental birds. As the data showed, the Monarch was unacceptable on sight alone at some time to all the experimental birds. The theory that models are unpalatable, and that their color pattern is a sign of unpalatability, is thus supported. Following a period of learning during which the experimental birds were given the Monarch, these birds did not even touch the Viceroy in many trials, and in no case did the birds eat a Viceroy. That the Viceroy is not as inherently unpalatable as the Monarch was shown by the control birds, and this will be discussed below. The highly significant difference in the treatment of the Viceroy by the two sets of birds is attributed to the experience of the experimental birds with the unpalatable Monarch, and to the subsequent association by these birds of the color pattern of the Viceroy with that of the Monarch. Under the conditions of the experiment, it has been shown that mimicry in the case of the Monarch and Viceroy is effective.

The status of the edibility of the Viceroy demands further consideration. According to Bates's idea (1862), the mimic was presumed to be a butterfly edible to vertebrate predators, especially to birds. Müller (1879), in dealing with a particular complex of closely related species of mimetic butterflies in South America (Heliconiidae) which had puzzled Bates, resolved the problem by suggesting that the entire complex of "models" and "mimics" was unpalatable to predators. The pooling of their numbers thus reduced the number of losses per species to the learning of inexperienced birds. In the present experiments, the status of the Viceroy as a Batesian or Müllerian mimic is not entirely evident, and it brings to mind Swynnerton's repeated emphasis on the range of edibility of butterflies to birds (1919). That the experimental birds in no instance regarded the Viceroys as edible is clear. The reaction to the Viceroy by E–1 and E–2 (fig. 1A) shows that those birds never killed that butterfly, although E–2 killed three Monarchs after the commencement of Viceroy trials. E–1 and E–2 were not discriminating the Viceroy from the Monarch as something different and edible. E–3 (fig. 1B) in trial 55 apparently did discriminate between its first Viceroy and Monarchs. Subsequent trials indicate either that the bird failed to maintain its discriminative ability, perhaps in part due to the design of the trials plus the adjustment for natural error, or that the bird found the Viceroy unpalatable, and established a general reaction to both Monarch and Viceroy as inedible. The fluctuating responses of E–4 (fig. 1B) to the Monarch and Viceroy show that the bird failed to remember on sight alone that the Monarch was inedible. The treatment of the Monarch and Viceroy was so consistent that it seems likely that this bird was not discriminating between the two, and perhaps found them both unpalatable. At any rate the Viceroy was not eaten in preference to the Monarch by E–4.

All four experimental birds showed identical, characteristic behavior, discussed above, to both the Monarch and Viceroy, which supports the idea that there was no discrimination of one from another. However, although the Monarch and Viceroy were treated in this generalized manner, a trial can be cited in which a bird clearly distinguished one Viceroy from Monarchs on the basis of a slight difference in color. For E–3, trial 62, there is reason to believe that the pecking of this Viceroy was due to its coloration which was somewhat more brown than previous or succeeding ones. The butterfly had been presented inadvertently, and the instant curiosity and pecking reaction of the bird to the slight color discrepancy was noted. The problem of discrimination *vs.* generalization of models and mimics by the Scrub Jays will be discussed in more detail in Part III of this series of papers. For the present, it can be said that although the data for the experimental birds show that the Viceroy was rejected by them, they do not clearly show why. For E–1 and E–2, the rejection of the Viceroy seems to be based on the fact that the birds learned that the Monarch was unpalatable, and could not discriminate the Viceroy from the Monarch. The data for E–3 and E–4 could be interpreted in the same way as for E–1 and E–2, or alternatively, that the Monarch and Viceroy were both found by experience to be unpalatable and were recognized as one or as separate color patterns, but in either case rejected.

The control birds gave further indication of the inherent edibility of the Viceroy. As the diagrams showed (fig. 2), the Viceroy was not as edible as the non-mimetic butterflies for C–3, and in part this was true for C–1 and C–4. C–2 regarded the Viceroy as an edible species. C–4 appears to have learned to eat the Viceroy. The reaction of C–3 to the Viceroy is comparable to the initial learning pattern of E–1 for the Monarch, but the Viceroy was more acceptable to C–3 toward the end of the experiment. It is possible that the control birds, particularly C–3, had some pre-capture experience with the Monarch, and associated the Viceroy with the Monarch, although no bird refused to peck its first Viceroy. The geographic distribution of the Viceroy precludes any experience with that exact color pattern in nature by these jays.

The learning patterns of E–3 and E–4 and of C–1, C–3, and C–4 could be interpreted to lend support to the hypothesis that the Viceroy is a classical Müllerian mimic. Additional evidence for this view is suggested by the reaction times of the control birds in seizing Viceroys, which were longer than those for seizing non-mimetic butterflies. However, rather than try to place the Viceroy in a rigid, all-or-none category which implies more than the data show, the Viceroy is here considered more edible than its model, the Monarch, but initially less edible (except to C–2) than the non-mimetic butterflies used in these experiments.

Summary

1) This paper is the first in a series of three which present experimental studies of mimicry in some North American butterflies.

2) The experiments were designed to study the effectiveness of mimicry in these butterflies with the use of eight Florida Scrub Jays (*Cyanocitta coerulescens coerulescens*) as caged predators. The butterflies were immobilized and their wings were folded together dorsally, so that only mimicry in the characters of the underside of the wings was being tested in these experiments.

3) The present experiments tested mimicry in the classic example of the Monarch (*Danaus plexippus*) and Viceroy (*Limenitis archippus archippus*).

4) The results of these experiments show that the non-mimetic butterflies, used in couplet with each model or mimic, were eaten in every trial by all birds.

5) The four experimental birds were given numerous trials with the model, the Monarch. The Monarch was not eaten by these birds in any trial, and in many trials was not touched, after initial learning had taken place.

6) After the experimental birds had been given more than 50 trials with the Monarch, the Viceroy was substituted for the Monarch at intervals. The Viceroy was never eaten by the four experimental birds, and in many trials was not even touched.

7) Characteristic behavior shown by the four experimental birds after the presentation of both Monarchs and Viceroys indicated no discrimination between the two species of butterflies.

8) The four control birds had no prior laboratory experience with the Monarch, and these birds ate the Viceroy in many trials.

9) A statistical analysis of the reaction to the Viceroy by the experimental and control birds indicated that the two groups did not react to the Viceroy in the same way. The difference in response is attributed to the prior laboratory experience of the experimental birds with the Monarch. The color pattern of the Viceroy was apparently associated with the complete inedibility and similar color pattern of the Monarch. Under the conditions of the experiment, mimicry has been shown to be effective.

10) The data indicate that the Viceroy is more edible than the Monarch, but less edible than the non-mimetic butterflies used in the experiments. In addition, the control birds took longer to seize Viceroys, on the average, than they took to seize the non-mimetic butterflies. Therefore the Viceroy is not termed either a Batesian or a Müllerian mimic in the classical sense.

11) Learning behavior and memory in the Scrub Jays were considered briefly. The records showed that three of the four experimental birds remembered to reject a Monarch and a Viceroy on sight alone, after a period of over two weeks had elapsed since their last experience with these butterflies.

Acknowledgments

I want to express my appreciation to Dr. C. L. Remington, under whom this work was carried out, for suggesting the problem of mimicry to me and for arranging my facilities at the Archbold Bio-

46 JANE VAN ZANDT BROWER

logical Station. I am indebted to Mr. Richard Archbold, Director of the Archbold Biological Station, where the experiments were conducted, for his generous cooperation and material assistance, and I wish to extend my thanks to the members of his staff for their help.

The suggestions of several persons were offered concerning the design and analysis of these experiments, and thanks are given to: Drs. C. I. Bliss, L. P. Brower, E. S. Deevey, Prof. G. E. Hutchinson, Drs. R. MacArthur, C. L. Remington, M. T. M. Rizki, and H. Seal. I am grateful to Drs. P. F. Bellinger, J. L. Brooks, E. B. Ford, F. R. S., Prof. G. E. Hutchinson, and Drs. C. L. Remington and P. M. Sheppard for reading and criticising the manuscript.

Dr. S. D. Ripley offered useful advice on the capture and care of the birds, and Drs. M. W. Sanderson and F. G. Werner identified the beetles.

Mr. S. A. Hessel very kindly helped with locating and collecting the *Viceroy hibernacula*.

The inestimable assistance of my husband, Lincoln P. Brower, in many phases of this work is gratefully acknowledged.

The experiments were conducted with the support of the Fanny Bullock Workman Scholarship offered by Wellesley College, and of a terminal, pre-doctoral National Science Foundation Fellowship. Summer field work in the preliminary stages of this investigation was supported by a Sigma Xi RESA Grant, and by a grant from the Higgins Fund of Yale University.

LITERATURE CITED

AMADON, DEAN. 1944. A preliminary life history study of the Florida Jay, *Cyanocitta c. coerulescens*. Amer. Mus. Nov. No. 1252.

BATES, H. W. 1862. Contributions to an insect fauna of the Amazon Valley. Lepidoptera: Heliconidae. Trans. Linn. Soc. Lond., 23: 495–566, 2 pls.

BENT, A. C. 1946. Life Histories of North American Jays, Crows, and Titmice. U. S. Nat. Mus. Bull. 191: 77–88.

CARPENTER, G. D. H. 1921. Experiments on the relative edibility of insects, with special reference to their coloration. Trans. Ent. Soc. Lond., 1921: 1–105.

——. 1942. Observations and experiments in Africa by the late C. M. F. Swynnerton on wild birds eating butterflies and the preferences shown. Proc. Linn. Soc. Lond., 154: 10–46.

DARLINGTON, P. J., JR. 1938. Experiments on mimicry in Cuba, with suggestions for further study. Trans. Roy. Ent. Soc. Lond., 87: 681–695, 1 pl.

DAVID, H. A., H. O. HARTLEY, AND E. S. PEARSON. 1954. The distribution of the ratio, in a single normal sample, of range to standard deviation. Biometrika, 41: 482–493.

DE RUITER, L. 1952. Some experiments on the camouflage of stick caterpillars. Behaviour, 4: 222–232.

FINN, FRANK. 1895. Contributions to the theory of warning colours and mimicry. I. Experiments with a Babbler (*Crateropus canorus*). Journ. Asiatic Soc. Bengal, 64: 344–356.

——. 1896. II. Experiments with a lizard *Calotes versicolor*). Journ. Asiatic Soc. Bengal, 65: 42–48.

——. 1897a. III. Experiments with a tupaia and a frog. Journ. Asiatic Soc. Bengal, 66: 528–533.

——. 1897b. IV. Experiments with various birds. Journ. Asiatic Soc. Bengal, 66: 613–668.

FORD, E. B. 1953. The genetics of polymorphism in the Lepidoptera. Advances in Genetics, 5: 43–87.

JONES, F. M. 1932. Insect coloration and the relative acceptability of insects to birds. Trans. Ent. Soc. Lond., 80: 345–386, 11 pls.

KLOTS, A. B. 1951. A Field Guide to the Butterflies. Houghton Mifflin Co., Boston.

LINDLEY, D. V., AND J. C. P. MILLER. 1953. Cambridge Elementary Statistical Tables. Cambridge Press.

MARSHALL, G. A. K. 1902. Five years' observations and experiments (1896–1901) on the bionomics of South African insects, chiefly . . . mimicry and warning colours. Trans. Ent. Soc. Lond., 1902: 287–584, 15 pls.

MORGAN, L. P. 1896. Habit and Instinct. Edward Arnold, London.

——. 1900. Animal Behavior. Edward Arnold, London.

MOSTLER, G. 1935. Beobachtungen zur Frage der Wespen-Mimikry. Zs. Morph. Oekol. Tiere, 29: 381–455.

MÜHLMANN, H. 1934. In Modellversuch künstlich erzeugte Mimikry und ihre Bedeutung für den "Nachahmer." Zs. Morph. Oekol. Tiere, 28: 259–296.

MÜLLER, FRITZ. 1879. (Trans. by R. Meldola.) *Ituna* and *Thyridia*; a remarkable case of mimicry in butterflies. Proc. Ent. Soc. Lond., **1879**: xx–xxix.

——. 1881. Bemerkenswerthe Fälle erworbener Aehnlichkeit bei Schmetterlingen. Kosmos. (Reviewed by A. R. Wallace, 1882. Nature, **26**: 86.)

POCOCK, R. I. 1911. On the palatability of some British insects, with notes on the significance of mimetic resemblances. Notes by E. B. Poulton. Proc. Zool. Soc. Lond., **1911**: 809–868.

POULTON, E. B. 1909. Mimicry in butterflies of North America. Ann. Ent. Soc. Amer., **2**: 203–242.

SADOVNIKOVA, MARY P. 1923. The study of the behavior of birds in the maze. Journ. Comp. Psych., **3**: 123–139.

SPRUNT, ALEXANDER, JR. 1954. Florida Bird Life. Coward-McCann Inc., New York, and Nat. Audubon Soc.

SWYNNERTON, C. M. F. 1915. A brief preliminary statement of a few of the results of 5 years' special testing of the theories of mimicry. Proc. Ent. Soc. Lond. **1915**: 32–44 (quote p. 42).

——. 1919. Experiments and observations bearing on the explanation of form and colouring, 1908–1913, Africa. Journ. Linn. Soc. Zool., **33**: 203–385.

THORPE, W. H. 1956. Learning and Instinct in Animals. Methuen and Co. Ltd., London.

WALSH, B. D., AND C. V. RILEY. 1869. Imitative butterflies. Amer. Entomologist, **1**: 189–193, 3 figs.

Relation of cue to consequence in avoidance learning

JOHN GARCIA AND ROBERT A. KOELLING

HARVARD MEDICAL SCHOOL AND MASSACHUSETTS GENERAL HOSPITAL

An audiovisual stimulus was made contingent upon the rat's licking at the water spout, thus making it analogous with a gustatory stimulus. When the audiovisual stimulus and the gustatory stimulus were paired with electric shock the avoidance reactions transferred to the audiovisual stimulus, but not the gustatory stimulus. Conversely, when both stimuli were paired with toxin or x-ray the avoidance reactions transferred to the gustatory stimulus, but not the audiovisual stimulus. Apparently stimuli are selected as cues dependent upon the nature of the subsequent reinforcer.

A great deal of evidence stemming from diverse sources suggests an inadequacy in the usual formulations concerning reinforcement. Barnett (1963) has described the "bait-shy" behavior of wild rats which have survived a poisoning attempt. These animals utilizing olfactory and gustatory cues, avoid the poison bait which previously made them ill. However, there is no evidence that they avoid the "place" of the poisoning.

In a recent volume (Haley & Snyder, 1964) several authors have discussed studies in which ionizing radiations were employed as a noxious stimulus to produce avoidance reactions in animals. Ionizing radiation like many poisons produces gastrointestinal disturbances and nausea. Strong aversions are readily established in animals when distinctively flavored fluids are conditionally paired with x-rays. Subsequently, the gustatory stimulus will depress fluid intake without radiation. In contrast, a distinctive environmental complex of auditory, visual, and tactual stimuli does not inhibit drinking even when the compound stimulus is associated with the identical radiation schedule. This differential effect has also been observed following ingestion of a toxin and the injection of a drug (Garcia & Koelling, 1965).

Apparently this differential effectiveness of cues is due either to the nature of the reinforcer, i.e., radiation or toxic effects, or to the peculiar relation which a gustatory stimulus has to the drinking response, i.e., gustatory stimulation occurs if and only if the animal licks the fluid. The environmental cues associated with a distinctive place are not as dependent upon a single response of the organism. Therefore, we made an auditory and visual stimulus dependent upon the animal's licking the water spout. Thus, in four experiments reported here "bright-noisy" water, as well as "tasty" water was conditionally paired with radiation, a toxin, immediate shock, and delayed shock, respectively, as reinforcers. Later the capacity of these response-controlled stimuli to inhibit drinking in the absence of reinforcement was tested.

Method

The apparatus was a light and sound shielded box (7 in. x 7 in. x 7 in.) with a drinking spout connected to an electronic drinkometer which counted each touch of the rat's tongue to the spout. "Bright-noisy" water was provided by connecting an incandescent lamp (5 watts) and a clicking relay into this circuit. "Tasty" water was provided by adding flavors to the drinking supply.

Each experimental group consisted of 10 rats (90 day old Sprague-Dawley males) maintained in individual cages without water, but with *Purina Laboratory chow ad libidum.*

The procedure was: A. One week of habituation to drinking in the apparatus without stimulation. B. Pretests to measure intake of bright-noisy water and tasty water prior to training. C. Acquisition training with: (1) reinforced trials where these stimuli were paired with reinforcement during drinking, (2) nonreinforced trials where rats drank water without stimuli or reinforcement. Training terminated when there was a reliable difference between water intake scores on reinforced and nonreinforced trials. D. Post-tests to measure intake of bright-noisy water and tasty water after training.

In the x-ray study an audiovisual group and a gustatory group were exposed to an identical radiation schedule. In the other studies reinforcement was contingent upon the rat's response. To insure that both the audiovisual and the gustatory stimuli received equivalent reinforcement, they were combined and simultaneously paired with the reinforcer during acquisition training. Therefore, one group serving as its own control and divided into equal subgroups, was tested in balanced order with an audiovisual and a gustatory test before and after training with these stimuli combined.

One 20-min. reinforced trial was administered every three days in the x-ray and lithium chloride studies. This prolonged intertrial interval was designed to allow sufficient time for the rats to recover from acute effects of treatment. On each interpolated day the animals received a 20-min. nonreinforced trial. They were post-tested two days after their last reinforced trial. The x-ray groups received a total of three reinforced trials, each with 54 r of filtered 250 kv x-rays delivered in 20 min. Sweet water (1 gm saccharin per liter) was the gustatory stimulus. The lithium chloride group had a total of five reinforced trials with toxic salty water (.12 M lithium chloride). Non-toxic salty water (.12 M sodium chloride) which rats cannot readily distinguish from the toxic solution was used in the gustatory tests (Nachman, 1963).

The immediate shock study was conducted on a more orthodox avoidance schedule. Tests and trials were 2 min. long. Each day for four consecutive acquisition days, animals were given two nonreinforced and two reinforced trials in an NRRN, RNNR pattern. A shock, the minimal current required to interrupt drinking (0.5 sec. at 0.08-0.20 ma), was delivered through a floor grid 2 sec. after the first lick at the spout.

The delayed shock study was conducted simultaneously with the lithium chloride on the same schedule. Non-toxic salty water was the gustatory stimulus. Shock reinforcement was delayed during first trials and gradually increased in intensity (.05 to .30 ma) in a schedule designed to produce a drinking pattern during the 20-min. period which resembled that of the corresponding animal drinking toxic salty water.

Results and Discussion

The results indicate that all reinforcers were effective in producing discrimination learning during the acquisition phase (see Fig. 1), but obvious differences occurred in the post-tests. The avoidance reactions produced by

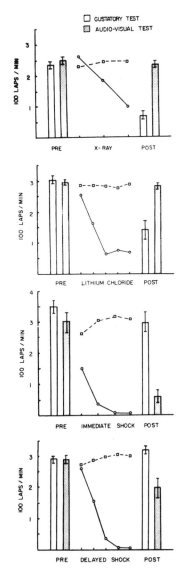

Fig. 1. The bars indicate water intake (± St. Error) during a gustatory test (a distinctive taste) and an audiovisual test (light and sound contingent upon licking) before and after conditional pairing with the reinforcers indicated. The curves illustrate mean intake during acquisition.

x-rays and lithium chloride are readily transferred to the gustatory stimulus but not to the audiovisual stimulus. The effect is more pronounced in the x-ray study, perhaps due to differences in dose. The x-ray animals received a constant dose while the lithium chloride rats drank a decreasing amount of the toxic solution during training. Nevertheless, the difference

between post-test scores is statistically significant in both experiments (p< 0.01 by ranks test).

Apparently when gustatory stimuli are paired with agents which produce nausea and gastric upset, they acquire secondary reinforcing properties which might be described as "conditioned nausea." Auditory and visual stimulation do not readily acquire similar properties even when they are contingent upon the licking response.

In contrast, the effect of both immediate and delayed shock to the paws is in the opposite direction. The avoidance reactions produced by electric shock to the paws transferred to the audiovisual stimulus but not to the gustatory stimulus. As one might expect the effect of delayed shocks was not as effective as shocks where the reinforcer immediately and consistently followed licking. Again, the difference between post-test intake scores is statistically significant in both studies (p< 0.01 by ranks test). Thus, when shock which produces peripheral pain is the reinforcer, "conditioned fear" properties are more readily acquired by auditory and visual stimuli than by gustatory stimuli.

It seems that given reinforcers are not equally effective for all classes of discriminable stimuli. The cues, which the animal selects from the welter of stimuli in the learning situation, appear to be related to the consequences of the subsequent reinforcer. Two speculations are offered: (1) Common elements in the time-intensity patterns of stimulation may facilitate a cross modal generalization from reinforcer to cue in one case and not in another. (2) More likely, natural selection may have favored mechanisms which associate gustatory and olfactory cues with internal discomfort since the chemical receptors sample the materials soon to be incorporated into the internal environment. Krechevsky (1933) postulated such a genetically coded hypothesis to account for the predispositions of rats to respond systematically to specific cues in an insoluble maze. The hypothesis of the sick rat, as for many of us under similar circumstances, would be, "It must have been something I ate."

References

Barnett, S. A. *The rat: a study in behavior.* Chicago: Aldine Press, 1963.

Garcia, J., & Koelling, R. A. A comparison of aversions induced by x-rays, toxins, and drugs in the rat. *Radiat. Res.,* in press, 1965.

Haley, T. J., & Snyder, R. S. (Eds.) *The response of the nervous system to ionizing radiation.* Boston: Little, Brown & Co., 1964.

Krechevsky, I. The hereditary nature of 'hypothesis'. *J. comp. Psychol.,* 1932, 16, 99-116.

Nachman, M. Learned aversion to the taste of lithium chloride and generalization to other salts. *J. comp. physiol. Psychol.,* 1963, 56, 343-349.

Note

1. This research stems from doctoral research carried out at Long Beach V. A. Hospital and supported by NIH No. RH00068. Thanks are extended to Professors B. F. Ritchie, D. Krech and E. R. Dempster, U. C. Berkeley, California.

Affectional Responses
in the Infant Monkey

Orphaned baby monkeys develop a strong and persistent
attachment to inanimate surrogate mothers.

Harry F. Harlow and Robert R. Zimmermann

Investigators from diverse behavioral fields have long recognized the strong attachment of the neonatal and infantile animal to its mother. Although this affectional behavior has been commonly observed, there is, outside the field of ethology, scant experimental evidence permitting identification of the factors critical to the formation of this bond. Lorenz (1) and others have stressed the importance of innate visual and auditory mechanisms which, through the process of imprinting, give rise to persisting following responses in the infant bird and fish. Imprinting behavior has been demonstrated successfully in a variety of avian species under controlled laboratory conditions, and this phenomenon has been investigated systematically in order to identify those variables which contribute to its development and maintenance [see, for example, Hinde, Thorpe, and Vince (2), Fabricius (3), Hess (4), Jaynes (5), and Moltz and Rosenblum (6)]. These studies represent the largest body of existent experimental evidence measuring the tie between infant and mother. At the mammalian level there is little or no systematic experimental evidence of this nature.

Observations on monkeys by Carpenter (7), Nolte (8), and Zuckermann (9) and on chimpanzees by Kohler (10) and by Yerkes and Tomilin (11) show that monkey and chimpanzee infants develop strong ties to their mothers and that these affectional attachments may persist for years. It is, of course, common knowledge that human infants form strong and persistent ties to their mothers.

Although students from diverse scientific fields recognize this abiding attachment, there is considerable disagreement about the nature of its development and its fundamental underlying mechanisms. A common theory among psychologists, sociologists, and anthropologists is that of learning based on drive reduction. This theory proposes that the infant's attachment to the mother results from the association of the mother's face and form with the alleviation of certain primary drive states, particularly hunger and thirst. Thus, through learning, affection becomes a self-supporting, derived drive (12). Psychoanalysts, on the other hand, have stressed the importance of various innate needs, such as a need to suck and orally possess the breast (2), or needs relating to contact, movement, temperature (13), and clinging to the mother (14).

The paucity of experimental evidence concerning the development of affectional responses has led these theorists to derive their basic hypotheses from deductions and intuitions based on observation and analysis of adult verbal reports. As a result, the available observational evidence is often forced into a preconceived theoretical framework. An exception to the above generalization is seen in the recent attempt by Bowlby (14) to analyze and integrate the available observational and experimental evidence derived from both human and subhuman infants. Bowlby has concluded that a theory of component instinctual responses, species specific, can best account for the infant's tie to the mother. He suggests that the species-specific responses for human beings (some of these responses are not strictly limited to human beings) include contact, clinging, sucking, crying, smiling, and following. He further emphasizes that these responses are manifested independently of primary drive reduction in human and subhuman infants.

The absence of experimental data which would allow a critical evaluation of any theory of affectional development can be attributed to several causes. The use of human infants as subjects has serious limitations, since it is not feasible to employ all the experimental controls which would permit a completely adequate analysis of the proposed variables. In addition, the limited response repertoire of the human neonate severely restricts the number of discrete or precise response categories that can be measured until a considerable age has been attained. Thus, critical variables go unmeasured and become lost or confounded among the complex physiological, psychological, and cultural factors which influence the developing human infant.

Moreover, the use of common laboratory animals also has serious limitations, for most of these animals have behavioral repertoires very different from those of the human being, and in many species these systems mature so rapidly that it is difficult to measure and assess their orderly development. On the other hand, subhuman primates, including the macaque monkey, are born at a state of maturity which makes it possible to begin precise measurements within the first few days of life. Furthermore, their postnatal maturational rate is slow enough to permit precise assessment of affectional variables and development.

Over a 3-year period prior to the beginning of the research program reported

The authors are on the staff of the primate laboratory, department of psychology, University of Wisconsin, Madison.

Fig. 1. Wire and cloth mother surrogates.

with, and are considered basic to, affection; these include nursing, clinging, and visual and auditory exploration.

In the course of raising these infants we observed that they all showed a strong attachment to the cheesecloth blankets which were used to cover the wire floors of their cages. Removal of these cloth blankets resulted in violent emotional behavior. These responses were not short-lived; indeed, the emotional disturbance lasted several days, as was indicated by the infant's refusal to work on the standard learning tests that were being conducted at the time. Similar observations had already been made by Foley (17) and by van Wagenen (16), who stressed the importance of adequate contact responses to the very survival of the neonatal macaque. Such observations suggested to us that contact was a true affectional variable and that it should be possible to trace and measure the development and importance of these responses. Indeed there seemed to be every reason to believe that one could manipulate all variables which have been considered critical to the development of the infant's attachment to a mother, or mother surrogate.

To attain control over maternal variables, we took the calculated risk of constructing and using inanimate mother surrogates rather than real mothers. The cloth mother that we used was a cylinder of wood covered with a sheath of terry cloth (18), and the wire mother was a hardware-cloth cylinder. Initially, sponge rubber was placed underneath the terry cloth sheath of the cloth mother surrogate, and a light bulb behind each mother surrogate provided radiant heat. For reasons of sanitation

here (15), some 60 infant macaque monkeys were separated from their mothers 6 to 12 hours after birth and raised at the primate laboratory of the University of Wisconsin. The success of the procedures developed to care for these neonates was demonstrated by the low mortality and by a gain in weight which was approximately 25 percent greater than that of infants raised by their own mothers. All credit for the success of this program belongs to van Wagenen (16), who had described the essential procedures in detail.

These first 3 years were spent in devising measures to assess the multiple

capabilities of the neonatal and infantile monkey. The studies which resulted have revealed that the development of perception, learning, manipulation, exploration, frustration, and timidity in the macaque monkey follows a course and sequence which is very similar to that in the human infant. The basic differences between the two species appear to be the advanced postnatal maturational status and the subsequent more rapid growth of the infant macaque. Probably the most important similarities between the two, in relation to the problem of affectional development, are characteristic responses that have been associated

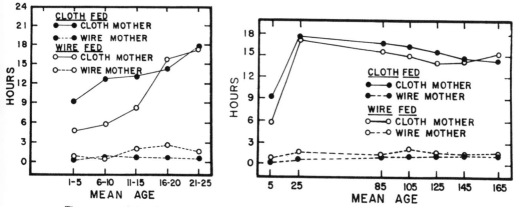

Time spent on cloth and wire mother surrogates. Fig. 2 (left). Short term. Fig. 3 (right). Long term.

and safety these two factors were eliminated in construction of the standard mothers, with no observable effect on the behavior of the infants. The two mothers were attached at a 45-degree angle to aluminum bases and were given different faces to assure uniqueness in the various test situations (Fig. 1). Bottle holders were installed in the upper middle part of the bodies to permit nursing. The mother was designed on the basis of previous experience with infant monkeys, which suggested that nursing in an upright or inclined position with something for the infant to clasp facilitated successful nursing and resulted in healthier infants (see *16*). Thus, both mothers provided the basic known requirements for adequate nursing, but the cloth mother provided an additional variable of contact comfort. That both of these surrogate mothers provided adequate nursing support is shown by the fact that the total ingestion of formula and the weight gain was normal for all infants fed on the surrogate mothers. The only consistent difference between the groups lay in the softer stools of the infants fed on the wire mother.

Development of Affectional Responses

The initial experiments on the development of affectional responses have already been reported (*19*) but will be briefly reviewed here, since subsequent experiments were derived from them. In the initial experiments, designed to evaluate the role of nursing on the development of affection, a cloth mother and a wire mother were placed in different cubicles attached to the infant's living cage. Eight newborn monkeys were placed in individual cages with the surrogates; for four infant monkeys the cloth mother lactated and the wire mother did not, and for the other four this condition was reversed.

The infants lived with their mother surrogates for a minimum of 165 days, and during this time they were tested in a variety of situations designed to measure the development of affectional responsiveness. Differential affectional responsiveness was initially measured in terms of mean hours per day spent on the cloth and on the wire mothers under two conditions of feeding, as shown in Fig. 2. Infants fed on the cloth mother and on the wire mother have highly similar scores after a short adaptation period (Fig. 3), and over a 165-day period both groups show a distinct pref-

Fig. 4. Typical fear stimulus.

erence for the cloth mother. The persistence of the differential responsiveness to the mothers for both groups of infants is evident, and the over-all differences between the groups fall short of statistical significance.

These data make it obvious that contact comfort is a variable of critical importance in the development of affectional responsiveness to the surrogate mother, and that nursing appears to play a negligible role. With increasing age and opportunity to learn, an infant fed from a lactating wire mother does not become more responsive to her, as would be predicted from a derived-drive theory, but instead becomes increasingly more responsive to its nonlactating cloth mother. These findings are at complete variance with a drive-reduction theory of affectional development.

The amount of time spent on the mother does not necessarily indicate an affectional attachment. It could merely reflect the fact that the cloth mother is a more comfortable sleeping platform or a more adequate source of warmth for the infant. However, three of the four infants nursed by the cloth mother and one of the four nursed by the wire mother left a gauze-covered heating pad that was on the floor of their cages during the first 14 days of life to spend up to 18 hours a day on the cloth mother. This suggests that differential heating or warmth is not a critical variable within the controlled temperature range of the laboratory.

Other tests demonstrate that the cloth mother is more than a convenient nest; indeed, they show that a bond develops between infant and cloth-mother surrogate that is almost unbelievably similar to the bond established between human mother and child. One highly definitive test measured the selective maternal responsiveness of the monkey infants under conditions of distress or fear.

Various fear-producing stimuli, such as the moving toy bear illustrated in Fig. 4, were presented to the infants in their home cages. The data on differential responses under both feeding conditions are given in Fig. 5. It is apparent that the cloth mother was highly preferred to the wire mother, and it is a fact that these differences were unrelated to feeding conditions—that is, nursing on the cloth or on the wire mother. Above and beyond these objective data are observations on the form of the infants' responses in this situation. In spite of their abject terror, the infant monkeys, after reaching the cloth mother and rubbing their bodies about hers, rapidly come to lose their fear of the frightening stimuli. Indeed, within a minute or two most of the babies were visually exploring the very thing which so shortly before had seemed an object of evil. The bravest of the babies would actually leave the mother and approach the fearful monsters, under, of course, the protective gaze of their mothers.

These data are highly similar, in terms of differential responsiveness, to the time scores previously mentioned and indicate the overwhelming importance of contact comfort. The results are so striking as to suggest that the primary function of nursing may be that of insuring frequent and intimate contact between mother and infant, thus facilitating the localization of the source of contact comfort. This interpretation finds some support in the test discussed above. In both situations the infants

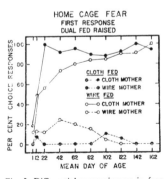

Fig. 5. Differential responsiveness in fear tests.

nursed by the cloth mother developed consistent responsiveness to the soft mother earlier in testing than did the infants nursed by the wire mother, and during this transient period the latter group was slightly more responsive to the wire mother than the former group. However, these early differences shortly disappeared.

Additional data have been obtained from two groups of four monkeys each which were raised with a single mother placed in a cubicle attached to the living-cage. Four of the infants were presented with a lactating wire mother and the other four were presented with a nonlactating cloth mother. The latter group was hand-fed from small nursing bottles for the first 30 days of life and then weaned to a cup. The development of responsiveness to the mothers was studied for 165 days; after this the individual mothers were removed from the cages and testing was continued to determine the strength and persistence of the affectional responses.

Figure 6 presents the mean time per day spent on the respective mothers over the 165-day test period, and Fig. 7 shows the percentage of responses to the mothers when a fear-producing stimulus was introduced into the home cage. These tests indicate that both groups of infants developed responsiveness to their mother surrogates. However, these measures did not reveal the differences in behavior that were displayed in the reactions to the mothers when the fear stimuli were presented. The infants raised on the cloth mother would rush to the mother and cling tightly to her. Following this initial response these infants would relax and either begin to manipulate the

mother or turn to gaze at the feared object without the slightest sign of apprehension. The infants raised on the wire mother, on the other hand, rushed away from the feared object toward their mother but did not cling to or embrace her. Instead, they would either clutch themselves and rock and vocalize for the remainder of the test or rub against the side of the cubicle. Contact with the cubicle or the mother did not reduce the emotionality produced by the introduction of the fear stimulus. These differences are revealed in emotionality scores, for behavior such as vocalization, crouching, rocking, and sucking, recorded during the test. Figure 8 shows the mean emotionality index for test sessions for the two experimental groups, the dual-mother groups, and a comparable control group raised under standard laboratory conditions. As can be seen, the infants raised with the single wire mother have the highest emotionality scores of all the groups, and the infants raised with the single cloth mother or with a cloth and wire mother have the lowest scores. It appears that the responses made by infants raised only with a wire mother were more in the nature of simple flight responses to the fear stimulus and that the presence of the mother surrogate had little effect in alleviating the fear.

During our initial experiments with the dual-mother conditions, responsiveness to the lactating wire mother in the fear tests decreased with age and opportunity to learn, while responsiveness to the nonlactating cloth mother increased. However, there was some indication of a slight increase in frequency of response to the wire mother for the first 30 to 60

days (see Fig. 5). These data suggest the possible hypothesis that nursing facilitated the contact of infant and mother during the early developmental periods.

The interpretation of all fear testing is complicated by the fact that all or most "fear" stimuli evoke many positive exploratory responses early in life and do not consistently evoke flight responses until the monkey is 40 to 50 days of age. Similar delayed maturation of visually induced fear responses has been reported for birds (3), chimpanzees (20), and human infants (21).

Because of apparent interactions between fearful and affectional developmental variables, a test was designed to trace the development of approach and avoidance responses in these infants. This test, described as the straight-alley test, was conducted in a wooden alley 8 feet long and 2 feet wide. One end of the alley contained a movable tray upon which appropriate stimuli were placed. The other end of the alley contained a box for hiding. Each test began with the monkey in a start box 1 foot in front of the hiding box; thus, the animal could maintain his original position, approach the stimulus tray as it moved toward him, or flee into the hiding box. The infants were presented with five stimuli in the course of five successive days. The stimuli included a standard cloth mother, a standard wire mother, a yellow cloth mother with the head removed, a blank tray, and a large black fear stimulus, shown in Fig. 9. The infants were tested at 5, 10, and 20 days of age, respectively, and then at 20-day intervals up to 160 days. Figure 10 shows the mean number of 15-second time

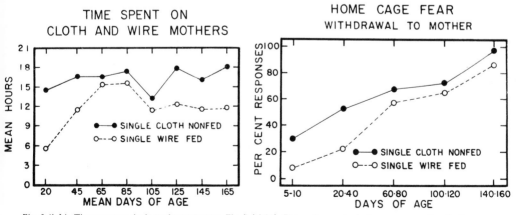

Fig. 6 (left). Time spent on single mother surrogates. Fig. 7 (right). Responsiveness to single surrogate mothers in fear tests.

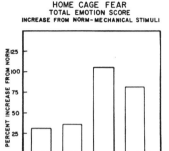

HOME CAGE FEAR
TOTAL EMOTION SCORE
INCREASE FROM NORM-MECHANICAL STIMULI

Fig. 8. Change in emotionality index in fear tests.

periods spent in contact with the appropriate mother during the 90-second tests for the two single-mother groups, and the responses to the cloth mother by four infants from the dual-mother group.

During the first 80 days of testing, all the groups showed an increase in response to the respective mother surrogates. The infants fed on the single wire mother, however, reached peak responsiveness at this age and then showed a consistent decline, followed by an actual avoidance of the wire mother. During test sessions 140 to 160, only one contact was made with the wire mother, and three of the four infants ran into the hiding box almost immediately and remained there for the entire test session. On the other hand, all of the infants raised with a cloth mother, whether or not they were nursed by her, showed a progressive increase in time spent in contact with their cloth mothers until approaches and contacts during the test sessions approached maximum scores.

The development of the response of flight from the wire mother by the group fed on the single wire mother is, of course, completely contrary to a derived-drive theory of affectional development. A comparison of this group with the group raised with a cloth mother gives some support to the hypothesis that feeding or nursing facilitates the early development of responses to the mother but that without the factor of contact comfort, these positive responses are not maintained.

The differential responsiveness to the cloth mother of infants raised with both mothers, the reduced emotionality of both the groups raised with cloth mothers in the home-cage fear tests, and the development of approach responses in the straight-alley test indi-

cate that the cloth mother provides a haven of safety and security for the frightened infant. The affectional response patterns found in the infant monkey are unlike tropistic or even complex reflex responses; they resemble instead the diverse and pervasive patterns of response exhibited by the human child in the complexity of situations involving child-mother relationships.

The role of the mother as a source of safety and security has been demonstrated experimentally for human infants by Arsenian (22). She placed children 11 to 30 months of age in a strange room containing toys and other play objects. Half of the children were accompanied into the room by a mother or a substitute mother (a familiar nursery attendant), while the other half entered the situation alone. The children in the first group (mother present) were much less emotional and participated much more fully in the play activity than those in the second group (mother absent). With repeated testing, the security score, a composite score of emotionality and play behavior, improved for the children who entered alone, but it still fell far below that for the children who were accompanied by their mothers. In subsequent tests, the children from the mother-present group were placed in the test room alone, and there was a drastic drop in the security scores. Contrariwise, the introduction of the mother raised the security scores of children in the other group.

We have performed a similar series of open-field experiments, comparing monkeys raised on mother surrogates with control monkeys raised in a wire cage containing a cheesecloth blanket from days 1 to 14 and no cloth blanket subsequently. The infants were introduced into the strange environment of the open field, which was a room measuring 6 by 6 by 6 feet, containing multiple stimuli known to elicit curiosity-manipulatory responses in baby monkeys. The infants raised with single mother surrogates were placed in this situation twice a week for 8 weeks, no mother surrogate being present during one of the weekly sessions and the appropriate mother surrogate (the kind which the experimental infant had always known) being present during the other sessions. Four infants raised with dual mother surrogates and four control infants were subjected to similar experimental sequences, the cloth mother being present on half of the occasions. The remaining four "dual-mother" in-

Fig. 9. Response to the fear stimulus in the straight-alley test.

fants were given repetitive tests to obtain information on the development of responsiveness to each of the dual mothers in this situation. A cloth blanket was always available as one of the stimuli throughout the sessions. It should be emphasized that the blanket could readily compete with the cloth mother as a contact stimulus, for it was standard laboratory procedure to wrap the infants in soft cloth whenever they were removed from their cages for testing, weighing, and other required laboratory activities.

As soon as they were placed in the test room, the infants raised with cloth mothers rushed to their mother surrogate when she was present and clutched her tenaciously, a response so strong that it can only be adequately depicted by motion pictures. Then, as had been observed in the fear tests in the home cage, they rapidly relaxed, showed no sign of apprehension, and began to demonstrate unequivocal positive responses of manipulating and climbing on the mother. After several sessions, the infants began to use the mother surrogate as a base of

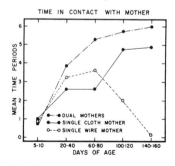

TIME IN CONTACT WITH MOTHER

Fig. 10. Responsiveness to mother surrogates in the straight-alley tests.

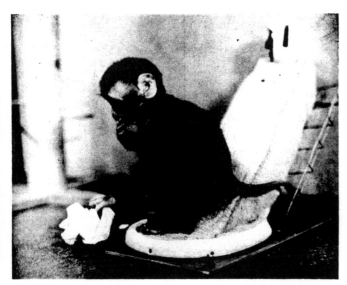

Fig. 11. Subsequent response to cloth mother and stimulus in the open-field test.

∩perations, leaving her to explore and handle a stimulus and then returning to her before going to a new plaything. Some of the infants even brought the stimuli to the mother, as shown in Fig. 11. The behavior of these infants changed radically in the absence of the mother. Emotional indices such as vocalization, crouching, rocking, and sucking increased sharply. Typical response patterns were either freezing in a crouched position, as illustrated in Fig. 12, or running around the room on the hind feet, clutching themselves with their arms. Though no quantitative evidence is available, contact and manipulation of objects was frantic and of short duration, as opposed to the playful type of manipulation observed when the mother was present.

In the presence of the mother, the behavior of the infants raised with sin-

gle wire mothers was both quantitatively and qualitatively different from that of the infants raised with cloth mothers. Not only did these infants spend little or no time contacting their mother surrogates but the presence of the mother did not reduce their emotionality. These differences are evident in the mean number of time periods spent in contact with the respective mothers, as shown in Fig. 13, and the composite emotional index for the two stimulus conditions depicted in Fig. 14. Although the infants raised with dual mothers spent considerably more time in contact with the cloth mother than did the infants raised with single cloth mothers, their emotional reactions to the presence and absence of the mother were highly similar, the composite emotional index being reduced by almost half when the mother was in the test situation. The infants raised with wire mothers were highly emotional under both conditions and actually showed a slight, though nonsignificant, increase in emotionality when the mother was present. Although some of the infants reared by a wire mother did contact her, their behavior was similar to that observed in the home-cage fear tests. They did not clutch and cling to their mother as did the infants with cloth mothers; instead, they sat on her lap and clutched themselves, or held their heads and bodies in their arms and engaged in convulsive jerking and rocking movements similar to the autistic behavior of deprived and institutionalized human children. The lack of exploratory and manipulatory behavior on the part of the infants reared with wire mothers, both in the presence and absence of the wire mother, was similar to that observed in the mother-absent condition for the infants raised with the cloth mothers, and such contact with objects as was made was of short duration and of an erratic and frantic nature. None of the infants raised with single wire mothers displayed the persistent and aggressive play behavior that was typical of many of the infants that were raised with cloth mothers.

The four control infants, raised without a mother surrogate, had approximately the same emotionality scores when the mother was absent that the other infants had in the same condition, but the control subjects' emotionality scores were significantly higher in the presence of the mother surrogate than in her absence. This result is not surprising, since recent evidence indicates that the cloth mother with the highly ornamental face is an effective fear stimulus

Fig. 12. Response in the open-field test in the absence of the mother surrogate.

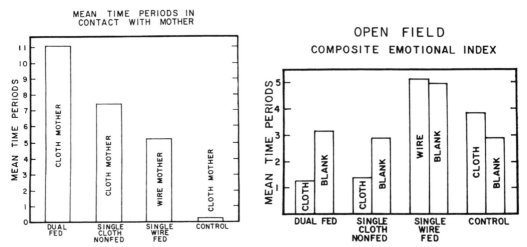

Fig. 13 (left). Responsiveness to mother surrogates in the open-field test. Fig. 14 (right). Emotionality index in testing with and without the mother surrogates.

for monkeys that have not been raised with her.

Further illustration of differential responsiveness to the two mother surrogates is found in the results of a series of developmental tests in the open-field situation, given to the remaining four "dual-mother" infants. These infants were placed in the test room with the cloth mother, the wire mother, and no mother present on successive occasions at various age levels. Figure 15 shows the mean number of time periods spent in contact with the respective mothers for two trials at each age level, and Fig. 16 reveals the composite emotion scores

for the three stimulus conditions during these same tests. The differential responsiveness to the cloth and wire mothers, as measured by contact time, is evident by 20 days of age, and this systematic difference continues throughout 140 days of age. Only small differences in emotionality under the various conditions are evident during the first 85 days of age, although the presence of the cloth mother does result in slightly lower scores from the 45th day onward. However, at 105 and 145 days of age there is a considerable difference for the three conditions, the emotionality scores for the wire-mother and blank conditions

showing a sharp increase. The heightened emotionality found under the wire-mother condition was mainly contributed by the two infants fed on the wire mother. The behavior of these two infants in the presence of the wire mother was similar to the behavior of the animals raised with a single wire mother. On the few occasions when contact with the wire mother was made, the infants did not attempt to cling to her; instead they would sit on her lap, clasp their heads and bodies, and rock back and forth.

In 1953 Butler (23) demonstrated that mature monkeys enclosed in a

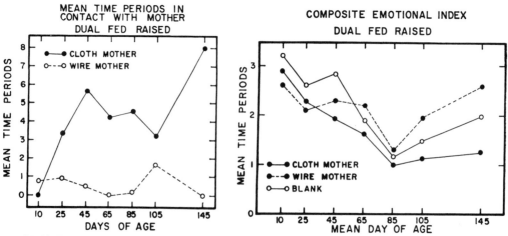

Fig. 15 (left). Differential responsiveness in the open-field test. Fig. 16 (right). Emotionality index under three conditions in the open-field test.

dimly lighted box would open and re-open a door for hours on end with no other motivation than that of looking outside the box. He also demonstrated that rhesus monkeys showed selectivity in rate and frequency of door-opening in response to stimuli of different degrees of attractiveness (24). We have utilized this characteristic of response selectivity on the part of the monkey to measure the strength of affectional responsiveness of the babies raised with mother surrogates in an infant version of the Butler box. The test sequence involves four repetitions of a test battery in which the four stimuli of cloth mother, wire mother, infant monkey, and empty box are presented for a 30-minute period on successive days. The first four subjects raised wth the dual mother surrogates and the eight infants raised with single mother surrogates were given a test sequence at 40 to 50 days of age, depending upon the availability of the apparatus. The data obtained from the three experimental groups and a comparable control group are presented in Fig. 17. Both groups of infants raised with cloth mothers showed approximately equal responsiveness to the cloth mother and to another infant monkey, and no greater responsiveness to the wire mother than to an empty box. Again, the results are independent of the kind of mother that lactated, cloth or wire. The infants raised with only a wire mother and those in the control group were more highly responsive to the monkey than to either of the mother surrogates. Furthermore, the former group showed a higher frequency of response to the empty box than to the wire mother.

In summary, the experimental analysis of the development of the infant monkey's attachment to an inanimate mother surrogate demonstrates the overwhelming importance of the variable of soft body contact that characterized the cloth mother, and this held true for the appearance, development, and maintenance of the infant-surrogate-mother tie. The results also indicate that, without the factor of contact comfort, only a weak attachment, if any, is formed. Finally, probably the most surprising finding is that nursing or feeding played either no role or a subordinate role in the development of affection as measured by contact time, responsiveness to fear, responsiveness to strangeness, and motivation to seek and see. No evidence was found indicating that nursing mediated the development of any of these responses, although there is evidence in-

dicating that feeding probably facilitated the early appearance and increased the early strength of affectional responsiveness. Certainly feeding, in contrast to contact comfort, is neither a necessary nor a sufficient condition for affectional development.

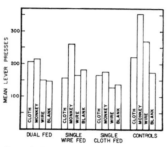

Fig. 17. Differential responses to visual exploration.

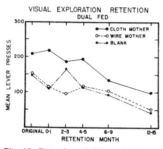

Fig. 18. Retention of differential visual-exploration responses.

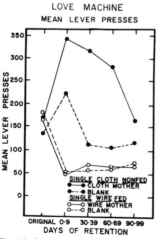

Fig. 19. Retention of differential visual-exploration responses by single-surrogate infants.

Retention of Affectional Responses

One of the outstanding characteristics of the infant's attachment to its mother is the persistence of the relationship over a period of years, even though the frequency of contact between infant and mother is reduced with increasing age. In order to test the persistence of the responsiveness of our "mother-surrogate" infants, the first four infant monkeys raised with dual mothers and all of the monkeys raised with single mothers were separated from their surrogates at 165 to 170 days of age. They were tested for affectional retention during the following 9 days, then at 30-day intervals during the following year. The results are of necessity incomplete, inasmuch as the entire mother-surrogate program was initiated less than 2 years ago, but enough evidence is available to indicate that the attachment formed to the cloth mother during the first 6 months of life is enduring and not easily forgotten.

Affectional retention as measured by the modified Butler box for the first 15 months of testing for four of the infants raised with two mothers is given in Fig. 18. Although there is considerable variability in the total response frequency from session to session, there is a consistent difference in the number of responses to the cloth mother as contrasted with responses to either the wire mother or the empty box, and there is no consistent difference between responses to the wire mother and to the empty box. The effects of contact comfort versus feeding are dramatically demonstrated in this test by the monkeys raised with either single cloth or wire mothers. Figure 19 shows the frequency of response to the appropriate mother surrogate and to the blank box during the preseparation period and the first 90 days of retention testing. Removal of the mother resulted in a doubling of the frequency of response to the cloth mother and more than tripled the difference between the responses to the cloth mother and those to the empty box for the infants that had lived with a single non-lactating cloth mother surrogate. The infants raised with a single lactating wire mother, on the other hand, not only failed to show any consistent preference for the wire mother but also showed a highly significant reduction in general level of responding. Although incomplete, the data from further retention testing indicate that the difference between these two groups persists for at least 5 months.

Affectional retention was also tested

428

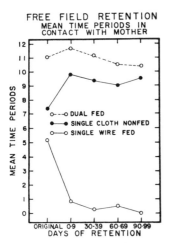

Fig. 20. Retention of responsiveness to mother surrogates in the open-field tests.

in the open field during the first 9 days after separation and then at 30-day intervals. Each test condition was run twice in each retention period. In the initial retention tests the behavior of the infants that had lived with cloth mothers differed slightly from that observed during the period preceding separation. These infants tended to spend more time in contact with the mother and less time exploring and manipulating the objects in the room. The behavior of the infants raised with single wire mothers, on the other hand, changed radically during the first retention sessions, and responses to the mother surrogate dropped almost to zero. Objective evidence for these differences are given in Fig. 20, which reveals the mean number of time periods spent in contact with the respective mothers. During the first retention test session, the infants raised with a single wire mother showed almost no responses to the mother surrogate they had always

known. Since the infants raised with both mothers were already approaching the maximum score in this measure, there was little room for improvement. The infants raised with a single nonlactating cloth mother, however, showed a consistent and significant increase in this measure during the first 90 days of retention. Evidence for the persistence of this responsiveness is given by the fact that after 15 months' separation from their mothers, the infants that had lived with cloth mothers spent an average of 8.75 out of 12 possible time periods in contact with the cloth mother during the test. The incomplete data for retention testing of the infants raised with only a lactating wire mother or a nonlactating cloth mother indicates that there is little or no change in the initial differences found between these two groups in this test over a period of 5 months. In the absence of the mother, the behavior of the infants raised with cloth mothers was similar in the initial retention tests to that during the preseparation tests, but with repeated testing they tended to show gradual adaptation to the open-field situation and, consequently, a reduction in their emotionality scores. Even with this over-all reduction in emotionality, these infants had consistently lower emotionality scores when the mother was present.

At the time of initiating the retention tests, an additional condition was introduced into the open-field test: the surrogate mother was placed in the center of the room and covered with a clear Plexiglas box. The animals raised with cloth mothers were initially disturbed and frustrated when their efforts to secure and contact the mother were blocked by the box. However, after several violent crashes into the plastic, the animals adapted to the situation and soon used the box as a place of orientation for exploratory and play behavior.

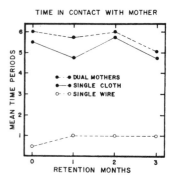

Fig. 22. Retention of responsiveness to mother surrogates in the straight-alley test.

In fact, several infants were much more active under these conditions than they were when the mother was available for direct contact. A comparison of the composite emotionality index of the babies raised with a single cloth or wire mother under the three conditions of no mother, surrogate mother, and surrogate-mother-box is presented in Fig. 21. The infants raised with a single cloth mother were consistently less emotional when they could contact the mother but also showed the effects of her visual presence, as their emotionality scores in the plastic box condition were definitely lower than their scores when the mother was absent. It appears that the infants gained considerable emotional security from the presence of the mother even though contact was denied.

In contrast, the animals raised with only lactating wire mothers did not show any significant or consistent trends during these retention sessions other than a general over-all reduction of emotionality, which may be attributed to a general adaptation, the result of repeated testing.

Affectional retention has also been measured in the straight-alley test mentioned earlier. During the preseparation tests it was found that the infants that had only wire mothers developed a general avoidance response to all of the stimuli in this test when they were about 100 days of age and made few, if any, responses to the wire mother during the final test sessions. In contrast, all the infants raised with a cloth mother responded positively to her. Maternal separation did not significantly change the behavior of any of the groups. The babies raised with just wire mothers continued to flee into the hiding booth in the presence of the wire mother, while all of the infants raised with

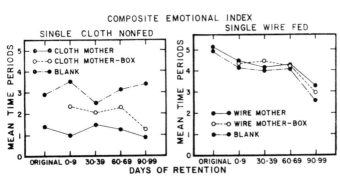

Fig. 21. Emotionality index under three conditions in the open-field retention tests.

Fig. 23. Typical response to cloth mother in the modified open-field test.

cloth mothers continued to respond positively to the cloth mother at approximately the same level as in the preseparation tests. The mean number of time periods spent in contact with the appropriate mother surrogates for the first 3 months of retention testing are given in Fig. 22. There is little, if any, waning of responsiveness to the cloth mother during these 3 months. There appeared to be some loss of responsiveness to the mother in this situation after 5 to 6 months of separation, but the test was discontinued at that time as the infants had outgrown the apparatus.

The retention data from these multiple tests demonstrate clearly the importance of body contact for the future maintenance of affectional responses. Whereas several of the measures in the preseparation period suggested that the infants raised with only a wire mother might have developed a weak attachment to her, all responsiveness disappeared in the first few days after the mother was withdrawn from the living-cage. Infants that had had the opportunity of living with a cloth mother showed the opposite effect and either became more responsive to the cloth mother or continued to respond to her at the same level.

These data indicate that once an affectional bond is formed it is maintained for a very considerable length of time with little reinforcement of the contact-comfort variable. The limited data available for infants that have been separated from their mother surrogates for a year suggest that these affectional re-

sponses show resistance to extinction similar to the resistance previously demonstrated for learned fears and learned pain. Such data are in keeping with common observation of human behavior.

It is true, however, that the infants raised with cloth mothers exhibit some absolute decrease in responsiveness with time in all of our major test situations. Such results would be obtained even if there were no true decrease in the strength of the affectional bond, because of familiarization and adaptation resulting from repeated testing. Therefore, at the end of 1 year of retention testing, new tests were introduced into the experimental program.

Our first new test was a modification of the open-field situation, in which basic principles of the home-cage fear test were incorporated. This particular choice was made partly because the latter test had to be discontinued when the mother surrogates were removed from the home cages.

For the new experiment a Masonite floor marked off in 6- by 12-inch rectangles was placed in the open-field chamber. Both mother surrogates were placed in the test room opposite a plastic start-box. Three fear stimuli, selected to produce differing degrees of emotionality, were placed in the center of the room directly in front of the start-box in successive test sessions. Eight trials were run under each stimulus condition, and in half of the trials the most direct path to the cloth mother was blocked by a large Plexiglas screen, illustrated in Fig. 23. Thus, in these trials

the infants were forced to approach and bypass the fear stimulus or the wire mother, or both, in order to reach the cloth mother. Following these 24 trials with the mothers present, one trial of each condition with both mothers absent was run, and this in turn was followed by two trials run under the most emotion-provoking condition: with a mechanical toy present and the direct path to the mother blocked.

We now have complete data for the first four infants raised with both a cloth and a wire mother. Even with this scanty information, the results are obvious. As would be predicted from our other measures, the emotionality scores for the three stimuli were significantly different and these same scores were increased greatly when the direct path to the mother was blocked. A highly significant preference was shown for the cloth mother under both conditions (direct and blocked path), although the presence of the block did increase the number of first responses to the wire mother from 3 to 10 percent. In all cases this was a transient response and the infants subsequently ran on to the cloth mother and clung tightly to her. Objective evidence for this overwhelming preference is indicated in Fig. 24, which shows the mean number of time periods spent in contact with the two mothers. After a number of trials, the infants would go first to the cloth mother and then, and only then, would go out to explore, manipulate, and even attack and destroy the fear stimuli. It was as if they believed that their mother would protect them, even at the cost of her life —little enough to ask in view of her condition.

The removal of the mother surrogates from the situation produced the predictable effect of doubling the emotionality index. In the absence of the mothers, the infants would often run to the Plexiglas partition which formerly had blocked their path to the mother, or they would crouch in the corner behind the block where the mother normally would have been. The return of the mothers in the final two trials of the test in which the most emotion-evoking situation was presented resulted in behavior near the normal level, as measured by the emotionality index and contacts with the cloth mother.

Our second test of this series was designed to replace the straight-alley test described above and provide more quantifiable data on responsiveness to fear stimuli. The test was conducted in an alley 8 feet long and 2 feet wide.

At one end of the alley and directly behind the monkeys' restraining chamber was a small stimulus chamber which contained a fear object. Each trial was initiated by raising an opaque sliding door which exposed the fear stimulus. Beginning at a point 18 inches from the restraining chamber, the alley was divided lengthwise by a partition; this provided the infant with the choice of entering one of two alleys.

The effects of all mother combinations were measured; these combinations included no mothers, two cloth mothers, two wire mothers, and a cloth and a wire mother. All mother conditions were counterbalanced by two distance conditions—distances of 24 and 78 inches, respectively, from the restraining chamber. This made it possible, for example, to provide the infant with the alternative of running to the cloth mother which was in close proximity to the fear stimulus or to the wire mother (or no mother) at a greater distance from the fear stimulus. Thus, it was possible to distinguish between running to the mother surrogate as an object of security, and generalized flight in response to a fear stimulus.

Again, the data available at this time are from the first four infants raised with cloth and wire mothers. Nevertheless, the evidence is quite conclusive. A highly significant preference is shown for the cloth mother as compared to the wire mother or to no mother, and this preference appears to be independent of the proximity of the mother to the fear stimulus. In the condition in which two cloth mothers are present, one 24 inches from the fear stimulus and the other 78 inches from it, there was a preference for the nearest mother, but the differences were not statistically significant. In two conditions in which no cloth mother was present and the infant had to choose between a wire mother and no mother, or between two empty chambers, the emotionality scores were almost twice those under the cloth-mother-present condition.

No differences were found in either of these tests that were related to previous conditions of feeding—that is, to whether the monkey had nursed on the cloth or on the wire mother.

The results of these two new tests, introduced after a full year's separation of mother surrogate and infant, are comparable to the results obtained during the preseparation period and the early retention testing. Preferential responses still favored the cloth as compared to the wire mother by as much as 85 to

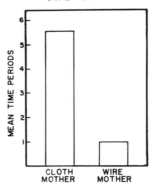

Fig. 24. Differential responsiveness in the modified open-field test.

90 percent, and the emotionality scores showed the typical 2:1 differential ratio with respect to mother-absent and mother-present conditions.

The researches presented here on the analysis of two affectional variables through the use of objective and observational techniques suggest a broad new field for the study of emotional development of infant animals. The analogous situations and results found in observations and study of human infants and of subprimates demonstrate the apparent face validity of our tests. The reliability of our observational techniques is indicated, for example, by the correlation coefficients computed for the composite emotional index in the open-field test. Four product-moment correlation coefficients, computed from four samples of 100 observations by five different pairs of independent observers over a period of more than a year, ranged from .87 to .89.

Additional Variables

Although the overwhelming importance of the contact variable has been clearly demonstrated in these experiments, there is reason to believe that other factors may contribute to the development of the affectional response pattern. We are currently conducting a series of new experiments to test some of these postulated variables.

For example, Bowlby (14) has suggested that one of the basic affectional variables in the primate order is not just contact but clinging contact. To test this hypothesis, four infant monkeys are being raised with the standard cloth mother and a flat inclined plane,

tightly covered with the same type of cloth. Thus, both objects contain the variable of contact with the soft cloth, but the shape of the mother tends to maximize the clinging variable, while the broad flat shape of the plane tends to minimize it. The preliminary results for differences in responsiveness to the cloth mother and responsiveness to the inclined plane under conditions that produce stress or fear or visual exploration suggest that clinging as well as contact is an affectional variable of considerable importance.

Experiments now in progress on the role of rocking motion in the development of attachment indicate that this may be a variable of measurable importance. One group of infants is being raised on rocking and stationary mothers and a second group, on rocking and stationary inclined planes. Both groups of infants show a small but consistent preference for the rocking object, as measured in average hours spent on the two objects.

Preliminary results for these three groups in the open-field test give additional evidence concerning the variable of clinging comfort. These data revealed that the infants raised with the standard cloth mother were more responsive to their mothers than the infants raised with inclined planes were to the planes.

The discovery of three variables of measurable importance to the formation and retention of affection is not surprising, and it is reasonable to assume that others will be demonstrated. The data so far obtained experimentally are in excellent concordance with the affectional variables named by Bowlby (14). We are now planning a series of studies to assess the effects of consistency and inconsistency with respect to the mother surrogates in relation to the clinical concept of rejection. The effects of early, intermediate, and late maternal deprivation and the generalization of the infant-surrogate attachment in social development are also being investigated. Indeed, the strength and stability of the monkeys' affectional responses to a mother surrogate are such that it should be practical to determine the neurological and biochemical variables that underlie love.

References and Notes

1. K. Lorenz, *Auk* 54, 245 (1937).
2. R. A. Hinde, W. H. Thorpe, M. A. Vince, *Behaviour* 9, 214 (1956).
3. E. Fabricius, *Acta Zool. Fennica* 68, 1 (1951).
4. E. H. Hess, *J. Comp. and Physiol. Psychol.*, in press.
5. J. Jaynes, *ibid.*, in press.
6. H. Moltz and L. Rosenblum, *ibid.* 51, 658 (1958).

7. C. R. Carpenter, *Comp. Psychol. Monograph No. 10* (1934), p. 1.
8. A. Nolte, *Z. Tierpsychol.* 12, 77 (1955).
9. S. Zuckerman, *Functional Affinities of Man, Monkeys and Apes* (Harcourt Brace, London, 1933).
10. W. Kohler, *The Mentality of Apes* (Humanities Press, New York, 1951).
11. R. M. Yerkes and M. I. Tomilin, *J. Comp. Psychol.* 20, 321 (1935).
12. J. Dollard and N. E. Miller, *Personality and Psychotherapy* (McGraw-Hill, New York, 1950), p. 133; P. H. Mussen and J. J. Conger, *Child Development and Personality* (Harper, New York, 1956), pp. 137, 138.
13. M. A. Ribble, *The Rights of Infants* (Columbia Univ. Press, New York, 1943); D. W. Winnicott, *Brit. J. Med. Psychol.* 21, 229 (1948).
14. J. Bowlby, *Intern. J. Psycho-Analysis* 39, part 5 (1958).
15. Support for the research presented in this article was provided through funds received from the graduate school of the University of Wisconsin; from grant M-772, National Institutes of Health; and from a Ford Foundation grant.
16. G. van Wagenen, in *The Care and Breeding of Laboratory Animals*, E. J. Farris, Ed. (Wiley, New York, 1950), p. 1.
17. J. P. Foley, Jr., *J. Genet. Psychol.* 45, 39 (1934).
18. We no longer make the cloth mother out of a block of wood. The cloth mother's body is simply that of the wire mother, covered by a terry-cloth sheath.
19. H. F. Harlow, *Am. Psychologist* 13, 673 (1958); —— and R. R. Zimmermann, *Proc. Am. Phil. Soc.* 102, 501 (1958).
20. D. O. Hebb, *The Organization of Behavior* (Wiley, New York, 1949), p. 241 ff.
21. A. T. Jersild and F. B. Holmes, *Child Develop. Monograph No. 20* (1935), p. 356.
22. J. M. Arsenian, *J. Abnormal Social Psychol.* 38, 225 (1943).
23. R. A. Butler, *J. Comp. and Physiol. Psychol.* 46, 95 (1953).
24. ——, *J. Exptl. Psychol.* 48, 19 (1954).

Neural and Hormonal Mechanisms of Behavior

Physiological Causes and Consequences

Elizabeth Adkins-Regan

Introduction

Ethology has long had as one of its goals explaining behavior in terms of underlying physiological mechanisms. In Tinbergen's (1963) four-part agenda for the field, mechanisms are referred to as "causation." This interest in underlying mechanisms has its origin in both philosophical and biological concerns. Philosophically, it is rooted in reductionism—the desire to account for behavior in terms of a lower level of organization, and the belief that such explanations are intellectually satisfying. Reductionism, with roots in logical positivism, is not unique to ethology, and in fact is the dominant or even exclusive rationale behind much of modern science. To the lasting benefit of the field, however, Tinbergen and other key figures had the wisdom to envision reductionism as one of several complementary levels of analysis in the study of behavior, thereby helping to immunize generations of students against the intellectual naiveté of pure reductionism. When ethologists study mechanisms, they tend to do so in a distinctive manner that incorporates or at least acknowledges other levels of analysis, as the papers in this section illustrate. Perhaps because behavior itself is richly complex, many ethologists (von Holst is a notable early example) eschew highly simplified forms of reductionism in which single components that can be isolated from one another are viewed as causing behavior in a strictly deterministic sense. Instead, these ethologists favor a systems approach in which mechanisms are sets of dynamically interacting processes (Fentress 1992).

The search for causal mechanisms is also driven by a compelling need to account for the two most striking characteristics of behavior: temporal variability and pattern. Behavior starts, stops, changes. It is unpredictable, at least at first glance. A rooster suddenly crows, for no apparent reason. A female hamster attracts a male to her burrow one day and aggressively drives him out the next. A leopard ignores the dead impala stored in a tree today but returns tomorrow to gorge itself. At the same time, behavior appears to consist of highly organized motor patterns that are predictable on a short time-scale. Picture, for example, a millipede crawling or a bird building an elaborate nest. Spontaneity, change, organization of movement—these are the qualities that make animals so different, so *animate*, compared with plants or pebbles. Ethologists seek to explain the variability of behavior over time—to discover the mechanisms and the intervening processes influencing the probability that a behavior will be performed under a given set of conditions—and to account for the organization of motor output.

Mechanism-oriented ethologists are of necessity multidisciplinary thinkers and researchers, simultaneously animal behaviorists and neurobiologists, endocrinologists, or sensory physiologists. During the initial decades of ethology, these other fields of study were in similarly early stages of development. For example, Ramón y Cahal's (1990) proposal that the neuron is the basic unit of the nervous system was not widely accepted until the early 1900s, and the term "hormone" was first used by Starling in 1905 (Carter 1974). Thus, the progress achieved by the authors of the classic papers in this section and by their forebears is all the more remarkable.

The Locus of Control: Exogenous or Endogenous? Central or Peripheral?

Attempts to explain what initiates, terminates, and organizes behavior have raised important questions about the source or locus of the factors responsible for behavioral change and patterning. Of particular importance are the relative roles of external stimuli and peripheral sensory and motor mechanisms as opposed to more internal, endogenous factors located in the central nervous system (note, for example, the title of von Holst's 1954 paper). From our late twentieth-century vantage point it is easy to see that this is ultimately a false dichotomy and that of course both kinds of factors must be operating together. But the significance of these central vs. peripheral debates for shaping research on mechanisms is unquestionable. Their conceptual roots can best be appreciated by recognizing that the dominant paradigm for thinking about behavioral and neural mechanisms during the early decades of the twentieth century assumed that Sherrington's stimulus-response reflex was the fundamental unit. Within this paradigm, each change in behavior is triggered by some external stimulus, patterned behavioral sequences consist of chains of successive reflexes, each providing the stimulus for the next, and therefore peripheral sensory and motor mechanisms are the source of behavioral change and organization. In animal psychology, this paradigm was manifest in behaviorism (also sometimes referred to as S-R, or stimulus-response psychology) which, in its most extreme form, denied both the existence of endogenous control mechanisms and also the validity of hypothesizing or conceptualizing any such processes.

Ethologists have long played an active role in challenging the hegemony of this stimulus-response paradigm; the idea that complex motor behavior was subject to considerable endogenous control was a major tenet of the early work of Lorenz, Tinbergen, and von Holst. Tinbergen (1950), in the paper included here, explicitly rejects the claim that behavior consists solely of reflexes, and argues that the spontaneity and variability of "instinctive" behavior make it no less respectable (only more difficult) for neural study than simpler reflex-like acts. Von Holst opens his paper by rejecting reflex-mechanism views of behavior. He is also responsible for some of the first experimental work pointing to a class of mechanisms that would later be called central pattern generators—mechanisms in the central nervous system that provide and maintain the patterned output commands to the muscles without requiring afferent input to coordinate the action. Subsequent work demonstrated that such patterned behaviors as walking, flight, and rhythmic sounds can be produced by neuronal networks isolated from sensory feedback, as in Wilson's (1961) important work on locust flight. These demonstrations of central pattern generators validated the early ethologists' assumption that complex sequences of movements could be wired into the nervous system as unitary chunks of behavior.

Biological Clocks

Ethologists were quick to appreciate that diurnal rhythms in behavior are one of the most salient and predictable forms of behavioral change, and to realize that such rhythms raise interesting questions about the locus of control. Many species are active only during the day or only at night. The dramatic change in light level (and often temperature as well) as the sun rises and sets provides possible ex-

ternal cues for diurnal patterning in behavioral change. This phenomenon then would appear at first glance to fit comfortably within the S-R paradigm. Experiments by ethologists and others, however, provided compelling evidence that the production and maintenance of these daily rhythms are not dependent on external cues, but rather are generated by an endogenous mechanism, a biological clock. Because of the dominance of the S-R paradigm, the proposal that daily rhythms are endogenously produced initially met with considerable skepticism and even outright denial. This skepticism was fueled by the difficulty of ensuring that all possible external physical cues for timing had been eliminated from the animal's environment.

The Aschoff paper included here was part of a 1960 symposium, the Cold Spring Harbor Symposium on Quantitative Biology (introduced by Bünning), devoted entirely to the subject of biological clocks. The symposium and its subsequent publication were landmark events, bringing rhythmic phenomena to the attention of the wider scientific community and convincing most of the remaining skeptics that the rhythms actually were endogenously generated. The critical evidence can be seen in Aschoff's experiments. When the obvious external cues are removed, the period of the animals' rhythms is not precisely 24 hours (thus the name "circadian rhythms"); that is, the period no longer corresponds to that of any known or probable physical cue. In addition, each individual subject's period is slightly different, which would not be the case if all were responding to some unknown external cue with a period slightly different from 24 hours. Aschoff's work is also important for clarifying the role of external cues: these cues function to synchronize or entrain the basic endogenous rhythms so that they are in register with the external world. He introduces the term "Zeitgeber" (time-giver) to refer to this role, and demonstrates experimentally the interplay ("coaction") between endogenous and exogenous factors. This paper also illustrates the value of engineering-based models of periodic variation for an understanding of behavioral rhythmicity in a wide array of animal species, making it clear why phrases such as "biological oscillators" and "animal chronometry" became widely used. The paper immediately following Aschoff's in the Cold Spring Harbor volume, by DeCoursey (1961), is a classic in its own right. The circadian activity rhythms of the flying squirrels in her experiments reveal a nearly perfect internal clock at work. Her paper addresses the issue of function by explaining why easy entrainment by light (compared with other possible Zeitgebers) would be adaptive for wild flying squirrels.

The papers by Aschoff and others from the symposium set an agenda for research on biological rhythms that would shape the field for the next decades (as revealed by the contents of a more recent volume edited by Aschoff [1981]). These early papers highlighted the most exciting and potentially fruitful topics, such as photoperiodism (control of season breeding by daylength), the role of circadian clocks in the measurement of daylength, and the intriguing possibility of endogenous monthly and annual rhythms. A particularly important insight for ethology was the realization that, if an animal has internal clocks that keep good time ("chronometers"), then it might be able to use the positions of the sun and stars to orient and navigate. This connection between internal clocks and celestial navigation inspired some of the most elegant and sophisticated experiments in all of

ethology (Emlen 1969; Keeton 1969; Kramer 1957; von Frisch and Lindauer 1956).

Others were eager to solve the mystery of the clock mechanism itself. Where is it? In the brain? What is it? A type of central pattern generator? How does it keep such exact time, even in animals whose body temperature varies widely? These questions stimulated much important research on the function and products of the pineal gland, the physiological and behavioral effects of pineal melatonin, and, more recently, the discovery that the suprachiasmatic nuclei in the hypothalamus are the locus of a clock whose integrity is essential for normal circadian, estrous, and seasonal rhythms (Rusak and Zucker 1979; Klein, Moore, and Reppert 1991).

Rhythmicity characterizes human life as well and, with the publication of Richter's book (1965), the importance of rhythms for an understanding of human behavior, health, and disease began to be appreciated. Women, like females of other Old World primate species, have a distinctive monthly physiological rhythm manifested in the menstrual cycle. McClintock's insight was the realization that, if other kinds of rhythms can become synchronized between individuals because of entrainment by a Zeitgeber, so too might menstrual cycles of women become synchronized as a result of exposure to an external cue. She demonstrated menstrual synchrony in humans for the first time by showing that the cycles of women living in a college dormitory became progressively more in phase as the academic year proceeded. This synchronization occurred only between close friends, as if the Zeitgeber is an interpersonal cue, rather than, say, the food in the dining hall. The resemblance to estrous synchrony in rodents and certain other mammals suggests that menstrual synchrony may be biologically significant (McClintock 1987). Menstrual synchrony continues to be a subject of lively discussion and debate with respect to the adaptive function of synchrony in mammalian ovulatory cycles and the identity of the Zeitgeber. Several laboratories have searched for a chemical cue that might be responsible for menstrual synchrony in humans; the most promising candidates are odors from the armpit (see, for example, Preti et al. 1986).

Social Influences on Hormonal Physiology

As the paper by McClintock implies, not only can changes in hormone levels cause behavioral change (the subject of the next section), but external cues also can alter internal hormonal secretion. When the external cues come from social interactions, behavioral change can cause hormonal change. This reversal of the usual direction of causality is the "other half" of the relationship between hormones and behavior. In spite of voluminous indirect evidence that external cues, including social cues, change gonadal hormone production, there was considerable reluctance by endocrinologists to acknowledge such phenomena. In contrast to the experience of the early biological rhythm researchers with resistance to the idea of endogenous control, here resistance was to the proposal of external control. The motivation for the skepticism, as so often happens in reductionistic science, was the absence of a known mechanism to account for the phenomenon. By what means could external stimuli change the hormonal activity of the gonads or other internal endocrine glands? Researchers already knew that the secretions of the anterior pituitary regulated gonadal status and

hormone secretion, but how on earth could the sight, sound, or smell of another animal (or even daylength, for that matter) have any effect on the gonads, which lie keep in the body and far from the brain? Or how could these external cues possibly affect the anterior pituitary, which, although close to the brain, is not innervated? Harris's (1955) discovery of the specialized portal circulatory system by which hypothalamic hormones regulate the anterior pituitary solved the problem. This mechanism of hypothalamic-pituitary communication made discoveries by animal behaviorists, such as those described in Lehrman's (1965) paper, credible in the eyes of physiologists. Knowledge of this mechanism of communication also encouraged serious interest in psychosomatic medicine (now more often called behavioral or psychological medicine) by the medical mainstream. It is no accident that Lehrman entitled one of the sections of his paper "psychosomatic regulation of the reproductive cycle."

Lehrman's paper reviews his experimental analyses of courtship and parental behavior during the reproductive cycle of ring doves. The paper includes his evidence that the presence of the male dove stimulates egg-laying by the female, and shows how, at each stage of the breeding cycle, social and hormonal interaction and feedback facilitate synchronization between the pair. Because male and female pigeons and doves share parental care in a closely coordinated manner, this physiological synchronization would appear to be highly adaptive. This paper illustrates nicely Lehrman's view of avian parental behavior not as reflexive, instinctive behavior rigidly controlled by hormones, but rather as flexible and learned interactions and negotiations between individuals. Parental behavior is mediated by responses, including hormonal responses to appropriate external stimuli. This emphasis on experiential, exogenous influences is rather different from the thinking of some of Lehrman's contemporaries, such as Klinghammer and Hess (1964), who emphasize innate factors. The resulting controversy (see Klopfer and Hailman 1972, for key references and a discussion) is but one of many manifestations of the difference in theoretical perspective between the integrative, epigenetic, environmental approach of some American ethologists (particularly those with intellectual connections to the American Museum of Natural History) and the European ethologists of the same period (see also Part 2).

McClintock's paper is another striking demonstration of the influence of external factors on gonadal hormone secretion, and is one of the first nonanecdotal extensions of this phenomenon to humans. Research such as McClintock's and Lehrman's has helped inspire several important lines of research central to ethology and to hormones and behavior. Silver's work (Silver and Norgren 1987) has shown how the circadian clock contributes to the synchronization of incubation by the male and female ring dove. Experiments by Cheng (1983) have added a new chapter to the ring dove story by showing that the proximate behavioral stimulus responsible for egg-laying by the female dove is not the male's courtship itself but the female's own behavior in response to the male. Crews (1980) successfully applied Lehrman's integrated approach to the study of hormones and behavior to a reptile, the green anole. Changes in circulating hormone levels in response to social stimuli are now known to occur commonly in a variety of vertebrates. In mammals, hormonal responses to conspecific stimuli can occur quite rapidly, within seconds or minutes, presenting a much more dynamic picture than the older view of hormonal change as sluggish. The behavioral interactions that occur when a male and female mouse first encounter one another are paralleled at the internal hormonal level (Bronson 1989) in a manner reminiscent of Lehrman's doves, but compressed in time. Odors are now known to be key elements of the social stimulus inducing hormonal change in rodents, one of many reasons why chemical communication is now an active subfield of ethology.

Behavioral Control by Hormones: Organization and Activation

It has long been known that removal of the testes reduces the sexual and aggressive behavior of male domestic animals. The discovery that glands such as the testes and adrenals produce hormones led to extensive investigation of the role of these hormones in the development and expression of adaptive behaviors such as courting, singing, mating, parenting, and fighting. A prototypical experiment uses the "remove-replace" design: the gonads or some other endocrine gland are removed, depriving the animal of the hormone of interest, and resulting in a decline or even cessation in behavior; afterwards, the animal is then given hormone replacement, restoring behavior to its original level. The identification and synthesis of the major gonadal sex hormones (the sex steroids) in the 1930s greatly improved the ability of investigators to control the identity and amount of hormone administered during the replacement phase. Experiments during the 1930s and 1940s used a wide array of species in addition to the common laboratory animals. The experiments of Noble at the American Museum of Natural History are particularly impressive with respect to phylogenetic diversity; his species repertoire included fish, frogs, lizards, gulls, herons, and chickens (see Young 1961, for a summary of these ground-breaking experiments in the comparative study of hormones and behavior).

Activational Effects of Hormones

Beach and Holz-Tucker's paper, while not the first to show an effect of hormones on behavior, is nonetheless an essential contribution. Their methods and the description of these methods are unusually careful, and a dose-response relationship between hormone administered and behavior stimulated is obtained. The paper raises interesting questions about hormones as a possible source of individual differences in sexual behavior that are still largely unanswered. Beach went on to become a major figure in the twentieth-century study of mechanisms of behavior. Beach's landmark book *Hormones and Behavior* (1948) defined the field, presented key concepts and questions to guide research, and helped attract students to the study of endocrine mechanisms of behavior. His other important contributions to ethology include: modeling and experimental analysis of sexual arousal in animals and humans; introduction of the concept of proceptivity (Beach 1976), an important departure from previously oversimplified notions of female sexual behavior; development of scenarios for the evolution of hormone-behavior relationships; and elucidation of the manner in which hormones and experience inter-

act to influence levels of adult behavior (see McGill et al. 1978, for further discussion and Beach's full bibliography). Interactions between hormones and social experience were also a central concern in Guhl's (1958, 1964) classic work on factors determining dominance rank in chickens.

Beach was an active contributor in the central vs. peripheral debate, reminding his colleagues who were taking the central side that there was no direct evidence that hormones acted on the central nervous system. Beach himself argued for the importance of peripheral loci of hormone action, and backed up his claim with experimental evidence that actions of androgens on penile morphology were an important determinant of masculine sexual behavior. Subsequent experiments by Fisher (1956) and Barfield (1969) in which small testosterone implants were placed directly into the brains of castrated male rats and roosters showed that hormones acting centrally alone could stimulate behavior, and it is now appreciated that the brain is not only a target for hormones (much as other organs such as the uterus or prostate are targets) but also produces hormones itself. In short, hormones act at multiple loci, some central, some peripheral. The peripheral zone is now known to include hormonal effects on the production of odors. The study of the interactions between hormones, pheromones, and behavior has greatly illuminated our understanding of the behavior of a wide array of species (see, for example, work by Signoret [1970] on pigs, by Stacey [1987] on goldfish, and by Vandenbergh [1988] on mice).

Thanks to foundations laid by Beach and others, behavioral endocrinology is now a thriving subfield of both ethology and physiology, with its own journal (*Hormones and Behavior,* initially coedited by Beach) and textbooks (Becker, Breedlove and Crews 1992; Brown 1994; Nelson 1995). Important technical advances have occurred since Beach and Holz-Tucker's paper, such as assays (especially radioimmunoassays) for accurately measuring hormones in small blood samples, methods for measuring hormone receptors in the nervous system and other tissues, and drugs that function as antihormones.

A particularly exciting development with roots in the ethological tradition is the emergence of field behavioral endocrinology. Much is being learned about how hormone levels actually change in wild species in relation to the different stages of their breeding cycles, generating annual rhythms of reproductive behavior (see, for example, the extensive review by Wingfield and Farner [1993] of the avian literature), how hormone levels respond to features of social organization such as territorial challenges or stressful dominance interactions (for example, Wingfield et al. 1990; Sapolsky 1993), and how hormones contribute to mating systems and reproductive strategies (for example, Thompson and Moore 1992; Wingfield et al. 1990). Some results have provided much-needed confirmation of conclusions derived from the laboratory (for example, in many wild birds copulation occurs mainly when sex hormone levels are elevated: Wingfield and Farner [1993]). Other results violate widely held assumptions (for example, the same territorial behavior that is highly correlated with testosterone during the breeding season can occur independently of testosterone outside the breeding season: Wingfield [1994]), raising important questions and stimulating new conceptual developments.

Organizational Effects of Hormones

By the 1950s, it was clear that gonadal and other hormones were major influ-ences on the expression of reproductive and aggressive behavior by adult animals, on the onset of adult reproductive behavior at puberty, and on the occurrence of sex differences in adult behavior. Furthermore, experimental embryologists such as Jost had shown that hormones produced during very early development (fetal or neonatal development) were a major determinant of adult reproductive anatomy (Jost 1985). During a critical period in early development, testicular hormones acted as organizers or inducers of male structures in genetic male mammals. It was Phoenix, Goy, Gerall, and Young's (1959) insight that the same might be true for reproductive behavior. In their experiments with guinea pigs, described in the paper included here, females born to mothers treated with a form of testosterone were simultaneously masculinized (showed more masculine sexual behavior than controls) and defeminized (showed less feminine sexual behavior than controls), effects which were permanent. These experiments were the first conclusive demonstration that hormones administered during a brief period in early development can permanently alter the responsiveness of the animal to adult hormones, as if the animal has been sex-reversed behaviorally. The realization that there might be ontogenetic parallels between the genitalia and the "neural tissues destined to mediate mating behavior in the adult" (see last sentence of Phoenix et al.'s abstract) was remarkable because very little was known at the time about neural tissues as targets of hormone action. The paper spells out what is still known as the organizational theory of hormone action, which distinguishes organizational effects (those that occur early in development, are limited to a critical period, and are permanent) from activational effects, such as those demonstrated by Beach and Holz-Tucker and Lehrman, in which behavior that already has been organized occurs in response to hormonal stimulation in adulthood. A review a few years later by Young, Goy, and Phoenix (1964) included some evidence that the same theory might account for adult sex differences in nonreproductive behavior, and might even apply to behavior that is not dependent on adult hormones for expression, such as play behavior by juvenile monkeys.

The organizational theory of hormone action generated much interest in studying hormonal contributions to the development of sex differences in behavior in a wide range of species. This theory remains, with important modifications, a major conceptual framework for understanding the development of sexual dimorphism in behavior and nervous systems. Furthermore, along with imprinting and song learning, sexual dif-ferentiation stands as one of the clearest examples in ethology of a critical period phenomenon and as an important model for the study of epigenetic processes in development. Several lines of research that descended from Phoenix et al.'s (1959) paper are relevant to modern ethology: explorations of the phylogenetic and behavioral generality of the organizational theory (for example, Adkins 1981), discoveries of sexual dimorphism in the brain and spinal cord underlying courtship and mating behavior that appear to be a product of sexual selection (Arnold and Gorski 1984; Kelley 1988), and challenges to the nature of the theory itself, which is deterministic and ignores the transactional nature of mammalian social development (Moore 1990). As ethologists learn more about mate choice and mating strategies, we also hope to see organization-like concepts extend into these important behavioral domains.

Neuroethology: The Evolutionary Approach to Behavioral Neuroscience

Neuroscience in all its forms has enjoyed an explosive increase in interest and attention in recent years. How does the ethological approach differ from other ways of studying the nervous system? Neuroethologists, with roots firmly planted in several disciplines, view the nervous system from a broad, integrated perspective. They make it their business to remember that neural mechanisms are themselves shaped by natural selection, just as behavior is. Neuroethologists also feel that behavior of biological significance in the animal's "real world" is most likely to reveal the functional organization of the nervous system (a concept discussed further in Part 5). To a neuroethologist, the behavior of the animal and the neural mechanisms for behavior should be studied together, rather than by different disciplines and people. Moreover, the behavior itself should be studied in natural or semi-natural environments whenever possible. Neuroethologists prefer to combine levels of analysis, particularly mechanism and function levels (and, more recently, ontogenetic as well—Stehouwer 1992). These researchers study well-defined behavioral units of likely adaptive significance, such as prey-catching responses or the production of courtship signals. The term "neuroethology" appeared relatively recently (first used in 1963 by Brown and Hunsperger and more fully articulated by Hoyle in 1984), but one can see clear precedents both in nomenclature (Tinbergen's "ethophysiology" and von Holst's "Verhaltens-physiologie" [comparative behavioral physiology], discussed by Ewert 1980) and in the research approach, exemplified in work by Von Holst (1954) and Dethier and Bodenstein (1958). Early work was somewhat primitive by modern standards, and ideas were seldom subjected to direct experimental tests. This lack of experimental rigor is understandable given the limited array of techniques available at the time and how little was known about the biochemistry or plasticity of the nervous system. Nevertheless, the intellectual creativity and breadth of vision of the early neuroethologists imparts lasting value to their work, and these qualities continue to contribute to neuroethology's distinctive flavor.

Neuroethologists search for relatively precise mechanisms or systems of mechanisms underlying species-specific behavior patterns, and so the choice of study animal is critical. The important criteria include: (a) ease of observation in the field; (b) reasonably normal behavior in the lab, even while the nervous system is manipulated (for example, during recording of neuronal activity); (c) one or more well-defined and easily recorded behavior patterns; and (d) nervous system accessibility. Invertebrates, especially arthropods such as insects, are often the choice; the papers by Dethier and Bodenstein and by von Holst (included in this volume) help show why. How convenient, for example, to be able to observe and interpret the feeding response of a blowfly head isolated from the body!

The paper by von Holst (1954) introduces and develops two key neural concepts: re-afference and efference copy. These concepts explain the ability of animals to execute visually guided movements correctly—to perceive a stable world while moving through it. Von Holst's use of re-afference and efference copy is a good example of a systems approach to causation, and was remarkably sophisticated for that era. These ideas had an important influence on subsequent research on visually guided behavior, such as Ewert's (1980) studies of prey catching by toads and Held and Hein's (1963) research on the role of active vs. passive movement in the development of visuo-motor coordination in cats. Appreciating the difference between active and passive movement is now seen as critical to understanding a variety of types of behavioral plasticity and learning, and it is possible to see remnants of von Holst's ideas in current thinking about the importance of active participation in song learning by birds and language learning by humans.

Tinbergen's (1950) paper was part of a collaborative symposium in 1949 between the Society of Experimental Biology and the Institute for the Study of Animal Behaviour (Cambridge, England). The symposium included papers by other notables of early ethology, such as von Holst, Lorenz, Koehler, and Baerends, and an entire section titled "Central and Peripheral Control of Behaviour Patterns." Tinbergen's contribution firmly established modeling as a major form of theorizing and conceptualization in ethology. His own model provided an ethological alternative to reflex models that could better explain the organized, integrated patterning of complex behavioral sequences. Tinbergen's model is explicitly hierarchical. Notice that the "highest" component is one triggered by internal factors, again showing his emphasis on endogenous control. Also, the greater complexity of his model was a significant improvement over Lorenz's "psychohydraulic" model, which could account only for the production of a single fixed action pattern instead of an organized sequence (and even then not very well). More recently, Dawkins (1976) and Fentress (1983) have provided theoretical and empirical support for the explanatory power of hierarchical models.

The von Holst and Tinbergen papers raise important issues about what is useful in a model. Tinbergen's was useful as a conceptual framework rather than a source of concrete, testable hypotheses. This model was never intended to be isomorphic with the nervous system itself—its components are behavioral, not physiological; while possible links between these components and neurophysiology are discussed, the components are not intended to correspond to particular parts of the brain, types of neurons, etc. (see Tinbergen 1950, p. 310). In spite of the name of the symposium (Physiological Mechanisms in Animal Behaviour), there is no actual physiology in Tinbergen's paper. Von Holst's model also was not intended to be isomorphic, but it was intended to generate testable hypotheses, making it unusual for its era. Whether or not isomorphism or direct experimental testability are essential features of neuroethological models depends on what these models are intended to accomplish. To a "pure" reductionist, a model must be isomorphic to be useful; to those primarily interested in analysis at the behavioral level, isomorphism is not essential so long as the model stimulates critical research that would not otherwise have been done.

The actual experiments conducted on neural mechanisms, however, have changed radically since the days of von Holst and Tinbergen, due primarily to the availability of sophisticated new techniques for manipulating the nervous system. Simultaneously, there has been a major infusion into theoretical work of concepts and models from engineering, especially computer science. In modern parlance, Tinbergen's model is based on serial processing, as were most models at that time and most accounts of nervous system organization and function. We now know that parallel processing is common in the nervous system (see, for example, Carr and Konishi 1990). Computer-based models (neural network, connectionist, computational, distributed control, etc.) raise the same issues as other machine-based models about what makes a good model. Are these newer models exciting? Irrelevant? Either way, these models remind us that the tradition of neural modeling begun by the early ethologists continues to provide insights on the neural mechanisms that support behavior (Heiligenberg 1991).

Many important advances in neuroethology have been made since the days of von Holst and Tinbergen. For example, there are now biochemical studies of the nervous system that reveal both the rich array of molecules underlying behavioral mechanisms and the functions of a few of these molecules. Study of the action of hormones and other neuromodulators on neural systems that function as central pattern generators has led to the development of more dynamic models for central pattern generators, helping to explain how the basic pattern is modified adaptively in the normal context (Harris-Warrick et al. 1992). A second example is the discovery that the function, biochemistry, and anatomy of neurons is much more plastic, much more subject to modification by external stimuli and experience (both in development and in adulthood) than previously realized (Easter et al. 1985; Greenough 1975; Kaas 1991). For instance, the cellular and gross anatomy of the songbird brain changes dramatically during the period of song learning, providing an excellent model system for investigating neural mechanisms of learned, adaptive behavior (Arnold et al. 1987; Nottebohm 1991). A third advance in the field of neuroethology is the recognition that the developmental approach (pioneered by the classic work of Sperry [1963] and others), like the comparative approach, can be a powerful tool in teasing apart the functional organization of the nervous system (see, for example, the special issue of the *Journal of Neurobiology* [1992, volume 23, no. 10] devoted to developmental neuroethology).

Neuroethology is now a full-fledged subfield of neuroscience and of ethology, with three textbooks (Camhi 1984; Ewert 1980; Guthrie 1980) and its own international society. A brief glance at these books or the programs of the society's meetings shows a continuing focus on the analysis of sensory and motor systems in controlling behavior, and a continuing interest in feature detectors, pacemakers, command neurons, and the like.

Motivation: A Central Concept or a Nineteenth Century Concept?

Neither "motivation" nor the related term "drive" is a kind of behavior or physiological mechanism. Instead, these terms refer to implied processes ("theoretical constructs" or, in more operational terms, "intervening variables") that account for a particular kind of variability in behavior. This variability is a change in the probability or intensity of a response in the absence of any obvious change in the environment. The prototypical example is feeding, which in many species depends importantly on time since the last meal, as if the animal gets hungrier (certainly we do) as the elapsed time increases. Thus motivation is not a physiological mechanism per se, but a behavioral mechanism in much the same way as innate releasing mechanisms, search images, imprinting, and sun compasses are behavioral mechanisms. Motivation is an internal process, something endogenous that can initiate and modify behavior, that can cause an animal to start searching for food and then to eat with greater relish.

Along with instinct, motivation has been a central concept in ethological theory. Early research on hormones and behavior often explicitly sought to understand "sex drive" (sexual motivation). Motivation ("action specific energy") also was key in Lorenz's psychohydraulic model, and Tinbergen's (1950) model placed motivation at the top of the hierarchy. Much creative experimental work attempted to understand what motivation is, how many different kinds there are, and how motivation effects a response appropriate to the environment and to the particular need state. As an internal process, motivation has figured importantly in several versions of the central vs. peripheral debate (Dethier and Bodenstein's 1958 study is a good example). Neural and hormonal substrates of motivation have been extensively explored, and the close links to these kinds of mechanisms explain why motivation is included in this section.

An important issue in the study of motivation has been whether there is a general component common to all motivated behavioral activities. Alternatively, each functional category of behavior could have its own separate motivational system and mechanisms. Looking for "brain centers" for motivation is one way to address this issue of general vs. specific mechanisms. Once again, some of the most important early work was conducted by von Holst. Together with von Saint Paul, von Holst was able to elicit in domestic fowl a surprisingly rich array of behavior patterns typical of highly motivated states by electrically stimulating the brain at a variety of loci (von Holst and von Saint Paul 1963). Results of this study revealed different drives, located in different places, that interact in complex ways. On the other hand, work by other researchers suggested that the hypothalamus was critical for a wide variety of motivated behaviors, including sex, aggression, feeding, drinking, thermoregulation, and flight from danger. These studies supported the concept of a central locus for a more general motivational state and ensured the status of the hypothalamus as a favorite structure for behavioral study for several decades.

Dethier and Bodenstein's (1958) goal was to extend to an invertebrate the study of variability in the tendency to feed: do blowflies have an endogenous central nervous system mechanism that causes the fly to feed when food-deprived? By means of meticulously described and executed experiments, the authors conclude that, unlike rats, which have both hunger and satiety centers, blowflies only have a satiety mechanism. Hunger is the absence of satiety ("hunger can be equated with absence of stimulating fluid in the foregut . . ." [p. 175]) and therefore hunger in the blowfly is not a drive-like central nervous system process. This paper, along with others in this volume, also serves as an important reminder of the virtues of "small science" by showing that research does not have to be technologically sophisticated or expensive to be exciting and worthwhile, a message also present in Dethier's delightful book *To Know a Fly* (1962, with foreword by Tinbergen).

As terms like "action specific energy," "psychohydraulic model" and "drive" reveal, classical ethological models of motivation were energy models. If motivational energy built up beyond a certain point, it "spilled" over into other, less appropriate behaviors (displacement). If two drive states were engaged simultaneously, conflicts resulted that were expressed through displays. Hinde's (1960) paper is reprinted here because it helped lay to rest this kind of theorizing in which physical energy and psychological drive were thought of as having the same properties. Its companion piece (Hinde 1959), critiqued "general drive" concepts; both papers are superb examples of conceptual analysis in ethology. Hinde reminds ethologists that primary concepts must not be used carelessly or unnecessarily. Rather, as in any discipline, these concepts must be carefully defined, their conceptual status acknowledged, and their use restricted to when the terms actually add something to our understanding of behavior. With the aid of Hinde's analysis, one can see that energy concepts of motivation bear a suspicious resemblance to nineteenth-century thinking that biological systems parallel mechanical systems by obeying all the laws of physics, including conservation of energy. While energy models may be appropriate as an initial approach to physiology (and were essential in the nineteenth-century refutation of vitalism), Hinde questions the usefulness of energy concepts for understanding behavior. The role that Hinde plays in this paper, as the critic who keeps other people intellectually honest, is a vital one for the health of any discipline (see Bateson [1991] for a recent summary of Hinde's many other important contributions to ethology).

Is motivation still a central concept in ethology, or did Hinde's essay and other critiques force it into the theoretical periphery? On the one hand, much important work continues to be produced by researchers whose explicit goal is to study motivation (see, for example, Satinoff and Teitelbaum 1983). Textbooks still include one or more chapters on motivation. McFarland's work (summarized in his 1985 textbook) has helped modernize the concept by formulating it in a more explicitly adaptive context. On the other hand, ethology journals of recent years reflect work in which motivation and drive figure much less prominently, both theoretically and empirically. As for textbooks, it is hard to fault Slater's (1985) conclusion that motivation is a difficult subject in contemporary ethology, or the statement by Toates and Birke (1982) that "'motivation' often means simply a convenient chapter heading for a discussion of hunger, thirst, and sex . . . with little or no attempt to present a synthesis or overall theoretical view" (pp. 191–92).

Instead of motivation and drive, more recent work emphasizes information processing, decision making, behavioral rules such as foraging rules, and other kinds of cognitive approaches to behav-

ior that tend to lack any affective component or process (see, for example, McFarland 1985). To those who agree with Beach that "motivation is the phlogiston of psychology" (Zucker 1983, p. 3), this change in emphasis is a welcome development. But to others, this emphasis is the late twentieth-century analogue of excessive behaviorism which, when combined with the excessive reductionism of sociobiology, does not bode well for the future vitality of ethology (see Bateson and Klopfer 1989). In causation as in the rest of the ethological agenda, it remains crucial to strike the proper balance between the need for objectivity and the need to retain the essence of the behavioral phenomena to be explained. Gaining a proper balance brings us back to the primary issue in Galef's historical introduction (Part 1). As mechanism-oriented ethologists enter the next century, they, too, must free themselves of unnecessary intellectual baggage that slows progress toward greater biological knowledge while they retain those concepts that are truly vital to an understanding of behavior.

Literature Cited

Adkins, E. K. 1981. Early organizational effects of hormones: An evolutionary perspective. Pp. 159–228 in *Neuroendocrinology of Reproduction: Physiology and Behavior*, N. T. Adler, ed. New York: Plenum Press.

Arnold, A. P., and R. A. Gorski. 1984. Gonadal steroid induction of structural sex differences in the central nervous system. *Annual Review of Neuroscience* 7:413–42.

Arnold, A. P., S. W. Bottjer, E. J. Nordeen, K. W. Nordeen, and D. R. Sengelaub. 1987. Hormones and critical periods in behavioral and neural development. Pp. 55–97 in *Imprinting and Cortical Plasticity*, J. P. Rauschecker and P. Marler, eds. New York: Wiley.

Aschoff, J. 1960. Exogenous and endogenous components in circadian rhythms. *Cold Spring Harbor Symposia in Quantitative Biology* 25:11–28.

Aschoff, J. 1981. *Handbook of Behavioral Neurobiology*.

vol. 4, *Biological Rhythms*. New York: Plenum Press.

Barfield, R. J. 1969. Activation of copulatory behavior by androgen implanted into the preoptic area of the male fowl. *Hormones and Behavior* 1:37–52.

Bateson, P., ed. 1991. *The Development and Integration of Behaviour: Essays in Honour of Robert Hinde*. Cambridge: Cambridge University Press.

Bateson, P. P. G., and P. H. Klopfer, eds. 1989. *Perspectives in Ethology*, vol. 8, *Whither Ethology?* New York: Plenum Press.

Beach, F. A. 1948. *Hormones and Behavior*. New York: Paul B. Hoeber.

Beach, F. A. 1976. Sexual attractivity, proceptivity, and receptivity in female mammals. *Hormones and Behavior* 7:105–38.

Beach, F. A., and A. M. Holz-Tucker. 1949. Effects of different concentrations of androgen upon sexual behavior in castrated male rats. *Journal of Comparative and Physiological Psychology* 42: 433–53.

Becker, J. B., S. M. Breedlove, and D. Crews. 1992. *Behavioral Endocrinology.* Cambridge: MIT Press.

Bronson, F. H. 1989. *Mammalian Reproductive Biology.* Chicago: University of Chicago Press.

Brown, J. L., and R. W. Hunsperger. 1963. Neuroethology and the motivation of agonistic behaviour. *Animal Behaviour* 11:439–48.

Brown, R. E. 1994. *An Introduction to Neuroendocrinology.* Cambridge: Cambridge University Press.

Camhi, J. M. 1984. *Neuroethology. Nerve Cells and the Natural Behavior of Animals.* Sunderland: Sinauer.

Carr, C. E., and M. Konishi. 1990. A circuit for detection of interaural time differences in the brain stem of the barn owl. *Journal of Neuroscience* 10: 3227–46.

Carter, C. S., ed. 1974. *Hormones and Sexual Behavior.* Stroudsburg: Dowden, Hutchison & Ross.

Cheng, M.-F. 1983. Behavioural "self-feedback" control of endocrine states. Pp. 408–21 in *Hormones and Behaviour in Higher Vertebrates,* J. Balthazart, E. Pröve and R. Gilles, eds. Berlin: Springer-Verlag.

Crews, D. 1980. Interrelationships among ecological, behavioral, and neuroendocrine processes in the reproductive cycle of *Anolis carolinensis* and other reptiles. Pp. 1–75 in *Advances in the Study of Behavior,* vol. 11, J. S. Rosenblatt, R. A. Hinde, C. Beer, and M.-C. Busnel, eds. New York: Academic Press.

Dawkins, R. 1976. Hierarchical organisation: a candidate principle for ethology. Pp. 7–54 in *Growing Points in Ethology,* P. P. G. Bateson and R. A. Hinde, eds. Cambridge: Cambridge University Press.

DeCoursey, P. 1961. Phase control of activity in a rodent. *Cold Spring Harbor Symposia on Quantitative Biology* 25:49–56.

Dethier, V. G. 1962. *To Know a Fly.* San Francisco: Holden Day.

Dethier, V. G., and D. Bodenstein. 1958. Hunger in the blowfly. *Zeitschrift für Tierpsychologie* 15: 129–40.

Emlen, S. T. 1969. Bird migration: influence of physiological state upon celestial orientation. *Science* 165:716–18.

Easter, S. S., D. Purves, P. Rakic, and N. C. Spitzer. 1985. The changing view of neural specificity. *Science* 230:507–11.

Ewert, J.-P. 1980. *Neuroethology. An Introduction to the Neurophysiological Fundamentals of Behavior.* Berlin: Springer-Verlag.

Fentress, J. C. 1983. Ethological models of hierarchy and patterning of species-specific behavior. Pp. 185–234, IN: *Handbook of Behavioral Neurobiology,* vol. 6, *Motivation,* E. Satinoff and P. Teitelbaum, eds. New York: Plenum Press.

Fentress, J. C. 1992. History of developmental neuroethology: early contributions from ethology. *Journal of Neurobiology* 23:1355–69.

Fisher, A. E. 1956. Maternal and sexual behavior induced by intracranial chemical stimulation. *Science* 124:228–29.

Greenough, W. T. 1975. Experiential modification of the developing brain. *American Scientist* 63: 37–46.

Guhl, A. M. 1958. The development of social organisation in the domestic chick. *Animal Behaviour* 6:92:111.

Guhl, A. M. 1964. Psychophysiological interrelation in the social behavior of chickens. *Psychological Bulletin* 61:277–85.

Guthrie, D. M. 1980. *Neuroethology: An Introduction.* New York: John Wiley and Sons.

Harris, G. W. 1955. *Neural Control of the Pituitary Gland.* London: Edward Arnold.

Harris-Warrick, R., E. Marder, A. I. Selverston, and M. Moulins, eds. 1992. *Dynamic Biological Networks.* Cambridge: MIT Press.

Heiligenberg, W. 1991. *Neural Nets in Electric Fish.* Cambridge: MIT Press.

Held, R., and A. Hein. 1963. Movement-produced stimulation in the development of visually guided behavior. *Journal of Comparative and Physiological Psychology* 56:872–76.

Hinde, R. A. 1959. Unitary drives. *Animal Behaviour* 7:130–141.

Hinde, R. A. 1960. Energy models of motivation. *Symposia of the Society for Experimental Biology* 14:199–213.

Hoyle, G. 1984. The scope of neuroethology. *Behavioral and Brain Sciences* 7:367–412.

Jost, A. 1985. Sexual organogenesis. Pp. 3–20 in *Handbook of Behavioral Neurobiology,* vol. 7: *Reproduction,* N. T. Adler, D. Pfaff and R. W. Goy, eds. New York: Plenum Press.

Kaas, J. H. 1991. Plasticity of sensory and motor maps in adult mammals. *Annual Review of Neuroscience* 14:137–67.

Keeton, W. T. 1969. Orientation by pigeons: is the sun necessary? *Science* 165:922–28.

Kelley, D. 1988. Sexually dimorphic behaviors. *Annual Review of Neuroscience* 11:225–51.

Klein, D. C., R. Y. Moore, and S. M. Reppert, eds. 1991. *Suprachiasmatic Nucleus: The Mind's Clock.* New York: Oxford University Press.

Klinghammer, E., and E. H. Hess. 1964. Parental feeding in ring doves (*Streptopelia roseogrisea*):

innate or learned? *Zeitschrift für Tierpsychologie* 21:338–47.

Klopfer, P. H., and J. P. Hailman. 1972. *Control and Development of Behavior: An Historical Sample from the Pens of Ethologists*. Reading: Addison-Wesley.

Kramer, G. 1957. Experiments on bird orientation and their interpretation. *Ibis* 99:196–227.

Lehrman, D. S. 1955. The physiological basis of parental feeding behavior in the ring dove (*Streptopelia risoria*). *Behaviour* 7:241–86.

Lehrman, D. S. 1965. Interaction between internal and external environments in the regulation of the reproductive cycle of the ring dove. Pp. 355–80 in *Sex and Behavior*, F. A. Beach, ed. New York: Wiley.

Lindauer, M. 1961. Time-compensated sun orientation in bees. *Cold Spring Harbor Symposia on Quantitative Biology* 24:371–78.

McClintock, M. K. 1971. Menstrual synchrony in humans. *Nature* 229:244–45.

McClintock, M. K. 1987. A functional approach to the behavioral endocrinology of rodents. Pp. 178–203 in *Psychobiology of Reproductive Behavior. An Evolutionary Perspective*, D. Crews, ed. Englewood Cliffs: Prentice-Hall.

McFarland, D. 1985. *Animal Behavior: Psychobiology, Ethology and Evolution*. Menlo Park: Benjamin/Cummings.

McGill, T. E., D. A. Dewsbury, and B. D. Sachs, eds. 1978. *Sex and Behavior: Status and Prospectus*. New York: Plenum Press.

Moore, C. L. 1990. Comparative development of vertebrate sexual behavior: Levels, cascades, and webs. Pp. 278–99 in *Contemporary Issues in Comparative Psychology*, D. Dewsbury, ed. Sunderland: Sinauer.

Nelson, R. J. 1995. *An Introduction to Behavioral Endocrinology*. Sunderland: Sinauer.

Nottebohm, F. 1991. Reassessing the mechanisms and origins of vocal learning in birds. *Trends in Neurosciences* 14:206–11.

Phoenix, C. H., R. W. Goy, A. A. Gerall, and W. C. Young. 1959. Organizing action of prenatally administered testosterone propionate on the tissues mediating mating behavior in the female guinea pig. *Endocrinology* 65:369–82.

Preti, G., W. B. Cutler, C. R. Garcia, G. R. Huggins, and H. J. Lawley. 1986. Human axillary secretions influence women's menstrual cycles: the role of donor extract of females. *Hormones and Behavior* 20:474–82.

Ramón y Cahal, S. 1990. *New Ideas on the Structure of the Nervous System in Man and Vertebrates*. Translated by N. Swanson and L. W. Swanson. Cambridge: MIT Press.

Richter, C. P. 1965. *Biological Clocks in Medicine and Psychiatry*. Springfield: Thomas.

Rusak, B., and I. Zucker. 1979. Neural regulation of circadian rhythms. *Physiological Reviews* 59:449–526.

Sapolsky, R. M. 1993. Endocrinology alfresco: psychoendocrine studies of wild baboons. *Recent Progress in Hormone Research* 48:437–68.

Satinoff, E., and P. Teitelbaum. 1983. *Handbook of Behavioral Neurobiology*, vol. 6, *Motivation*. New York: Plenum Press.

Signoret, J. P. 1970. Reproductive behaviour of pigs. *Journal of Reproduction and Fertility*, Supplement 11:105–17.

Silver, R. and R. B. Norgren. 1987. Circadian rhythms in avian reproduction. Pp. 120–48 in *Psychobiology of Reproductive Behavior. An Evolutionary Perspective*, D. Crews, ed. Englewood Cliffs: Prentice-Hall.

Slater, P. J. B. 1985. *An Introduction to Ethology*. Cambridge: Cambridge University Press.

Sperry, R. W. 1963. Chemoaffinity in the orderly growth of nerve fiber patterns and connections. *Proceedings of the National Academy of Sciences* (USA) 50:703–10.

Stacey, N. E. 1987. Roles of hormones and pheromones in fish reproductive behavior. Pp. 28–60 in *Psychobiology of Reproductive Behavior. An Evolutionary Perspective*, D. Crews, ed. Englewood Cliffs: Prentice-Hall.

Stehouwer, D. J. 1992. The emergence of developmental neuroethology. *Journal of Neurobiology* 23:1353–54.

Thompson, C. W., and M. C. Moore. 1992. Behavioral and hormonal correlates of alternative reproductive strategies in a polygynous lizard: tests of the relative plasticity and challenge hypotheses. *Hormones and Behavior* 26:568–85.

Tinbergen, N. 1950. The hierarchical organization of nervous mechanisms underlying instinctive behavior. *Symposia of the Society for Experimental Biology* 4:305–12.

Tinbergen, N. 1963. On aims and methods of ethology. *Zeitschrift für Tierpsychologie* 20:410–33.

Toates, F. M., and L. I. A. Birke. 1982. Motivation: a new perspective on some old ideas. Pp. 241–42 in *Perspectives in Ethology*, vol. 5. *Ontogeny*, P. P. G. Bateson and P. H. Klopfer, eds. New York: Plenum Press.

Vandenbergh, J. G. 1988. Pheromones and mammalian reproduction. Pp. 1679–96, in *The Physiology of Reproduction*, E. Knobil and J. Neill et al., eds. New York: Raven Press.

von Frisch, K., and M. Lindauer. 1956. The "language" and orientation of the honey bee. *Annual Review of Entomology* 1:45–58.

von Holst, E. 1954. Relations between the central nervous system and the peripheral organs. *British Journal of Animal Behaviour* 2:89–94.

von Holst, E., and U. von Saint Paul. 1963. On the functional organisation of drives. *Animal Behaviour* 11:1–20.

Wilson, D. M. 1961. The central nervous control of flight in a locust. *Journal of Experimental Biology* 38:471–90.

Wingfield, J. C. 1994. Control of territorial aggression in a changing environment. *Psychoneuroendocrinology* 19:709:21.

Wingfield, J. C., and D. S. Farner. 1993. Endocrinology of reproduction in wild species. Pp. 164–328 in *Avian Biology*, vol. 9, D. S. Farner, J. R. King and K. C. Parkes, eds. London: Academic Press.

Wingfield, J. C., R. E. Hegner, A. M. Dufty, and G. F. Ball. 1990. The "challenge hypothesis": theoretical implications for patterns of testosterone secretion, mating systems, and breeding strategies. *American Naturalist* 136:829–46.

Young, W. C. 1961. The hormones and mating behavior. Pp. 1173–1239 in *Sex and Internal Secretions*, vol. 2, third edition, W. C. Young, ed. Baltimore: Williams & Wilkins.

Young, W. C., R. W. Goy, and C. H. Phoenix. 1964. Hormones and sexual behavior. *Science* 143:212–18.

Zucker, I. 1983. Motivation, biological clocks, and temporal organization of behavior. Pp. 3–22 in *Handbook of Behavioral Neurobiology*, vol. 6 *Motivation*, E. Satinoff and P. Teitelbaum, eds. New York: Plenum Press.

THE HIERARCHICAL ORGANIZATION OF NERVOUS MECHANISMS UNDERLYING INSTINCTIVE BEHAVIOUR

By N. TINBERGEN
University of Leiden

As long as workers on instinct focused their attention on to the fact that instinctive behaviour as a whole is often highly variable and plastic, while the only constant aspect was considered to be the end towards which it is directed, there was little hope of finding nervous mechanisms underlying the behaviour.

Although some writers called attention to the relatively rigid and simple innate activities that do exist, the majority of writers considered these rigid activities to be of minor importance; they were either considered to be no 'true' instinctive activities, or it was attempted to show that even they were not rigid but 'variable'. In pointing out this variability the need for sharply distinguishing between intrinsic variability and effects due to rigid responsiveness to variable external factors was rarely felt.

We begin to realize now that the fact that an animal may use various behaviour patterns in turn in order to attain one special end does not necessarily mean that this plastic behaviour is not dependent on nervous mechanisms, but that it might also, and certainly does, mean that the underlying nervous mechanisms are much more complicated than was expected before. The notion that a relatively limited number of mechanisms are responsible for plastic behaviour gains in probability when one obvious fact is better realized, viz. that in each case of plastic behaviour, however complicated it may be, there are many more ways in which an animal could, yet does not, try to attain the end than there are methods which it does use. One instance may make the point clear. When one of the three eggs of a gull is taken out of the nest and put at a foot's distance in front of the sitting gull, the bird will 'retrieve' it. This it does by bringing the bill behind the egg, and shovelling it back while carefully balancing it on the relatively narrow edge of the lower mandible. When watching the repeated clumsy attempts of the bird, which have often to be repeated quite a number of times before they meet with success, one is impressed by the bird's obvious 'stupidity'. Why does it not shovel the egg back by one sweep of the extended wing? Or with one of its webbed feet? The answer is, of course, the effectors are there all right, effectors much better

306   THE HIERARCHICAL ORGANIZATION OF NERVOUS

suited to this task than the bill, but the animal simply does not have the central nervous mechanisms to employ these effectors in this situation. In psychological terms, it does not occur to the bird to retrieve the egg in a more intelligent way.

This recognition of the probability that an animal may have a limited (although admittedly sometimes great) number of nervous mechanisms in the service of the attainment of one biological end is an important step in the study of underlying nervous mechanisms.

Another, even more important, step has been the recognition that the nervous mechanisms are organized in hierarchical systems. The first suggestion of such a hierarchical organization is given by the fact that the relatively simple movements are often elements, building stones, of the more complicated movements.

Two sciences, originally working in isolation from each other, have contributed toward our knowledge of central nervous hierarchy: ethology and neurophysiology.

I will approach the problem by ethological study first. Here the principle was first discovered in the three-spined stickleback.

The male three-spined stickleback hibernates in the sea or in deep fresh water. With the awakening of the reproductive instinct, it migrates towards shallow fresh water. A rise in temperature and visual stimulation by a special type of vegetation make it settle on a 'territory'. To this territory all its reproductive activities are confined. It fights off other males, it builds a nest; when this is finished, it courts females which spawn in the nest, after which the male fertilizes the eggs. The eggs and young are guarded by the male.

All these activities are partly dependent on internal impulses and partly upon external stimuli. The responsiveness to the external stimuli, such as, for instance, those releasing fighting, or building, is, however, dependent on the activation of the reproductive instinct as a whole, which is primary. While a temperature and a visual stimulus (vegetation) determine whether the fish will settle and begin with fighting, building, etc., these subordinated activities are dependent on special additional stimuli. Fighting, for instance, is released by the visual stimulus situation 'male in nuptial markings intruding', nest building by visual stimuli from certain plants.

Now fighting may consist of very different movements; there are at least five different methods of fighting. Which type will be shown depends on additional, still more specific, stimuli from the intruding male. Biting will evoke biting in return, escape evokes pursuit, etc.

This type of evidence suggests a hierarchical system in the central nervous system, including partial systems at different levels of integration.

The idea of well-defined central nervous mechanisms is further rendered probable in view of the fact that, at each level, special causal factors evoke special responses. This is most obvious at the lowest level; the relatively simple and stereotyped motor response 'biting' is released by a simple tactile stimulus.

At the higher levels, equally simple stimuli ('sign stimuli') have an equally specific, though less clearly recognizable, effect. This effect, it is true, cannot be described as a special motor response, yet it is a narrowing down of the potential responses in the sense that it increases the animal's readiness to perform a special limited set of responses, at the same time decreasing its readiness to perform all other types of activity. Thus fighting as a whole (that is to say, the readiness to respond with one of the types of fighting) is released by the stimulus 'red male intruding'.

This can mean but one thing: this hierarchical system is a system of nervous 'centres', the higher centres controlling a number of centres of a next lower level, each of these in their turn controlling a number of lower centres, etc.

In order to understand what these 'centres' (which are, provisionally, to be thought of as functionally characterized systems) are actually doing, it is necessary to analyse the motor responses a little further.

As Lorenz (1937), partially referring to the studies of Craig (1918), has shown, instinctive behaviour often consists of two successive parts of very different kinds. An animal, in which an instinctive urge or drive is activated, starts 'random', 'exploratory' or 'seeking' behaviour. When this type of behaviour is closely observed, it is found that it is typically purposive in the sense of McDougall. Further, it is continued until the animal comes into a situation that provides the sign stimuli necessary to release the motor response of one of the centres of the lowest level. To mention an instance: a peregrine falcon in which the hunting drive becomes active, searches for prey until it is found. The sight of the prey releases the motor response of catching, killing, eating, which is a chain of simple, relatively rigid, responses. Or, a female rat runs through a maze, keeps searching, until it finds its young, which releases the maternal motor responses. Or again, an animal migrates toward the end of the day to a quiet place, where it can sleep (Holzapfel, 1940).

The final motor response, bringing the striving of the animal to an end, is called the consummatory act (Craig, 1918), the preparatory searching behaviour: appetitive behaviour. Whereas the analysis of the consummatory act has made a good start already, that of the much more complicated appetitive behaviour has not yet proceeded very far.

Baerends (1941), who has analysed the behaviour of the digger wasp,

Ammophila campestris, along these lines, has succeeded in carrying the analysis one step further. I will take as an example the hunting behaviour of a peregrine falcon, because this happens to be familiar to me. What I said above on the hunting behaviour of this species was purposely simplified.

A peregrine falcon in which the hunting urge becomes active leaves its perch and begins to roam about its hunting territory, which may measure several, up to at least ten, miles in diameter. It flies more or less aimlessly around, on the look-out for potential prey, perhaps purposely visiting special locations where it has met with success before. However, it will depend upon the type of prey sighted what type of hunting it will show. It may take a starling, a lapwing or a teal out of a flock, it may pick up a weak gull or lapwing from the water, it may even take a small mammal from the ground. Now the important point is, that the sight of one of these prey will not immediately call forth the consummatory act, but it releases another, more special type of appetitive behaviour. For instance, upon the sight of a flock of teals, the falcon will not at once try to catch one of them, but it will make sham attacks and continue them until one of the teals fails to keep up with the quickly manœuvring flock and gets isolated. This stimulates the falcon to the final swoop, which is a still more specific type of purposive, appetitive behaviour, and only if the swoop brings it into a favourable position will the falcon catch the prey, which is a real consummatory act.

This shows that activation of the centre of the highest level results in appetitive behaviour of a generalized kind. This is carried on until a new stimulus with a more restricted effect releases a subordinated type of appetitive behaviour. This again is continued until the next stimulus releases a still more restricted type of appetitive behaviour, and this is carried on until the consummatory act is released.

Next we have to consider the part played by the external stimuli. Certain facts, not to be entered upon here, point to the conclusion that the releasing stimuli are not calling forth the response in the way of a reflex, however involved this reflex response might be. We must rather assume that they merely remove a block and thus provide free passage for the motor impulses coming from the activated centre. The instinctive centres seem to be in a state of readiness, they are constantly being loaded from within, but their discharge is prevented by a block. If there were no such block, continuous simultaneous discharge of all centres and, as a consequence, chaotic movement would be the result. The adequate sign stimuli act upon a reflex-like 'innate releasing mechanism' (I.R.M.), and this mechanism, upon stimulation, removes the block.

Each centre, on each level, has such a block with a corresponding I.R.M.

MECHANISMS UNDERLYING INSTINCTIVE BEHAVIOUR   309

As long as this block is not removed by stimulation of the I.R.M., the centre cannot 'get rid' of its motor impulses.

This hypothesis fits in very well with what scanty knowledge we have of so-called displacement activities. In short, these seem to be the result of accumulated motivational 'tension' in a centre which cannot be discharged because the block is not removed. The impulses under certain circumstances find an outlet through neighbouring centres. Lack of space does not per-

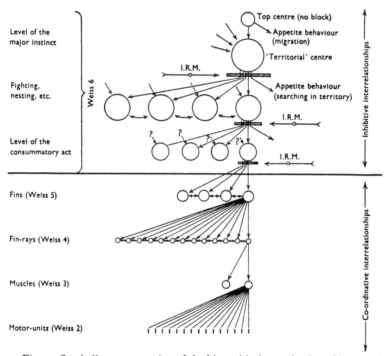

Fig. 1. Symbolic representation of the hierarchical organization of 'centres' playing a part in instinctive behaviour. Explanation in text.

mit to elaborate this further; I may refer to Kortlandt (1940), Tinbergen, (1940), Tinbergen & van Iersel (1947) and my forthcoming book on instinct.

I have attempted to develop a graphic model of the organization of such an instinctive centre. It is based upon the instinct with which I happen to be familiar, the reproductive instinct of the male three-spined stickleback (Fig. 1). 'Centres' are represented by circles, causal factors by arrows, blocks by shaded rectangles, innate releasing mechanisms by schematic pictures of a nerve cell, and motor paths by drawn lines.

The centre on the highest level has no corresponding block. This seems to be typical for the highest levels, as indicated by the fact that activation of a major drive always results in (the highest type of) appetitive behaviour.

If there were blocks, even at this highest level, the animal would have no possibility of getting rid of impulses, which, so far as we know, must lead to neurosis.

In the case of the stickleback, day-lengthening, through various endocrine glands increasing the level of gonadal hormones in the blood, activates the highest centre in early spring. The impulses generated by this centre can travel one of two ways. They can travel toward the centre of the next lower level, but as this centre is blocked at the distal end, these impulses cannot go further. The other possible way is through the appetitive behaviour. The appetitive behaviour belonging to this highest centre is migration from the sea, or deep fresh water, to shallow fresh water. This migration is carried on until the sign stimuli adequate to activate the I.R.M. of the next lower centre are met with. These stimuli are, as said before, a sudden rise in temperature, and visual stimuli provided by the vegetation. These stimuli remove the block, and thereby provide free passage to the impulses. The result is that the animal stops migrating. This state of affairs leads to the hypothesis that the threshold for the activation of the highest appetitive behaviour must be higher than that for the activation of the lower type of appetitive behaviour.

On the level of the lower centre, the 'territorial centre', the situation is similar to the one we have just seen. Here again the impulses can flow to the next lower level of centres, those of fighting, or building, etc., or they may flow to the appetitive behaviour. As long as the blocks of the fighting, building, etc., centres are not removed by the adequate sign stimuli, the animal will perform the appetitive behaviour. This is a type of restless swimming all over the territory, while the animal is on the look-out for rivals and for nesting materials. This is carried on until, for instance, an intruding male appears. This removes the block of the fighting centre; all impulses flow to this centre, and the animal stops its aimless wandering over the territory and attacks. The type of attack, that is to say, the activation of one of the five next lower centres, is decided by the behaviour of the intruder; his movements finally provide the stimuli which call forth (or rather enable the fish to perform) one special type of fighting which is a consummatory act.

So much for the interpretation of the ethological evidence. Though this is not the place to dwell on the neurophysiology of instinct, it should be mentioned that there are certain neurophysiological data which lend support to the views presented above.

First, the existence of internal, intrinsic central nervous mechanisms responsible for and *controlling co-ordinated motor patterns of an order of complexity of the consummatory acts* has been made highly probable by the work of von Holst, Gray & Lissmann, and Weiss. Moreover, their work shows that the relatively low centres, of the level of the consummatory act and locomotion, have their anatomical basis somewhere in the spinal cord.

Further, it is of the greatest interest that Hess & Brügger (1943) discovered that certain behaviour patterns, which according to his descriptions must be considered the result of activation of the highest centres, can be called forth by direct electrical stimulation of certain strictly localized centres in the hypothalamic region of the brain. In cats, Hess succeeded in calling forth sleep, fighting and eating with the appetitive behaviour patterns belonging to these activities. This work, when seen in connexion with the work on spinal mechanisms, suggests the possibility of parallelization and mutual fertilization of ethology and neurophysiology and opens wide perspectives.

Still, in another way it is possible to link up ethological work with neurophysiology. After the above views on the hierarchical organization of instinct were developed, I read Weiss's paper on self-differentiation of central nervous patterns (1940). In this paper, Weiss emphasizes the hierarchical organization of central nervous mechanisms, but his considerations apply to levels lower than the consummatory act. In fact, the highest level of his system of six levels, called by him the level of the behaviour of the animal as a whole, is equivalent to the whole complex of centres we have been considering here. In other words, his highest level is again a complex of several levels, and our system, being an analysis of what Weiss considered the highest level, can be fitted without any trouble into his system. Weiss's analysis of the levels below that of the consummatory act shows interesting parallels to von Holst's work on co-ordination in fish.

In my figure I have indicated a number of horizontal double-headed arrows between centres of the same level. These represent interrelations. In the levels above the heavy line these interrelations must be supposed to exist on the ground that the centres of the same level mutually suppress each other's activities. Further, the existence of displacement activities suggests something of the kind. Below the solid line, the interrelations are not of an inhibitive type, but they are responsible for the co-ordinative phenomena. Especially the phenomena of superposition and the magnet effect, discovered and analysed by von Holst, must be based on some kind of nervous interrelationships.

These considerations, however provisional they may be, enable us to see the problems of instinct much more clearly than twenty years ago. Most of the older writers considered the directiveness of instinctive activities as one of the main characteristics of an instinct. Instincts were distinguished on the ground of their being directed to special ends: escape, feeding, reproduction, etc.

At present we are in a position to formulate neurophysiological arguments in favour of the distinction of several instincts. Although it is too early to give a definition of 'an instinct', we are justified in concluding that in any definition of an instinct its neurophysiological foundation will have to be mentioned just as well as the end toward which it is directed. The hierarchical structure will have to be mentioned, the dependence of internal (motivational) and external factors, and the innate and self-differentiating character will have to be stressed.

The fact that neurophysiological, or, to put it in more general terms, objectivistic research no longer confines itself to the relatively lower levels but, for the first time, embarks upon a study of the levels above the consummatory act, creates a peculiar and new situation. Objectivistic research begins to discover the functioning of mechanisms in activities that, as introspection teaches us, are accompanied by subjective phenomena, viz. the instinctive emotions in McDougall's sense. No doubt, there will be psychologists who will see this situation as a conflict. In my opinion, however, results of the type discussed here must not necessarily be in conflict with psychological results obtained by means of introspection. It will be of the greatest importance for future ethological research to clarify the relations between these two types of research.

REFERENCES

BAERENDS, G. P. (1941). Fortpflanzungsverhalten und Orientierung der Grabwespe *Ammophila campestris* Jur. *Tijdschr. Ent.* **84**, 68–275.

CRAIG, W. (1918). Appetites and aversions as constituents of instincts. *Biol. Bull. Woods Hole,* **34**, 91–107.

GRAY, J. (1939). Aspects of animal locomotion. *Proc. Roy. Soc.* B, **128**, 28–62.

HESS, W. R. & BRÜGGER, M. (1943). Das subkortikale Zentrum der affektiven Abwehrreaktion. *Helv. Physiol. Acta,* **1**, 33–52.

HOLST, E. VON (1937). Vom Wesen der Ordnung im Zentralnervensystem. *Naturwissenschaften,* **25**, 625–31, 641–7.

HOLZAPFEL, M. (1940). Triebbedingte Ruhezustände als Ziel von Appetenzverhalten. *Naturwissenschaften,* **28**, 273–80.

KORTLANDT, A. (1940). Wechselwirkung zwischen Instinkten. *Arch. néerl. Zool.* **4**, 442–520.

LISSMANN, H. W. (1946). The neurological basis of the locomotory rhythm in the spinal dogfish (*Scyllium canicula, Acabthias vulgaris*). II. The effect of deafferentation. *J. Exp. Biol.* **23**, 162–76.

LORENZ, K. (1937). Ueber die Bildung des Instinktbegriffs. *Naturwissenschaften,* **25**, 289–300, 307–18, 324–31.

TINBERGEN, N. (1940). Die Uebersprungbewegung. *Z. Tierpsychol.* **4**, 1–40.

TINBERGEN, N. & IERSEL, J. J. A. VAN (1947). 'Displacement reactions' in the three-spined stickleback. *Behaviour,* **1**, 56–63.

WEISS, P. (1940). Self-differentiation of the basic patterns of coordination. *Comp. Psychol. Monogr.* **17**, 1–96.

WEISS, P. (1941). Autonomous versus reflexogenous activity of the central nervous system. *Proc. Amer. Phil. Soc.* **84**, 53–64.

Relations Between the Central Nervous System and the Peripheral Organs[*]

By E. von HOLST

Max-Planck-Institut, Wilhelmshaven

The relation of the Central Nervous System (CNS) to the peripheral senses and muscular movement is an old and much discussed problem. Here we are at the heart of the physiology of behaviour, and in comparison to that which is not known, our present knowledge is very meagre and vague ! Under these circumstances, our knowledge and conceptions are dependent upon the method which happens to be popular at the moment. In this field, the method which has played the greatest role consists of, first, artificially inactivating the CNS and then, through peripheral stimulation, evoking a particular response. On this basis, the CNS is often held to be only a reflex-mechanism, yet we know today that this view is one-sided. In order to be in co-ordinated activity, the CNS often needs a minimum of stimulation or loading by afferent impulses: the conception of chain-reflex-co-ordination has been recognised almost everywhere as being incorrect. Isolated, that is de-afferented, parts of the nervous system show continued electrical activity. One can therefore say that, as a rule, deafferented ganglion cells, under otherwise normal conditions, possess "automaticity."

These facts allow us to regard the function of the peripheral senses from a new viewpoint. The classical reflex-concept assumes that the peripheral stimulus initiates the central nervous activity. Since we now know that this supposed cause is often unnecessary, it is possible to start from the CNS. We can ask the question, what effect is produced on the sensory-receptors by the motor impulses which initiate a muscular movement? Thus, we look from the opposite direction, not from the outside inward, but from the centre to periphery. You will quickly see that in this manner we shall come upon new problems and experimentally verifiable hypotheses.

In order to make myself clear, I should like first to explain a few terms. The whole of the

impulses which are produced by whatever stimuli in whatever receptors I shall term *afference*, and in contradistinction to this I shall call the whole of the motor impulses *efference*. Efference can only be present when ganglion cells are active; afference, on the contrary, can have two quite different sources: first, stimuli produced by muscular activity, which I shall call re-afference: second, stimuli produced by external factors, which I shall call *ex*-afference. Re-afference is the necessary afferent reflexion caused by every motor impulse: ex-afference is independent of motor impulses.

Here are some examples: when I turn my eyes, the image present on the retina moves over the retina. The stimuli so produced in the optic nerve constitute a re-afference, for this is the necessary result of my eye movement. If I shake my head, a re-afference necessarily is produced by the labrynth. If, on the other hand, I stand on a railway platform looking straight at a train when it starts to move, the moving image on the retina of my unmoving eye produces an ex-afference: likewise, when I lie in a tossing ship, the impulses of my labyrinth will constitute an ex-afference. If I shake the branch of a tree, various receptors of my skin and joints produce a re-afference, but if I place my hand on a branch shaken by the wind, the stimuli of the same receptors produce an ex-afference. We can see that this distinction has nothing to do with the difference between the so called proprio- and extero-receptors. The *same* receptor can serve both the re- and the ex-afference. The CNS, must, however, possess the ability to distinguish one from the other. This distinction is indispensable for every organism, since it must correctly perceive its environment at rest and in movement, and stimuli resulting from its own movements must not be interpreted as movements of the environment. I want to describe experiments which show how the CNS distinguishes between ex-afference and re-afference.

When one rotates a striped cylinder around a

[*] Lecture delivered at the Zoological Laboratory, Cambridge, on 30th October, 1953.

THE BRITISH JOURNAL OF ANIMAL BEHAVIOUR

quietly sitting insect, for instance the fly Eristalis, the animal turns itself in the same sense (Fig. 1a). This is a well-known optomotor—"reflex." As soon as the animal moves itself, for instance, "spontaneously" (or stimulated by a smell), one observes that it turns itself unhindered by the stripes of the stationary cylinder. We must ask ourselves why the animal at every turn is not turned back by his optomotor "reflex," since the movement of the image on the retina is the same as in the first

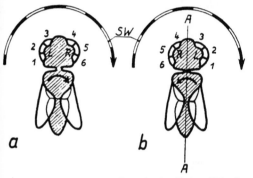

Fig. 1. Insect (Eristalis) in striped cylinder (SW). L, R = left, right eyes; a = head in normal, b = in turned position.

case, when the cylinder moved and the animal was stationary. A possible answer according to the reflex - theory is that in locomotion the optomotor-"reflex" is inhibited or "blocked." But we shall see that this answer is incorrect. It is possible, as has been shown by my colleague Mittelstaedt, to turn the head of the insect through 180° about the long axis (Fig. 1b A-A): then the head is fixed to the thorax, so that the two eyes are effectively interchanged and the order of the visual elements is reversed. The unmoving animal now responds, when the cylinder turns to the right, by turning itself to the left, as is to be expected from the reversed position of the eyes. If it is indeed the case that in spontaneous (or otherwise caused) locomotion the optomotor-reflex is "blocked," the animal should move *unhindered* in the stationary cylinder. But the opposite is the case: once the insect begins to move, it spins rapidly to right or left in small circles until it is exhausted. We have observed the same behaviour with fishes, whose eyes have been turned 180° about the optic axis. But we have found this behaviour only in patterned optical surround-

ings; in optically homogeneous surroundings the animal moves normally. This indicates that the optomotor-"reflex" is not "blocked" in locomotion, but on the contrary, the associated re-afference plays an important role. Exactly what that role is will be made clearer by the next example.

If a vertebrate is turned over on its side by external forces, the well-known righting "reflexes" are initiated by the ex-afference of the labyrinth. But, just as in my first example, every animal is able to take up any position without righting reflexes being produced by the re-afference of the labyrinth. Again, it has been believed that the reflexes were "blocked" during position changing; and, again, we can show that this is not the case.

The righting reflexes, as is well-known, are released by the statoliths in the labyrinths, which, when the head is tilted, produce a shearing force on the underlying sensory organ, as we have found in fishes. One can increase this mechanical force which the statoliths exert on the sense organs, through the addition of a constant centrifugal force. We have built for this purpose a small revolving laboratory, capable of more than doubling the gravitational force. In this manner the statolith is made heavier, and the corresponding shearing stimuli produced by every tilting of the head are quantitatively increased. If one records the tilting of free swimming fish under these conditions, one finds that the degree of tilting becomes proportionally less, the heavier the statoliths are made. (For the method of measurement see v. Holst u. Mittelstaedt, 1950). If the statoliths are removed, then the behaviour of the fish is the same under normal and centrifugal conditions. We see, therefore, that the re-afference of the labyrinth is not "blocked", but has a quantitative effect upon the *degree* of tilting, and, indeed, the greater the re-afference, the smaller the degree of the movement. One can say that the CNS "measures" the degree of movement by the magnitude of the re-afference thereby released.

Thus we have learned two facts: if the form of the re-afference is reversed, as in the first example, than the initiated movement is increased progressively. Secondly, if the re-afference keeps its normal form but is increased, as in the second example, the initiated movement is correspondingly decreased. These facts allow us to formulate a hypothesis about the

THE BRITISH JOURNAL OF ANIMAL BEHAVIOUR

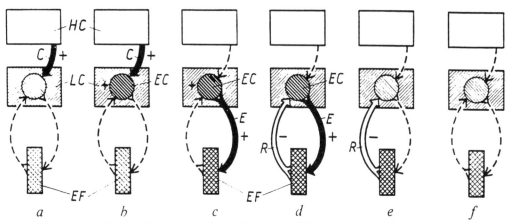

Fig. 2. Illustration of the re-afference principle; see explanation in text.

mechanism here involved. We shall propose that the efference leaves an "image" of itself somewhere in the CNS, to which the re-afference of this movement compares as the negative of a photograph compares to its print: so that, when superimposed, the image disappears. Figure 2 illustrates this in a number of subsequent steps. A motor impulse, a "command" C (Fig. 2a), from a higher centre HC causes a specific activation in a lower centre LC (Fig. 2b), which is the stimulus-situation giving rise to a specific efference E (Fig. 2c) to the effector EF (i.e. a muscle, a joint, or the whole organism). This central simulus situation, the "image" of the efference, may be called "efference copy," EC. The effector, activated by the efference, produces a re-afference R, which returns to the lower centre, nullifying the efference copy by superposition (Fig. 2d-f). Because of the complementary action of these two components we can arbitrarily designate the whole efferent part of this process as plus (−, dark coloured) and the afferent part as minus (—, white coloured). When the efference copy and the re-afference exactly compensate one another, nothing further happens. When, however, the afference is too small or lacking, then a − difference will remain or when the re-afference is too great, a — difference will remain. This difference will have definite effects, according to the particular organisation of the system, The difference can either influence the movement itself, or for instance, ascend to a higher centre and produce a perception.

Let us first consider the simple situation of Fig. 2. The initiated movement will continue,

until the re-afference exactly nullifies the efference copy. Then we must predict the following: first, if through external influence the re-afference is increased, then the initiated movement will end prematurely. We have already seen that this is the case in the fish labyrinth experiment with the centrifuge. Secondly (Fig. 3a), if the re-afference is inverted, that is changed from − to − , there will be no nullification, but summation (Fig. 3b) and the movement will progressively increase, as we have

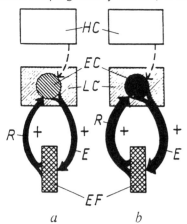

Fig. 3. Illustration of the experiment with the eyes in turned position (Fig. 1); see explanation in text.

already seen in the experiment with the inverted eyes*. Thirdly (Fig. 4), in the case where the re-afference is lacking (for instance, due to the

* This is the so called "positive feed-back."

destruction of the afferent pathways) the in-itiated movement will not be increased, as in the second case, but will continue until something else limits it. This behaviour occurs widely and

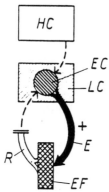

Fig. 4. Illustration of the experiment with interrupted afference; see explanation in text.

can be seen particularly well in fish without labyrinths in optically homogeneous surround-ings. Every turning or tilt leads to circling or summersaulting. Also, in the human disease Tabes dorsalis, where the dorsal roots are destroyed, the well known exaggerated, ataxic movements of the limbs indicate that the same mechanism is involved. Therefore, contrary to the chain-reflex theory, the stimulus, originating with every movement, that is the re-afference, produces not an augmenting, excitatory, but a

limiting, effect on the movement. Only those forms of locomotion, such as the swimming of fish, which do not require a constant adjust-ment to the surrounding medium, proceed just as before after de-afferentation. These movements are automatically co-ordinated in the CNS and therefore require no limiting re-afference (v. Holst, Lissmann).

With this simple scheme we are able to understand a number of previously unexplained types of behaviour. The most hypothetical part of this theory is the postulated efference copy: this "image" in the CNS, produced by the "command" and matched by the re-afference. I am going to present direct proof of the exist-ence of this phenomenon. For this purpose I choose two human examples, in which the difference between the efference copy and re-afference is transmitted to a higher centre and produces a perception. My first example is concerned with the already mentioned human eye movement.

A re-afference from the actively moving eye can have two sources: firstly, movement of the image across the retina and secondly, impulses from the sensory cells of the eye muscles. The former results in a conscious perception; the latter is of no importance for the following consideration. Consider my eye mechanically fixed and the muscle receptors narcotised (Fig. 5a). When I want to turn my eye to the right, an efference E and, according to the theory, an efference-copy EC is produced, but

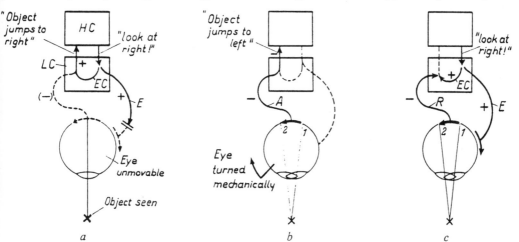

Fig. 5. Illustration of the experiments with human eye; explanation in text (for the letters compare the text of Fig. 2).

92

the immovable eye does not produce any re-afference. The efference-copy will not be nullified, but transmitted to higher centres and could produce a perception. It is possible to predict the exact form of this perception (v. Holst und Mittelstaedt, 1950). The perception, if I want to turn my eye to the right, must be that "the surroundings have jumped to the right." This is indeed the case! It has been known for many years from people with paralysed eye muscles and it has been established exactly from the experiments of Kornmuller on himself that every intended but unfulfilled eye movement results in the perception of a quantitative movement of the surroundings in the same direction. Since here *nothing* happens on the afferent pathways, this false perception *can* only result from the activity, originated by the intention of the eye movement, being returned to higher centres. This is another way of saying that the unmatched efference-copy causes the perception.

Now, we make a simple experiment and turn the paralysed eye mechanically to the right (Fig. 5b). In this case both the motor intention and also the efference-copy are lacking, but the image moves across the retina and afference A is transmitted, unmatched by an efference-copy, to higher centres and produces, as is known, the perception that "the surroundings move to the left." This is also a false perception. If now we combine the first case with the second, that is, if my eye is moved mechanically at the same time I intend this movement—which is the same as *voluntarily* moving a *normal* eye—then in fact these two complementary effects just mentioned are produced: firstly, the perception of the returning "command" causing a jump of the surroundings to the right and, secondly, an image-motion on the retina producing a jump of the surroundings in the opposite direction. These two phenomena, the efference-copy and the re-afference, now compensate each other (Fig. 5c); and as a result *no moving* of the surroundings is perceived. The surroundings appear stationary during this normal eye movement, and *this* perception is *physically correct*. As we have already seen, the correct perception results from two opposite and false perceptions which cancel each other. Thus, we understand a phenomenon with which Psychology has been concerned for many years, that is, the perception of the surroundings as nearly stationary during eye movements ("Raumkonstanz").

Now we come to the second example, visual accommodation. The eye is focussed for distant vision when at rest, since the elastic lens is flattened by its zonal fibres. For near-accommodation a circular muscle, working against these fibres, allows the lens to round up. We should also like to apply our theory to this system. If the accommodation apparatus is narcotised, (for instance by atropine), that is, the eye is permanently accommodated for distant vision, than an intention for near-accommodation will start a motor-impulse, which cannot be nullified by any re-afference and, therefore, must return to a higher centre, where it can produce a perception. This is indeed the case. All objects in the visual field become small, and this false perception is called "micropsia." The same phenomenon must exist with a normal eye, if we imprint an after-image of a distant cross on the retina and then look upon a near surface. Since the after-image remains the same size and sharpness on the retina, it must appear very small on the near surface, because again only the "command" for accommodation returns to the centre of perception. This is also the case, as one can easily convince onself. These false perceptions appear, although the *peripheral stimulus-situation* is *un*altered. If, on the contrary, the *accommodation* of a normal eye is *un*altered, that is, if we look first at a small and then at a large cross at the *same* distance, then naturally the changed afference will be transmitted to the centre of perception and we see the second cross to be larger. Now we combine this last case with the first, that is, we observe with a *normal* eye a cross, moving from a distant point nearer to the eye. This initiates the accommodation-impulse, which returning, tells us "the cross is becoming smaller"; but at the same time the enlargement of the retinal image states, "the cross is becoming larger." The two cancel one another out, with the result that we perceive the cross to be of *constant* size. Again, the correct perception is the result of two opposite false perceptions; and, further, we come to an understanding of a phenomenon, long discussed in Psychology, the "GrossenKonstanz der Seh-Dinge" (Hering), which means that we see the objects to be nearly the same size irrespective of their distance from us.

I could present still further examples from man and from lower and higher animals which would show what role the re-afference plays in general in behaviour. It serves either to limit the magnitude of movement or to insure the constancy of the perceived surroundings during movement, and so makes possible the distinction between real and apparent motion of objects. The first step in both of these functional mechanisms is the comparison of the re-afference with the efference-copy.

In conclusion, permit me a few general considerations. I have attempted to show through the example of this central nervous mechanism, that it is possible in the field of the Physiology of Behaviour to avoid formulating "theories," which are only generalised descriptions of observations; rather should we follow the example of the exact sciences, namely, that a theory must exactly predict what will happen under defined conditions, so that one can by experiment verify or disprove it. Thus one avoids the error of false generalisation, which often occurs in central nervous physiology. For this reason I would like to emphasize that the principle of re-afference is only *one* of *many* central nervous mechanisms. There exists a large number of other mechanisms with other modes of function, and of these we know as yet very little. We recognise fragments of some of them and call them "reflexes"; but this term denotes fragments of very different mechanisms. I believe the whole Central-Nervous System is a "hierarchical system" of such different func-

tional parts, a concept which you find also in Tinbergen's book "The Study of Instinct."

One final point. I have spoken of neither electrical spikes, nor nerve pathways, nor anatomical centres, in which particular functions might be localised. In the realm of behavioural analysis these things are indeed of secondary interest. The functional schemata, constructed in order to illustrate definite causal relationships, are quite abstract, although the consequences they predict are concrete and experimentally verifiable. The physiologist who fully understands such a causal system is still unable to deduce where the cell elements which perform this function are located, or how they operate. Such questions are dealt with at another level of investigation, where the electrophysiologist works and develops his own terminology. It is useful and justifiable for every level of investigation to have its own language, but we must expect, that, with a greater advancement of our knowledge, it will be easy to translate one such language into another. Until such a time, each field must develop along its own lines, unhindered by the many possibilities for misinterpretation.

REFERENCE

v. Holtst u. Mittelstaedt (1950). *Naturwissenchaften*, 464-476.

Exogenous and Endogenous Components in Circadian Rhythms

Jürgen Aschoff

Max-Planck-Institut für Verhaltensphysiologie, Erling-Andechs/Obb., Germany

PREFACE

The main topic of the symposium "Biological Clocks" says nothing about the types of clocks we will be concerned with. It is an open question whether the clocks run continuously or stop after one revolution and have to be started anew. Also the term "clock" does not imply that one revolution is finished in 24 hours. Watches are instruments to measure time. Organisms have to measure time-spans of quite different lengths and for different purposes. Each measurement of speed, for instance, needs timing and mostly within the limits of milliseconds. One can expect, therefore, that organisms possess several clocks with perhaps extremely different periods [1]. There is no need for these clocks to run continuously; for some purposes it would be sufficient if the clock were started always at the beginning of timing (principle of a sandglass, stop watch). "Biological clocks" therefore is a more general concept than what will be mainly discussed at the symposium. On the other hand, the term "circadian," as introduced by Halberg [2] and used in 45 percent of all titles concerning the special clock-problem, comprises two different things: a) the average period of the clock is about 24 hours; b) the clock runs continuously.

At present the circadian clock can be studied only by measuring the periodic course of one or more functions in an organism. Still, nobody can say how these observable functions are related to the clock, how well respectively "the clock" is represented by one function. To answer this question, it may be in the future still more necessary than now to measure several functions simultaneously in one organism, and to observe how they are synchronized or become desynchronized during certain experimental conditions. Nevertheless, many circadian problems can be solved if only one function is followed. It may be that to answer some special questions one or the other function is more useful. But with respect to the general mechanism of the clock, all functions are nearly equivalent. Scientists who work on eosinophiles or on locomotor activity are surely not nearer to, but may be also not so much further from, the clock than those who train animals to feed at definite times of day or to run always toward the same quarter. It will be more important—at least with regard to the problem of synchronization—if we can demonstrate that the results of certain circadian experiments are always equal, independent from whatever functions may have been selected (compare Figs. 17 and 18).

An organism, executing under natural conditions a day-night periodicity, does not necessarily possess a circadian one. The periodicity can be purely exogenous. In that case, environment is the real and only cause of the rhythm which ceases in artificial constant conditions. Contrary to this, the circadian rhythms are endogenous, that means: they are caused in the organism itself. The periodic environment operates only as a synchronizing agent. Periodic factors of the environment, which are able to synchronize a circadian periodicity, have been designed as Zeitgeber [3, 4] or synchronizer [5]. Entraining agent, time-giver, time-cue, and others are used as analogues. As it is not yet decided what term may in general be accepted by the English literature, the German "Zeitgeber" will be used here throughout. The question, whether a biological periodicity is endogenous (circadian), and whether an environmental factor is a Zeitgeber, has to be tested in each single case by certain experiments. Some experiments of this type will be discussed here. So far as they have been carried out in our own laboratory, they are to a great extent a consequence of the most stimulating visit Pittendrigh paid last year to Germany. The results have been obtained by collaboration with Iwan Diehl, Ursula Gerecke, and Rütger Wever.

I. ENDOGENOUS COMPONENTS

I.1. The Problem of Endogenous and Exogenous

To establish a periodicity as an endogenous one, it is necessary to exclude all possible Zeitgeber. Therefore, the organism to be tested must be transferred to what is called "constant conditions." But a priori it is not clear what sort of environmental factors should be considered as potential Zeitgeber. In keeping constant all variables of the environment most commonly controlled, as light, temperature, and so on, three results of such an experiment are possible:

a) The periodicity ceases suddenly or damps out within few periods. This result is neither a convincing proof against "endogenous" nor for "exogenous." It could be the case that a periodicity,

11

although endogenous, becomes unobservable by the tested function because the environmental factors chosen for constancy are too unfavorable (environment too hot, too cool, too bright, too dark, and so on; compare chapter II.3).

b) The periodicity continues and that with a period of exactly 24 hours. In this case, one can not exclude that an overlooked or unknown periodic factor of the environment was effective as Zeitgeber. All experiments in constant conditions, the results of which show an unaltered frequency and phase of the organism, do not prove an endogenous periodicity and still less that this periodicity was inherited.

c) The periodicity continues, but with a frequency deviating by a certain, more or less constant, amount from that of earth-rotation. If there is no other periodicity in the environment (perhaps a tidal one), with which the organism is in synchrony, then the periodicity is really endogenous.

This spontaneous frequency or, to use Pittendrigh's phrase [6, 7], the free running period, is the only convincing evidence of an endogenous (circadian) periodicity. Thereby, one has to mention that a frequency may not be ascertained without measuring several periods. Also, in studying the circadian clock, at least 5 to 7 periods should be registrated before speaking of a spontaneous frequency.

The free running period we can observe in an organism is, of course, nothing like a physical constant. Organisms as open systems are always correlated to the environment. Behavior and function are results of an "inside-outside" coaction. Therefore, we can not expect that the spontaneous frequency of an organism will be the same in all conditions. In a constant environment the period may depend as well on the functional state of the organism—breeding time or anoestrus for instance—as on the special environmental conditions, as intensity of illumination, temperature, and so on. That the organism, although living in a constant environment, behaves rhythmically, we call "spontaneous" and "endogenous." The actual value, however, of the rhythm, the frequency, is determined by all circumstantial conditions—external as well as internal. These statements have already been made by Pfeffer [8].

The term "spontaneous frequency" means that the periodicity arises in the organism spontaneously without external periodic stimuli. Thus "spontaneous" applies to neurophysiological usage. Spontaneous rhythm in neurophysiology means: rhythmic output from the living system in contrast to continuous constant input from the environment. (Without input— no Life!) Heart beat and the rhythmic impulses from the respiratory center are spontaneous. Both centers operate rhythmically even if the cellular milieu is kept constant. The frequency of the impulses depends on all conditioning circumstances, e.g., on constant temperature or on constant CO_2-tension. It is well known that the sensitivity, or better responsiveness, of such a system to an irritation changes periodically. To describe such a rhythm in terms of "phase-shift," of "autophasing," or of "continuously resetting by CO_2-shocks" would be quite unusual.

Discussing the problem of "spontaneous," a few words seem to be necessary with regard to a hypothesis suggested by F. A. Brown. Even in 1957 he emphasized, "that organisms in constant conditions may retain unaltered phase relationships with the external physical cycles even for periods of month," and, "that the clock system maintains its regular frequencies through some kind of an external pacemaking signal which continues to be effective under what is usually deemed 'constant conditions' " [9]. Confronted with such results, everybody would agree that Zeitgeber must have been operating in these experiments, although they are not necessarily "still unknown external factors" [10]. During the last two years, Brown has put into his hypothesis the free running period deviating from 24 hours [11]. He describes the spontaneous frequency as an effect of "autophasing," whereby it is not always clear whether this autophasing is the result of external time-cues or of a constant environment [12]. Both explanations presume a periodic changing sensitivity of the organism. If time-cues were operating, as a consequence of the varying sensitivity, the organism should be in synchrony with the cues; deviating periods are out of the range of entrainment, and the time-cues behave in these cases as a constant environment with respect to the organism. If, on the other hand, there are no time-cues, but real constant conditions, then the coaction of a rhythm (with a changing sensitivity) with such a constant environment is again in correspondence with the definition of "spontaneous" given above. One last example: The frequency of an electronic circuit depends on inductance and capacity. The system oscillates continuously and spontaneously if the inevitable losses of energy are replenished by an anode-voltage via feed-back. If one of the parameters of the circuit is made sensitive to light by use of a photocell, the frequency of the system will become a function of the intensity of illumination. Nevertheless, the oscillation remains spontaneous (endogenous); only the actual value of the frequency is determined by internal as well as by external factors. It is evident that such a system reaches once during each period a point of highest sensitivity against light. It would be unusual, however, to describe the oscillation as an effect of "autophasing," as "resetting the phase of the oscillation," or as an "exogenous" periodicity. The consequence would be that one frequency of a system would be called "frequency" and another one "resetting."

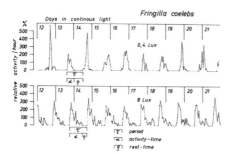

FIGURE 1. Activity of a male chaffinch. Constant illumination of two intensities, 0.4 lux (above) and 8 lux (below). Ordinate: Continuously registrated activity per hour.

I.2. THE SPONTANEOUS FREQUENCY

As a simple example of what is called a spontaneous frequency, Fig. 1 shows the activity of a caged chaffinch in constant conditions—constant temperature, continuous feeding, isolation from noises, and continuous illumination with an intensity of 0.4 lux (above) and 8 lux (below). The diagram explains two things: a) With an intensity of 0.4 lux the period is longer than with 8 lux; b) Within one period two fractions can clearly be divided: activity-time, when the bird is awake and jumps around in its cage, and rest-time, when the bird sleeps [13]. Offering an intensity of 0.4 lux, activity-time is shorter and rest-time longer than offering 8 lux. Both the fractions of the period will be discussed later on. First frequency requires the main interest. The length of each period is easiest to derive from the sharp onsets of activity. But also peaks of activity or the ends of activity-time could be used as reference points between which the period is measured. Without a special examination one can not predict which point of the biological period may be more important or useful in studying the clock. It has been shown, however, in the results of several experiments that, in measuring activity, the onsets scatter less around the average period than other points (compare Fig. 2) [14]. In the following discussion often the onsets, but also sometimes the peaks of activity, will be used for reference.

The accuracy with which an organism keeps its frequency is easy to demonstrate by marking the successive daily delays or advances of onsets with respect to local time in a diagram. In Fig. 2 the sequence of days as the independent variable is drawn on the abscissa. The ordinate represents hours of onsets and ends of activity of a chaffinch, given in Central European time for each single day. Under the conditions of alternating 12 hours of light and 12 hours of darkness (LD 12:12), the activity-time of the bird is strongly fixed to the light-time (= time of illumination) [15].

In continuous illumination with an intensity of 0.4 lux, activity starts and ends each day a little later. After 30 days of continuous illumination, the bird becomes newly synchronized with LD (12:12) during 12 days and then again exposed to continuous illumination, but now with an intensity of 120 lux. In this high intensity of LL, activity starts each day a little earlier. The diagram shows: a) The onsets of activity offer a more precise measurement of period and thus also of frequency than the ends of activity; b) By a continuous illumination with an intensity of 0.4 lux, activity-time is shorter than in LL with 8 lux; c) A strong artificial, nearly unnatural Zeitgeber (compare Fig. 14) with alternating 12 hours of bright light and 12 hours of total darkness, catches the free running frequency suddenly; also, after continuous illumination is reestablished, the spontaneous frequency starts at once.

A certain intensity of constant illumination given, the spontaneous frequency varies with each organism tested, and also in one organism the frequency can change without obvious causes. The onsets of activity of 4 chaffinches in LL are drawn in Fig. 3. They have been measured in 4 different intensities of illumination during at least 20 days. Although the frequencies vary inter- and intraindividually, each intensity of illumination is characterized by a certain frequency, averaged on all 4 birds. This average spontaneous frequency is lowest in 0.4 lux and highest in 120 lux.

I.3. FREQUENCY, DEPENDING ON LIGHT INTENSITY

Since long ago it has been known that the spontaneous frequency depends on the intensity of illumination [16, 3]. This fact has been confirmed recently in several organisms. In general, the frequency varies in a linear scale with the logarithm of light intensity. A survey on all results so far published is presented in

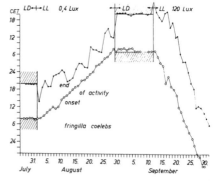

FIGURE 2. Onset and end of activity of a chaffinch in alternating 12 hours of light and 12 hours of darkness (LD, 12:12) or in continuous illumination (LL) of two intensities (0.4 lux and 120 lux). Ordinate: Daily onset and end of activity at Central European time.

14 ASCHOFF

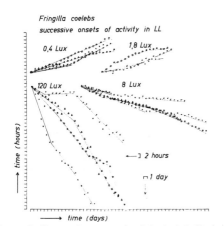

Fringilla coelebs
successive onsets of activity in LL

0,4 Lux *1,8 Lux*

120 Lux *8 Lux*

←⌐ 2 hours

⌐1 day

→ *time (days)*

FIGURE 3. Consecutive onsets of activity in 4 chaffinches during at least 20 days. Continuous illumination of 4 different intensities. Delay of onsets with 0.4 and 1.8 lux, advance of onsets with 8 and 120 lux. Notice different time-scale on ordinate and abscissa.

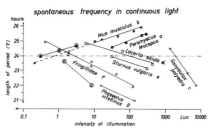

spontaneous frequency in continuous light

FIGURE 4. Spontaneous (circadian) period of divers organisms in constant environment, depending on intensity of illumination. Each point represents the average of several individuals and periods. *a)* Activity, Johnson [16]; *b)* Activity, ● old and ◉ new experiments; *c)* Luminescence, Hastings and Sweeney [17]; *d)* Activity, Hoffmann [18]; *e)* Activity; *f)* Activity, ○ first and ◉ second series of experiments; *g)* Leaf movements, Pfeffer [19.]

Fig. 4. Each point of the diagram represents the average of several organisms (with the exception of *Phaseolus*), which in each intensity of illumination have been measured during several periods. The figures concerning starlings, finches (1 and 2 series), and mice (2 series) have not yet been published in detail. The results of the first and the second experimental series on mice are close to one another. Contrary to this, finches perform a much higher frequency in the second experiment than in the first. Such a variation of sensitivity to light is by no means astonishing; for instance, it could be expected especially as a consequence of seasonal changes in the functional state of the organism. That the straight line representing a species in Fig. 4, should always have the

same slope and position in the diagram by repeated experiments seems to be unlikely. Each intervening treatment of an organism (of whatever kind) may change the spontaneous frequency from one test-experiment to the next. Also quite clearly an organism in *LL* will have different frequencies in an empty cage and in a cage wherein is offered a dark corner to move in. The figures concerning *Phaseolus* are correctly arranged in their relative positions; the absolute value of light intensity used by Pfeffer could only be guessed.

The results of divers authors as they are collected in Fig. 4 suggest the hypothesis that in *LL* with increasing intensity of illumination, light-active animals (finches, starlings, lizards) increase their spontaneous frequency, while dark-active animals (housemice, white-footed mice) decrease it. This rule, when mentioned first, was formulated unsatisfactorily [20, 21]. Discussion with Rawson and his students at Madison contributed to the final version [22]. Until now no results seem to be published which contradict the rule in principle; nevertheless, exceptions may be expected. Details of the rule will be discussed in chapter I.5.

The frequency of an oscillating system is measured as the reciprocal of period. Beside this, the system has within one single period what may be called a speed. This speed can be described in terms of an angular velocity of a rotating vector. Only a true sinusoidal oscillation has a constant speed throughout the whole cycle. In all other oscillations the speed varies within one period. That means: in a circadian oscillation there are fractions of the period during which the system runs slower than in the average and other fractions during which it runs faster. It seems reasonable to assume that in *LD* this varying speed is co-ordinated with the Zeitgeber in some way. Accepting this, the data presented in Fig. 4 suggest that in light active animals increasing intensity of light (or the change from *D* to *L*) tends to accelerate the speed of the system and that decreasing intensity (the change from *L* to *D*) tends to slow it down. The rate of change in speed depends, of course, on the phase when the organism is exposed to higher or lower light-intensities. Moreover, if one looks at an animal in constant conditions, one could expect that the varying speed of the system is correlated to the two fractions of the period: activity-time and rest-time. If one assumes that activity-time coincides with a speed higher than in the average and rest-time with a lower one, a testable prediction can be made: In lengthening the activity-time—i.e., the time during which the system runs faster than in the average—the period must become shorter. This has been shown in many cases. Differences in the speed of the clock have still not been measured directly. But there are two observations which perhaps can be taken as indirect evidence. First, in the hierarchy of rhythms [23, 24] the 2-hour rhythms of activity ("bursts" of activity) are most striking. It has been shown that these

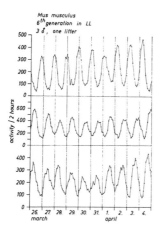

FIGURE 5. Activity of three mice in *LL*. One litter, born November 11th, of the 6th generation, raised in continuous illumination. Original figures smoothed, using a running average of 5 points (compare Fig. 6, left column, F_5).

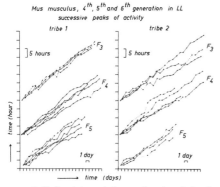

FIGURE 6. Daily delay of the peaks of activity in 6 litters of mice in *LL*. F_3-, F_4-, and F_5-generation of two strains, raised in continuous illumination of about 100 lux.

I.4. INHERITANCE OF PERIODICITY

If an organism has been shown to possess an endogenous (circadian) periodicity, the question arises whether the periodicity was acquired by the individual during ontogeny or has become part of hereditary disposition during phylogeny. It is self-evident that only the potentiality to behave rhythmically can be inherited. The realization of such a potentiality depends, as in all other functions, on several circumstantial conditions. There is no need for an inherited periodicity to be in full function directly after birth. If the periodicity unfolds only in the growing organism step by step (as in man), this does not necessarily mean that the periodicity has been impressed upon the organism by the environment. All organs or centers of co-ordination necessary for the periodicity have to be fully grown up before an overt periodicity can operate even if it is inherited [26, 27].

Animals which are raised in constant conditions from very early states of development and which, nevertheless, show a spontaneous frequency can be said to have an innate periodicity. Successful experiments of such kind have been carried out in chickens [26] and lizards [28] by raising them from the egg on in continuous illumination or darkness. Also, lizards which grew up in an artificial 18-hour or 36-hour day showed, when tested in constant darkness, natural periods similar to those of normally growing animals and independent from the foregoing lighting regimen [29]. Experiments in which the animals are raised in constant conditions for several generations give no proof for an innate periodicity as long as the period remains 24 hours; the possible effect of a Zeitgeber can not be excluded. If, on the other hand, mammals raised from birth on in *LL* or *DD* develop a spontaneous frequency, one can object that the nursing mothers may have impressed their rhythm upon the young [30]. It may be, however, interesting enough to note the fact that 6 generations of mice raised in *LL* kept all the same spontaneous frequency. The activity of three litter-mates of such a F_5-generation is shown in Fig. 5. The spontaneous frequencies of the 4th, 5th, and 6th *LL*-generation in two strains are in good accordance with each other; this is demonstrated in Fig. 6. Here the successive peaks of activity are marked for all individuals of six litters during at least 20 days. All periods are within the limits of 25.0 and 25.6 hours.

I.5. THREE IMPORTANT PARAMETERS

In constant conditions, spontaneous frequency and activity-time are two parameters of the oscillation easy to measure. A third one, more difficult to ascertain but perhaps equally important, is given by the level around which the system oscillates. The level is not identical with the amplitude. An electronic circuit can oscillate with exactly the same amplitude around

bursts have a higher frequency during activity-time (in *LL* as well as in *LD*) than during rest-time [13]. Second, animals using the sun as a compass for orientation have to compute the sun-movement with an average angular velocity of 15° per hour. Unpublished experiments indicate that some light-active fish compute more than 15° per hour during light-time and less during dark-time [25]. The question, whether these observations corroborate the hypothesis of an interaction between the Zeitgeber and a varying speed in the oscillating system or not, can not be discussed here in detail.

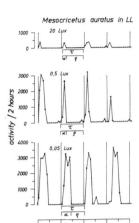

FIGURE 7. Activity of a hamster, measured in a running wheel in *LL* with three intensities of illumination. Increasing rest-time (ρ), decreasing activity-time (α), and amount of total activity with increasing light intensity. Nearly no change of period (τ).

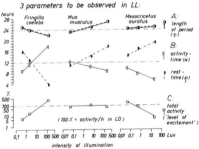

FIGURE 8. Three circadian parameters, depending on intensity of illumination in constant environmental conditions. *A*) Length of spontaneous period; *B*) Ratio of activity-time to rest-time; *C*) Total amount of activity per unit-time (100% = average activity per hour in *LD*). Each point represents the average of 3 (finch and mouse) or 2 (hamster) individuals, measured during at least 6 periods each.

a low or a high level. The two states differ in the amount of energy they spend per unit time. Applying this to biology, one can call the average level around which the system oscillates "level of excitement." Body temperature can oscillate around 37° as an average as well as around 38° with exactly the same amplitude. In this case the average temperature would be the level. The total amount of activity performed by the animal per unit time may be (although imperfect) a measure of this level. Likewise, or even better, one can probably use the metabolism of energy (oxygen consumption) as a measurement for the level of excitement. Evidently,

TABLE 1. PERIOD (τ), ACTIVITY-TIME (α) AND TOTAL ACTIVITY (A) IN STARLINGS DURING CONSTANT ILLUMINATION OF THREE INTENSITIES

LL, Lux	τ, hours	α, hours	A, counts/30 min
0.6	24.3	9.3	117
7.5	23.5	14.7	300
65.0	22.5	16.5	750

neither activity nor oxygen-consumption is a precise measurement of "the level." There are many cases in which the same oxygen consumption coincides with different frequencies. The effective power may play a role. On the other hand, it is clear that the level is an important parameter in each oscillating system. Although there is no clear answer at this time, it seems reasonable not to exclude the question: What does "level" mean in the circadian oscillation of an organism? The terms "light-active" and "dark-active" are meaningful only in the way that light increases activity (or the level) in light-active animals and decreases it in dark-active animals. Without such a causal connection between intensity of illumination and level of excitement (or activity) there would be no separation between light- and dark-active animals. One of the primary reactions of the organism to light is that the level of excitement increases or decreases.

In many of the animals so far studied, level of excitement, spontaneous frequency, and α:ρ-ratio (activity-time:rest-time) seem to be correlated. Comparing the three parameters in a dark-active animal in *LL* with three intensities of illumination (Fig. 7), one notices: Higher intensities of light are combined with a smaller amount of total activity, with a shorter activity-time, and with a (only little) longer period. The reverse applies to light-active animals. Indeed, all three parameters are not always equally dependent on the intensity of illumination. It may be that in one organism total activity (= level), in another one period remains nearly constant. In each case, however, at least two parameters seem to follow the rule which is demonstrated in Fig. 8 with respect to one light-active and two dark-active species. Each point of the diagram is the average of measurements on three animals (two hamsters only) during at least 7 days. Quite the same results as shown in Fig. 8 with finches Hoffmann [18] could obtain in experiments on starlings (Table 1). Concerning Fig. 8, a few more remarks may be useful. In finches and mice, the α:ρ-ratio is 1:1 when the period is 24 hours. Perhaps from these facts one can derive that also in hamsters the period will be shorter than 24 hours if in very low intensities of illumination the α:ρ-ratio becomes 1:1. Further it is remarkable that in mice there is no change in the total amount of activity. This coincides with the observation that mice easily alternate between light- and dark-activity and that they always have two peaks of activity, the main one during dark-time and the second one during light-time. Other conclusions will be referred to later (compare Fig. 15).

The results as presented in Fig. 8 and in Table 1 encourage us to formulate the circadian rule as follows: In light-active animals spontaneous frequency, $\alpha{:}\rho$-ratio, and total activity increase with increasing intensity of continuous illumination. The opposite applies to dark-active animals. The rule makes no statements about the absolute values of the spontaneous frequency. The original concept that in light-active animals the period should be shorter than 24 hours in LL and longer in DD—as it is in finches and starlings—must not necessarily be generalized. It may be that there are species whose natural period is always longer or shorter than 24 hours (as the hamster seems to indicate). At any rate, theoretically there is no need for crossing the line which represents a free running period of 24 hours. Synchronization and phase-control will also be possible—at least in a model as shown by Wever [31]—if an organism has periods only longer or shorter than 24 hours within the whole range from lowest to highest intensities of illumination. This does not mean that such a possibility must be realized in nature. We rather expect that normally the free running period in LL has a trend from less to more than 24 hours with increasing or decreasing light intensity.

There is one other conclusion which may be derived from the results represented in Fig. 8. If level of excitement and spontaneous frequency are in any way related to each other, then one could expect changes of frequency also if the level is changed by any other means than by light. Temperature has a noteworthy effect on the level as measured by activity. Several experiments have been carried out to study the influence of temperature on the spontaneous frequency. The

TABLE 2. Period (τ) and Activity-Time (α = Time during Which Luminescence Is Above a Threshold) in *Gonyaulax* in Constant Conditions with 5 Levels of Constant Temperature (T)

T, C.°	τ, hours	α, hours
16.0°	22.8	22.8
19.0°	23.0	16.4
22.0°	25.3	14.7
23.0°	25.7	8.7
26.8°	26.5	6.5

TABLE 3. Period (τ) and Activity-Time (α) in Lizards in Constant Conditions with 3 Levels of Constant Temperature (T)

T, C.°	τ, hours	α, hours
16°	25.25	5.0
25°	24.34	7.3
35°	24.19	12.7

results seem to corroborate the hypothesis: Organisms which become excited (or more active) by an increasing temperature shorten their period simultaneously; organisms whose activity is depressed by a higher temperature lengthen their period (Fig. 9). The period of 5 organisms mentioned in Fig. 9 has been measured exactly. In the case of *Avena* it was only noted that the period at 27° with respect to 17° "was unaffected, or, if a slight shortening occurred at the higher temperature, it was not more than one hour in 24" [37]. Activity was measured quantitatively only in a few organisms; but there is no doubt in any case about the direction into which the level of excitement (or the amount of activity) is changed with increasing temperature (dotted lines in Fig. 9). It stands to reason that the hypothetical relation as indicated in the diagrams of Fig. 9 needs further examination. That the hypothesis may be right is in addition supported by two of the 6 organisms. In these two cases, the $\alpha{:}\rho$-ratio varies with increasing temperature toward a direction one would expect from the rule: a) With respect to *Gonyaulax*, this can be computed from data published by Hastings and Sweeney, using their Fig. 1 [32]. If one takes cipher 1 on the ordinate scale as a threshold of luminescence above which all figures are included in the "activity-time," one gets the results shown in Table 2; b) Less doubtful are figures Hoffmann could get in measuring the activity of lizards (Table 3). In accordance with the rule, *Gonyaulax* shortens activity-time and lizards lengthen it with increasing temperature. Clearly, this test is only meaningful in a range of temperature wherein the animals increase their activity with increasing temperature. Only in this range the organism may be called "warm-active" as it is "light-

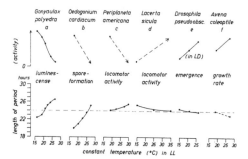

FIGURE 9. Total amount of "activity" and spontaneous period in constant environmental conditions, depending on temperature. *a*) Hastings and Sweeney [31]; *b*) Bühnemann [33]; *c*) Bünning [34]; *d*) Hoffmann [35]; *e*) Pittendrigh [36]; *f*) Ball and Dyke [37].

active" in the sense of increasing activity with increasing intensity of illumination. In both cases there is an "optimum" above which activity decreases; then the organism becomes dark-active and cool-active respectively. Light-active animals will mostly show "warm-activity" and dark-active animals "cool-activity" in testing the animals within the range of intensities in which they normally live. *Gonyaulax* as an exception seems to be light-active and cool-active together. The same may apply to *Saintpaulia ionontha*, as shown by Went, because in a cyclic environment *Saintpaulia* prefer the cool fraction of the temperature cycle coinciding with the light fraction of the light-dark-cycle. Therefore one should expect that *Saintpaulia* would change its spontaneous frequency (if there is one) in opposite directions by increasing light-intensity and by increasing temperature.

I.6. ACCURACY OF THE CLOCK

If in constant conditions the clock does not run with a period of exactly 24 hours, this may not be "a measurement of inaccuracy of the clock" [38]. Accuracy is measured by how precisely the clock keeps a certain frequency. For this purpose one has to measure how much each single period deviates from the average of all measured periods. Usually this measurement is given as standard deviation. To compare the accuracies of several frequencies, it is preferable to express the standard deviation in per cent of the mean value. This has been done in Fig. 10 with period, activity-time, and rest-time of 4 chaffinches. The activity of the birds was measured in constant illumination with 4 different intensities. Each point of the diagram represents the standard deviation in one frequency of one bird during at least 10 days. In some cases, if the spontaneous frequency has changed from one to another value during the same illumination (compare Fig. 3), the standard deviation has been computed for both frequencies. Each point in the upper-most diagram corresponds with a point in the middle (activity-time) and in the lower-most (rest-time).

At first sight it is evident from Fig. 10 that the accuracy of the clock varies with the frequency and that there is a difference of accuracy between activity-time and rest-time. Judging the diagram concerning period, one can say that the clock runs the less accurately the shorter the period; most precise is the clock at a spontaneous period which corresponds to that of earth-rotation. Whether it becomes again less accurate toward longer periods (that means: at a low frequency in very dim light) is still not cleared up, but indicated by few points on the right end of the diagram. Activity-time (diagram in the middle of Fig. 10) remains nearly equally precise within a large range of frequencies; however, accuracy here becomes also worse with increasing length of activity-time (which means increasing frequency). Thirdly, the rest-time

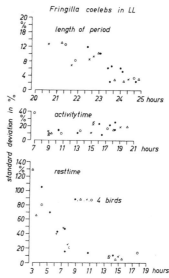

FIGURE 10. Standard deviation from mean value of period, activity-time and rest-time, depending on spontaneous frequency in *LL* with varying intensity of illumination. Each point represents the average of several periods in one bird.

becomes quickly less and less precise the more rest-time shortens (with other words: the faster the clock runs). Above all, in the average of all 4 birds and all frequencies, activity-time has a smaller standard deviation than rest-time. Regarding this, one other observation may be important. Analyzing each single period with respect to activity-time and rest-time, one finds a strong negative correlation between these two components. Whenever activity-time is longer than in the average, the following rest-time is shorter and vice versa. From these two facts one could conclude that the clock regulates its period especially by keeping constant activity-time and by correcting errors via rest-time. But this explanation, although thinkable, is only one of several possibilities. If only the onset of activity is more precisely fixed than the end (as for instance the firing potential as compared with the cutoff potential of a brush discharge in an electronic circuit), the same negative correlation between activity- and rest-time would happen. In this case, the clock will keep its accuracy by correcting period instead of activity-time, which seems to be more likely.

II. EXOGENOUS COMPONENTS

II.1. GENERAL REMARKS

The periodical factors of the environment, which we call Zeitgeber, have several effects:

a) Zeitgeber synchronize a circadian periodicity

with the environment;

b) Zeitgeber synchronize several individuals of one species and keep them in phase;

c) Zeitgeber synchronize several circadian clocks in an organism, if it has more than one;

d) Zeitgeber synchronize by determining phase;

e) Zeitgeber influence the circadian pattern.

Statement a) is right by definition. Statement b) implies that all individuals have the same sensitivity to the Zeitgeber; if this were not so, the individuals would not be in phase (compare chapter II,4), and one would not find what one calls a species-pattern (compare chapter II.6). Statement c) tackles the question of endodiurnal organization as Pittendrigh calls it. Most important is statement d) as a consequence of statement a); it is related to, but not identical with, the last statement e) (compare again II.4 and II.6).

With respect to rhythms, synchronization means that at least two rhythms are in synchrony with each other. Therefore, only such factors of the environment which pass off periodically can operate as a Zeitgeber. One single event can influence the phase of a rhythm; therefore, a group of rhythms running out of phase with each other can be pushed into phase for a short time by one single event. But a single event can never synchronize continuously. Therefore, we may not agree with such a statement as: single isolated events are "an important class of time-givers" [39]. Only repeated single events can operate as a Zeitgeber. The first more detailed definition of Zeitgeber says: "Es kann sich ebenso gut um diskontinuierliche (z. B. ein Tonsignal alle 24 Stunden) wie um kontinuierliche periodische Vorgaenge handeln (z. B. taeglicher Temperaturgang) bzw. um den regelmaessig wiederkehrenden Uebergang eines Zustandes in einen anderen (Hell-Dunkel-Wechsel)" [4]. That means: A Zeitgeber can be represented by a) one short signal each 24 hours, b) a continuously changing factor such as the daily course of temperature, or c) alternating steady state conditions, e.g., alternating light and darkness. These are three thinkable types of Zeitgeber. Whether they all are effective in practice has to be tested.

The question, which parameter of a Zeitgeber acts as the really efficient one, is more difficult to answer. Two possibilities which have to be considered have been discussed together with the definition of Zeitgeber cited above. The difference between both of them is easy to explain by using alternating light and darkness as a Zeitgeber. Case one: The steady state itself is effective, that means: the whole light-time and/or the whole dark-time. Case two: Only the transitions from one to the other state are effective, that means: under natural conditions dawn and dusk. Combinations of both cases may also be possible. Following a suggestion made by Wever [31], the first type may be called proportional effect and the second type differential effect (Fig. 11). Proportional effect means: The effect

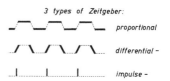

FIGURE 11. Simplified scheme of 3 possible types of effective Zeitgeber components.

is proportional to the reaction which the stimulus of the Zeitgeber in its steady state causes in the organism; the effect continues steadily as long as the intensity of the stimulus is unchanged. Differential effect means: the effect depends only on changing factors of the environment; their steady state has no effect. A special type of Zeitgeber not realized in nature is that of repeated single events. They may be called impulse Zeitgeber. Impulse Zeitgeber are a borderline case of Zeitgeber with two differential effects, so far as the change from one steady state to another one is followed by the reverse change to the original state within an extremely short time.

To become effective, a Zeitgeber must influence the phase of the organism. There are two possibilities to shift the phase of an oscillation. a) Each sudden downward or upward step of the level results in a phase-shift. The amount of the shift depends on the phase at which the level has been changed and on the difference between the two levels. In at least two phases there will be no phase-shift from changing the level. b) Each change of speed for a given fraction of period causes a phase-shift; the amount of the shift depends on the rate of changed speed and how long it remains changed. The question now arises by what means the proportional and differential stimuli of a *LD*-Zeitgeber shift the phase of an organism. As mentioned above, the proportional effect may be described as the influence on speed. The longer the system runs faster or slower—i.e., the longer light-time or dark-time—the greater the advance or the delay respectively. On the other hand, the differential Zeitgeber is effective during a nearly immeasurable short time in an artificial lighting regimen. It would be meaningless to speak of an altered angular velocity during this very short time. The effect rather immediately is a phase shift. The amount of the shift depends on the value of the differential coefficient. Whether only one or both of these types of Zeitgeber also influence the level of excitement can so far not be derived from our knowledge. How important this question may be with respect to all problems of phase-control will be discussed by Wever [31].

Until now there is no decision possible whether the natural Zeitgeber operate proportionally or differentially. Probability seems to indicate that in most of all organisms both types of Zeitgeber are cooperating. Nevertheless, it is quite possible that one species follows mainly the differential principle while another one (e.g., mice?) is synchronized more proportionally (compare II.4). Theoretically, an impulse Zeitgeber should also be possible. Curiously enough, this case never has been tested experimentally. A few observations seem to contradict it. If light-time or dark-time of an artificial LD-regimen with a total period of 24 hours is shortened extremely—that means to less than one hour—some organisms lose synchronization and run with their own frequency. These results may not necessarily prove that an impulse Zeitgeber is ineffective on principle. One has rather to test whether in these cases the differential impulse was not only too weak and thus unable to counteract the strong influence on speed of the extremely long light-time (or dark-time). Obviously, an impulse Zeitgeber with a 24-hour period must be made the stronger the more the other conditions (e.g., intensity of light during L and length of L) push off the free running period of the organism from 24 hours (compare II.2).

To become effective, a Zeitgeber must send out stimuli, and to become synchronized, the organism must be sensitive to these stimuli. A periodically varying sensitivity or responsiveness is one of the inevitable characteristics of each self-sustained oscillation. An excised frog heart is often used to demonstrate the rhythmically varying threshold to electric stimulation. Also, diurnal changes of responsiveness have been shown, mostly by using photoperiodic effects as an indicator [39a], but also true sensitivity-curves by measuring thresholds in men [40]. How important, however, such a rhythm may be with regard to phase-control became only evident in the excellent experiments wherein phase-shifts resulting from single light-signals have been measured [14, 17, 39, 41, 42]. The results demonstrate directly the circadian periodicity of responsiveness. Similar facts have been found out by using temperature shocks [44, 45].

The circadian response-curves, when measured by using extremely short flashes of light as perturbations, are the most clear expression of the differential effect of a Zeitgeber. Clearly enough, proportional effects of a Zeitgeber can not be derived from such a curve. If, on the other hand, light-signals have been used with a measurable duration of time, the resulting phase-shift is the combination of two differential and one proportional stimuli. It is very difficult to decide whether the phase-shift was caused by one or both of the differential stimuli, by the proportional stimulus, or by all three together. In increasing the duration of the light-signal the proportional effect must become greater with respect to that of the two differential stimuli. That means: The greater the distance between light-on and light-off the clearer the proportional effect if there is one. Therefore LL shows—as mentioned above—the proportional effect of L without any additional differ-

ential stimuli (Fig. 4). Considering once more Fig. 9, one can guess that temperature only in *Gonyaulax* has a remarkable proportional effect but not in the five other organisms. A few difficulties may be briefly mentioned. If proportional effects have been proven by measuring τ in different steady state intensities of LL, this does not necessarily mean that also in LD the proportional stimulus becomes effective; there may be something like a "blocking" effect from the preceding differential stimulus. If, on the other hand, τ does not change with increasing intensities of illumination in LL, this does not necessarily mean that there is no Zeitgeber-effect; if different light-intensities are combined with different levels, alternating changes between two intensities are differentially effective.

II.2. Testing a Zeitgeber

An environmental periodicity may be called a Zeitgeber only if the organism synchronized with the environment possesses a circadian periodicity. Studying the mechanism of Zeitgeber therefore presumes organisms whose spontaneous frequency has been demonstrated. In testing whether an environmental periodicity operates as a Zeitgeber on such an organism or not, three types of experiments may be considered:

a) Catching the free running clock. In constant conditions the spontaneous frequency of an organism is measured; afterward the Zeitgeber to be tested is added for several periods and then again the conditions are kept constant. An example of such an experiment is given in Fig. 2.

b) Phase-shift. The periodical factor of the environment the organism already is synchronized with has to be shifted suddenly by a certain phase-angle. If the Zeitgeber is fully effective, the organism should follow the shift within a few periods and should regain the original phase-relation to the Zeitgeber. An experiment in this line carried out with finches in LD (12:12, 400 lux: 0.4 lux) is shown in Figs. 12 and 13. The 12-hour shift is done once by doubling light-time and once by doubling dark-time. Within about 4 periods, the birds are always resynchronized. During synchronizing the amplitude is damped (Fig. 12). The phase-angle-difference between organism and Zeitgeber becomes evident in Fig. 13. In this diagram again the sequence of days (as the independent variable) is represented by the abscissa; onsets and ends of activity are given on the ordinate in Central European time. Better than with words, the picture explains the mechanism of shifting phase: By doubling light-time, the period of the bird is shortened, by doubling dark-time, it is lengthened. The opposite to this one would expect in dark-active animals.

c) Varying frequency. By use of an effective Zeitge-

EXOGENOUS AND ENDOGENOUS COMPONENTS   21

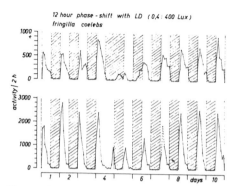

FIGURE 12. Activity of chaffinch in *LD* (12:12). Phase-shift of the Zeitgeber for 12 hours by doubling light-time (below) or dark-time (above).

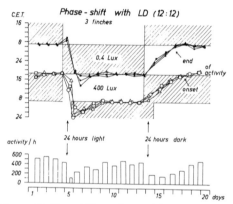

FIGURE 13. Phase-shift experiment with *LD* (12:12, 400 lux: 0.4 lux). Onset and end of activity of three chaffinches, given for each consecutive day (abscissa) on the ordinate-scale in Central European time.

ber it must be possible to entrain an organism to different frequencies within certain limits. Examples of such experiments are presented in Figs. 15 and 17.

The original definition of Zeitgeber says that all environmental factors may be considered "to whose stimuli an organism is sensitive" [4]. Recently some authors incline more toward the opinion that Zeitgeber are more specific. Periodicities of temperature and of illumination are generally accepted as Zeitgeber. On the other hand, few experiments have been published indicating that a periodic feeding or irritation by noise may be no Zeitgeber [14, 39]. But there are doubts whether these results are fully convincing. From experiments just carried out at Heidelberg we must conclude that birds running with their spontaneous frequency in *LL* (0.4 lux) can be synchronized with 24

hours if they are fed for 12 hours and kept without food for the following 12 hours. The same seems to be true if an unspecific noise for 12 hours alternates with 12 hours of silence. The experiments are still in progress; therefore, a final answer is impossible. At all events, these unspecific Zeitgeber if ever effective become ineffective if the continuous illumination is increased to higher intensities. As mentioned above, a relatively weak 24-hour Zeitgeber can not catch a free running period deviating too far from 24 hours.

Obviously, not all possible Zeitgeber are equally effective. Light and temperature are the main parameters of the environment to which the organisms have been adapted. The "strength" of a Zeitgeber depends on several elements, some of which may be mentioned here: sensitivity of the organism to the stimulus represented by the Zeitgeber; time-ratio and intensity-ratio of the two fractions of the Zeitgeber; absolute value of stimulus-intensity during the whole period of the Zeitgeber (e.g., intensity of illumination as well during light-time as during dark-time); general circumstances (e.g., degree of constant temperature in *LD*, functional state of the organism, and so on).

II.3. MASKING FACTORS AND ZEITGEBER

Proportional and differential Zeitgeber are only effective if the respective amplitude is big enough. In *LD* there must be a certain difference in intensity of illumination between *L* and *D*. Moreover, not only the difference, but also the absolute values, of intensity are important. An amplitude of 200 lux may represent an excellent Zeitgeber, if *D* means "total darkness." The same amplitude operating between 4 lux and 204 lux may be on the borderline to become ineffective (compare Fig. 16). With respect to this, it may be right to carry out experiments in *LD* with a *D* of "total darkness." But in studying some special qualities of the Zeitgeber-mechanism, e.g., phase-control, this may be wrong. In nature, night never means a darkness as we can perform it in the laboratory. Experiments with an "absolute" dark *D* can therefore be misleading, especially if optical orientated animals are used. An example is presented in Fig. 14, published in an earlier paper of our own [46]. A greenfinch, kept in an artificial 24-hour day with 10, 8, 6, or 4 hours of light, starts and finishes its activity exactly with "light-on" and "light-off." But if there had been an illumination with an intensity of only 0.1 lux during dark-time, quite another result may have happened, (compare Figs. 16 and 19). Figure 14 is shown here to stress that certain (sometimes overlooked) experimental conditions can obscure the real Zeitgeber-mechanism. We may call them masking conditions [47]. Each experiment stays as a question, and the organism should be free to answer as it inclines; this was not the case in the experiment drawn in Fig. 14. To use other words: Especially in experiments with Zeitgeber, the conditions should

22   ASCHOFF

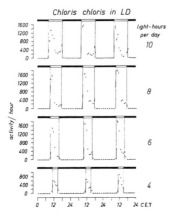

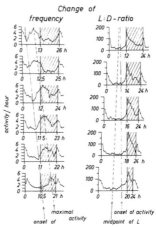

FIGURE 14. Activity of a greenfinch in LD (24-hour period) with 10, 8, 6 and 4 hours of light alternating with total darkness. Misleading experiment (compare text). (From Z. Tierpsychol. *12* (1955) Parey-Verlag. Berlin.)

not be masking (inhibitive) but as "permissive" as possible.

II.4. THE PHASE-ANGLE

Synchronization takes place by determining the phase. That means: An organism synchronized with a Zeitgeber keeps a definite phase-angle-difference to the Zeitgeber. In order to measure the phase-angle-difference, one must select reference points as well in the period of the Zeitgeber as in that of the organism. For this purpose several possibilities are on hand which may be discussed in the light of Fig. 15. The diagrams of the two columns show the activity of mice in LD, each curve averaged on three individuals and several periods. The right side diagrams are the results of experiments [48] in which the LD-period was always 24 hours; the $L:D$-ratio has been varied within the limits 12:12 and 20:4. The left column shows results of experiments in LD with a period varying from 26 to 21 hours; L and D have been kept on equal lengths.

The right column shall be discussed first. Two main reference points of the Zeitgeber are "dawn" and "dusk." The midpoints, however, of light-time or dark-time as representatives of both time spans could also be used. The biological period offers the peaks of activity or the minima (= onsets) for reference. In both cases it is not clear a priori which reference points are more meaningful or more important for phase-control. In examining the diagrams, it is evident that the peaks keep the same phase-angle-difference with respect to dawn and dusk in all conditions, but a changing one with respect to the midpoints. The onsets are not as clear. But also here, the dashed line running parallel to dawn, seems to fit the onsets with the exception of only the lower-most diagram. Is there

FIGURE 15. Pattern of activity of mice in LD, averaged on 3 individuals and at least 6 periods in each diagram. Left column: Period of the Zeitgeber varying from 26 to 21 hours; $L:D$-ratio constant (50% of period each). Right column: Varying $L:D$-ratio of the Zeitgeber, period remaining constant (24 hours).

any possible conclusion from these facts? If dawn and dusk both had a differential effect on the organism, one would expect that the phase should be bound to a point between them. The result would be that onsets as well as peaks of activity in the right column of Fig. 15 would be kept parallel to the midpoints. If, on the other hand, in this case the Zeitgeber would be a proportional one, the increasing influence on angular velocity of the increasing light-time should be compensated by a phase-shift of the organism with respect to light- and dark-time. Therefore, in the case of a proportionally operating Zeitgeber, the organism would change its phase-angle-difference to the midpoints. As the mice do this, whether one watches peaks or onsets, we may predict that LD in mice operates more as a proportional than as a differential Zeitgeber. On the other hand, it is evident that both dawn and dusk are followed by an increasing activity; that may be rather a differential effect of the Zeitgeber on pattern.

In varying the frequency of the Zeitgeber (left column of Fig. 15), there is a clear and big change in the phase-angle-difference of the organism whatever may be used for reference. If the period of the Zeitgeber is longer than 24 hours, the phase of the organism is advanced; if the period is shorter, the phase is delayed. In an artificial 26-hour day, activity starts at the very beginning of light-time; the mice are completely light-active. In the 21-hour day, the onsets have been shifted toward the beginning of dark-time; the mice are dark-active. There is one more interesting fact. In the 26-hour day as well as in the 21-hour day, the mice perform nearly the same amount of total activity. That means: in mice, total activity is nearly independent of whether the mice are active in light or in darkness. This fact, corresponding with the results of experiments in *LL* of different light-intensities (Fig. 8), indicates again that phase-control in mice is done mostly by influencing angular velocity instead of level of excitement [31.] This may be taken as one more evidence for *LD* operating mainly as a proportional Zeitgeber on mice.

Phase does not depend only on the *L:D*-ratio or on *LD*-frequency. Intensity of illumination in *L* and *D* is also important. As shown in Fig. 13, activity-time of chaffinches in *LD* (12:12) may be identical with the light-time, even if *D* does not mean "total darkness" but dim illumination with an intensity of 0.4 lux. One gets quite another picture if the intensity of illumination during *D* is increased to 4 lux (Fig. 16, upper diagram). Activity now starts in the middle of dark-time; it ends as in the experiment shown in Fig. 13 together with light-off. Under the influence of 4 lux during *D* instead of 0.4 lux, the chaffinch keeps quite another phase-angle-difference to the Zeitgeber using onset of activity for reference; activity-time is now longer by 50 per cent. That an alternating *LD* with 200 lux during *L* and 4 lux during *D* is efficient as a Zeitgeber is shown by the phase-shift executed with the Zeitgeber. After doubling light-time, two chaffinches are resynchronized within few periods and keep then the same phase-angle to the Zeitgeber as before. A third bird behaves in a different way (Fig. 16, lower diagram). This bird is remarkable from the beginning, because instead of being synchronized, it runs during the first 4 days of the experiment with a spontaneous period of about 16 hours through the Zeitgeber and is caught not earlier than on the fifth day. But even then, its

phase is quite abnormal: activity starts when the illumination changes from 200 lux to 4 lux, and activity ends with the beginning of bright light. After the Zeitgeber has been shifted, the bird is still synchronized; but now, the phase-angle-difference between bird and Zeitgeber is reversed by 180°. The diagrams of Fig. 16 are an example of the "strength" of a Zeitgeber depending on the absolute value of stimulus-intensity in both of its time-fractions. Using a Zeitgeber on the borderline between being effective or ineffective, synchronization may still be possible but phase-control becomes ambiguous.

One more conclusion seems possible by observing the diagrams in Fig. 16. As other experiments have shown, the free running period of a chaffinch is shorter than 24 hours in *LL* with an intensity of 200 lux as well as with an intensity of 4 lux. If the *LD*-Zeitgeber would operate on birds only as a proportional one—that means: only by influencing angular velocity—one could not expect synchronization with 24 hours by an alternating illumination with intensities of 200 lux and 4 lux. Therefore, *LD* can not operate as a pure proportional Zeitgeber on birds; there must be a more or less strong differential effect.

II.5. Comparison of Phases in Different Organisms

From the experiments described in the foregoing chapter one concludes that phase is a function of at least three parameters of a Zeitgeber: *L:D*-ratio, *LD*-frequency, and intensity of illumination in *L* and in *D*. (This applies also to other Zeitgeber respectively, but only the relationships in *LD* are clarified.) It may be interesting enough to survey phase-relation of different organisms in different experimental conditions. Evidently it could be that each species reacts differently if one changes one or more of the three parameters of a Zeitgeber mentioned above. Surprisingly enough this is not the case as a survey on literature shows. All organisms so far tested behave the same in experiments with varying *LD*-ratio or *LD*-frequency. Figures 17 and 18 contain 10 examples each, representing phase-control in both types of experiments. In varying frequency (Fig. 17) all organisms shift their phases toward the same direction: with high frequencies the phases are delayed; with low frequencies they are advanced with respect to the phase-relation at a 24-hour period. If frequency remains unaltered and only the *L:D*-ratio is varied (Fig. 18), the resulting phase-shifts remain far less. In some organisms phase tends to stay more closely to the midpoints, in others more to dawn or dusk. In general, in light-active animals phase stays more or less parallel to "light-on," in dark-active animals parallel to "dark-on." As mentioned above, this is no strong indication of a differential effect, but rather more of a proportional one. (The figures for

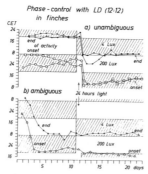

FIGURE 16. Chaffinches in *LD* (12:12) with 200 lux in *L* and 4 lux in *D*. Onset and end of activity in Central European time on the ordinate. Phase-shift of the Zeitgeber on the 11th day by doubling light-time. Upper diagram: Effective Zeitgeber in two finches, unambiguous phase-control. Lower diagram: Catching the spontaneous frequency of one bird by a weak Zeitgeber with ambiguous phase-control.

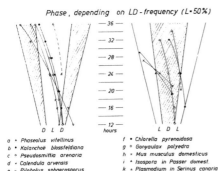

Phase, depending on LD-frequency (L·50%)

a ∘ Phaseolus vitellinus	f ∘ Chlorella pyrenoidosa
b ∘ Kalanchoë blossfeldiana	g ∘ Gonyaulax polyedra
c ∘ Pseudosmittia arenaria	h ∘ Mus musculus domesticus
d ∘ Calendula arvensis	i ∘ Isospora in Passer domest.
e ∘ Pilobolus sphaerosporus	k ∘ Plasmodium in Serinus canaria

FIGURE 17. Phase in *LD*, depending on *LD*-frequency (*L* and *D* 50% of period each). *a)* Leaf movements, Pfeffer [19]; *b)* Opening of flowers, Bünsow [43]; *c)* Emergence, Remmert [50]; *d)* Flowers, Stoppel [51]; *e)* Spore formation, Uebelmesser [52]; *f)* Cell division, Lorenzen [53]; *g)* Luminescence, Hastings and Sweeney [54]; *h)* Maximal activity, Tribukait [49]; *i)* Isospora reproduction, Boughton [55]; *k)* Plasmodium reproduction, Boyd [56].

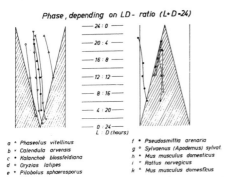

Phase, depending on LD- ratio (L+D ·24)

a ∘ Phaseolus vitellinus	f ∘ Pseudosmittia arenaria
b ∘ Calendula arvensis	g ∘ Sylvaenus (Apodemus) sylvat.
c ∘ Kalanchoë blossfeldiana	h ∘ Mus musculus domesticus
d ∘ Oryzias latipes	i ∘ Rattus norvegicus
e ∘ Pilobolus sphaerosporus	k ∘ Mus musculus domesticus

FIGURE 18. Phase in *LD*, depending on *L*:*D*-ratio (*L* + *D* = 24 hours). *a)* Leaf movements, Pfeffer [19]; *b)* Opening of flowers, Stoppel [51]; *c)* Flowers, Bünsow [43]; *d)* Oviposition, Egami [57]; *e)* Spore formation, Uebelmesser [52]; *f)* Emergence, Remmert [50]; *g)* Peak of activity, Miller [58]; *h)*, *i)* and *k)* Peak of activity.

Sylvaenus are derived from very sketchy data and therefore not conclusive.)

The differences in the amount of phase-shift in the two experiments as shown in Figs. 17 and 18 may be partly the result of one clear fact: In varying the frequency of the Zeitgeber, the two effective parameters of the Zeitgeber are moved in the same direction. If frequency remains constant and *L*:*D*-ratio varies, the differential and proportional parameters of the Zeitgeber are moved in opposite directions. Therefore, the Zeitgeber effects are summated in the first case, but are conflicting in the other one. The possible consequence of the second fact on some problems of

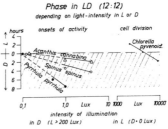

Phase in LD (12:12)
depending on light-intensity in L or D

FIGURE 19. Phase in *LD*, depending on intensity of illumination in *L* or *D*. Ordinate: Hours before or after "light-on." Figures for *Chlorella* from Pirson and Lorenzen [59].

seasonal photoperiodism will be discussed elsewhere [31].

The influence of intensity of illumination on phase has not been studied in detail. A few results seem to indicate that also here divers organisms behave similarly. In the left diagram of Fig. 19, the onsets of activity of three different birds are marked on the ordinate; hour '0' corresponds with "light-on." In all experiments, the intensity of illumination during *L* was the same (200 lux); the very low intensity during *D* was varied. In all three species, activity starts the earlier, the brighter the light in *D*. The results may be compared with experiments carried out on *Chlorella* [59]. In this case, dark-time was kept absolutely dark while the intensity of illumination during light-time was varied. In these experiments, the phase of the organism was also advanced with an increasing intensity of illumination.

II.6. Coaction of Endogenous and Exogenous Components

As mentioned in chapter II.1, and in relation to Fig. 15, Zeitgeber do not only synchronize but can also influence the pattern of circadian periodicity. More important than these external influences are the endogenous ones. The special structure of an organism determines whether it becomes in *LD* light- or dark-active, and causes, moreover, some single events during the total period. Two examples may illustrate such species and individual patterns [60]. Differences in circadian pattern have been known for a long time. In Fig. 20 the patterns of three mainly dark-active species are drawn. In these diagrams, variations in the total amount of activity have been excluded by computing all figures as per cent of the daily mean. Each curve represents the average of one individual on several days. The three curves of the three individuals in one diagram are close to each other in their respective shapes, thus providing a clear species pattern. The pattern of one species is clearly separated from those of the two others.

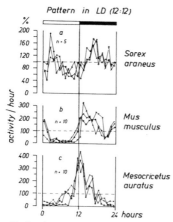

FIGURE 20. Species-pattern of activity in *LD* (12:12). Three individuals in each species. Each pattern-diagram represents the average for 5 days (*Sorex*) or 10 days (*Mus* and *Mesocricetus*) of one individual. Figures for *Sorex* from Crowcroft [61].

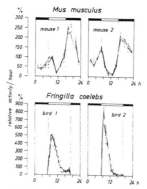

FIGURE 21. Individual pattern of activity in two mice (*LD* 9:15) and two greenfinches (*LD* 12:12). Five consecutive days superimposed in each diagram. (100% of ordinate = average activity per 24 hours.)

Although generally several individuals in one species have similar patterns, it is nevertheless possible to work out individual differences. In Fig. 21 the individual patterns of two mice and two birds are shown, all measured in artificial *LD*. The pattern of one individual remains nearly unaltered during 5 consecutive days whose activity-figures are superimposed in one diagram. The differences between the patterns of two individuals of one species are statistically significant [60]. Inter-individual differences in sensitivity to light may be one of the elements responsible for the special individual pattern; they also cause the individually different time-lags between dawn or dusk, and onset of activity

(individual phase-angle-differences) [39, 62, 63]. Obviously, the individual as well as the species pattern will not remain constant for too long a time. It is known that the pattern varies as the organism grows up, and also in the adult we may expect variations of the pattern at least from season to season [23]. The migrating birds offer a good example in alternating between light- and dark-activity. Nevertheless, we can speak of a species-pattern as well as of an individual pattern so long as the conditions (inside and outside the organism) remain comparable.

REFERENCES

1. ASCHOFF, J. 1959. Zeitliche Strukturen biologischer Vorgaenge. Nova acta Leopoldina, *21:* 147–177.
2. HALBERG, F. 1959. Physiologic 24-hour periodicity in human beings and mice, the lighting regimen and daily routine. Pp. 803–878. *Photoperiodism and Related Phenomena in Plants and Animals*, ed. Withrow. Washington: A.A.A.S.
3. ASCHOFF, J. 1951. Die 24-Stunden-Periodik der Maus unter konstanten Umgebungsbedingungen. Naturwiss., *38:* 506–507.
4. ———. 1954. Zeitgeber der tierischen Tagesperiodik. Naturwiss., *41:* 49–56.
5. HALBERG, F., M. B. VISSCHER, and J. J. BITTNER. 1954. Relation of visual factors to eosinophil rhythm in mice. Am. J. Physiol., *179:* 229–235.
6. BRUCE, V. G., and C. S. PITTENDRIGH. 1957. Endogenous rhythms in insects and microorganisms. Am. Natural., *91:* 179–195.
7. PITTENDRIGH, C. S. 1958. Perspectives in the study of biological clocks. Pp. 239–268. *Symposium on Perspectives in Marine Biology*. Berkeley: Univ. of California Press.
8. PFEFFER, W. 1909. Untersuchungen ueber die Enstehung der Schlafbewegungen der Blattorgane. Abhandl. saechs. Akad. Wiss., math.-physik. Kl., *30:* 257–272.
9. BROWN, F. A., JR. 1957. Biological chronometry. Am. Natural., *91:* 129–133.
10. ———. 1957. The rhythmic nature of life. Pp. 287–304. *Advances in Invertebrate Physiology*. Univ. Oregon Publ.
11. ———. 1959. Living clocks. Science, *130:* 1535–1544.
12. WEBB, H. M., and F. A. BROWN, JR. 1959. Timing long-cycle physiological rhythms. Physiol. Rev., *39:* 127–158.
13. ASCHOFF, J., and J. MEYER-LOHMANN. 1954. Die Schubfolge der lokomotorischen Aktivitaet bei Nagern. Pflügers Arch., *260:* 81–86.
14. RAWSON, K. 1956. Homing behavior and endogenous rhythms. Ph.D. Thesis, Harvard Univ.
15. ASCHOFF, J., and J. MEYER-LOHMANN. 1954. Die 24-Stunden-Periodik von Nagern im natuerlichen und kuenstlichen Belichtungswechsel. Z. Tierpsychol., *11:* 476–484.
16. JOHNSON, M. 1939. Effect of continuous light on periodic spontaneous activity of white-footed mice (*Peromyscus*). J. Exper. Zool., *82:* 315–328.
17. HASTINGS, J. W., and B. M. SWEENEY. 1958. A persistent diurnal rhythm of luminescence in *Gonyaulax polyedra*. Biol. Bull., *115:* 440–458.
18. HOFFMANN, K. 1960. Versuche zur Analyse der Tagesperiodik I. Der Einfluss der Lichtintensitaet. Z. vergl. Physiol. (in press)

19. PFEFFER, W. 1915. Beitraege zur Kenntnis der Entstehung der Schlafbewegungen. Abhandl. saechs. Akad. Wiss., math.-physik. Kl., *34:* 1–154.

20. ASCHOFF, J. 1952. Frequenzaenderungen der Aktivitaetsperiodik bei Maeusen im Dauerlicht und Dauerdunkel. Pflügers Arch., *255:* 197–203.

21. ——. 1958. Tierische Periodik unter dem Einfluss von Zeitgebern. Z. Tierpsychol., *15:* 1–30.

22. ——. 1959. Periodik licht- und dunkelaktiver Tiere unter konstanten Umgebungsbedingungen. Pflügers Arch., *270:* 9.

23. ——. 1957. Aktivitaetsmuster der Tagesperiodik. Naturwiss., *44:* 361–367.

24. PITTENDRIGH, C. S., and V. G. BRUCE. 1957. An oscillator model for biological clocks. Pp. 75–109. *Rhythmic and Synthetic Processes in Growth.* Princeton Univ. Press.

25. SCHWASSMANN, H., and W. BRAEMER. 1960. Personal communication.

26. ASCHOFF, J., and J. MEYER-LOHMANN. 1954. Angeborene 24-Stunden-Periodik beim Kuecken. Pflügers Arch., *260:* 170–176.

27. HELLBRÜGGE, TH. 1960. The development of circadian rhythms in infants. Cold Spring Harbor Symp. on Quant. Biol. Vol. 25.

28. HOFFMANN, K. 1957. Angeborene Tagesperiodik bei Eidechsen. Naturwiss., *44:* 359–360.

29. ——. 1959. Die Aktivitaetsperiodik von im 18- und 36-Stunden-Tag erbrueteten Eidechsen. Z. vergl. Physiol., *42:* 422–432.

30. ASCHOFF, J. 1955. Tagesperiodik von Maeusestaemmen unter konstanten Umgebungsbedingungen. Pflügers Arch., *262:* 51–59.

31. WEVER, R. 1960. Possibilities of phase-control, demonstrated by an electronic model. Cold Spring Harbor Symp. on Quant. Biol. Vol. 25.

32. HASTINGS, J. W., and B. M. SWEENEY. 1957. On the mechanism of temperature independence in a biological clock. Proc. Nat. Acad. Sci., *43:* 804–811.

33. BÜHNEMANN, F. 1955. Das endodiurnale System der Oedogonium-zelle III. Ueber den Temperatureinfluss. Z. Naturforschg., *10b:* 305–310.

34. BÜNNING, E. 1958. Ueber den Temperatureinfluss auf die endogene Tagesrhythmuk, besonders bei *Periplaneta americana.* Biol. Zbl., *77:* 141–152.

35. HOFFMANN, K. 1957. Ueber den Einfluss der Temperatur auf die Tagesperiodik bei einem Poikilothermen. Naturwiss., *44:* 358–359.

36. PITTENDRIGH, C. S. 1954. On temperature independence in the clock system controlling emergence time in *Drosophila.* Proc. Nat. Acad. Sci., *40:* 1018–1029.

37. BALL, N. G., and I. J. DYKE. 1954. An endogenous 24-hour rhythm in the growth rate of the *Avena* coleoptile. J. Exper. Bot., *5:* 421–433.

38. HASTINGS, J. W. 1959. Unicellular clocks. Ann. Rev. Microbiol., *13:* 297–312.

39. DECOURSEY, P. J. 1959. Daily activity rhythms in the flying squirrel, *Glaucomys volans.* Ph. D. Thesis, Univ. of Wisconsin.

39a. ASCHOFF, J. 1955. Jahresperiodik der Fortpflanzung bei Warmbluetern. Studium generale *8:* 742–776

40. GEHRKE, K. 1956. Tageszeitliche Unterschiede der Gehoersempfindlichkeit. Dr. Dissertation, Univ. Goettingen.

41. PITTENDRIGH, C. S., V. G. BRUCE, and P. KAUS. 1958. On the significance of transients in daily rhythms. Proc. Nat. Acad. Sci., *44:* 965–973.

42. BRUCE, V. G., and C. S. PITTENDRIGH. 1958. Resetting the *Euglena* clock with a single light stimulus. Am. Natural., *92:* 295–306.

43. BÜNSOW, R. 1953. Endogene Tagesrhythmik und Photoperiodismus bei *Kalanchoë blossfeldiana.* Planta, *42:* 220–252.

44. STEPHENS, G. C. 1957. Influence of temperature fluctuations on the diurnal melanophore rhythm of the fiddler crab *Uca.* Physiol. Zool., *30:* 55–69.

45. BÜNNING, E. 1959. Zur Analyse des Zeitsinnes bei *Periplaneta americana.* Z. Naturforschg., *14b:* 1–4

46. ASCHOFF, J., and J. MEYER-LOHMANN. 1955. Die Aktivitaet gekaefigter Gruenfinken im 24-Stunden-Tag bei unterschiedlich langer Lichtzeit mit und ohne Daemmerung. Z. Tierpsychol., *12:* 254–265.

47. FRY, F. E. J. 1947. Effects of the environment on animal activity. Univ. Toronto Stud. Biol. Ser., *55:* 1–62.

48. ASCHOFF, J., and J. MEYER-LOHMANN. 1955. Die Aktivitaetsperiodik von Nagern im kuenstlichen 24-Stunden-Tag mit 6 bis 20 Stunden Lichtzeit. Z. vergl. Physiol., *37:* 107–117.

49. TRIBUKAIT, B. 1956. Die Aktivitaetsperiodik der weissen Maus im Kunsttag von 16 bis 29 Stunden Laenge. Z. vergl. Physiol., *38:* 479–490.

50. REMMERT, H. 1955. Untersuchungen ueber das tageszeitlich gebundene Schluepfen von Pseudosmittia arenaria. Z. vergl. Physiol., *37:* 338–354.

51. STOPPEL, R. 1910. Ueber den Einfluss des Lichtes auf das Oeffnen und Schliessen einiger Blueten. Z. f. Bot., *2:* 369–453.

52. UEBELMESSER, E. R. 1954. Ueber den endonomen Tagesrhythmus der Sporangientraegerbildung von *Pilobolus.* Arch. Mikrobiol., *20:* 1–33.

53. LORENZEN, H. 1957. Synchrone Zellteilung von *Chlorella* bei verschiedenem Licht-Dunkel-Wechsel. Flora, *144:* 473–496.

54. HASTINGS, J. W., and B. M. SWEENEY. 1959. The Gonyaulax clock. Pp. 567–584. *Photoperiodism and Related Phenomena in Plants and Animals,* ed. Withrow. Washington: A.A.A.S.

55. BOUGHTON, D. C., F. O. ATCHLEY, and L. C. ESKRIDGE. 1935. Experimental modification of the diurnal oocyst-production of the sparrow coccidium. J. Exper. Zool., *70:* 55–74.

56. BOYD, G. H. 1929. Experimental modification of the reproductive activity of *Plasmodium cathemerium.* J. Exper. Zool., *54:* 111–126.

57. EGAMI, N. 1954. Effect of artificial photoperiodicity on time of oviposition in the fish Oryzias latpes. Annot. Zool. Jap., *27:* 57–62

58. MILLER, R. S. 1955. Activity rhythms in the wood mouse, *Apodemus sylvaticus,* and the bank vole, *Clethryonomys glareolus.* Proc. Zool. Soc. London, *125:* 505–519.

59. PIRSON, A., and H. LORENZEN. 1958. Ein endogener Zeitfaktor bei der Teilung von *Chlorella.* Z. f. Bot., *46:* 53–66.

60. ASCHOFF, J., and K. HONMA. 1959. Art- und Individualmuster der Tagesperiodik. Z. vergl. Physiol., *42:* 383–392.

61. CROWCROFT, P. 1954. The daily cycle of activity in British shrews. Proc. Zool. Soc. London, *123:* 715–729.

62. DUNNETT, G. E., and R. A. HINDE. 1953. The winter roosting and awakening behaviour of captive great tits. Brit. J. Animal. Beh., *1:* 91–95.

63. SCHEER, G. 1952. Beobachtungen und Untersuchungen ueber die Abhaengigkeit des Fruehgesanges

der Voegel von inneren und aeusseren Faktoren. Biol. Abhandl., Heft *3/4:* 1–68.

DISCUSSION

BÜNNING: It is certainly interesting from an ecological standpoint to know in which way the length of the periods is depending on continuous white light or on continuous darkness. But for the physiological analysis it is perhaps more important to know that continuous light has quite different effects, depending on the light quality. In *Phaseolus* and in several other plants the periods in continuous red light are by 2–5 hours longer than in continuous darkness, whereas in continuous far-red light these are shorter than in darkness. White light, thus, means in certain cases an interaction of 2 antagonistic effects.

ASCHOFF: I agree fully that different wave-lengths will have different influences on the clock-system; also it may be more important to study quality of light instead of intensity. On the other hand, as I in my talk was mostly concerned with features of the free running oscillation and with its entrainment by Zeitgeber, it seemed reasonable to me to refer mainly to all the several experiments with different intensities of white light. Under natural conditions, white light is the main Zeitgeber, and one can use it to study general clock-mechanisms.

CLOUDSLEY-THOMPSON: If the change in phase-length depending on light intensity has an adaptive function in altering the time of the onset of activity with changing season, it would seem that the clock may not only be regulated in this way, but may also be synchronized by a Zeitgeber each day. My work on terrestrial anthropods suggests that under one set of constant condition the free-running period may be almost exactly 24 hours, yet it may be less or more than this under another. Why, therefore, do you say that an exact 24-hour period is unlikely to be found? In your figures of the times of the onset of activity in the finch, the curve slopes upwards or downwards according to the light intensity. Surely, if the correct intermediate intensity was selected, it would be horizontal.

ASCHOFF: Full agreement, that by using the right constant conditions the spontaneous period may be 24 hours. This never has been refuted.

ENRIGHT: Dr. Aschoff has, in his paper, indicated that he is not seriously concerned with amplitude in activity cycles; but yet he has emphasized what he has called "α" and "ρ" i.e., the proportion of the cycle during which the organism is active above some zero level. I would like to suggest that α-ρ measurements may, in fact, be only a reflection of amplitude, no less, but no more reliable, than amplitude as a characteristic of a given rhythm.

My own work on activity rhythms has produced, under some conditions, a discontinuous, peaked graph such as shown in Dr. Aschoff's graphs for finches, for example. By changing experimental conditions—not with regard to temperature or light intensity, but, for example, simply by depriving the animals of a sand substrate, a factor which would presumably have no bearing on the "rate of energy flow through the system"—the activity data can be transformed to a continuous function with, in fact, no zero level.

I would suggest, then, that light, or substrate, or the like, can markedly affect amplitude of an activity rhythm by determining the zero, or threshold level, below which cyclic processes continue but are not evidenced by overt activity; that the position of this threshold level may be shifted upward or downward by a whole array of factors from both the recent past history of the organism, and the experimental conditions; and that this shifting of threshold would also result in changed values of α and ρ. If the threshold is shifted upward enough, there will be zero amplitude, and zero α; shifting downward enough there will be continuous activity and a zero ρ.

I bring this point up not to discount the significance of α and ρ as characteristics of a given cycle, but rather to emphasize what I believe is the very real importance of amplitude measurements of the cycles; to suggest that α and ρ may simply be different aspects of a truncated continuous cycle.

ASCHOFF: I agree that α and ρ are also measurements of the amplitude so long as the period is constant. If, on the other hand, amplitude remains constant, the $\alpha = \rho$ ratio is only a function of the average level. The threshold above and below which we measure activity-time and rest-time is quite clearly arbitrary. This threshold may be zero as well as any other value of the ordinate on which we draw the observed activity; as well it may be the surface of a layer of sand wherein the organism vanishes during rest-time and where from it arises with beginning activity-time. The main idea is not to shift the threshold—which clearly would change the $\alpha = \rho$ ratio—but to keep the threshold at a definite value of the ordinate. If then $\alpha = \rho$ changes without a change in amplitude (upper + lower width), the level must have been changed. The whole problem is easy to understand if one draws up a curve of body temperature and uses 37° as a threshold. The upward step of the level by one degree during fever (without change in amplitude, as is often the case) will bring an increase in $\alpha : \rho$ ratio.

KALMUS: *LD* means lethal dose in toxicology and is the measurement of insecticidal action. One might perhaps avoid such abbreviations like *LD* 50 per cent in describing charges of illumination.

ASCHOFF: *L* and *D* for light and dark are now so commonly used that it seems rather difficult to avoid using them.

KALMUS: Deviation from an exact 24-hour period in constant conditions should not be used as the sole criterion for endogenouity. Continuation of an inverted or phase-shifted rhythm in constant conditions is another important criterion.

ASCHOFF: Exactly these types of experiments have been used to make proof of an exogenous periodicity. Hypothetical explanations for the continuation of inverted rhythms as exogenous ones are, for instance, given by F. A. Brown. Serious explanations may be based on the possibility of semistable phase control. To avoid such criticisms, only the spontaneous period deviating from 24 hours should be taken as proof for endogenous.

RAWSON: What evidence is there to substantiate the concept of an *angular velocity* within a one-day period? It appears to me that most of our evidence is based on the measurement of *one event per day* such as peak of activity or onset of activity but that we do not have many time measures by organisms throughout the day. There are a few measures of time within the day, such as those of sun compass orientation, where the "clock" is read during the day. Until this kind of evidence is analyzed we can say nothing about *angular velocity* of the rhythm or even whether the expression "angular velocity" is a useful term describing the circadian rhythm of an organism.

ASCHOFF: The statements with regard to angular velocity have been started only from arguments which are implicit in an oscillating system. The first question was whether one can assume or not that the varying speed of the system is correlated to *LD* or activity- and rest-time respectively. The next question is by what means we perhaps can measure the speed; few possibilities are discussed in the paper.

SOLLBERGER: I want to point out that the amplitude may often rise with the level, just indicating an increased load on a constantly running system. In that case only a change in the relation between amplitude and level ("amplitude line") signifies a real change in "angular velocity." Perhaps the latter may preferably be called average velocity instead.

244

NATURE VOL. 229 JANUARY 22 1971

Menstrual Synchrony and Suppression

by

MARTHA K. McCLINTOCK

Department of Psychology, Harvard University, Cambridge, Massachusetts 02138

Synchrony and suppression among a group of women living together in a college dormitory suggest that social interaction can have a strong effect on the menstrual cycle.

STUDIES of the influence of pheromones on the oestrous cycles of mice[1-4], and of crowding on variables such as adrenalin production in mice and other species[5] have suggested that social grouping can influence the balance of the endocrine system. Although there has been little direct investigation with humans, anecdotal and indirect observations have indicated that social groupings influence some aspects of the menstrual cycle. Menstrual synchrony is often reported by all-female living groups and by mothers, daughters and sisters who are living together. For example, the distribution of onsets of seven female lifeguards was scattered at the beginning of the summer, but after 3 months spent together the onset of all seven cycles fell within a 4 day period.

Indirect support is given by the investigation of Collet et al.[6] on the effect of age on menstrual cycle patterning. A higher percentage of anovulatory cycles were reported for college age women than for older women. Although Collet et al. attributed this to a maturational factor, it is interesting that most of the college aged women attended all female schools. Considering the parallel with the Lee–Boot effect in mice[1] (groups consisting only of females become pseudopregnant or anoestrous), it seems possible that an interpersonal factor is operating together with the maturational factor.

Subjects were 135 females aged 17–22 yr—all residents of a dormitory in a suburban women's college. The dormitory in which they resided has four main corridors each with approximately twenty-five girls living in single and double rooms. Six smaller living areas, separated from the main corridors by at least one door, each house approximately eight girls in single rooms.

Three times during the academic year, each subject was asked when her last and second to last menstrual periods had begun; thus the date of onset was determined for all cycles between late September and early April. The average duration of menstruation and presence of dysmenorrhoea were noted. In addition, subjects estimated how many times each week they were in the company of males and listed by room number the girls ($N \leq 10$) with whom they spent the most time, indicating which two of these they saw most often.

The date of menstrual onset was compared for room mates and closest friends, for close friend groups and for living groups. Two people qualified as "closest friends" only if both had indicated that they saw each other most often. While menstrual cycle timing in women using birth control pills was individually invariant, these women were still included in the analysis, because their influence on the menstrual cycles of the others was unknown. For room mates and closest friends, the difference between the date of onset in October for one arbitrarily chosen member of the pair and the closest date of onset for the other was calculated. This difference was compared with a difference for March calculated in a similar way, but with one change : instead of choosing the closest onset dates for the pair, both onsets for March were chosen to follow

the initial October onset by an equal number of cycles. For example, if onset 6 occurred on March 10 for the first member of the pair, and onsets 5 and 6 for the other member occurred on March 1 and March 29 respectively, then the March 10 and March 29 dates were used to calculate the difference in onset. This procedure was used to minimize chance coincidences that did not result from a trend towards synchrony.

The Wilcoxon matched-pairs signed-ranks test[7] was used to test for a significant decrease in the difference between onset dates of room mates and closest friends. This test utilizes both the direction and magnitude of change in differences and is therefore a relatively powerful test.

There was a significant increase in synchronization (that is, a decrease in the difference between onset dates) among room mates ($P \leq 0.0007$), among closest friends ($P \leq 0.003$) and among room mates and closest friends combined ($P \leq 0.0003$). The increase in synchrony for room mates did not differ significantly from the increase for closest friends. The increase in synchrony was further substantiated by non-overlapping confidence intervals, calculated for the median difference on onset dates[8] (Table 1).

This synchrony might be due to some factor other than time spent with an individual; Koford[9] has attributed synchrony of the breeding season in *Macaca mulata* on Cayo Santiago to common seasonal changes in available food. The fact that the subjects generally eat as a dormitory group in a common dining room might be a significant factor in creating synchrony. A similar life pattern and common, repeated stress periods might also effect synchrony. Subjects were therefore randomly paired and tested for synchrony within the dormitory as a whole, but no significant trend (N.S., $P \leq 0.8$) was found, and the confidence intervals for the median difference in onset date overlapped completely.

Group synchrony was also investigated and the data were analysed to verify that the decrease in difference between onset dates was a true measure of synchrony. All subjects were divided into fifteen groups of close friends ($5 \leq N \leq 10$), using the lists of close friends made by each subject. During the interview, it was stressed to each subject that her list of "close friends" should include the people she saw most often and with whom she spent the most time, not necessarily those with whom she felt the closest. But because there is usually some overlap, the term "close friends" was adopted. Only subjects who mutually listed each other were included in a group.

A mean onset date (μ_t) was determined for each group in October, late November, January, late February and April. As before, the onset dates (X_t) being compared, each followed the October onset (X_1) by an equal number of cycles. The mean individual difference from the group onset mean

$$\frac{\sum\limits_{i}^{n}(X_t - \mu_t)}{n}$$

was determined for each group and compared across time in two ways. First, a linear rank method, designed by Page[10] to

Table 1 Confidence Intervals (> 0.99, in days) for the Median Difference in Onset Date between Members of the Pair

	October	March
Close friends and room mates		
N=66	$7 < M < 10$	$3 < M < 7$
Random pairs		
N=33	$6 < M < 14$	$5 < M < 15$

NATURE VOL. 229 JANUARY 22 1971

test ordered hypotheses for multiple treatments, showed a significant decrease in individual differences from the group onset mean for close friend groups ($P \leq 0.001$). Second, a graph of this decrease as a function of time (Fig. 1) indicated that the greatest decrease occurred in the first 4 months with little subsequent change. This asymptotic relation indicated that the decrease in difference between onset dates was indeed an increase in synchrony for close friend groups.

Usually those who considered themselves close friends lived together. Because this was not always the case, however, subjects were divided into thirteen living groups ($5 \leq N \leq 12$), solely on the basis of arrangement of rooms, to test the importance of geographic location. When grouped in this way, there was no significant increase in synchrony within groups.

Dewan[11] has suggested that the menstrual cycles of monkeys around the equator are synchronized because each cycle is locked in phase with the Moon. As the production by the pineal gland of a substance which inhibits the action of luteinizing hormone is suppressed by light, the continuous light of nights with a full Moon would facilitate ovulation across a group of monkeys and induce synchrony. This suggests that the synchrony in close friend groups and among room mates comes from a common light–dark pattern, perhaps with common stress periods in which the subjects may stay up for a large part of the night. It would be expected that if synchrony arose from common light–dark cycles, room mates would exhibit a more significant amount of synchrony than do closest friends. The opposite trend was found, however, although it was not significant (room mates, $P \leq 0.007$; closest friends, $P \leq 0.003$). It does not seem likely therefore that a photo-periodic effect is a significant cause of synchrony. This is further supported by the lack of significant synchrony in random pairings in the dormitory.

Paralleling the Whitten effect in mice[3] (in which suppression of oestrus in groups of females can be released by the introduction of a male pheromone) synchrony may result from a pheromonal interaction of suppression among close friend groups, followed by a periodic release due to the presence of males on the weekend. However, this would be insufficient to explain the synchrony which occurred among room mates and close friends, but did not occur throughout the dormitory.

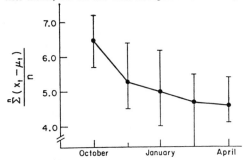

Fig. 1 The median individual mean difference from the group onset mean $\left(\frac{\left(\Sigma X_t - \mu_t \right)}{n} \text{ days} \right)$ as a function of time. The asymptotic relation and non-overlapping confidence intervals[8] for the medians in October and late February, and October and April (> 0.99), indicate an increase in synchrony for close friend groups.

Some additional pheromonal effect among individuals of the group of females would be necessary. Perhaps at least one female pheromone affects the timing of other female menstrual cycles.

Another possible source of synchrony might be the awareness of menstrual cycles among friends. A sample taken from the dormitory, however, indicated that 47% were not conscious

Table 2 Mean Cycle Lengths and Duration of Menstruation

Estimated exposure to male (days/week)	Length of cycle (days)	Duration (days)
0–2 N=56	30.0 ± 3.9	5.0 ± 1.1
3–7 N=31	28.5 ± 2.9	4.8 ± 1.2
P	≤ 0.03	N.S. ≤ 0.2

of their friends' menstrual cycles, and, of the 53% who were, 48% (25% of the total) were only vaguely aware.

The significant factor in synchrony, then, is that the individuals of the group spend time together. Whether the mechanism underlying this phenomenon is pheromonal, mediated by awareness or some other process is a question which still remains open for speculation and investigation.

Subjects were divided into two groups: those who estimated that they spent time with males, once, twice or no times per week (N=42), and those who estimated that they spent time with males three or more times per week (N=33). Borderline cases and those taking birth control pills were discarded. After testing for homogeneity of variance, the mean cycle length and duration of menstruation was compared using Student's t test. Those who estimated seeing males less than three times per week experienced significantly ($P \geq 0.03$) longer cycles than those of the other group whose mean cycle length corresponded with national norms (approximately 28 days)[12]. There was no significant difference in duration of menstruation itself ($P \geq 0.2$, Table 2).

The possibility that the results were confounded by a maturational factor was tested, as subjects included members of the freshman, sophomore, junior and senior classes. The subjects were regrouped and compared according to class: underclassmen were compared with upperclassmen. There was no significant difference in cycle length (underclassmen 29.6 ± 5.6 days; upperclassmen 29.9 ± 5.7 days).

Exposure to males may not be the significant factor. It may be, for example, that those with longer cycles are less likely to spend time with males. However, many subjects spontaneously indicated that they became more regular and had shorter cycles when they dated more often. For example, one subject reported that she had a cycle length of 6 months until she began to see males more frequently. Her cycle length then shortened to 4.5 weeks. Then, when she stopped seeing males as often, her cycle lengthened again. Whether this is due to a pheromone mechanism similar to the Lee–Boot effect in mice[1] has yet to be determined.

Although this is a preliminary study, the evidence for synchrony and suppression of the menstrual cycle is quite strong, indicating that in humans there is some interpersonal physiological process which affects the menstrual cycle.

I thank Professor Patricia Sampson and Monty Slatkin for help in preparing the manuscript.

Received July 28, 1970.

[1] Van der Lee, S., and Boot, L. M., *Acta Physiol. Pharmacol. Neerl.*, 5, 213 (1956).
[2] Whitten, W. K., *J. Endocrinol.*, 18, 102 (1959).
[3] Whitten, W. K., *Science*, 16, 584 (1968).
[4] Parkes, A. S., and Bruce, H. M., *J. Reprod. Fertil.*, 4, 303 (1962).
[5] Thiessen, D., *Texas Rep. Biol. Med.*, 22, 266 (1964); Leiderman, P. H., and Shapiro, D., *Psychobiological Approaches to Social Behavior* (Stanford University Press, 1964).
[6] Collet, M. E., Wertenberger, G. E., and Fiske, V. M., *Fertil. Steril.*, 5, 437 (1954).
[7] Siegal, S., *Nonparametric Statistics for the Behavioral Sciences* (McGraw-Hill, New York, 1956).
[8] Nair, K. R., *Indian J. Statistics*, 4, 551 (1940).
[9] Koford, C. B., in *Primate Behavior; Field Studies of Monkeys and Apes* (edit. by Devore, I.) (Holt, Rinehart and Winston, New York, 1965).
[10] Page, E. B., *Amer. Stat. Assoc. J.*, 58, 216 (1963).
[11] Dewan, E. M., *Science Tech.*, 20 (1969).
[12] Turner, C. D., *General Endocrinology* (Saunders, Philadelphia, 1965).

15

INTERACTION BETWEEN INTERNAL AND EXTERNAL ENVIRONMENTS IN THE REGULATION OF THE REPRODUCTIVE CYCLE OF THE RING DOVE

Daniel S. Lehrman

MAJOR FEATURES OF SEXUAL AND PARENTAL BEHAVIOR

The ring dove (*Streptopelia risoria*), a small relative of the domestic pigeon, is unusually well suited for psychobiological investigation. It breeds very well in captivity, exhibits behavior patterns which do not appear to have degenerated from the dove wild type in spite of hundreds of generations of domestication, and has been the subject of attention by investigators who have provided a good background of useful information both about its reproductive behavior (Craig, 1909) and about its reproductive physiology (Riddle, 1937).

If a male and female of this species, both of which have had recent breeding experience, but not with each other, are placed together in a breeding cage containing a nest bowl and a supply of nesting material, a regular and predictable sequence of changes in behavior can be observed. The principal activity observed in the first day or so is a characteristic pattern of courtship behavior, in which the male bows and coos at the female (Figure 1). After some hours, the birds indicate their selection of a nest site by crouching in a concave place (e.g., the inside of a nest bowl) and uttering a characteristic coo. Once the site

The research program described in this chapter is supported by research grant MH-02271 and by Research Career Award MH-K6-16,621, both from the National Institute of Mental Health. Grateful acknowledgment is made to the following graduate students who are participating in this work: Mr. Philip N. Brody, Mr. Carl J. Erickson, Miss Miriam Friedman, Mr. Barry R. Komisaruk, Miss Sheila Miller, and Mrs. Rochelle P. Wortis.

355

356   MAJOR FEATURES OF SEXUAL AND PARENTAL BEHAVIOR

(a)

(b)

Figure 1. Male ring dove (a) bowing and (b) cooing to female, initial courtship behavior

has been selected, both the male and the female actively participate in building a nest there. The male characteristically gathers most of the nesting material and carries it to the female, who spends most of her time standing on the nest, where she takes the nesting material from the male and builds it into the nest. The male may do some of the nest building and the female some of the collecting, but the pattern is predominantly as I have described it.

After several days of nest-building activity, a fairly abrupt change occurs in the behavior of the female. She becomes noticeably more attached to the nest, and more difficult to dislodge from it; it is possible to pick her up while she is standing on the nest and to lift her, in which case she may actually lift the partly built nest clutched in her claws! This kind of behavior usually indicates that the female is about to lay an egg. She lays the first of her two eggs at about 5:00 P.M. and the second egg at about 9:00 A.M. on the second day following. At some time between the laying of the first and second eggs, the female begins to sit on the eggs. Usually on the day when the second egg is laid, the male also begins to sit. The birds take turns in incubating the eggs, the male sitting for about six hours in the middle of the day, the female continuously for the remaining 18 hours. The eggs hatch after about 14 days, and for several days following hatching the parents continue to sit on the young, following approximately the same schedule as that on which they had incubated the eggs. At first the parents feed the young by regurgitating to them a substance produced by the lining epithelium of their crops ("crop milk") (Patel, 1936). The young leave the nest at 10 to 12 days of age, but continue to beg for food from the parents. Starting when the young are 14 to 15 days old, the parents gradually become less and less willing to respond to the begging of the young, while the squabs gradually develop the ability to peck for food from the ground. When the young are about 20 days old, the bowing-coo courtship behavior of the adult male increases in frequency, the parents begin to build a new nest, lay new eggs, and the cycle is repeated. This cycle—of courtship, nest building, egg laying, incubation, and care of the young—lasts 6 to 7 weeks, and may continue throughout the year (at least in our laboratory, where there are no seasonal changes in temperature or length of the day).

Now, these regular, highly predictable changes in behavior are not merely casual or superficial changes in the birds' preoccupations. They represent striking changes in the overall atmosphere of the events going on in the breeding cage. At its appropriate stage each of the behaviors I have mentioned represents the *predominant* activity of the animals at the time. Further, these changes in the behavior are not solely responses to external stimuli. The birds do not build the nest merely because the nesting material is available. In fact nesting material may be in the cage throughout the cycle, but nest-building behavior will still be concentrated at the stage that I have described as the nest-building stage. Similarly, reactions to the eggs and to the young appear to occur only at appropriate stages in the cycle.

These cyclic changes in behavior thus represent, at least in part, changes in the internal condition of the animals, rather than merely changes in their external situation. In fact, the changes in behavior are associated with equally striking and equally pervasive changes in the anatomy and physiological state of the birds. For example, when she is first introduced into the cage, the oviduct of a female dove may weigh about 800 milligrams. Eight or nine days later, when she lays her first egg, the same oviduct may weigh on the order of 4000 milligrams, an increase of some 400 per cent. The crops of both male and female weigh about 900 milligrams when the birds are introduced into the cage. When they start to sit on the eggs ten days later, the crop will weigh about the same. But two weeks later, when the eggs hatch, the crops may weigh about 3000 milligrams, an increase of about 250 per cent. Equally striking changes in the condition of the ovary, the weight of the testes, the length of the gut, the weight of the liver, the histology of the pituitary gland, etc., are correlated with the behavioral cycle (Schooley and Riddle, 1938).

Now, if a male dove or a female dove is placed *alone* in such a cage, no such cycle of behavioral or anatomical changes takes place. A female dove alone in a cage does not lay two eggs every six or seven weeks; she lays no eggs at all. A male dove alone in a cage never shows any interest in nesting material, in eggs, or in young.

The cycle of psychobiological changes which I have described is therefore one which occurs more or less synchronously in each member of a pair of doves living together, but will not occur independently in either of the pair. The subject of the present chapter is our analysis of the origin of this cycle, and of the mechanisms by which it is regulated.

EXTERNAL STIMULI AND THE INDUCTION OF INCUBATION BEHAVIOR

Let us first consider the problem of what makes these birds capable of sitting on eggs. In a preliminary experiment (Lehrman, 1958a), we introduced pairs of ring doves into breeding cages each containing a nest bowl in which was a nest and two eggs. Each member of the pair had been in isolation for from three to five weeks since its last breeding experience. Under these conditions birds do not sit on the eggs until after 5 to 7 days, during which they first court, then exhibit nest-building behavior (Figure 2, Group 1).

We considered the possibility that the latency (5 to 7 days) which preceded the onset of incubation behavior might represent the amount of time required for the birds to become habituated to a strange cage, and to recover from the disturbance of being handled. Pairs of birds were therefore placed in cages containing no nest bowl or nesting material and with an opaque partition separating the male and the female. After the birds had spent seven days in such a cage, the opaque partition was removed, and a nest containing eggs was simultaneously added. Now, if the 5-to-7 day latency period in the original group was due to the trauma of being handled and to the fact that the birds

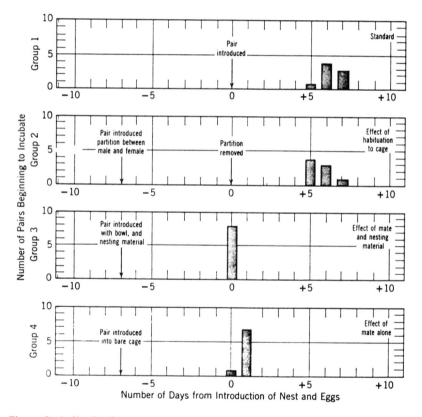

Figure 2. A distribution of latencies of incubation behavior under different conditions of association with mate and availability of nesting material. Nest and eggs introduced at 9 A.M. of Day 0. (From Lehrman, 1958a.)

were in a strange cage, the birds in this group (Figure 2, Group 2) should sit immediately when the partition is removed. However, they did not sit until from 5 to 7 days after removal of the partition.

Another possibility is that stimuli coming from the eggs induce the birds to become interested in sitting on eggs, but that it takes from 5 to 7 days for this effect to reach a threshold value at which the behavior actually changes to incubation. In order to test this possibility, we placed pairs of birds in cages each containing an empty bowl and a supply of nesting material, but no eggs. After seven days the nest bowl was removed with its partially or fully built nest, and replaced with a new nest bowl containing a nest with two eggs. Under this treatment all birds sat on the eggs almost immediately (Figure 2, Group 3). It appears that the stimuli which cause our subjects to change from birds that are not interested in sitting on eggs to birds that are interested in sitting on eggs come, not from the eggs, but from some combination of the presence of the mate and of the nesting situation (nest bowl and nesting material).

In an attempt to separate the effects of the presence of the mate from those of the presence of the nesting situation, we introduced pairs of birds into cages containing no nest bowl or nesting material, and then, after seven days, introduced a nest containing two eggs. These birds did not sit immediately; but neither did they wait 5 to 7 days. Instead the birds sat after approximately one day, during which they engaged in intensive nest-building behavior (Figure 2, Group 4). A fifth group of males and females placed alone in cages containing nest and eggs exhibited no incubation behavior at all.

It appears from these experiments that at least two changes in condition are induced by the treatments we have described. First, stimuli arising from association with a mate cause them to change from birds interested primarily in courtship to birds interested primarily in nest building. Secondly, when the birds are interested in nest building, stimuli arising from the presence of the nest bowl and nesting material appear to facilitate the transition from interest in building a nest to interest in sitting on eggs.

A further demonstration of the effect of external stimuli derived from the mate and/or the nesting material on the development of incubation behavior is found in a second group of experiments (Lehrman, Brody, and Wortis, 1961). We placed pairs of birds in breeding cages for varying numbers of days, in some cases with, and in other cases without, nest bowl and nesting material. Then we introduced a nest and eggs into the cage, and allowed the birds three hours to sit. If neither bird sat within three hours, the test was scored as negative, and both birds were sacrificed for autopsy. If either bird sat within three hours, that bird was removed, and the other bird was given an additional three hours to sit. This experiment therefore provided evidence showing, independently for the male and for the female, the development of readiness to incubate as a function of the number of days with the mate, with or without the opportunity to build a nest. Figures 3 and 4 show the results of this experiment for the females and for the males respectively.

It is apparent from these graphs that association with the mate gradually brings the birds into a condition of readiness to incubate, and that this effect is greatly enhanced by the presence of nesting material. The effect of exposure to the nesting situation is not to stimulate the

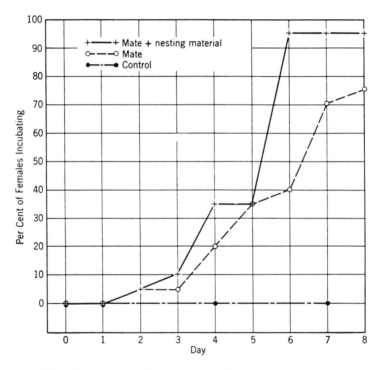

Figure 3. Effect of association with a mate or with a mate plus nesting material on the development of incubation behavior in female ring doves. Each point is derived from tests of 20 birds. No individual bird is represented in more than one point. The abscissa represents the duration of the subject's association with a mate or with a mate plus nesting material, or (for the control group), the time spent alone in the test cage, just before being tested for response to eggs. "Day 0" means that the bird was tested immediately on being placed in the cage. (From Lehrman, Brody, and Wortis, 1961.)

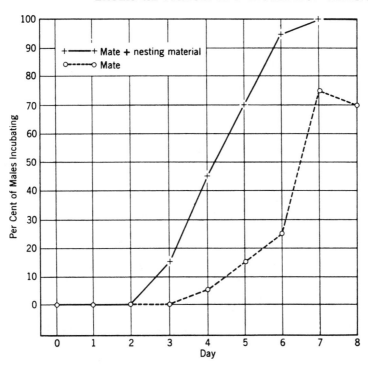

Figure 4. Effect of association with a mate or with a mate plus nesting material on the development of incubation behavior in male ring doves. See the caption for Figure 3.

onset of readiness to incubate in an all-or-none way. Rather, it has an effect upon the development of readiness to incubate which is synergistic with the effect of stimulation provided by the mate (Hinde and Warren, 1959).

It will be noted that by the third day the presence of nesting material makes a difference in the number of males incubating, but a comparable effect is not seen in females until the sixth day. In this connection it is helpful to recall that when the nest is built the male gathers nesting material and carries it to the female. It may be that the condition of the female is affected by the presence or absence of nesting material, starting about the sixth day, because the behavior of the male toward her has been different, depending upon the presence or absence of nesting material, starting at about the third day.

It is clear from this group of experiments that external stimuli normally associated with the breeding situation play an important role in inducing a state of readiness to incubate. We may now inquire into the nature of this state.

HORMONAL INDUCTION OF INCUBATION

We next attempted to induce incubation behavior by injecting hormones into the birds, instead of subjecting them to particular types of external stimulation. For this experiment (Lehrman, 1958b), we treated birds just as we had in the experiments illustrated in Figure 2, except that some of the birds were injected with hormones while they were in the isolation cages, starting one week before they were due to be placed in pairs in the test cages. The results are shown in Figure 5.

Group 1 of Figure 5 is a replication of Group 1 of Figure 2. Here again, birds placed in pairs in a cage containing a nest and eggs sat on the eggs only after an interval of 4 to 7 days, during which they first courted, and then built a nest. When doves were injected with progesterone while in the isolation cages, they sat upon the eggs almost immediately on their introduction into the test cage (Figure 5, Group 2). When the injected substance was an estrogen (diethylstilbestrol), the effect on most birds was to make them sit on the eggs after a latency period of 1 to 3 days during which they engaged in intensive nest-building behavior (Figure 5, Group 3). Testosterone had no effect upon incubation behavior.

Since prolactin has been reported to induce incubation behavior in the domestic chicken (Riddle, Bates, and Lahr, 1935), we have also attempted to induce ring doves to sit on eggs by injecting this hormone (Lehrman and Brody, 1961). We find that prolactin is not nearly as effective as progesterone in inducing incubation behavior, even at dosage levels which induce full development of the crop (the growth of which is under the control of prolactin). For example, when the total dosage of prolactin was 400 I.U., only 40 per cent of the birds sat on eggs, even though their average crop weight was about 3000 milligrams (compared with about 900 milligrams for uninjected control birds). Injection of 10 I.U. of this hormone caused significant increases in crop weight (to 1200 milligrams), but elicited no increase in frequency of incubation behavior, as compared with the performance of control birds. Since, in a normal breeding cycle the crop begins to increase in weight only *after* incubation begins, it seems most unlikely that prolactin plays a role in the initiation of normal incubation behavior in this species.

It appears, therefore, that estrogen may be associated with nest-building behavior, and that progesterone is probably the hormonal initiator of incubation behavior.

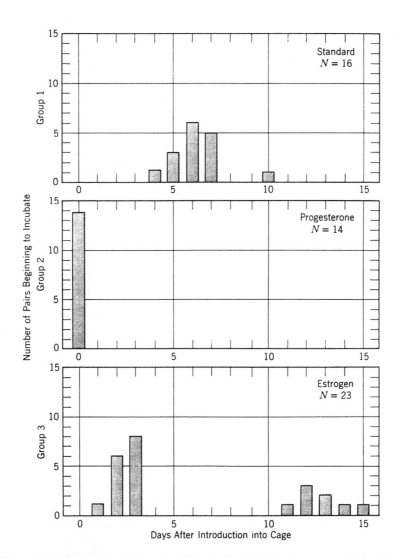

Figure 5. A distribution of latencies of incubation behavior in groups of ring doves subjected to different hormone pretreatments. Birds introduced into experimental cages between 9 A.M. and 9:30 A.M. on Day 0. N = number of pairs. (From Lehrman, 1958b.)

EXTERNAL STIMULI AND GONADOTROPIN SECRETION

The foregoing experiments strongly suggest the conclusions that stimuli coming from the mate and from the nesting situation induce readiness to incubate, and that this readiness depends on gonadal hormones. It seems logical to assume that these external stimuli must influence readiness to incubate via stimulation of the secretion of gonad-stimulating hormones by the birds' pituitary glands. Figure 6 shows the percentage of female doves that ovulated after varying periods of association with a male, with and without nesting material. It can be seen that the effect on ovulation is almost exactly coincident with the effect of these stimuli on the onset of incubation behavior in the female, as shown in Figure 3. In individual females there is a very high degree of association between the occurrence or nonoccurrence of ovulation and that of incubation behavior (chi square = 200.4, $p <$ .001).

We carried out these observations because of our interest in the

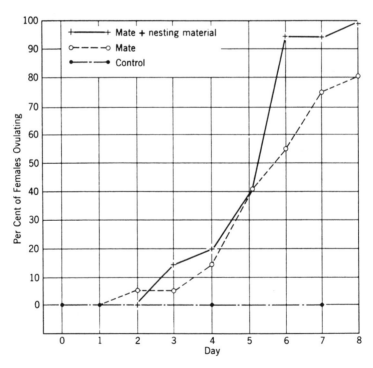

Figure 6. Effect of association with a mate or with a mate plus nesting material on the occurrence of ovulation in female ring doves. See the caption for Figure 3.

coincidence of ovulation and of the onset of readiness to incubate. However, it is not surprising that external stimuli provided by the male can induce ovarian activity in female doves. Harper (1904) observed that ovarian development in the domestic pigeon followed introduction of a male into her cage; and Craig (1911) showed, for the ring dove, that ovulation occurred after association with the male without the necessity of actual copulation. Matthews (1939) showed that a female domestic pigeon can be induced to lay an egg in reaction to visual and auditory stimuli provided by a male from whom she is physically separated.

Mr. Carl Erickson and I have verified this finding in the ring dove, and are presently investigating some of the conditions and properties of this effect. Female ring doves with previous breeding experience are placed in a cage containing similarly experienced male doves, from which they are separated by a glass divider. During the seven days following their introduction into the cages, the oviduct weights of the females rise from a mean of approximately 900 milligrams to approximately 3100 milligrams. If, on the other hand, the "stimulus" males have been castrated two months before the beginning of the experiment, then the oviduct weights rise to only 1800 milligrams. This represents a striking and significant depression of the stimulating effects of the males. Apparently the stimulating effect when the male is present involves not merely the presence of a second bird on the other side of the glass plate, but rests, at least in part, on some (presumably) behavioral characteristics of the male; characteristics which are dependent on the presence of male hormones (Carpenter, 1933). This investigation is continuing.

It is clear from these data that stimuli from the mate (and from the nesting situation) actually stimulate the secretion of gonadotrophic hormones by the female's pituitary gland. It should not be presumed, however, that secretion of each of the pituitary products occurs quite independently as a response to a different external stimulus. The fact is that certain functional relationships within the endocrine system are equally important in some cases. A good example is seen in the process of ovulation. This process is partly set in motion by stimuli from the mate, and further facilitated by stimuli arising from the nesting situation. However, once it has begun the process may continue to completion even if the initiating stimuli are withdrawn soon after ovarian activity has been set in motion, so to speak. Figure 7 shows the result of an experiment (Lehrman, Wortis, and Brody, 1961), in which female ring doves were placed in cages with males, with or without nest bowl and nesting material, and left there for seven days, although

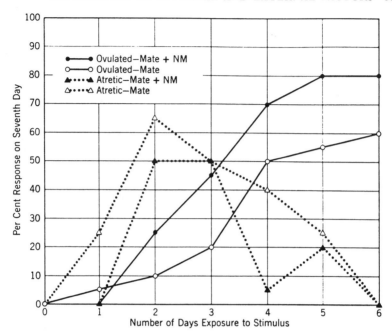

Figure 7. Number of birds ovulating (solid lines) and number of birds developing atretic follicles (dotted lines) by the seventh day, as a function of duration of exposure to a male (light lines) or to a male and nesting material (heavy lines) during the first six days. Atretic follicles are follicles whi.. ..ave begun to develop as a result of stimulation by the male, then degenerated after stimulation was withdrawn. (From Lehrman, Wortis, and Brody, 1961.)

the male (and the nesting material) were removed from the cage at various times after their introduction.

The longer the male was left with the female from the beginning of the 7-day test period, the more likely the female was to go on to ovulate by the end of the 7-day period, after the male had been removed. However, once ovarian development was well under way, the continued presence of the male was not necessary. In this situation the effects of external stimuli clearly are interactive with those of physiological events within the animal; events probably similar to those underlying cyclic and "spontaneous" ovulation.

EXTERNAL STIMULI AND PROLACTIN SECRETION

Since growth of the crop occurs only under the stimulation of prolactin (Riddle and Bates, 1939), increases in weight of the crop may be used as an indicator of the occurrence of prolactin secretion. It will be recalled that the crop increases in weight by about 250 per cent during the 14-day period when the birds are sitting on the eggs. Patel (1936) found that the crops of domestic pigeons would not increase in weight if the birds were removed from the breeding cages at the time the eggs were laid, and thus deprived of the opportunity to sit on eggs. If a male pigeon is removed from the cage while the birds are incubating, the female will continue to sit most of the time. Patel removed male pigeons from the breeding cages just after the eggs were laid, but placed the male in a cage immediately adjacent to the breeding cage, so that he could see his mate sitting on the eggs. Under these conditions, the crops of the male pigeons grew just as if they were sitting on the eggs themselves. Obviously, visual stimuli, and perhaps other external stimuli associated with the incubation situation, induced the birds' pituitary glands to secrete prolactin.

Saeki and Tanabe (1955) found that the pituitary glands of domestic chickens contained large amounts of prolactin as long as the hens were allowed to sit on their eggs. Apparently this phenomenon is fairly general among birds, even including species which do not feed their young by regurgitating a crop secretion.

Working with ring doves in our laboratory, Miss Miriam Friedman has verified Patel's finding that the crop of the male will continue to develop if he is separated from the nest, the eggs, and the incubating female, by a glass plate. The plate is inserted as a cage divider several days after the beginning of incubation. If, on the other hand, the male and female are thus separated from the time of their introduction into the cage, the female will usually build a nest and lay and incubate the (sterile) eggs, but males examined after the female has been incubating for 13 days show *no* crop development. Further studies are now being made of the preconditions for the induction of prolactin secretion by stimulation of this kind.

EFFECTS OF PROLACTIN ON DOVE BEHAVIOR

I have indicated that prolactin is relatively ineffective in inducing incubation behavior in doves, but that progesterone appears to be capable of producing this effect. However, since prolactin appears to be secreted *during* the incubation period, Mr. Philip Brody and I have investigated the question of whether prolactin is capable of *maintaining* incubation behavior which has already been induced by other means.

We tested male and female doves for incubation responses by introducing pairs into a cage containing an artificial nest with two eggs. If they have been in isolation cages for several weeks since their last breeding, birds thus tested never sit on eggs. On the other hand, if they are placed in pairs, allowed to court, build a nest, and lay eggs, and then tested by being moved in (reassorted) pairs to strange cages, all birds sit on the eggs. Thus, under our present methods of testing, ring doves are invariably ready to incubate on the day that the second egg is laid. Similar tests after seven days of incubation and after 12 days of incubation (of the normal 14-day incubation period) yield similar results; i.e., all birds incubate.

Now, if the birds are removed from the breeding cages on the day that the second egg is laid and placed in individual isolation cages, then recombined into pairs after seven days of such isolation and tested for incubation behavior, only 20 per cent of the birds incubate. After 12 days of such isolation, the number of birds ready to incubate is decreased to five per cent. This means that readiness to incubate disappears when the birds are not exposed to the eggs or the incubation situation.

If prolactin (400 I.U., divided evenly) is injected daily for seven days in birds which have been isolated from the day the second egg is laid, 80 per cent of these birds will be found ready to incubate at the end of the treatment period. If this treatment is spread over 12 days, 72.5 per cent of the birds sit. This is to be compared with the figures of only 40 per cent of the birds which sit after seven days of treatment starting when they have been in isolation for several weeks, and only five per cent positive responses when the same amount of prolactin is divided into 12 daily doses, starting when the birds have been in isolation for several weeks.

These data suggest that, even though prolactin is incapable of initiating incubation behavior in birds treated when they are in "sexual" condition, this hormone is capable of *maintaining* incubation behavior in birds which are already (through the influence of progesterone) at the beginning of their incubation period.

One result of the secretion of prolactin during the incubation period is that the crop is well developed at the time the egg hatches and is therefore able to produce crop milk, which the parents regurgitate to the young. I have therefore investigated the effect of prolactin upon the birds' readiness to feed young (Lehrman, 1955).

Twelve adult ring doves with previous breeding experience, kept in isolation cages since their last breeding, were each injected with 450 I.U. prolactin over a 7-day period. Each of these birds, and each of 12 control animals similarly treated except that they were not given any prolactin, were placed, one bird at a time, in a cage with a 7-day-old squab. Ten of the 12 prolactin-injected birds fed the squabs, while none of the controls did so.

It thus appears that the prolactin which is secreted starting when the birds are sitting on their eggs contributes to the ability of ring doves to show parental feeding behavior in response to stimuli from the squabs.

"PSYCHOSOMATIC" REGULATION OF THE REPRODUCTIVE CYCLE

We are now able to give a tentative answer to the problem raised at the beginning of this chapter. How does the reproductive cycle originate, and how is it regulated, in pairs of ring doves, when it appears not to occur independently in either the male or the female?

The succession of behavioral changes constituting the cycle seems to depend upon changes in endocrine secretion. Courtship, nest building, incubation, and feeding of the young can all be induced, in ring doves and in some other birds, by the injection of hormones equivalent to those secreted by the birds' own glands at appropriate stages in the cycle. The regulation of the behavioral cycle is, therefore, the regulation of the activity of the endocrine systems of the birds. The experiments summarized here point to the conclusion that changes in the activity of the endocrine system are induced or facilitated by stimuli coming from various aspects of the environment at different stages of the breeding cycle. Thus, participation in courtship appears to induce the secretion of hormones which facilitate the building of a nest; participation in nest building under these conditions contributes stimulation of the secretion of the hormone(s) which induce the birds to sit on eggs. Stimulation arising from the act of sitting on the eggs induces the further secretion of a hormone which (a) induces the birds to continue incubating, and (b) helps bring the birds into a condition of readiness to feed the young when they hatch.

The regulation of the reproductive cycle of the ring dove therefore appears to depend, at least in part, on a double set of reciprocal interrelationships. First, the effects of hormones on behavior are paralleled by effects of external stimuli on the secretion of hormones. This includes stimuli arising from the behavior of the animal and of its mate. Second, there is a complicated reciprocal relationship between the effects of the presence and behavior of one mate upon the endocrine system of the other bird. Stimuli from bird A produce endocrine changes in bird B. These physiological changes evoke alterations in the behavior of B. Bird B's new behavior in turn influences hormonal functioning in A with resultant changes in A's behavior. This is a rather involved way of saying that the cycle appears to originate, and to be synchronized in male and female, because the behavior of the male and of the female, and the products of this behavior (nest, eggs, etc.), provide stimulation which influences the endocrine systems of the birds in ways which in turn contribute to the sequential changes in their behavior (Lehrman, 1959).

Such a conception of the regulation of the reproductive cycle obviously implies that some functions of the pituitary gland must be controlled in large part by the nervous system. This view is entirely consistent with recent work upon the regulation of pituitary activity. It has become increasingly clear in recent years that the secretion of several hormones by the anterior pituitary is regulated by the hypothalamus, via a neurosecretory link which we need not describe here (Harris, 1955).* Hypothalamic regulation of the pituitary gland is evident not only in the way in which the reproductive cycle of the ring dove is regulated by stimuli from the mate and nesting situation, but in a wide variety of other phenomena. Ovulation in rabbits occurs in response to the stimulation of copulation. The development of the reproductive systems of birds takes place in response to the increasing length of the day in the spring. The timing of ovulation and of estrus in rats depends partly upon stimuli arising from the daily cycle of sunrise and sunset. The production of milk by the mammalian mammary gland is a response to stimuli provided by the sucking young, etc. (Lehrman, 1961).

If the hypothalamus controls and regulates the activity of the pituitary gland, it is not surprising that the secretion of pituitary and gonadal hormones can occur in response to environmental stimuli of the kind which we ordinarily associate with behavioral responses, and that this fact should play a substantial role in the normal regulation of the breeding cycle.

SOME UNRESOLVED PROBLEMS

The foregoing discussion has brought together the results of a number of published and unpublished studies into a general description of the psychobiology of the reproductive behavior cycle in the ring dove. I should now like to call attention to a number of unresolved problems, on some of which we have preliminary, but far from conclusive, data.

In all of the above descriptions, I have been unavoidably vague about the problem of physiological changes in the male during the period of courtship and of the transition to incubation behavior. By contrast, we have been able to correlate oviduct growth, and ovula-

* This subject is discussed in Chapter 20.

tion, with the change in behavior from nest building to incubation. We have not succeeded in finding reliable anatomic differences between males who have been kept in isolation and those which have been with a female for several days; or between males which will, and those which will not, incubate eggs. There is, however, little doubt that relevant physiological changes do take place in the male during this time, since it is clear that the readiness of males to sit on eggs develops as a consequence of their association with the female, during the time when the female's readiness to incubate is also developing, and when she exhibits anatomical indications of increased ovarian activity. Our studies of the cycle in the male are continuing.

We have considered the question of whether the observed changes in the female reproductive system are induced merely by the *presence* of the male, or whether they must be paced by sequential changes in the behavior of the male, dependent upon changes in his own physiological condition; changes which are synchronized with those in the female. Preliminary experiments have consisted of allowing females to stay for seven days with a sequence of males, changed daily. In this situation, we have compared the physiological and behavioral changes in different groups of birds in which each of the males introduced on successive days has (*a*) come directly from several weeks of isolation, (*b*) come directly from having spent seven days with another female, (*c*) come directly from having spent the same number of days with another female as the experimental female had spent with a succession of males. Observations of the male's behavior in this situation strongly suggest that the behavior of the male depends partly upon his own stage in the reproductive cycle, and partly upon his response to the behavior of the female, which is more closely tied to her own stage in the reproductive cycle. This conclusion must be regarded as quite tentative until we are able to return to this experiment, with the lavish expenditure of birds, cage facilities, and observer time which it entails.

I have spoken of progesterone as a hormone involved in the onset of incubation behavior, because this behavior occurs in association with ovulation, and because it is able to induce incubation behavior when injected. However, Riddle and Lahr (1944) found that DCA was just as effective as progesterone. We have not tried this hormone, but I want to introduce a note of caution at this point, particularly with respect to the male. It is true that Lofts and Marshall (1959) have presented evidence from paper chromatography that the avian testis contains progesterone during later (i.e., postnuptial) stages in the reproductive cycle. However, decisions about the role of progesterone and DCA in the incubation behavior of male doves ought to wait for chemical analysis of changes in the blood during the cycle.

All experiments described in this paper have been performed under a constant photoperiod of 14 hours light, 10 hours dark. The normal situation in temperate zone birds is, of course, that the development of the reproductive system and of reproductive activity in the spring is stimulated by the increasing length of the day, and that the decline of breeding activity in the fall and winter is some combined function of the decreasing length of the day and of the refractoriness of the birds' endocrine system to light stimulation (Farner, 1955; Marshall, 1959). In natural situations the kinds of influences with which we have been dealing, such as stimulation by the mate, by nesting situations, by the eggs, etc., serve to regulate the timing and sequence of events, and the repetitions of short breeding cycles, *within* the overall breeding season (Lehrman, 1959). Since our birds live constantly under conditions of a long daylength, their endocrine systems were already at some unspecified and as yet unanalyzed level of activity at the *beginning* of all of our treatments, and during the periods which, for purposes of our experiments, we regarded as inactive periods. We have lately begun studying the effects of various daylengths on endocrine activity in the ring dove, with a view to analyzing the interactions between environmental stimulation resulting from increasing daylengths, and stimulation by the mate and other aspects of the environment, in the regulation of the reproductive cycle. This analysis must advance considerably beyond its present point before we can make confident statements about the role played by these various processes in the regulation of normal avian breeding cycles.

Studies of local effects of hormones in the brain, with a view to the analysis of the neural mechanisms of instinctive behavior, and of the modes of hormonal action on behavior; studies of the role of experience in the development of hormone-induced behavior patterns (Lehrman and Wortis, 1960; Lehrman, 1962); studies of behavioral development; and studies of various other aspects of neural and endocrine mechanisms of behavior now under way in our laboratory, are necessary for a broad understanding of the psychobiology of reproductive behavior in the ring dove. The data and the unresolved problems which I have selected for discussion here are intended to illustrate some of the directions in which analysis of the regulation of the reproductive cycle in this species contributes to various problems of comparative psychology, physiology, and ecology.

Questions and Group Discussion

MR. G. BATESON: In connection with your brief reference to effects of external stimulation on the estrous cycle, do the females in a colony stay "in step" with one another?

DR. LEHRMAN: Not as far as I know; but we have no specific data on this point.

DR. LEVINE: It has been observed in any number of laboratories that female mice in the same cage tend to synchronize their cycles.

DR. LEHRMAN: Female mice placed in a room with no male mice tend to come into estrus at very irregularly distributed times. The regularity of the four-day cycle will not be established for a long time. In contrast, if females are moved to a room containing male mice, they will establish regular estrous cycles very quickly. In fact, they need not be in the room with male mice. They can be in a room in which male mice lived until yesterday. It seems that the stimuli affecting the regulation of the estrous cycle are olfactory.

A female mouse that has recently copulated with a fertile male mouse will become pregnant; but if this female is caged with a strange male mouse within 24 hours after she has copulated with the first male, pregnancy may be prevented or interrupted. The implantation of the egg is blocked; and the reliability of the blocking effect is increased if the male is of a different strain from that of the male with which the female was previously mated. If she is exposed to a sibling of the first male, pregnancy blockage is much less likely to occur. The important stimulus is olfactory, for blockage occurs if the second male is kept in an adjoining cage and does not have physical contact with the female.

MR. G. BATESON: Is there any evidence to show that the relation of external stimuli to endocrine processes in birds is subject to conditioning?

DR. LEHRMAN: Yes, in a very general way. A female ring dove that has been isolated from other ring doves and reared by human keepers will lay an egg if the keeper strokes her head for a few minutes every day for several days. She is less likely to lay an egg in response to being placed with a male ring dove. Now that means that somewhere, somehow the gonadotrophin-secreting parts of the pituitary gland have become connected to environmental stimuli by a process which includes some form of conditioning, but I am not prepared to say anything about the details involved.

DR. TINBERGEN: Did you show that experienced doves react differently than ones without experience?

DR. LEHRMAN: If you inject experienced ring doves with prolactin, they will go and feed a young dove offered to them. Inexperienced doves will not. They act extremely uncomfortable, in a way that suggests that if they would only go and feed the young they would feel better, but they don't do it.

DR. ROSENBLATT: Rats continue to lactate for long periods of time if they are periodically given foster young of nursing age. Is there a parallel for the duration of the prolactin secretion in ring doves?

DR. LEHRMAN: You can make a ring dove secrete prolactin a little longer than it would normally by substituting younger young, but the effect is not nearly so dramatic.

DR. ROSENBLATT: It drops off?

DR. LEHRMAN: Very easily; if you remove the young as soon as they are hatched, then the crop will very quickly go down to its resting state, and the female will lay new eggs at least two weeks earlier than she would have otherwise. By the time the new eggs are laid the crop will be in a nonsecreting condition.

DR. PEIRCE: I would like to ask how stimuli from the male induce changes in the female. Is there a possibility that there is some process that need not directly involve the hypothalamus but that leads the animal to do something which *then* leads to the change of endocrine conditions?

DR. LEHRMAN: That is an extremely good question. When a male dove secretes prolactin as a result of seeing a female sitting on her nest, he acts as if he himself wishes to sit on a place where there are no eggs, and he may spend some time sitting there, acting as though he were paying attention to nonexisting eggs. It is entirely possible that that behavior provides part of the stimulation that causes the pituitary to secrete prolactin.

I wouldn't be prepared to say that the hormone could be secreted entirely in response to the visual stimulus. We should probably include changes in autonomically mediated conditions of tension in the animal's body as an immediate response to the stimulus, which tensions, in turn, might provide stimulation for the hormone secretion. There just isn't any evidence on this point at present.

DR. BEACH: Dr. Lehrman mentioned cases of coition-induced ovulation in the cat, rabbit, etc. Many female mammals can ovulate without this stimulus. There is a rather interesting intermediate case represented by the rat. Without copulation the female ovulates, but she does not secrete progesterone. If she is mated with a sterile male she will go into a condition of pseudopregnancy and won't cycle for about two weeks. This is also true of the hamster.

I am convinced that this ties in with the fact that in the rat and the hamster the male achieves a number of successive intromissions before he ejaculates. We just recently completed the first step in an experiment dealing with my hypothesis. Jim Wilson, one of my assistants, has tested two groups of female rats. He first established the fact that all of them were fertile. Then in the second mating, he arranged the situation for one group so that the male would ejaculate after four or fewer intromissions. In the other group the male took the normal number of intromissions to ejaculate, i.e., an average of 10. Only one female in the first group became pregnant. All females in the second group were impregnated.

DR. LEHRMAN: How did you arrange the difference between these?

DR. BEACH: We took the males that we wanted to have a small number of intromissions, put them with nonexperimental females and allowed them to have several intromissions but not to ejaculate, and then immediately placed them with the experimental females.

MRS. JOHNSON: Was the sperm count high enough on this ejaculation?

DR. BEACH: The sperm count should have been the same, the males had the same amount of stimulation.

DR. DIAMOND: Are there any good cases of evidence of inducing human ovulation?

DR. MASTERS: Ovulation is certainly not in any sense a regularly reproducible phenomenon. The same individual may not respond similarly the second time she is stimulated.

The catch in the human is this. Either you have got to find the egg (which is a difficult chore), or you have to establish a pregnancy to prove ovulation. There have been a few pregnancies produced by direct stimulation of ovulation. Recently we're really beginning to get purified pituitary extract so far as the human is concerned. Prior to the last four or five years what we had to work with has been a little less than sensational in terms of extract purity.

DR. TINBERGEN: You said something to the effect that brooding affects brooding, and this is a positive feedback?

DR. LEHRMAN: I think so.

DR. TINBERGEN: Now it provides a negative feedback for the fertilization of eggs, we have agreed on that.

DR. LEHRMAN: Yes.

DR. TINBERGEN: But as I understand positive feedback, that is something that happens as a consequence of a process which then accelerates a process, and I think continuation of brooding is resupplying stimuli which, I think, is not a positive feedback; or am I misunderstanding you?

DR. LEHRMAN: You are. I am considering the feedback not between two things within the animal, but rather between the egg and the pituitary gland. The pituitary gland produces prolactin, which makes the bird want to sit. It sits. Stimuli coming from the egg, as a result of its sitting makes the gland continue to secrete prolactin that makes it want to sit more.

The amount of prolactin secreted keeps going up and up and up during the incubation period. It does not reach a level at which you have a negative kind of feedback, and I think you will agree that birds are less easy to scare away from the nest later in the incubation period. They sit tighter later.

DR. PRIBRAM: I think it is dangerous to think of this in terms of positive feedback. It may be a negative feedback, but not a completely balancing one.

DR. LEHRMAN: Prolactin level increases as a result of stimulation from the egg. The increased prolactin causes increased contact with the egg; increased contact with the eggs causes higher prolactin level. That is a special case of negative feedback with a positive sign.

DR. PRIBRAM: I wouldn't say it is positive or negative. It is a feedback, we all agree on that.

To determine whether it is positive or negative would be very difficult. We are talking about a mathematical problem, and I prefer at this stage to say that it could be either.

DR. DE GROOT: Gemzell and Heijkenskjöld (1957) have studied feedback mechanisms involved in the production of ACTH. The level of circulating ACTH influences the rate of synthesis of this trophic hormone by the anterior pituitary. When a concentration gets to a certain level it exerts a damping effect upon the pituitary.

DR. PRIBRAM: Yes. That is another thing you might find in positive feedbacks, tremendous oscillations. I think it is a mistake to think that whenever you get an accruing function that it necessarily involves a positive feedback.

DR. ROSENBLATT: Would you say that when you have a decreasing function it is difficult to say that it is a negative feedback?

DR. PRIBRAM: The characteristics of a negative feedback are much better known. It does not tend to break up the system, so whenever you have a system that persists for any length of time, the chances are it is a negative rather than a positive feedback.

DR. GINSBURG: Isn't the pay dirt here in the kind of evidence that Dr. DeGroot was summarizing? He suggests that there are very specialized functions of the hypothalamus with respect to controlling the pituitary. Now, is there any neurochemical or neuroanatomical evidence to support this very reasonable view?

DR. LEHRMAN: There is some evidence from brain stimulation. Direct stimulation of the hypothalamus will cause LH secretion. In Sawyer's laboratory I was shown evidence that stimulation of the hypothalamus at one locus causes the manufacture of prolactin in the pituitary gland, but stimulation at another locus is required for the prolactin to be released from the pituitary gland.

DR. DE GROOT: The very final piece of evidence in this theoretical chain still has to be demonstrated in my opinion. It has yet to be shown that within the blood or the portal system there are various substances or just one substance which when taken from the portal blood of one animal and injected into another animal will cause release of such specific trophic hormones.

DR. MICHAEL: I have a vested interest in this because later I hope to present some data concerning actual transport in the portal vessels.

DR. LEHRMAN: The evidence already available is really impressive. If we interrupt the connection between the hypothalamus and the pituitary gland, we interrupt gonadotrophin stimulation, and the rate at which it becomes re-established is consistent, not with the rate of regrowth of nerve connections, but with the rate of regrowth of blood connections, especially in individual cases. Thus individual differences between operated animals in which this hypothalamic control becomes re-established and those in which it does not, are correlated with the individual rate at which the blood vessels regenerated. That is Harris and Jacobson's experiment.

DR. LEVINE: Even more interesting evidence is available. Harris and others have been taking extracts of median eminence and hypothalamus, injecting them directly in the pituitary and producing ovulation.

REFERENCES

Carpenter, C. R. (1933). Psychobiological studies of social behavior in Aves. I. The effect of complete and incomplete gonadectomy on the primary sexual activity of the male pigeon. *J. comp. Psychol.*, **16** (1), 25–94.

Craig, W. (1909). The expression of emotion in the pigeons. I. The blond ring doves. *J. comp. Neurol.*, **19**, 29–80.

———— (1911). Oviposition induced by the male in pigeons. *J. Morph.*, **22**, 299–305.

Farner, D. (1955). The annual stimulus for migration. In A. Wolfson (Ed.),

INTERACTION OF INTERNAL AND EXTERNAL FACTORS 379

Recent Studies in Avian Biology. Urbana: University of Illinois Press, 198–237.

Gemsell, C. A. and Heijkenskjöld, F. (1957). *Acta Endocrinol.,* **24,** 249–254.

Harper, E. H. (1904). The fertilization and early development of the pigeon's egg. *Amer. J. Anat.,* **130,** 421–432.

Harris, G. W. (1955). *Neural Control of the Pituitary Gland.* London: Edward Arnold and Company.

Hinde, R. A. and Warren, R. P. (1959). The effect of nest building on later reproductive behavior in domestic canaries. *Anim. Behav.,* **7,** 35–41.

Lehrman, D. S. (1955). The physiological basis of parental feeding behavior in the ring dove (*Streptopelia risoria*). *Behaviour,* **7,** 241–286.

———— (1958a). Induction of broodiness by participation in courtship and nest-building in the ring dove (*Streptopelia risoria*). *J. comp. physiol. Psychol.,* **51,** 32–36.

———— (1958b). Effect of female sex hormones on incubation behavior in the ring dove (*Streptopelia risoria*). *J. comp. physiol. Psychol.,* **51,** 142–145.

———— (1959). Hormonal responses to external stimuli in birds. *Ibis,* **101,** 478–496.

———— (1961). Hormonal regulation of parental behavior in birds and infra-human mammals. In W. C. Young (Ed.), *Sex and Internal Secretions.* Baltimore: Williams and Wilkins, 1268–1382.

———— (1962). Interaction of hormonal and experiential influences on development of behavior. In E. L. Bliss (Ed.), *Roots of Behavior.* New York: Harper, 142–156.

Lehrman, D. S. and Brody, P. (1961). Does prolactin induce incubation behaviour in the ring dove? *J. Endocrinol.,* **22,** 269–275.

Lehrman, D. S., Brody, P., and Wortis, R. P. (1961). The presence of the mate and of nesting material as stimuli for the development of incubation behavior and for gonadotropin secretion in the ring dove (*Streptopelia risoria*). *Endocrinology,* **68,** 507–516.

Lehrman, D. S. and Wortis, R. P. (1960). Previous breeding experience and hormone-induced incubation behavior in the ring dove. *Sci.,* **132,** 1667–1668.

Lehrman, D. S., Wortis, R. P., and Brody, P. (1961). Gonadotropin secretion in response to external stimuli of varying duration in the ring dove (*Streptopelia risoria*). *Proc. Soc. exp. Biol. and Med.,* **106,** 298–300.

Lofts, B. and Marshall, A. J. (1959). The post-nuptial occurrence of progestins in the seminiferous tubules of birds. *J. Endocrinol.,* **19,** 16–21.

Marshall, A. J. (1959). Internal and environmental control of breeding. *Ibis,* **101,** 456–478.

Matthews, L. H. (1939). Visual stimulation and ovulation in pigeons. *Proc. Roy. Soc.* London, **126B,** 557–560.

Patel, M. D. (1936). The physiology of the formation of "pigeon's milk." *Physiol. Zoöl.,* **9,** 129–152.

Riddle, O. (1937). Physiological responses to prolactin. *Cold Spr. Harb. Symp. quant. Biol.,* **5,** 218–228.

Riddle, O. and Bates, R. W. (1939). The preparation, assay and actions of lactogenic hormone. In E. Allen, C. H. Danforth. and E. A. Doisy (Eds.), *Sex and Internal Secretions* (2nd Ed.). Baltimore: Williams and Wilkins, 1088–1117.

Riddle, O., Bates, R. W., and Lahr, E. L. (1935). Prolactin induces broodiness in fowl. *Am. J. Physiol.,* **111,** 352–360.

Riddle, O. and Lahr, E. L. (1944). On broodiness of ring doves following implants of certain steroid hormones. *Endocrinology,* **35,** 255–260.

380 REFERENCES

Saeki, Y. and Tanabe, Y. (1954). Changes in prolactin potency of the pituitary of the hen during nesting and rearing in her broody period. *Bull. nat. Inst. agric. Sci.*, Tokyo, Ser. G., 8, 101–109.

——— (1955). Changes in prolactin content of fowl pituitary during broody periods and some experiments on the induction of broodiness. *Poult. Sci.*, 34, 909–919.

Schooley, J. P. and Riddle, O. (1938). The morphological basis of pituitary function in pigeons. *Am. J. Anat.*, 62, 313–350.

EFFECTS OF DIFFERENT CONCENTRATIONS OF ANDROGEN UPON SEXUAL BEHAVIOR IN CASTRATED MALE RATS[1]

FRANK A. BEACH, *Yale University*

AND

A. MARIE HOLZ-TUCKER, *American Museum of Natural History*

Received October 11, 1948

Several investigators have employed testosterone propionate to evoke copulatory behavior in castrated male animals (Shapiro, 1937; Moore and Price, 1938; Stone, 1939). The majority of studies have dealt with the rat, and the dosages have varied from 100 to 1250 micrograms of hormone per day (1 microgram = .001 mg). It is well established that large amounts of androgen will maintain or restore normal sexual behavior in the adult castrate but we still do not know just how much hormone is required to achieve this result.

The present experiment was conducted to measure the sexual behavior of castrated male rats receiving various amounts of testosterone propionate by daily injection. We hoped to define the "adequate maintenance dose" which would keep behavior exactly at preoperative levels, and in addition to observe the behavioral effects of holding the hormone level below or above the minimal concentration needed for maintenance.

PROCEDURE

Subjects for the investigation were male rats, 90 to 100 days of age at the beginning of the tests. The animals were drawn from a colony which for ten years has been inbred to the extent of avoiding the introduction of any new stock, although no systematic plan of sibling crosses or back crosses has been followed. The strain was derived from a cross between wild *Rattus norvegicus* and tame albino rats from the Wistar Institute. For five or six years preceding this experiment males used as sires were individuals chosen for their willingness to mate promptly and vigorously. If one assumes that such characteristics are determined in part by heredity it follows that our practice may have resulted in a gradual increase in the sexual excitability of the strain.

Selection of the experimental animals was preceded by a series of preliminary sex tests in which males were placed singly with a female in heat and observed for the execution of mating responses. Individuals which failed to show any copulatory activity after two or three tests were discarded. This procedure eliminated sexually-sluggish animals and perhaps some others which were emotionally disturbed by the general testing situation.

The regular sex tests began after 52 suitably active males had been selected. These tests were conducted once each week in a quiet room with adequate ventilation and lighting. The circular observation cage was 34 inches in diameter, 30 inches high, and had no cover.

[1] This investigation was supported by a grant from the Committee for Research in Problems of Sex, National Research Council. The experiment was conducted while the senior author held the Chairmanship of the Department of Animal Behavior at the American Museum of Natural History. The junior author conducted most of the tests, performed the operations, and made the hormone injections. The senior author's responsibility included planning the experiment, interpretation of the data and preparation of the manuscript.

Stimulus animals were spayed females which had been brought into heat by the administration of ovarian hormones (Beach, 1942). Each day the injected females first were tested with non-experimental males of known sexual vigor. Only those individuals that displayed normal heat behavior in response to the mating attempts of the "indicator male" were employed in the experimental tests.

A male was placed in the observation cage and allowed a three-minute adaptation period and then a sexually-receptive female was quietly deposited in the center of the cage. Each test lasted for 10 minutes from the time of the first complete or incomplete copulation by the experimental male. In a complete copulation intromission is achieved. Incomplete copulations involve mounting the female and executing pelvic thrusts but insertion is lacking. Ejaculation is not involved in either type of response. If mounting responses did not occur within 10 minutes after the introduction of the female, the test was terminated and scored as negative. The behavioral items noted and the various measures employed will be described in the presentation of experimental results.

At the conclusion of the sixth preoperative test all males were castrated, and hormone injections were begun 48 hours later.

TABLE 1

Schedule of androgen treatment after castration

GROUP	N	MICROGRAMS OF TESTOSTERONE PROPIONATE PER DAY	
		First post operative period (9 weeks)	Second post operative period (10 weeks)
I	11	0	1
II	10	25	75
III	10	50	0
IV	10	100	omitted
V	11	500	0

The operated animals were divided into five experimental groups which were equated as closely as possible in terms of the average frequency of copulations per preoperative test. Each male was injected subcutaneously once every 24 hours with .2 cc. of sesame oil. The concentration of testosterone propionate contained in this amount of oil ranged from 25 to 500 micrograms for various groups and a control group received plain oil with no hormone.[2]

Tests were continued for a period of nine weeks, at which time the hormone dosages were changed. A final ten-week period concluded the experimental tests. The amounts of androgen administered daily to males in each group are shown in table 1.

Animals in Group IV received 100 micrograms per day during the first nine weeks after operation and 5 micrograms daily for the remainder of the experiment. These rats showed no response to the smaller dose. Their behavior was quite similar to that of castrates in Groups III and V who were given no hormone in the second post operative period. However, Group I exhibited definite increases in sexual activity in response to daily injections of 1 microgram of testosterone propionate. This suggests that treatment at a high dose level may render animals insensitive to much lower concentrations but our data are insufficient to establish the point. We do not here report the scores of Group IV rats during the second post operative period because they were in no way different from those of Groups III and V and because inclusion of their records would have obscured an otherwise clear

[2] The hormone preparations used in this study were generously supplied by Dr. Edward Henderson of Schering Corporation, Bloomfield, N. J.

relationship between hormone dosage and frequency of sexual responses. It is felt that exclusion of these data is justified by the fact that Group IV was the only group in which androgen concentration was reduced but not brought to zero,—a procedure which evidently yields results quite different from (1) increasing the dosage or (2) withdrawing the hormone altogether.

All animals were sacrificed within 24 hours after the final test. Completeness of testis removal was checked by macroscopic inspection and samples of seminal vesicle tissue were removed for histological study.

RESULTS

The experimental results will be discussed in terms of the various behavioral items studied.

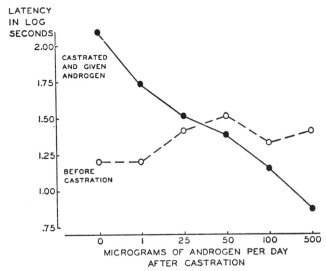

FIG. 1. AVERAGE DELAY PRECEDING THE FIRST SEXUAL MOUNT IN PRE- AND POST-OPERATIVE TESTS

Changes in Latency

Latency is defined as the number of seconds elapsing between the time that the receptive female was placed in the observation cage and the time that the male executed his first sexual mount (complete or incomplete copulation). This measure proved to be a very sensitive indicator of the amount of hormone administered to castrated males.

In order to normalize the distributions, raw scores have been translated into logarithms. Figure 1 is a graphic representation of the average log of latencies for each group before and after castration. Some differences existed between the averages for the various groups in preoperative tests, and after castration the average latency was directly proportional to the size of the hormone dosage. During the first preoperative test many animals showed long latencies which seemed to reflect emotional disturbance due to the strangeness of the environment. Therefore the scores in this test were not used in computing group averages. Immediately after operation the latency scores tended to be quite variable, but they were stabilized after the castrates had been receiving hormone for a week or ten days. Accordingly the records of the first postoperative test have been omitted from these particular calculations. Preoperative averages are based upon the results of tests 2 to 6 and the values representing postoperative performances of all groups are based upon nine tests conducted from 2 to 10 weeks after castration. In addition we have inserted the average score for Group I during the second post operative period at which time these rats were receiving 1 microgram of testosterone propionate per twenty-four hours.

In order to determine the significance of changes in latency we have compared the average log latencies of each group before and after castration. Preoperative means are based upon scores made in the last three tests before castration, and post operative means are based upon the records of six successive tests conducted

TABLE 2
Mean log latency

GROUP	N	NORMAL (TESTS 4-6)	CASTRATE (TESTS 10-15)	DIFFERENCE	PROBABILITY	MICROGAMS OF ANDROGEN PER DAY
I	7	1.14	2.26	+1.12	<.01	0
I	8	1.22	1.66*	+0.44	.03	1
II	10	1.33	1.53	+0.20	.04	25
III	10	1.47	1.34	−0.13	.11	50
IV	10	1.28	1.15	−0.13	.06	100
V	11	1.30	0.78	−0.52	<.01	500

* Based on tests 19 to 25.

after the animals had been receiving androgen for one month. In addition there are included the records of Group I during the last six tests in the second postoperative period when 1 microgram per day was being injected. The post operative change in mean values for each group was evaluated in terms of Student's *t* test, and using Snedecor's tables we have calculated the probability that the differences could have occurred by chance.

Table 2 presents the results of these comparisons. Castration followed by no treatment resulted in an average increase in lateny which would be expected to occur by chance less than 1 time in 100 and is therefore clearly significant in the statistical sense. Apparently 50 micrograms of testosterone propionate per day was adequate to maintain latency scores at preoperative levels. Lower doses were associated with latencies which, while shorter than would be expected without any replacement therapy, were significantly longer than normal. One hundred micrograms of androgen per twenty-four hours produced a decrease in latency scores which was significant at the 6 per cent level of confidence, and under the influence of 500 micrograms of hormone per day castrated male rats showed latencies that were definitely shorter than normal.

Data collected during the second post operative period corroborate these conclusions. Males in Group II were given 25 micrograms during the first and 75 micrograms during the second post operative period. The increase in dosage occasioned a shortening of the average log latency which was significant at below the 1 per cent level. In fact, under the influence of 75 micrograms per day these animals showed latencies shorter than they had displayed before castration (significant at the 8 per cent level).

Males in Groups III and V received 50 and 500 micrograms of androgen respectively during the first post operative period and no hormone in the second post operative period. The withdrawal of hormone was followed by increases in log latencies which were highly significant in both cases.

Changes in the copulatory response

The rat's copulatory response includes five distinct elements. (a) The male mounts the receptive female from the rear, clasping his forelegs around her sides in the lumbar region. (b) The male's forelegs are pressed downward and backward and then brought forward again in a series of very rapid "palpation" movements which help stimulate the female to elevate the perineum and thus expose the genitalia. (c) Concomitantly with the forelimb palpations, the male executes a series of short pelvic thrusts which bring the penis into contact with the genitalia of the female. (d) After several such preliminary pelvic movements the male accomplishes the brief intromission by means of a single, deep thrust. (e) Insertion usually is maintained for only a fraction of a second and the male dismounts abruptly with a vigorous backward lunge which often carries him half a foot or more away from the female. This pattern is clearly recognizable under ordinary circumstances. Ejaculation does not occur with each copulation, and when it does take place the behavior is different as subsequent description will reveal.

Per cent of each group copulating in each test: The first noticeable post operational change in sexual behavior was the complete disappearance of the copulatory response from the behavior of some males in some tests. We shall first consider the results on a "present-or-absent" basis, paying no attention to the frequency of copulations during tests in which such behavior occurred.

The trends represented in figure 2 are based upon 3-point moving averages. Results of tests 1 and 2 are not shown, but the percentage of copulators increased for all groups in the first few preoperative tests and then tended to stabilize. This probably reflects gradual adaptation to the testing situation as well as some general "conditioning" effect. In the interest of legibility we have refrained from reproducing the curves for Groups III and IV which received 50 and 100 micrograms per day respectively. The performance of Group IV males was very similar to that of Group V and the curve for Group III falls midway between those of Groups II and IV.

During the first post operative period the records of the several groups fell in a definite order which corresponded to the magnitude of hormone dosages. Because of inter-group differences in preoperative performance, comparisons between behavior in the first and second periods of the experiment must be drawn

438   FRANK A. BEACH AND A. MARIE HOLZ-TUCKER

with considerable caution. However, comparing each group with itself in the two periods we may tentatively conclude that daily injections of 50, 100, or 500 micrograms sufficed to maintain the proportion of copulators at preoperative levels if not to increase it somewhat.

Administration of 25 micrograms per day resulted in a decrease in the per cent of a group copulating in each test. Control injections of sesame oil had no recognizable effect, and the proportion of copulators in Group I progressively decreased until only approximately 10 per cent of the animals were responding in any given test. It should be stated that this did not reflect the persistence of copulatory behavior in a single member of the group. It was not always the same individual who copulated in successive post operative tests. Various males executed the

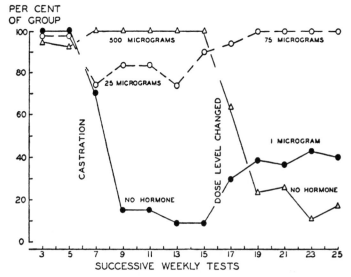

FIG. 2. PROPORTIONS OF THREE EXPERIMENTAL GROUPS SHOWING THE
COPULATORY RESPONSE

response at different times, and the post castrational decline in behavior was not a regularly progressive change in individual animals although it was for the group as a whole. For example, one male in Group I exhibited no sexual activity for the first seven weeks after castration and then, in the eighth postoperative test, displayed 3 complete and 20 incomplete copulations. Another rat's scores were negative until the fifth test after castration at which time copulation and ejaculation occurred. Similar irregularities appeared in the individual records of males in Groups III and V during the second post operative period when androgen treatment was discontinued.

Hormone dosages were changed nine weeks after castration and subsequent alterations in behavior fully supported the foregoing conclusions. An increase from 25 to 75 micrograms in the daily dosage for Group II occasioned a prompt rise in the proportion of copulators per test, and within three weeks the record of this group was equal to its own preoperative scores. Groups III and V were deprived of androgen in the second post operative period and the change in their performance was comparable to that of Group I during the first period after castration when plain oil was injected. The scores for Group I during the second post operative period show that as little as 1 microgram of testosterone propionate per day was sufficient to increase the percentage of copulators.

Frequency of copulations in positive tests: Thus far we have paid no attention to the number of times an animal copulated in a particular test. As long as the response occurred once the individual was scored as positive. We are now ready to consider the effects of castration and subsequent androgen administration upon copulatory frequency during those tests in which such behavior appeared.

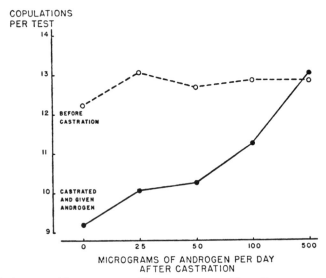

FIG. 3. AVERAGE FREQUENCY OF COPULATIONS PER TEST BEFORE AND AFTER CASTRATION

The average copulation frequency tended to increase in all groups during the first 2 or 3 preoperative tests and then to level off at approximately 12 to 13 copulations per test. During the last 4 tests before operation no major changes occurred in the group averages. Following castration the scores for all groups declined progressively. The drop from preoperative levels was quite marked in Groups II and III which received 25 and 50 micrograms of androgen respectively, and somewhat less pronounced in Groups IV and V receiving 100 and 500 micrograms. The average scores for Group I (no hormone) were extremely variable, due to the small number of males copulating, but obviously tended to decrease sharply.

In order to determine the statistical significance of these changes we have compared the average copulatory frequencies for each group during the last 3 preoperative tests with the mean scores during tests 10 to 15 which took place 4 to 9 weeks after castration. The comparisons are graphically represented in figure 3.

It is plain that in every group save one the average frequency of copulation was lower after castration, and the extent of the decrease was inversely proportional to the amount of androgen administered. Only those males which received 500 micrograms per day maintained their behavior at preoperative levels.

Application of the *t* test to differences between pre- and post operative means revealed that the difference was not statistically significant in the case of Group I, presumably because of the small N (5) and the high test-to-test variability shown by these untreated castrates. The decreases were significant, however, in the case of Groups II, III and IV (at the 1 per cent level for II and III and the 3 per cent level for IV).

During the second post operative period when Groups III and V were deprived of hormone the average frequency of copulations per positive test decreased markedly for both groups, but because relatively few males copulated at all the group averages were too variable to justify further statistical analysis. Group I males which had been receiving plain oil during the first post operative period showed no increase in copulation frequency when they were given 1 microgram of androgen per day although, as pointed out earlier, this treatment did produce an increase in the proportion of the group displaying the copulatory response at least once in a test.

During the first period after operation males in Group II were given 25 micrograms of hormone per day and copulated much less frequently during positive tests than they had before castration. In the second post operative period when the dosage was increased to 75 micrograms copulation frequency gradually increased until the group was performing as well as it had prior to operation. During the last 5 tests under the influence of 75 micrograms these animals copulated an average of 13.5 times per positive test which was slightly higher than their preoperative average. The average increase consequent to raised hormone dosage amounted to 3.4 copulations per rat per test and this difference proved to be significant at below the 1 per cent level of confidence.

It seems paradoxical that 100 micrograms of testosterone propionate per day was insufficient to maintain normal copulatory frequency during the first 9 weeks after castration in the case of Group IV and that 75 micrograms constituted an adequate dosage at a later stage in the experiment for Group II. It is possible that the long period of treatment at a relatively low dose level somehow sensitized the males in Group II so that the increase to 75 micrograms was peculiarly effective. At the present state of knowledge it seems useless to speculate further.

Changes in the incomplete copulatory response

The male rat's incomplete copulatory response consists of mounting the female and displaying pelvic thrusts and palpating movements of the forelimbs. There is no intromission, and when he dismounts the male slides weakly off the female's back instead of throwing himself backward with the vigorous lunge which terminates the complete copulatory response. The difference between complete and incomplete copulations usually is quite obvious to the trained observer although in perhaps 5 per cent of the cases there may be some question as to the occurrence of intromission.

Per cent of each group showing incomplete copulations in each test. Castration without androgen replacement resulted in a decrease in the number of animals exhibiting the incomplete copulatory response. Groups I, III and V were injected with plain oil during either the first or the second post operative period and in all groups the proportion of tests in which this response occurred decreased by approximately 50 per cent from precastrate levels.

Castrates receiving 1, 25, or 50 micrograms of testosterone propionate per day showed incomplete copulatory reactions in about the same proportion of tests as they had before the operation. A daily dose of 100 micrograms caused some increase in the number of tests in which such behavior appeared; and 500 micrograms produced an even more marked rise. These results are summarized in table 3.

TABLE 3

Proportion of tests in which incomplete copulation occurred

GROUP	N	AVERAGE PERCENTAGE OF TESTS POSITIVE FOR GROUP*			MICROGRAMS OF ANDROGEN PER DAY
		Normal (tests 4–6)	Castrate (tests 10–15)	Difference	
I	11	48	46†	−2	1
II	10	53	62	+9	25
III	10	40	45	+5	50
IV	10	47	65	+18	100
V	11	54	82	+28	500
			(tests 20–25)		
I	11	48	21‡	−27	0
III	10	40	15	−25	0
V	11	54	30	−24	0

* A test is "positive" if the response in question occurs at least once.
† Refers to tests 20–25 for this one group.
‡ Refers to tests 10–15 for this one group.

It should be held in mind that we are considering here merely the proportion of tests in which incomplete copulation occurred one or more times. The question as to the number of responses executed per test is dealt with below.

Frequency of incomplete copulations in positive tests. There was a marked tendency for untreated castrates or those receiving small amounts of androgen to display an increase in the average frequency of incomplete copulations during those tests in which such behavior appeared. Mean values presented in table 4 show that when they were injected with bland oil animals in Groups I, III and V displayed more of these responses in positive tests than they had prior to castration. Administration of 1 or 25 micrograms per day was accompanied by a statistically significant rise in the average number of incomplete copulations per positive test (Groups I and II). Castrates receiving 50, 100, or 500 micrograms during the first post operative period exhibited no significant change in the average frequency of incomplete copulations. Furthermore, when the dosage for males in Group II was raised from 25 to 75 micrograms per day the average frequency of this response decreased from 5.2 to 2.7 per test.

The data shown in tables 3 and 4 suggest that lack of testis hormone caused castrated male rats to show incomplete copulation in fewer tests but to execute the response a greater number of times during the tests in which it did occur. Administration of 1 or 25 micrograms of testosterone propionate per day induced no marked change in the proportion of tests in which incomplete copulations appeared, but the frequency of responses per positive test was increased by these dosages. This particular reaction was comparatively unaffected by castration if 50 micrograms of androgen were injected daily after the operation. Finally, large doses of 100 or 500 micrograms produced an increase in the number of tests in

TABLE 4

Average number of incomplete copulations per positive test*

GROUP	N	NORMAL (TESTS 4–6)	CASTRATE (TESTS 10–15)	DIFFERENCE	PROBABILITY	MICROGRAMS OF ANDROGEN PER DAY
I	7	1.7	9.4†	+7.7	<.01	1
II	10	1.8	5.2	+3.4	.02	25
III	8	2.1	2.5	+0.4	.25	50
IV	9	3.2	2.7	−0.5	.32	100
V	8	2.9	3.7	+0.8	.35	500
			(TESTS 20–25)			
I	5	2.1	8.7‡	+6.6	.05	0
III	5	2.4	13.8	+11.4	.01	0
V	5	2.0	4.9	+2.9	.07	0

* A positive test is one in which the response in question occurred at least once.
† Refers to tests 20–25 for this one group.
‡ Refers to tests 10–15 for this one group.

which the behavior was present but did not change the frequency of incomplete copulations per test.

Changes in the ejaculatory response

When a male rat ejaculates the event is clearly reflected in the overt behavior. At first the pattern progresses in the manner described for the copulatory response, but when insertion is achieved the male does not release the female immediately. Instead he maintains the mating clasp with his forelegs and prolongs intromission, pressing tightly against the female's hind quarters. After several seconds the clasp is relaxed and the male raises his forelimbs rather slowly, rearing upwards and backwards and often coming to rest in a semi-sitting position. Occasionally instead of releasing the female in this manner the male may grip her more tightly and fall slowly to one side, pulling her with him. In either event the difference from a copulation without ejaculation is clear-cut and unmistakable.

In the present report the terms "ejaculation" or "ejaculatory pattern" refer exclusively to this sequence of overt responses and not to the occurrence of emission. Under certain circumstances male rats may display all of the outward signs of ejaculation even though they are incapable of emitting seminal fluid (e.g. after removal of the accessory sex glands), but in our records an animal was credited with an ejaculation each time the behavioral pattern appeared regardless of the presence or absence of a vaginal plug.

Analysis of ejaculatory frequency in normal animals. Before discussing the effects of castration and androgen treatment upon the ejaculatory response it will be profitable to consider briefly several aspects of this element in the sexual performance of the normal male. The data to be described bear directly upon

TABLE 5

Comparison of average scores of rats that ejaculated once with those that ejaculated twice in the sixth preoperative test

ITEM FOR COMPARISON	17 MALES THAT EJACULATED TWICE	29 MALES THAT EJACULATED ONCE	DIFFERENCE	PROBABILITY
Copulations preceding first ejaculation	8.5	12.2	+3.7	< .01
Seconds per copulation before first ejaculation	18.6	31.3	+12.7	< .01
Seconds to attain first ejaculation after mating began	152.4	349.2	+196.8	.05
Seconds latency	24.9	36.5	+11.6	†
Seconds recovery after first ejaculation	261.5	270.7*	+9.2	†

* Only 8 of the 29 males in this group recovered sufficiently to resume copulating before the test was terminated.

† Not significant.

questions of individual differences in sexual ability and possess considerably significance in relation to problems of sexual impotence.

The data upon which the following analysis is based were obtained during the last preoperative test for all rats. During this test a few animals failed to copulate, 5 copulated without ejaculating, 29 ejaculated once and 17 ejaculated twice. Disregarding those males which did not copulate or copulated without ejaculating we have attempted to discover such differences as might have existed between the 17 rats that ejaculated twice and the 29 males that did so only once.

The first and most obvious possibility was that a difference in general health was involved, but a comparison of body weights revealed no significant difference. Males ejaculating once weighed an average of 316 grams and the average weight of those ejaculating twice was 327 grams with a great deal of overlap between the two distributions.

Other possible explanations were analyzed and the results are summarized in table 5. Rats that ejaculated twice in a time-limited test were animals that reached the first ejaculation after fewer intromissions than did those individuals who ejaculated only once in the same period. "Double-ejaculators" allowed less time to elapse between preejaculatory intromissions with the result that the average number of seconds per copulation was significantly lower, and, as a natural consequence of these two differences, "double ejaculators" achieved their first ejaculation earlier in the test than did "single ejaculators".

Males which were going to ejaculate twice tended to initiate sexual activity with less delay than rats that were going to ejaculate once, but this difference in latency scores was not statistically significant. Following the first ejaculation the "double ejaculators" resumed their copulatory activity after an average delay of about $4\frac{1}{2}$ minutes. Eight "single ejaculators" recovered in approximately the same length of time but failed to ejaculate again even though they completed as many copulations after the first ejaculation as did the other males. Twenty-one "single ejaculators" showed no copulatory behavior after the initial ejaculation. In their case ejaculation occurred so late in the test that not enough time remained for recovery and the renewal of sexual activity.

TABLE 6

Comparison of activity preceding first and the second ejaculations for 17 rats that ejaculated twice during the sixth preoperative test

ITEM FOR COMPARISON	FIRST EJACULATION	SECOND EJACULATION	DIFFERENCE	PROBABILITY
Copulations preceding ejaculation....	8.5	4.6	−3.9	<.01
Seconds preceding ejaculation after mating begins......................	152.4	97.7	−54.7	<.01
Seconds per copulation................	18.6	21.5	+2.9	†

† Not significant.

One additional item of interest was seen in the fact that when a second ejaculation occurred it seemed to do so after fewer copulations and in less time than had been necessary for the first ejaculation. These conclusions are based upon data summarized in table 6.

Per cent of each group ejaculating in successive tests. Most of the sexually-active male rats of our strain will ejaculate at least once in the majority of a series of weekly tests, but not every male will ejaculate on every test. Figure 4, based on a 3-point moving average, shows the proportion of Groups I, II and V which achieved ejaculation at least once in successive tests. The curves for Groups III and IV fall between those of II and V but have been omitted so that the figure can be read easily.

There was an obvious relationship between the amount of androgen administered and the proportion of a group ejaculating. Comparison of figures 2 and 4 reveals that more hormone was necessary to evoke ejaculation than to call forth copulation.

Frequency of ejaculation in positive tests. Having seen the effects of castration and hormone treatment upon the presence or absence of the ejaculatory response, we are now in a position to consider the frequency of ejaculations during those tests in which this reaction occurred at least once. Most rats which continued to ejaculate under the influence of very small amounts of androgen tended to do so only once within the 10-minute test. For example, before castration males in Group I ejaculated an average of 1.7 times per positive test. When these animals were castrated and injected with 1 microgram of testosterone propionate per 24 hours ejaculation occasionally occurred, but the average frequency was 1.0 per positive test. The difference between the means was significant at the 3 per cent level.

Animals in Group II also displayed a drop in the frequency of ejaculations when they were castrated and given daily injections of 25 micrograms, but the change was slight and the difference between the pre- and post operative means

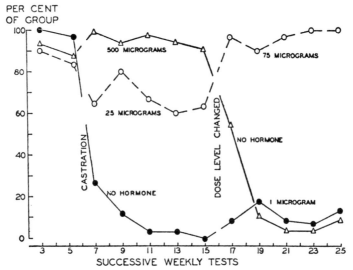

FIG. 4. PROPORTIONS OF THREE EXPERIMENTAL GROUPS EJACULATING AT LEAST ONCE DURING THE TEST

not statistically significant. Rats receiving 50 or 100 micrograms of androgen per day tended to ejaculate more frequently after castration although the increases were not marked. However, males in Group V ejaculated an average of 1.4 times per positive test before castration and 1.7 times after operation while receiving 500 micrograms of hormone per day. This rise in ejaculatory frequency was highly significant (<.01).

None of the animals ejaculated more than twice in the same 10-minute test and changes in ejaculatory frequency during positive tests can therefore be illustrated by showing the proportions of the various groups which ejaculated twice in each test. This scheme has been followed in preparing figure 5 which is based upon a 3-point moving average. Curves for groups III and IV are not shown but they fitted in with the trends revealed by the other three groups. Group IV males received 100 micrograms and their scores fell approximately

446   FRANK A. BEACH AND A. MARIE HOLZ-TUCKER

midway between the 25 and 500 microgram groups. Group III with 50 micrograms showed two ejaculations about as frequently as rats receiving 25 micrograms.

There was some tendency for the percentage of a group ejaculating twice to increase in successive tests before operation. After castration if no androgen was given the ability to ejaculate twice in the same test was soon lost by all males. Administration of 25 or 50 micrograms of hormone per day maintained the proportion of "double ejaculators" at levels roughly comparable to those attained in the last preoperative tests. Higher dosages, particularly 500 micrograms per

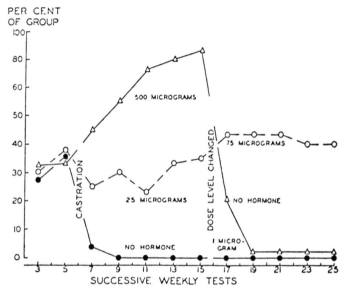

Fig. 5. Proportions of Three Experimental Groups Ejaculating Twice During the Test

day, brought about a striking increase in the number of males capable of attaining two ejaculations within the time-limited test.

Number of copulations preceding ejaculation. Normal male rats display considerable individual variation with respect to the number of intromissions which are necessary to evoke ejaculation, and at the same time most individuals show a remarkable constancy in this function from one test to another. Some males regularly achieve the first ejaculation in each test after 5, 6 or 7 copulations whereas others just as reliably take 13, 14 or 15 intromissions to attain the ejaculatory peak.

Table 7 shows the effects of castration and subsequent androgen therapy on this particular aspect of the male's sexual behavior. Animals receiving 1 microgram per day or no hormone at all ejaculated so rarely that there are no reliable data concerning the number of copulations necessary for ejaculation in such cases. However, the values incorporated in table 7 strongly suggest that castrated rats receiving 25, 50 or 100 micrograms of testosterone propionate per day tended to ejaculate after fewer copulations than they had before operation. In contrast, males injected with 500 micrograms per day showed no significant change in the frequency of copulations preceding the first ejaculation.

These mean scores suggest that relatively large amounts of androgen were necessary to maintain the preejaculatory copulation score at normal levels, and the performance of males in Group II during the three experimental periods supported this conclusion. Prior to operation these rats ejaculated after an average of 11.0 copulations. After castration while they were receiving 25 micrograms of testosterone propionate per day the mean number of copulations preceding ejaculation fell to 9.3. Then, when the daily dosage was increased to 75 micrograms the score rose again to an average value of 11.1. The increase occasioned by the higher dose level was highly significant (p = <.01).[3]

Time necessary to achieve ejaculation. The fact that castrated males receiving less than 500 micrograms of androgen per day tended to ejaculate after fewer intromissions during the first postoperative period than they had before gonad-

TABLE 7

Average number of copulations preceding the first ejaculation

GROUP	N	NORMAL (TESTS 4–6)	CASTRATE (TESTS 10–15)	DIFFERENCE	PROBABILITY	MICROGRAMS OF ANDROGEN PER DAY
II	10	11.0	9.3	−1.7	.05	25
III	10	11.1	8.5	−2.6	<.01	50
IV	10	11.7	8.7	−3.0	<.01	100
V	11	9.9	9.2	−0.7	.25	500

ectomy was an unexpected finding, and we proceeded to determine whether these rats ejaculated earlier in the tests after castration.

Basing our calculations upon the number of seconds intervening between the first sexual mount (complete or incomplete copulation) and the occurrence of the first ejaculation we found that rats receiving plain oil or 1 microgram of androgen invariably took longer to ejaculate than they had in preoperative tests. Group II males ejaculated in an average of 265 seconds before operation and 281 seconds after castration while they were receiving 25 micrograms of hormone per day. When the dosage was raised to 75 micrograms the average time needed to achieve ejaculation decreased to 257 seconds. Rats receiving 50 or 100 micrograms per day ejaculated more quickly in post operative tests than they had before castration; and under the influence of 500 micrograms of androgen per 24 hours castrated males displayed the most marked decrease in the number of seconds elapsing before ejaculation (255 seconds as normal and 211 as castrate, significant at the 5 per cent level).

Castrates receiving small doses of androgen tended to ejaculate after fewer

[3] Here, as at several other points in this report, it appears that Group II males responded more strongly to 75 micrograms than did Group IV animals to 100 micrograms. We are unable to explain this phenomenon but call attention to the fact that rats in Group II received 25 micrograms per day for 9 weeks before the higher concentration was employed.

448   FRANK A. BEACH AND A. MAEIE HOLZ-TUCKER

copulations than normal, but it took them longer to reach the point of ejaculating. This suggests that the delay between copulations must have increased when the androgen level was low. Gonadectomized males injected with very large amounts of hormone ejaculated sooner than they had before operation, but the number of preejaculatory intromissions was unchanged. Here a shortened intercopulatory interval is indicated.

These indications are borne out by values presented in table 8 which shows the mean recovery period in seconds for each group before and after castration. Post operative changes in the amount of time elapsing between copulations were directly proportional to the amount of androgen given daily to the castrated animals. Only very large doses were sufficient to maintain copulation at normal speeds. The significance of all the differences shown in table 8 cannot be demonstrated statistically, but the regularity of progression in the Difference column argues against the belief that this is a chance effect. For Groups II, III, IV and

TABLE 8

Average number of seconds intervening between copulations which precede the first ejaculation

GROUP	N	NORMAL (TESTS 4–6)	CASTRATE (TESTS 10–15)	DIFFERENCE	MICROGRAMS OF ANDROGEN PER DAY
I	3	18.3	32.6	+14.3	1
II	10	24.1	30.2	+6.1	25
III	10	25.3	30.9	+5.6	50
IV	10	26.0	29.0	+3.0	100
V	11	25.7	22.9	−2.8	500
III	2	26.0	39.8	+13.8	0
V	2	21.7	44.8	+23.1	0

V in which the N was large enough to permit application of the *t* test the probabilities were .07, .02, .05 and .11 respectively. Finally, in Group II the mean intercopulatory interval increased by 6.1 seconds while the castrates were receiving 25 micrograms of androgen per day, and then decreased by almost the same amount when the dosage was raised to 75 micrograms.

Number of incomplete copulations preceding ejaculation. We have already seen that the frequency of incomplete copulations increased in castrated males given relatively small amounts of androgen. In the case of these animals if we combine the incomplete and the complete copulations preceding the first ejaculation we arrive at the following tentative conclusions. Copulations with intromission occurred less frequently after castration and in their place there appeared more of the incomplete copulatory responses. The increase in incomplete reactions equalled or exceeded the decrease in complete ones with the result that under the influence of low hormone doses gonadectomized males actually mounted the female at least as frequently to achieve ejaculation as they had before operation or as did other castrates which received larger amounts of androgen.

Changes in the recovery period after ejaculation

In male rats of our experimental strain ejaculation is followed by several minutes of relative inactivity during which the animal appears refractory to sexual arousal. The total refractory period can be thought of as including an "absolute" and a "relative" phase. During the absolute phase the male seems incapable of copulatory performance. No amount of stimulation will evoke mounting behavior and very little attention is paid to the female. During the relative phase of the refractory period sexual responsiveness is gradually regained. This period may be shortened if the female spontaneously resumes the exhibition of heat responses. It seems as though the copulatory threshold becomes progressively lower and can be crossed before it reaches normal levels if the stimulation is sufficiently intense.

We have defined the post ejaculatory refractory period as the number of seconds intervening between an ejaculation and the occurrence of the next complete or incomplete copulation. Inspection of the experimental records revealed that in the first test after castration the length of this period was greatly increased unless at least 75 micrograms of testosterone propionate were supplied each day. If no androgen was injected the refractory period was very long in tests given one week after gonadectomy or cessation of hormone treatment. In such cases, however, ejaculation was soon eradicated and consequently there are not enough data on duration of the refractory period to permit statistical treatment. A similar situation obtains in the case of Group I rats which received 1 microgram per day during the second period after operation.

Males in Groups II to V ejaculated with sufficient frequency after operation to permit quantitative analysis of changes in the post ejaculatory refractory period. Prior to gonadectomy the average recovery periods for these 4 groups varied from 251 to 280 seconds. During the first postoperative period the group averages ranged from 229 to 309 seconds. The longest mean refractory period was that of Group II whose members received 25 micrograms of hormone per 24 hours, and the shortest was that of rats in Group V which were given 500 micrograms daily. Animals treated with 50 micrograms showed a slight increase in average recovery time and those receiving 100 micrograms a minor decrease. The average times for Group II in the three periods of experimentation were as follows: 281 seconds before operation, 309 seconds after castration while receiving 25 micrograms per day and 256 seconds after castration while receiving 75 micrograms per day.

In order to measure the statistical significance of changes in the post ejaculatory refractory period the raw scores were transformed into logarithms and differences between the pre- and postoperative averages for each group were evaluated by means of the t test. The results appear in table 9. It will be seen that the increased mean refractory periods shown by animals receiving 25 or 50 micrograms of androgen per day were statistically significant. The shortened recovery times for Groups IV and V were not significant although of course the possibility of a true difference is not disproven.

When the dosage for Group II was increased from 25 to 75 micrograms the mean log refractory period decreased from 2.49 to 2.41 and the change was significant at the 2 per cent level.

Morphological changes and general health

Condition of seminal vesicles: Within 24 hours after the final test all animals were sacrificed and samples of seminal vesicle tissue were removed from several members of each group for sectioning and histological study. It was regarded as undesirable to subject the rats to abdominal operation at the conclusion of the first postoperative period and therefore no information is available concerning

TABLE 9

Average duration of recovery period after first ejaculation (in log seconds)

GROUP	N	NORMAL (TESTS 4–6)	CASTRATED (TESTS 10–15)	DIFFERENCE	PROBABILITY	MICROGRAMS OF ANDROGEN PER DAY
II	5	2.40	2.49	+.09	<.01	25
III	7	2.39	2.45	+.06	.02	50
IV	6	2.42	2.37	−.05	.14	100
V	11	2.38	2.34	−.04	.13	500

TABLE 10

Average body weights

GROUP	N	NORMAL WEEKS 4–6	CASTRATE WEEKS 10–15	CASTRATE WEEKS 20–25	PER CENT INCREASE OVER PREOPERATIVE WEIGHT First postoperative period	PER CENT INCREASE OVER PREOPERATIVE WEIGHT Second postoperative period	MICROGRAMS OF ANDROGEN PER DAY First postoperative period	MICROGRAMS OF ANDROGEN PER DAY Second postoperative period
I	11	300	344	350	15	2	0	1
II	10	297	339	367	14	8	25	75
III	10	306	366	376	19	3	50	0
IV	10	318	369	382	16	—	100	—
V	11	304	349	375	15	7	500	0

the response of accessory glands to hormone dosages used during the first 9 weeks after castration.

Males in Groups III and V were given 50 and 500 micrograms per day respectively in the first period and plain sesame oil during the second period after castration. At the end of the second period the seminal vesicles were typical of the long-term castrate. Gross size was markedly reduced and there was no evidence of secretory activity. Much of the epithelium was totally desquamated and such cells as remained covering the connective tissue were very low with quite small, spherical nuclei.

Group I males received no hormone for 9 weeks after operation and then were give 1 microgram per day for the next 10 weeks. This amount of testosterone propionate exerted no observable effect upon the seminal vesicles, and the cytological picture in these animals was exactly the same as that described for Groups III and V after 10 weeks without androgen. This finding is particularly interesting in view of the fact that 1 microgram per day did produce a distinct increase in some aspects of the sexual behavior.

Daily injections of 75 micrograms of hormone maintained normal seminal vesicles in males of Group II. In their gross and microscopic aspects the accessories of these animals were indistinguishable from those of intact rats.

General health: The health of the experimental males was checked by observation and by recording bodily weight once each week. There was no indication that either the operation or the hormone treatment had any adverse effects. All animals gained weight during the first 9 weeks after castration and group averages increased from 14 to 19 per cent with no apparent relationship between magnitude of increase and size of hormone dosage. During the second postoperative period 41 rats gained a little more weight and 11 lost an average of 1 to 15 grams. These minor decreases were shown by some rats in every group and probably reflect the fact that the animals were getting to an age at which the weight curve tends to level off. Average weights are shown in table 10.

DISCUSSION

In interpreting the foregoing results it is necessary to hold in mind certain limitations of the conditions under which they were obtained. First, the experimental population was a highly selected group drawn from an inbred strain. Second, our own results show that the behavioral effects of a given concentration of androgen tend to vary depending upon previous hormonal treatment. Castrated rats which have never been given androgen show an improvement in some aspects of their sexual performance when they are injected with 1 microgram per day; yet five times this amount has no appreciable effect upon the behavior of castrates previously treated with 100 micrograms. Finally, although it is probable that other forms of androgen would facilitate sexual performance in the castrated male, it is entirely possible that the magnitude of such an effect is a function of the form of the hormone and the method of its administration.

Within these limits the results of our experiment are reasonably consistent and clear cut. It has long been known that sexual behavior can be maintained in castrated rats by the administration of androgen but it can now be added that if the androgen is testosterone propionate, and if it is administered in the form of daily injections, the amount needed to hold performance at or near preoperative levels is approximately 50 to 75 micrograms. This is considerably less than the amounts employed in most of the earlier studies. We have also obtained new evidence concerning the effects of hormonal concentrations falling well above or below this "maintenance level."

In a very general way it is correct to state that the strength of the "sex drive" in a castrated male rat is roughly proportional to the amount of exogenous testicular hormone which is injected. But any such generalization must be followed by the qualification that "sex drive" is a meaningful concept only when it is operationally defined, and that when this is done the strength of the drive tends

452   FRANK A. BEACH AND A. MARIE HOLZ-TUCKER

to vary according to the behavioral criterion selected to measure it. For example, if we choose the frequency of intromissions per 10 minutes as *the* criterion our results indicate that sex drive decreases after castration unless more than 100 micrograms of androgen are given each day. In contrast it would be entirely permissible to state that sex drive will be measured in terms of the number of intromissions necessary to cause a sexual climax, or the speed of sexual arousal as measured by initial latency scores. In either instance the minimum androgen concentration sufficient to maintain normal sex drive would be considerably below 100 micrograms.

We are impressed with the need for an unambiguous, operational definition of sex drive,—a definition based upon actual mating performance rather than some less direct type of behavior such as the tendency to cross an electrically charged grid to reach the receptive female. Present data are insufficient to establish the point but they contain some suggestion that initial latency, time to achieve the first ejaculation, and duration of the post ejaculatory recovery period may be positively correlated. If such a correlation were high enough it would be entirely feasible to devise a combined score which would incorporate these measures and would serve as a valid and reliable index to sex drive in the individual male. When this can be done it seems likely that there will prove to be a fairly high positive relationship between sex drive and concentration of exogenous androgen administered to the castrated male.

A word of caution should be addressed to the reader who is unfamiliar with endocrinological data. The results herein reported *do not* signify that individual differences in the sexual responsiveness and potency of *unoperated* male rats are due to differences in levels of endogenous hormone. This particular problem cannot be approached until there is available some sensitive and reliable method of determining the amount of hormone present in the blood. Nothing of this sort can be done as yet with an animal as small as the rat. There are many potential sources of variability in sexual excitability and differences between intact males are quite possibly entirely independent of any hormonal basis.

The results of this experiment tell us nothing with respect to the locus or specific nature of hormonal action. They are, however, harmonious with the general hypothesis that one important function of the hormone is to lower the threshold in those nervous mechanisms specifically concerned with the mediation of sexual responses. For instance, the reduced initial latencies characteristic of castrated males receiving high doses of androgen suggest that such animals reach the threshold of arousal necessary to mounting activity with less preliminary stimulation than do other males whose blood contains less male hormone. Similarly, the abbreviated period of sexual refractoriness after an ejaculation is suggestive of a lowered threshold and a consequent reduction in the duration or intensity of the liminal stimulus.

SUMMARY

Fifty-two male rats were observed in six weekly mating tests with receptive females and then castrated. Sex tests were continued after operation while the males received daily injections of testosterone propionate. The amount of androgen administered to different groups varied from 1 to 500 micrograms per day. A control group was treated with plain sesame oil.

The amount of hormone necessary to maintain sexual performance at or near preoperative levels varied somewhat depending upon the behavioral criterion selected as a measure. In the main, however, 50 to 75 micrograms per day represented a maintenance dose.

Castrated rats receiving no hormone and those injected with less than 50 micrograms were less likely to show any sexual responses toward the female than were normal animals or other castrates given higher concentrations of male hormone. When they did display copulatory reactions the low dose castrates were slow in initiating sexual contact, and mounting activity was apt to be preceded by relatively protracted periods of inattention or investigation of the estrous female. Such mating behavior as did occur consisted to a large extent of copulatory attempts which failed because of lack of intromission. Successful intromissions often were widely spaced in time. Ejaculation rarely was achieved but when it did appear it usually occurred after fewer intromissions and was followed by a longer period of sexual inactivity than is the rule in normal rats or in castrates receiving larger amounts of hormone.

Castrated male rats injected with 100 or 500 micrograms of testosterone propionate per day exhibited sexual behavior that was equal or superior to that shown prior to operation. They were more likely to copulate in every test than they had been before gonadectomy. At the same time the occurrence of incomplete copulations increased. Castrates receiving these larger doses of hormone tended to initiate sexual relations after shorter delays than they had before operation. The frequency of intromissions in a 10-minute test was not increased but multiple ejaculations occurred more frequently, and the first sexual climax was reached at an earlier point in the test. Finally, the castrates supplied with large amounts of androgen tended to recover more rapidly from the sexually depressing effects of an ejaculation.

REFERENCES

1. BEACH, F. A.: *Hormones and Behavior*, New York: Paul B. Hoeber, Inc., 1948.
2. MOORE, C. R. AND PRICE, D.: Some effects of testosterone and testosterone propionate in the rat. *Anat. Rec.*, 1938, **71**, 59–78.
3. SHAPIRO, H. A. Effect of testosterone propionate upon mating. *Nature*, 1937, **139**, 588–589.
4. STONE, C. P.: Copulatory activity in adult male rats following castration and injections of testosterone propionate. *Endocrinol.*, 1939, **24**, 165–174.

ORGANIZING ACTION OF PRENATALLY ADMINISTERED TESTOSTERONE PROPIONATE ON THE TISSUES MEDIATING MATING BEHAVIOR IN THE FEMALE GUINEA PIG[1]

CHARLES H. PHOENIX, ROBERT W. GOY, ARNOLD A. GERALL
AND WILLIAM C. YOUNG

Department of Anatomy, University of Kansas, Lawrence, Kansas

ABSTRACT

The sexual behavior of male and female guinea pigs from mothers receiving testosterone propionate during most of pregnancy was studied after the attainment of adulthood. As a part of the investigation, the responsiveness of the females to estradiol benzoate and progesterone and to testosterone propionate was determined.

The larger quantities of testosterone propionate produced hermaphrodites having external genitalia indistinguishable macroscopically from those of newborn males. Gonadectomized animals of this type were used for tests of their responsiveness to estradiol benzoate and progesterone and to testosterone propionate. The capacity to display lordosis following administration of estrogen and progesterone was greatly reduced. Male-like mounting behavior, on the other hand, was displayed by many of these animals even when lordosis could not be elicited. Suppression of the capacity for displaying lordosis was achieved with a quantity of androgen less than that required for masculinization of the external genitalia.

The hermaphrodites receiving testosterone propionate as adults displayed an amount of mounting behavior which approached that displayed by the castrated injected males receiving the same hormone.

The data are uniform in demonstrating that an androgen administered prenatally has an organizing action on the tissues mediating mating behavior in the sense of producing a responsiveness to exogenous hormones which differs from that of normal adult females.

No structural abnormalities were apparent in the male siblings and their behavior was essentially normal.

The results are believed to justify the conclusion that the prenatal period is a time when fetal morphogenic substances have an organizing or "differentiating" action on the neural tissues mediating mating behavior. During adulthood the hormones are activational.

Attention is directed to the parallel nature of the relationship, on the one hand, between androgens and the differentiation of the genital tracts, and on the other, between androgens and the organization of the neural tissues destined to mediate mating behavior in the adult.

Received February 9, 1959.

[1] This investigation was supported by research grant M-504 (C6) from the National Institute of Mental Health, Public Health Service.

369

INVESTIGATORS interested in reproductive behavior have demonstrated that one role of the gonadal hormones in adult male and female mammals is to bring to expression the patterns of behavior previously organized or determined by genetical and experiential factors (1, 2, 3, 4, 5). The hypothesis that these hormones have an organizing action in the sense of patterning the responses an individual gives to such substances has long been rejected (6, 7, 8, 9, 10). As far as the adult is concerned, this conclusion seems well founded. Female hormone, instead of feminizing castrated male rats as Kun (11) claimed, increased their activity as males (6). Male and female guinea pigs gonadectomized the day of birth, and a female rat with a congenital absence of the ovaries, displayed normal patterns of behavior when injected with the appropriate hormones as adults (3, 9, 12).

Unexplored since the studies of Dantchakoff (13, 14, 15), Raynaud (16) and Wilson, Young and Hamilton (17), is the possibility that androgens or estrogens reaching animals during the prenatal period might have an organizing action that would be reflected by the character of adult sexual behavior. If the existence of such an action were revealed, it would 1) extend our knowledge of the role of the gonadal hormones in the regulation of sexual behavior by providing information bearing on the action of these hormones or related substances during the prenatal period, 2) be suggestive evidence that the relationship between the neural tissues mediating mating behavior and the morphogenic fetal hormones parallels that between the genital tissues and the same hormones, and 3) direct attention to a possible origin of behavioral differences between the sexes which is *ipso facto* important for psychologic and psychiatric theory (18). Although comprehensive experiments have not yet been performed, initial investigations with an androgen have yielded effects which are so much more in line with current thought in the area of gonadal hormones and sexual differentiation (19, 20, 21, 22) than the earlier experiments on behavior, that the results are summarized here.

MATERIALS AND METHODS

(the production of hermaphrodites)

Most of the experimental animals were born to mothers which had received intramuscular injections of testosterone propionate[2,3] during much of gestation. One group was composed of females in which there were no visible abnormalities of the external

---

[2] Testosterone propionate (Perandren propionate) was supplied by Ciba Pharmaceutical Products, Inc.

[3] The injections were made by Mr. Myron D. Tedford, a Public Health Service Predoctoral Fellow, who is using these and other animals treated similarly for a study of the structural changes in the gonads, genital tracts, and external genitalia, and the course of gestation. We are indebted to him for supplying us with the animals whose behavior was investigated.

genitalia. These are referred to as the *unmodified females*. Their mothers were given an initial injection of 1 mg. of testosterone propionate some time between day 10 and day 27 after conception and 1 mg. every third or fourth day thereafter until the end of pregnancy.

The larger group was composed of females in which the external genitalia at the time of birth were indistinguishable macroscopically from those of their male siblings and untreated males. These animals are designated *hermaphrodites*. Laparotomy was necessary in order to distinguish these genetical females from males; it was performed within the first week after birth. Their mothers received an initial injection of 5 mg. of testosterone propionate on day 10, 15, 18, or 24 of the gestation period and 1 mg. daily thereafter until day 68.

Control animals were females and males from untreated mothers from the same stock as the experimental animals.

All these animals, i.e. the unmodified females, the hermaphrodites, their male siblings, and the control females and males, were used in four experiments designed to test the effects of testosterone propionate received prenatally on the responsiveness of the animals as adults to male and female hormones.

EXPERIMENTAL

Experiment I. *The behavior of gonadectomized adult unmodified females and hermaphrodites injected with estradiol benzoate and progesterone.*

Subjects

Fourteen females from untreated mothers.

Fourteen unmodified females.

Nine hermaphrodites.

Eight males from untreated mothers.

Except for four unmodified females gonadectomized when they were 45 days old, all the unmodified females and hermaphrodites were gonadectomized at 80 to 150 days of age. No data from the laboratory indicate that the response to exogenously administered sex hormones is influenced by age at the time of gonadectomy. The eight males were castrated before they were 21 days old.

Tests

After gonadectomy, when the animals were 90 to 160 days old, tests were made of the responsiveness to 1.66, 3.32, and 6.64 μg. of subcutaneously injected estradiol benzoate followed 36 hours later by 0.2 mg. of progesterone.[4] Observations were continuous for 12 hours, beginning immediately after the injection of progesterone. Following the procedure of Goy and Young (4) hourly checks were made for the occurrence of the lordosis reflex in response to fingering. Individual records were kept of this measure of behavior and of the frequency of male-like mounting.

In three tests the control females, hermaphrodites, and males were ob-

---

[4] Estradiol benzoate (Progynon-B) and progesterone (Proluton) were supplied by the Schering Corporation.

served for the occurrence of mounting in the absence of exogenous hormone. The unmodified females were given one such test.

The means and medians of the measures of behavior for which data were obtained were calculated from the individual averages and they are based on the data from the animals which responded to the hormones. For purposes of statistical analysis, maximum values (12 hours) for latency and 0 values for all other measures were arbitrarily assigned to the individuals failing to respond.

Results

The data bearing on all the measures of the estrous response except mounting are summarized in Table 1. The lower values for the per cent of

TABLE 1. DURATION OF HEAT AND LORDOSIS IN GONADECTOMIZED GUINEA PIGS GIVEN DIFFERENT AMOUNTS OF ESTRADIOL AND 0.2 MG. OF PROGESTERONE

Subjects	Tests* N	Per cent of tests positive for estrus	Mean latency in hours	Mean duration of heat in hours	Median duration of max. lord. in seconds
			1.66 μg.		
Control females	19	89	5.7	5.7	11.5
Unmodified females	20	65	6.5	2.8	8.5
Hermaphrodites	9	22	8.5	2.5	2.0
Castrated males	8	38	6.0	1.2	2.0
			3.32 μg.		
Control females	33	94	4.4	7.3	12.3
Unmodified females	38	68	5.6	2.8	5.1
Hermaphrodites	18	22	8.0	2.0	3.0
Castrated males	16	31	4.5	3.2	2.7
			6.64 μg.		
Control females	28	96	3.7	7.2	9.3
Unmodified females	22	77	5.8	3.3	6.0
Hermaphrodites	18	22	9.2	2.0	2.0
Castrated males	16	0	—	—	—

* All the animals were given one or more tests at each level of hormone.

tests positive for estrus, the mean duration of heat, and the median duration of the maximum lordosis were conspicuous effects of the treatment given prenatally and the differences among the groups are highly significant (P <.001). Among the two groups of experimental females and the castrated males, the low gutteral growl which is so characteristically a part of the pattern of lordosis in normal females, was commonly, and in some individuals always, lacking. Had the estimation of the duration of maximum lordosis been based only on complete responses, the differences among the groups would have been even greater.

Variations in medians for the duration of maximum lordosis were not systematically related to quantity of estradiol given prior to the tests. The analysis, therefore, was based on the medians of individual averages over all dosages. These medians were 11.3, 6.5, 2.3, and 2.5 seconds for control females, unmodified females, hermaphrodites, and castrated males, respectively. The median of the unmodified females, which most closely resembles that of the control females, is significantly different ($U = 22$, $P < .002$) from that of the controls.

Other differences also are indicative of the changes that were induced. Per cent response and duration of heat tended to increase in the control groups as the quantity of injected estradiol was increased. Latency which is related inversely to the duration of heat (4) decreased. Among the experimental groups (unmodified females, hermaphrodites, and castrated males), similar relationships were seen only in the unmodified females.

In general the suppression of the capacity to display lordosis was proportional to the quantity of androgen injected prenatally. Amounts in-

TABLE 2. THE QUANTITY OF MOUNTING WITH AND WITHOUT
ESTRADIOL AND PROGESTERONE

Subjects	Without hormone	With hormone*
	Mean number of mounts	Mean number of mounts
Control females	0	10.7
Unmodified females	0	8.8
Hermaphrodites	4.4	5.6
Castrated males	11.8	16.7

* Variation in the amount of mounting was not related to the quantity of estradiol. The means therefore are based on the averages for each individual whether the dosage was 1.66, 3.32, or 6.64 µg. of the hormone.

sufficient to alter external genital structures resulted in disturbances in the lordosis in only 50% of the animals, but the larger amounts that produced the hermaphrodites affected the lordosis in all. Within each group the effect on lordosis was not related to the quantity of androgen received prenatally. Among unmodified females, even siblings differed, one showing complete suppression of the lordosis and the other responding normally. The findings demonstrate that suppression of the capacity for displaying lordosis does not depend on masculinization of the external genitalia; clearly less androgen was required for the former than for the latter.

Additional evidence for the masculinizing effect of the prenatally administered androgen is provided by the data on the male-like mounting displayed by each group (Table 2). When estradiol and progesterone were injected all groups displayed mounting, and the differences among the groups are not statistically significant. In contrast, on tests when no hormones were given, the hermaphrodites and castrated males were the only animals that mounted.

The interval from the beginning of the test to the display of mounting differed among the groups. Of the males which mounted, all did so at least once during the first hour. Of the 7 hermaphrodites which mounted, 5 or

71% mounted at least once during the first hour, but only 1 normal female (7%) and 1 unmodified female (7%) mounted this early in the test. The modal time for the onset of mounting was the 1st hour for the castrated males and hermaphrodites and the 6th and 7th hours for the control females and unmodified females, respectively. In this respect the hermaphrodites closely resembled the castrated males and seem to have been masculinized. The latency of mounting in the unmodified females was not different from that in the control females.

In one way the mounting performance of the unmodified females did differ from that of the controls. More unmodified than control females displayed mounting on tests after injections when the lordosis reflex could not be obtained. Of 8 unmodified females which failed to show lordosis, 6 or 75% mounted. Because of the small number of control females which failed to display lordosis after injection, older data on normal females from the same genetical stock are used for comparison. These data combined with those from the present study reveal that of 38 normal females failing to display lordosis after injection with comparable amounts of estradiol and progesterone only 4 or 10.5% mounted. The difference between the proportions of control females and unmodified females displaying mounting in the absence of lordosis is significant (C.R. = 4.02, P < .001). Inasmuch as mounting was displayed spontaneously by the hermaphrodites, it was not possible with the animals available to determine the extent to which this behavior was being shown in response to the estradiol and progesterone.

Conclusions

1. Prenatally administered testosterone propionate suppressed the capacity for displaying lordosis following gonadectomy and the injection of estradiol and progesterone. The effect was manifested either by an absence of lordosis or by a marked abnormality in its character when it was displayed.

2. Suppression of the capacity for displaying lordosis was achieved with a smaller quantity of the androgen than was necessary for the gross modification of the external genitalia.

3. The capacity to display male-like mounting was not suppressed.

4. Quantities of testosterone propionate sufficient to suppress lordosis and masculinize the genitalia also reduced the interval before mounting behavior was displayed.

Experiment II. *Permanence of the effects of prenatally administered androgen.*

Subjects

Group 1. Three hermaphrodites used in the previous experiment.

Group 2. Seven unmodified females used in the previous experiment.

Group 3. Eight control females used in the previous experiment.

Group 4. Six hermaphrodites injected with 500 μg. of testosterone propionate per 100 gm. body weight per day from birth to 80 days of age.

Group 5. Six normal females injected with the same amount of testosterone propionate from birth to 80 days of age.

Group 6. Five mothers of hermaphrodites injected with testosterone propionate during pregnancy as described in Materials and Methods.

Group 7. Eight untreated females comparable in age with those injected with testosterone propionate during pregnancy.

The animals in Groups 1 through 5 were gonadectomized when they were 80 to 150 days of age, those in Groups 6 and 7 when they were 1.5 to 3 years old. The operations on the animals in Group 6 were performed approximately 10 months after the last injection of testosterone propionate.

Tests

All the animals received 3.32 μg. of estradiol benzoate followed 36 hours later with 0.2 mg. of progesterone. The tests were similar to those given the hermaphrodites, unmodified females, and controls in Experiment I. The number, however, differed for each group and is shown in the description of the results. The values reported in the tables and the statistical treatment of the data were determined by the methods described in Experiment I.

Results

The behavior of the 3 hermaphrodites, the 7 unmodified females, and the 8 control females is summarized in Table 3 and compared with that dis-

TABLE 3. BEHAVIORAL RESPONSES TO 3.32 μG. OF ESTRADIOL AND
0.2 MG. OF PROGESTERONE

		Tests at 6–9 months of age	Tests at 11–12 months of age
Hermaphrodites (Group 1)	Per cent response	33.0	0
	Latency to heat in hours	7.5	—
	Duration of heat in hours	2.5	—
	Median maximum lordosis in seconds	2.0	—
	Mean number of mounts	3.0	45.2
Unmodified females (Group 2)	Per cent response	55.0	71.0
	Latency to heat in hours	6.3	7.5
	Duration of heat in hours	2.2	2.3
	Median maximum lordosis in seconds	4.0	5.8
	Mean number of mounts	8.7	17.5
Normal females (Group 3)	Per cent response	95.0	94.0
	Latency to heat in hours	4.4	6.1
	Duration of heat in hours	7.2	4.5
	Median maximum lordosis in seconds	10.0	10.2
	Mean number of mounts	9.9	9.6

TABLE 4. PER CENT RESPONSE, DURATION OF HEAT, AND MAXIMUM LORDOSIS AFTER CESSATION OF TREATMENT WITH TESTOSTERONE PROPIONATE FROM BIRTH TO 80 DAYS OF AGE

		Approximate age in days at time of test			
		90	140	160	175
Hermaphrodites (Group 4)	Per cent response	0	0	0	0
	Mean duration of heat in hours	—	—	—	—
	Median maximum lordosis in seconds	—	—	—	—
Females (Group 5)	Per cent response	0	84	66	66
	Mean duration of heat in hours	0	4.6	1.7	3.7
	Median maximum lordosis in seconds	0	9.0	5.5	9.5

played during the earlier tests when the animals were 6 months old. The results reported in Table 3 are based on at least 2 tests of each individual at each age level. No significant change occurred in the hermaphrodites and unmodified females for per cent response, latency to heat, duration of heat, and the duration of maximum lordosis. The normal females, however, showed a significant decrease in the duration of heat ($T = 0$, $P = .01$), reflecting perhaps a decrease in responsiveness to the hormones as the animals aged. The increase in mounting is significant for the unmodified females ($T = 0$, $P = .02$). The 3 hermaphrodites displayed increased mounting behavior, but the increase could not be evaluated statistically. Of the normal females, 3 showed increases, 3 a decrease, and 2 remained the same.

The contrast between the effects of prenatal and postnatal treatment is revealed by the results obtained from the animals treated neonatally (Groups 4 and 5) and from those treated during pregnancy (Group 6). During the period after withdrawal of the testosterone propionate, 5 of the 6 normal females which had been injected for 80 days after birth regained the ability to display lordosis, whereas the hermaphrodites did not (Table 4). The effects of the postnatally administered androgen on the mounting behavior displayed by the animals in the two groups were complex and their presentation is being postponed until a further discussion can be given. The females treated with testosterone propionate while pregnant (Group 6) did not, like their "daughters," lose the capacity to display lordosis. Comparison of their behavior in response to estradiol and progesterone in five tests with that of untreated females of the same age (Group 7) (Table 5), revealed that the differences between the groups are not significant for latency, duration of heat, and mounting.

TABLE 5. BEHAVIOR OF NORMAL FEMALES TREATED WITH TESTOSTERONE PROPIONATE FOR 50 DAYS DURING PREGNANCY AND TESTED 10 MONTHS LATER

	Per cent response	Latency of heat in hours	Duration of heat in hours	Mean no. of mounts
Treated females (Group 6)	84	6.7	4.2	17.8
Untreated females (Group 7)	62	7.6	3.2	8.1

Conclusions

1. The suppression of the capacity for displaying the feminine components of the sexual behavior pattern which followed the administration of testosterone propionate prenatally appears to have been permanent.

2. Amounts of testosterone propionate which were effective prenatally had no conspicuous lasting effects when administered postnatally.

Experiment III. *The behavior of gonadectomized hermaphrodites in response to testosterone propionate.*

Subjects

Five hermaphrodites gonadectomized between 86 and 112 days of age.
Five normal females gonadectomized between 80 and 106 days of age.
Eight normal males castrated before 21 days of age.
When the animals were approximately 180 days old all received 2.5 mg. of testosterone propionate daily for 16 consecutive days.

Tests

A sexual behavior test was given the day before the first injection. Additional tests were given on days 1 and 2 of the injection period, and every other day thereafter until each animal had received 9 tests. The ninth test was given the day of the sixteenth injection.

Results

The median value for mounting by the hermaphrodites and females in the single test prior to the injection of testosterone propionate was 0. For the males the median was 5.5.

The remaining data are summarized in Table 6. They demonstrate the masculinizing effect of prenatally administered testosterone propionate on the female. Castrated males and hermaphrodites obtained the highest sexual behavior scores, the control females the lowest. The overall difference in scores was significant ($P \sim .02$). The differences between the castrated males and hermaphrodites were not significant, whereas both groups differed significantly from the control females ($P = .05$). The overall difference in the number of tests to the first display of mounting was significant ($P < .01$). As with the sexual behavior scores, the difference between males

TABLE 6. MASCULINE BEHAVIOR IN GONADECTOMIZED ADULT ANIMALS
INJECTED WITH TESTOSTERONE PROPIONATE

Group	Mean sexual behavior score	Mean mounts per test	Median number of tests to the first display of mounting	Median mg. of t.p. prior to the display of mounting
Spayed untreated females	2.1	5.8	7.0	30.0
Spayed hermaphrodites	3.6	15.4	3.0	10.0
Males castrated prepuberally	5.0	20.5	1.5	3.8

and hermaphrodites was not significant, but both groups differed significantly from the control females (P = .02). There was a significant overall difference (P <.01) in the amount of testosterone propionate required before the first appearance of mounting. Again, the hermaphrodites resembled the castrated males in that there was no significant difference between these two groups, but both groups displayed mounting with significantly less hormone (P = .02) than the control females.

Conclusions

1. Adult hermaphrodites gonadectomized and injected with testosterone propionate were more responsive to this hormone than gonadectomized normal females.

2. The earlier appearance and greater strength of masculine behavior by the hermaphrodites given testosterone propionate are believed to be effects of the prenatally administered testosterone propionate on the tissues mediating masculine behavior and therefore to be expressions of its organizing action.

Experiment IV. *The behavior of adult male siblings of the hermaphrodites.*

Subjects

Five males from untreated mothers.

Five males born to mothers receiving testosterone propionate during pregnancy. No hormone was administered after birth.

Five males born to mothers receiving testosterone propionate during pregnancy. These animals received 500 μg. of the hormone per 100 gm. body weight daily beginning 1 to 3 days after birth and continuing 80 to 90 days.

Tests

Five tests were given when the animals were 11 months old. In a test the subject was placed with a receptive female of approximately the same size, and the frequency of the display of selected measures of behavior was recorded for a maximum of 10 minutes. These measures included sniffing and nibbling, nuzzling, abortive mounting, mounting, intromissions, and ejaculation. A description of the measures and the method for computing scores are given by Valenstein, Riss and Young (23).

Results

The mean scores are summarized in Table 7. It is clear that any effect of the exogenous testosterone propionate was slight. There was no evidence of suppression of the capacity to display masculine behavior, if anything, the animals receiving the hormone prenatally achieved higher scores than the controls.

TABLE 7. MEAN SEXUAL BEHAVIOR SCORES OBTAINED BY THE
THREE GROUPS OF ADULT MALES

Groups*	Tests				
	I	II	III	IV	V
Untreated	6.9	6.6	9.2	7.2	10.4
Testosterone propionate prenatally	10.4	9.3	9.1	9.3	12.2
Testosterone propionate prenatally and postnatally	10.9	11.2	7.3	11.1	9.4

* Difference among the groups not significant; $F = 1.30$; $df = 2, 12$.

Conclusion

The sexual behavior of adult males which had received testosterone propionate prenatally was not significantly different from that of untreated controls.

DISCUSSION

The data from the four experiments summarized in the preceding sections support the hypothesis that androgenic substances received prenatally have an organizing action on the tissues mediating mating behavior in the sense of altering permanently the responses females normally give as adults. This possibility was suggested by the work of Dantchakoff (13, 14, 15), Raynaud (16), and Wilson, Young and Hamilton (17). Probably, however, because interest in the role of gonadal hormones in the regulation of mating behavior was concentrated so largely on the neonatal individual and adult, the suggestion was never incorporated in our theories of hormonal action. This step may now be taken, but when what has been learned from the present investigation is related to what has long been known with respect to the action of androgens on the genital tracts, a concept much broader than that suggested by the older studies is revealed.

The embryonic and fetal periods, when the genital tracts are exposed to the influence of as yet unidentified morphogenic substances (19, 20, 21, 22, 24), are periods of differentiation. The adult period, when the genital tracts are target organs of the gonadal hormones, is a period of functional response as measured by cyclic growth, secretion, and motility. The response depends on whether Müllerian or Wolffian duct derivatives have developed, and although generally specific for hormones of the corresponding sex, it is not completely specific (25). For the neural tissues mediating mating behavior, corresponding relationships seem to exist. The embryonic and fetal periods are periods of organization or "differentiation" in the direction of masculinization or feminization. Adulthood, when gonadal hormones are being secreted, is a period of activation; neural tissues are the target organs and mating behavior is brought to expression. Like the genital tracts, the neural tissues mediating mating behavior respond to androgens or to estrogens depending on the sex of the individual, but again the specificity is not complete (26, 27).

An extension of this analogy is suggested by the work done on the embryonic differentiation of the genital tracts, particularly that by Burns and Jost and summarized in their reviews (20, 21, 22). It will be recalled from the data reported in the present study that testosterone propionate administered prenatally affected the behavior of the male but slightly, whereas the effects on the female were profound. Not only was there a heightened responsiveness to the male hormone as revealed by the stronger masculine behavior displayed when testosterone propionate was given, but there was a suppression of the capacity to display the feminine components in response to treatment with an estrogen and progesterone. In studies of the genital tracts there were no effects on the male except for a slight acceleration in the development of the prostate and seminal vesicle and an increase in the size of the penis (28). Within the female, the Wolffian duct system was stimulated (13, 14, 15, 28, 29), and locally, when a fetal testis was implanted into a female fetus (20, 21), there was an interruption of the Müllerian duct on that side. What has not been seen when an exogenous androgen was administered, except by Greene and Ivy (30) in some of their rats, is a suppression or inhibition of the Müllerian duct system corresponding to the suppression of the capacity for displaying the feminine component of behavior.

The failure to detect a corresponding suppressing action on the Müllerian duct does not exclude the possibilities 1) that such an effect will be found, and 2) that the suppressing action is in the nature of a reduction in the responsiveness of the genital tract to estrogens rather than in the inhibition of its development. Such an effect was encountered in rats given testosterone propionate prenatally (17) when it was found that uterine as well as behavioral responses to estrogen and progesterone were suppressed.

A final suggestion with respect to the analogy we have postulated arises from a comparison of our results with those reported by Dantchakoff and Raynaud. These investigators stressed the increased responsiveness of their masculinized guinea pigs and mice to exogenous androgens, and seemed to regard the change as the expression of an inherent bisexuality. The possibility that there might have been a suppression of the capacity to respond as females and therefore an inequality of potential does not seem to have been considered. Like Dantchakoff (13, 14, 15), Raynaud (16), and many others (9, 31, 32, 33, 34, 35), the existence of a bisexuality is assumed. We suggest, however, that in the adult this bisexuality is unequal in the neural tissues as it is in the case of the genital tissues. The capacity exists for giving behavioral responses of the opposite sex, but it is variable and, in most mammals that have been studied and in many lower vertebrates as well, it is elicited only with difficulty (27). Structurally, the situation is similar. Vestiges of the genital tracts of the opposite sex persist and are responsive to gonadal hormones (36, 37), but except perhaps in rare instances, equivalence of organs and responses in a single individual is not seen (36, 37, 38, 39).

The concept of a correspondence between the action of gonadal hormones on genital tissues and neural tissues contains much that is new and its full scope is not yet clear. The possibility must be considered that the masculinity or femininity of an animal's behavior beyond that which is purely sexual has developed in response to certain hormonal substances within the embryo and fetus.

Thus far the permanence of the effect achieved when testosterone propionate was received prenatally has not been achieved when the same hormones were administered to adults or to newborn individuals. The dependence of this "permanence" on the action of the hormone during a possible critical period must be ascertained.

The nature of the modifications produced by prenatally administered testosterone propionate on the tissues mediating mating behavior and on the genital tract is challenging. Embryologists interested in the latter have looked for a structural retardation of the Müllerian duct derivatives culminating in their absence, except perhaps for vestigial structures found in any normal male. Neurologists or psychologists interested in the effects of the androgen on neural tissues would hardly think of alterations so drastic. Instead, a more subtle change reflected in function rather than in visible structure would be presumed.

Involved in this suggestion is the view that behavior may be treated as a dependent variable and therefore that we may speak of shaping the behavior by hormone administration just as the psychologist speaks of shaping behavior by manipulating the external environment. An assumption seldom made explicit is that modification of behavior follows an alteration in the structure or function of the neural correlates of the behavior. We are assuming that testosterone or some metabolite acts on those central nervous tissues in which patterns of sexual behavior are organized. We are not prepared to suggest whether the site of action is general or localized.

REFERENCES

1. BEACH, F. A.: *J. Genet. Psychol.* **60**: 121. 1942.
2. ZIMBARDO, P. G.: *J. Comp. & Physiol. Psychol.* **51**: 764. 1958.
3. VALENSTEIN, E. S., W. RISS AND W. C. YOUNG: *J. Comp. & Physiol. Psychol.* **48**: 397. 1955.
4. GOY, R. W. AND W. C. YOUNG: *Psychosom. Med.* **19**: 144. 1957.
5. ROSENBLATT, J. S. AND L. R. ARONSON: *Animal Behav.* **6**: 171. 1958.
6. BALL, J.: *J. Comp. Psychol.* **24**: 135. 1937.
7. BALL, J.: *J. Comp. Psychol.* **28**: 273. 1939.
8. BEACH, F. A.: *Endocrinology* **29**: 409. 1941.

382   PHOENIX, GOY, GERALL AND YOUNG   *Volume 65*

9. BEACH, F. A.: *Anat. Rec.* **92**: 289. 1945.
10. RISS, W., E. S. VALENSTEIN, J. SINKS AND W. C. YOUNG: *Endocrinology* **57**: 139. 1955.
11. KUN, H.: *Endokrinologie* **13**: 311. 1934.
12. WILSON, J. G. AND W. C. YOUNG: *Endocrinology* **29**: 779. 1941.
13. DANTCHAKOFF, V.: *Compt. rend. Acad. sci.* **206**: 945. 1938.
14. DANTCHAKOFF, V.: *Compt. rend. soc. Biol.* **127**: 1255. 1938.
15. DANTCHAKOFF, V.: *Compt. rend. soc. Biol.* **127**: 1259. 1938.
16. RAYNAUD, A.: *Bull. Biol. France et Belgique* **72**: 297. 1938.
17. WILSON, J. G., W. C. YOUNG AND J. B. HAMILTON: *Yale J. Biol. & Med.* **13**: 189. 1940.
18. HAMPSON, J. L. AND J. G. HAMPSON: Allen's Sex and Internal Secretions, ed. by W. C. Young, Baltimore, Williams & Wilkins. In press.
19. JOST, A.: *Arch. Anat. micro. et Morph. exper.* **36**: 271. 1947.
20. JOST, A.: *Rec. Prog. Hormone Res.* **8**: 379. 1953.
21. JOST, A.: Gestation. Transactions of the Third Conference, ed. by C. A. Villee, New York, Josiah Macy, Jr. Foundation, 129. 1957.
22. BURNS, R. K.: Allen's Sex and Internal Secretions, ed. by W. C. Young, Baltimore, Williams & Wilkins. In press.
23. VALENSTEIN, E. S., W. RISS AND W. C. YOUNG: *J. Comp. & Physiol. Psychol.* **47**: 162. 1954.
24. HOLYOKE, E. A. AND B. A. BEBER: *Science* **128**: 1082. 1958.
25. BURROWS, H.: Biological Actions of Sex Hormones, Cambridge, Cambridge University Press. 1949.
26. ANTLIFF, H. R. AND W. C. YOUNG: *Endocrinology* **59**: 74. 1956.
27. YOUNG, W. C.: Allen's Sex and Internal Secretions, ed. by W. C. Young, Baltimore, Williams & Wilkins. In press.
28. GREENE, R. R.: *Biol. Symposia* **9**: 105. 1942.
29. TURNER, C. D.: *J. Morphol.* **65**: 353. 1939.
30. GREENE, R. R. AND A. C. IVY: *Science* **86**: 200. 1937.
31. STEINACH, E.: *Zentrabl. Physiol.* **27**: 717. 1913.
32. STEINACH, E.: *Arch. f. Entwcklngsmechn. d. Organ.* **42**: 307. 1916.
33. LIPSCHÜTZ, A.: The Internal Secretions of the Sex Glands. Baltimore, Williams & Wilkins. 1924.
34. BEACH, F. A.: *J. Comp. Psychol.* **36**: 169. 1942.
35. BEACH, F. A.: *Physiol. Zool.* **18**: 390. 1945.
36. MAHONEY, J. J.: *J. Exper. Zool.* **90**: 413. 1942.
37. PRICE, D.: *Anat. Rec.* **82**: 93. 1942.
38. BURNS, R. K.: *Contr. Embryology*, Carnegie Institution of Washington, **31**: 147 1945.
39. BURNS, R. K. *Am. J. Anat.*, **98**: 35. 1956.

Division of Biology, University of Pennsylvania, Philadelphia 4, Pa. and The United States Public Health Service, Gerontology Branch, Baltimore City Hospitals, Md.

Hunger in the Blowfly *)

by V. G. Dethier and Dietrich Bodenstein

With 2 figures

Eingegangen am 11. Februar 1958

Introduction

Although psychologists and ethologists have directed a great deal of attention to the study of motivation and drive in vertebrates and are now seeking a physiological basis for these concepts, no penetrating studies of the problem have been undertaken with the invertebrates. Because of the earlier position which the invertebrates occupy on the evolutionary scale and their relatively simpler behavioral organization it would seem that an investigation of drive and motivation in these animals should prove rewarding.

The two kinds of behavior which have lent themselves most readily to analyses of drive and motivation in the vertebrates have been feeding and sexual activity. Since the physiology of feeding is understood better in insects than in other invertebrates, and better than the physiology of sex, feeding is the activity of choice in the study contemplated. Most adult insects are periodic feeders; they eat for a while, then stop and remain trophically inactive for varying periods of time. Some knowledge of the identity of the physiological factors which start and stop feeding, and the conditions under which they are operative, is available. The question of what constitutes hunger and satiety remains unanswered. An analysis of these two states would appear to be a logical initial step toward the larger problem of drive. That is the purpose of this communication.

One of the earliest observed correlates of food deprivation in insects was change in threshold of response to such acceptable substances as sugars. As early as 1922 MINNICH had shown that the concentration of sucrose which was acceptable to the red admiral butterfly, *Pyrameis atalanta* L., and elicited extension of the coiled proboscis, was low if the butterfly had been deprived of food for some time but was elevated as a consequence of feeding. KUNZE (1927) and von FRISCH (1927) obtained results of a similar nature when they experimented with honey bees. Such changes in the response thresholds for taste substances, coincidental with changes in the duration of time since the last feeding, have since been reported by a great number of investigators for a variety of insects (cf. e. g., HASLINGER 1935). A striking feature of the change is its magnitude. In the blowfly *Phormia regina* Meigen a ten-million-fold increase in the acceptance threshold for sucrose has been demonstrated to occur (DETHIER and RHOADES 1954). Changes of the order of one-hundred-thousand-fold have been reported for other species (ANDERSON 1932), and ten-thousand-fold changes are quite usual (cf., e. g., MINNICH 1929).

Another characteristic of the change in threshold is the dependence of the rate of change upon the nature of the food previously ingested. EVANS and DETHIER (1957) have shown, for example, that the fall in glucose threshold of the blowfly after a meal of 1 M fucose is very rapid, requiring only twenty-five hours to regain its basal level; while after 1 M mannose feeding a period of one hundred hours is required. Although after the feeding of any particular sugar there may be variations in the rate of decline depending upon which set of receptors (i. e., oral or tarsal) is tested (MINNICH 1922a, 1922b; HASLINGER 1935), the comparative rates with different sugars ingested bear a constant relation to one another. This fact plus the fact that the rate of threshold change depends

---

*) This work was aided by a grant from The National Science Foundation.

9

130   V. G. DETHIER and DIETRICH BODENSTEIN

upon the identity of the sugar ingested and not upon which sugar is employed as a test
stimulus suggests that threshold change is causally related to feeding rather than to
events occurring in the peripheral sensory system.

All of these changes in behavioral threshold to acceptable taste substances bear a
predictable relation to food deprivation and satiety; consequently, it was reasonable to
assume that an analysis of the regulation of threshold would yield information relative
to the problems of hunger and feeding. It had been shown that feeding by a hungry fly
was initiated by adequate sensory input from mouth or leg receptors and that it continued only as long as the sensory input continued effectively (DETHIER 1955). When
receptors became adapted, feeding ceased temporarily until disadaptation occurred
(DETHIER, EVANS, and RHOADES 1956; EVANS and DETHIER 1957). However, the ultimate
cessation of feeding for any given meal could not be ascribed to sensory adaptation.
Prolonged elevation of threshold and prolonged feeding inactivity had to be attributed
to some other factor. The following were eliminated as regulatory factors: blood-sugar
levels, stored glycogen depletion, and crop contents (EVANS and DETHIER 1957, HUDSON
1958). It was concluded that some region of the gut other than the crop was the critical
part for threshold regulation. The experiments reported here were designed to reveal the
mechanisms of threshold regulation and its bearing on hunger, satiation, and feeding activity.

Materials and Methods

The black blowfly, *Phormia regina* Meigen, was the animal employed in all experiments. All the animals were operated upon without anaesthesia. In order to fix them in
the desired position for the various operations, they were held on the operating table by
small strips of plasticene. The table consisted of a black wax plate into which depressions

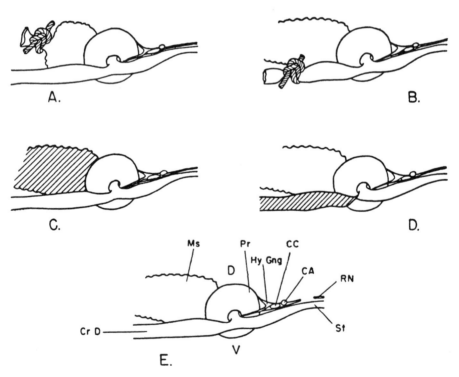

Fig. 1. Diagrams of the region of the foregut of *Phormia regina*. A. Point at which midgut
was ligated. B. Point at which crop duct was ligated. C. Most anterior advance of fluid
injected into the midgut via the anus. D. Most anterior advance of fluid in crop duct when
crop valve is closed. E. Point at which recurrent nerve was cut. Ms = mesenteron (midgut),
Pr = proventriculus, CrD = crop duct, St = stomodaeum (foregut), HyGng = hypocerebral ganglion, CC = corpus cardiacum, CA = corpus allatum, RN = recurrent nerve,
D = dorsal, V = ventral

of different shapes were cut to aid in bedding and holding the animals in the desired position. The operations were performed under a dissecting microscope. Sometimes, as in the case of the cutting of the recurrent nerve, a rather high magnification (50×) was necessary. Watchmaker's forceps, sharpened to a fine point, and iridectomy scissors in which the blades had been thinned, narrowed, and finely pointed by appropriate grinding were satisfactory instruments.

Severing of the Recurrent Nerve • The procedure for cutting the recurrent nerve, perhaps the most difficult operation performed, was as follows: a shallow depression in the shape of the fly's body (thorax and abdomen) was cut into the wax. The depth of the depression was about two-thirds of the dorsal-ventral width of the thorax. In front of this depression a second smaller one was made to hold the fly's head. Thus a narrow wax bridge separated the two depressions. The fly was placed dorsal side up into the large depression and held in place by a strip of plasticene laid across the anterior portion of the thorax. The head of the fly was then carefully manipulated into the second small depression so that the neck stretched across the wax bridge. The head was held down by a plasticene strip. In this manner the dorsal neck skin was brought into clear view. A drop of physiological solution was placed on the neck; it covered part of the thorax and the head. This drop not only kept moist the tissues that were to be exposed, but also acted as a lens and thus aided considerably in the performance of the next steps. With the drop in place, the skin of the neck was cut along the midline with an iridectomy scissors. The cut had to extend posteriorly to the point where the neck meets the thorax. The wound was then widened by pulling the cut skin apart with two pairs of forceps. This exposed the longitudinal neck muscles that extend dorsally close below the skin. Some of these were severed but care had to be taken not to destroy too many of the lateral neck muscles. With some of the uppermost dorsal neck muscles cut, the foregut and the two large lateral tracheal trunks come into clear view. Near the prothorax one can also see the anterior border of the proventriculus (Fig. 1 E). Somewhat left and slightly in front of the proventriculus, the allatum-cardiacum complex is located; it is rather easy to distinguish from other tissues by the light bluish sheen of the cardiacum. The recurrent nerve which passes from the anterior part of the cardiacum forward, is as yet not visible. The cardiacum adheres laterally to the left tracheal trunk. With the fine tip of the forceps, this adhesion was broken. As soon as this is done, the recurrent nerve becomes visible. By holding the nerve close to the cardiacum with forceps, it can be lifted up slightly and then cut. This finished the operation. The wound border was now pressed together with forceps, the physiological solution removed and the fly freed. After a successful operation, the freed animal behaves normally and if given food, it feeds almost immediately.

Operations on the Intestinal Tract • In order to ligate the midgut (Fig. 1, A) behind the proventriculus, the fly was placed dorsal side up into a depression of the wax plate and held in position by two plasticene strips — one across the abdomen and the other across the head. With iridectomy scissors, a small wedge of muscle tissue was removed from the prothorax. The cut extended anteriorly almost to the neck, but left intact a small chitin ridge close to the neck. Posteriorly, the cut extended to about one-half of the length of the thorax. The broadest part of the tissue wedge to be removed was the dorsal surface of the thorax. The sloped cuts on each side met in the midline, just above the gut. Removal of this cut wedge exposes the anterior portion of the midgut and shows clearly the junction of the midgut and the proventriculus. A drop of physiological solution was placed in the wound. The cut-out tissue wedge was also placed in a drop of physiological solution to keep it from drying. With forceps, the midgut could now be lifted up and cut. If ligation was desired, the gut was lifted up with one pair of forceps, while at the same time a fine silk thread, held with the second pair of forceps, was pushed below the gut from one side to the other. The diameter of the silk thread used corresponded roughly to that of a human hair. Silk has however the advantage of being more flexible than hair. The two ends of the thread were then manipulated into a sling which was tightened just behind the proventriculus. After the knot was tight, the ends of the thread were cut. If it was desired to cut the midgut, a second ligature had to be applied. The gut was then cut between the two ligatures. Now, the cut-out tissue wedge was replaced and the wound thus closed. Any excess of blood and physiological solution that was pressed from the wound by the replacement of the cut-out piece was carefully soaked up by filter paper. The fly was now ready for the test. Animals thus operated upon are, of course, unable to fly because the various incisions have severed the flight muscles.

To ligate or remove the crop, the fly was held on the operating table, ventral side up, by two plasticene strips, one crossing the mid-thorax, and the other the posterior end of the abdomen. The crop lies somewhat to the right side, close to the midline, on the border of

132   V. G. Dethier and Dietrich Bodenstein

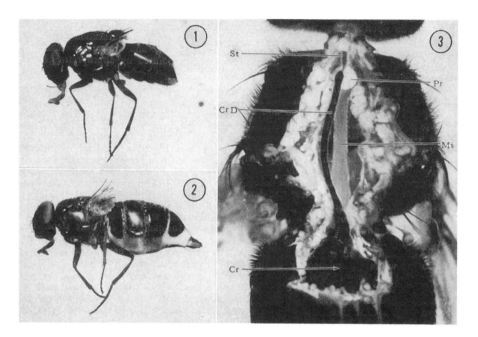

Fig. 2. 1. Normal fly which has fed to repletion. 2. Hyperphagic fly in which the recurrent nerve has been cut. 3. Ventral view of fly showing the anterior region of the alimentary canal. St = stomodaeum (foregut), Pr = proventriculus, CrD = crop duct, Ms = mesenteron (midgut), Cr = crop

thorax and abdomen (Fig. 2, Cr.). At this site, a small incision was made into the skin of the abdomen and the crop carefully pulled out. A silk loop was slipped over the exposed crop and its duct tied off at its entrance into the crop (Fig. 1 B). The crop was then pushed back through the wound into the abdomen and the animal released. If the crop was to be removed, a second ligature was tied around the duct and the duct cut between the two constrictions.

The action of the stomach valves and the route ingested food takes were observed in the living fly. For this, the fly was placed on the operating table ventral side up and held by sealing its wings to the wax table by a heated needle. As much of the ventral half of the thorax as was necessary to uncover the intestinal tract was then removed and physiological solution dropped into the gaping opening (Fig. 2). Flies prepared in this manner live for several hours and take in food which is offered them. By offering a sugar solution containing a dye (carmine or fuchsin), one can follow the course of the ingested food visually with ease.

Injection Experiments • All injection experiments were performed with an injection apparatus used for *Drosophila* work (Bodenstein 50). In the transfer of blood from fed to unfed individuals, physiological solution was injected first into the abdominal cavity of fed flies. The injected solution was allowed to mix with the blood of the host for several minutes. It was then withdrawn, and the mixture of physiological solution and blood thus obtained was injected into unfed individuals.

For certain experiments it was necessary to inject sugar solution via the anus into the gut. To this end, the fly was sealed by its wings to the wax table, ventral side up, and then injected. This injection procedure proved to be rather difficult. The wall of the hindgut, close to the anus, is very thin and the needle often pierced the gut wall; this resulted in forcing the injected fluid into the abdominal cavity instead of into the gut. The success of the injection was therefore carefully checked in each case by opening the animal after testing its threshold responses. Since a colored solution was used for the injection, it could easily be determined whether or not the injected fluid was in the gut or in the abdomen. In successful cases, the hind and midgut were filled with the colored fluid (Fig. 1 C). The proventricular valve blocked the solution from passing into the proventriculus. Only such positive cases were used in the evaluation of the experiments.

In some instances sugar solution was also injected directly into the midgut through a prepared opening in the thorax, as was used for the midgut ligation experiments.

Threshold Determination • Measurements of the lowest concentration of sucrose which would elicit extension of the proboscis by the fly were made by stimulating the chemoreceptors of the tarsi with doubled concentrations of sucrose beginning with 0.001 M and presented consecutively. For this purpose all flies were affixed by the wings to paraffin-tipped applicator sticks and permitted to drink water ad libitum so that there would be no proboscis extension to water.

Results

Blood Transfusion • As EVANS and DETHIER (1957) had demonstrated, changes in the levels of the normal blood sugar, trehalose, and of various sugars which entered the blood by way of the midgut or by direct injection had no measurable effect on the threshold of response to sugars. The possibility still remained, however, that feeding caused liberation into the blood of a humoral agent which effected threshold change. To test this possibility blood was transfused from flies which had been fed two hours previously to eighty-one flies which had been starved for twenty-four hours. When the thresholds of transfused flies were measured immediately after transfusion and at intervals over a subsequent two hour period, eighty per cent of the flies had a threshold which did not differ from the previously measured starvation level. There is a possibility that the quantity of blood injected or the effect of dilution with saline might have obscured the action of a hormone, but barring this it can be concluded that feeding did not cause the liberation into the blood of any humoral substances contributing to a rise in threshold. This conclusion is strengthened by the observation that an isolated head, removed from a fully fed fly with a high threshold, will feed.

Midgut Involvement • Earlier work had eliminated the crop as the regulatory region of the alimentary canal but had not provided direct proof of the role of any other region. The two types of complementary experiments which were conducted to supply the missing information consisted of elimination of the midgut and hindgut and of loading the mid- and hindgut.

Five unfed flies, which exhibited low acceptance thresholds, received ligatures immediately behind the proventriculus (Fig. 1). A post-operative check of threshold immediately after ligation indicated that the operation had had no untoward effect on behavior. The flies were then fed to satiation on 1 M sucrose. Threshold was now measured periodically over a four-hour period. Up to three hours after feeding the threshold remained high and only then began to fall. Identical results were obtained if the flies were fed 1 M sucrose prior to ligation. Furthermore, no differences were observed between the effects of ligation alone and ligation plus complete section of the midgut. From these experiments it was concluded that the presence of sugar in the mid- and hindgut was not necessary to maintain threshold elevation for a period longer than that accounted for by sensory adaptation.

The complementary experiments consisted of loading the mid and hindgut. Of sixty-four starved flies which were given enemas of 1 M glucose, thirty-one were shown on subsequent post-mortem examination to have had the gut completely loaded from the rectum forward to the cardiac valve. While the flies were still living, periodic tests of threshold over a four hour period revealed no change in the initial low pre-operative value. In other words, loading the gut by enema did not result in a rise in threshold.

In a comparable series of experiments 2 M glucose was injected through the wall of the midgut of twenty-five starved flies, and water was similarly injected in eight flies. Again in eighty per cent of the cases there was no rise in threshold over that of the starved condition. In a sham operation the thres-

134   V. G. Dethier and Dietrich Bodenstein

hold remained the same in seventy-six per cent of the cases. The results of these experiments are in agreement with the results obtained with ligation and sectioning in showing that the presence of sugar in the mid- and hindgut does not cause a rise in threshold.

Foregut Involvement • If the gut was indeed involved in threshold regulation, the only portion remaining for consideration was the small area of foregut extending from the head to and including the proventriculus (Fig. 1). With one exception, attempts to inject sugar into this delicate region without damaging nerve tissue were unsuccessful. In the single successful case there was a thirty-fold rise in threshold. While little reliance can be placed on a single case, the result was suggestive. Accordingly, an attempt was made to approach the problem by denervating this particular region of the foregut.

Recurrent Nerve Section • An examination of Fig. 1 will show that the gut is innervated by the recurrent nerve, which with its branches and ganglia constitutes the stomatogastric system, the analogue of the vertebrate autonomic system. The recurrent nerve passes from the frontal ganglion posteriorly along the dorsal wall of the oesophagus to the vicinity of the junction of the crop duct. Here it is connected with the hypocerebral ganglion, gives off nerves to the endocrine glands, and sends fibers to the crop and proventriculus. Experiments in which the crop had been removed had demonstrated that section of this branch of the recurrent nerve was without effect as far as threshold regulation was concerned (Evans and Dethier 1957). The experiments involving removal of the midgut as reported above demonstrated that section of the proventricular branch of the recurrent nerve at a point posterior to the proventriculus was also without effect. There remained only to section the recurrent nerve anterior to the hypocerebral ganglion in the hope thereby of removing any innervation of the foregut. This difficult operation was performed successfully on twenty-five starved flies (Fig. 1 E, RN).

The results were decisive and spectacular. Section of the nerve trunk produced a complete hyperphagia. Flies began to feed in the normal fashion and ceased, as was to be expected, when the oral receptors became adapted; however, as soon as disadaptation occurred vigorous feeding was resumed. Tests revealed that the normal prolonged rise in threshold failed to appear. If a hyperphagic fly was fed on 1 M sucrose and disadaptation of the receptors allowed to occur, the fly then showed a threshold of response to 0.1 M sucrose. As a consequence of the absence of threshold rise the operated flies fed continuously as long as they lived and became enormously bloated (Fig. 2). Death finally ensued probably as a result of starvation, because in the absence of stomatogastric innervation the valves regulating the passage of food in the midgut failed to operate properly, and food never left the swollen crop.

Discussion

The picture which emerges of the feeding activity of the fly is as follows. When an unfed fly encounters food (e. g., sugar), contact chemoreceptors on the tarsi are stimulated. This sensory input results in extension of the proboscis. Extension brings the chemosensory hairs on the aboral surface of the labellum into contact with the sugar. In response to this stimulation the labellar lobes open, thus bringing the receptors (interpseudotrachael papillae) on the oral surface into contact with the sugar. Stimulation of these receptors, as well as of the labellar hairs, initiates sucking. Feeding is thus initiated and driven by input from oral receptors (cf. Dethier 1955). At the beginning of feeding the threshold of response (assuming a starved fly) is at its lowest level and, while there is as yet no direct evidence that this level equals the absolute threshold of the receptors themselves — as MINNICH (1929) and von FRISCH (1935) postulated —, it seems reasonably certain that it represents the minimal amount of *summed* sensory input that can activate the motor system under any circumstances.

As imbibition continues, its rate is determined by the intensity of sensory input (EVANS and DETHIER 1957). The greater the intensity, that is, the concentration of the stimulus, the more rapidly the muscles of the cibarial pump contract. The duration of imbibition depends upon the time required for a high level of adaptation of the oral receptors (loc. cit.), and varies, of course, with the concentration of the stimulus. For example, duration varies from 90 sec. at 2 M sucrose to 36 sec. at 0.25 M sucrose. When adaptation attains a sufficiently high level, feeding ceases. There is now no effective sensory input, and the threshold of response is high.

The sugar solution which is imbibed goes initially to both the midgut and crop. When the midgut is filled, the cardiac valve closes, and the continued influx of sugar is directed into the crop. Waves of peristalsis passing posteriorly in the crop duct constantly press the fluid back into the crop itself. When feeding has ceased, the crop duct closes. The region of the foregut now contains sugar residue from more anterior regions of the oesophagus and preoral cavity.

When the sugar in the midgut is utilized, the crop valve opens momentarily, peristalsis in the duct is reversed so that a slug of fluid is driven energetically into the foregut region, the crop valve closes, a wave of peristalsis in the foregut drives the fluid toward the proventriculus where the cardiac valve opens briefly to permit passage to the midgut.

Although disadaptation of the oral receptors has by this time set in, the threshold of response remains elevated, and feeding is blocked. Threshold elevation can only be due to impulses passing up the recurrent nerve and inhibiting sensory input from the mouth. This inhibition, as manifest by high threshold, continues as long as there is sugar in the foregut.

When the recurrent nerve is cut, feeding resumes every time disadaptation sets in so that a fly feeds almost continuously in an intermittent fashion. In these hyperphagic flies the crop valve remains open and the cardiac valve closed, except in occasional cases where extreme hydrostatic pressure apparently forces some liquid through into the midgut. The enormous distension characteristic of hyperphagic flies is reminiscent of the honey ants (*Myrmecocystus horti-deorum*) which under normal circumstances are hyperphagic. It would be interesting to speculate upon the evolutionary development of these castes in terms of possible functional changes in the stomatogastric nervous system.

The fine synchronization of valves and peristalsis of the normal fly is disrupted in the hyperphagic fly. It would appear that emptying of the midgut somehow signals the crop to deliver food to the foregut and that the cardiac valve then opens in response to some neural message which is not relayed when the recurrent nerve is cut. The complete sequence of events in this interesting region of the gut invites further investigation.

It would appear that hunger in the fly can be equated with absence of stimulating fluid in the foregut, in other words, absence of inhibition. As early as 1929 MINNICH has postulated that starvation could be conceived of as eliminating inhibition in the central nervous system. This is now proved to be the case. At the moment there is no conclusive evidence of hunger "drive" in the sense of positive input from external receptors, internal receptors, or endogenous centers within the central nervous system. As long as there is sensory input from oral or leg receptors and no recurrent inhibition, the fly feeds. If the head is removed from a fly which has fed to repletion and possesses a high acceptance threshold, the isolated head will show a low threshold and will feed until sensory adaptation occurs. This experiment confirms the conclusion that high threshold and satiety occur as a result of neural information received directly from regions posterior to the head, and that no humoral factor in the blood is necessary for this purpose.

On the other hand, low threshold alone is not sufficient to insure feeding. The fly must encounter food. Naturally its chances are enhanced if it increases its activity. Experiments being conducted in this laboratory by Evans and Barton Browne do indeed show that as the period of food deprivation increases a fly becomes increasingly active. The increase in activity is correlated with a drop in threshold. Further experiments are in progress to ascertain whether or not the regulation of threshold and of activity are under unitary control.

It is of considerable interest that the consummation of feeding is brought about, neither by the fulfillment of a metabolic need nor by the fulfillment of any motor pattern. If a fly is fed the non-metabolizable sugar fucose to repletion, its acceptance threshold remains elevated and feeding is terminated. Such a fly is "metabolically hungry" but "behaviorally satiated". It will eventually starve to death through failure to feed even in the presence of enough glucose to keep it alive.

The fulfillment of a motor pattern (e. g., swallowing) is not in itself a consummatory act. It cannot terminate feeding as seen from the fact that flies with recurrent nerve sections indulge in prolonged sucking and swallowing without a cessation of feeding ensuing. Consummation is brought about by inhibition originating somewhere in the foregut region. In other words, feeding is controlled by a sensory feedback mechanism. As soon as internal receptors are stimulated, they block at some internuncial level the flow of information from external receptors which drive feeding. The exact nature of this feedback mechanism is not yet completely understood. The receptors have not yet been identified. It is not known whether the adequate stimulus is chemical, osmotic, or mechanical.

The exact mode of operation of the feedback requires clarification. It has been shown, for example, that threshold does not remain maximally elevated as long as there is the least bit of sugar in the crop. Instead, it falls off gradually in a manner characteristic of the specific sugar imbibed (Evans and Dethier 1957). It is known, moreover, that the concentration of the crop contents remains unaltered from beginning to end. How then is the message from the recurrent nerve graded in such a way as to inhibit less and less effectively as the crop progressively empties? One possibility which remains open for investigation is that the frequency of crop delivery to the midgut varies with time after feeding and that the degree of recurrent inhibition is a function of frequency.

It was pointed out earlier that the characteristic slope of threshold decline with increasing deprivation depended upon the nature of the food which had been ingested. This fact can be explained in terms of rate of utilization of the specific food. If different carbohydrates are absorbed from the midgut at different rates, then these rates would determine the frequency

Hunger in the Blowfly   137

of demand upon the crop, hence, the rate of crop delivery to the foregut region. If the hypothesis that frequency of stimulation in the foregut determines the effectiveness of inhibition is correct, the differences in rate of threshold decline after feeding on different sugars are understandable.

In comparison with the rat the fly apparently possesses a much simpler mechanism for the regulation of feeding although more work with the fly may reveal that this simplicity is an illusion. In the rat there are two neural centers known to regulate feeding. It has been shown (see BROBECK 1946 for summary) that a bilateral lesion in the region of the ventromedial nuclei of the hypothalamus causes a pronounced increase in food consumption. Lesions in the lateral hypothalamus, on the other hand, produce hypophagia or aphagia (ANAND and BROBECK 1951: TEITELBAUM and STELLAR 1954). Thus there appears to be in the rat a "feeding" center and a "satiety" center. ANLIKER and MAYER (1957) have presented evidence in support of the contention that the latter receives information from the bloodstream via glucoreceptors. Destruction of the satiety center acts to release the feeding center from the inhibitory influence of the satiety center with resulting hyperphagia. There is yet no evidence of a feeding center in the fly if by feeding center one means a center whose endogenous activity initiates feeding. If by feeding center one means a region of internuncials which processes information received from peripheral sense organs, then the fly does possess one. On the other hand, no way has been found so far to produce experimental hypophagia or aphagia.

In the rat it seems possible to separate experimentally factors controlling food intake and those controlling hunger considered as a drive, or in other words, to separate the control of satiation from hunger drive reduction. MILLER, BAILEY, and STEVENSON (1950) were able to show that when rats were trained to press a bar to obtain food, the normals would work harder than rats (hyperphagics) with lesions in the ventral hypothalamus. Yet normally the hyperphagics ate more. Accordingly, it appears that the lesions had different effects than hunger and that the mechanism of hunger and that governing food consumption may be different.

Furthermore, MILLER and KESSEN (1956) have shown that the volume of food taken with increased deprivation increased to a maximum by six hours after which there was no further increase, that stomach contractions paralleled the intake curve, but that work performed and the amount of quinine which would be tolerated in the food continued to increase. Thus it was clear that intake leveled off but that hunger increased (up to 54 hours).

In the fly the evidence presently available suggests that the factors governing intake (satiation) and possible factors responsible for that which MILLER terms "hunger drive" are one and the same. Our experiments to date indicate that hunger is merely the absence of satiation. The usual simpler types of experiments employed to separate the factors of satiation and hunger cannot be applied successfully to the fly. For example, an increase in the viscosity of a solution makes it more difficult for a fly to imbibe, and the rate of sucking decreases (EVANS and DETHIER 1957). One might ask whether flies will work harder, that is, longer, to imbibe a viscous solution as they become hungrier. However, since feeding is driven by input from oral receptors and ceases when they adapt, the amount of solution taken in will not be a function of the degree of hunger but rather of the rate of sensory adaptation (which controls satiation).

Similarly, although both the fly and the rat will tolerate greater quantitites of an adulterant (e. g., quinine or salt) as they become hungrier, the mechanisms probably are different. In the rat the amount of quinine tolerated seems to be a measure of hunger (MILLER 1957). In the fly the amount tolerated actually reflects a change in the behavioral threshold to sugar, which has been shown to be a function of recurrent nerve inhibition, that is, the mechanism controlling satiation. The observation that insects will tolerate greater concentrations of unacceptable compounds mixed with sugar as the period of deprivation lengthens does not really indicate that rejection threshold increases with starvation. Rejection threshold is the lowest concentration of a substance which will prevent acceptance of the material (sugar) with which it is mixed. As the concentration (hence stimulating effectiveness) of sugar is increased, the amount of adulterant required to prevent response must likewise be increased (DETHIER 1955). As an insect is starved its sensitivity to sugar increases; therefore, more salt is required to prevent the response. HASLINGER (1935) measured the rejection threshold of the fly *Calliphora* for HCl during starvation by presenting HCl in fructose the concentration of which was varied so as to be just three times the threshold for fructose on each day of testing. Under these conditions no real change in rejection threshold for HCl was found, nor for salts, sugar alcohols, or quinine. For the honey bee von FRISCH (1934) found no change in salt threshold but an increase for HCl and quinine. Thus, although the relation of rejection to food deprivation is somewhat conflicting and closer scrutiny of the phenomenon is required, the data of HASLINGER strongly suggest that changes in rejection really mean changes in acceptance threshold and are, therefore, associated with the mechanism controlling satiety.

In the rat, according to MILLER (1957), there are three possible sources of drive reduction: the performance of the consummatory response, the taste of the food, the presence of food in the stomach. MILLER and KESSEN (1952) showed that the first two are certainly not the sole factors. Clearly in the fly the performance of the consummatory response is not a factor. The "taste" is, and the foregut (= gastric factor?) is only insofar as it inhibits "taste".

Changes in behavior associated with starvation must be considered as resulting from the usual stimuli which impinge upon the external receptors having a different effect in the central nervous system due to modifying influences from internal receptors, chemical or osmotic changes in the internal milieu, or endogenous activity within the central nervous system. There seems to be little point in talking about "drive" and "motivation" unless one is referring to behavior resulting from one or more of the mechanisms referred to above. And it is manifestly more informative to talk in terms of mechanisms. It is certainly true in the fly that changes in the threshold of response are attributable to the modifying (in this case, inhibitory) effect which internal stimuli ultimately produce upon peripheral sensory input to the brain. Hunger, aside from being a synonym for food deprivation, is the absence of this inhibition.

Presumably the question of the presence of spontaneous activity in the central nervous system which activates the fly to seek food and to feed may be approached further by studying feeding as a function of some deterring action which does not operate through the chemosensory system. But until the nature of the relation between threshold and general activity is clarified, no function of activity can be used as a tool to separate the sensory factors which regulate feeding from possible internal factors.

Summary

When an unfed fly encounters food, contact chemoreceptors on the tarsi are stimulated. As a result of this sensory input the proboscis is extended. Extension brings chemosensory hairs on the labellum into contact with the food. In response to this stimulation the labellar lobes open, thus bringing the receptors on the oral surface into contact with the food. Stimulation of either set of receptors on the proboscis initiates sucking.

As feeding continues its rate is determined by the intensity of oral sensory input. The duration of feeding depends upon the time required for a high level of adaptation of oral receptors to occur, and it varies with the concentration of the stimulus. When adaptation attains a sufficiently high level feeding ceases.

At the beginning of feeding the threshold of response is at its lowest level. At the cessation of feeding the threshold is very high. Even after disadaptation has set in the threshold remains elevated for many hours. The duration of threshold elevation varies with the kind of food eaten. None of the following factors are involved in threshold regulation: blood sugar level, stored glycogen depletion, crop content, midgut content, humoral agents in the blood. Threshold is regulated by information originating in the foregut and passing by way of the recurrent nerve to the brain where it inhibits the effect of sensory input from oral receptors. When the recurrent nerve is cut, inhibition no longer occurs, and feeding is continuous until the fly dies. In these hyperphagic flies the threshold of response to sugar is never elevated.

Hunger can be equated with absence of stimulating fluid in the foregut, i.e., absence of inhibitory impulses carried by the recurrent nerve. At the moment there is no conclusive evidence of hunger "drive" in the sense of positive input from external or internal receptors or from endogenous centers within the central nervous system.

Consummation of feeding is brought about neither by the fulfillment of a metabolic need nor of a motor pattern. If a fly is fed the non-metabolizable sugar fucose, its threshold of response is elevated and feeding is terminated. Such a fly is "metabolically hungry" but "behaviorally satiated". Flies with recurrent nerve sections suck and swallow indefinitely; hence, the motor pattern itself is not a consummatory act.

In the fly there is no evidence of a "feeding center" if by feeding center is meant a higher center whose endogenous activity drives feeding.

Zusammenfassung

Wenn Futter die Geschmacksorgane an den Tastern einer Fliege berührt, streckt sie ihren Rüssel aus; dadurch kommen die geschmacksempfindlichen Haare des Rüssels in Kontakt mit der Nahrung. Auf diesen Reiz hin öffnen sich die Labellen und geben die Chemoreceptoren der Mundöffnung frei. Reizung aller Sinnesorgane auf den Labellen löst das Saugen aus, dessen Geschwindigkeit vom Reizungsgrade der Mundsinnesorgane abhängt. Das Saugen hält an, bis der Adaptationsgrad der Mundsinnesorgane einen hohen Wert erreicht hat, was wiederum von der Konzentration der Nahrung abhängt. Ist der kritische Wert erreicht, dann hört die Fliege auf zu saugen.

Der Schwellenwert der Nahrungsaufnahme liegt also zu Beginn ganz tief und beim Aufhören der Reaktion ganz hoch. So bleibt er stundenlang, auch wenn die sensorische Adaptation schon abklingt. Wie lange die Schwelle

140   V. G. Dethier and Dietrich Bodenstein: Hunger in the Blowfly

hoch bleibt, das hängt mit von der Art der Nahrung ab, nicht jedoch vom
Blutzuckerspiegel, dem Abbau des Glykogenspeichers, der Füllung des
Kropfes oder des Mitteldarmes, noch auch vom Hormongehalt der Haemo-
lymphe. Die Schwellenhöhe wird vielmehr eingestellt durch Informationen im
Vorderdarm, die über den *n. recurrens* zum Oberschlundganglion gehen, wo
sie die von den Mundsinnesorganen einlaufende Erregung hemmen. Zer-
schneidet man diesen Nerv, so entfällt die Hemmung, und die Fliege saugt
ununterbrochen bis zu ihrem Tode. Die Konzentrationsschwelle für Zucker-
lösungen hebt sich bei diesen Dauertrinkern nie.

Hunger ist hier Fehlen von reizender Flüssigkeit im Vorderdarm und
somit von hemmenden Impulsen im *n. recurrens*. Es besteht bisher kein
Anlaß, den Hunger Trieb zu nennen, wenn man darunter einen positiven
Erregungsbetrag versteht, gleich ob er von äußeren oder inneren Sinnes-
organen oder einem Zentrum her einströmt.

Hier hängt das Einstellen der Nahrungsaufnahme also weder von der Be-
friedigung eines Stoffwechselbedarfs noch auch vom Ablauf einer Bewegung ab.
Wenn die Fliege unverdaulichen Zucker (Fukose) trinkt, hebt sich der Schwellenwert genau so wie bei verdaulicher Kost, und sie saugt nicht weiter; sie ist
stoffwechsel-physiologisch hungrig, aber verhaltensmäßig satt. Die Fliege mit
zerschnittenem rücklaufendem Nerv saugt und schluckt, bis sie platzt; also ist
diese Bewegungsweise auch keine triebverzehrende Endhandlung.

Auch für ein „Saugzentrum" spricht nichts, wenn man sich darunter
ein höheres Nervenzentrum vorstellt, dessen endogene Aktivität zum Saugen
antreibt.

References

Anand, B. K., and J. R. Brobeck (1951): Localization of a "feeding center" in the
hypothalamus of the rat. Proc. Soc. Exp. Biol. and Med., 77, 323—324 • Anderson, A. L.
(1932): The sensitivity of the legs of common butterflies to sugars. J. exp. Zool., 63,
235—259 • Anliker, J., and J. Mayer (1957): The regulation of food intake. Some experiments relating behavioral, metabolic and morphologic aspects. In: Symposium on Nutrition
and Behavior. Nutrition Symposium Series No. 14, 46—51, The National Vitamin Foundation, Inc. N. Y. • Bodenstein, D.: The postembryonic development of Drosophila. In
"Biology of Drosophila", pp. 275—367 (ed. M. Demerec). John Wiley and Son, N. Y 1950 •
Brobeck, J. R. (1946): Mechanism of the development of obesity in animals with hypothalamic lesions. Physiol. Revs., 26, 541—559 • Dethier, V.G. (1955): The physiology
and histology of the contract chemoreceptors of the blowfly. Quart. Rev. Biol., 30, 348—371 •
Dethier, V. G. and M. V. Rhoades (1954): Sugar preference aversion functions for the
blowfly. J. exper. Zool. 126, 177—204 • Dethier, V. G., D. R. Evans and M. V. Rhoades
(1956): Some factors controlling the ingestions of carbohydrates by the blowfly. Biol. Bull.
Woods Hole, 111, 204—222. • Evans, D. R., and V. G. Dethier (1957): The
regulation of taste thresholds for sugars in the blowfly. J. Insect Physiol., 1, 3—17 •
von Frisch, K. (1927): Versuche über den Geschmackssinn der Bienen. Naturwiss., 15,
321—327; (1935): Über den Geschmackssinn der Biene. Z. vgl. Phys. 21, 1—156 • Haslinger,
F. (1935): Über den Geschmackssinn von Calliphora erythrocephala Meigen und über die
Verwertung von Zuckern und Zuckeralkoholen durch diese Fliege. Z. vgl. Phys., 22,
614—639 • Hudson, A. (1958): The effect of flight on the taste threshold and carbohydrate
utilization of Phormia regina Meigen. J. Insect Physiol., (in press) • Kunze, G. (1927):
Einige Versuche über den Geschmackssinn der Honigbiene Zool. Jb. Abt. Allg. Zool. u.
Physiol., 44, 287—314 • Miller, N. E. (1957): Experiments on motivation. Science, 126,
1271—1278 • Miller, N. E., C. J. Bailey, J. A. F. Stevenson (1950): Decreased "hunger"
but increased food intake resulting from hypothalamic lesions. Science, 112, 256—259 •
Miller, N. E., and G. L. Kessen (1952): Reward effects of food via stomach fistula compared with those of food via mouth. J. Comp. Physiol. Psychol., 45, 555—564 • Minnich,
D. E. (1922a): The chemical sensitivity of the tarsi of the red admiral butterfly, Pyrameis
atalanta L. J. exp. Zool. 35, 57—81; (1922b): A quantitative study of tarsal sensitivity to
solutions of saccharose, in the red admiral butterfly, Pyrameis atalanta L. J. exp. Zool. 36,
445—457; (1929): The chemical sensitivity of the legs of the blowfly, Calliphora vomitoria L.,
to various sugars. Z. vgl. Phys. 11, 1—55 • Teitelbaum, P., and E. Stellar (1954): Recovery
from the failure to eat produced by hypothalamic lesions. Science, 120, 894—895.

ENERGY MODELS OF MOTIVATION

By R. A. HINDE

Field Station for the Study of Animal Behaviour, Cambridge University
Department of Zoology

INTRODUCTION

The problem of motivation is central to the understanding of behaviour. Why, in the absence of learning and fatigue, does the response to a constant stimulus change from time to time? To what is the apparent spontaneity of behaviour due? This paper is concerned with one type of model which has been developed to help answer such questions—namely that in which changes in the organism's activity are ascribed to changes in the quantity or distribution of an entity comparable to physical, chemical or electrical energy.

Such models have been developed by theoreticians with widely differing backgrounds, interests and aims, and the frameworks of ideas built round them diverge in many respects; but in each case the energy treatment of motivation is a central theme (cf. Carthy, 1951; Kennedy, 1954). They have had a great influence on psychological thought, and although they are unlikely to continue to be useful, it is instructive to examine their nature, their achievements and their limitations.

THE MODELS

The four models or theories to be discussed here are those of Freud, McDougall, Lorenz and Tinbergen. They are only four of many in which energy concepts are used, but in them the energy analogy is made explicit in terms of a mechanical model, instead of being merely implied by a 'drive' variable which is supposed to energize behaviour. The models were designed to account for many features of behaviour in addition to the phenomena of motivation, and here it will be necessary to extract only those aspects relevant to the present theme.

In the psycho-analytic model (Freud, 1932, 1940) the id is pictured as a chaos of instinctive energies which are supposed to originate from some source of stimulation within the body. Their control is in the hands of the ego, which permits, postpones or denies their satisfaction. In this the ego may be dominated by the super-ego. The energy with which Freud was particularly concerned—the sexual energy or libido—is supposed not to require immediate discharge. It can be postponed, repressed, sublimated, and so on. The source of this energy lies in different erogenous zones as the individual develops, being successively oral, anal and phallic, and it is in relation to these changes that the individual develops his responses to the external world.

200   ENERGY MODELS OF MOTIVATION

The instinctual energy is supposed to undergo various vicissitudes, discussions of which often imply that it can be stored, or that it can flow like a fluid. It may become attached to objects represented by mental structures or processes (libidinal cathexes) and later withdrawn from them in a manner that Freud (1940) likened to protoplasmic pseudopodia: it has also been compared with an electric charge. Thus some of the characteristics of the energy depend on its quantitative distribution.

McDougall (1913) envisaged energy liberated on the afferent side of the nervous system, and held back by 'sluice gates'. If the stimuli necessary to open the gates are not forthcoming, the energy 'bubbles over' among the motor mechanisms to produce appetitive behaviour. On receipt of appropriate stimuli, one of the gates opens, and the afferent channels of this instinct become the principal outlet for all available free energy. Later (1923) he used a rather more complex analogy in which each instinct was pictured as a chamber in which gas is constantly liberated. The gas can escape via pipes leading to the executive organs when the appropriate lock(s) is opened. The gas is supposed to drive the motor mechanisms, just as an electric motor is driven by electrical energy.

The models of Lorenz and Tinbergen have much in common with McDougall's. Lorenz's 'reaction specific energy' was earlier (1937) thought of as a gas constantly being pumped into a container, and later (e.g. 1950) as a liquid in a reservoir. In the latter case it is supposed that the reservoir can discharge through a spring-loaded valve at the bottom. The valve is opened in part by the hydrostatic pressure in the reservoir, and in part by a weight on a scale pan which represents the external stimulus. As the reservoir discharges, the hydrostatic pressure on the valve decreases, and thus a greater weight is necessary to open the valve again.

Tinbergen (1951) pictured a heirarchy of nervous centres, each of which has the properties of a Lorenzian reservoir. Each centre can be loaded with 'motivational impulses' from a superordinated centre and/or other sources. Until the appropriate stimulus is given the outflow is blocked and the animal can show only appetitive behaviour: when the block is removed the impulses can flow into the subordinate centre or be discharged in action.

It is important to emphasize again that the theories of these authors have little in common except for the energy model of motivation—they were devised for different purposes, and the more recent authors have been at pains to emphasize their differences from the earlier ones. For instance, for McDougall the most important feature of instinct was the 'conative-affective core', while for Lorenz it was the stereotyped motor pattern. Furthermore, the models differ greatly in the precision with which they are defined. The Freudian model is a loose one: its flexibility is perhaps necessary in view of the great range of behavioural and mental phenomena it comprehends, but makes it very difficult to test. The other models are more tightly defined, but differ, as we shall see, in their supposed relations to the nervous system.

In spite of such differences, all these models share the idea of a substance, capable of energizing behaviour, held back in a container and subsequently released in action.† In the discussion which follows, I shall be little concerned with the other details of the models, or with the ways in which the theories based on them differ. Furthermore, I shall disregard the niceties of terminology, lumping instinctual energy, psychophysical energy, action specific energy and motivational impulses together as, for present purposes, basically similar concepts.

REALITY STATUS OF THE MODELS

Until recently, students of the more complex types of behaviour could get little help from physiology, and had to fashion their concepts without reference to the properties of the nervous system. Many, indeed, advocated this course from preference, either on grounds of expediency, suggesting that knowledge of the nervous system was still too primitive and might be misleading, or on principle, claiming that physiology and behaviour were distinct levels of discourse. At present the models and theories used in attempts to understand, explain or predict behaviour range from those whose nature is such that physiological data are irrelevant (Skinner, 1938) to those which consist of a forthright attempt to describe psychological data in physiological terms (Hebb, 1947, 1955). The former type may be applicable over a wide range of phenomena, but only at a limited range of analytical levels: the latter may point the way to analysis at lower levels, but their expectancy of life depends on their compatibility with the phenomena found there.

The originators of all the models discussed here regard them as having some relation to structures in the nervous system, but vary in the emphasis which they lay on this. Tinbergen, although freely emphasizing the hypothetical status of his model, clearly regards his 'centres' as neural structures, and his 'motivational impulses' as related to nerve impulses. He speaks of his hierarchical scheme as a 'graphic picture of the nervous mechanisms involved'. McDougall likewise regards the relationship between model and nervous system as a close one, for he localizes the 'sluice gates' in the optic thalamus. Lorenz, on the other hand, usually treated his model in an 'as if' fashion—he did not suggest that we should look for reservoirs in the body. He did, however, bring forward physiological evidence in its support—quoting, for instance, Sherrington's work (1906) on spinal contrast, and von Holst's (1936) work on endogenous rhythms in fishes; and he sometimes uses such terms as 'central nervous impulse flow' as a synonym for 'reaction specific energy'. His use of physiological evidence was, however, *post hoc*—the model was based on behavioural data and the physiological evidence came later.

---

† It will be clear that in some respects the postulated entity has the properties of a material substance, rather than energy. However it is on the 'energy' properties of flowing and 'doing work' that the models primarily depend.

202   ENERGY MODELS OF MOTIVATION

Freud's model developed from physiology, in particular from a sensory-excitation–motor-discharge picture of nervous function, and its basic postulates are almost a direct translation of such ideas into psychological terms—excitation into mental energy, the discharges of excitation into pleasure, and so on (Peters, 1958). However, Freudian theory developed far beyond these primitive notions, and then bore little or no relation to physiology, even though the instincts were supposed to have an ultimately physiological source.

Thus two of these models (Tinbergen and McDougall) had explicitly physiological implications; that of Lorenz was usually used in an 'as if' fashion; and Freud's, although it had physiological roots, became divorced from any supposed structures or functions in the nervous system. However, as we shall see, all have been influenced by the covert introduction of existence postulates concerning the explanatory concepts used.

THE RELATION OF BEHAVIOURAL ENERGY
TO PHYSICAL ENERGY

In these theories the concept of energy, earlier acquired by the physical sciences from everyday observation of behaviour, is reclaimed for use in its original context. In accounting for the organism's changing responsiveness, the theorist is concerned with its capacity for doing work and an energy concept seems an obvious choice. The use of such a concept, however, brings with it the temptation of ascribing to the behavioural energy the various properties of physical energy. Thus it may be said to flow from point to point, or to exist in more than one form (bound or free, in Freudian theory). It is of fundamental importance to the theorist to recognize that such properties are additional postulates in the behaviour theory: because behavioural energy is postulated to account for the activity of organisms, it *need* share no other properties with the energy postulated to account for the movement of matter. The distinction is particularly important in that students of behaviour, while using a concept of behavioural energy to explain the behaviour of the whole animal, may simultaneously be concerned in establishing bridgeheads with physiologists, who use energy in a manner closely similar to the physicists.

Freud ascribed many of the properties of physical energy to mental energy, which was stored, flowed, discharged, and so on, but he did recognize the importance of distinguishing between them. Thus he wrote (1940) 'We have no data which enable us to come nearer to a knowledge of it [mental energy] by analogy with other forms of energy'. The use of the phrase 'other forms' is revealing, and finds an echo in the work of the psychoanalyst Colby (1955), who discussed this question in detail and elaborated an even more complex energy model. Colby regards mental energy as a postulated form of energy in addition to mechanical, thermal, electrical and chemical, and states that it 'does not disobey' the principles formulated for other forms of energy, though conversion into these other forms is not 'yet' possible. It appears that for him psychic energy would be expected to obey the Laws of Thermodynamics but for the fact that organisms are open systems. Colby, however, is clearly ambivalent on this issue, for elsewhere he emphasizes that psychic energy is not mechanical, thermal, chemical or electric, and writes 'Perhaps we have no right to speak of energy at all'. Another analyst (Kubie, 1947), with a highly sophisticated attitude to energy concepts, draws a sharp distinction between psychic and physical energies, although he does invest them with similar properties:

'It is therefore scientifically necessary to keep clearly in mind the fact that the psychodynamics dealt with in psychoanalysis refer to something which is loosely analogous to, but still very far from, the exacter field of thermodynamics. These psychodynamics deal with an effort to estimate (a) the sources of energy, (b) the kinds and quantities of energy, (c) the transformations of energy, one into another, and (d) the distributions of energy. But the "energy" referred to here means not what is intended by the physicist, but *simply apparent intensities of feelings and impulses, or in psychoanalytic terms "the libido".*'

McDougall was less reserved than Freud on this issue, and clearly regarded his 'psycho-physical energy' as a form of physical energy. 'We are naturally inclined to suppose that it is a case of conversion of potential energy, stored in the tissues in chemical form, into the free or active form, kinetic or electric or what not; and probably this view is correct.' He further suggested that there is a positive correlation between the flow of energy and the 'felt strength' of the impulse.

With Lorenz and Tinbergen this question was concealed. Lorenz's reaction specific energy, pictured as a liquid, was clearly only distantly related to physical energy. Since the model was primarily an 'as if' one, the question of conversion to physical energy did not arise. Tinbergen's motivational impulses, although supposedly related to physical energy exchanges in the nervous system, were not regarded as physical energy sums themselves. In spite of this, a number of properties of physical energy came to be ascribed to them—thus they could be 'discharged in action,' 'stored,' 'released,' 'flow,' 'spark over' and so on (Hinde, 1956). This has undoubtedly influenced the course of research (see below, p. 206).

Thus although, in the four schemes at present under consideration, the relation between behavioural energy and physical energy was not particularly close, some properties of the latter insinuated themselves almost unnoticed into the behavioural theories. Physical and behavioural energy have in fact often been confused in theories of motivation: for instance, Brown (1953) uses as evidence for an energizing function of drives the 'marked disproportionality between the energy content of a stimulus and the energy expended in response'.

The transposition of properties from physical to behavioural energy could be helpful, suggesting new questions which open further avenues of research or co-ordinating previously unrelated facts. However, their presence is a danger, and likely to lead to sterile endeavour, if their nature is not recognized, and if they are introduced not as stated postulates but as known properties of physical energy which are therefore without further thought ascribed to a hypothetical behavioural energy. Some examples of confusions which have arisen in this way are discussed below.

NUMBER OF FORMS OF ENERGY POSTULATED

An important issue in these models is the number of forms of energy postulated. The behaviour of an organism is diverse: is a different form of energy to be postulated for each type of behaviour, or is there only one form producing behaviour differing according to the structure within which it acts?

Freud, in an early model, postulated two basic types of energy—sexual energy (libido) and energy pertaining to the self-preservative instincts.† Later, sexual and self-preservative were grouped together as 'Life' instincts, in contrast to the 'Death' instincts, which become manifest when directed outwards as the instinct of destruction.‡ Within each major group of instincts are recognized component instincts differing in their source, aim, object and in their quantitative distribution: but it is not always clear whether these differences are thought to lie in the nature of the energy, or in the structure within which the energy acts. Sometimes Freud stated that the instincts differ primarily because of the differing quantities of excitation accompanying them (Freud, 1915), but later he often implied differences of quality in the energy itself (Freud, 1924). Colby (1955), who, as we have seen, associated the concept of behavioural energy closely with physical energy, emphasized that all energy must be neutral, its 'aim', etc. being acquired only when it acts through a structure.

McDougall is ambivalent on the number of sources of energy, stating the possibilities on the one hand that the instincts each have their own energy, and on the other that they all draw on a common supply: he inclines towards the second view.

Lonrenz is concerned primarily with limited sequences of behaviour, and not with a synthesizing model of the behaviour of the whole organism. It was therefore sufficient for him to talk about action specific energy. From

---

† If the difference between the energies of the sexual and self-preservative instincts is one of kind, then the dichotomy is distasteful to biologists, for there is no reason for supposing that different examples of sexual (in a broad sense) behaviour on the one hand, and self-preservative behaviour on the other, classified together on functional grounds, have anything causally in common. Lloyd Morgan (1912) made a similar dichotomy in his top level instincts of self-preservation and race maintenance, which also involves a confusion of functional and causal categories.

‡ There remains, by implication, a third type of energy—Ego—which may oppose both sexual and destructive instincts. Freud thought of the Ego as deriving energy from the Id, but there are divergent views on this. Indeed there are numerous variants on this theme—some psychoanalysts, for example, postulate a neutral energy.

his earlier writings it was not clear whether the specificity of this energy was supposed to be due to its nature or to the structure within which it acts, but later, because of the occurrence of displacement activities (see below, p. 206), he concluded that the specificity was due to the structure. Furthermore, Lorenz did not suppose that the reservoirs for functionally related activities were fed from a common source: he (1937) emphatically opposed McDougall's view of superordinated instincts which employ motor mechanisms as means to an end, regarding such instincts merely as functional categories. Rather Lorenz emphasized the individuality of each type of response, and ascribes to the external situation the integration of discrete responses into functional units.

Tinbergen's scheme differs from that of Lorenz in this matter, for the motivational impulses were supposed to descend the hierarchical system of nervous centres: each such system constituted an 'instinct'. The impulses were thus general at least to all the activities of one hierarchical system, and, since he regarded (1952) 'sparking over' from one system to another as possible, perhaps to all. Since Tinbergen suggests that each activity is supplied by motivational impulses both from the superordinated centre and from its own specific source, his scheme combines features from those of McDougall and Lorenz.

The importance of this question for energy theorists is emphasized by Thorpe (1956). Most behaviour is directive in the sense that variable means are used to a constant end. If all behaviour depends on one type of energy, then the directiveness must be a consequence of the structure in which the energy acts: since the motor patterns used to achieve a certain goal may be diverse, this seems to demand a fantastic complexity of structure. Thorpe therefore prefers to think of there being an element of directiveness in the drive itself—and thus prefers Lorenz's model, in which the energy is specific to the action, to Tinbergen's scheme, in which the motivational impulses flow down a hierarchical scheme containing a limited number of channels.

UTILITY AND LEVEL OF APPLICABILITY OF ENERGY MODELS

These energy models of motivation were developed with a minimum of reference to physiological data—they were intended for the understanding or prediction of behaviour from behavioural data. It is therefore on this level that, in the first instance, they must be assessed. It has been said that it does not help to ascribe feeding behaviour to a feeding drive, or to feeding specific energy, any more than it helps to postulate a locomotive force to explain the progress of a railway engine. This type of criticism is based on a misunderstanding. Although the mere postulation of a locomotive force may be of little use, there is a level—that of classical dynamics—in which the language of forces and so on helps a great deal in predicting the behaviour of railway engines. We can say, for instance, what will happen if the engine meets a stationary truck on the line (see also McDougall, 1923).

In a similar way, energy models have been surprisingly successful. The Freudian energy model not only accounts for the more general properties of motivated behaviour, such as its apparent spontaneity and persistence, but also for the manner in which instincts can change their aim (displacement in the psychoanalytical, not the ethological, sense) and the way in which component instincts can replace each other. Similarly, Lorenz's reservoir analogy and Tinbergen's hierarchy comprehend the relation between the threshold of stimulation necessary to elicit a response and the time since it was last released; the occurrence of appetitive behaviour, responses to normally inadequate stimuli, and ultimately vacuum activities (i.e. responses in the absence of the appropriate stimuli) if the releasing stimuli are withheld; the variations in intensity at which instinctive activities appear; the initial 'warming up' phase and the after-response shown by many responses, and many other aspects of changing responsiveness. Further, differences between the characteristics of response patterns can be related in an 'as if' fashion to differences in the dimensions of the reservoir. These models are thus of value in illustrating diverse properties of behaviour in a simple manner, and can be used for explanation and exposition. In addition, the analytical study of behaviour forms only a first stage in its understanding—the products of the analysis must be re-synthesized so that the relations between them can be understood: for this such models can be an invaluable aid.

However, we have seen that some of the models purport to go further than this—they are not just 'as if' models of the mechanisms underlying the behaviour, but representations of those mechanisms themselves. They must thus be assessed also by their compatibility with our knowledge of the nervous system. Indeed even the 'as if' models must ultimately be assessed in this way, for only if the model is close to the original will the questions it poses be relevant, and only then will it continue to be of service as analysis of the original proceeds. To take an example from physics, the ray theory of light, used originally for explanations of shadow-casting, etc., suggested questions (e.g. 'What is it that travels?') which paved the way for the corpuscular and wave theories. Although the latter is essential in some contexts (such as the explanation of diffraction), the ray theory retains its usefulness, for a treatment of shadow-casting in terms of the wave theory would be unnecessarily clumsy. Further, the two theories remain compatible with, and translatable into, each other (Toulmin, 1953). For similar reasons, it is important to assess these energy models of motivation not only at the behavioural level, but also in terms of their compatibility with lower ones. Although a model must not resemble the original too closely, or it will lose just those properties of simplicity and manipulability which makes it useful, it must approximate to it, or the questions it suggests will be irrelevant.

DIFFICULTIES AND DANGERS

In the following paragraphs we shall consider some of the difficulties and dangers inherent in the use of an energy model of motivation. These arise in part from misunderstandings of the nature of the model, and in part from incompatibilities between the properties of the model and those of the original.

(i) *Confusion between behavioural energy and physical energy*

We have already seen that behavioural energy, postulated to account for changes in activity, need share no properties with physical energy. Not only is there no necessary reason why it should be treated as an entity with any of the properties of physical energy, but the question of its convertibility into physical energy is a dangerous red herring. The way in which the properties of the model may be confused with those of the original have been discussed for Freudian theory by Meehl & McCorquodale (1948). Concepts like libido or super-ego may be introduced initially as intervening variables without material properties, but such properties have a way of creeping into discussion without being made explicit. Thus Meehl & McCorquodale point out that libido may be introduced as a term for the 'set of sexual needs' or 'basic strivings,' but subsequently puzzling phenomena are explained in terms of properties of libido, such that it flows, is dammed up, converted, regresses to earlier channels, and so on. Such properties are introduced surreptitiously as occasion demands, and involve a transition from admissible intervening variables, which carry no existence postulates, to hypothetical constructs which require the existence of decidedly improbable entities and processes.

Such difficulties are especially likely to occur when a model which purports to be close to the original, like that of Tinbergen, develops out of an 'as if' model, like that of Lorenz. This case has been discussed elsewhere (Hinde, 1956). To quote but one example, ethologists have called behaviour patterns which appear out of their functional context 'displacement activities'. These activities usually appear when there is reason to think that one or more types of motivation are strong, but unable to find expression in action: instead, the animal shows a displacement activity, which seems to be irrelevant. Thus when a chaffinch has conflicting tendencies to approach and avoid a food dish, it may show preening behaviour. Such irrelevant activities were explained on the energy model by supposing that the thwarted energy 'sparked over' into the displacement activity—sparking over being a property of (electrical) energy which was imputed to the behavioural energy. This idea hindered an analytical study of the causal factors underlying displacement behaviour. Thus it has recently become apparent that many displacement activities are not so causally irrelevant as they appear to be, for those factors which elicit the behaviour in its normal functional context are also present when it appears

208   ENERGY MODELS OF MOTIVATION

as a displacement activity. For example, some displacement activities appear to be due to autonomic activity aroused as a consequence of fear-provoking stimuli or other aspects of the situation. The displacement activity may consist of the autonomic response itself (e.g. feather postures in birds) or of a somatic response to stimuli consequent upon autonomic activity (Andrew, 1956; Morris, 1956). In other cases the displacement behaviour consists of a response to factors continuously present, which was previously inhibited through the greater priority of the incompatible behaviour patterns which are in conflict (van Iersel & Bol, 1958; Rowell, 1959). Of course it remains possible that the intensity of the apparently irrelevant behaviour is influenced by factors not specific to it, including those associated with the conflicting tendencies (see also Hinde, 1959).

Similarly in psychoanalytic theory we find not only that within one category of instincts (e.g. sexual) the constituent instincts can change their aim, but also that 'they can replace one another—the energy of one instinct passing over to another' (Freud, 1940). Explanations of this type may be useful at a descriptive level, but are misleading as analysis proceeds.

(ii) *The distinction between the accumulation and release of energy*

In all energy models, the energy is supposed to build up and subsequently to be released in action. McDougall, Lorenz and Tinbergen, all of whom were influenced by Wallace Craig, compare the releasing stimulus to a key which opens a lock. This apparent dichotomy between releasing and motivating effects is a property of the model, and may not be relevant to the mechanisms underlying behaviour. Although many factors appear to have both a motivating and a releasing effect on the responses they affect—they appear both to cause an increase in responsiveness, and to elicit the response—this does not necessarily imply that two distinct processes are at work. For example, if a given input increased the probability of a certain pattern of neural firing, it might appear in behaviour both that the responsiveness was increased and that the behaviour was elicited.

This sort of difficulty is the more likely to arise, the more precisely the model is portrayed. Thus McDougall, who did not work out his model in such detail as Lorenz and Tinbergen, implied that motivation and release were in fact one process when he wrote 'The evoking of the instinctive action, the opening of the door of the instinct on perception of its specific object, increases the urgency of the appetite'.

(iii) *Implications about the cessation of activity*

In all these theories, the cessation of activity is ascribed to the discharge of energy—the behavioural energy flows away as a consequence of performance. Influenced by the analogy with physical energy, Freud held that the main function of the nervous system is to reduce excitation to the lowest possible

ENERGY MODELS OF MOTIVATION   209

level. McDougall, Lorenz and Tinbergen imply a similar view, and the two latter emphasize that it is the performance of more or less stereotyped motor patterns which involve the discharge of the energy.†

This view of the cessation of activities comes naturally from models in which the properties of physical energy are imputed to behavioural energy. It is, however, also supported by another type of argument, also involving a *non sequitur*. Much behaviour is related to an increase in stimulation. Therefore, it might be argued, all activity is due to an increase in stimulation, and cessation of activity is related to a decrease. On an energy model, stimulation may increase the energy, and thus decrease in activity is related to a decrease in energy.

Such a view is incompatible with the data now available on two grounds. First, cessation of activity may be due to the animal encountering a 'goal' stimulus situation, and not to the performance of an activity. If this goal stimulus situation is encountered abnormally early, the behaviour which normally leads to it may not appear at all. McDougall recognized this, and indeed defined his instincts in terms of the goals which brought about a cessation of activity. This, however, made it necessary for him to be rather inconsistent about his energy model. While the energy was supposed to drive the motor mechanisms, it was apparently not consumed in action, but could flow back to other reservoirs or to the general source. The more precisely described Lorenz/Tinbergen models, on the other hand, do not allow for reduction in activity by consummatory stimuli: reduction in responsiveness occurs only through the discharge of energy in action. These models are misleading because they are too simple—energy flow is supposed to control not only what happens between stimulus and response, but also the drop in responsiveness when the response is given. In practice, these may be due to quite different aspects of the mechanisms underlying behaviour: for instance the energy model leaves no room for inhibition (Kennedy, 1954). Further, even if the cessation of activity is in some sense due to the performance, many different processes may be involved: the mechanism is not a unitary one, as the energy model implies (see below).

Secondly, if activity is due to the accumulation of energy and cessation to its discharge, the organisms should come to rest when the energy level is minimal. In fact, much behaviour serves the function of bringing the animal into conditions of increased stimulation. This has been shown dramatically with

---

† In doing so they did not imply that behavioural energy is converted into physical energy—thus Tinbergen (1952) suggests that even sleep is an activity, in which, presumably, behavioural energy is discharged.

The view that a fall in responsiveness is normally due to the performance of a stereotyped activity is not a necessary consequence of the use of energy models—Freud and McDougall did not suggest that energy could be discharged only in this way—but their use makes such errors more likely. Lorenz and Tinbergen were apparently also influenced by over-generalizing from the observation that performance of some stereotyped activities leads to a fall in responsiveness, to the conclusion that all falls in responsiveness are due to such activities.

humans subjected to acute sensory deprivation—the experimental conditions are intolerable in spite of the considerable financial reward offered (Bexton, Heron & Scott, 1954). Energy theories are in difficulty over accounting for such 'reactions to deficit' (Lashley, 1938).

(iv) *Unitary nature of explanation*

In these energy models, each type of behaviour is related to the flow of energy. Increase in strength of the behaviour is due to an increased flow of energy, decrease to a diminished flow. The strength of behaviour is thus related to a single mechanism. It is, however, apparent that changes in responsiveness to a constant stimulus may be due to many different processes in the nervous system and in the body as a whole—for instance, the changes consequent upon performance may affect one response or many, may or may not be specific to the stimulus, and may have recovery periods varying from seconds to months. Energy models, by lumping together diverse processes which affect the strength of behaviour, can lead to an over-simplification of the mechanisms underlying it, and distract attention from the complexities of the behaviour itself. Similarly, energy models are in difficulty with the almost cyclic short-term waxing and waning of such activities as the response of chaffinches to owls, the song of many birds, and so on.

Kubie (1947) has emphasized this point with reference to the psychoanalytic model. Changes in behaviour are referred to quantitative changes in energy distribution, but in fact so many variables are involved (repression, displacement, substitution, etc.) that it is not justifiable to make easy guesses about what varied to produce a given state. Similar difficulties in relation to other models have been discussed by Hinde (1959).

Precht (1952) has elaborated the Lorenzian model to allow for some complication of this sort. Analysing the changes in strength of the hunting behaviour of spiders, he distinguishes between 'drive' which depends on deprivation, and 'excitatory level', which is a function of non-release of the eating pattern. The distinction is an important one, but it may be doubted whether the elaborate hydraulic system which he produced is really an aid to further analysis.

Tinbergen's model translated the Lorenzian reservoir into nervous 'centres'. Changes in response strength are ascribed to the loading of these centres. Now for many types of behaviour it is indeed possible to identify *loci* in the diencephalon whose ablation leads to the disappearance of the behaviour, whose stimulation leads to its elicitation, and where hormones or solutions produce appropriate effects on behaviour. There is, however, no evidence that 'energy' is accumulated in such centres, nor that response strength depends solely on their state. Indeed the strength of any response depends on many structures, neural and non-neural, and there is no character by character correspondence between such postulated centres and any structure in the brain.

Although the greatest attraction of these energy models is their simplicity—a relatively simple mechanical model accounting for diverse properties of behaviour—there is a danger in this, for one property of the model may correspond to more than one character of the original. This difficulty has in fact arisen in many behaviour systems irrespective of whether they use an energy model of motivation. Thus a single drive variable is sometimes used not only with reference to changes in responsiveness to a constant stimulus, but also to spontaneity, temporal persistence of the effects of the stimuli, after-responses (i.e. the persistence of activities after the stimulus is removed), the temporal grouping of functionally related activities, and so on. As discussed elsewhere (Hinde, 1959), there is no *a priori* reason why these diverse characters of behaviour should depend on a single feature of the underlying mechanism: an over-simple model may hinder analysis.

(v) *Independence of activities*

Another difficulty which arises from the use of energy models, though by no means peculiar to them, is due to the emphasis laid on the independence of different activities. Lorenz & Tinbergen (1938) write 'If ever we may say that only part of an organism is involved in a reaction, it may confidently be said of instinctive action'. Activities are interpreted as due to energies acting in specific structures, and not as responses of the organisms as a whole. Both types of attitude carry disadvantages, but an over-emphasis on the independence of activities leads to a neglect of, for instance, sensory, metabolic or temperamental factors which affect many activities.

IS AN ENERGY CONCEPT NECESSARY?

We have seen that these energy models will account for diverse properties of behaviour, but that they meet with serious difficulties when the behaviour is analysed more closely. They have also been strangely sterile in leading to bridgeheads with physiology. These shortcomings of energy models have been emphasized by a number of other writers (e.g. Kubie, 1947; Deutsch, 1953; Bowlby, 1958). Energy concepts are useful in descriptions of changes in behaviour, but are they necessary? Colby states that 'a dynamic psychology must conceive of psychic activities as the product of forces, and forces involve energy sums. It is thus quite necessary that metapsychology have some sort of energy theory'. Is this really so?

Kubie (1947) has pointed out that psychological phenomena are the product of an interplay of diverse factors. A rearrangement of these factors can alter the pattern of behaviour without any change in hypothetical stores of energy. Such a view is in harmony with the known facts about the functioning of the nervous system. The central nervous system is not normally inert, having to be prodded into activity by specific stimuli external to it. Rather it is in a state of continuous activity—a state supported primarily by the non-specific effects of stimuli acting through the brainstem reticular system. Factors such as stimuli and hormones which affect specific patterns of behaviour are to be thought of as controlling this activity, of increasing the probability of one pattern rather than another. Changes in strength or threshold can thus be thought of as changes in the probability of one pattern of activity rather than another, and not as changes in the level of energy in a specific neural mechanism. This involves some return to a 'telephone exchange' theory of behaviour, but with emphasis on the non-specific input necessary to keep the switch mechanism active, and with switches which are not all-or-none, but determine the probability of one pattern rather than another. Furthermore, switching does not depend solely on external stimuli—i.e. we are not concerned with a purely reflexological model. This is not the place to pursue this view further: it suffices to say that it seems possible and preferable to formulate behaviour theories in which concepts of energy, and of drives which energize behaviour, have no role.

SUMMARY

1. Phenomena of motivation have often been explained in terms of an energy model.

2. The energy models used by Freud, McDougall, Lorenz and Tinbergen are outlined briefly.

3. The extent to which these models are considered by their authors to correspond with structures in the nervous system is discussed.

4. The relation between physical energy and the postulated behavioural energies are examined.

5. The number of forms of energy postulated by each author is discussed.

6. These models have had considerable success in discussions of the behaviour of the whole animal.

7. They have, however, certain grave disadvantages. In particular, these arise from a confusion between the properties of physical and behavioural energy, and from attempts to explain multiple processes in terms of simple unitary mechanisms.

8. It seems doubtful whether an energy concept is in fact necessary at all.

ACKNOWLEDGEMENTS

I am grateful to Drs J. W. L. Beament, John Bowlby, Charles Kaufman and W. H. Thorpe for their comments on the manuscript.

REFERENCES

ANDREW, R. J. (1956). Some remarks on conflict situations, with special reference to *Emberiza* spp. *Brit. J. Anim. Behav.* **4**, 41–45.

BEXTON, W. H., HERON, W. & SCOTT, T. H. (1954). Effects of decreased variation in the sensory environment. *Canad. J. Psychol.* **8**, 70–76.

ENERGY MODELS OF MOTIVATION 213

Bowlby, J. (1958). The nature of the child's tie to his mother. *Internat. J. Psychoanalysis*, 39.

Brown, J. S. (1953). Problems presented by the concept of acquired drives. In *Current Theory and Research in Motivation*, a symposium. Univ. of Nebraska.

Carthy, J. D. (1951). Instinct. *New Biology*, 10, 95–105.

Colby, K. M. (1955). *Energy and Structure in Psychoanalysis*. New York: Ronald Press Co.

Deutsch, J. A. (1953). A new type of behaviour theory. *Brit. J. Psychol.* 44, 304–317.

Freud, S. (1915). Instincts and their vicissitudes. *Collected Papers*, Vol. IV.

Freud, S. (1923). *The Ego and the Id*. London: Hogarth Press, 1947.

Freud, S. (1924). The economic problem in masochism. *Collected Papers*, Vol. II, XXII.

Freud, S. (1932). *New Introductory Lectures on Psychoanalysis*. London: Hogarth Press, 1946.

Freud, S. (1940). *An Outline of Psychoanalysis*. New York. London: Hogarth Press, 1949.

Hebb, D. O. (1947). *The Organization of Behaviour*. New York: Wiley.

Hebb, D. O. (1955). Drives and the C.N.S. (Conceptual Nervous System). *Psych. Rev.* 62, 243–254.

Hinde, R. A. (1956). Ethological models and the concept of drive. *Brit. J. Philos. Sci.* 6, 321–331.

Hinde, R. A. (1959). Unitary drives. *Anim. Behav.* 7, 130–141.

Hinde, R. A. (1960). (in press; to be inserted).

Holst, E. von (1936). Versuche zur Theorie der relativen Koordination. *Pflüg. Arch. ges. Physiol.* 237, 93–121.

Iersel, J. J. A. van & Bol, A. C. (1958). Preening of two tern species. A study on displacement. *Behaviour* 13, 1–89.

Kennedy, J. S. (1954). Is modern ethology objective? *Brit. J. Anim. Behav.* 2, 12–19.

Kubie, L. S. (1947). The fallacious use of quantitative concepts in dynamic psychology. *Psychoanalytic Quart.* 16, 507–518.

Lashley, K. S. (1938). Experimental analysis of instinctive behaviour. *Psychol. Rev.* 45, 445–471.

Lloyd Morgan, C. (1912). *Instinct and Experience*. London: Methuen.

Lorenz, K. (1937). Uber die Bildung des Instinktbegriffes. *Naturwiss* 25, 289–300, 307–318, 324–331.

Lorenz, K. (1950). The comparative method in studying innate behaviour patterns. *Sym. Soc. Exp. Biol.* IV, 221–268.

Lorenz, K. & Tinbergen, N. (1938). Taxis und Instinkthandlung in der Eirollbewegung der Graugans. *Z. Tierpsychol.* 2, 1–29. Translated in Schiller, C. H. (1957). *Instinctive Behaviour*. London: Methuen.

McDougall, W. (1913). The sources and direction of psychophysical energy. *Amer. J. Insanity*. Not consulted. Quoted in McDougall (1923).

McDougall, W. (1923). *An Outline of Psychology*. London: Methuen.

Meehl, P. E. & McCorquodale, K. (1948). On a distinction between hypothetical constructs and intervening variables. *Psych. Rev.* 55, 95–107.

Morris, D. (1956). The feather postures of birds and the problem of the origin of social signals. *Behaviour* 9, 75–113.

Peters, R. S. (1958). *The Concept of Motivation*. London: Routledge and Kegan Paul.

Precht, H. (1952). Über das angeborene Verhalten von Tieren. Versuche an Springspinnen. *Z. Tierpsychol.* 9, 207–230.

Rowell, C. H. F. (1959). The occurrence of grooming in the behaviour of Chaffinches in approach-avoidance conflict situations, and its bearing on the concept of 'displacement activity'. *Ph.D. Thesis*. Cambridge.

Sherrington, C. S. (1906). *Integrative Action of the Nervous System*. New York: Scribner.

Skinner, B. F. (1938). *The Behaviour of Organisms*. New York: Appleton Century.

Thorpe, W. H. (1956). *Learning and Instinct in Animals*. London: Methuen.

Tinbergen, N. (1951). *The Study of Instinct*. Oxford.

Tinbergen, N. (1952). Derived activities: their causation, biological significance, origin and emancipation during evolution. *Quart. Rev. Biol.* 27, 1–32.

Toulmin, S. E. (1953). *The Philosophy of Science*. London: Hutchinson.

5

Sensory Processes, Orientation, and Communication

Biology of the *Umwelt*

Fred C. Dyer and H. Jane Brockmann

Introduction

Of the many influential authors whose work had to be excluded from this volume to keep the length within bounds, perhaps none was more difficult to omit than Jakob von Uexküll. His concept of the *Umwelt*, set forth principally in long monographs (e.g., 1934/1957), was of profound importance in the development of ethology. Lorenz (1958/1971, p. 277) himself attested to the influence of von Uexküll's ideas on ethology by stating that "this young science owes more to his teaching than to any other school of behaviour study." Many of the phenomena and operational concepts that gave focus to classical ethology were first described, or at least anticipated, in the writings of von Uexküll. These concepts, commonly attributed to other founders of ethology, such as Lorenz (1935; see this volume) or Tinbergen (1951), include sign stimuli, innate re- leasing mechanisms, stimulus filtering, and search images. Perhaps more important, a component of the *Umwelt* concept, that animals inhabit unique, species-specific perceptual worlds, has become one of the core principles of the modern study of animal behavior. The papers collected in this section of the book represent the areas of ethology most influenced by von Uexküll, and so it seemed appropriate to organize this introduction around the *Umwelt* concept. (For details about von Uexküll's broader scientific contributions, see Thorpe 1979 and Lorenz 1958/1971.)

At the outset it should be said that, unlike other founders of ethology, von Uexküll was no Darwinian. He viewed species-specific specializations of behavior and morphology not as the result of organic evolution but as evidence of a vitalistic "pre-established harmony" (Lorenz 1958, 1971, p. 276) between each species and its environment. As emphasized by Lorenz, however, the approach von Uexküll advocated was hardly mystical; it was both rigorously mechanistic and clearly focused on understanding both the physiological causation and the survival value of behavior.

A glance at a variety of introductory texts in animal behavior shows that the *Umwelt* concept is often identified solely with the observation that animals of different species live in distinct sensory worlds. In fact, von Uexküll's conception of the *Umwelt* was more comprehensive (see Lorenz 1958/1971, p. 274). An animal's *Umwelt* encompasses both its perceptual world (*Merkwelt*), or all of the environmental features to which it responds; and its "effector world" (*Wirkwelt*), or all of the actions that are appropriate for responding to different environmental features. Equally important, the perceptual cues, or "releasers" (von Uexküll's term), and responses associated with a particular stimulus, are integrated into what von Uexküll called a "functional cycle," which he took to be a fundamental unit of behavior. The key insight here is that mechanisms by which animals perceive the world are linked physiologically to the mechanisms whereby the animal responds to the world.

The paradigmatic illustration by von Uexküll (1934, 1957) of an *Umwelt* in action was a tick "hunting" for a mammalian host. The tick's highly impoverished *Umwelt* consists only of three releasers and three associated responses at this stage in its life cycle: butyric acid, emitted by all mammals, which induces the tick to drop from its perch; the mechanical stimulation produced by the host's fur, which releases crawling; and warmth (of the host's skin), which induces the tick to

bore in for its meal. Simple though the tick's *Umwelt* may be, "the very poverty of this world guarantees the unfailing certainty of her actions, and security is more important than wealth" (von Uexküll 1934, 1957, p. 12).

The *Umwelten* of other animals may be richer and may include learned responses to the environment (von Uexküll [1934, 1957] cites both spatial learning and search images). But the tick teaches a valuable lesson, one that is a central theme in von Uexküll's writing: we can appreciate how and why an animal does what it does only if we view it as the subject as well as the object of processes that influence its behavior. Progress in understanding these processes comes from imagining what it might be like to be the animal, not only possessing its sensory apparatus but also being attuned, both in perception and in response, to the objects and relationships in the outside world that are most relevant to its survival.

The papers collected in this section all illustrate the liberating influence of this perspective for understanding animal behavior. Freed from the anthropomorphic assumption that animals perceive the world in much the same way as we do, early ethologists uncovered sometimes astounding capabilities of animals to detect and respond to environmental features that we can detect only with specially designed instruments. Furthermore, acknowledging the specific capabilities of animals, and the general possibility that animals have capabilities that we do not, has repeatedly proved essential for understanding the mechanisms and evolution of behavior. This perspective has proved particularly important in the three areas of research included here: the biology of the senses, orientation, and communication.

Sensory Worlds

The study of the sensory capabilities of animals by ethologists can be divided into two related problems. First, what physical features of the environment can be detected by a given animal, and how do the sensory organs transduce useful information from a given feature? Second, of the features that potentially can be detected, what are the actual stimuli that elicit behavior in a given context, and how do sensory systems extract relevant stimuli from irrelevant (but nevertheless detectable) background noise?

A major figure in studies on the first of these problems was Karl von Frisch, who discovered much about the *Umwelt* of the honeybee (although he did not use this term and was not obviously influenced by von Uexküll). Among von Frisch's earliest discoveries was his demonstration that honeybees have a well-developed ability to discriminate colors (von Frisch 1914). Also, the bees' visual spectrum is shifted relative to ours, excluding red but including ultraviolet (see also von Frisch 1950).

The discovery of color vision in an insect challenged a widely held dogma of the times, but it required only a modest realignment of thinking about the sensory capabilities of animals. Subsequent work by von Frisch and others forced biologists to acknowledge the possibility of sensory abilities that previously would have been thought ludicrous. One such example was von Frisch's (1949) discovery (see von Frisch and Lindauer 1956, reprinted here) that bees can see the polarized light of blue sky and use it for orientation. This ability turns out to be shared by a wide variety of invertebrates and many vertebrates (Waterman 1981), although not, of course, by humans.

Just a few years earlier, Griffin and Galambos (1941), in a paper reprinted here, published their equally revolutionary finding that vespertilionid bats could navigate in total darkness by emitting high-frequency sound pulses and extracting information from the returning echos. As Griffin (1958) later recounted, this discovery solved a mystery that extended back at least to the eighteenth century. In 1793, Lazzaro Spallanzani (who was best known for disproving the possibility of spontaneous generation) performed experiments that implicated a role for sound in bat navigation, but he could not fathom how this could be since no sounds could be heard. Griffin et al. (1960) subsequently showed that some bats use echolocation not only to avoid obstacles but also to detect, pursue, and capture flying insect prey. Given the challenges inherent in localizing and pursuing prey, while at the same time avoiding obstacles and ignoring sounds produced by other bats, it seems clear that bats live in an auditory world that must be as rich to them as our visual world is to us. Over the years, an extensive literature has developed exploring the neural basis of bat echolocation (see Suga [1990] for an overview), as well as the adaptive design of echolocation systems for performance in different environments (reviewed by Neuweiler 1989). Also, in one of the more compelling sequels to the story begun by Griffin and Galambos, Roeder and Treat (1961, reprinted here) found that noctuid moths can detect the ultrasonic cries of predatory bats, and can undertake various evasive maneuvers when under attack (see May 1991 for a review of similar abilities in other species). Thus, the *Umwelten* of predator and prey interlock in a grim ballet set to a score that only the dancers can hear.

The chemical senses of animals also have been dissected to reveal olfactory worlds of impressive richness and specialization. Odors are, of course, widely used by animals to detect and orient to sources of food or to mates, to avoid sources of danger, or to communicate with conspecifics in the context of reproduction, social behavior, and territorial defense. Odors also are used for dramatic feats of orientation: a notable example is the ability of salmon to use, among other cues, learned odors to return for spawning to the stream in which they hatched (Hasler 1960, reprinted here). Among the most intensively studied roles of chemical cues is in intraspecific communication. In the late 1950s, chemicals used for interspecific communication came to be called pheromones. The review of pheromone biology by Wilson and Bossert (1963), reprinted here, was seminal in the development of research on both the communicative functions of odors and the sensory processes whereby odors influence behavior.

The human *Umwelt* is dominated by sights, sounds, and smells, and yet it is difficult to imagine what it might be like to have the vision of bees, the hearing of bats, or the olfactory sensitivity of salmon. Even more challenging is to imagine being able to extract useful information from infrared images, as do pit vipers (Newman and Hartline 1982); from electrical fields, as do sharks (Kalmijn 1971) and weakly electric fish (Lissmann, 1951, 1958); or from magnetic fields, as do pigeons (Keeton 1974), newts (Phillips 1986), sea turtles (Lohmann 1992), and a wide variety of other species. It is both inspiring and a bit unsettling to consider the difficulty of discovering sensory abilities that are so counter to our everyday experience.

As we suggested earlier, identifying the major modalities through which an animal detects the world around it provides only a partial description of its sensory world. An additional problem, noted by von Uexküll and others (e.g., Lashley 1938; Tinbergen 1951; and Marler 1961a), arises from the observation that animals often respond preferentially to a small subset of the stimuli that their sensory organs seem equipped to detect. Thus, a process of "filtering" appears to select the stimuli that actually elicit behavior.

In many cases, this selective responsiveness to certain stimuli is the stimuli that matter to the animal. Examples include the tick's sensitivity to butyric acid and to no other chemical (von Uexküll 1934/1957); the sensitivity of the noctuid moth's ear only to ultrasonic frequencies, which are produced mainly by bats (Roeder and Treat 1961); and the pheromone receptors of many insects and mammals. Also, Wehner (1987) beautifully reviews several examples of sensory filtering in which the spatial configuration of a receptor or receptor array is matched to the spatial features of the stimulus that the sensory organ is designed to detect.

Stimulus filtering may also occur at a central level in the nervous system. Many of the most famous sign stimuli worked out by Tinbergen and colleagues (see Tinbergen 1951) comprise specific configurations of visual features whose salience is apparently not a consequence of selective sensitivity but of selective responsiveness. Examples abound in the classical ethological literature: the red belly of the male stickleback, which elicits attacks by other males (ter Pelkwijk and Tinbergen 1937); the red spot on the end of the herring gull's lower mandible, which elicits begging by chicks (Tinbergen and Perdeck 1950); and the features that induce the greylag goose to roll a displaced egg back into its nest (Lorenz and Tinbergen 1938). A more recent example that has been studied physiologically is the mechanism by which a toad's brain recognizes an object as potential prey (Ewert 1987): this mechanism consists of a set of neurons that respond selectively to visual stimuli of just the right shape, size, and speed of movement.

Early ethologists emphasized the innateness of such filtering mechanisms, but the classic studies by Hailman (1967) on the pecking responses of gull chicks established that experience can modify the configurations of stimuli that elicit a behavior. Ewert's work on toads also showed that the responsiveness of prey-detection neurons can be modified by experience. Indeed, associative learning generally can be regarded as a filtering mechanism that selects the particular features of an animal's environment that are worth responding to.

Orientation

Interest by ethologists in orientation is based in part on their recognition of the integral role that orientation plays in behavior, and in part on the sense of wonder induced by the impressive navigational abilities of some species. The study of orientation addresses two broad complementary issues. One concerns the features of the environment that provide the animal with information about its position and direction of movement. Answering this question involves understanding the sensory abilities of animals, and has been aided enormously by the realization that animals live in distinct sensory worlds. Indeed, many previously unknown sensory abilities, including several mentioned above, have been discovered through studies of orientation.

The other major question in studies of orientation concerns how an animal uses the information provided by a particular feature of the environment to complete a given task. A major impetus for this question was the early realization that even quite simple animals (Jennings 1906) do not always react to a given orientation cue in the consistent or stereotyped way assumed by various "tropism" (Loeb 1900) or "reflex" (James 1901) theories. Instead, an animal may modulate its responses to a cue when exposed to other stimuli, when competing demands arise, or when its internal state changes with age, season, or stage in a behavioral sequence. To take a simple example, a nesting insect may respond to the same light gradient with positive phototaxis if embarking on a foraging trip, but with negative phototaxis if returning to the nest. Much early work in animal orientation dealt with such simple behavior, attempting to classify the general types of response (e.g., taxes, kineses) involved in orientation to stimuli, and the circumstances under which animals change their responses (see Fraenkel and Gunn 1940).

At the opposite extreme, many species carry out feats of orientation that appear to require the intricate coordination of mechanisms on multiple levels, including sensory, perceptual, and hormonal processes, endogenous rhythms, and learning. Three papers reprinted here deal with organisms that have served as models for investigating the control of such complex feats of orientation: honeybees (von Frisch and Lindauer 1956), salmonid fishes (Hasler 1960), and passerine birds (Emlen 1969). In each case the animals orient themselves over long distances (relative to their body sizes), using various reference cues both to determine their position relative to their goal and to maintain their course.

One of the most important ingredients for success in these sorts of studies has been the realization that animals may use several distinct cues for orientation. Moreover, an animal's responses to different cues may be organized either redundantly or by the context in which they are used (Able 1980). These insights have important implications for the design of experiments to test the relevance of particular stimuli for orientation by a given species. Specifically, the ability of an animal to orient in the absence of a particular cue is not positive evidence that this cue cannot be used, since the animal might have access to a backup system that relies on other cues. Moreover, an inability to orient in the presence of a given cue might simply mean that the animal has not been observed in the behavioral or motivational context in which it normally uses the cue. Redundant organization of orientation systems is illustrated by honey bees (von Frisch and Lindauer) and homing pigeons (Keeton 1974). In both species, a demonstrably sufficient reference for compass orientation, the disk of the sun, is not necessary for orientation under all circumstances. Deprived of the sun, bees can rely upon landmarks to move around their familiar

foraging range, while displaced pigeons can use the earth's magnetic field to set and maintain a homeward course (and then use landmarks when in familiar areas). The importance of internal state is illustrated by Emlen's (1969) discovery of season changes in the responsiveness of migrating birds to specific stimuli. Hasler's (1960) paper shows how a migrating salmon may use a succession of stimuli along its trip inland to the spawning stream. The same point is beautifully illustrated in recent studies of the seaward orientation of hatching sea turtles (summarized by Lohmann 1992). Hatchlings at each point in their rush toward open ocean respond preferentially to the stimuli that are appropriate for the stage that they have reached: light gradients, landmarks, and slope when on the beach; wave movements when in the water near the shore; magnetic and celestial compass references when offshore.

The study of orientation continues to be a thriving enterprise. For reviews of the literature on avian orientation, see papers in Berthold (1991); for invertebrates, see Wehner (1981, 1984) and Dyer (1994).

Communication

The study of communication has always required adopting von Uexküll's perspective of animals as the subjects of their world. Specifically, the observer at least must attempt to examine an interaction between two animals from the points of view of both the sender and the receiver of a signal (Slater 1983). The investigations of communication by early ethologists also were explicitly guided by von Uexküll's conceptual framework, focusing on interconnected releasers and responses. In such classic studies as Tinbergen's (see 1951 for review) investigations of the courtship rituals of male and female sticklebacks, the signals emitted

by one member of a communicating pair were viewed as releasers of the behavior of its partner, which in turn released the next behavior in the sequence. With this approach, studies of communication were placed squarely into the same explanatory framework that applied to behavior with no communicative function.

Studies of communication generally try to account for the specific features of communicative traits and for the diversity of signals produced by different species and by individuals of the same species in different contexts. Most early work focused on two major questions. One concerned the historical origins of traits that currently function as signals. The other dealt with the information communicated by the signals. Regarding the historical origins of signals, an important insight was that many signals evolved through a process of "ritualization" from behavioral or physiological traits that originally had a non-signaling function (Huxley 1914, Tinbergen 1952). The significance of this point is that the current form of the signal could be related to the ancestral behavioral context in which it acquired its communicative function. To take some well-known examples from mammals: threat displays, such as a dog's baring its teeth, could have evolved from so-called "intention movements" associated with biting during fights; vocalizations, raising of hackles, and scent-marking with urine or with glandular secretions could have evolved from autonomic responses produced during the course of contests (Hinde 1970). Another possible example of ritualization occurred in the evolution of the honey bee dance language, one of the most astonishing behaviors discovered during the history of ethology. As discussed in the article by von Frisch and Lindauer (1956), reprinted here, the orientation of the waggle dance (which correlates with the direction to food) could have evolved from an "intention movement" (the dancer aiming her body in preparation for a flight to the food) which came to be interpreted by other bees in the nest as a signal of the direction to fly to find the food. See Dyer (1991) for more recent insights into the evolution of the dance language.

Early work on the information communicated by signals focused on why a trait might have acquired meaning as a signal, and therefore why it was ritualized to enhance that meaning. The 1950s were an important decade for the development of ideas about these issues, and the article by Marler (1961b), reprinted

here, represents an important synthesis of the progress that was made. Marler categorized signals in two dimensions. The first dimension classifies signals according to their referents: e.g., the species, sex, and identity of the signaler, the signaler's motivational state, and objects or events in the environment. The second dimension classifies signals according to their structural attributes: e.g., whether graded or discrete, conspicuous or subtle, stereotyped or variable, simple or complex. Interpreting the design of particular signals thus is taken to be a matter of understanding why given attributes are appropriate for communicating about particular referents. This interpretation is linked to the study of the phylogenetic history of signals, because the current structure of a signal presumably is non-arbitrarily related to the behavioral and motivational context in which the signal evolved (see the introduction to Part 6, by Arnold and Brockmann).

Beginning in the 1960s, new approaches to the study of communication have developed. These approaches have expanded, and in some cases overturned, the insights developed by early workers about the forces guiding the ritualization process. The most dominant of these new approaches (reviewed by Krebs 1991) emphasizes the coevolution between the signaler and the receiver of the signal. This approach acknowledges that (a) it is the receiver's response to the signal that exerts the most important selective pressure on the signal, and (b) neither the signaler nor the receiver should be expected to do anything that is not in its own best genetic interests. This approach arose with the development of gene-centered thinking about the adaptive significance of traits (Hamilton 1964, this volume, Part 6) and game-theoretical analyses of interactions among animals (Maynard Smith and Price 1973, this volume, Part 6). These ideas led to the perspective that communication is best viewed as the more or less selfish attempts by signalers to manipulate the behavior of the receivers to the advantage of the signalers, and of equally selfish attempts by receivers to avoid being deceived or manipulated to their disadvantage.

The critical issue separating the coevolutionary view and what Dawkins and Krebs (1978) call the "classical" view actually is fairly narrow, albeit important. The issue is the possibility that some signals were designed to communicate the motivational state of the signaler. Consider the alternative interpretations that can be given for the common observation that some signals are varied in a graded fashion according to the apparent level of excitement of the signaler, whereas others are emitted with a "typical intensity" (Morris 1957) regardless of the presumed motivational level of the signaler. As represented by Marler (1961b), the classical interpretation of these patterns was that there are contexts in which a signaler might need to convey gradations in its motivational states (e.g., degree of readiness to attack or to flee), and other contexts in which the exact motivational state is not important to communicate, or when the need for an unambiguous, all-or-none signal is paramount. By the coevolutionary view, stereotyped signals that carry little or no information about the signaler's motivations or intentions are actually to be expected under some conflict situations, when an opponent could turn such information to the signaler's disadvantage. Graded signals, on the other hand, at least in situations of conflict, are not viewed as conveying particular motivational levels. Instead they are assumed to reflect the escalation of effort between competitors to intimidate each other and to assess each other's ability, progressing from displays that are relatively inexpensive to produce to ones that are relatively costly. See Krebs (1991) for a recent overview of these issues.

Another flourishing approach to the study of communication, which complements those already discussed, has addressed the design of signals in relation to the environmental conditions under which the communication takes place. Here the emphasis is on the physical factors that influence the propagation of the signal and the preservation of information that it contains. The paper by Wilson and Bossert (1963), dealing with the design of chemical communication systems, was one of the earliest attempts to address this problem. Other studies have explored the design of acoustic (Morton 1975; Wiley and Richards 1982) and visual (Lythgoe 1979) signals for communication in different environments. Excellent general reviews are provided by Gerhardt (1983), Hopkins (1983), and Endler (1992).

It is fitting to close by noting that these new approaches to animal communication, to the extent that they have produced new insights into the hows and whys of animal behavior, may have succeeded because they heeded von Uexküll's advice to view the world from the animal's perspective. This is most obvious in regard to the physical properties of signals, where we can view the animal's communication system as an extension of its sensory world. It is equally true that the coevolutionary view of animal communication systems requires the ethologist to view the animal, as did von Uexküll, as the subject of its *Umwelt*, and to imagine what it would be like to be the one at the center of that world.

Literature Cited

Able, K. P. 1980. Mechanisms of orientation, navigation, and homing. Pp. 284–373 in *Animal Migration, Orientation, and Navigation*, S. A. Gauthreaux, Jr., ed. New York: Academic Press.

Berthold, P., ed. 1991. *Orientation in Birds*. Basel: Birkhäuser Verlag.

Dawkins, R., and J. R. Krebs. 1978. Animal signals: information or manipulation? Pp. 282–309 in *Behavioural Ecology: An Evolutionary Approach*, 1st ed., J. R. Krebs and N. B. Davies, eds. Oxford: Blackwell.

Dyer, F. C. 1991. Comparative studies of dance communication: analysis of phylogeny and function. Pp. 177–98, in *Diversity in the genus* Apis, D. R. Smith, ed. Boulder: Westview Press.

Dyer, F. C. 1994. Spatial cognition and navigation in insects. Pp. 66–98 in *Behavioral Mechanisms in Evolutionary Ecology*, L. A. Real, ed. Chicago: University of Chicago Press.

Emlen, S. T. 1969. Bird migration: influence of physiological state upon celestial orientation. *Science* 165:716–18.

Endler, J. A. 1992. Signals, signal conditions, and the direction of evolution. *The American Naturalist* 139 (Supplement): 125–53.

Ewert, J.-P. 1987. Neuroethology of releasing mechanisms: prey-catching in toads. *Behavioral and Brain Sciences* 10:337–406.

Fraenkel, G. S., and D. L. Gunn. 1940. *The Orientation of Animals*. Oxford; Oxford University Press.

Frisch, K. von. 1914. *Der Farbensinn und Formensinn der Bienen*. Jena: G. Fischer.

Frisch, K. von. 1949. Die Polarisation des Himmelslichtes also orientierender Faktor bei den Tänzen der Bienen. *Experientia (Basel)* 5:142–48.

Frisch, K. von. 1950. *Bees: Their Vision, Chemical Senses, and Language*. Ithaca: Cornell University Press.

Frisch, K. von, and M. Lindauer. 1956. The language and orientation of the honey bees. *Annual Review of Entomology* 1:45–58.

Gerhardt, H. C. 1983. Communication and the environment. Pp. 82–113 in *Animal Behaviour*. Vol. 21, *Communication*, T. R. Halliday and P. J. B. Slater, eds. Oxford: Blackwell.

Griffin, D. R. 1958. *Listening in the Dark*. New Haven: Yale University Press.

Griffin, D. R., and R. Galambos. 1941. The sensory basis of obstacle avoidance by flying bats. *Journal of Experimental Zoology* 86:481–506.

Griffin, D. R., F. A. Webster, and C. R. Michael. 1960. The echolocation of flying insects by bats. *Animal Behaviour* 8:141–54.

Hailman, J. P. 1967. The ontogeny of an instinct: the pecking response in chicks of the laughing gull (*Larus atricilla* L.) and related species. *Behaviour*, Supplement 15.

Hamilton, W. D. 1964. The genetical theory of social behaviour (I and II). *Journal of Theoretical Biology* 7:1–16, 17–32.

Hasler, A. D. 1960. Guideposts of migrating fishes. *Science* 132:785–92.

Hinde, R. A. 1970. *Animal Behavior: A Synthesis of Ethology and Comparative Psychology*, 2d ed. New York: McGraw-Hill.

Hopkins, C. D. 1983. Sensory mechanisms in animal communication. Pp. 114–55 in *Animal Behaviour*. Vol. 2, *Communication* (T. R. Halliday and P. J. B. Slater, eds. Oxford: Blackwell.

Huxley, J. S. 1914. The courtship habits of the great crested grebe (*Podiceps cristatus*); with an addition to the theory of sexual selection. *Proceedings of the Zoological Society of London* 35:491–562.

James, W. 1901. *Principles of Psychology*. New York: Henry Holt.

Jennings, H. S. 1906. *Behavior of the Lower Organisms*. Bloomington: Indiana University Press.

Kalmijn, A. J. 1971. The electric sense of sharks and rays. *Journal of Experimental Biology* 55:371–83.

Keeton, W. T. 1974. The orientational and navigational basis of homing in birds. Pp. 47–132 in *Advances in the Study of Behavior*, vol. 5, D. S. Lehrman, J. S. Rosenblatt, R. A. Hinde, and E. Shaw, eds. New York: Academic Press.

Krebs, J. R. 1991. Animal communication: ideas derived from Tinbergen's activities. Pp. 60–73 in *The Tinbergen Legacy*, M. S. Dawkins, T. R. Halliday, and R. Dawkins, eds. London: Chapman and Hall.

Lashley, K. 1938. Experimental analysis of instinctive behavior. *Psychological Review* 45:445–71.

Lissmann, H. W. 1951. Continuous electrical signals from the tail of a fish, Gymnarchus niloticus Cuv. *Nature*, 199:88–89.

Lissmann, H. W. 1958. On the function and evolution of electric organs in fish. *Journal of Experimental Biology*. 35:156–91.

Loeb, J. 1900. *Comparative Physiology of the Brain and Comparative Psychology*. New York: G. P. Putman's Sons.

Lohmann, K. J. 1992. How sea turtles navigate. *Scientific American* 266:100–107.

Lorenz, K. Z. 1935. Der Kumpan in der *Umwelt des Vogels*. *Journal für Ornithologie* 83:137–213. (Translated a Companionship in bird life, in C. H. Schiller, ed., *Instinctive Behavior*, 83–128.

New York: International Universities Press [1857].)

Lorenz, K. Z. 1958. Methods of approach to the problems of behaviour. *The Harvey Lectures 1958–59.* New York: Academic Press. (Reprinted in K. Z. Lorenz, *Studies of Animal and Human Behaviour,* vol. 2, 246–80. Cambridge: Harvard University Press, [1971].)

Lorenz, K. Z., and N. Tinbergen. 1938. Taxis und Instinkthandlung in der Eirollbewegung der Graugans. *Zeitschrift für Tierpsychologie* 2:1–29. (Translated as Taxis and instinctive action in the egg-retrieving behavior of the Greylag Goose, in C. H. Schiller, ed., *Instinctive Behavior,* 176–208. New York: International Universities Press [1957].)

Lythgoe, J. N. 1979. *The Ecology of Vision.* Oxford: Clarendon Press.

Marler, P. 1961a. The filtering of external stimuli during instinctive behaviour. Pp. 150–66 in *Current Problems in Animal Behaviour,* W. H. Thorpe and O. L. Zangwill, eds. Cambridge: Cambridge University Press.

Marler, P. 1961b. The logical analysis of animal communication. *Journal of Theoretical Biology* 1: 295–317.

May, M. 1991. Aerial defense tactics of flying insects. *The American Scientist* 79:316–29.

Maynard Smith, J., and G. R. Price. 1973. The logic of animal conflict. *Nature* 246:15–18.

Morris, D. 1957. "Typical intensity" and its relation to the problem of ritualisation. *Behaviour* 11: 1–12.

Morton, E. S. 1975. Ecological sources of selection on avian sounds. *The American Naturalist* 109: 17–34.

Neuweiler, G. 1989. Foraging ecology and audition in echolocating bats. *Trends in Ecology and Evolution* 4:160–66.

Newman, E. A., and P. H. Hartline. 1982. The infrared "vision" of snakes. *Scientific American* 246– 116–24.

Pelkwijk, J. J. ter, and N. Tinbergen. 1937. Eine reizbiologisch Analyse einiger Verhaltensweisen von *Gasterosteus aculeatus* L. *Zeitschrift für Tierpsychologie* 1:193–200.

Phillips, J. B. 1986. Two magnetoreception pathways in a migratory salamander. *Science* 233: 765–67.

Roeder, K. D., and A. E. Treat. 1961. The detection and evasion of bats by moths. *American Scientist* 49:135–148.

Slater, P. J. B. 1983. The study of communication. Pp. 9–42 in *Animal Behaviour,* vol. 2: *Communication.* T. R. Halliday and P. J. B. Slater, eds. Oxford: Blackwell.

Suga, N. 1990. Biosonar and neural computation in bats. *Scientific American* 262:60–68.

Thorpe, W. H. 1979. *The Origins and Rise of Ethology.* London: Heinemann.

Tinbergen, N. 1951. *The Study of Instinct.* Oxford: Oxford University Press.

Tinbergen, N. 1952. Derived activities; their causation, biological significance, origin, and emancipation during evolution. *Quarterly Review of Biology* 27:1–32.

Tinbergen, N., and A. C. Perdeck. 1950. On the stimulus situation releasing the begging response in the newly hatched herring gull chick. *Behaviour* 3:1–39.

Uexküll, Jakob von. 1934. *Streifzüge durch die Umwelten von Tieren und Menschen.* Berlin: Springer-Verlag. (Translated as: A stroll through the worlds of animals and men, in C. H. Schiller, ed., *Instinctive Behavior,* pp. 5–80. New York: International Universities Press [1957]).

Waterman, T. H. 1981. Polarization sensitivity. Pp. 281–469 in *Handbook of Sensory Physiology,* vol. 7/6B, H. Autrum, ed. Berlin: Springer-Verlag.

Wehner, R. 1981. Spatial vision in arthropods. Pp. 287–616 in *Handbook of Sensory Physiology,* vol. 7/6B, H. Autrum, ed. Berlin: Springer-Verlag.

Wehner, R. 1984. Astronavigation in insects. *Annual Review of Entomology* 29:277–98.

Wehner, R. 1987. "Matched filters"—neural models of the external world. *Journal of Comparative Physiology A* 161:511–31.

Wiley, R. H., and D. G. Richards. 1982. Adaptations for acoustic communication in birds: sound transmission and signal detection. Pp. 131–181 in *Acoustic Communication in Birds,* vol. 1, D. E. Kroodsma and E. H. Miller, eds. New York: Academic Press.

Wilson, E. O., and W. H. Bossert. 1963. Chemical communication among animals. *Recent Progress in Hormone Research* 19:673–716.

THE "LANGUAGE" AND ORIENTATION
OF THE HONEY BEE[1]

By K. von Frisch and M. Lindauer

Zoologisches Institut, Universität München, München, Germany

We have been asked to report upon the latest results of our study of the "language" and orientation of bees. The knowledge we had up to 1950 has been presented in a series of lectures entitled "Bees" and given by the senior author on the occasion of an invitational tour of the United States. These lectures have been edited in book form by von Frisch (11), and a bibliography up to the year 1950 is to be found in this booklet. The present review is concerned primarily with developments since that time.

The "language" of the bees is not a verbal one. It depends on the senses of touch and smell. Their words are rhythmic movements and scents. Before going into details we should like to restate the means by which the bees communicate with each other.

A bee informs her hive companions of a new, rich food source (nectar, sugar solution, or pollen) found in the close vicinity of the hive by means of a round dance (see Fig. 2a). Those of the bees which are in a co-operative mood follow the returned bee closely during the dance on the honeycomb and learn that there is a profitable source of food. By the special form of dance, which gives no indication of direction, they are taught that it is to be found close to the hive. The specific scent of the visited plant sticking to the body of the dancer, to the pollen, or to the nectar tells them for which kind of flower to search.

If the food is farther away than 50 to 100 m., the round dance is replaced by the waggle dance (Fig. 2c). With its help the hive mates following the dancer are informed not only of the existence of a promising source of food and of its characteristic scent, they even learn at what distance and in which direction from the hive to seek it. The distance is taught by the measure (rhythm) of the dance [von Frisch (11)]. The direction is given in relation to the position of the sun. If the feeding place is to be found exactly in the same direction in which the sun lies at the time, the dancer indicates this on the vertical honeycomb of the hive by orientating the straight wagging of the dance vertically upwards. If the feeding place lies at an angle of 60° to the left of the sun, the straight part within the dance will also be carried out at an angle of 60° to the left of the vertical, and so forth. They transpose thus the angle to the sun, that has to be made while flying to the food source, into the field of the gravity when dancing on the honeycomb [von Frisch (11, Fig. 44, p. 77)]. The hive mates which have accompanied the dance fly out and will find the source of food according to the communications they have received.

[1] The survey of the literature pertaining to this review was completed in April, 1955.

46   FRISCH AND LINDAUER

The food gathering bees dance after each flight, as long as it is advantageous to draw further gatherers to the exploited feeding place. As soon as the food becomes scarce, they stop dancing, although they continue collecting, and from then on this worker group does not increase anymore. By these means the principle of relations between supply and demand is applied to the behaviour of bees.

Certain objections have been raised to our use of the word "language" to describe these means of communication [Révész (37)]. It is evident that the mental principles of communication between bees are quite different from those of the human language. In order to underline this difference we write the word "language" in quotation marks. There is, however, no doubt that the language of the bees is on a higher level than the means of communication among birds and mammals with the exception of man. Warning or attracting calls and mating songs among birds express only the animal's motivation. This can convey itself to other members of the species. The dances of bees, however, transmits the knowledge of significant facts [von Frisch (14, 17)].

THE INDICATION OF DISTANCE

With increasing distance between feeding place and hive the number of straight runs within the waggle dance decreases, the rhythm of the dance becomes slower. At the same time the number of abdominal waggles during the straight run increases. It cannot be determined from the beginning whether the number of straight runs in a given time, or the number of waggles, indicate the distances. The latter are so quick that one can only approximately estimate their number by direct observation. Relying on such data [von Frisch (9, p. 20)] Haldane & Spurway (20) suggest that the number of waggles is the principal means by which distance is conveyed. We think, however, that before coming to this conclusion we must count the mean number of waggles per distance as accurately as possible. For this reason, we have made slow-motion pictures of dances at varying distances from the feeding place. The evaluation of this work has not yet been completed.

Two important observations have been made which might answer the problem of how a bee is able to estimate the distance she has flown. If the bee has a head wind on its outward journey to the feeding place she indicates in her dance a greater distance than when there is no wind. Similarly she indicates a shorter distance than normal when she had the wind in her favour [von Frisch (10, p. 15)]. If she has to fly uphill on her way to the feeding place she will, when dancing in the hive convey a greater distance than when she flies the same length on the level. And if she has to fly downhill she will communicate a shorter distance still [Heran & Wanke (23)]. It seems, therefore, that the time or energy spent on the flight from hive to feeding place[2] are very important, instead of the actual distance. Probably the main factor in the evaluation of distance is the energy spent on the flight; the experiments concerning this problem, however, are not yet finished.

[2] All given data on the dance concern the flight to the feeding place, none the flight homewards to the hive.

LANGUAGE OF THE HONEY BEE   47

The following experiment shows how accurately the dancer's indication of distance is understood and followed by the other bees. Several bees individually numbered with coloured dots are, at a certain distance from the hive, fed with a sugar solution to which a scent, for example lavender oil, was added. Upon their return to the hive, the numbered gatherers dance, and during the dance the hive companions smell the lavender and search for this specific odour when flying out. On the direct line from the hive to the feeding place and even further away "scent plates" are placed at various distances. They emit the scent, but offer no food. The bees looking out for this odour at the indicated distance are attracted by the scent plates if they come near them. They fly around them and finally alight on them and are thereupon counted by an observer. In Figure 1 we show the results of two such experi-

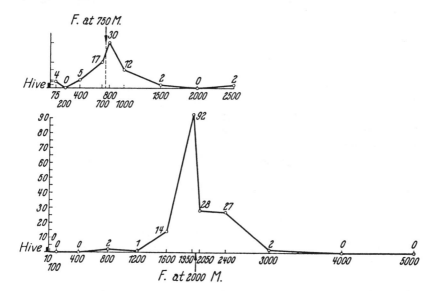

FIG. 1. Experiments concerning the reaction to the indication of distance. The distances of the scent plates from the hive are indicated on the abscissa in meters. The numbers on the points of the graph correspond with the number of bees which have come to each scent plate.

(Above) Trial of June 27, 1949. Distance between feeding place (F) and hive is 750 m. Duration of the experiment: 90 min.

(Below) Trial of July 20, 1952. Distance to feeding place is 2000 m. Duration of experiment: 3 hr.

ments. In the first trial the feeding place was at a distance of 750 m. and in the second at 2000 m. from the hive. The data given for the different points of the graph indicate how many of the searching bees came to each scent plate during the time of observation. The graphs show that the indication of distance given by the dancer is quite accurate and has been well understood by the other bees [von Frisch (13, 17)].

48   FRISCH AND LINDAUER

THE INDICATION OF DIRECTION

Normally the bees dance on the vertical honeycomb. In the dark hive they show the direction to be flown in order to indicate the feeding place to their companions by transposing the angle to the sun at which they flew from hive to food into the vertical plane. Sometimes, however, it happens that the returning bee dances outside the hive on the horizontal landing plank, in the daylight. In that case she points directly towards the goal during the straight wagging runs of the dance [von Frisch (11, Fig. 50, p. 87)]. This kind of communication of direction is the simpler one and more easily understood, as many insects possess a light-compass-reaction. They can register the position of the sun during their excursions or flights and keep thus a straight line of movement always maintaining the same angle to the sun. During her flight the bee also memorises the position of the sun and upon her return to the hive she maintains on the horizontal surface the same memorised angle to the sun, when performing the straight run of the dance. Thus she points directly towards the feeding place during the wagging dance.

Based on these observations we presumed that the dancing on a horizontal surface and in daylight was the more primitive and phylogenetically older form of indication of direction. It was, therefore, important to investigate the behaviour of other species of bees such as *Apis florea* Fabricius the dwarf honey bee. Its behaviour is comparatively primitive. The nest consists of one honeycomb only and possesses no nest cover. The honeycomb is built in the open air, suspended from a bush, the top of the comb being attached to a branch. The upper end of the honeycomb forms a broadened horizontal platform. The homecoming bees go immediately onto this platform where they convey, by a horizontal wagging dance in the same manner as our honey bees, what they found and whereabouts. They indicate the direction to be flown according to the position of the sun or the polarized light rays of the blue sky. If the honeycomb is turned around its horizontal or its vertical axis, the dancers walk always to the top in order to find a horizontal surface for dancing [Lindauer (unpublished observations)]. The hypothesis that the dancing on a horizontal surface is the original and more primitive means of communication among bees is considerably strengthened by these observations.

It is rather difficult to conceive how in their social evolution the bees have acquired the possibility of transposing their orientation to the sun into an orientation to gravity for the vertical dance in the dark hive. But some new observations show that among other insects such translations are also to be found, although biologically they appear to be pointless.

Vowles (41) working in a dark room on a horizontal plane forced ants into a motion of flight. As long as a lamp burnt the ants always maintained the same angle to the light source while in flight (light-compass-movement). When the light was turned out, and the surface put in a vertical position, the ant would run at approximately the same angle to the field of gravity as it formerly had taken to the light on the horizontal plane. Birukow (6) found

LANGUAGE OF THE HONEY BEE   49

a similar behaviour in the dung beetle, *Geotrupes silvaticus* Panz., in *Coccinella septempunctata* Linn., and in *Melasoma populi* Linn. The conformity with the behaviour of the bees, however, is not complete. The ants do indeed maintain the same angle to gravity when on a vertical surface as they had kept to the light but with four possibilities of choice: they run at the same angle either upwards to the right or to the left, or downwards to the right or to the left. *Geotrupes* translates correctly not only the angle but even the direction to the light into the direction to gravity, with the difference that it identifies the downward direction with the direction to the sun and the upwards direction with the one leading away from the sun. The bees, on the other hand, direct their waggle dance to the zenith when they fly towards the sun and to the nadir when they fly away from the light. This latter behaviour can also be found in some other beetles (*Coccinella* and *Melasoma*). But when these ran, e.g., 20° to the left towards the light, they can in the dark and on a vertical plane run either 20° to the left or the right against the field of gravity [Teuckhoff, after Birukow (6)]. There is little doubt that with further research other insects will be found which transpose exactly in the same way as the bees. This, however, is not so very important. The essential fact is that there seems to exist a primary capacity of the nervous centers to transpose a given angle to the light into an action of the sense organ of gravity as soon as the optical stimulus ceases to act [Birukow (6)]. Rothschild (38) comes in a more philosophic way to the same conclusion. Therefore we need not suppose that on a conference of the workers union the bees agreed to adopt a certain key reaction in order to transpose the angle to the light into the angle to gravity. The fact that during their evolution they were able to make such a correct and unmistakable use of their primary faculty of transposition is miraculous enough.

We tried in vain to find an expression for "upwards" and for "downwards" in the "language" of the bees. We carried a hive to a deep valley setting it up in the middle of the steep slope. Then two feeding places were put out, both being in exactly the same direction from the hive and at the same distance, but one being on the valley floor, the other exactly above it on a bridge. The dancing bees were unable to inform their hive mates whether they meant them to fly to the upper or the lower feeding place. In another experiment the feeding place was exactly above the opening platform of the hive on a broadcasting tower. The dancers could not convey the vertical direction of flight to the other bees and performed round dances thus sending their companions out into the surrounding meadows instead of upwards to their own feeding place [von Frisch, Heran & Lindauer (18)].

"Dialects" in the Bees Language

The bees belonging to the *carnica* race (*Apis mellifera* var. *carnica* Polm.) perform round dances (Fig. 2a) when the feeding place is closer than 50 to 100 m. from the hive. Baltzer and Tschumi found that Swiss bees execute another form of dancing, the sickle dance (Fig. 2b), when the feeding place is in the neighbourhood of the hive. Whereas the round dance of *A. m. carnica*

50   FRISCH AND LINDAUER

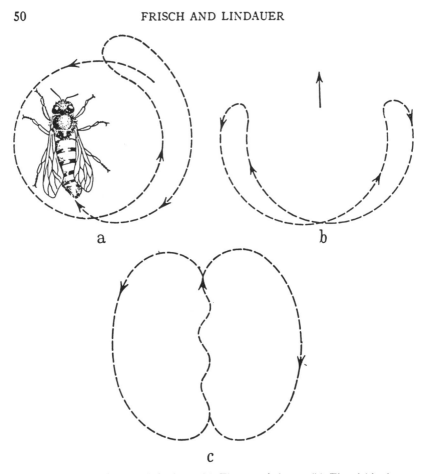

a   b

c

FIG. 2. The dance figures of the bees. (a) The round dance. (b) The sickle dance.
(c) The waggle dance.

does not indicate the direction for feeding places in the immediate neighbourhood this sickle dance shows the direction to be flown. The axis of the
danced semicircle (arrow in Fig. 2b) gives the appropriate direction to be
followed, exactly in the same way as the straight part indicates this in the
true waggle dance. According to Tschumi (40) the Swiss bees perform the
sickle dance for as short a distance as 10 m. The same behaviour has been
observed among Dutch bees by Hein (21).

This latter form of dancing seems to be a special character of the subspecies. In populations crossbred from the dark *A. m. carnica* and the yellow
striped Italian bee (*Apis mellifera* var. *ligustica* Spin.) representatives of the
dark race were to be found in the same hive with bees having the yellow
abdominal segments of the Italian subspecies. Upon their return from a food
source 10 m. distant the dark *carnica* bees performed mainly round dances,

LANGUAGE OF THE HONEY BEE   51

whereas the yellow bees of the same population performed sickle dances [von Frisch (13)]. Further experiments with such cross-breeds have, however, shown that the form of the dance is not always coupled with the pigmentation of the dancer in this way. The pure Italian breed differs clearly however from the pure *carnica* colonies because of the large number of sickle dances performed at short feeding distances.[3] Even the indication of distance reveals variations between these subspecies. At a given distance the Italian bees dance slower than the bees of *A. m. carnica* [Baltzer (2); von Frisch (13) and other unpublished experiments].

More important differences than those among the subspecies of our common honey bee have been found among the various species of *Apis* which Lindauer (unpublished data) was able to study on Ceylon.

Apis indica Fabr., closely related to *Apis mellifera* and often interpreted as one of its subspecies, indicates distance and direction in the same way as our honey bee but possesses the pecularity that sickle dances indicating direction begin at a distance of 0.5 m. between the hive and source of food and that well directed waggle dances are performed as soon as the distance is above 2 m. The dance rhythm, for the indication of distance is slower than in *Apis mellifera*, although otherwise the graphs obtained from distance indication follow a very similar course. The dances of *A. dorsata* Fabr., the giant honey bee, differ from those of *A. m. carnica* only in that the direction-indicating waggle dances begin at as short a distance as 3 m. and that the dance rhythm for a given distance is slower. We have already mentioned that *A. florea* uses a horizontal surface for its dances. We would, however, like to add here that they perform round dances up to a distance of 5 m. between hive and food. As soon as the distance is longer they switch over to the waggle dance.

The Dances of Swarming Bees

The goal which the dancers indicate is normally a rich source of food. But during the search for a new nesting site exactly the same dances are carried out to convey information of good housing possibilities. Not only the daily bread but also favourable breeding conditions are highly important for the survival of a colony of bees.

A short time after a swarm has left the mother hive and formed a cluster in its neighbourhood, some bees may be seen dancing on the cluster. Now they are not food gatherers but scouts which have found a nesting site and indicate their discovery to the clustered, waiting colony. By close observation of these dances it is possible to predict the direction in which the swarm will finally fly and over what distance, sometimes several kilometers. On three occasions merely through the reading of the scout's dances we succeeded in finding, even before the colony moved to it, the new, hardly visible site.

At the beginning the scouts indicate several nesting sites to the cluster,

[3] The colonies in which Tschumi and Hein observed the sickle dances were, as was discovered later, crossbred with Italian bees.

52   FRISCH AND LINDAUER

as they search everywhere at the same time. Each scout advertises his own
discovery. Before the cluster moves into a new nest an agreement is, however, found, and the bees will fly to the best of all the proposed nesting sites.
If on a small island or in a wide plain, with few or no natural sites, artificial
nesting places are offered to a swarm, we are able to detect which qualities of
a nesting site are favourably and which are adversely appraised by the scout-
ing bees. Among other qualities the distance between nesting site and
mother hive, the size of the nest cavity, its protection against wind, and its
thermal insulation seem to be very important in their choice of the nesting
place.

Those of the scouts which find an especially favourable site, dance much
longer and more energetically than those which find a place of lesser value.
In this way the attention of the clustered bees is mainly drawn to the best
proposition. The bees that take part in the dances fly out and inspect the
proposed nesting site. These, on their return to the cluster, also dance and
make further propaganda for it.

We must add that those scouts that find an inferior site change their mind
if other bees having found a better place dance very energetically near them.
Attracted by these vigorous dancers they follow the latter and inspect the
better place. Upon their return these also will dance in favour of the new goal.

We were also able to observe which of the bees became scouts. They are
levied among the experienced field bees, which change their activity during
the swarming time. These seasoned bees stop gathering when all the cells of
the hive contain sealed brood, nectar, and pollen. Every apiarist knows that
on such occasions the bees remain lazily and inactively in the hive and cluster
in the form of a beard at the flight hole. He interprets his observation correctly
as a sign for an approaching swarm emission. Under such circumstances,
when there is no need for other activities some of the old bees take the initia-
tive and reconnoiter new nesting sites. At this point the preparations for the
swarming have begun. Thus the preliminary activities for swarming begin
early, several days before the swarm leaves the hive [Lindauer (27, 28, 30)].

The Perception of the Sun Through a Clouded Sky

It is much more important for bees than for man to know always where
the sun stands. This knowledge is indispensable for their own orientation
and for the communication with their hive mates concerning feeding possibilities. In consequence they are much better equipped for such an ac-
complishment than we human beings.

They are even able to detect the position of the sun through the entirely
covered sky. This can easily be shown if we put an experimental hive into a
horizontal position and give the bees a view of the covered sky through a
glass window. On the horizontal honeycomb the bees dance correctly. Their
waggle lines point directly towards the goal, but only if they can see the point
in the clouded sky behind which the sun is hidden. Their dance lacks orientation if they can not look in this direction, even if the other parts of the sky
are visible.

LANGUAGE OF THE HONEY BEE   53

This ability of detection is not attributable to the perception of infrared rays. To our great astonishment experiments with filters have shown that the ultraviolet rays between 3000 and 4000 Å are decisive for the bees' view of the sun when it is concealed behind the clouds [von Frisch (16); Heran (22)].

THE PERCEPTION OF POLARIZED LIGHT

If the sun is hidden behind a forest, a mountain, a house or even below the horizon, the bees are still able to orientate themselves and to convey correctly their findings under the condition that the sky, or only a small spot of it, is clear and blue. The light coming out of the blue sky is partially polarized, and its plane of vibration is in direct correlation with the position of the sun. The bee's eye is able to recognise the plane of vibration of polarized light and can in consequence detect the position of the unseen sun if a spot of blue sky is visible [see von Frisch (11, pp. 88–109)].

Autrum suggests that each ommatidium of the compound eye is able to analyse polarized light by its eight radially arranged sensory cells [see von Frisch (11, Fig. 55, p. 98)]. He strongly supported his hypothesis by carrying out electrophysiological experiments on eyes of bees and flies. The magnitude of the exposure potentials depends on the light intensity, if an ommatidium of a cut off head[4] is exposed. The rotation of a polarizer put in front of the eye does not influence the magnitude of the potentials. In consequence the ommatidium does not possess a common polarizer. If this were the case maxima and minima would appear. Autrum found, however, that the exposure of an ommatidium with polarized light gives clearly higher potentials than the exposure with unpolarized light of the same intensity. This result corresponds very well with his hypothesis. In polarized light the intensity of the light becomes much stronger within the plane of its vibration when compared with unpolarized natural light of the same fundamental intensity. As a result of this fact the effective intensity becomes higher in those sensory cells of an ommatidium which are the most sensitive to the light of the given plane of vibration. It is irrelevant that the other sensory cells of the same ommatidium contribute a lower potential because the highest potential within the lot is decisive for the entire effect [Autrum & Stumpf (1)].

Optical research on the effect of polarization carried out with eyes of bees and flies have produced results which agree with these findings. Light arriving vertically on the ommatidium is not doubly refracted either by the corneal lens or by the crystalline cone. When the visual cells of an ommatidium are investigated in the direction of their axis, the rhabdomeres, however, show double refraction. Theoretically it is to be supposed that one of the two directions of vibration of double refracted light is suppressed in the visual

---

[4] The head was put on a fine steel pin (indifferent electrode). The different electrode was a second pin stuck into the eye. The exposure of a single ommatidium was made possible by a source of light the angle of vision of which under the given conditions was considerably smaller for the eye than the angle of aperture of the ommatidium.

54   FRISCH AND LINDAUER

cell. The small size of the elements in question has not made it possible to prove this [Menzer & Stockhammer (31)]. It has been proved, however, that for flies (*Calliphora et al.*) the planes of vibration of the speedier rays are radially arranged and correspond thus with the position of the visual cells themselves.

Reactions to the direction of vibration of polarized light have, since their discovery among bees, been proved among many other Arthropoda: such as ants [Schifferer, see von Frisch (12, p. 220); Carthy (7); Vowles (41)], among flies [Autrum & Stumpf (1) for *Calliphora;* Wellington (45) for *Sarcophaga;* Stephens, Fingerman & Brown (39) for *Drosophila*], among beetles [Birukow (5) for *Geotrupes*], among larvae of Lepidoptera and Hymenoptera [Wellington, Sullivan & Green (46)], among larvae of Diptera [Baylor & Smith (3)], among crustaceans [Kerz (24) for *Eupagurus*], among *Amphipoda* [Pardi & Papi (33, 35, 36) for *Talitrus*], among several *Cladocera* [Baylor & Smith (3); Eckert (8)], among spiders for the wolf spider *Arctosa* [Papi (32)], among mites for *Hydrachna* [Baylor & Smith (3)] and finally in *Agalena* [Görner (unpublished data)] and in *Limulus* [Waterman (42, 43, 44)].

Some authors think that the reactions to polarized light, which they were able to observe, can be explained by a double refraction in the dioptric system. Such an explanation is however only possible if the effective light rays arrive obliquely (to the longitudinal axis of the ommatidium) on the eye [see Berger & Segal (4); Stephens, Fingerman & Brown (39); Baylor & Smith (3)]. It is possible that the reactions to polarized light of those Arthropoda which have been examined are not all attributable to the same type of function. For bees and flies, at least, we cannot agree with the above mentioned supposition.

The results obtained by Autrum & Stumpf, by Menzer & Stockhammer, and by Stockhammer indicate that there is a double refraction within the visual cells of each ommatidium. Furthermore the bees are able to analyse the direction of vibration of polarized light extremely well, even when the experimental conditions are such that we cannot assume the participation of obliquely arriving rays.

The Importance of the Sun as a Means of Orientation

It is not difficult to conceive that the sun can be used as a compass when the bees take a short flight. By flying at a constant angle to the sun they are able to keep on a straight course. Furthermore we have seen that they use the angle to the sun in order to indicate the direction of a goal to their hive mates. Even this is easily comprehensible for short periods of activity.

During recent years we found that bees are capable of using the sun for orientation over periods of many hours during which the sun's position continually changes. They seem to be acquainted with the movement of the sun and to possess a well developed time sense which enables them to calculate the change of the solar position and to correct accordingly their direction of flight. The proof that they possess this faculty was first obtained by a dis-

LANGUAGE OF THE HONEY BEE   55

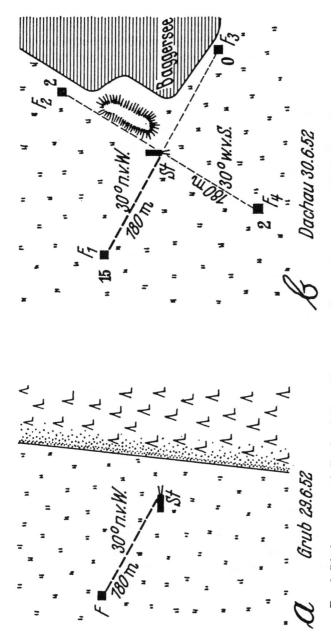

Fig. 3. Displacement test indicating ability of honey bee to allow for change in solar position. (a) Position of hive (St) and feeding place (F) on afternoon of June 29, 1952 near Grub. The marked bees were fed during this afternoon. (b) New position of hive and feeding places (F¹ to F⁴) on the following morning (June 30) near Dachau. The numbers to the left of each feeding place indicate the numbers of previously marked bees collected at each feeding place this morning. Every arriving bee is immediately caught so that the other bees at the hive cannot be informed of the change in the situation.

56   FRISCH AND LINDAUER

placement test [von Frisch (12); von Frisch & Lindauer (19)]. As an example we show here the method and its results.

An experimental hive was brought into a region entirely unknown to the colony. On the afternoon of June 29, 1952 we opened the flight hole and fed some of the numbered and thus individually known bees until the evening. The feeding place was 180 m. from the hive in the direction 30° north of west (Fig. 3a). On the following morning the hive was taken to another entirely unknown landscape of different structure. The bees, therefore, found no familiar landmarks for their orientation during flight. The sun stood in the east, whereas during the preceding afternoon it had been in the west. Furthermore the hive had a different position. Thus the position of the flight hole could not help in the orientation of the bees. All around the hive at a distance of 180 m. we put up feeding plates. Notwithstanding the different landscape, the different position of hive and sun and in spite of the different time, the known bees, with few exceptions, flew to the feeding place 30° north of west which they had found and visited during the preceding afternoon. The other feeding places were almost completely ignored (Fig. 3b). In their search for the previously known feeding place in this unknown territory the bees can only have been helped by the sun. In order to use the sun as a means of orientation bees must be able to calculate the solar movements.

This capability is, however, shown in a much more surprising fashion by the following observation. Gathering bees do not dance for a longer time than 1 or 2 min. They start afterwards for a new flight. Some of the scouts become nonstop-dancers, when they find a nesting site for a preparing swarm, performing for hours their orientated waggle dance. During that time they slowly correct the direction of their dance according to the change of the sun's position. Even when they dance in a hive, set up in a closed room, so that they can see neither sun nor sky, they change their dance accordingly. They hereby demonstrate most impressively their knowledge of the solar position for every hour of the day and this even in a completely closed room [von Frisch (15); Lindauer (29)].

There is no doubt that among the Arthropoda the honey bees with their complicated instincts are on a very high rank of evolution. It is, however, remarkable that the use of the sun as a compass in correlation with a perfect time sense apparently does not necessitate a very high development of the mental faculties, because the same means of orientation is known to exist among crustaceans and spiders. *Talitrus saltator*, a crustacean which belongs to the Amphipoda, lives on the seashore. If by a storm or a human hand this animal is carried out of the region of moist sand which it inhabits and is put down on dry land, it makes his way hastily back to the shore. Even if it is carried many miles inland it takes exactly the same direction which would normally bring it from dry sand back to the moist sand of its habitat. This occurs only under the condition that it is given the opportunity of detecting the solar position, either directly or with the help of the polarized light from the blue sky. It finds this definite direction at every hour of the day even if

LANGUAGE OF THE HONEY BEE   57

it has previously been kept for hours in a dark room [Pardi & Papi (33, 35, 36)]. The same means of orientation has been found in *Tylos latreillii*, a crustacean belonging to the Isopoda, which is also an inhabitant of the moist sand of beaches [Pardi (34)] and in the wolf spider *Arctosa perita*, which lives on the shores of the sea and lakes [Papi (32)]. It is surmisable that the sun serves also as a compass for the long distance migrations of insects such as butterflies. This method of orientation, is however, not only known among the Arthropoda but even among birds. Birds can use the sun in connection with the time of day as a guide for their migrations [Kramer (25, 26)].

THE RELATIVE IMPORTANCE OF LANDMARKS AND THE SOLAR COMPASS

It has been known for a long time that bees use conspicuous landmarks such as trees, rocks, houses, etc. for their orientation in the landscape. But as we have seen, they orientate themselves as well by the position of the sun and by polarized light. In order to find out which of the two, the terrestrial or the celestial clue, is the more important for the bee, we set them in competition in the following experiment.

We trained bees to find a feeding place 180 meters to the south of their hive. The arrangement was such that on their way from hive to food source the bees flew along the edge of a forest running from north to south. On the following day we transplanted the colony in a new unknown landscape with a similar edge of forest, this time running from east to west. The bees did not look for the feeding place in the trained direction to the south of the hive, but flew along the edge of the forest 180 m. to the west. The very conspicuous landmark won the competition over the solar compass. In another set of conditions, however, when the edge of the forest was 200 m. lateral to the course to be flown, the forest being at an angle of vision of 3 to 4°, after their transplantation the bees followed in the majority of cases the clue given by the celestial compass.

In the same way as an adjacent forest edge, a shore line or a road is a superiour means of orientation when compared to solar orientation. When geographic characters are used it appears to be most important that the course to be flown follow a continuous and unbroken landmark. Correspondingly experiments have shown that a single high tree or an isolated cluster of bushes were not instrumental in deviating the bees from their accustomed direction of flight after a transplantation into a new landscape [von Frisch & Lindauer (19)].

Our method of investigation is time consuming but it gives us the possibility of judging the relative importance of the different means of orientation in bees.

LITERATURE CITED

1. Autrum, H., and Stumpf, H., *Z. Naturforsch.*, 5b, 116–22 (1950)
2. Baltzer, F., *Archiv Julius Klaus-Stift. Vererbungsforsch., Sozianthropol. u. Rassenhy.*, 27, 197–206 (1952)
3. Baylor, E. R., and Smith, F. E., *Am. Naturalist*, 87, 97–101 (1953)
4. Berger, P., and Segal, M. I., *Compt. rend.*, 234, 1308–10 (1952)

58   FRISCH AND LINDAUER

5. Birukow, G., *Naturwissenschaften*, **40**, 611–12 (1953)
6. Birukow, G., *Z. vergleich. Physiol.*, **36**, 176–211 (1954)
7. Carthy, J. D., *Behaviour*, **3**, 275–303 (1951)
8. Eckert, B., *Českoslov. Biol.*, **2**, 76–80 (1953)
9. von Frisch, K., *Österr. zool. Z.*, **1**, 1–48 (1946)
10. von Frisch, K., *Naturwissenschaften*, **35**, 12–23, 38–43 (1948)
11. von Frisch, K., *Bees—their Vision, Chemical Senses and Language* (Cornell University Press, Ithaca, N. Y., 119 pp., 1950)
12. von Frisch, K., *Experientia*, **6**, 210–21 (1950)
13. von Frisch, K., *Naturwissenschaften*, **38**, 105–12 (1951)
14. von Frisch, K., *Psychol. Rundschau*, **4**, 235–36 (1953)
15. von Frisch, K., *Verhandl. deut. Zool. Ges. Freiburg 1952*, 58–72 (Leipzig, Germany, 1953)
16. von Frisch, K., *Sitzber. bayr. Akad. Wiss., Math.-naturw. Kl.*, 197–99 (München, Germany, 1954)
17. von Frisch, K., *Festrede bayr. Akad. Wiss.* (Verlag bayr. Akad. Wiss., München, Germany, 1955)
18. von Frisch, K., Heran, G., and Lindauer, M., *Z. vergleich. Physiol.*, **35**, 219–45 (1953)
19. von Frisch, K., and Lindauer, M., *Naturwissenschaften*, **41**, 245–53 (1954)
20. Haldane, J. B. S., and Spurway, H., *Insectes Sociaux (Paris)*, **1**, 247–83 (1954)
21. Hein, G., *Experientia*, **6**, 142 (1950)
22. Heran, H., *Z. vergleich. Physiol.*, **34**, 179–206 (1952)
23. Heran, H., and Wanke, L., *Z. vergleich. Physiol.*, **34**, 383–93 (1952)
24. Kerz, M., *Experientia*, **6**, 427 (1950)
25. Kramer, G., *Ibis*, **94**, 265–85 (1952)
26. Kramer, G., *Verhandl. deut. Zool. Ges. Freiburg 1952*, 72–84 (Leipzig, Germany, 1953)
27. Lindauer, M., *Maturwissenschaften*, **38**, 509–13 (1951)
28. Lindauer, M., *Naturwissenschaften*, **40**, 379–85 (1953)
29. Lindauer, M., *Naturwissenschaften*, **41**, 506–7 (1954)
30. Lindauer, M., *Z. vergleich. Physiol.*, **37**, 263–324 (1955)
31. Menzer, G., and Stockhammer, K., *Naturwissenschaften*, **38**, 190–91 (1951)
32. Papi, F., *Z. vergleich. Physiol.*, **37**, 230–33 (1955)
33. Papi, F., and Pardi, L., *Z. vergleich. Physiol.*, **35**, 490–518 (1953)
34. Pardi, L., *Z. Tierpsychol.*, **11**, 175–81 (1954)
35. Pardi, L., and Papi, F., *Naturwissenschaften* , **39**, 262–63 (1952)
36. Pardi, L., and Papi, F., *Z. vergleich. Physiol.*, **35**, 459–89 (1953)
37. Révész, G., *Psychol. Rundschau*, **4**, 81–83 (1953)
38. Rothschild, F. S., *Schweiz. Z. Psychol. u. ihre Anwendungen*, **12**, 177–99 (1953)
39. Stephens, G. C., Fingerman, M., and Brown, F. A., *Ann. Entomol. Soc. Amer.*, **46**, 75–83 (1953)
40. Tschumi, P., *Schweiz. Bienenzeitung*, 129–34 (1950)
41. Vowles, D. M., *J. Exptl. Biol.*, **31**, 341–55 (1954)
42. Waterman, T. H., *Science*, **111**, 252–54 (1950)
43. Waterman, T. H., *Proc. Natl. Acad. Sci. U. S.*, **40**, 258–62 (1954)
44. Waterman, T. H., and Wiersma, C. A. G., *J. Exptl. Zool.*, **126**, 59–85 (1954)
45. Wellington, W. G., *Nature*, **172**, 1177 (1953)
46. Wellington, W. G., Sullivan, C. R., and Green, G. W., *Can. J. Zool.*, **29**, 330–51 (1951)

THE SENSORY BASIS OF OBSTACLE AVOIDANCE BY FLYING BATS

DONALD R. GRIFFIN AND ROBERT GALAMBOS

Biological Laboratories, Harvard University, Cambridge, Massachusetts

Anyone who has ever watched a bat in flight has been impressed by its agility and the quickness with which it dodges any solid object in its path. Most bats are nocturnal and can fly through dark forests or long tortuous caves where their eyes must be useless for lack of light. Spallanzani (1794) blinded bats and found that they could fly as skillfully as before. Since these classic experiments, many attempts have been made to learn what sense organ enabled sightless flying bats to avoid collisions with obstacles and with each other.

Shortly after Spallanzani's experiments Jurine (1798) found that covering the ears of bats impaired their ability to avoid obstacles. Spallanzani himself confirmed these experiments towards the end of his life (see Senebier, 1807). This early auditory explanation was opposed by Cuvier (1800), who believed that the obstacles were detected by means of tactile receptors on the bat's skin, especially the skin of the wing membranes. Cuvier's wing membrane tactile theory has been accepted almost without question by many naturalists such as Godman (1831), Bell (1874), Flower and Lydekker (1881), Millais ('04), Whitaker ('06), Seton ('09) and Barrett-Hamilton ('11). Indeed the original experiments of Jurine and Spallanzani seem to have been forgotten, and as far as we know they have not been cited in any paper since 1874.

Rollinat and Trouessart ('00) covered the ears of several bats and observed considerable impairment of the flight in five out of six such experiments which they describe in detail.

481

482   DONALD R. GRIFFIN AND ROBERT GALAMBOS

They therefore concluded that the ear was the most important sense organ involved. Hahn ('08) made an extensive series of tests in which the percentage of hits was recorded as bats flew between vertical metal wires. Eyes and ears were covered, external ears and tragi were cut off and the whole skin including the wing membrane was coated with a thick layer of vaseline. Of these treatments, only the plugging of the ears increased the number of hits to any great extent. Hahn therefore concluded "obstacles are perceived chiefly through sense organs located in the internal ear. Perception is probably due to condensation of the atmosphere between the moving animal and the obstacle it is approaching." Since a flying bat seldom makes any audible sound, the sense of hearing as such was not considered capable of playing any part in obstacle avoidance.

However, Whitaker ('06) had pointed out that bats have very acute hearing and seem to be especially sensitive to high-pitched sounds. And Hartridge ('20), a physiologist, set forth clearly the auditory hypothesis of obstacle avoidance as follows: "I suggest that bats during flight emit a short wave-length note and that this sound is reflected from objects in the vicinity. The reflected sound gives the bat information concerning its surroundings . . . It is probable that if a bat made use of short wave-length sound it would be able to estimate the position in space of an object ahead of it with considerable accuracy." Several years later, Pierce and Griffin ('38) found experimentally that bats did emit a fairly intense supersonic sound with a frequency of forty-five to fifty thousand vibrations per second. This sound is of too high a frequency or too short a wave-length to be heard by the human ear, which is insensitive to sounds consisting of more than about twenty thousand vibrations per second.

This discovery opened new possibilities for the auditory theory, and in 1939 and 1940 we attempted to test and develop this theory more fully. It had been amply demonstrated that covering the ears of bats would cause them to strike obstacles, but we wished to have adequate experimental controls, not previously available, to establish with certainty that this impairment was not due to injury or irritation by the ear plug. Furthermore the auditory theory predicts that covering the mouth should prevent emission of sounds and hence avoidance of obstacles. Total elimination of either hearing or sound production should result in complete inability to perceive obstacles. The bat should then strike them just as often as an inert object of the same size would strike if thrown aimlessly at the same obstacles. In man, both ears are required for accurate auditory localization of sound sources, so that covering one of a bat's ears should seriously impair its flight. Experiments were designed to explore all these points and the results are described below.

Preliminary tests were made by one of us (DRG) during the summer of 1939 while he was a research fellow at the Edmund Niles Huyck Preserve at Rensselaerville, New York. We are much indebted to the officers of the Preserve for this opportunity and to Mrs. F. S. Greene of Rensselaerville for the use of the building in which the experiments were conducted. The remaining work was done at the Physical Laboratories of Harvard University. Apparatus for producing, detecting and recording supersonic sounds was indispensable for this work and it was placed at our disposal by Professor G. W. Pierce. We are correspondingly grateful to Professor Pierce for his generosity both with laboratory facilities and constructive advice. Mr. James Brewster of the Harvard Film Service took motion pictures of flying bats which were extremely useful in analysing the bats' behavior. Finally we are much indebted to Professor A. C. Redfield for his guidance and criticism and his reading of this manuscript.

Elsewhere we plan to present, (1) a detailed analysis of the sounds produced by bats and a discussion of the auditory theory of obstacle avoidance, (2) data on the sensitivity of the bat's ear to supersonic sounds, and (3) a complete history of the development of this problem.

484   DONALD R. GRIFFIN AND ROBERT GALAMBOS

METHODS

The ability of bats to avoid obstacles was measured by making them fly through a standardized barrier of metal wires, much like those used by Hahn ('08). Such tests were made with 144 normal bats and with several whose eyes, ears and mouths had been covered. None of these impairments injured the bats in any way, and at the end of each experiment the animals could be restored to normal. The terms "blind," "deaf," and "gagged" will be used in this paper to describe bats with the corresponding organs temporarily impaired by methods described in detail below.

Metal wires about 1 mm. in diameter were suspended vertically from the ceiling at intervals of 30 cm. ($\pm$ 2.5 cm.) in a row across the middle of a small room. The room used at Rensselaerville, New York, with bats numbers 1 to 28 was approximately 9 feet square and 7 feet high, while the room at the Physical Laboratories of Harvard University used for the remaining experiments was a sound-proof chamber 4.5 m. by 3.3 m. and 3.3 m. high. The walls and ceiling of the room at Rensselaerville were wood, and those of the Laboratory room were thick rock wool padding covered by cheesecloth, a surface which had been designed to reflect as little sound as possible.

Only one bat was tested at a time, and as it flew back and forth through this barrier a count was kept of the number of hits and misses. When the bat touched a wire there was an easily audible sound; and even when a wing tip brushed lightly against it, a wire would continue to vibrate perceptibly after the bat had passed. If a bat struck a wire and turned back or fell back into the half of the room from which it first approached the barrier, the collision was not counted. Nor did we count cases in which the bat turned back from the barrier without striking it. This simplified the observations and yielded data quite satisfactory for comparing bats under different conditions. We could see no indication that when we impaired sense organs, we altered the frequency with which a bat would turn back after it had once struck a wire.

No trial was counted in which the bat passed between the wall and the vertical wire nearest to it, or struck the ceiling or the floor while passing through the barrier. Thus in all recorded cases the bat was flying through a simple barrier of vertical wires at 30 cm. intervals uncomplicated by other obstructions. Hahn's data were taken in a slightly different manner and are not strictly comparable with ours.

Bats nos. 1–28 were caught at summer colonies in buildings at Rensselaerville, New York, and were used the same day they were caught or the next day. Nos. 29–46 had been taken while hibernating in various caves in western New England, and they were kept in hibernation in a laboratory cold room at about 5.6° C. When needed for an experiment they would awaken in a few minutes if brought to room temperature, and when returned to the cold room they soon relapsed into hibernation again. Nos. 47 and 48 were caught in a summer colony in Massachusetts and were used the following day.

The criterion selected for judging ability to avoid obstacles was the per cent of trials (as defined above) in which the bat touched a wire. Thus a low percentage of hits (referred to below as "score") means good avoidance, and a high score indicates little or no ability to dodge the wires. Like Hahn ('08), we found that our barrier was so difficult that no bat avoided it completely. One or two especially skillful individuals could pass through the wires thirty or forty times without a hit, but this was very exceptional. Therefore we were interested in the difference between the scores of normal, blind, deaf and gagged bats. Often only fifty trials, or occasionally even fewer, could be obtained in each series without tiring the bat to the point that his score increased greatly. In many cases, however, series of 200 or more trials were possible.

Objectivity in drawing conclusions from the data necessitated a statistical test of the significance of these differences, and the Chi squared test described by Fisher ('25, p. 84) was selected as the best available for our data. The statistical function Chi squared (χ^2) was obtained from the equation:

$$\chi^2 = \frac{(ad - bc)^2 \, (a + b + c + d)}{(a + b) \, (c + d) \, (a + c) \, (b + d)}$$

where

a = number of hits of bats in one condition,
b = number of misses of bats in one condition,
c = number of hits of bats in a second condition,
d = number of misses of bats in a second condition.

By means of tables derived from the equation for the "normal curve" of probability, one can compute for any value of χ^2 the probability P that the difference between the two scores is significant, i.e., the probability that the two scores were obtained under basically different conditions. This probability of significance of the difference is the real criterion by which the experiment should be judged. A probability of 0.95 means that the chances are 19:1 that the difference is significant; P = 0.98 means 98:2 or 49:1 and P = 0.99 means 99:1. For most purposes, 19:1 is considered as proving a real difference, but 49:1 or 99:1 are sometimes used as conservative standards. In the tables containing our data, both χ^2 and P are listed for all comparisons of bats' scores under different conditions.

OBSTACLE AVOIDANCE OF NORMAL BATS

There was much individual variation in the normal bat's ability to avoid the barrier of wires described in the preceding section. During our experiments, each of the 144 bats passed through the barrier fifty or more times; and the distribution of their scores is shown in table 1. A bat which is tired, in poor health or incompletely revived from hibernation will fly in a very clumsy fashion. Obviously we are dealing with a very refined skill which reaches its full development only under the best of conditions. Even the bats listed in table 1 represent a selection of the most active individuals. When experimentally impairing sense organs we naturally chose bats whose normal performance was the best obtainable.

TABLE 1
Obstacle avoidance of normal bats

SCORE PER CENT HITS	MYOTIS L LUCIFUGUS	MYOTIS KEENII SEPTEN- TRIONALIS	MYOTIS SODALIS	PIPISTRELLUS SUBFLAVUS	EPTESICUS FUSCUS
5.1 to 10.0	..	1	..	..	..
10.1 to 15.0	1	1	..	..	..
15.1 to 20.0	5	3	1	..	..
20.1 to 25.0	13	3	..	..	1
25.1 to 30.0	18	..	..	..	..
30.1 to 35.0	24	..	..	..	..
35.1 to 40.0	25	1	..	1	..
40.1 to 45.0	14	..	..	..	1
45.1 to 50.0	14	..	..	..	..
50.1 to 55.0	7	..	..	..	..
55.1 to 60.0	6	..	..	..	..
60.1 to 65.0	2	..	..	..	2
Total	129	9	1	1	4
Average per cent hits	35	20	..	..	48

Otherwise a treatment which interfered with the ability might not have been recognized as such because the bat's skill was impaired even in the control series. Therefore the scores recorded for the experimental animals "intact" (all senses unimpaired) are often considerably below the average for normal bats.

Eptesicus f. fuscus is a larger bat than the others with a wingspread of about 29 cm. and its flight seems much clumsier with fewer turns and swallow-like maneuvers. Hahn ('08) noted that Eptesicus f. fuscus and Corynorhinus macrotus seemed less agile fliers than Myotis or Pipistrellus. Borell ('37), Van Tyne ('33) and Osgood ('35) describe methods of collecting bats with wires and bird nets, and it is interesting to note that the smaller insectivorous bats of the family Vespertilionidae seem less likely to collide with such obstacles than such bats as Tadarida and Artibeus.

OBSTACLE AVOIDANCE OF BATS WITH THEIR EYES COVERED

Many experimenters from Spallanzani in 1793 to Hahn in 1908 have shown that blinded bats can fly and avoid obstacles as well as intact controls. In the course of our experiments with other sense organs, many bats' eyes were covered with dark blue collodion mimeograph correction fluid. This material dries opaque in 5 to 10 minutes if not too thick. It gave no evidence of injuring the bats, and could be peeled off or washed away with ether at the end of an experiment.

There is very little difference in the flight of normal and blind bats. Hahn observed that some bats flew more slowly and cautiously when deprived of their eyesight; and in the vicinity of obstacles there seemed to us to be somewhat more of the fluttering and hovering which Eisentraut ('37) describes as "Rüttelflug" in contrast to the normal flapping flight or "Ruderflug." However, most bats with their eyes covered flew through the wires with that same elaborate agility at which all observers have marvelled. Slow motion moving pictures show that when the bat approached our barrier, it would often modify its flight in some appropriate manner such as gliding through with wings partly folded, or banking to 90° in a sharp turn with wings spread vertically. The results of our obstacle avoidance tests with blinded bats are shown in table 2. The average score of intact bats was 30 per cent and of blinded bats 24 per cent. No statistical test is needed to prove that such differences are the result of chance variations. Whatever sense may warn the bat of wires and walls ahead, it certainly is not the eye.

Hahn noted a few cases in which a bat did seem to be guiding its flight by eyesight alone. These bats, which included Myotis l. lucifugus, Myotis keenii septentrionalis and one Corynorhinus macrotus, flew repeatedly against glass windows and wire screens. We have never observed such behavior in normal or blinded bats. Furthermore, eyesight is of little use to animals whose hearing has been impaired, as will be shown below (p. 492 and table 5).

Several bats listed in table 2 and in Hahn's data showed marked improvement in obstacle avoidance when their eyes were covered. This phenomenon, if it be more than accidental, might result from increased attention and care in detecting the wires by means of some very efficient non-visual sense organ.

OBSTACLE AVOIDANCE BY BATS   489

TABLE 2

Obstacle avoidance of bats with eyes covered
Experiments with Myotis l. lucifugus

NUMBER	INTACT		EYES COVERED	
	No. of trials	Per cent hits	No. of trials	Per cent hits
2	73	30	57	32
3	170	38	100	34
4	69	38	50	38
5	430	15	100	18
7	100	47	180	23
8	120	25	50	18
10	245	32	100	40
11	100	33	30	7
12	230	44	70	46
14	160	29	50	38
17	50	56	40	23
18	50	32	150	37
19	50	26	50	18
21	78	27	50	38
22	64	28	50	34
24	150	19	50	18
25	110	21	30	26
26	120	33	100	23
27	100	33	100	29
28	106	33	100	24
30	29	59	30	63
32	53	32	125	43
34	98	16	108	17
35	50	48	65	22
36	53	28	50	32
44	50	24	50	32
45	60	23	61	10
46	53	32	20	15
Total trials and weighted average	3021	30	2016	24

OBSTACLE AVOIDANCE OF BATS WITH THEIR EARS COVERED

The hearing of twenty-six bats was impaired by the various methods described below and the Chi squared test was used to determine the statistical significance of the differences between these bats' scores when normal and when deaf. Tables 3 and 4 show the actual scores. In all cases where we are certain that the external meatus was completely closed, the bat's flight became strikingly abnormal and its ability to

TABLE 3
Obstacle avoidance of bats (M. lucifugus) intact and with ears covered

NUMBER	INTACT		EYES COVERED		χ²	PROBABILITY OF SIGNIFICANCE
	Trials	Per cent hits	Trials	Per cent hits		
1	70	30	150	79 (b)	48.7	0.99 – 1.00
1	100	38 (c)	150	79 (b)	42.4	0.99 – 1.00
9	58	30	280	44 (a)	4.44	0.95 – 0.98
10	245	32	50	62 (a)	15.8	0.99 – 1.00
12	210	46	50	64 (a)	5.14	0.95 – 0.98
14	160	29	100	69 (a)	40.5	0.99 – 1.00
15	115	34	50	70 (a)	18.4	0.99 – 1.00
16	90	33	50	68 (a)	15.5	0.99 – 1.00
20	57	23	377	67 (a)	40.3	0.99 – 1.00
23	102	32	100	67 (a)	25.6	0.99 – 1.00
29	30	30 (d)	30	50 (e)	2.5	not significant
45	60	23	20	80 (e)	20.5	0.99 – 1.00
Total trials and weighted average	1297	34	1407	65		

(a) ears sewed   (b) ears covered with cellodion   (c) holes opened in cellodio
(d) glass tubes in ears open   (e) glass tubes in ears closed

TABLE 4
Obstacle avoidance of bats with eyes covered and with ears covered

NUMBER	SPECIES	EYES COVERED		EYES AND EARS COVERED		χ²	PROBABILITY OF SIGNIFICANCE
		Trials	Per cent hits	Trials	Per cent hits		
7	M. lucifugus	180	23	180	70 (a)	77.3	0.99 – 1.00
8	"	50	18	90	63 (a)	26.5	0.99 – 1.00
14	"	50	38	50	72 (a)	11.7	0.99 – 1.00
18	"	150	37	70	63 (a)	13.2	0.99 – 1.00
19	"	50	18	50	68 (a)	25.4	0.99 – 1.00
27	"	100	16	50	92 (a)	79.1	0.99 – 1.00
28	"	100	24	50	82 (a)	46.1	0.99 – 1.00
29	"	114	37 (d)	120	58 (e)	10.8	0.99 – 1.00
31	Eptesicus fuscus	30	63 (d)	30	87 (e)	4.35	0.95 – 0.98
34	M. sodalis	108	17	60	50 (a)	21.0	0.99 – 1.00
40	Pipistrellus	50	24 (d)	50	60 (e)	13.3	0.99 – 1.00
42	M. keenii	100	24 (d)	50	62 (e)	20.8	0.99 – 1.00
43	M. keenii	130	23 (d)	110	47 (e)	15.6	0.99 – 1.00
44	M. lucifugus	60	45 (d)	110	66 (e)	7.35	0.99 – 1.00
45	M. lucifugus	92	41 (d)	82	83 (e)	31.6	0.99 – 1.00
46	M. keenii	60	42 (d)	30	73 (e)	8.04	0.99 – 1.00
47	M. lucifugus	54	30	54	70 (b)	17.9	0.99 – 1.00
Totals and weighted average		1478	29	1236	67		

(a) ears sewed   (b) ears covered with cellodion   (d) glass tubes in ears open
(e) glass tubes in ears closed

avoid obstacles was drastically impaired. Whereas twelve bats tested had an average avoidance score of 34 per cent when intact, the average when deaf was 65 per cent. Seventeen others averaged 29 per cent blind and 67 per cent when blind and deaf. In the great majority of cases, there was a probability of significance greater than 0.99. The two bats (nos. 29 and 46) which did not show a significant difference had become fatigued after a very few trials.

A deaf bat is very reluctant to fly at all. Proddings which would frighten a normal or a blind bat into a hurried flight produce no response at all in the deaf animal. Indeed such bats seem totally preoccupied in vigorous scratching of the collodion covering their ears. One must ordinarily drop such a bat in mid air before it begins to fly, and even then its flight is extremely slow and hesitant. Some deaf bats fly in a straight line until they meet a wall or barrier of wires, or they may drop to the floor and resume without delay their frantic scratching. Others will fly in small circles 5 or 6 feet in diameter. Usually they form the habit of circling consistently in one direction in trial after trial, and it is difficult to make them pass the barrier at all. The experimenter must then release the bat at such an angle to the barrier of wires that its circling will cause it to pass through.

A deaf bat seldom shows any change in flight as it approaches or hits the wires. Conspicuously absent is the normal bat's agile dodging and gliding through the barrier with partly folded wings. Usually the deaf bat is using "Rüttelflug" (Eisentraut, '37) almost entirely and his speed is far below normal. As will be described elsewhere in detail normal bats regularly change the rate of their supersonic note as they approach the barrier of wires. When they fail to do so, they are much more likely to strike the wires. Bats with ears plugged never show any such alteration in their supersonic calls.

On approaching a solid wall, a deaf bat's helplessness is particularly apparent. A head-on collision almost invariably results (when the ear canals are blocked completely); and the bat usually tries to continue flying forward even after the collision. It bumps its nose, bounces backwards off the wall and then flies on again, repeating the process doggedly in evident bewilderment until the loss of altitude at each collision brings it to the floor. The cheesecloth walls of the laboratory room were fortunately soft enough so that a bat was not injured in any way even by a long series of such head-on collisions.

Such repeated collisions with the wall after the initial bump were far less frequent in the room at Rensselaerville than in the padded laboratory room. The wooden walls of the former room may have reflected enough of the sound emitted by the bat to penetrate the less efficient ear plugs used in these earlier experiments. The slight reflection from the cheese-cloth-rock wool walls, although detected by the intact bat, failed to pass through the plugged ear canal sufficiently to warn the deaf bat of the wall ahead. It seemed as though the bat continued to fly forward as long as it heard no warning sound reflection even though its nose had already bumped the wall. Such a bat was in the same dilemma as a bird which flutters helplessly against a glass window; and seemed just as incapable of adjusting itseslf to the unprecedented situation.

Several normal bats bumped head-on into a single sheet of cheesecloth stretched across a doorway. This was not a consistent occurrence, but it suggests that such surfaces which reflect very little sound may deceive a normal bat just as a window pane deceives a bird.

Covering the eyes of a deaf bat makes little difference to its flight or obstacle avoidance as shown by the similarity of the scores in tables 3 and 4, and also by table 5. Perhaps it is only large obstacles which are avoided by visual means.

Ear plugging experiments have been criticized on the grounds that the poor flight and obstacle avoidance is due to injury or irritation arising from the ear plug. This is rendered unlikely by the experiments of Rollinat and Trouessart ('00) and Hahn ('08), in which they cut off the entire

OBSTACLE AVOIDANCE BY BATS   493

TABLE 5

The effects of covering the eyes of bats (M. lucifugus) which are already deafened

NUMBER	ONLY EARS COVERED		EYES AND EARS COVERED		χ^2	PROBABILITY OF SIGNIFICANCE
	Trials	Per cent hits	Trials	Per cent hits		
14	100	69 (a)	50	72 (a)	0.143	not significant
23	50	62 (a)	50	76 (a)	2.30	not significant
29	30	50 (b)	120	58 (b)	0.678	not significant
45	20	80 (b)	82	83 (b)	0.090	not significant
Totals and weighted average	200	65.5	302	70.1	0.04	not significant

(a) ears sewed   (b) ear tubes closed

external ear and tragus. This drastic mutilation did not impair the bats' ability to avoid obstacles, and it is difficult to see why the mild irritation of an ear plug should do so. Furthermore, most of the deafened bats returned to their normal level of obstacle avoidance scores when the ear plugs were removed. Table 6 shows the results in several typical cases where bats were tested intact or blind after the trials in the deafened condition.

In our experiments bats numbers 6, 7 to 16, 18 to 20, 23, 27, 28, and 34 had the cavity of the external ear filled with a tightly compressed wad of cotton and the various flaps of

TABLE 6

Obstacle avoidance of bats (M. lucifugus) before and after ears were covered

NUMBER	INTACT		EARS COVERED		INTACT (LATER)	
	Trials	Per cent hits	Trials	Per cent hits	Trials	Per cent hits
6	100	24	50	70 (a)	150	35
14	110	27	50	72 (a)	50	32
15	65	34	50	70 (a)	50	34
16	60	38	50	68 (a)	30	23
23	52	35	50	76 (a)	50	28
	ONLY EYES COVERED		EYES AND EARS COVERED		ONLY EYES COVERED (LATER)	
7	120	22	180	69 (a)	60	25
8	50	18	90	63 (a)	20	20
18	100	31	70	63 (a)	50	48
27	50	4	50	92 (a)	50	28

(a) ears sewed

494   DONALD R. GRIFFIN AND ROBERT GALAMBOS

the pinna were sewed together over this wad so that it was
securely held in place. The folded pinna itself thus served
to close the ear in addition to the compressed cotton. The
edge of the pinna is poorly supplied with blood vessels and
never bled when perforated by the needle. The bats did not
even seem to resent this sewing process any more than they
resented merely being held in the hand. The control experiments whose protocols are presented below showed that this
slight injury did not in itself interfere with the bats' obstacle
avoidance.

	NUMBER OF TRIALS	PER CENT HITS
Bat No. 5 (Myotis l. lucifugus)		
Intact	110	16
Needle passed through pinna as in sewing up ears, but thread pulled out	100	15
Bat No. 11 (Myotis l. lucifugus)		
Intact	100	33
Eyes covered with collodion	30	7
Ears sewed, but left open, no cotton	40	40
Ears sewed together across top of head	20	20

Bats numbers 1, 47, and 48 had their ears covered with
the same collodion which was used to cover bats' eyes. The
fact that this collodion was not of itself detrimental to
obstacle avoidance is shown by the following experiment:

Bat No. 1 (Myotis l. lucifugus)

CONDITION	NUMBER OF TRIALS	PER CENT HITS
Intact	70	30
Both ears plugged with collodion	50	82
Plugs left in place, but small holes opened through them with a needle	100	38
Fresh collodion applied to top of head	20	40

To provide a final answer to the objection that flight impairment of deaf bats results from injury, we inserted small
glass tubes into the ears. When these tubes were opened
and closed, any injury as irritation remained constant. The
tubes used were 4 to 8 mm. long and 2 to 3 mm. in diameter,
with a flange increasing the diameter at one end by about
1 mm. This flange was pushed into the slight enlargement
of the meatus which lies just inside the base of the pinna.

OBSTACLE AVOIDANCE BY BATS   495

A thread was then passed around the pinna and through
the outer layers of the skin just at the base of the opening
of the external ear. This loop, drawn tight and knotted, held
the tube tightly in place and pressed the base of the pinna
around it so that there was practically no communication
between the ear canal and the outside except through the bore
of the tube. Any remaining chinks were closed by a thin
layer of collodion spread around the tube and the base of
the pinna.

When it was desired to deafen the bat, plugs of tight-fitting
thread dipped in collodion were inserted into these tubes,
and removed again when we wished the bat to hear. At
the end of any series the bat could be restored completely
to a normal condition by cutting and pulling out the threads,
removing the glass tubes and washing off the collodion with
ether.

Bats numbers 21, 31, 40, 42, 43, 44, 45, and 46 flew quite
well with such tubes in their ears but open, while these
same bats with ear tubes closed showed the impaired flight
and high percentage of hits characteristic of all deafened
bats (see tables 3 and 4). The probabilities of significance
of these differences were all 0.99 or greater, except in two
cases where sufficient trials could not be made because of
fatigue. To be sure, the bats with open tubes in their ears
yielded somewhat higher scores than when their ears were
free of all obstructions, but this is scarcely surprising in
view of the drastic alterations in size and shape of the meatus.
The following protocol is typical of several bats tested alternately with ear tubes open and closed:

Bat No. 42, Myotis keenii septentrionalis

CONDITION	TRIALS	PER CENT HITS
Blind, ear tubes closed	10	60
" " " open	30	26
" " " closed	40	63
" " " open	50	22
" " " closed	strikes all objects in its path, very reluctant to fly at all	
" " " open	20	25

496   DONALD R. GRIFFIN AND ROBERT GALAMBOS

Any injury or irritation was constant, and we can say definitely that it was the blocking of the air passage to the ear drum which seriously impaired the bats' ability to avoid the wires.

OBSTACLE AVOIDANCE OF BATS WITH ONE EAR COVERED

With one ear covered a bat strikes our type of wire barriers almost as frequently as when it is completely deaf (see tables 7 and 8). The difference between scores with one ear and with both ears covered has, in most cases, a very low probability of significance. Yet two bats showed P's of 0.95 and 0.98, and if the χ^2 test is applied to the series as a whole, $P = 0.99$ to 1.00 indicating a significant but very slight difference in ability to avoid obstacles.

A few bats refused to fly at all when one ear was covered. Usually such bats land normally on the wall instead of bumping it. They often turned back from the wires or wall while totally deaf bats practically never did so. This indicates that one ear is sufficient to inform the animal of the general proximity of a large obstacle, but that fine discriminations are almost impossible.

TABLE 7

Obstacle avoidance of bats (M. lucifugus) with one ear covered

NUMBER	INTACT		ONE EAR COVERED		χ^2	PROBABILITY OF SIGNIFICANCE
	Trials	Per cent hits	Trials	Per cent hits		
15	115	34	50	56 (a)	7.03	0.99 – 1.00
16	90	33	50	45 (a)	2.02	not significant
	ONLY EYES COVERED		EYES AND ONE EAR COVERED			
7	180	23	50	66 (a)	33.4	0.99 – 1.00
11	90	24	10	80 (a)	13.2	0.99 – 1.00
12	70	46	30	73 (a)	6.44	0.98 – 0.99
14	50	38	100	55 (a)	3.86	0.95 – 0.98
18	150	37	50	58 (a)	7.01	0.99 – 1.00
19	50	18	50	56 (a)	15.6	0.99 – 1.00
Totals and weighted average	590	30	390	59		

(a) ears sewed

OBSTACLE AVOIDANCE BY BATS 497

TABLE 8

Obstacle avoidance of bats (M. lucifugus) with one ear covered and with both ears covered

NUMBER	ONE EAR COVERED		BOTH EARS COVERED		χ^2	PROBABILITY OF SIGNIFICANCE
	Trials	Per cent hits	Trials	Per cent hits		
12	30	73 (a)	50	64 (a)	0.73	not significant
14	100	55 (a)	50	72 (a)	3.97	0.95 – 0.98
15	50	56 (a)	50	70 (a)	2.15	not significant
16	60	54 (a)	50	68 (a)	5.84	0.98 – 0.99
	EYES AND ONE EAR COVERED		EYES AND BOTH EARS COVERED			
7	50	66 (a)	180	69 (a)	0.15	not significant
12	20	80 (a)	30	80 (a)	0	" "
13	50	68 (a)	156	73 (a)	0.483	" "
18	50	58 (a)	70	63 (a)	0.289-	" "
19	50	56 (a)	50	68 (a)	1.54	" "
45	15	73 (b)	82	83 (c)	0.77	" "
47	50	78 (d)	50	76 (d)	0.06	" "
48	35	71 (d)	35	60 (d)	1.03	" "
Totals and weighted average	560	62	853	71	13.7*	0.99 – 1.00*

(a) ears sewed   (b) ear tubes open   (c) ear tubes closed

(d) ears covered with cellodion   (*) χ^2 and Probability for whole series

Most theories explaining the localization of sound sources assume that the judgment is based on differences of phase, of intensity, or of time of first arrival of the sound waves at the two ears. At first thought, therefore, it might seem that the bat with only one ear closed should have no ability to localize sound sources; and the slight improvement shown by the bat with one ear freed might be used to argue against the auditory theory. Actually, however, the difference in scores is so slight that it might well result if the bat merely knew it was in the vicinity of the barrier and decreased the average wingspread or turned so as to approach the barrier more nearly at right angles. The chief difference between "one ear" bats and deaf ones is that the former turn away from obstacles and land normally on walls. These bats may be warned of the obstacle simply by the fact that the reflected sound, although heard with only one ear, nevertheless becomes louder as the bat flies forwards. Or the animal

498   DONALD R. GRIFFIN AND ROBERT GALAMBOS

might be able to accomplish a rudimentary form of localization by turning his head and noting the change in intensity of the reflected sound. This explanation assumes, to be sure, that the bat is capable of a fairly complex reaction to slight differences in sound sensations, but this same assumption is implicit in the auditory theory in any form.

THE OBSTACLE AVOIDANCE OF BATS WITH THEIR MOUTHS COVERED

If the auditory theory be correct, it should be possible to impair obstacle avoidance by preventing the bat from emitting sound. If a loop of thread is tied around the bat's snout, so that the mouth is held closed, and collodion is then applied all around the lips, production of both audible and supersonic sound is reduced to negligible proportions. Care must obviously be taken to leave the nostrils free, but if this precaution is observed, the bat flies well and is able to breathe sufficiently even during prolonged flight.

Gagged bats flew in the same clumsy, hesitant and bewildered manner as deaf bats. Table 9 shows that this treatment raised the scores of six bats from an average of 38 per cent to 65 per cent. The probability of significance of this difference is in every case greater than 0.99; the impairment is just as

TABLE 9

Obstacle avoidance of bats (M. lucifugus) with mouth covered

NUMBER	INTACT		EYES AND MOUTH COVERED		χ^2	PROBABILITY OF SIGNIFICANCE
	Trials	Per cent hits	Trials	Per cent hits		
32	54	33	112	70	19.0	0.99 – 1.00
37	40	28	42	31 (a)	0.12	not significant
39	50	28	113	60	14.4	0.99 – 1.00
41	52	46	65	68	5.49	0.98 – 0.99
	ONLY EYES COVERED		EYES AND MOUTH COVERED			
32	155	46	112	70	15.1	0.99 – 1.00
33	31	10	40	75	30.0	0.99 – 1.00
41	60	42	65	68	8.40	0.99 – 1.00
Totals and weighted average	442	38	549	65		

(a) made considerable supersonic sound in spite of mouth covering.

OBSTACLE AVOIDANCE BY BATS   **499**

serious as that which results from covering the ears. In one case, No. 37, the difference was not significant, and this was the only bat which continued to make supersonic notes easily detectable with the amplifier even after we had covered its mouth with as much collodion as we could apply. Perhaps this vociferous animal produced so much sound that a serviceable intensity escaped through the nostrils.

Several other bats in the course of the experiments would succeed in scratching a small hole in the collodion covering their lips, and when this happened the drop in their obstacle avoidance scores always occurred at the same moment that we could first detect the production of supersonics. A very slight crack in the collodion would enable the bat to dodge the wires.

Since only a negligible amount of air could have reached the lungs through such a hole, we believe that covering the bats' mouth impairs its flight, not by interfering with respiration, but by preventing the emission of supersonic cries.

OTHER RECEPTORS

Cuvier (1800) proposed the theory that bats detect obstacles by tactile sensitivity of their wing membranes. The origin and history of this and other theories will be discussed elsewhere, but it is significant to note here that Cuvier's theory was accepted by most naturalists for more than 100 years, in spite of experiments recorded by Spallanzani (1794), Rollinat and Trouessart ('00) and Hahn ('08), in which the wing membranes were covered with varnish or vaseline without seriously impairing the bat's ability to avoid obstacles.

One of our preliminary experiments is of interest in this connection; its protocol follows:

Bat No. 49 (Myotis l. lucifugus)

CONDITION	NUMBER OF TRIALS	PER CENT HITS
Intact	22	9
Underside of wings covered with collodion	40	38
Eyes and ears also covered	40	53
Eyes and ears covered, wings free of all collodion	30	67
Intact	20	5

Many observers, including Hahn ('08), Allen ('39) and the writers have noted that bats often fly to the sources of drafts, such as cracks in doors or open windows. We can find no records of experiments testing this ability which may well be tactile.

While covering the wings of this bat seems to have produced some impairment, it is equally apparent that wing membrane sensitivity was of no help to the bat when deaf. Evidence is presented below that deaf or gagged bats strike as often as they would be expected to do if they had no knowledge of the wires, and thus it seems very likely that tactile sensitivity does not play any important part in the detection of obstacles of this sort. The helpless manner in which a deaf bat bumps repeatedly into the wall makes it equally doubtful that even a large solid obstruction is perceived by means of any tactile sense.

Blatchley (1896), while discussing the habits of bats in an Indiana cave, records that when "a door was put in the opening . . . through which they had been wont to pass in numbers, they flew blindly against it and were killed by thousands." In our experience no normal bat was ever even stunned by a full speed collision with a wooden wall, and we should like to see whether Blatchley's observation could be repeated. It could only be explained, apparently, by assuming that the bats formed a habit of flying full speed through this part of the cave and disregarded the sensory cues which would ordinarily have warned them of the door.

This raises an interesting question concerning the role of the non-auditory receptors of the inner ear. Anaesthesia of the inner ear has disastrous results on the flight of bats, as Rollinat and Trouessart reported in 1900. We filled the external ear canals of one Myotis l. lucifugus with a solution of "Procaine." After a few minutes it was released and proceeded to fly off balance, crashing into the walls and wires at every opportunity. The speed of this bat's flight was much greater than that of any other we observed. The organs of equilibrium must have been affected for the animal flew with its body held at all sorts of angles.

Bats might have a well-developed sense of the direction of their motion and acceleration, arising from the saccule, utricle and semi-circular canals. They could then fly by a sort of dead reckoning when thoroughly familiar with a cave, making certain turns at definite intervals which they remembered from countless previous flights through the same passage-ways. This would explain their failure to avoid the newly erected door where Blatchley reports such wholesale destruction. One might deduce that bats flying repeatedly through our experimental rooms, or Hahn's, would have learned while intact the approximate position of the barrier. Even though gagged or deaf, they might, as they neared the barrier, alter their flight by decreasing their wingspan or changing from Ruderflug to Rüttelflug (Eisentraut, '37), tending thus to lower their avoidance scores. We noted no clear tendency for the bats to use so crafty a device, but here may lie a parallel explanation for the fact that gagged and deafened bats gave slightly lower scores than the chance average computed below.

In a few of our experiments, bats were already blind when first released in the experimental room, so that they could not possibly have had any memory of the location of the wires. Yet they avoided just as well as other normal or blinded bats. This makes it obvious that dead reckoning by means of labyrinth memory could not account in any important degree for the flying skill of sightless bats.

AVOIDANCE BY CHANCE AND THE COMPLETENESS OF IMPAIRMENT

It is of interest to determine what score a bat might be expected to make if he simply flew through the wires with just the same wing beats as though the air were free of all obstructions. Hahn ('08) attempted to do this by preparing a cotton ball "with one diameter equal to the expanse of the bat and the other slightly smaller to compensate for the upward stroke of the bat's wing when the distance from tip

502   DONALD R. GRIFFIN AND ROBERT GALAMBOS

to tip was somewhat less.'' He threw this ball at the barrier an unstated number of times and it struck 82 per cent of the "trials.''

High speed motion pictures of flying bats have now made it possible to estimate their average wingspread more accurately. A series of such pictures, which were taken by Dr. H. E. Edgerton at the Massachusetts Institute of Technology, show that the minimum wingspread at the bottom of the stroke may be as little as 25 per cent of the maximum spread. The average wingspread of Myotis l. lucifugus as shown by these motion pictures was about 60 per cent of the maximum. Eisentraut ('37) gives diagrams of the positions of the wing-tips of a horseshoe bat (Rhinolophus hipposideros) throughout its stroke, and the minimum is 35 per cent of the maximum, with an average of 65 per cent.

The average maximum wingspread of 89 Myotis l. lucifugus was 24.7 cm., and the average spread during flight should therefore be about 60 per cent of 24.7 or 14.9 cm. Thus if the bats always flew directly towards the wires, one would expect that they would strike 5.9/12 or 49 per cent of the time. Actually, of course, they may approach from any angle, and if the angle is sufficiently oblique, there would be no chance of their missing. There is, however, some tendency for bats to approach the wires at approximately right angles more often than at extremely oblique angles. If we disregard this fact and assume that the bat is equally likely to approach from any angle, we can compute the probability that it will hit a wire. This probability will be the expected average score of a bat flying with no knowledge of the wires' existence. This computed score will be somewhat too high because of the unmeasured tendency for bats to approach at an angle of about 90° more often than at oblique angles.

Let W = average wingspread of bat
   S = horizontal distance between the wires
   P = probability that bat will strike if it makes no effort to avoid
   θ = projection on a horizontal plane of the angle at which the bat approaches the barrier of wires. (In order to simplify the calculations, θ will be expressed in radians.)

If the bat flies directly at the wires, $\Theta = \frac{\pi}{2}$, and $P = \frac{W}{S}$, and in general:

$$P = \frac{W}{S \sin \Theta} \quad \text{when} \quad \frac{\pi}{2} \geq \Theta \geq \sin^{-1} \frac{W}{S}$$

If we assume that Θ is equally likely to be any angle from 0 to $\frac{\pi}{2}$, we can find an average value for P, which will be called $P_0 \rightarrow \frac{\pi}{2}$.

It is convenient to consider two sorts of values for Θ, those $< \sin^{-1} \frac{W}{S}$ for which $P = 1$, and those between $\sin^{-1} \frac{W}{S}$ and $\frac{\pi}{2}$ for which P varies between 1 and $\frac{W}{S}$.

$$P_0 \rightarrow \frac{\pi}{2} = \frac{(P_0 \rightarrow \sin^{-1} \frac{W}{S})\,(\sin^{-1} \frac{W}{S}) + (P_{\sin^{-1} \frac{W}{S}} \rightarrow \frac{\pi}{2})\,(\frac{\pi}{2} - \sin^{-1} \frac{W}{S})}{\pi/2}$$

$$P_0 \rightarrow \sin^{-1} \frac{W}{S} = 1$$

$$P_{\sin^{-1} \frac{W}{S}} \rightarrow \frac{\pi}{2} = \frac{\displaystyle\int_{\sin^{-1} \frac{W}{S}}^{\pi/2} \frac{W}{S \sin \Theta} \, d\Theta}{\frac{\pi}{2} - \sin^{-1} \frac{W}{S}} =$$

$$P_0 \rightarrow \frac{\pi}{2} = \frac{\sin^{-1} \frac{W}{S} + \frac{W}{S}\,[\log \tan \frac{\pi}{4} - \log \tan \frac{1}{2} \sin^{-1} \frac{W}{S}]}{\pi/2}$$

here $W = 5.9$ inches
        $S = 12$ inches

$$\frac{W}{S} = 0.49$$

$$P_0 \rightarrow \frac{\pi}{2} = \frac{0.51 + (0.49)\,(1.40)}{1.57}$$

$$= 0.762$$

Thus the probability of striking wires $= 0.762$, and the chance average score $= 76.2$ per cent hits.

Or, translating this conclusion from mathematics into English, the bat should strike 76 per cent of the trials if he were completely ignorant of the wires' existence. The actual scores

504   DONALD R. GRIFFIN AND ROBERT GALAMBOS

of twenty-two deaf bats and six gagged animals were 66 per cent and 65 per cent respsectively.[1]

The calculated score is undoubtedly too high because the bat tends to fly parallel to the long axis of the room and very seldom approaches the wires at angles of less than about 30° If we were to make the not unreasonable assumption that the angle of approach might have any value between 30° and 90°, $\text{Po} \rightarrow \sin^{-1}\frac{w}{s}$ could be neglected (for $\sin^{-1} 0.49 =$ almost exactly 30°). Then P would equal about 65 per cent. Even without this qualification, the difference between chance average and the scores of deaf and gagged bats is of very little significance. Seventy-six per cent is a very conservative value for the chance score, and if any cues besides reflected sound are used to detect obstacles, they enable the bat to lower his percentage of hits only to 65 per cent. This evidence supports the hypothesis that other senses than hearing play a very minor role, if any, in obstacle avoidance by flying bats.

SUMMARY AND CONCLUSIONS

1. Obstacle avoidance scores were obtained for bats in the form of per cent hits in passing through a standardized barrier of metal wires. The eyes, ears and mouth were covered without permanent injury to the bats in order to determine the effects of these treatments on their ability to avoid obstacles.

2. Normal bats varied from scores of 7 per cent to 62 per cent, the average for Myotis l. lucifugus being about 35 per cent.

3. Bats with their eyes covered could not be distinguished from normal bats by their flight or obstacle avoidance.

4. Covering both of a bat's ears results in great reluctance to fly, frequent bumping of the walls, and an average obstacle

---

[1] Hahn's scores for deaf bats were about the same as ours, but he did not have motion pictures to show how much the maximum wingspread exceeds the average and was thus led to the incorrect conclusion that the chance score would be 82 per cent and that "there is therefore some avoidance even when the ears are stopped".

OBSTACLE AVOIDANCE BY BATS 505

avoidance score for twenty-nine bats of 66 per cent. Control tests showed that this effect was not due to injury attending the plugging of the ears.

5. Bats with one ear covered often turned away from the walls and the barrier of wires. Most of them could land normally, but they avoided wires very little better than bats with both ears closed.

6. When bats' mouths are covered so as to prevent the production of supersonic notes, their obstacle avoidance was impaired just as much as that of deafened bats (average score, 65 per cent).

7. The scores of bats with either ears or mouth covered are practically as high as the score computed for a bat flying through the barrier with no knowledge of its existence.

8. No other sense than hearing is necessary for essentially normal obstacle avoidance.

9. All this evidence points conclusively towards Hartridge's auditory theory in a slightly modified form which may be stated as follows: Flying bats detect obstacles in their path by (1) emitting supersonic notes, (2) hearing these sound waves when reflected back to them by the obstacles, and (3) detecting the position of the obstacle by localizing the source of this reflected sound. This localization is presumably accomplished binaurally by some auditory mechanism, similar in principle to that used by other mammals for sounds of ordinary frequencies.

LITERATURE CITED

ALLEN, G. M. 1939 Bats. Cambridge, p. 136.
BARRETT-HAMILTON, G. E. H. 1911 A History of British Mammals. London, p. 41.
BELL, THOMAS 1874 A History of British Quadrupeds. London, pp. 5–6.
BLATCHLEY, W. S. 1896 Twenty-first annual report of Indiana Dept. Geology and Natural Resources, p. 21.
BORELL, A. E. 1937 A new method of collecting bats. J. Mammalogy, vol. 18, pp. 478–480.
CUVIER, GEORGES 1800 Leçons d'Anatomie Comparée, T. ii, p. 581.
EISENTRAUT, M. 1937 Die deutschen Fledermäuse. Zentralbl. f. Kleintierkunde u. Peltztierkunde, vol. 13 (4), pp. 29–44.

506   DONALD R. GRIFFIN AND ROBERT GALAMBOS

FISHER, R. A. 1932 Statistical methods for research workers. Edinburgh, p. 84.

FLOWER, W. H., AND R. LYDEKKER 1891 Mammals, living and extinct. London, p. 645.

GODMAN, J. D. 1831 American Natural History, 2nd ed. Philadelphia, pp. 57–58.

HAHN, W. L. 1908 Some habits and sensory adaptations of cave-inhabiting bats. Biol. Bull., vol. 15, pp. 165–193.

HARTRIDGE, H. 1920 The avoidance of objects by bats in their flight. J. Physiol., vol. 54, pp. 54–57.

JURINE, LOUIS 1798 Experiments on bats deprived of sight by M. deJurine; (translated) from the Journal de Physique for 1798. Philosophical Magazine, vol. 1, pp. 136–140.

MILLAIS, J. G. 1904 Mammals of Great Britain and Ireland. London, vol. 1, pp. 96–97.

OSGOOD, F. L. 1935 Notes on bat netting. J. Mammalogy, vol. 16, p. 228.

PIERCE, G. W., AND D. R. GRIFFIN 1938 Experimental determination of supersonic notes emitted by bats. J. Mammalogy, vol. 19, pp. 454–455.

ROLLINAT, R., AND E. TROUESSART 1900 Sur le sens de la direction chez chauves-souris. Comptes Rend. Soc. Biol., vol. 52, pp. 604–607.

SENEBIER, J. 1807 Rapports de l'Air. 3 vol. Geneva. See vol. 2, p. 102 ff.

SETON, E. T. 1909 Life histories of northern animals. New York, pp. 1154–1158.

SPALLANZANI, LAZZARO 1793–94. Lettere sopra il sospetto d'un nuovo senso nei Pipistrelli; to be found in ''Le Opere di Lazzaro Spallanzani'' vol. 3 of the 5 vol. edition published at Milan in 1932 by Ulrico Hoepli for Reale Accademia D'Italia.

VAN TYNE, J. 1933 The trammel net as a means of collecting bats. J. Mammalogy, vol. 14, pp. 145–146.

WHITAKER, A. 1906 The development of the senses in bats. The Naturalist, London 1906, pp. 145–151.

Fig. 1. External opening of the right ear in *Agrotis ypsilon*. The external surface of the tympanic membrane faces obliquely backwards and outwards into the cavity below arrow. The body of the moth is about ³/₄ inch in length.

AMERICAN SCIENTIST

SUMMER   JUNE 1961

THE DETECTION AND EVASION OF BATS BY MOTHS

By KENNETH D. ROEDER and ASHER E. TREAT[1]

A CENTRAL objective of a large segment of biological and psychological research is to provide a physiological basis for behavior. The first step toward this objective is analytic, and consists of determining the structure and function of neural components after they have been isolated from their connections with the rest of the nervous system. There has been much progress in this direction, and it is now possible to describe in terms of input and output performance the operation of many isolated sense cells, neurons, and muscle fibers, even though the principles of their internal operation are mostly not understood.

The next step, the synthetic process of assembling this information on isolated neural components and relating it to the behavior of the intact animal, is hampered by two kinds of difficulty. The first appears to be methodological, but is somewhat hard to define. When one regards the evergrowing literature on the unit performance of sense cells, nerve cells, and muscle fibers, it is to experience that sense of dismay first encountered at a tender age when the springs, gears, and screws of one's first watch were strewn upon the table. The *modus operandi* of analysis or taking apart seems to come naturally, and the problems encountered are essentially technical in nature. Synthesis or the derivation of a system from its components seems to lack the *a priori* logic of analysis.

The second general difficulty is technical, and stems from the fact that even the simplest behavior of the higher animals and man is accompanied by the simultaneous activity of millions of sense cells, nerve cells, muscle fibers, and glands. Even if it were possible to register the traffic of nervous and chemical information generated and received by each and all of these neural elements during the behavior, it is doubtful whether the record would provide a meaningful description of the action.

Even though these problems cannot be solved directly at the present

[1] Much of the experimental work reported in this paper was made possible by Grant E-947 from the U. S. Public Health Service.

135

time, they become less formidable if the behavior selected for study is simple and stereotyped, and only a small number of nerve cells are concerned in its execution. These conditions are partly fulfilled by the sensory mechanisms whereby certain nocturnal moths detect the approach of insectivorous bats.

Echolocation and Countermeasure

Bats detect obstacles in complete darkness by emitting a sequence of high-pitched cries or chirps and locating the source of the echoes. As Griffin (1958) and others have shown, this form of Sonar is unbelievably precise. By means of it, insectivorous bats locate and track flying moths, mosquitoes, and small flies (Griffin, et al., 1960). North American bats,

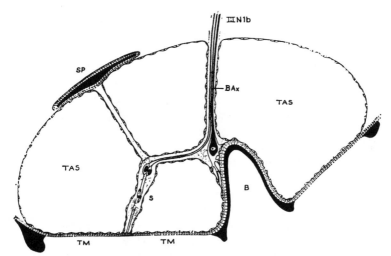

FIG. 2. Diagram of the tympanic organ of a noctuid moth. The sensillum (*S*) contains the pair of acoustic receptors or A cells. The A nerve fibers are joined by that of the B cell (*BAX*) to form the tympanic nerve (*IIINIB*). *TAS*, tympanic air sac; *B* and *SP*, skeletal supports; *TM*, tympanic membrane. (After Treat and Roeder, 1959.)

such as *Myotis lucifugus* and *Eptesicus fuscus*, emit chirps about 10 times a second when they are cruising in the open. Each chirp lasts from 10 to 15 milliseconds (msec) with an initial frequency of 80 kilocycles (kc) dropping about one octave in pitch toward its end (see Figure 5).

The frequencies in these chirps are ultrasonic, that is, inaudible to human ears, which cannot detect tones much above 15–18 kc. The higher frequencies used by bats make possible more discrete echoes from smaller objects. The chirps can be rendered audible by detecting them with a special microphone and rectifying the ultrasonic component. They then can be heard through headphones as a series of clicks. These clicks fuse into what Griffin has called a "buzz" when the bat is chasing an insect or avoiding an obstacle.

Several families of moths (in particular the owlet moths or *Noctuidae*) have evolved countermeasures enabling them to detect the chirps of bats. A pair of ultrasonic ears is found near the "waist" of the moth between thorax and abdomen (Figure 1). An extremely thin eardrum or tympanic membrane is directed obliquely backward and outward into the recess (dark area) found at this point.

Internal to the eardrum is an air-filled cavity that is spanned by a thin strand of tissue running from the center of the eardrum to a skeletal support (Figure 2). This tissue contains the sound-detecting apparatus, consisting of two acoustic sense cells (A cells). A single nerve fiber arises from each A cell and passes close to the skeletal support, where the pair is joined by a third nerve fiber arising from a large cell (B cell) in the membranes covering the support. The three fibers continue their course to the central nervous system of the moth as the tympanic nerve.

The traffic of nerve impulses passing over the three fibers from A cells and B cell to the nervous system of the moth can be followed if a fine metal electrode is placed under the tympanic nerve. Another electrode is placed in inactive tissue nearby. As each impulse passes the site of the active electrode it can be detected as a small action potential lasting about 1 msec. Since the tympanic nerve contains only three nerve fibers, it is not difficult to distinguish and to read out the respective reports to the nervous system from the pair of A cells and the B cell. A similar experiment in a mammal is practically meaningless since the auditory nerve contains about 50,000 nerve fibers.

This method of detection shows that the A cells transmit organized patterns of impulses over their fibers only when the ear is exposed to sound (Roeder and Treat, 1957). The B cell transmits a regular and continuous succession of impulses that can usually be distinguished from the A impulses by their greater height. The B impulses are completely unaffected by acoustic stimulation, and change in frequency only when the skeletal framework and membranes lining the ear are subjected to steady mechanical distortion (Treat and Roeder, 1959). The B cell behaves in a manner similar to receptors found in other parts of the body that convey information about mechanical stress on joints, muscles, and skeleton. The role of such a receptor in the ear of a moth is unknown.

In the absence of sound, the A cells discharge irregularly spaced and relatively infrequent impulses (Figure 3A). A continuous pure tone of low intensity elicits a more regular succession of more frequent impulses in one of the A fibers (Figure 3B). The other fiber is not yet affected. Any slight increase in the intensity of the tone causes a corresponding increase in the impulse frequency of the active fiber. When the intensity of the tone is increased to about 10-fold that producing a detectable response in the more sensitive A fiber, the second A fiber begins to respond in like manner. Its action potentials are superimposed on those of the first (Figures 3C and D) by the method of recording, but actually reach the central nervous system over their own pathway. This experiment reveals two of the ways in which the moth ear codes sound intensity. It is like an instrument having a graded fine adjustment (the

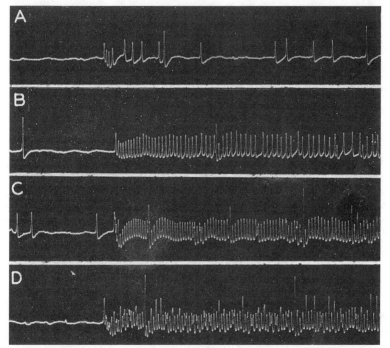

Fig. 3. Tympanic nerve response in *Prodenia eridania* to a pure tone of 40 kc. The occasional large spikes originate in the B cell. (*A*) Response to a sound intensity close to the threshold of the sensitive A cell. (*B*) Intensity 7 db above that in (*A*). (*C*) Intensity 15 db above that in (*A*). (*D*) Intensity 23 db above that in (*A*). The less sensitive A cell discharges occasionally in (*C*), and frequently in (*D*), as indicated by the double peaks. Time line 100 msec. (From Roeder, 1959.)

intensity-frequency relation) and a coarse adjustment of two steps (the pair of A cells). Other ways of coding intensity will appear later.

The moth ear responds in this manner to tones from 3 kc to well over 100 kc but there is no evidence that it is capable of discriminating between tones of different frequency. It is most sensitive near the middle of its range, that is, to frequencies such as those contained in bat chirps.

In Figure 3 it will be noticed that, in each of the recordings, the intervals between the successive impulses increase as the pure tone stimulus continues. In terms of the nerve code outlined above, the A

THE DETECTION AND EVASION OF BATS BY MOTHS   139

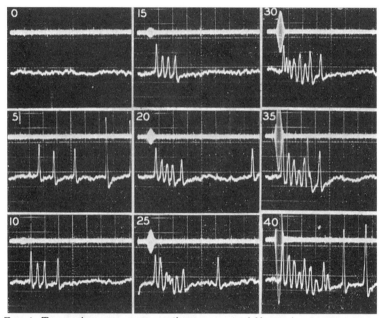

FIG. 4. Tympanic nerve responses (lower traces) of *Noctua* (= *Amathes*) *c-nigrum* to a 70 kc. sound pulse recorded simultaneously by a Granith microphone (upper traces). The numbers indicate the intensity of the sound pulse in decibels above a reference level (0). The threshold of the sensitive A cell lies between 0 and 5 db. The large spikes appearing in some of the records are from the B cell. The less sensitive A cell responds in the 25 db recording. Vertical lines, 4 msec. apart.

cells report that the sound is declining in intensity with time, although in fact it was kept constant. This adaptation to a constant stimulus occurs in most receptors registering changes in the outside world. In terms of our own experience, the impact of our surroundings would be shocking and unbearable if it were not distorted in this manner by sense organs. The brilliance of a lighted room entered after dark would continue to be blinding and the noise of a jet engine would remain unbearable. However, the A cells of the moth's ear adapt very rapidly to a continuous tone, and their full effectiveness as pulse detectors is revealed only when they are exposed to short tone pulses similar to bat chirps.

In the experiment illustrated in Figure 4 a tone pulse of 3 msec duration was generated at regular intervals. It is similar to a bat chirp except for its regularity and the absence of frequency modulation. A microphone (upper trace) and moth ear (lower trace) were placed within range, and the intensity of the stimulus pulse was adjusted so that it just produced a detectable response in the most sensitive A fiber (0 db). The intensity was then increased by 5 decibel[1] (db) steps as each

[1] The decibel (db) notation expresses relative sound pressures. An intensity of 20 db is 10-fold that of the reference level (0 db), a 40 db sound is 100-fold the reference level.

recording was made. It will be seen that the microphone begins to detect the sound pulse when it is about 10 db above the threshold of the most sensitive A cell in the moth's ear. As before, the increase in frequency of A impulses is evident if the 5 and 10 db records are compared, and a response of the less sensitive A cell appears first in the 25 db record where the extra peaks of its action potentials overlap those of the more sensitive A unit. In addition to these two ways of coding intensity, two more can now be recognized. If the interval between detection of the sound by the microphone and by the moth ear is compared at different sound intensities, it will be noticed that the tympanic nerve response occurs earlier and earlier on the horizontal time axis. In other words, the latency of the response decreases with increasing loudness. Also, the sense cells are seen to discharge impulses for some

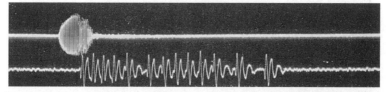

FIG. 5. The cry of a flying bat (*Myotis*) recorded by a Granith microphone (upper trace) and the A cells (lower trace) of a noctuid moth (*Agroperina dubitans*). The A spikes shown in the lower trace have been distorted in form by the recording technique. Time line, 10 msec. Made in collaboration with Dr. Fred Webster in his laboratory.

time after the sound has ceased, and this after-discharge becomes longer with increasing sound intensity.

The Detection of Bats

These experiments with artificial sounds suggest how the moth ear might be expected to respond to a bat cry. A few laboratory observations were made with captured bats. In one of these, in collaboration with Dr. Fred Webster, the cries of a flying bat were picked up simultaneously by a moth ear and a microphone, and recorded on high-speed magnetic tape (Figure 5). Interesting though they were, these experiments served mainly to show that the full potentialities of the moth ear as a bat detector could not be realized within the confines of a laboratory, and efforts were made to transport the necessary equipment to a spot where bats were flying and feeding under natural conditions.

Finally, about 300 pounds of equipment was uprooted from the laboratory and reassembled at dusk of a July evening on a quiet hillside in the Berkshires of western Massachusetts. Moths attracted to a light provided experimental material. The insect subject was pinned on cork so that one of its ears had an unrestricted sound field, and with the help of a microscope its tympanic nerve was exposed and placed on electrodes. After amplification, the action potentials were displayed on

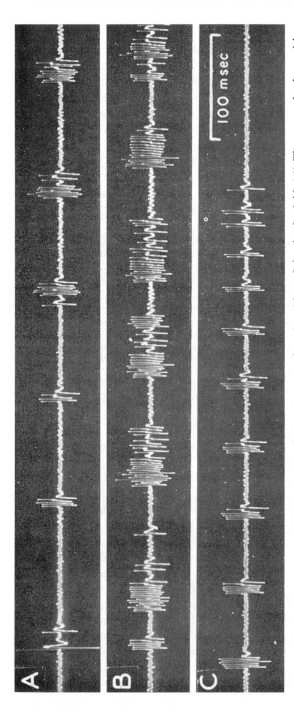

Fig. 6. Tympanic responses of *Noctua* (=*Amathes*) *c-nigrum* to the cries of bats flying in the field. (*A*) The approach of a cruising bat emitting pulses at about 10 per second. (*B*) Tympanic response to the original cry and its echo made by a bat cruising nearby. (*C*) A "buzz." Time line, 100 msec. (From Roeder and Treat, 1961.)

an oscilloscope. They were also made audible as a series of clicks by means of headphones connected to the amplifier, and were stored on magnetic tape for later study.

It was dark before all was ready, but bats immediately revealed their presence to the moth ear by short trains of nerve impulses that recurred about 10 times a second (Figure 6A). The approach of a cruising bat from maximum range was coded as a progressive increase in the number and frequency of impulses in each train, first from one and then from both A fibers. It was not long before we learned to read something of the movements of the bats from these neural signals. Long trains, sometimes with two frequency peaks, suggested the chirps of nearby bats that echoed from the wall of a neighboring house (Figure 6B). An increase in the repetition rate of the trains coupled with a decrease in the number of impulses in each train signified a "buzz" as the bat attacked some flying insect in the darkness (Figure 6C).

All of this was inaudible and invisible to our unaided senses. With a powerful floodlight near the nerve preparation we were able to see bats flying within a radius of 20 feet, and some attacks on flying insects could then be both seen and also "heard" through the "buzz" as coded by the moth's typanic nerve. However, most of the sounds detected by the moth ear were made by bats maneuvering well out of range of the light. A rough measure of the sensitivity of the moth ear to bat chirps was obtained at dusk on another occasion when the bats could still be seen. The A cells first detected an approaching bat flying at an altitude of more than 20 feet and at a horizontal distance of over 100 feet from the moth—a performance that betters that of the most sensitive microphones.

Direction

Since differences in sound intensity are coded by the tympanic nerve in at least four different ways, the horizontal bearing of a bat might be derived from a comparison of the nerve responses to the same chirp in the right and left ears. A difference in right and left responses might be expected only if each ear had directional properties, that is, a lower threshold to sounds coming from a particular direction relative to the moth's axis.

Directional sensitivity was measured in an open area where echoes were minimal. A source of clicks of constant intensity was placed on radii to the moth at 45 degree intervals. The source was moved in and out on each radius until a standard tympanic nerve response was obtained, and the distance from the moth noted. Horizontal distances along 8 radii were combined to make a polar plot of sensitivity (Roeder and Treat, 1961). The plot showed that, although there was little

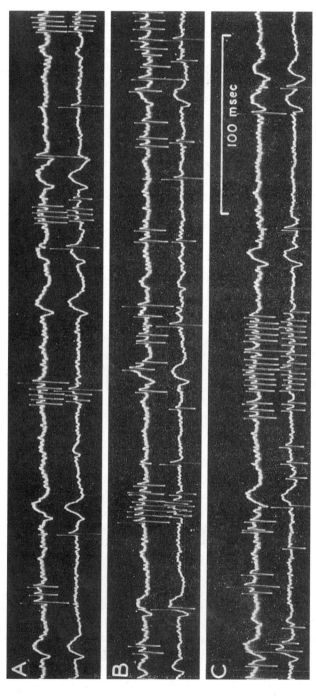

100 msec

Fig. 7. Binaural tympanic responses of *Feltia sp.* to the cries of red bats flying in the field. The electrocardiogram of the moth also appears on both channels as slow waves. B impulses (large spikes) appear regularly in the records from both tympanic nerves. (*A*) An approaching bat. Differential response is marked at first (response latency, number of spikes) but has practically disappeared in the final train. (*B*) A "buzz" registered mainly by one ear. (*C*) A "buzz" registered by both ears. Time line, 100 msec.

difference in sensitivity fore and aft, a click on the side nearest the ear at about 90 degrees relative to the moth's longitudinal axis was audible at about twice the distance of a similarly placed click on the far side.

This led to further field experiments in the presence of flying bats. The tympanic nerve responses from both ears of a moth were recorded simultaneously on separate tracks of a stereophonic magnetic tape. The tape was subsequently re-played into a two-channel oscilloscope and the traces photographed (Figure 7). In the upper record (A) the increase in number of impulses in each succeeding train suggests the approach of a bat. When the signals from right and left ears are compared, it is evident that the greatest difference exists when the signal is faintest, the first response of the series occurring in one ear only. When both ears respond, the differential nature of the binaural response can be seen first as a difference in the number of spikes generated in right and left ears, second in the differential spike frequency, and third in the latency of the response, which is greater on that side generating fewer spikes. It is also evident that, as the sound intensity increases (presumably due to the approach of the bat), the differential becomes less until the responses of right and left ears become almost identical. In another experiment, it was found that the tympanic nerve response saturates, i.e., becomes maximal, when the sound intensity is about 40 db (100-fold) above threshold. From this it can be concluded that the moth's nervous system receives information that would enable it to determine whether a distant bat was to the right or left, but if the bat was at close quarters this information would not be available. In Figure 7C the "buzz" was picked up by one ear only, presumably because during this part of its performance the chirps of a bat are much less intense.

It is tempting to estimate just how close the bat must be before the moth fails to get information on its location. If it is assumed that a bat is first detected at 100 feet and approaches on a straight path at right angles to the moth's course while making chirps of constant loudness, the differential tympanic nerve response would diminish throughout the approach and disappear completely when the bat was 15 to 20 feet away. However, we have not yet determined how much of the information that we are able to read out of its auditory mechanism is actually utilized by the moth in its normal behavior.

The Evasive Behavior of Moths

Although the evasive behavior of moths in the presence of bats must have been witnessed hundreds of times, it is hard to find an adequate account of the maneuvers of either party. The contest normally takes place in darkness, and, even when it is illuminated by a floodlight, the action is too fast and complex to be appreciated by the eye. The flight path of the bat and its ability to intercept and capture its prey have

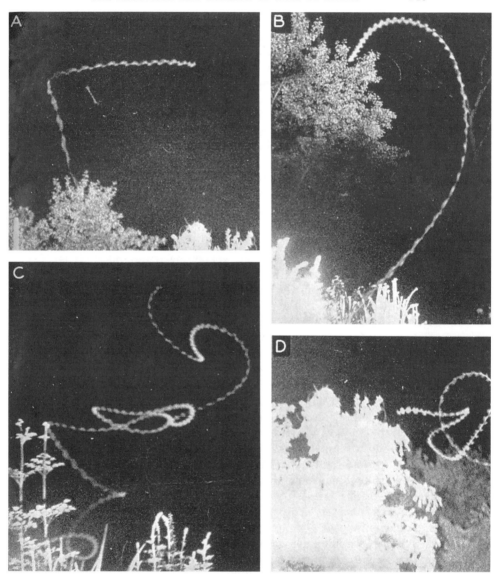

Fɪɢ. 8. Flight tracks registered by various moths just before, and immediately following, exposure to a series of simulated bat cries. The dotted appearance of the tracks is due to the individual wingbeats of the moth. The beginning of each track appears in each photograph, and the moth finally flies out of the field.

been studied by Griffin (1958) and his students. More recently, Webster (in press) has shown by means of high-speed sound motion pictures that bats become adept at using echoes to plot an interception course with an object moving in a simple ballistic trajectory. Many people have noted the seemingly erratic dives and turns made by moths when bats are near, and similar behavior has been described when moths are exposed to artificial sources of ultrasound (Schaller and Timm, 1950; Treat, 1955).

In an effort to learn more about the behavior of moths under field conditions their flight was tracked photographically as they reacted to a series of ultrasonic pulses simulating bat cries. The sounds were generated by the equipment used in the experiment shown in Figure 4. The pulses were similar in form to those shown, although longer in duration (6 msec). Each pulse ranged from 50 to 70 kc with a rise and fall time of about 1 msec. Pulse sequences up to 50 per second could be released on closure of a switch. The sounds were emitted by a plane-surfaced condenser loud-speaker mounted so as to project a fairly directional beam over an open area of lawn and shrubs illuminated by a 250-watt floodlight.

The observer sat behind the sound generator and floodlight, holding in one hand the cable release of a 35 mm camera set on "bulb," and in the other the switch controlling the onset of the sound-pulse sequence. Many moths and other insects flew out of the darkness into this floodlight arena. A number were attracted directly to the light and were disregarded. Many others moved across the arena at various angles but without marked deviation toward the light. When one of these appeared to be in line with the loudspeaker the camera shutter was opened and the sound pulses turned on.

Some of the tracks registered by the camera as the illuminated moths moved against the night sky are shown in Figure 8. Many insects, including some moths, showed no change in flight pattern when they encountered the sound. In others, the changes in flight path were dramatic in their abruptness and bewildering in their variety. The simplest, and also one of the commonest reactions was a sharp power dive into the grass (Figure 8A, B). Sometimes the dive was not completed and the insect flew off at high speed close to the ground. Almost as frequently the dive was prefaced or combined with a series of tight turns, climbs, and loops (Figure 8C, D).

It is not known whether these maneuvers are selected in some random manner from the repertoire of individual moths, or whether they are characteristics of different species. However, Webster (in press) has shown that bats soon learn to plot an interception course with food propelled through the air in a simple ballistic trajectory. The random behavior elicited by simulated bat cries in the natural moth population seems to be a natural answer to this predictive ability in bats, while the sharpness of the turns must certainly tax the maneuverability of the heavier predator.

The reacting moths shown in Figure 8 were mostly within 25 feet of the camera and sound source, and were exposed to an unknown but probably high sound intensity. Under these circumstances, the evasive behavior appeared to be completely unorientated relative to the sound source, as might be predicted from the binaural tympanic nerve recordings. In some instances, moths flying at a greater distance or only on the edge of the sound beam appeared to turn away from the area and fly off at high speed. This must be checked in future experiments.

The Survival Value of Evasion

In spite of the evidence that the moth ear is an excellent bat detector, and that acoustic stimulation releases erratic flight patterns, one may well ask whether this behavior really protects moths from attack by bats.

This question has been answered (Roeder and Treat, in press) by observing with a floodlight 402 field encounters between moths and feeding bats. In each encounter we recorded the presence or absence of evasive maneuvers by the moth, and the outcome, that is, whether it was captured by the bat or managed to escape. From the pooled data we determined the ratio of the percentage of nonreactors surviving attack to the ratio of reactors surviving attack. Thus computed, the selective advantage of evasive action was 40 per cent, meaning that for every 100 reacting moths that survived, there were only 60 surviving nonreactors.

This figure is very high when compared with similar estimates of survival value for other biological characteristics. It seems more than adequate to account for the evolution of the moth's ear through natural selection even if the detection of bats turns out to be its only function.

Conclusion

As with most investigations, this work raises more questions than it has answered. The role of the B cell remains completely obscure. There is no evidence to connect it with the auditory function even though it is located in the ear, and its regular impulse discharge is a characteristic feature of the tympanic nerve activity of many species of moth (Treat and Roeder, 1959. See also Figure 7). The manner in which the A cells transduce sound waves recurring 100,000 times a second into the much slower succession of nerve impulses remains a mystery, and the synaptic mechanisms whereby information from the A fibers is translated into action by the nervous system of the moth, await investigation.

During the field experiments it was noticed that many other natural sounds initiated impulses in the A fibers. These included the rustling of leaves, the chirp of tree and field crickets, and, in one instance, ultrasonic components in the wingbeat sounds made by another moth. Occasionally, the A fibers discharged regularly as if detecting a rhythmic sound, though none was audible to the observers and its source (if any) remains a mystery. There is no evidence that these identified and unidentified sounds are important in the life of a moth, yet it must be said that a moth can detect them, and a careful study of moth behavior in their presence would be of value.

Several families of moths lack ears and show no response to ultrasonic stimuli. Some of these, such as the sphinx or hawk moths and the larger saturniid moths, are probably too much of a mouthful for the average bat, and might find no survival advantage in a warning device. Others are of the same size and general habits as the noctuids and might be expected to suffer attacks by bats. Included in this group are some common pests such as the tent caterpillar. It will be interesting to learn whether these forms owe their success in survival to some structural or behavioral countermeasure that compensates for the lack of a tympanic organ.

In spite of these unanswered questions, we believe that some progress has been made in putting together the sensory information received by an animal, and relating this to what the animal does. That this has been possible in moths is only because of the small number of channels through which acoustic information reaches the nervous system in these insects. Further examples of this favorable situation have been described in other insects, and still others are waiting to be explored.

LITERATURE

1. Griffin, Donald R. *Listening in the Dark.* Yale University Press, New Haven, 1958.
2. Griffin, D. R., F. A. Webster, and C. R. Michael. The Echolocation of Flying insects by Bats. *Animal Behaviour, 8,* 141–154, 1960.
3. Roeder K. D. A Physiological Approach to the Relation between Prey and Predator. *Smithsonian Miscellaneous Collections, 137,* 287–306, 1959.
4. Roeder, K. D. and A. E. Treat. Ultrasonic Reception by the Tympanic Organ of Noctuid Moths. *Journal of Experimental Zoology, 134,* 127–158, 1957.
5. Roeder, K. D. and A. E. Treat. The Detection of Bat Cries by Moths. *Sensory Communication* (ed. W. Rosenblith), M.I.T. Technology Press, 1961.
6. Roeder, K. D. and A. E. Treat. The Acoustic Detection of Bats by Moths. *Proceedings of the XI (1960) International Entomological Congress,* in press.
7. Schaller, F. and C. Timm. Das Hörvermögen der Nachtschmetterlinge. *Zeitschrift für Vergleichende Physiologie, 32,* 468–481, 1950.
8. Treat, A. E. The Response to Sound of Certain Lepidoptera. *Annals of the Entomological Society of America, 48,* 272–284, 1955.
9. Treat, A. E. and K. D. Roeder. A Nervous Element of Unknown Function in the Tympanic Organs of Moths. *Journal of Insect Physiology, 3,* 262–270, 1959.

Guideposts of Migrating Fishes

New findings have added to our knowledge of how fish use olfactory and visual cues to find their way home.

Arthur D. Hasler

Homing in migrating fishes such as the salmon may be defined as a behavior pattern in which an animal spends its early life in one locality and subsequently returns to this locality, after undertaking migratory journeys of long or short duration to areas where the environment is drastically different. Fishes inhabiting smaller bodies of water, such as a stream pool, a pond, or a lake, return readily to a home territory when displaced (1), but the homing ability is of far less magnitude in these fishes than in salmon.

The object of this article is to describe and appraise the recent research on homing in fishes, to point up the gaps in knowledge, and to mention a few problems that should be pursued to help clarify differences of opinion. Emphasis is placed on the salmon, for which we have more factual information than for any other group.

There is comparatively good evidence for homing in the Atlantic salmon (*Salmo salar*), the five species of Pacific salmon (*Oncorhyncus kisutch, O. nerka, O. gorbuscha, O. keta, O. tschawytscha*), and the steelhead or rainbow trout (*Salmo gairdnerii*), whose returning migrations constitute a spectacular movement of large numbers of the species as they return to their parent stream to spawn.

In short, salmon spawn in freshwater streams and spend several years (two to seven, depending on the species), at sea, until they reach sexual maturity. Generation after generation, families of salmon return to the same riverlet so consistently that populations in streams that are not far apart follow distinctly separate lines of evolution. During a spawning movement into a river system the majority of fish swim upstream (Fig. 1) until they locate their home creek, where they spawn. In the genus *Oncorhyncus* the adults die after spawning, while in *Salmo*, spawnings are observed in subsequent years.

For a detailed and comprehensive review of migration of fishes, Scheuring's (2) splendid monograph is recommended. For relatively recent accounts of the mechanisms of homing, Scheer (3) and Hasler (4) may be consulted. The eel's life story has been reviewed thoroughly in the excellent treatise by Bertin (5), whose volume is available in both French and English, and the life history of the Atlantic salmon was summarized well by Jones (6).

Homing ability may be classified into three principal types (7): (i) The ability of an animal to find home by relying on local landmarks within familiar territory and the use of exploration in unfamiliar areas; (ii) the ability to maintain a constant compass direction in unfamiliar territory [this type is called *Richtungsfinden (-fliegen)* by Kramer (8)]; and (iii) the ability to head for home from unknown territory by true navigation.

Let us review briefly some of the details of homing in the salmon and eel.

The most overwhelming evidence for the precision of homing in salmon is that of Clemens *et al.* (9). They marked 469,326 fingerlings in a stream of the Fraser River system and recovered almost 11,000 when the salmon returned there from the sea. Although traps were placed on nearby tributaries, all of the marked fish were captured in the stream of their origin. There were no strays. Other workers report some straying, an aspect of variation; this is as it should be in a biological system. Nevertheless, one cannot but marvel at the accuracy of the majority, for in this drive to return home there appear to be advantages which are of distinct value for survival.

Seemingly, more spectacular than the migration of the salmon is that of the eel, according to Schmidt (10), who could trace the migration of the elvers of *Anguilla* from a spawning ground near the Bahama Islands to North America and Europe. After extensive study of the distribution of *Leptocephalus* larvae of the two species of Atlantic eel, he concluded that the adults migrate long distances at sea to spawn and that the young find their way back to their respective continents. Most recently some inadequacies of Schmidt's explanation have been treated by Tucker (11), who has suggested an alternative hypothesis—that there is only one species of Atlantic eel and that the European eel population is derived from American spawn which becomes transported via the Gulf Stream to Europe. He supplies arguments as to why the adult European eel could not migrate so far. D'Ancona (12), in a brief note, dis-

The author is professor of zoology at the University of Wisconsin, Madison. This article is based on a paper by the author which will appear in *Ergebnisse der Biologie*, vol. 22.

Fig. 1. Salmon ascending falls en route to their home stream to spawn. [U.S. Fish and Wildlife Service]

agrees with Tucker on a number of points, including his main thesis regarding the effect of temperature on the differentiation in the numbers of myomeres of the two groups of elvers as they are transported to North America and Europe, respectively. D'Ancona concludes that this new interpretation of Schmidt's results "requires so many new hypotheses that it is too difficult to accept it on the basis of present knowledge." It is not the object of this article to appraise the Schmidt-Tucker theories in detail; the true explanation can come principally from recaptures of marked adult eels through intensive hauls at sea made in an attempt to fill in the gaps in our knowledge of the oceanic routes of migration. This is an extremely difficult task and may not be accomplished for a long time. There is need for new questions to be posed and for new tests and experiments to be designed to answer them. If the European eel does not cross the Atlantic to spawn, which Tucker claims, are there yet undiscovered spawning grounds nearer to the European continent? Work needs to be done to determine if there is passive transport, or if there is an inherited migratory behavior pattern which makes possible the return to Europe, or if there is active searching. We greatly need more records of adult eels recaptured at sea to unravel the riddle of the eel's life history.

In spite of the impressive evidence

of homing in salmon, we still lack a complete record of the migratory path at sea. So far, only a few salmon have been marked in the home stream, caught and remarked at sea, and recaptured in the home stream (13).

Recently the Oregon State Game Commission has released a report of remarkable series of recaptures worthy of special attention.

"April 1958: Steelhead fingerling (probably 6 to 8 inches long) released from Alsea River hatchery on the central Oregon coast; marked by removing both ventral fins and the adipose fin."

"September 5, 1958: Captured 75 miles southeast of Geese Island, which is southwest of Kodiak Island, Alaska; fish was 365 mm. long when marked with a 'spaghetti' tag."

"February 5, 1960: Fish was recovered at Alsea River hatchery; 558 mm."

A. C. DeLacy of the University of Washington states (14): "This was one of only 59 tagged on the high seas and . . . two others of the group have been recovered in Washington—one in the Samish River (near Bellingham) and the other at Chehalis on the Washington coast. Of course these last two were not fin clipped but they do suggest that the Alsea return does not represent a freak migration."

Each year, through the self-sacrificing and strenuous efforts of fishery biologists from Japan, Canada, and the United States in the wide expanses

of the Pacific Ocean (15), the gaps in our knowledge of the routes at sea are being filled in. In summarizing the reports of the International North Pacific Fisheries Commission, W. F. Thompson of the University of Washington comments (16): "The red salmon of Bristol Bay feeds in great numbers around Attu Island which is 1200 miles from Bristol Bay. We have a great number of tags from that district retaken in our own North American estuaries, particularly Bristol Bay, but not the Gulf of Alaska. Pink salmon spawning in Kamchatka have been tagged all along the Aleutian Islands to a distance of 1200 miles from their own streams. The chum salmon returning to the Okhotsk Sea have been tagged 1700 miles from their home streams. The same is true of chum salmon spawning in Hokkaido. King and steelhead have been tagged in the vicinity of Adak Island and have returned to the Columbia River system and to Washington streams showing a migration of approximately 2500 miles; but Kings and steelheads are taken in very small numbers along the Aleutians. Their main migration is from the waters of southeastern Alaska southward as far as the Columbia River" (Fig. 2).

Donaldson and Allen's (17) recent work nicely confirms the findings of earlier work, that homing is not due to genetic factors but is rather the result of some sort of "imprinting" by environmental factors. They switched salmon stock and let the eggs develop in different waters. The fish returned, as adults, after a sojourn at sea, to the waters in which they had been hatched and in which they had lived as fry, not to the parent stream. In fact, Donaldson has now, in this way, built up a run of salmon to a newly built hatchery at the University of Washington on Lake Union.

A Working Hypothesis on Homing of Salmon

While the details of the routes of salmon are still not fully known, it appears justifiable to accept the home stream behavior as a working base. A leading question is: What cues and mechanisms guide these anadromous fishes in their migration? There are two major parts to this question which require an explanation. (i) How do the salmon recognize the main river as well as the home tributary? (ii)

How do they negotiate the route at sea, without visual landmarks?

In an attempt to answer the first question Wisby and I (18) have proposed the hypothesis that young salmon are "imprinted" by or "conditioned" to an organic odor of the home stream during their early fingerling period. To state this in simplest terms, we have proposed that the mature salmon, upon reaching the mouth of the home river, would be stimulated to enter because of a characteristic odor. (That is, each stream acquires a different odor which may be derived from a community of plants or their decomposition products in the stream or in the drainage basin.) Subsequently, the salmon would swim constantly upstream, responding positively to the water current. It would reject tributary after tributary until it began to detect traces of the home stream, to which it would respond positively, owing to the early "imprinting." On the way it might make faulty choices but would continue the search after backtracking— a behavior pattern which is frequently observed (19).

I would like here to object to the often stated hypothesis that an animal "follows a gradient," because, if an animal stayed in the gradient, it would soon become adapted to the odor (20) and would be incapable of responding to it. I think it more likely, therefore, that the behavior during the ascent

of the river system is analogous to that of a dog following an odor track. The dog does not stay exactly on the track but progresses along it by crisscrossing, responding to the presence and absence of the scent as it follows its prey. According to this theory, a salmon returning to its parent stream reacts differently to the odor of that stream than to that of any other. In order for a salmon to return to its home stream there must be some possibility of a differential reaction, something more than a simple response to a repellent or an attractant. The guiding odor must remain constant from year to year and have meaning only for those salmon which were conditioned to it during their fresh-water sojourn.

This theory presents three distinct problems. (i) Do streams have characteristic odors to which fish can react? If so, what is the nature of the odor? (ii) Can salmon detect and discriminate between such odors if they do exist? (iii) Can salmon retain odor impressions from youth to maturity?

In order to answer the first question, Wisby and I trained a group of blunt-nosed minnows to discriminate between the waters of two chemically different Wisconsin creeks, one in a quarzitic drainage basin, the other in a dolomitic one (18). That scent-perceiving organs were the sole means of discrimination in these tests was proved by destroying the olfactory

tissue of trained fishes; after this they no longer responded to the training odors. Chemical analysis of the stream waters indicated that the major difference between them was in the total organic fraction. Experimental evidence to substantiate this was obtained by separating the water into various fractions and then presenting these to trained fishes. The fishes trained previously to react to natural water did not react to the redissolved inorganic ash, or to the distillate or residue of water fractionated at 100°C. However, they recognized the distillate, but not the residue, of water fractionated by vacuum distillation at 25°C—a strong indication that the odorous stimulant is a volatile organic substance.

A test was conducted of the retentive capacities of the trained minnow, and it was determined that even this fish, which is not specialized for long migrations, could differentiate between the odors for a comparatively long period after the cessation of training. Learned behavior was found to be retained longer by young fish than by old ones. Numerous examples are known where animals retain imprinting, or early learning, through to adulthood — circumstantial evidence that odor imprinting can be retained for long periods (21).

The method of training which had been used with such success with the minnows was then applied to salmon

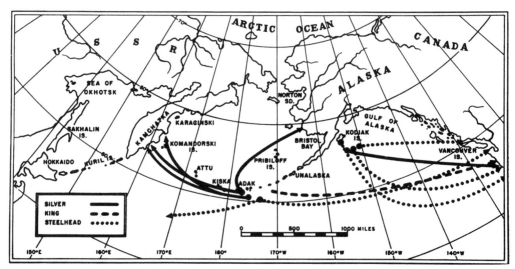

Fig. 2. Generalized distribution of recaptures of king and silver salmon and steelhead trout tagged at sea from 1956 to 1960. The routes shown are the shortest distances between the marking and the recapture points. [International North Pacific Fisheries Commission]

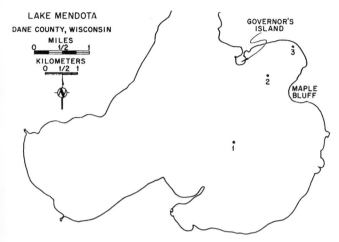

LAKE MENDOTA

DANE COUNTY, WISCONSIN

MILES
0   1/2   1

KILOMETERS
0   1/2   1

GOVERNOR'S
ISLAND

MAPLE
BLUFF

Fig. 3. Map of Lake Mendota. Releases of white bass were made at stations 1, 2, and 3.

fingerlings. After a short period of training it was evident that these fish, too, could discriminate between the odors of water from the two Wisconsin creeks.

Pursuing further this concept of the recognition, by odor, of a home stream, we (22) captured sexually ripe coho salmon (*Oncorhyncus kisutch*) at two branches of the Issaquah River in Washington and returned them downstream, below the fork, to make the run and selection of stream again. We plugged the nasal sac of half of the 302 specimens with cotton. The great majority of those recaptured from the other group had again selected the stream of their first choice, while the fish with plugged nasal sacs returned in a nearly random fashion. This field experiment is indicative of the important role that a functional olfactory system has in the orientation of salmon returning to their home stream.

In an effort to further test this hypothesis we proposed to employ an artificial substance to which salmon fry could be conditioned in a hatchery and which could then be used to decoy them, upon their return, into riverlets downstream from the site of conditioning. Moreover, if it proved practical, it might be used to direct fish into rehabilitated streams, formerly dammed or polluted, or to salvage a run which would not be able to pass a newly constructed power dam. Such an odor must be neither a repellent nor an attractant for the unconditioned salmon.

One of my students (23) has suggested that the compound morpholine

might fit the requirements for this trial. It is soluble in water, thus permitting accurate dilutions; it is detectable in extremely low concentrations, thus making the treatment of large volumes of water feasible; and it is chemically stable under stream conditions. Furthermore, at these low concentrations, it is neither an attractant nor a repellent for unconditioned salmon and thus should influence the behavior of only those salmon that have been previously conditioned to it. Field tests may now be conducted to determine whether salmon fry and fingerlings which have been conditioned to morpholine in a hatchery can be decoyed to a hatchery located on a stream other than that of their birth upon their return to fresh water as migrants. A field experiment of this kind to determine the nature of imprinting is of the highest importance relative to our hypothesis. At least it would be instructive to determine if the "memory" of an artificial odor could be superimposed upon that of the natural odor of the stream. That specific carbon dioxide concentrations in the rivers may serve as guide-posts for migrating fishes has been proposed (24). If fishes were found to be attracted or repelled by substances such as carbon dioxide, that would not signify that they were responding to it in homing. Indeed, their homing behavior would seem to preclude the possibility that they are so attracted or repelled. If this were the case, salmon might be expected to follow a chemical track regardless of their origin.

Recently, Teichmann (25) has ex-

amined the olfactory acuity of the eel with respect to pure chemicals and has found it to be remarkably high. Concentrations of 3×10^{-20} of β-phenylethylalcohol were detected by young eels conditioned by training to these chemicals. If 1 milliliter of the compound were diluted to this concentration, the volume of liquid would be of an order of magnitude 58 times the volume of the Lake of Constance (the Bodensee). Teichmann (25) computed that the amount of the chemical in the eel's olfactory sac at this concentration would be as little as two or three molecules.

If and how, in migration, eels use this sense is not known; however, Creutzberg (26) suggests that elvers use the olfactory sense in discriminating between the water of ebb and flow tide and are hence able to take advantage of the transport of the flow tide in bringing them from sea to fresh water. In laboratory tests he found that elvers were not able to discriminate between ebb and flow waters after the water had been filtered through charcoal. The odor can be presumed to be organic in nature.

Contemplation of the various roles that olfaction might play in eel migrations is provocative, and the opportunity to discover new mechanisms awaits the imaginative and industrious investigator.

In the ocean, it seems to me, odor might play a role principally by giving the fish a sign stimulus for home recognition. If the fish were swimming within a water mass, it would have no sense of being displaced, as the mass moved, unless there were fixed visual or tactile features in the environment; (compare the experiences of balloonists in a cloud). On the other hand, in the place of contact of two water masses there might be differences in salinity, dissolved gases, and odor (4) that possibly could be perceived. Unpublished data from our laboratory have convinced us that the minnow can smell the difference between water from Georges Bank and samples from the Sargasso Sea. When two water masses met there might be a sliding of one over the other—a "shear effect"— that would help the fish to sense that it had reached the edge of a water mass, stimulating it either to enter the new mass or stay in the water mass in which it had been swimming.

The sensing of salinity, gases, or odors at any one place at sea would appear to me to be signals for recognizing, for example, an oceanic spawning site once the fish had arrived rather than cues for directional orientation.

In this article I may have overstressed the olfactory sense and its importance in migration, but I hasten to acknowledge the many limitations of our results and the need for much more work to fill in the gaps in our knowledge. As I contemplate the role of olfaction in salmon migration I believe it to be, above all, useful to a salmon within a stream system, far less useful in the open sea.

Open Sea Orientation

While home-finding in a stream system may depend upon the recognition of an odor and on other yet undiscovered guideposts, it seemed to me that the olfactory hypothesis was inadequate to explain the movements of salmon in the ocean. Certainly other cues are used.

Since it was abundantly clear that a Wisconsin team would have difficulty in conducting field studies on salmon at sea, I then asked myself the question: Is there a fresh-water fish which must find its spawning ground from open water? If so, then the mechanism of open-lake orientation in such a species might give us a clue as to how the salmon accomplishes these feats at sea.

Our initial attack on this problem consisted of study of a less complex type of homing than that in salmon. For a number of years my co-workers and I (27) have studied the natural history of the white bass (*Roccus chrysops* Raf.) in Lake Mendota, Wisconsin; we have been able to locate only two major spawning sites in the entire lake, and these are of very limited area. These spawning grounds, Maple Bluff and Governor's Island, are both on the north shore of the lake and are separated by a distance of 1.6 kilometers (Fig. 3). Here the white bass congregate at spawning time in late May and early June, when temperatures range from 16° to 24°C.

During three different spawning seasons (1955, 1956, and 1957) white bass were captured in fyke nets, marked with numbered disk tags, and transported in open tanks to the different release stations in the lake for daytime releases.

From the start, we observed that a large percentage of the displaced fish returned and were recaptured in nets; moreover, as the observations accumulated we were impressed with the high percentages of recapture (89 to 96 percent) among the Maple Bluff and Governor's Island spawners. They returned to their original spawning site from a release point located 2.4 kilometers from the spawning grounds, in a lake having an area of 39.4 square kilometers and a shoreline of 32.4 kilometers. For tagged fish released on the spawning ground without being displaced the percentages for recapture and for time lapse between release and recapture were of the same order of magnitude as those for fish that were displaced. This would indicate, therefore, that there was an almost complete return of the displaced fish to a spawning ground, it being assumed that mortality of the two groups was similar, that the efficiency of the net was constant, and that the nondisplaced fish remained as a catchable portion of the fish in the spawning grounds.

Subsequently, the "take-off" direction of displaced white bass was observed. To the fish was attached a plastic float (Fig. 4), which could be followed as the fish towed it along. More refined methods of tracking are currently being tested in our laboratory. The releases reported by my co-workers and me in 1958 (27), and many additional releases since then, continue to convince us that the course taken upon release is generally north, toward the spawning grounds, on sunny days. On cloudy days, however, the fish swim randomly. It would appear that this tendency to take off toward the north serves the purpose of bringing the fish promptly to shore in the general vicinity of the spawning areas. Once there, they appear to locate their specific spawning areas by other cues. Orientation of type (ii) discussed above appears, then, to characterize the white bass of Lake Mendota.

How might this mechanism of orientation be explained? While we were investigating the field aspects of this problem, the classical studies of von Frisch (28) and Kramer (29) on sun orientation had been published. I considered it important to explore the possibility that a sun-compass mechanism could be helpful to fish in open-water migration. To expose this notion to laboratory and field tests, new methods and apparatus had to be developed. In addition, we had to determine what fish were most suitable for this type of laboratory experiment.

The principal method which we developed to test for sun orientation relied upon an escape, or cover-seeking, response, which was used for scoring. The fish were tested under the open sky in a specially designed tank (Fig. 5) (27). During training the fish usually attempted to seek cover and found it

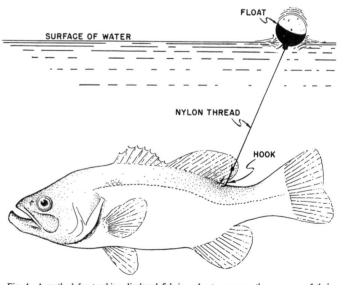

Fig. 4. A method for tracking displaced fish in order to measure the accuracy of their direction of take-off.

in only one of the 16 small compartments, within 360°, the others having
been covered by a metal band. The
arrangement of the small containers
was such that they could not be seen
by the fish from its starting point in
the middle of the large tank. Training
tests were conducted at frequent intervals. In this process the fish was removed from the small container and
placed in a cage in the center of the
tank. Upon release it was given a
small electric shock to frighten it. This
resulted in the fish's seeking cover
again in the small open container,
which was always in the same compass
direction.

Tests were then conducted with all
16 of the boxes open and available to
the fish. When it was determined that
the fish had learned the location of the
box which lay in the same compass
direction, the critical tests were begun.
The tests were made between 0800 and 0900
hours and 1500 and 1600 hours CST.
All 16 small containers were available
for entry, but usually the fish chose
those which lay in the compass direction in which it had been trained to
seek shelter.

The trained fish were disoriented
when tested under completely overcast skies, at times when the experimenter could not detect the presence
of the sun. This demonstrates that the
sun was the fish's point of reference
and that the fish had learned to seek
cover in the same direction at different times of day—that is, to allow
for the movement of the sun (see
Fig. 6).

The crucial and definitive test was
then conducted—namely, substitution
of an artificial "sun," indoors, for the
actual sun. A sun-compass fish responded as though it were responding
to the real sun out-of-doors at that
time of day, choosing a hiding box at
the appropriate angle to the artificial
sun. Hence, the existence of an orientation rhythm which is associated with
the so-called "biological clock" has
been established.

It now becomes imperative that
field studies be made of migrating
salmon at sea. Sexually mature salmon
should be captured in purse seines
near the mouth of a home river,
equipped with tracers similar to those
described above for the white bass, and
displaced several kilometers out to
sea, and their "take-off" and swimming direction should then be charted.

Future Research on
Open-Sea Navigation

From field and laboratory evidence,
compiled from four different species
of fresh-water fish, it is clear that many
fish possess a sun-compass mechanism,
and of importance here are the results
of our unpublished laboratory tests
on young silver salmon, which show
that they too have this mechanism for
orientation.

We cannot stop here, because a
salmon with a sun-compass mechanism
needs augmentive sensory and physiological abilities to accomplish the
migratory feat it appears to perform.

If it could be shown that salmon do
swim in a constant compass direction
when the sun shines and that they
cannot do so under a completely cloud-
covered sky, the evidence would be
clear. On the other hand, negative
results in both cases could be due to
excessive handling, but positive results, with and without the sun, would
point toward orienting mechanisms
other than our hypothetical sun compass.

If salmon could maintain a compass
course in the open sea, even a slight
drift with currents would displace
them to such great distances that they
would be driven to shore hundreds of
kilometers from their home-river system (see Fig. 2).

Possible Influence of Sun Altitude

Perhaps there are other cues which
are perceived in navigation. Celestial
cues may only be the only significant
ones.

In seeking an understanding of the
mechanism of the sun compass in
animals, the application of laboratory
methods has been extremely illuminating. Nevertheless, it must be kept in
mind that these methods have not
provided us with an explanation of
navigational ability.

The results from field tests have
established navigational ability in birds
(30), but such ability requires, in addition to the sun-compass type of
orientation, some other factors, such
as those that enable homing pigeons,
for example, to find their home loft
after being released at distant, unfamiliar places. For salmon, comparable field tests have not been made,
even though oceanic migration appears
to demand some type of navigational
ability.

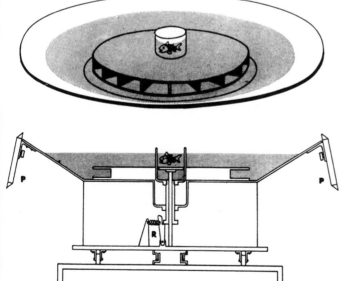

Fig. 5. Tank for training fish to a compass direction. (Top) View from above, showing
the hidden boxes. (Bottom) Side view, showing periscopes (*P*) for indirect observation
and the release lever (*R*) used to permit the cage to be recessed by remote control
when the fish is released.

The take-off directions of the white bass when released out of sight of shore in the middle of a large lake (27) support the view that at least one species of migrating fish uses its ability to maintain a compass direction when the sun is visible. This sun-compass mechanism in fish appears to be quite similar to that found in invertebrates (28) and in birds (29). Other characteristics of the sun-compass mechanism, as judged from laboratory tests, lead to further fruitful avenues of exploration into the nature of the physiological mechanisms involved.

Braemer (31) simulated drastic longitudinal displacements with trained fish by shifting their "time sense." He did this by conditioning them to light cycles which were delayed or advanced from those of their normal day. When these fish were tested out of doors, he found that the compass direction which they now indicated deviated by a definite angle from the previous direction. This deviation was roughly equivalent to the amount of shift in time (or longitude). When testing these fish during the noon hour, he observed that the zenith position of the sun could be correlated with a partial reorientation which resulted in choices that were in the compass direction of take-off of their training period.

This result suggests that the altitude of the sun plays a role in orientation.

In order to shed further light on this point Schwassmann and I (32) flew some sunfish (Lepomus cyanellus) that had been trained in orientation to the sun for several weeks at Madison, Wisconsin (43°N), to the equator at Belém, Brazil (1°S), where we tested them out of doors in our circular sun-orientation maze (Fig. 5). These fish did not adapt to this new and radically different daily sun movement but continued to make the compensation for the azimuth curve of the sun that would have been "correct" for Madison. One of these sunfish, when flown to Montevideo (30°S) and retrained briefly under the sun (which appears to move counterclockwise in the Southern Hemisphere) continued to make the adjustment that would have been correct for the Northern Hemisphere. What would have been the response of a sexually mature salmon that had moved from one latitude to the other at a normal and gradual rate?

Under the equinoctial sun at the equator a rapid deterioration of the oriented behavior of the displaced

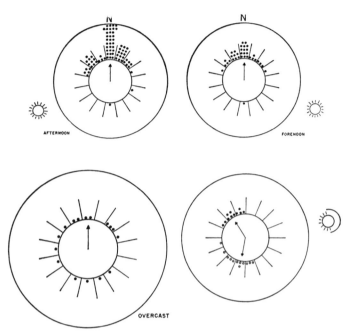

Fig. 6. Scores of a fish trained to north. Scores of the fish (top left) tested in the afternoon, with 16 possible choices; (top right) tested in the forenoon, with 16 possible choices; (bottom left) tested under completely overcast skies on two different days; (bottom right) tested with an artificial light, the angle of altitude being the same as for the sun (solid circles, scores in the forenoon; open circles, scores in the afternoon).

fish occurred, and all the fish showed an increasing tendency to maintain one and the same angle to the azimuth position of the sun throughout the day. Because of the somewhat preliminary nature of this experiment, no definite conclusion can be drawn as to the probable cause of this behavior.

It is clear that other types of studies must be undertaken in order to solve the riddle of the salmon's migratory accomplishments. Certainly we must be boldly imaginative in approaching the problem, by going to sea with new techniques and logical hypotheses to be tested and returning to the laboratory to elaborate on the mechanisms used by the fish to find its way.

Is it possible that the salmon, upon reaching sexual maturity, accumulates a certain temperature budget within a water mass in which it is swimming, as it obeys direction from its sun compass? Its subsequent deviation from this course might take place only when this budget was exceeded. At this point the fish might take another angle to the sun in order to correct its course and keep from being drifted too far off course. This is no doubt too bold

a hypothesis, but it is this type of thinking which will be required as we take the next step of the many which will be necessary before the complete behavior pattern is deciphered.

Adult salmon are known to move during the night in the sea. In order to take advantage of this activity the gill-net fishermen set their nets at night. Clifford Barnes of the University of Washington (33) observed salmon migrating at right angles to his oceanographic research vessel, which was on course at night in the northeastern Pacific. Because of a luminescent sea, this school of large salmon could be seen clearly. The fish swam on a fairly straight course until they were out of sight. We need to know more about the directed movements of salmon at night. We badly need a technique for tracking them, in order to gain knowledge of their night activities at sea. The work on night migrating birds (34) and the possible influence of star patterns as beacons for orientation suggest new points of reference for fish.

The tuna constitute another group of fishes which make long migrations in the open sea. Transpacific and transatlantic (east-west) migration of marked and recaptured tuna are on record. Field data are accumulating rapidly, owing to the efforts of Japanese and North American workers as they add knowledge about the migratory routes. Soon, enough information will have accumulated so that experimental work can be planned. We have here opportunities of unlimited potentiality, exciting to contemplate.

Summary

The pertinent literature regarding the mechanisms employed by sexually mature salmon in order to return from the sea to their home stream or rivulet is the subject of this article. Emphasis is placed on the Pacific salmon as reported in the American literature. Although the migrations of the eel are alluded to briefly, the new theories of odor and sun orientation and their possible role in homing of salmon are reviewed in somewhat greater detail.

In salmon two main phases of homing are stressed: (i) the finding of the home stream or rivulet after the main river has been reached and (ii) migration from long distances at sea to the rivers. The oceanic migration seems to demand an ability to navigate—that is, to return home from a distant place in an oriented fashion and not by random searching.

In the river system, it is proposed, a young salmon becomes "imprinted" or conditioned to the odor of its parent stream. After three to five years at sea and during the return river migration, it swims against the current, rejecting all odors, until it arrives at the home tributary to which it has been initially conditioned. Circumstantial evidence is cited in support of this hypothesis: (i) from training experiments in the laboratory, in which it was shown that fish can discriminate the scents of streams and that a fish may retain the odor of a stream in its "memory"; (ii) from field studies, which showed that after normal salmon and salmon with plugged nasal sacs had been displaced, the latter were unable to return accurately to the home stream.

Additional experiments are suggested for definitive tests of the theory.

Recent studies by Dutch and German researchers on the olfactory sense of eels points up the importance of this sense in orientation, particularly as an aid for locating a river inlet.

At sea, it is proposed, the salmon may possess, among other capabilities, a sun-compass mechanism for orientation, as first described by von Frisch for bees and by Kramer for birds. Recent work has shown that a pelagic American lake fish employs the sun in order to strike a compass course when displaced from its spawning ground near shore. Laboratory experiments prove that several fresh-water fish, in addition to young salmon, have a sun-compass mechanism, but true navigation in fish has not yet been demonstrated by field studies.

Some of the unsolved problems are stressed here, and suggestions are made for future observations and for experiments which should be performed in future investigations of the remarkable feat of homing in the salmon. An increased effort, by the International North Pacific Fisheries Commission to capture and mark thousands of Pacific salmon on the high seas is filling the gaps in our knowledge about the routes of salmon at sea. Comparable efforts are needed in order to obtain records of the oceanic movements of eels, tuna, and other fish which migrate long distances.

References and Notes

1. A. D. Hasler and W. J. Wisby, *Ecology* **39**, 289 (1958); S. D. Gerking, *Biol. Revs.* **34**, 221 (1959).
2. L. Scheuring, *Ergebn. Biol.* **5**, pt. 1, 405 (1929); *ibid.* **6**, pt. 2, 4 (1930).
3. B. T. Scheer, *Quart. Rev. Biol.* **14**, 408 (1939).
4. A. D. Hasler, *J. Fisheries Research Board Can.* **11**, 107 (1954).
5. L. Bertin, *Eels: A Biological Study* (Cleaver-Hume, London, 1956), p. 192 [*Les Anguilles* (Payot, Paris, 1942), p. 218].
6. J. W. Jones, *The Salmon* (Collins, London, 1959), p. 192.
7. D. R. Griffin, *Am. Scientist* **41**, 209 (1953).
8. G. Kramer, *J. Ornithol.* **94**, 201 (1953).
9. W. A. Clemens, R. E. Foerster, A. L. Pritchard, in "Migration and Conservation of Salmon," *Publ. Am. Assoc. Advance. Sci.* No. 8 (1939), pp. 51–59.
10. J. Schmidt, *Rapp. Conseil Exploration Mer* **5**, 137, 267 (1906)
11. D. W. Tucker, *Nature* **183**, 495 (1959).
12. U. D'Ancona, *ibid.* **183**, 1405 (1959).
13. A. G. Huntsman, *Science* **95**, 381 (1942); A. L. Pritchard, *Fisheries Research Board Can. Progr. Repts. Atlantic Coast Stas.* No. 57 (1943), pp. 8–11; A. A. Blair, *J. Fisheries Research Board Can.* **13**, 225 (1956).
14. A. C. DeLacy, personal communication.
15. *Intern. North Pacific Fisheries Comm.*, Vancouver, B.C., *Ann. Repts.* (1956–1958); A. C. Hartt, *Univ. Wash. Fisheries Research Inst. Circ. No. 106* (1959).
16. W. F. Thompson, personal communication.
17. L. R. Donaldson and G. H. Allen, *Trans. Am. Fisheries Soc.* **87** (1957), 13 (1958).
18. A. D. Hasler and W. J. Wisby, *Am. Naturalist* **85**, 223 (1951).
19. W. E. Ricker and A. Robertson, *Can. Field-Naturalist* **49**, 132 (1935); W. Wickett, *Fisheries Research Board Can. Progr. Repts. Atlantic Coast Stas. No. 11* (1958), pp. 18–19.
20. T. J. Walker and A. D. Hasler, *Physiol. Zoöl.* **22**, 45 (1949).
21. W. H. Thorpe and F. G. W. Jones, *Proc. Roy. Soc.* (London) **B124**, 56 (1938).
22. J. Wisby and A. D. Hasler, *J. Fisheries Research Board Can.* **11**, 472 (1954).
23. W. J. Wisby, thesis, University of Wisconsin (1952).
24. E. B. Powers and R. T. Clark, *Ecology* **24**, 109 (1943); G. B. Collins, *U.S. Fish Wildlife Serv. Fishery Bull. No. 52* (1952), pp. 375–396.
25. H. Teichmann, *Naturwissenschaften* **44**, 242 (1957).
26. F. Creutzberg, *Nature* **184**, 1961 (1959).
27. A. D. Hasler, R. M. Horrall, W. J. Wisby, W. Braemer, *Limnol. Oceanog.* **3**, 353 (1958).
28. K. von Frisch, *Experientia* **6**, 210 (1950).
29. G. Kramer, *Naturwissenschaften* **37**, 188 (1950).
30. G. V. T. Matthews, *Bird Navigation* (Cambridge Univ. Press, Cambridge, England, 1955), p. 141.
31. W. Braemer (working in our laboratory), *Verhandl. deut. zool. Ges.*, in press.
32. A. D. Hasler and H. O. Schwassmann, *Cold Spring Harbor Symposia Quant. Biol.*, in press.
33. C. Barnes, personal communication.
34. F. Sauer, *Z. Tierpsychol.* **14**, 29 (1957); F. C. Bellrose, *Wilson Bull.* **70** (1958), p. 20.

Chemical Communication among Animals

EDWARD O. WILSON AND WILLIAM H. BOSSERT

The Biological Laboratories, Harvard University, Cambridge, Massachusetts

I. Introduction

The study of communication among animals by means of the transfer of chemical substances has expanded in the past few years into a new discipline, attracting the cooperative attention of a growing corps of animal behaviorists, neurophysiologists, endocrinologists, and organic chemists. The subject is sufficiently developed to bear at this time not only a general review, but a start at quantitative theoretical analysis.

Karlson *et al.* (66, 67) offered the term *pheromone* for the somewhat awkward, self-contradictory *ectohormone* of Bethe (7), and restricted it to embrace the set of substances used in intraspecific communication. These authors made a useful distinction between olfactorily and orally acting pheromones, with the assumption that most of the former cause immediate behavioral responses; i.e., are "chemical releasers" in the terminology of the ethologist, whereas the latter produce their primary effects on the endocrine and reproductive systems. The best examples of orally acting pheromones are the caste-determining and reproduction-inhibiting substances of the social insects. Here the principal question before us concerns the physiological mechanisms through which the pheromones act. Lüscher's work (79, 80, and contained references) has proved that caste inhibition is effected via the endocrine system, but from this and similar studies on honey bees (20, 21) and ants (8), it is still not known whether the endocrine system is controlled via the nervous system activated by gustatory chemoreception or directly by absorbed pheromone molecules. In the case of the honey bee, Butler (20) has shown that at least one inhibitory substance distinct from the orally ingested "queen substance" (9-ketodec-2-enoic acid) is perceived olfactorily. Loher (78) has established the presence in adult gregarious-phase, male locusts (*Schistocerca gregaria*) of a volatile epidermal secretion which accelerates the maturation of young locusts. The substance is perceived at least in part by antennal olfaction. The picture is further complicated by the recent discovery of olfactorily acting pheromones which alter reproductive physiology in mice without inducing immediate behavioral responses (91).

The above considerations require a somewhat new way of viewing pheromone action. We propose to distinguish the *releaser effect,* involving the

674   EDWARD O. WILSON AND WILLIAM H. BOSSERT

classical stimulus-response mediated wholly by the central nervous system, from the *primer effect,* in which the endocrine and reproductive (and possibly other) systems are altered physiologically (see Fig. 1). In the latter effect, the body is in a true sense "primed" for new biological activity. The alteration, which of course requires more time than the releaser effect, may lead in a few cases to behavioral responses induced *in vacuo* by internal stimuli. But, as a rule, new external stimuli following

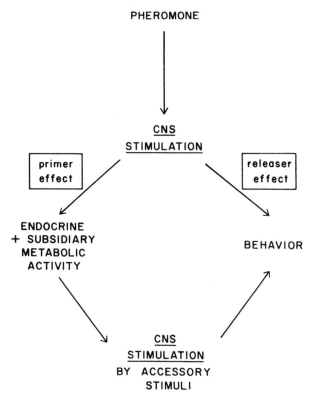

FIG. 1. The pathways of influence of pheromone action. The pheromone may be the primary stimulus causing a quick behavioral response (*releaser effect*), or it may act more slowly and indirectly by altering the physiology and "priming" the animal for a different behavioral repertory (*primer effect*). In the latter case behavior is usually induced not by the pheromone, but by accessary external stimuli, such as those received during courtship or parent-young interaction. The CNS-endocrine-reproductive activity pathway occurs in mammals. Other pathways, such as the shorter endocrine-to-reproductive activity pathway, or even a direct alteration of reproductive physiology by the pheromone, are possibilities not yet exemplified.

the priming are required to release the altered behavior patterns. It is quite possible for the same pheromone to be both a primer and a releaser; e.g., 9-ketodec-2-enoic acid, which inhibits worker ovary development (20), inhibits queen-cell building by workers and attracts males during the nuptial flight (48). Our rather limited knowledge of the primer pheromones has been explained competently by the authors cited in the previous paragraph. The following account will deal mostly with the problems of communication by the releaser effect.

II. General Characteristics of Chemical Transmission

It is conventional to emphasize the limitations of a chemical communication system (82), but this has been largely due to our long-standing ignorance of the subject. In the discussion of certain characteristics of the system to follow later, it will be seen how some are remarkably advantageous. In another section it will be further shown that information in a simple pheromone system can be multiplied by relatively slight physiological and behavioral modifications.

A. Minimum Potential Thresholds of Response

Certain recent estimates of response threshold concentrations are surprisingly low. Schutz (101) found that the alarm substances of some fish species are effective in dilutions up to one volumetric part in 2×10^{11}; a fragment of skin weighing 2 μg., and thought to contain on the order of 10^{-3} μg. of the *Schreckstoff*, was enough to disperse a school of minnows (*Phoxinus laevis*) from a feeding place. In field tests Jacobson *et al.* (63) recorded responses of the male gypsy moths (*Porthetria dispar*) to as little as 10^{-7} μg. of pure sex attractant. But even this minute quantity is probably far above the true threshold. Butenandt and Hecker (16, 17) obtained positive behavioral responses from male silkworm moths (*Bombyx mori*) from samples of pure female natural sex attractant diluted to 10^{-10} μg. per cubic centimeter (cm.^{3}) of petroleum ether and presented in a film on the tip of glass rods 1 cm. from the male's antennae. The material, solvent plus solute, on the glass rod is given as about 0.01 cm^3. It can be therefore roughly estimated that only 2600 molecules were presented. If these diffused through a sphere of 1-cm. radius, the concentration at the edge of the sphere can be shown to be no more than 192 molecules per cubic centimeter of air (12). Very likely the real threshold concentration is lower. Butenandt and Hecker (17) have synthesized a geometric isomer of the attractant which is effective in the same test when diluted to 10^{-12}. Probably no more than one or several molecules of this substance suffices to activate the moth. This seems more plausible when the lowest recorded figures for vertebrates are considered: 1700 molecules of butyric acid per cubic centimeter of air for the Alsatian dog (87, 88); and 1770 molecules of α-phenylethanol per cubic centimeter of water for the eel *Anguillula anguillula* (109).

One may reasonably conclude from these estimates that as few as several molecules of an appropriate substance are enough to fire a chemoreceptor cell. In fact, Neuhaus (87, 88) has deduced that just one molecule of butyric acid is enough to induce a response in the Alsatian dog. DeVries and Stuiver (40) have shown that at most 8 molecules of secondary butyl mercaptan are needed to fire one human olfactory cell, and as few as 40 are needed to cause a human subject to sense the odor consciously.

The great potential efficiency of olfactory communication can be illustrated by further considering the case of the *Bombyx mori* sex attractant. Butenandt and his associates (17) obtained 12 mg. of bombykol esters from the equivalent of 250,000 *Bombyx* females, a yield quite comparable to the 20 mg. of attractant obtained from 500,000 females of *Porthetria* by Jacobson *et al.* In the case of *Porthetria* the efficiency of yield very likely exceeded 50% (5). Thus a female *Bombyx* contained at a given instant in time at least 10^{-2} μg., or about 10^{14} molecules, and probably an amount not much greater than this. If 200 molecules is taken as the threshold quantity to release the male response (it is possibly less), we can see that the minute trace of attractant in a single average female is still enough to draw a response from over 10^{11} males, certainly far more than exist at any one time. Looked at in a somewhat more realistic way, this amount of attractant can be drawn out by air transport to form an active ellipsoidal zone several kilometers long, which is a reasonable approximation to what the females of many moth species accomplish in nature (*vide infra*).

B. POTENTIAL SIGNAL DIVERSITY

As the number of carbon atoms in an organic molecule is increased, the number of possible kinds of hydrocarbons increases in an irregular but approximately exponential manner, as illustrated in Table I. With the addition of oxygen atoms to produce alcohols, aldehydes, ketones, etc., the number increases. When we add to this list the stereoisomers, the increase in potential molecular species with molecular weight is even greater. Both vertebrates and insects can distinguish some, but not all, isomeric forms. Thus, iso-α-irone (*trans*-2,6-methyl-6α-ionine) and neoirone (*cis*-2,6-methyl-6α-ionone) are perceptibly different to humans, as are geraniol and nerol. Even some optical isomers are distinguishable to the human nose (84). Geometric isomers can also be distinguished by insects. Steiner *et al.* (106) have recently shown that the *trans* form of siglure (*sec*-butyl ester of 6-methyl-3-cyclohexene-1-carboxylic acid), a new synthetic lure of the Mediterranean fruit fly (*Ceratitis capitata*) is more than twice as effective as the *cis* form. Similar results were obtained by Beroza *et al.* (6) with the hydrochlorinated derivative medlure. Butenandt and Hecker (17) synthesized two geometric isomers of hexadecadienol which

TABLE I

The Number of Possible Structural Isomeric Hydrocarbons of the Methane Series as a Function of the Number of Carbon Atoms[a]

Carbon content	No. of isomers	Carbon content	No. of isomers
1	1	11	159
		12	355
2	1	.	.
		.	.
3	1	.	.
4	2	15	4347
		.	.
5	3	.	.
		20	366,319
6	5	.	.
7	9	.	.
		.	.
8	18	30	4.1×10^9
		.	.
9	35	.	.
		.	.
10	75	40	6.2×10^{13}

[a] From Henze and Blair (61).

become behaviorally effective only at concentrations 10^{12} and 10^{13} times greater than the third, most effective, isomer. These early findings on insects, while based on synthetic compounds, strongly suggest the possibility of the existence of related phenomena in natural pheromone differentiation. It is necessary to conclude that the number of possible odorant molecules that can be manufactured is, in evolutionary time, effectively unlimited; that is, the potential number must be far greater than the number that can be used.

How many different odorants can a single organism distinguish? On the basis of subjective experience, according to Hainer *et al.* (58), some experts estimate that humans can recognize at least 10,000. Since our species is notably microsmatic, it is likely that other animals, at least many other mammal species, can do better. When the odorants are presented in mixtures, much as other types of signals are presented syntactically in communication, the number of signifiable discrete messages increases to a great but unknown extent. There seems to be no effective limit to the chemical "vocabulary" that a species, through evolution, could come to employ other than the one imposed by the ability of its central nervous system to process information.

With these considerations in mind, it becomes easier to accept the extreme cases of olfactory acuity reported in the literature. Kalmus (64), in measuring the almost legendary tracking ability of hunting dogs, proved that these breeds can separate the odors of identical twins, and Schmid (99) showed that German police dogs can even distinguish trails made by another single dog if one trail is 30 minutes older than the other. Hasler and his associates (review, 59) have presented convincing evidence supporting their hypothesis that migratory salmon home to the stream of their birth because they are imprinted to its specific odor or combination of odors. Fish of at least some social species are able to recognize the odor of their own school in opposition to alien schools of the same species (53). Female mice distinguish the odors left in soiled cages by their mates from those left by alien males, and this discrimination has a profound effect on their reproductive physiology (91). The ability of social insects to separate instantly members of the same colony from strangers belonging to the same species is well known. Recent work has shown that dietary differences alone are adequate to differentiate the colony odors of honey bees (65), whereas either dietary differences or differences in chemical composition of the nest walls are adequate in the ants (70). The role of diet in odor differentiation is probably a key one in other animal groups as well. Geographic variation in the chemical composition of the beaver's castoreum is in large part a reflection of dietary variation (71). Even some of the variation in human body odor, especially in perspiration, is due to dietary differences (1), a fact that might explain the unexpected ability of dogs to separate identical twins.

C. FADE-OUT TIME

The obvious design-feature which in abstract separates chemical communication from systems in other modalities is the inherently long fade-out time. In Fig. 2, we have plotted fade-out time of chemical signals released as puffs in still air as a function of amount of material in the puff, thereshold concentration, and the diffusion coefficient. To cite one example from a

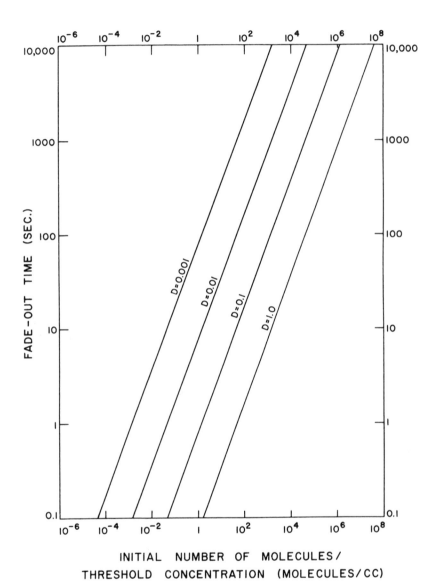

FIG. 2. The time required for a pheromone, released as a puff in still air, to diffuse to the extent that the concentration is everywhere below threshold level. *D*, diffusion coefficient of the substance in air.

case history discussed elsewhere in this article, the sex attractant contained in one female gypsy moth, 10^{-2} μg., would require about 8×10^7 seconds, or 930 days, to diffuse everywhere below threshold level; i.e., fade out, if it were released altogether in one puff in perfectly still air. One gram of the same substance would require over ten million years! Even if the diffusion coefficient were to be greatly increased by air turbulence, say to 100 cm.^{2}/second, the contents of a female gypsy moth would still require 8×10^4 seconds, or about one day, to fade out, and 1 gm. over 10,000 years. Dissipation is further slowed when, as is usually the case, the pheromone is emitted in liquid form and must be evaporated first; and even more when the substance is partly absorbed into the substrate.

The consequences of slow fade-out are not hard to visualize. The rate of signal transmission must be low. Hence the system is poor for communicating rapid changes in position and mood of an animal, for which visual and acoustic signals are superbly adapted. Hence pheromones do not appear to be used, for instance, in the maintenance of dominance hierarchies. This limitation has been somewhat overcome by the use of high threshold concentration values, sensory adaptation, and dissipation by air (or water) currents, as will be seen in the description of case histories in a later section. The slow fade-out of chemical signals can be a positive asset, however. Chiefly, it can allow an animal to broadcast a signal over a long period of time and over a relatively great distance, at an absolute minimum of energy expenditure. In this combination of values, chemical communication is unique. As a consequence, it seems logical to find that it is very commonly used in territorial marking and trial laying.

D. Limitations in Pattern Designing

Patterning could conceivably occur (1) spatially, in the laying down of odor "shapes"; or (2) temporally, through modulation in pheromone emission or variation in its sequence and duration. But the slowness of pheromone transmission and fade-out and their great irregularity due to turbulence, combine to make these phenomena difficult to achieve. Spatial patterning was suggested in the chemical communication of ants by Forel, who postulated a special "topochemical sense" by which the workers discerned the shape of odor spots in the chemical trails and other crucial parts of the odor environment (43). Recent work on the odor trail (25, 81, 117) has shown that the trials are indeed laid as spots or streaks, occasionally with distinct shapes that conceivably could be used in determining the direction in which the trail was laid. But although humans can sometimes make out shapes visually, the ants cannot do so by chemoreception, since they zigzag back and forth across the trails in too erratic a fashion. Instead, they use other cues, probably visual, to gauge the direction of the trails with respect to the nest. The trails themselves lead them along by virtue of their linear shape but provide no further information. Tracking dogs crisscross trails in a similar exploratory manner, and it is this irregularity which probably prevents them from adapting to the odor and losing the trail altogether (59). There is also to our knowledge no evidence of information on specific odor shapes being utilized by insects or mammals at scent posts.

The difficulties in discerning temporal patterning are also considerable. These can be visualized by considering the problems of odor modulation, as illustrated in Fig. 3. This hypothetical case, using the emission and threshold values of the *Porthetria* sex attractant, shows nicely that in still air periodic pulses can carry information only over what are, in relative terms, extremely short distances. Since any pattern of emission can be decomposed into sine waves, the result applies to temporal patterns of all sorts. Recent analysis (W. H. Bossert, unpublished) shows that the maximum rate of information transfer over distances of a meter or more by temporal patterning must be less than 0.1 bit/sec. In a moderate wind, however, this means of communication would be practical over greater distances. The limiting rate of information transfer over a range of several meters on the downwind direction exceeds 100 bits/sec.

Something that approaches true temporal patterning appears to occur in the chemotactic orientation of aggregating slime-mold amoebae. The amoebae are attracted to each other by an unidentified pheromone referred to as "acrasin." Shaffer (105) was unable to find any differences in the amounts of acrasin secreted by the centers and by different parts of the affluent streams of amoebae. A simple diffusion hypothesis creates a paradox: it would appear that in order to progress further toward the centers the amoebic streams would have to move to some extent down gradients of their own making. But in fact, there is also secreted a substance of large molecular weight which inactivates the acrasin. According to the "relay hypothesis" of Shaffer, the interaction of acrasin and antiacrasin results in a series of directional impulses, following one another in concentric manner. The amoebae then are able to orient with respect to a series of small ephemeral gradients, which are generated and followed in a relay fashion. The recent research on this subject has been reviewed by Bonner in his book on the biology of slime molds (11).

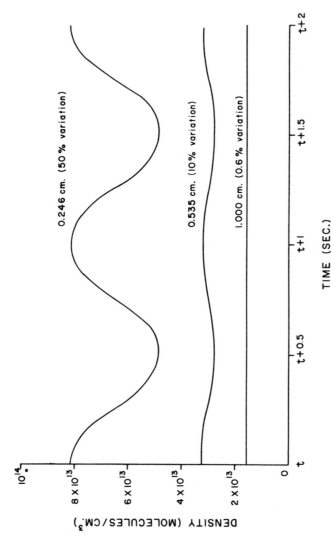

FIG. 3. A hypothetical case of modulated rate of pheromone emission, to illustrate the short range over which communication by this means can be effected. The maximum rate of emission is set at 10^{13} molecules per second, approximately that of the gypsy moth, *Porthetria dispar*. The true pattern of emission of this species is unknown. The emission in this example is arbitrarily made periodic at one cycle per second, thus approximating the rate of extrusion of the scent glands described for same British moth species (see 68). The emission is further made sinusoidal and the coefficient of diffusion set at 0.1, both arbitrary but reasonable features. Note that by human standards of density discrimination (see text) the pattern is lost less than 0.32 cm. from the source.

E. Limitations in Quantitative Communication

It would be difficult to encode and transmit information on quantity by means of frequency modulation, for the reasons just given in the preceding section.

Quantity may be transmitted more readily by relating it, in an analogue fashion, to the amount of pheromone emitted and hence the concentration perceived by the receiver organism. Hainer *et al.* (58) have provided some interesting estimates of concentration perception in humans that allow us to form an intuitive notion of the limits of quantitative communication. For several odorants checked, the ratio of threshold concentration to the lowest concentration that gives the maximum perception is about 1:190,000. Concentration perception follows the "Weber-Fechner law": the magnitude of nervous input increases as a function of the logarithm of stimulus intensity; in humans, the increase in concentration required to make a just noticeable difference is about 52%. Thus, there are about 30 distinguishable levels of concentration. If these various concentration levels were used in a chemical communication code, the receiver organism would have to stand very near the emission source to overcome the damping effects of diffusion and turbulent transport mentioned earlier. Furthermore, the rapid adaptation characteristic of odor perception would present an acute problem. However, there is no doubt that variation in concentration is sometimes used in chemical communication. In some species of ants, e.g., *Pogonomyrmex badius*, the alarm substances act as an attractant in the lowest perceptible amounts. At a concentration approximately ten times that of the attraction threshold, the worker ants switch to true alarm behavior, and if exposed to higher concentrations for a longer period of time they switch to digging behavior. The possibility exists that more than one pheromone exists, but the shift from alarm to digging behavior has also been induced by other, pure odorants (116, 118). The number of fire ant workers (*Solenopsis saevissima*) responding to recruitment trails is an approximately linear function of the amount of attractant substance in the trails (117).

III. The Optimal Range of Molecular Weights

There are at least two prima facie reasons why pheromones should not be expected to be of very low molecular weight. First, the number of kinds of compounds is restricted. Fewer than 300 alkanes, alkenes, monohydroxy alcohols, ketones, and aldehydes contain five carbon atoms or less; and fewer than 100 of these substances contain four carbon atoms or less. Further, only a fraction of these substances could be manufactured by any given taxon. But the number of molecular species is sharply increased with increase in molecular size. Hence, in order to achieve diversity of signals, of the sort needed in species specificity, substances with at least five carbon atoms are likely, and with even more carbon members molecular diversity probably ceases to be limiting.

A second condition making the choice of very small molecules less likely is their lowered stimulative efficiency. Dethier (36, 38) has summarized empirical information on chemical correlates of stimulative efficiency in insects in the following way: the efficiency is proportional to chain length, molecular weight, and boiling point; and inversely proportional to water solubility. Cook (cited by Dethier, 38) has shown that the optimum concentrations in a moderately large series of alcohols and esters with respect to attractiveness to flies is related to the boiling points as follows: the logarithm of the optimum concentration is equal to 13.78 minus 7.1 times the logarithm of the boiling point in degrees Celsius. Boiling point is of course approximately proportional to molecular weight in a given homologous series. Thus isoamyl butyrate has a boiling point 2.5 times higher than that of ethyl acetate and a molecular weight 1.8 times greater, but an optimal concentration only 0.001 times as high. Dethier (38) states that in insects threshold concentration is proportional to optimal concentration. Of course, chemoreceptors of a given species could and must be modified to alter threshold concentrations to chemical substances crucial to the species' biology. But the generalizations just cited suggest again that, at least for insects, molecules containing at least five carbon atoms are basically advantageous.

We would therefore like to predict on purely logical grounds that in the great majority of cases of chemical communication, where pheromones with relatively high specificity and stimulative efficiency are required, the pheromones will contain at least five carbon members and molecular weights of at least 65. One additional premise can be added on the basis of preliminary empirical information: almost all known pheromones contain at least one oxygen atom. Therefore the predicted minimum molecular weight can be adjusted to 80.

It is necessary in passing to eliminate one possible counterargument that has strong intuitive appeal. One might expect that the smallest molecules would have the highest diffusion coefficients and hence provide the fastest transmission, in accordance with Graham's law. This is true, but to a degree that scarcely matters. In the short list of substances for which diffusion (D) coefficients in air have been experimentally determined (62), D varies from 0.634 (hydrogen) to 0.089 (ether vapor, molecular weight 74.12); extrapolating crudely from these values, it can be estimated that D of gases of molecular weight 200 are still greater than 0.01. Elsewhere (12) we have shown that variation of even one order of magnitude in D fails to add significantly to the already great variance among chemical communication systems. For instance, it does not affect in any way the distance in still air over which the pheromone can act. The true diffusion coefficient becomes almost negligible in significance when turbulent transport is involved.

In the factors so far considered, and eliminating D in still air from consideration, it would appear that the larger the molecule the better. Of course there must be a point of diminishing return due to other factors. As molecular size is increased in any homologous series, the energy cost of biosynthesis and intracellular transport per molecule increases. The number of molecules per unit volume of reservoir space decreases. The vapor pressure and hence volatility decreases, eventually to the point where gaseous transport becomes impracticable. It would be impossible at this time to deduce the upper limit of molecular weight of pheromones due to those factors. Intuitively we would expect it to be in the vicinity of 300 or 400, since in that weight range further advantages due to diversity and stimulative efficiency are greatly diminished. In fact, the pheromone with the largest molecular size known is gyplure ($C_{18}H_{34}O_3$), the *Porthetria* sex attractant, with molecular weight 298.

In summary, we would like to suggest that the great majority of pheromones can be expected to contain between 5 and 20 carbon members and have molecular weights between 80 and 300. Perhaps the lower ends of these optimal ranges will be raised somewhat as our knowledge increases. Also, it might be expected that sex attractants in which specificity and effectiveness have a higher premium, will as a rule have greater molecular size; and this has so far held up in the limited roster of identified substances. Of five known or suspected sex attractants (Fig. 6) four have molecular weights between 225 and 298. One, the honey bee queen substance, which serves only secondarily as an attractant, has a molecular weight of 184. The most effective synthetic attractants of flies, which act in some ways like sex attractants, have molecular weights around 200 (55). Of seven alarm and recruitment substances known in the social insects, six have molecular weights from 114 and 194, and one (dendrolasin) has a weight of 218 (Fig. 5).

IV. Functions and Properties of Releaser Pheromones

This discussion will be concerned with the releaser effect, as distinguished in the Introduction. Early classifications of releaser stimuli have tended to use functional categories. Dethier *et al.* (39) have made a useful distinction between (1) true attractants, which cause movement toward the source, and (2) arrestants, which merely cause cessation of locomotion. They have listed as further categories of chemical stimuli (3) locomotor stimulants, which cause more rapid dispersal through increased undirected locomotion; (4) repellents, which cause oriented movements away from the source; (5) stimulants of specific acts such as feeding, mating, etc. (the "releasers" of ethological terminology); (6) deterrents, the inhibitors of specific behavioral acts. These categories are logically exhaustive and refer to chemical stimuli as a whole instead of just pheromones. Marler (83), in adapting Charles Morris' linguistic theory (85) to ethology, distinguishes the signal functions of "identifiors," which direct to a place in time and space; "designators" and "prescriptors" which dispose the receiver toward specific response sequences; and "appraisors" that provide opportunities for choosing among signal sources. Under this scheme of "pragmatics" one releaser pheromone can serve two or more functions. For example, the sex attractants of moths serve both as identifiors and designators, the latter function being achieved at higher concentrations. In a complex gradient created by two or more females, the pheromone can serve as an appraisor.

Both of these recent classification systems should succeed in serving as more than nomenclatural guides to intuitively obvious ideas. They are also potentially useful in stimulating the careful examination of behavioral function of individual pheromones. For the moment, however, we wish to go a step beyond and take up specific biological functions of pheromones. It is our purpose to propose and support the following theory: *that the chemical properties of the pheromone, the emission rate, and the response threshold concentration have been adjusted in evolution to maximize efficiency with respect to a specific function.* This requires the examination of several specific cases within a narrowly defined function. Other functional categories will be listed for which we lack comparable information.

A. TRAIL MARKING

Chemical traits that attract or orient or do both are widespread in the animal kingdom. In some cases they serve as a long-lasting guideline to some fixed object. For instance, chemical cues, probably a trail, appear to be provided by the moth ear mite *Myrmonyssus phalaenodectes* to the initially infested ear. Subsequently, invading mites proceed to this ear and thus bilateral infestations are avoided (111). In some trail-following ants, such as the fungus-growing species of *Atta*, the trails lead to persistent food sources and are themselves persistent. Army ants lay "exploratory trails" as their foraging columns advance (118). Other ant species, as well as certain meliponid bees (75), lay "recruitment trails," which serve to draw out sister workers from the nests to newly discovered food sources or nest sites.

Although several analyses are underway, no trail substance has yet been chemically identified. The abdominal Nassanoff gland secretion of honey bees, which is released into the air to guide workers to food finds and new nest sites (97), has recently been identified as geraniol ($C_{10}H_{18}O$) by Boch and Shearer (10). It is interesting to note that several years previously Kullenberg (69) had correctly guessed this identity on the basis of smell alone.

We have chosen the recruitment trail of the fire ant *Solenopsis saevissima* for detailed communication analysis. When workers find food or a superior nest site they return to the nest, laying a trail substance through their extruded stings. The substance is an attractant, which induces the workers to leave the nest and follow the trail outward. On glass at room temperature the trail fades out in an average of 104 seconds; that is, its concentration drops below threshold level in that time. This relatively quick fade-out is important in that it allows the colony as a whole to regulate the amount of attractant in the trail and hence the number of workers using it. At the same time it imposes a restriction in accuracy of following, so that only about four bits of information are transmitted with respect to compass direction and two bits with respect to distance. Both aspects have been treated in some detail in an earlier paper (117).

In analyzing this case further, the "goals" of a short-lived recruitment trail were intuited as follows. The effective width attained through diffusion must not be great, or otherwise the orientating function of the trail will be impaired. In order to achieve greater accuracy the ants probably would employ the gradient of concentration to stay near the center of the trail. In such a circumstance, it is important that the gradient of concentration near the effective boundary of the diffusing trail substance be large so as to strongly correct departing individuals. Finally, the fade-out must be moderately quick, in order to prevent excessive response, as well as confusion due to overlapping trails. But not too quick, for then the effective range of the trail will be shortened to ineffectiveness.

In a mathematical analysis to be developed more completely elsewhere (12), we have constructed a model of trail formation in still air. The maximum length of a trail at any time is

$$X_{max} = \frac{Q}{2K\pi D} \tag{1}$$

when Q is the rate at which the pheromone is laid in molecules per second; K is the threshold density, in molecules per cubic centimeter, below which the ants cannot detect the trail; and D is the diffusion coefficient. The maximum radius of the cross section of the trail is computed as

$$r_{X_{max}} = \sqrt{\frac{2Q}{eK\pi u}} \qquad (2)$$

where u is the rate of locomotion of the ant in centimeters per second. From laboratory tests X_{max} was determined to be about 42 cm., and $r_{X_{max}}$, 10 mm. From Eq. (1) D was estimated to be 0.00646 cm.^{2}/sec. This value is probably less than the actual diffusion coefficient, because it is improbable that the material is initially all in the gaseous state as assumed in the model. Since the substance is quite volatile, however, the model is still acceptably accurate, with the value of D reduced by a factor due to evaporation time. From Eq. (2) the ratio of emission rate to threshold density (Q/K) is computed as 1.69 cm.^{3}/sec. Knowing the volume of the Dufour's gland reservoir, the source of the trail substance, and the probable limits of emission rate as fractions of this volume, we estimated the reservoir to contain between 10^{16} and 6×10^{16} molecules, the trail to contain between 10^{14} and 6×10^{15} molecules per centimeter in a straight line distance, and Q to be between 4×10^{13} and 2.4×10^{15} molecules per second. The range of K corresponding to this estimate is from 2.36×10^{13} to 1.42×10^{15} molecules per cubic centimeter. The widths and persistence of the fire ant trail at various times can be interpolated as shown in Fig. 4.

The intuitively expected design features are thus upheld. The persistence of the trail is, as expected, sufficient to permit trails of adequate length, but the fade-out, about 100 seconds, is rapid enough to prevent confusion of trails (while permitting quantitative control in mass communication). This results from a proper balance of Q, which increases the persistence for higher values, and K and D, which reduce the persistence. In the case of *Solenopsis saevissima*, Q is moderate, low enough to permit adequately long trails yet large enough to maintain the persistence in the longer trails. The diffusion rate is low, serving both to keep the persistence high and the trail narrow. K, the threshold concentration, is unexpectedly high, especially in comparison with that of other kinds of pheromones (see below). This of course ensures that a new trail will be recognized in high background levels from old trails. In addition, the high value of K and the corresponding value of the ratio Q/K is in striking adherence to another principal design feature, the maintenance of a large gradient at the trail boundary.

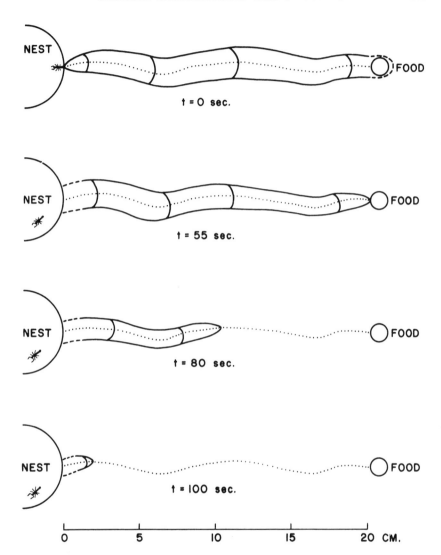

FIG. 4.   The form of the odor trail of *Solenopsis saevissima* laid on glass. As the trail substance diffuses from its line of application on the surface, it forms a semiellipsoidal zone within which the pheromone is above threshold concentration. The times shown here are given from the moment a worker reaches the nest after laying a trail in a straight or only slightly wavering line from a food source 20 cm. away. The trail is shown as continuous. In nature it is irregularly segmental, but the dimensions and fade-out times remain nearly the same.

The gradient is increased for large values of Q and small values of D. For a given Q and D, the gradient is at a maximum where

$$\frac{Q}{K} = e\pi DX \tag{3}$$

In this case

$$\frac{Q}{K} = e\pi \,(0.00646)X.$$

It would be most advantageous to maximize the gradient near the fade-out point of a long trial. The maximum length of an active trail was estimated as 42 cm. If X is set as 42-cm., then $Q/K = 2.31$ cm.^{3}/sec. By the earlier procedure Q/K was computed to be 1.69 cm.^{3}/sec. It is this reasonably close correspondence especially which gives confidence in the validity of the model.

B. ALARM

This category includes a variety of specific responses related only by the threatening nature of the stimuli that induced them. Fish and tadpoles of the toad *Bufo* scatter from the vicinity of the epidermal alarm substances (41, 101). Ants always react to alarm substances by excited, circular or zigzag motion. In the immediate nest vicinity the workers of most species posture aggressively and run toward the stimulus source, but those of other species characteristically flee. A few actually attack the stimulus source, even though it is a sister worker (14, 116, 118). Aggressiveness increases with the size and health of the colony. As a rule, ant alarm substances are attractants at low concentrations and release aggressive or retreat behavior at higher concentrations (118). Termites lay trails to the area of the threatening stimulus, whether it is an intruding insect or a breach in the nest wall. In the lower termites, trail laying appears to serve this function exclusively: only in later evolutionary stages has it come to serve in recruitment to food sources as well (107). When honey bees sting they emit a "pleasantly smelling scent" from unicellular glands next to the quadrate plate of the sting; this substance is responsible for the well-known tendency of other bees to sting in the same area (51).

The pioneering work of Pavan and his associates on the chemistry of exocrine secretions of ants (92–95), together with more recent behavioral studies utilizing the pure products (118, 119), have resulted in the identification of some of the first natural alarm substances. They have so far proved to be mostly terpenes with molecular weights between 100 and 200 (Fig. 5). An exception is the furan dendrolasin. Iridomyrmecin, a complex lactone secreted by *Iridomyrmex humilis,* has proved behaviorally inert (120), and it is our guess that the related substances isoiridomyrmecin, iridodial, and dolichodial will also prove to be behaviorally unimportant. The main function of these anal gland substances is probably defensive, as Pavan (92) originally suggested for iridomyrmecin.

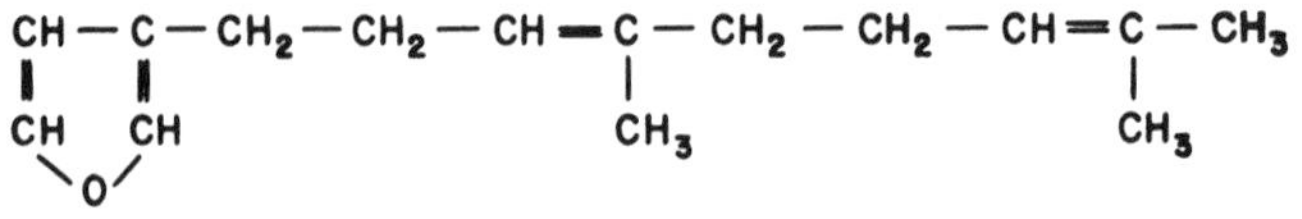

CH — C — CH$_2$ — CH$_2$ — CH = C — CH$_2$ — CH$_2$ — CH = C — CH$_3$

CH　CH　　　　　　　　CH$_3$　　　　　　　　CH$_3$

＼O／

(I) DENDROLASIN

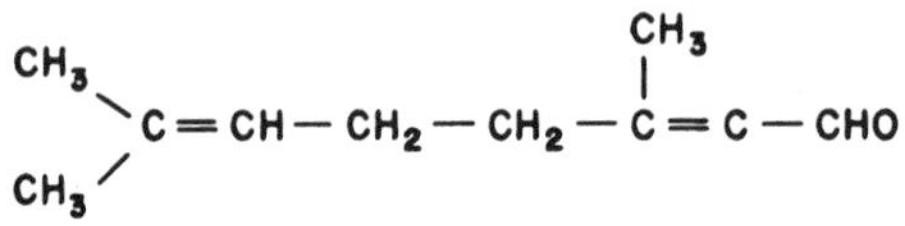

CH$_3$
＼　　　　　　　　　　CH$_3$
　C = CH — CH$_2$ — CH$_2$ — C = C — CHO　　(II) CITRAL
／
CH$_3$

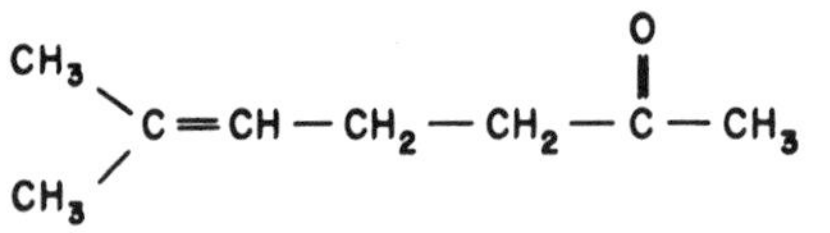

CH$_3$
＼　　　　　　　　　O
　C = CH — CH$_2$ — CH$_2$ — C — CH$_3$　　(III) METHYLHEPTENONE
／
CH$_3$

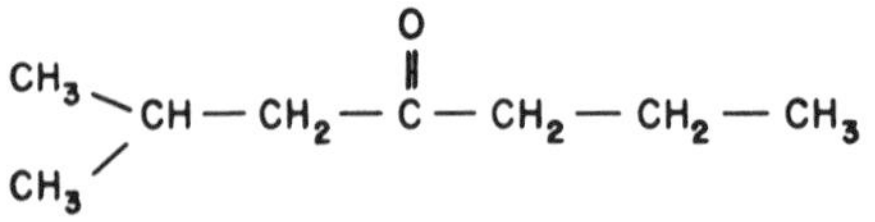

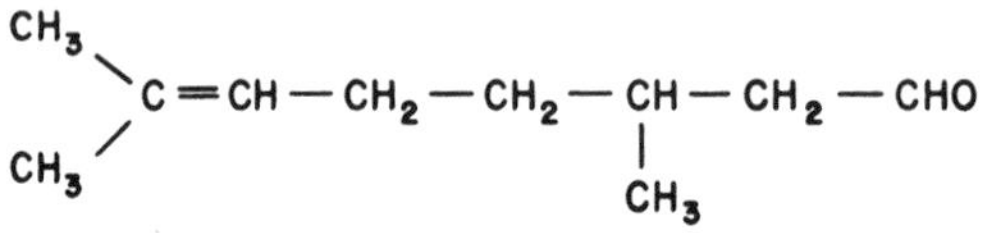

CH$_3$　　　　　　O
＼　　　　　　　‖
　CH — CH$_2$ — C — CH$_2$ — CH$_2$ — CH$_3$　　(IV) PROPYL ISOBUTYL KETONE
／
CH$_3$

CH$_3$
＼
　C = CH — CH$_2$ — CH$_2$ — CH — CH$_2$ — CHO　(V) CITRONELLAL
／
CH$_3$　　　　　　　　　　　CH$_3$

O
‖
CH$_3$ — CH$_2$ — CH$_2$ — CH$_2$ — CH$_2$ — C — CH$_3$　(VI) 2-HEPTANONE

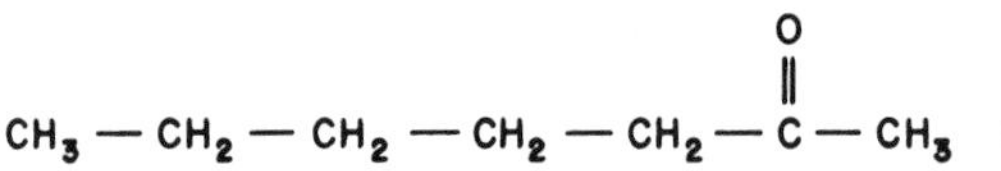

FIG. 5. Alarm substances of ants. For anatomical location of glandular sources mentioned here see Fig. 8. (I) Dendrolasin: *Lasius (Dendrolasius) fuliginosus,* mandibular gland (93). (II) Citral: *Atta rubropilosa,* mandibular gland (19); *Acanthomyops claviger,* minor constituent, mandibular gland (29, 50). (III) Methylheptenone: widespread in Dolichoderinae (*Diceratoclinea, Iridomyrmex, Tapinoma*), known to be alarm substance in *Tapinoma,* anal gland (28, 95, 119). (IV) Propyl isobutyl ketone: *Tapinoma nigerrimum,* active at least on *T. sessile,* anal gland (95, 119). (V) Citronellal: one of two isomers shown here, *Acanthomyops claviger,* major constituent, mandibular gland (29, 50). (VI) 2-Heptanone: *Iridomyrmex pruinosus,* anal gland (9).

We have chosen the alarm substance of *Pogonomyrmex badius,* secreted by the mandibular glands but as yet chemically unidentified, for experimental analysis. In a procedure to be described in detail elsewhere (12), we crushed the heads of workers in long glass tubes containing resting sister workers in still air and measured the onset times and distances of responses. The release of a puff of the pheromone was simulated by crushing the heads of workers and removing them from the experimental chamber after brief intervals. Estimates of the diffusion of coefficient in six trials ranged from 0.33 to 0.80, a parameter undoubtedly raised by unavoidable air turbulence caused by movement of the ants. For 3-second exposures, the estimates of the ratio of pheromone molecules released to threshold concentration per cubic centimeter (i.e., Q/K ratio) were 635–918. The Q/K ratio for entire contents of a worker head were estimated to lie between only 939 and 1800, thus indicating considerable volatility, in accord with the subjective impression received during dissection of the glands. To estimate Q and K separately, it was assumed that all of the mandibular gland reservoir is filled by the alarm substance and that the density of the unknown pheromone (or pheromones) is the same as water. (From information on other alarm substances, these assumptions are believed to introduce an error of less than 30%.) The head of a worker then contains about 6.26×10^{16} molecules, and K is about 4.47×10^{13} molecules per cubic centimeter of air.

These are of course no more than crude predictions, probably safe only to the nearest order of magnitude. Even at this low level of accuracy, however, they allow us to describe the design features objectively. In a three-dimensional situation in really still air, it can be predicted that the crushing of a worker head will cause a spread of the pheromone through a maximum sphere of attractiveness with radius of about 6 cm. and this will be attained in 13 seconds. In about 35 seconds the signal will have faded out completely. These estimates correspond reasonably well with experimental results obtained in an earlier study (116). The response at lower concentrations is simple attraction. It has been observed that at a concentration approximately 10 times the attraction threshold, true alarm behavior is induced. The maximum range of the latter effect will be about 3 cm., and the necessary concentration will disappear in about 8 seconds. When a worker is merely alarmed and not crushed, Q will undoubtedly be much less, and the range and duration of the dual signals will be shorter. In any case K is seen to be high, and Q/K moderate (actually, a little higher than that of the trail pheromone but far below that of the sex attractant to be described next). The system seems very well designed for the intuitive "goals" of an alarm system in ant colonies: single workers can spread swift alarm over a short distance for a short period of time. If the danger is local, the signal quickly fades and the bulk of the colony is undisturbed. If it is persistent, increasing numbers of workers become involved and the signal spreads. When the danger stimuli cease, the signal quickly fades.

C. Sexual Stimulation

Sex pheromones release one or both of two responses: attraction and sexual behavior. In many cases the former occurs at low concentration and the latter at higher concentration as the responder draws near the source. Sex pheromones are widespread through the animal kingdom. They are especially common among the insects (35, 55), but they are also known to occur in Crustacea (22, 115), fish (108), salamanders (89), snakes (90), and mammals (4, 13, 23, 76, 91). In the lower animal phyla sex recognition and attraction is almost certainly primarily by means of pheromones, but the subject has received little analysis (26, 45). Opportunities for discovery abound in this area.

The identification of the first insect sex attractants, those of the moths *Porthetria* and *Bombyx*, was a remarkable achievement of two independent teams of researchers. Chemists and entomologists of the United States Department of Agriculture worked intermittently over a period of thirty years on *Porthetria* (63), while Butenandt and his associates in Germany required twenty years to solve the problem in *Bombyx* (18). In both cases hundreds of thousands of female moths were extracted, each yielding on the order of 10^{-2} μg. of pure attractant. The structural formulas are given in Fig. 6. A similar effort by Wharton and Wharton (114) over a ten-year period to identify the sex attractant of the cockroach *Periplaneta americana* has not yet met with success.[1] Gary (48) has identified the queen substance of honey bees as one of a medley of sex attractants. The stimulative efficiency, however, is far below that of the moth substances, since 0.1 mg. of the pure synthetic substance was required per assay tube in the open air to get a response. Numerous artificial sex lures effective on a wide variety of insects have been discovered by screening methods in connection with control work. These have been described in detail in reviews by Dethier (35) and most recently by Green *et al.* (55). Their stimulative efficiency varies widely, and in some cases they induce aberrant responses. The chemical relation of

---

[1] *Note added in proof:* Jacobson *et al.* (63a) reported the identification of the *Periplaneta americana* attractant as 2,2-dimethyl-3-isopropylidenecyclopropyl propionate, with a molecular weight of 182.

694   EDWARD O. WILSON AND WILLIAM H. BOSSERT

these synthetic substances to the true sex attractants is unknown. Kullenberg (69) made guesses as to the identity of hymenopteran sex attractants
on the basis of smell tests of crushed individuals. In many cases he was

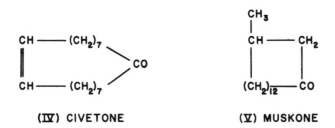

$$CH_3 - (CH_2)_2 - CH = CH - CH = CH - (CH_2)_8 - CH_2 - OH$$

(I) BOMBYKOL

$$CH_3 - (CH_2)_5 - CH - CH_2 - CH = CH - (CH_2)_5 - CH_2 - OH$$
$$\underset{\underset{O}{\overset{\|}{C} - CH_3}}{\overset{|}{O}}$$

(II) GYPLURE

$$CH_3 - \overset{\overset{\textstyle O}{\|}}{C} - (CH_2)_5 - CH = CH - \overset{\overset{\textstyle O}{\|}}{C} - OH$$

(III) HONEYBEE QUEEN SUBSTANCE

(IV) CIVETONE   (V) MUSKONE

FIG. 6. Known and possible sex pheromones. (I) "Bombykol," hexadeca-10,12-dien-
1-ol: sex lure of the moth *Bombyx mori*. Of four geometric isomers based on the diene
groupings that were synthesized and tested by Butenandt and Hecker (17), only one
(10-*trans*,12-*cis*) proved as active as natural Bombykol. (II) Gyplure, dextrorotary
10-acetoxyhexadec-*cis*-7-en-1-ol: sex lure of the moth *Porthetria dispar*. The *dl*-form of
the substance was also synthesized by Jacobson *et al.* (63) but found not to differ from
the *d*-form in effectiveness. (III) "Queen substance," 9-ketodec-*trans*-2-enoic acid: a
secondary (?) sex attractant of the honey bee (21, 48). (IV) Civetone: secretion of
para-anal glandular pouch of the civet *Viverra zibetha*, which evidently functions in
defense, but possibly also as a sex pheromone or territorial marker or both (13, 54, 71,
121). (V) Muskone: secreted in the preputial glands of the musk deer, *Moschus mos-
chiferus*, which probably functions either as a sex pheromone or territorial marker or
both (13, 71).

successful in attracting males with synthetic lures containing the predicted chemicals.

Several musk substances of mammals have been identified (see Fig. 6 and review by Lederer, 71) but, surprisingly, have been the subject of very little behavioral study under natural conditions. Recent authors have suggested that the musks function variously in defense, territorial marking, or as sex pheromones (13, 54, 71, 121). In reality, their individual functions may prove to be combinations of these, as Bourlière (13) has suggested. The structural resemblance of civetone to the odorant steroid $\triangle^{16}$-androstenol-3, if the latter's cyclic structure is partially opened, has been remarked by Lederer (71). This possibility of a relation between hormonal steroids and sexual pheromones in mammals is an intriguing one. The primer effect of the odor of male mice remains in soiled cages after the males have been removed (91) and could conceivably be in the urine. Odorous urine steroids have been identified: some of the characteristic "urine odor" of humans, for instance, is imparted by $\triangle^2$-androstenone-17. Beach and Gilmore (4) noted that sexually active male dogs spend more time investigating the urine of receptive females than that from nonreceptive females. LeMagnen (74) discovered that adult male rats were attracted to the odor of estrous females, while Carr and Pender (24) found that male rats can discriminate the odors of urine from estrous versus diestrous females. Carr and Caul (23) were successful in training female rats to distinguish the odors from normal versus castrated males. It is conceivable that even humans may respond in some way to pheromones. LeMagnen (72, 73) reported that the odor of the lactone of 14-hydroxytetradecanoic acid ("exaltolide") is perceived clearly only by sexually mature females, and most clearly by these about the time of ovulation. A male subject became more sensitive following an injection of estrogen. A parallel difference in sensitivity to the odor of old urine was noted. Any conclusions drawn from these fragmentary reports must of course be highly speculative. The subject of olfactorily determined behavior in mammals has remained in general remarkably fallow, considering its potential significance.

The case of *Porthetria dispar* has been chosen here for critical examination of the mechanism of pheromone transport. The mathematical analysis of wind transport of a pheromone, the method of *P. dispar* and most other insects using sex pheromones, is complicated and has been undertaken in a separate paper on methods of measurement of pheromone communication (12). Using the rather crude "maximum distance" measures taken by Collins and Potts (30) over which the female *P. dispar* can draw males, we concluded that a group of 10 to 15 females together have a maximum

696   EDWARD O. WILSON AND WILLIAM H. BOSSERT

"reach" of between 3.72 and 16.09 kilometers. This estimate is in full accord with the results of numerous field studies made on other moth species. As Schwinck (104) has shown in experiments with several moth genera, including *Porthetria*, orientation is possible over great distances and in the absence of a detectable gradient because the males become positively

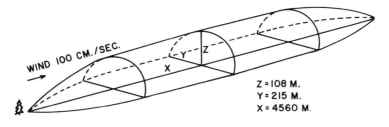

Z = 108 M.
Y = 215 M.
X = 4560 M.

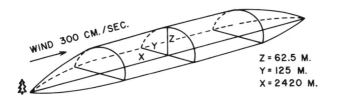

Z = 62.5 M.
Y = 125 M.
X = 2420 M.

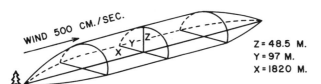

Z = 48.5 M.
Y = 97 M.
X = 1820 M.

FIG. 7. Predicted active spaces downwind from "calling" females of the moth *Porthetria dispar*, at various wind velocities. Further explanation in the text.

anemotactic when stimulated by the pheromone. By flying upwind they approach the stimulus source. The ratio of pheromone emission rate in molecules per second over threshold concentration in molecules per cubic centimeter of air, i.e., Q/K, was estimated to lie between 1.87×10^{10} cm.^{3}/ sec. and 3.03×10^{11} cm.^{3}/sec. In Fig. 7, we have illustrated the expected shape of the active semiellipsoid space downwind from the female and the effect of wind velocity on the length of the long axis. One unexpected result is that in moderate to heavy winds the "reach" of the female is inversely proportional to wind velocity. From the data of Butenandt on *Bombyx mori* discussed previously, K of this species can be expected to be in the vicinity of no more than 200 molecules per cubic centimeter of air in still air. Let us for the moment assume that the parameter is the same for *Porthetria dispar*. Jacobson *et al.* obtained an average of about 4×10^{-8} gm. of gyplure per female in their extraction procedure. Even if the extraction procedure is at least 50% efficient, which is believed to be the case (5), a female is likely to have no more than 8×10^{-8} gm., or 8.18×10^{13} to 1.616×10^{14} molecules, at a given instant. It would follow that the female therefore releases about one-eighth to one-fourth of the quantity in her reservoir each second. If this is to be rejected as impossible (which should not be done out of hand), then Q must be higher or K lower than given here, and Q/K even higher than the very high estimate already given.

The minimum figure of 1.87×10^{10} cm.^{3}/sec. is in marked contrast to the estimates for the trail and alarm substances, which are on the order of 1 and between 10^1 and 10^3, respectively. The significance of these design features seems clear enough. They allow a female to call over great distances for long periods of time with a minimum expenditure of energy. This can be achieved because, unlike the "goals" of the trail and alarm systems, a long fade-out time can be tolerated or even sought, and K can therefore be adjusted toward the minimal figure.

D. Aggregation

Pheromones exist which serve solely as attractants and do not release more complex behavior. Consequently, they promote simple aggregation. It has long been known, for instance, that many aposematic insect species form dense, conspicuous clusters (31), but the mechanisms by which such aggregations are established and maintained were unknown. Now Eisner and Kafatos (42) have been able to prove that in the beetle *Lycus loripes* the males produce a volatile attractant which is responsible for the aggregation. Wilson (117) has demonstrated the presence of a volatile substance, probably carbon dioxide, that attracts and promotes clustering in workers of the ant *Solenopsis saevissima*. Barnhart and Chadwick (2) discovered that house flies, *Musca domestica*, visiting a bait contribute to it a "fly-factor" substance that increases its attractiveness to other flies. The presence of other aggregation pheromones can probably be sought with profit among the many fascinating cases of simple aggregation in insects and other animals.

698   EDWARD O. WILSON AND WILLIAM H. BOSSERT

E. Nonterritorial Dispersal

The intraspecific use of repellents, locomotory stimulants, or deterrents, to use the terminology of Dethier et al. (39), should result in increased species dispersal. Adults of the flour beetle *Tribolium confusum* aggregate at low population density, apparently distribute themselves randomly at intermediate density, and distribute uniformly at high density (86). The last effect is evidently due to the secretion of quinones, which act as repellents above a certain concentration. Loconti and Roth (77) have shown that these substances are produced by thoracic and abdominal glands and in the case of *T. castaneum* consist primarily of 2-ethyl-1,4-benzoquinone and 2-methyl-1,4-benzoquinone. A third (trace) substance, 2-methoxy-1,4-benzoquinone, is not effective as a repellent. A similar effect may be responsible for the dispersal of young *Zinaria* millipedes, which secrete a substance, apparently hydrocyanic gas, during dispersal from their initial aggregations, but the evidence is anecdotal (121, 122). Dispersal can also be achieved by a reproductive deterrent, as shown by Salt (98). Females of the parasitic wasp *Trichogramma evanescens* receive chemical cues from eggs already parasitized by other females and avoid ovipositing on them. The discrimination has been referred to as "perfect" by Salt, and further experimental data make its adaptive significance clear: when host eggs were heavily superparasitized in laboratory studies the parasites were either dwarfed, imperfectly developed, or wholly inviable.

F. Territory and Home Range Marking

The use of chemical trails and scent parts to mark territories and home ranges is widespread in the mammals. The subject has received lengthy treatment in the reviews of Hediger (60), Grassé et al. (54), Bourlière (13), and Wynne-Edwards (121). The odorants are either passed in excrement or else are secreted by special exocrine glands located, according to taxon, in one or more of a variety of body sites: pedal, carpal, tarsal, metatarsal, preorbital, occipital, caudal, preputial, anal, etc. Our knowledge of the chemistry of these substances is very limited. The best review is that of Lederer (71). The Crocodilia also release scent, at least in the mating season. The male American alligator emits jets of musk that can on occasion be smelled for miles (R. L. Ditmars, quoted from Wynne-Edwards, 121). In the South American alligatorids *Caiman sclerops* and *C. latirostris*, musk glands located near the anus yield 4–35 gm. of a mixture composed of about 80% fatty acids (including isovaleric and palmitic), some cholesterol, some cetyl alcohol, and about 4% "yacarol," which is principally *d*-citronellol

(71). The males of bumble bees and some solitary apidae mark their flight paths with scent. Male bumble bees (*Bombus*) daub vegetation which scent spots placed at intervals in a circuit, around which they travel hour after hour and day after day (46, 56, 57, 47). The scenting usually takes place once in the morning but is sometimes reinforced during the day. Kullenberg (69) noted that the scent spots of *Bombus hortorum* smell very much like hydroxy-citronellal, and found that dummy marks of synthetic hydroxycitronellal were sufficient to reorient males under natural conditions.

The principal failing of behavioral observations on "territorial" phero-mones in general is that they are usually inadequate to prove what the true functions are. Thus, Free and Butler (47) assume that the *Bombus* scent posts form networks that attract females and ensure their swift cap-ture. Wynne-Edwards (121) suggests that in addition they serve to disperse the males and ultimately to control the population density. Actually, the field observations of Frank (46), Haas (56, 57), and Kullenberg (69) are inconclusive in this respect. With reference to mammals, scent posts may serve as territorial markers, as orientation signals in undefended home range, or as sex attractants, singly or in combination. The relative weights of these functions have not been the subject of critical analysis (13).

V. The Increase of Information

The limited capacity of exocrine glands to release multiple pheromones separately, together with the slow fade-out of chemical signals, results in considerable restriction of the potential rate of information transmission. Special devices, however, can be imagined which might increase the rate. These include (1) the use of multiple exocrine glands to produce independent pheromones; (2) the production of medleys of pheromones by the same gland, with various independent or combinatorial effects; (3) the use of the same pheromone in different environmental contexts to produce different effects; (4) the shaping of variable behavioral responses to differing concentrations of the same pheromone. All these modifications are in fact to be found some-where in the animal kingdom.

The first device has been employed to a certain degree by several animal groups. Kullenberg (69), for example, found that in the females of many of the aculeate Hymenoptera he studied, simple attractants were released from the head and sexual excitants from the abdomen. But the highest known development of a multiple glandular system for this purpose is to be found in the ants. As shown in Fig. 5 and 8, species in the various subfamilies and tribes have modified different exocrine glands for the production of a variety of species-specific alarm and trail substances. Pavan's gland is of

700   EDWARD O. WILSON AND WILLIAM H. BOSSERT

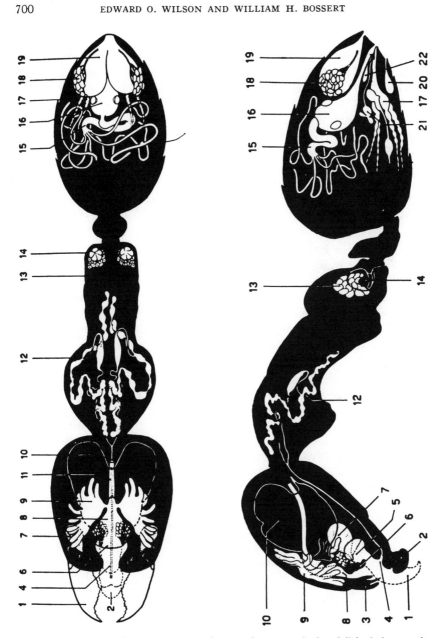

FIG. 8. Exocrine glandular system of a worker ant of the dolichoderine species *Iridomyrmex humilis*, which can be taken as representative of the ants as a whole (from Pavan and Ronchetti, 94). The glands known or suspected of producing pheromones are

special interest in this connection, since it was evolved *de novo* in the Dolichoderinae as the site of trail substance production. Other pheromones of unknown glandular origin, induce grooming behavior, transport, and food exchange (117). The exocrine glandular system of ants is an extraordinarily well-developed one. It is a reasonable guess that many of the glands of hitherto unknown function will prove to be pheromone producers; e.g., the metapleural gland (no. *13* in Fig. 8), which is an organ peculiar to the ants. The exploration of the "chemical codes" of the ants and other social insects can be said to be in its earliest stages. For details, see the recent reviews of Karlson and Butenandt (66) and Wilson (118).

Medleys of pheromones are also known. In ants combinations of alarm substances in the same glandular reservoir are common (Fig. 5), although it is not known whether these have different behavioral effects. The mandibular gland of the honey bee queen produces both queen substance, which inhibits worker ovary development and the tendency by workers to build queen brood cells, as well as a worker attractant (20, 21). At least one other worker attractant is secreted elsewhere on the queen's body. An extraordinarily rich mixture of chemicals is found in the castoreum of the beaver (71). About 45 substances have already been identified, including a surprisingly diverse array of alcohols, phenols, ketones, organic acids, and esters, as well as salicylaldehyde and castoramine ($C_{15}H_{23}O_2N$). The communicative function of these chemicals, if any, is unknown. Besides producing separate behavioral effects through independent action, it is conceivable that pheromones can act synergistically; that is, the attractiveness of mixtures might exceed the sum of those exhibited by the individual ingredients. Several cases of synergistic effects of mixtures of synthetic lures on insects are known (55). For example, a mixture of geraniol and eugenol is more attractive to the Japanese beetle than either chemical alone.

---

the following: (*7*) mandibular gland, alarm substances, *Atta, Pogonomyrmex, Acanthomyops* (19, 29, 50, 116); (*12*) labial gland, queen and larval foodstuffs, *Formica* (52); (*16*) hindgut, probably containing glandular tissue, trail substance, *Lasius, Paratrechina* (25, 118); (*18*) anal glands and (*19*) anal gland reservoirs, alarm substance, subfamily Dolichoderinae only (*Diceratoclinea, Monacis, Iridomyrmex, Tapinoma, Liometopum,* etc.) (9, 119, 120); (*20*) Pavan's gland (*organo ventrale*), trail substance, Dolichoderinae only (119); (*21*) "true" poison glands and reservoir, trail substance, tribe Attini (*Atta, Acromyrmex*) (9); (*22*) Dufour's gland, trail substance, *Solenopsis, Pheidole* (118). Number *17* is the ovaries; *13* and *14* mark the metapleural glands and their reservoirs, characteristic of almost all the ants but of unknown function. Other exocrine glands shown in the head function at least in part in the secretion of digestive enzymes.

An example of a pheromone which is used in different contexts to produce various effects is the trail substance of the fire ant *Solenopsis saevissima* (117). This pheromone acts in most circumstances as a simple attractant. As such it is employed variously in the recruitment of workers to new food sources, in the organization of colony emigration to new nest sites, and in reinforcement of the cephalic alarm substance when the nest is disturbed. By itself it provides no information as to the context. Instead, the ultimate response is determined by the context stimuli; i.e., other stimuli received from the environment in the vicinity of the point of emission. The pheromone is not always an attractant, however. When released experimentally in the interior of the nest, it causes workers to leave the nest and explore outside in what appears to be typical foraging behavior (120). Another example of a pheromone whose effect is context-determined is the queen substance of the honey bee, which serves its simultaneous multiple functions in worker control within the hive and as a sex attractant during the nuptial flight.

Finally, we have already mentioned that the same pheromone can have both qualitatively and quantitatively variable effects at different concentrations. In the insects alarm and sex substances are typically simple attractants at low concentrations and releasers of complex behavioral patterns at higher concentrations. Furthermore, continued exposure to higher concentrations of mandibular gland secretion causes workers and *Pogonomyrmex badius* to switch from alarm to digging behavior (116).

VI. Interspecific Communication

There is no reason to doubt that the essential characteristics of intraspecific chemical communication thus far described are also true of interspecific communication. The problems are different but closely parallel. To take the extreme case, there are systems in which species specificity and stimulative efficiency of a very high order are required, so that we can expect to find dramatic behavioral effects produced across species by the transmission of a limited set of chemicals. This is precisely the case in the phagostimulants that induce postspecificity in feeding, in both herbivorous and carnivorous forms. Herbivorous specificity, especially in insects, is relatively well understood and has been the subject of recent reviews by Dethier (37), Thorsteinson (110), and Fraenkel (44). Some cases of carnivorous-prey specificity mediated by chemical cues are also known. To cite one extreme example from the ants, workers of certain species of the tribe Dacetini capture and feed on nothing but entomobryomorph and symphypleonan springtails (15). Other potential prey are antennated and then rejected.

Species specificity in chemical communication is also commonplace in symbiosis. An example which has been the subject of a classical experimental demonstration by Welsh (112, 113) is the relation between bivalve mollusks and parasitic water mites of the genus *Unionicola*. The *Unionicola* are positively phototactic in the absence of mollusks, but when extracts of the gills of the host species are added to the surrounding water, the mites abruptly become negatively phototactic. Extracts from nonhost species of mollusks fail to induce the response. The adaptive significance of this taxis reversal seems clear enough: positive photataxis aids dispersal, and negative photataxis brings the mites to the bottom, where the clams are located. Although such decisive behavioral assays are hard to find, evidence has continued to accumulate to indicate the widespread occurrence of parallel phenomena in all forms of symbiosis (32, 33, 34). But, as Caullery (27) and Davenport (32) have emphasized, the experimental study of these and other "symbiosis-effective stimuli" in the great array of symbiotic animal groups has hardly begun.

Chemical cues may also be used by prey species in the recognition and avoidance of predators. George (49) has analyzed a remarkable kind of communication between the North American pickerels *Esox americanus* and *E. niger* and one of their prey species, the mosquito fish *Gambusia partruelis*. When a strange fish enters the vicinity of the *Gambusia*, the latter swim cautiously toward it, while orienting visually to avoid the "attack cone" or conic area in front of the head of the intruder. When the odor of *Esox* is detected, the *Gambusia* undergo rapidly the following changes: the eyes darken, a conspicuous suborbital pigment band appears, the fins are held more erect, the body becomes straighter and the posture more rigid, the caudal sweep of the tail fin is diminished, and the fish tend to move upward and remain near the surface film. While they remain in this alerted state, the *Gambusia* are easily induced to give the jump response, an erratic skipping along and through the water surface, at this time they are less likely to swim downward when a shadow passes over. George showed that the chemical releaser can be passed through a Büchner funnel, and that among a variety of freshwater fish tested, it is peculiar to *Esox*. In laboratory tests, the *Gambusia* exhibiting the response were preyed on less than those failing to show it. Thus the *Gambusia* appear to have developed an elaborate response specifically related to predation by *Esox*, which is one of their principal enemies in nature. But there may be more to the story. George found that the *Gambusia* color change does not communicate the alert state to other *Gambusia* in the absence of the *Esox* odorant. He proposed the novel hypothesis that the *Gambusia* use this as a means of communicating back to the *Esox* the fact, potentially important to *Gambusia* and *Esox* alike, that the *Gambusia* have been alerted and hence are more difficult to catch.

VII. Pheromones as Interspecific Isolating Mechanisms

It is tempting to analyze examples of chemical communication abstractly and in isolation, a procedure we have been more or less required to follow in the present paper. However, full understanding of a biological phenomenon is gained only when it is viewed as a part of the whole biology of the species, and beyond that, of the ecological community in which the species lives. The importance of this generalization, so often neglected by physiologists, can be illustrated forcefully by considering the particular case of the function of sex pheromones in moths.

Since sex pheromones serve both as lures and releasers of precopulatory behavior, it is tempting to predict from evolutionary theory that these substances are species specific with reference to species that have their nuptial flights in the same localities at the same times. Specifically, the sex pheromones appear especially liable to diversification in evolution so as to function as "intrinsic isolating mechanisms" in the process of species formation. Yet experimental evidence so far seems to indicate a failure of this device among closely related sympatric moth species. Barth (3) found that distantly related European pyralid species were not interactive, but that more closely related species were interactive to various degrees. Schwinck (102, 103) noted good interactivity between the lymantriids *Porthetria dispar* and *P. monacha*. Using the important new technique of recording olfactory potentials from the male antennae (electroantennograms, EAG) Schneider (100) found that seven saturniid species were mostly interactive, although none were interactive with the bombycid *Bombyx mori*. Schneider concluded that chemical differences among the saturniid sex attractants "might be slight or not present at all," and that they "would not be a very effective tool to prevent the species from interbreeding."

What, then, does prevent the sympatric species from interbreeding? The truth is that the sex attractants are more significant than these experimental studies alone suggest. They play an important role in a complex of independently acting isolating mechanisms which in the aggregate are quite effective. The nature of this complex can be partly deduced from the important but all-but-forgotten study by Rau and Rau (96) on sex attraction in the four species of North American saturniids *Hyalophora cecropia*, *Samia* (*Platysamia*) *cynthia*, *Antheraea* (*Telea*) *polyphemus*, and *Callosamia promethea*. *Hyalophora cecropia* is partially isolated from *S. cynthia*

CHEMICAL COMMUNICATION AMONG ANIMALS   705

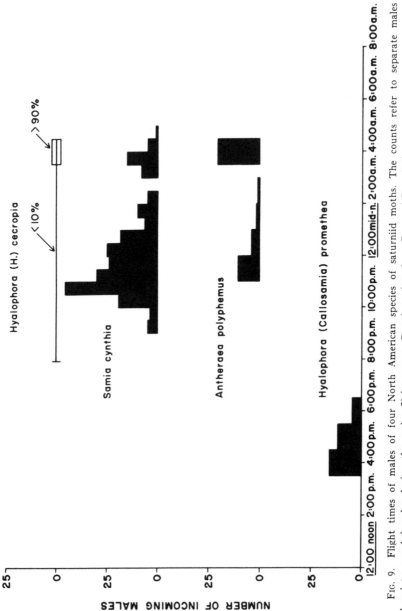

Fig. 9. Flight times of males of four North American species of saturniid moths. The counts refer to separate males lured to caged females during the entire flight season. Based on data from Rau and Rau (96).

706   EDWARD O. WILSON AND WILLIAM H. BOSSERT

and *C. promethea* by differences in the seasons of nuptial flight. In the Raus' study, this species emerged mostly in May, whereas *S. cynthia* and *C. promethea* appeared mostly in June. *Antheraea polyphemus* had an overlapping emergence period, late May and early June. As shown in Fig. 9, the species are further isolated from each other by the fact that the males fly only during brief periods each day that differ among the species. The flight times are relatively invariable, being apparently based on endogenous circadian rhythms mostly independent of weather. The sex attractants also have a key role. Schneider (100) recorded a negative EAG (not differing from control EAG) in the *cecropia-promethea* test. Other combinations of the four North American species gave either positive EAG's or were not

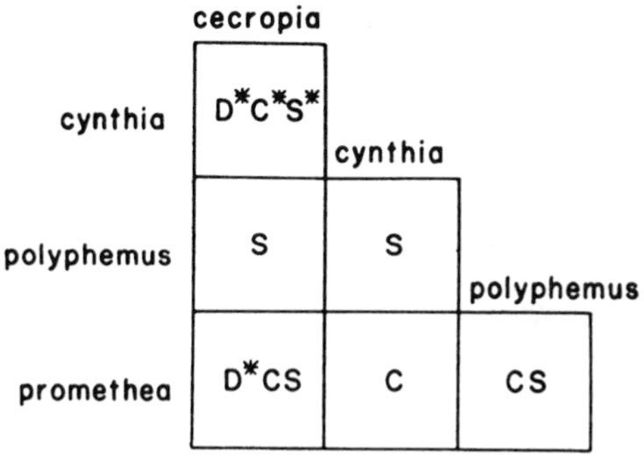

Fig. 10. Reproductive isolating mechanisms known to separate the four saturniid species. *D*, difference in season of flight; *C*, difference in daily flight time; *S*, species-specific sex attractant. An asterisk indicates that the isolation is incomplete. Based on data from Schneider (100) and Rau and Rau (96).

tested. However, as this author made clear, the EAG method does not permit the distinction between strong and weak effects, at least in the Saturniidae. The full behavioral significance cannot therefore be deduced. Field studies are more meaningful. Rau and Rau (96) observed that *A. polyphemus* females never attracted males of the other species, or vice versa. *Samia cynthia* females attracted *H. cecropia* males, but only in the absence of *H. cecropia* females. Furthermore, only young *H. cecropia* males were attracted, and these left without settling on the *S. cynthia* cages. The information of sexual isolation among the four species is summarized in Fig. 10.

ACKNOWLEDGMENTS

The authors are grateful to Professor John H. Law and Mr. Neil B. Todd, of Harvard University, for reading parts of the manuscript and offering many helpful suggestions. Dr. Morton Beroza, Dr. Murray S. Blum, Dr. Dietrich Schneider, and Mrs. Katherine

S. Ralls supplied additional useful information during the course of the study. The figures were prepared by Mrs. Joyce S. Todd. The original research reported in the paper was supported by a grant from the United States National Science Foundation.

REFERENCES

1. Barail, L. C. 1946. *Rayon Textile Monthly* **27**, 663; 1947. **28**, 93, 439, 496; cited by Moncrieff (84).
2. Barnhart, C. S., and Chadwick, L. E. 1953. *Science* **117**, 104.
3. Barth, R. 1937. *Zool. Jahrb. Abt. Zool. Physiol.* **58**, 297.
4. Beach, F. A., and Gilmore, J. 1949. *J. Mammal.* **30**, 391.
5. Beroza, M. 1962. Personal communication.
6. Beroza, M., Green, N., Gertler, S. I., Steiner, L. F., and Miyashita, D. H. 1961. *J. Agr. Food Chem.* **9**, 361.
7. Bethe, A. 1932. *Naturwissenschaften* **20**, 177.
8. Bier, K. 1958. *Ergeb. Biol.* **20**, 97.
9. Blum, M. S. 1962. Personal communication.
10. Boch, R., and Shearer, D. A. 1962. *Nature* **194**, 704.
11. Bonner, J. T. 1959. "The Cellular Slime Molds." Princeton Univ. Press, Princeton, New Jersey.
12. Bossert, W. H., and Wilson, E. O. *J. Theoret. Biol.* (in press).
13. Bourlière, F. 1956. "The Natural History of Mammals." Knopf, New York.
14. Brown, W. L. 1960. *Psyche* **67**, 24.
15. Brown, W. L., and Wilson, E. O. 1959. *Quart. Rev. Biol.* **34**, 278.
16. Butenandt, A. 1959. *Naturwissenschaften* **15**, 461.
17. Butenandt, A., and Hecker, E. 1961. *Angew. Chem.* **73**, 349.
18. Butenandt, A., Beckmann, R., Stamm, D., and Hecker, E. 1959. *Z. Naturforsch.* **14b**, 283.
19. Butenandt, A., Linzen, B., and Lindauer, M. 1959. *Arch. anat. microscop. morphol. exptl.* **48**, 13.
20. Butler, C. G. 1961. *J. Insect Physiol.* **7**, 258.
21. Butler, C. G., Callow, R. K., and Johnston, N. 1961. *Proc. Roy. Soc.* **B155**, 417.
22. Carlisle, D. B. 1959. *J. Marine Biol. Assoc. United Kingdom* **38**, 481.
23. Carr, W. J., and Caul, W. F. 1962. *Animal Behaviour* **10**, 20.
24. Carr, W. J., and Pender, B. 1962. Cited in Carr and Caul (23).
25. Carthy, J. D. 1951. *Behaviour* **3**, 304.
26. Carthy, J. D. 1958. "An Introduction to the Behavior of Invertebrates." Allen & Unwin, London.
27. Caullery, M. 1950. "Parasitism and Symbiosis." Sidgwick & Jackson, London.
28. Cavill, G. W. K., and Hinterberger, H. 1961. *Verhandl. 11th Intern. Kongr. Entomol., Vienna, 1960* **3**, 53.
29. Chadma, M. S., Eisner, T., Monro, A., and Meinwald, J. 1962. *J. Insect Physiol.* **8**, 175.
30. Collins, C. W., and Potts, S. F. 1932. *U.S. Dept. Agr. Tech. Bull.* **336**, 1.
31. Cott, H. B. 1957. "Adaptive Coloration of Animals." Methuen, London.
32. Davenport, D. 1955. *Quart. Rev. Biol.* **30**, 29.
33. Davenport, D., and Norris, K. S. 1958. *Biol. Bull.* **115**, 397.
34. Davenport, D., Camougis, G., and Hickok, J. F. 1960. *Animal Behaviour* **8**, 209.

708   EDWARD O. WILSON AND WILLIAM H. BOSSERT

35. Dethier, V. G. 1947. "Chemical Insect Attractants and Repellents." McGraw-Hill (Blakiston), New York.
36. Dethier, V. G. 1953. *In* "Insect Physiology" (K. D. Roeder, ed.), p. 544. Wiley, New York.
37. Dethier, V. G. 1954. *Evolution* **8**, 33.
38. Dethier, V. G. 1954. *Ann. N.Y. Acad. Sci.* **58**, 139.
39. Dethier, V. G., Browne, L. B., and Smith, C. N. 1960. *J. Econ. Entomol,* **53**, 134.
40. De Vries, H., and Stuiver, M. 1961. *In* "Sensory Communication" (W. A. Rosenblith, ed.), p. 159. M.I.T. Press, Cambridge, Massachusetts, and Wiley, New York.
41. Eibl-Eibesfeldt, I. 1949. *Experientia* **5**, 236.
42. Eisner, T., and Kafatos, F. C. 1962. *Psyche* **69**, 53.
43. Forel, A. 1929. "The Social World of the Ants," Vol. 1. A. and C. Boni, New York. (Translation by C. K. Ogden of "Le Monde Social des Fourmis.")
44. Fraenkel, G. S. 1959. *Science* **129**, 1466.
45. Fraenkel, G. S., and Gunn, D. L. 1961. "The Orientation of Animals." Dover, New York (annotated ed.).
46. Frank, A. 1941. *Z. vergleich. Physiol.* **28**, 467.
47. Free, J. B., and Butler, C. G. 1959. "Bumblebees." Collins, London.
48. Gary, N. E. 1962. *Science* **136**, 773.
49. George, C. 1960. Behavioral interaction of the pickerel and the mosquitofish. Ph.D. thesis, Harvard Univ., Cambridge, Massachusetts.
50. Ghent, R. L. 1961. Adaptive refinements in the chemical defense mechanisms of certain Formicinae. Ph.D. thesis, Cornell Univ., Ithaca, New York.
51. Ghent, R. L., and Gary, N. E. 1962. *Psyche* **69**, 1.
52. Gösswald, K., and Kloft, W. 1960. *Zool. Beiträge* **5**, 519.
53. Goz, H. 1954. *Z. vergleich. Physiol.* **29**, 1.
54. Grassé, P.-P. (ed.) 1955. *In* "Traité de Zool." Vol. 17, p. 194 (plus various parts of the same volume by other authors). Maisson, Paris.
55. Green, N., Beroza, M., and Hall, S. M. 1960. *In* "Advances in Pest Control Research," Vol. 3, p. 129. Interscience, New York.
56. Haas, A. 1946. *Z. Naturforsch.* **1**, 596.
57. Haas, A. 1952. *Naturwissenschaften* **39**, 484.
58. Hainer, R. M., Emslie, A. G., and Jacobson, A. 1954. *Ann. N.Y. Acad. Sci.* **58**, 158.
59. Hasler, A. D. 1960. *Ergeb. Biol.* **23**, 94.
60. Hediger, H. 1951. "Observations sur la psychologie animale dans les parcs nationaux du Congo Belge." *Explorat. Parcs Natl. Congo Belge* **1**, 1.
61. Henze, H. R., and Blair, C. M. 1931. *J. Am. Chem. Soc.* **53**, 3077.
62. Hodgman, C. D., Weast, R. C., and Selby, S. M., eds. 1961. "Handbook of Chemistry and Physics," 43rd ed., p. 2229. Chem. Rubber Publ., Cleveland, Ohio
63. Jacobson, M., Beroza, M., and Jones, W. A. 1960. *Science* **132**, 1011.
63a. Jacobson, M., Beroza, M., and Yamamoto, R. T. 1963. *Science* **139**, 48.
64. Kalmus, H. 1955. *Brit. J. Animal Behaviour* **3**, 25.
65. Kalmus, H., and Ribbands, C. R. 1952. *Proc. Roy. Soc.* **B140**, 50.
66. Karlson, P., and Butenandt, A. 1959. *Ann. Rev. Entomol.* **4**, 39.
67. Karlson, P., and Lüscher, M. 1959. *Nature* **183**, 55.
68. Kettlewell, H. B. D. 1946. *Entomologist* **79**, 8.

69. Kullenberg, B. 1956. *Zool. Bidrag Uppsala* **31**, 253.
70. Lange, R. 1960. *Z. Tierpsychol.* **17**, 389.
71. Lederer, E. 1950. *Fortschr. Chem. org. Naturstoffe* **6**, 87.
72. LeMagnen, J. 1948. *Compt. rend. acad. sci.* **226**, 694.
73. LeMagnen, J. 1950. *Compt. rend. acad. sci.* **230**, 1367.
74. LeMagnen, J. 1952. *Arch. sci. physiol.* **6**, 295.
75. Lindauer, M. 1961. "Communication among Social Bees." Harvard Univ. Press, Cambridge, Massachusetts.
76. Lipkow, J. 1954. *Z. Morphol. Ökol. Tiere* **42**, 333.
77. Loconti, J. D., and Roth, L. M. 1953. *Ann. Entomol. Soc. Am.* **46**, 281.
78. Loher, W. 1960. *Proc. Roy. Soc.* **B153**, 380.
79. Lüscher, M. 1960. *Ann. N.Y. Acad. Sci.* **89**, 549.
80. Lüscher, M., and Springhetti, A. 1960. *J. Insect Physiol.* **5**, 190.
81. Macgregor, E. C. 1948. *Behaviour* **1**, 267.
82. Marler, P. 1959. *In* "Darwin's Biological Work: Some Aspects Reconsidered" (P. R. Bell, ed.), p. 150. Cambridge Univ., London.
83. Marler, P. 1961. *J. Theoret. Biol.* **1**, 295.
84. Moncrieff, R. W. 1954. *Ann. N.Y. Acad. Sci.* **58**, 73.
85. Morris, C. W. 1946. "Signs, Language and Behavior." Prentice-Hall, Englewood Cliffs, New Jersey.
86. Naylor, A. F. 1959. *Ecology* **40**, 453.
87. Neuhaus, W. 1955. *Z. vergleich. Physiol.* **37**, 234.
88. Neuhaus, W. 1957. *Z. vergleich. Physiol.* **40**, 65.
89. Noble, G. K. 1931. "The Biology of the Amphibia." McGraw-Hill, New York.
90. Noble, G. K. 1937. *Bull. Am. Museum Nat. Hist.* **73**, 673.
91. Parkes, A. S., and Bruce, H. M. 1961. *Science* **134**, 1049.
92. Pavan, M. 1950. *Ricerca sci.* **20**, 1853.
93. Pavan, M. 1961. *Atti accad. nazl. ital. entomol. rend.* **8**, 228.
94. Pavan, M., and Ronchetti, G. 1955. *Atti soc. ital. sci. nat. e museo civico storia nat. Milano* **94**, 379.
95. Pavan, M., and Trave, R. 1958. *Insectes sociaux* **5**, 299.
96. Rau, P., and Rau, N. L. 1929. *Trans. Acad. Sci. St. Louis* **26**, 82.
97. Ribbands, R. 1953. "The Behaviour and Social Life of Honeybees." Bee Research Assoc. Ltd., London.
98. Salt, G. 1936. *J. Exptl. Biol.* **13**, 363.
99. Schmid, B. 1935. *Z. vergleich. Physiol.* **22**, 524.
100. Schneider, D. 1962. *J. Insect Physiol.* **8**, 15.
101. Schutz, F. 1956. *Z. vergleich. Physiol.* **38**, 84.
102. Schwinck, I. 1955. *Z. vergleich. Physiol.* **37**, 439.
103. Schwinck, I. 1955. *Z. angew. Entomol.* **37**, 349.
104. Schwinck, I. 1958. *Proc. 10th Intern. Congr. Entomol., Montreal, 1956* **2**, 577.
105. Schaffer, B. M. 1957. *Am. Naturalist* **91**, 19.
106. Steiner, L. F., Mitchell, W. C., Green, N., and Beroza, M. 1958. *J. Econ Entomol.* **51**, 921.
107. Stuart, A. 1960. Experimental studies on communication in termites. Ph.D. thesis, Harvard Univ., Cambridge, Massachusetts.
108. Tavolga, W. N. 1956. *Zoologica* **11**, 49.
109. Teichmann, H. 1957. *Naturwissenschaften* **44**, 242.

710   EDWARD O. WILSON AND WILLIAM H. BOSSERT

110. Thorsteinson, A. J. 1955. *Can. Entomologist* **87**, 49.
111. Treat, A. E. 1961. *Verhdl. 11th Intern. Kongr. Entomol. Vienna 1960* **1**, 619.
112. Welsh, J. H. 1930. *Biol. Bull.* **59**, 165.
113. Welsh, J. H. 1931. *Biol. Bull.* **61**, 497.
114. Wharton, M. L., and Wharton, D. R. A. 1957. *J. Insect Physiol.* **1**, 229.
115. Williamson, D. I. 1954. *Dove Marine Lab. Ref.* **3**, 49.
116. Wilson, E. O. 1958. *Psyche* **65**, 41.
117. Wilson, E. O. 1962. *Animal Behaviour* **10**, 134.
118. Wilson, E. O. 1963. *Ann. Rev. Entomol.* **8**.
119. Wilson, E. O., and Pavan, M. 1959. *Psyche* **66**, 70.
120. Wilson, E. O. 1962. Unpublished observations.
121. Wynne-Edwards, V. C. 1962. "Animal Dispersion in Relation to Social Behavior." Oliver & Boyd, London.
122. Young, F. N. 1958. *Proc. Indiana Acad. Sci.* **67**, 171.

DISCUSSION

J. J. Christian: I would like to congratulate Dr. Wilson for this most intriguing presentation. He made the statement that odors did not play a role in the establishment of dominance hierarchies, and certainly it is true that visual and acoustic stimuli do. However, he also stated that these olfactory substances played a role in marking out territories and home ranges. It seems to me that odors should participate in the establishment of, and even more the maintenance of, dominance hierarchies. I would like to know what the evidence is, at least in mammals, that they do not.

E. O. Wilson: The evidence is strictly negative. I made that conjecture on the basis of the fact that we had gotten a fairly complete understanding of the structure of dominance hierarchies in vertebrates by observing visual and acoustic signals, and hence there has not been any compelling need to look for other communication systems. In many vertebrate societies, as for instance a rhesus clan, relatively complex signals must be presented with great rapidity, since the animal is often required to express change of mood in a matter of seconds. We have established that chemical transmission is both inherently poor in patterning and slow in fade-out. It seems more likely, however, that "status" in a society could be signaled persistently by some substance. I know of no information on this subject. As Dr. Christian infers, territoriality (i.e., defense of part of the home range) is nevertheless linked with dominance hierarchies. The responses or the signals (acoustical and visual) that are used to maintain territories may be the same as those used to maintain dominance hierarchies. For instance, in *Anolis* lizards, one can shift lizards from solitary territorial maintenance to dominance hierarchies by putting them together artificially in groups, and then overlord lizards come to dominate the group who maintain their position by the same signals used in nature to maintain territoriality.

J. Robbins: You used the amusing term "smelling up" the substance. Is there any information whether these substances are chemically transformed during their action?

E. O. Wilson: This has not been investigated.

K. J. Ryan: It seems to me that there are two aspects to this. One is the production of the pheromone, and the other is the ability to respond to it, and I wonder in the course of development whether there is any dichotomy, or whether these two abilities coincide in the development of the animals or the insects.

E. O. Wilson: You mean in the course of evolution?

K. J. Ryan: Not only in the course of evolution, but in the development from immature to mature animals.

E. O. Wilson: You mean, for instance, that an individual in the course of normal development might become sensitive to a pheromone before it began to produce it. Is that correct?

K. J. Ryan: Yes. I was thinking of it particularly in the reproductive sense—what influences these might have on maturation.

E. O. Wilson: Apart from cases of sexual and parent-young communication, where transmission of particular pheromones is one-way, I cannot recall examples of your dichotomy. But the subject of behavioral ontogeny with reference to pheromones is terra incognita.

E. H. Frieden: This very interesting work emphasizes the intriguing and important problem of the mechanisms involved in the activation of physiological systems which respond to the presence of very small numbers of molecules. Many of the participants here will no doubt recall the estimates made by Selig Hecht and others that the simultaneous entrance of only a few light quanta suffice for stimulation of the rods of the retina. The exquisite sensitivity of the receptor brings up another problem, namely, the effect of the signal upon the signaler. To press the visual analogy a bit further, it is as though an organism that can respond to light of extremely low intensity were to carry around a powerful searchlight, bright enough to blind itself as well as any of its fellows in the immediate neighborhood. To minimize the response of the signaler to its own signal, does not some sort of quenching mechanism have to be provided?

With respect to possible mechanisms of response to substances present in very low concentrations, Passoneau and Lowry have recently reported that it is possible to measure extremely small concentrations of pyridine nucleotides by the technique of "enzymatic cycling," in which a small amount of a nucleotide is used to couple two other reactions by alternate cycles of oxidation and reduction. "Amplification" factors of considerable magnitude may thus be realized. It is conceivable that some of the substances you have been talking about may operate in a similar fashion.

E. O. Wilson: It was J. B. S. Haldane and H. Spurway (*Insectes sociaux* **1,** 247, 1954) who pointed out most forcefully that most animal communication consists of an invitation to imitate—a "follow me." This is certainly true in many cases of chemical communication, where the animal is going through behavior at the moment it is releasing a pheromone which induces the same behavior. Of course, this is not true where transmission is across sexes or across castes in the case of social insects, where the emitter is insensitive to that particular substance in the first place. But in the case of alarm substances, to take an example, the individual who is emitting the pheromone is both in the center of the active zone and showing high-intensity alarm. Also, in trail laying, the layer follows its own trail. It comes into the nest laying a trail; then it turns about and goes out along its own trail with about the same amount of accuracy as other follower ants, while laying a second trail. In my information analyses of single trails, I had to remove the incoming ant as soon as it got into the nest in order to avoid confusion. I have seen single ants lay as many as four trails, one upon the other.

R. O. Greep: As a follow up of the matter of "follow me" substances, I would like to be permitted to ask a practical question. I wonder whether you can tell us something about the molecular weight and composition of Chanel No. 5. Are the active ingredients of perfumes thought to be similar to some of the natural signals?

E. O. Wilson: My best source of information for that was Lederer's 1950 paper (71) on the chemistry of perfumes. This review indicates, and it is generally well known, that many of the bases of commercial perfumes are the animal musks. Civetone and muskone were identified by biochemists in the course of research on perfumes. I don't know what the behavioral connection is and I have often wondered myself why animal musk is used. What role can we really attribute to the influences of culture? Even if the role were almost total, say 99% by some operational measure we might devise, why did man start using animal musk in the first place?

J. F. Grattan: Recently I had an opportunity to listen to a fascinating paper delivered by Dr. John E. Amoore of the Botany Department of the University of Edinburgh, Scotland, titled: "Stereochemical Theory of Olfaction." As pointed out by Dr. Amoore, man can differentiate from one another seven primary odor groups. Further, an intriguing theory has been proposed which nicely explains human perception of odors. According to this theory, man's tissue receptor sites contain 7 separate and distinct geometric patterns, each of which is expressly available to accept but one of the seven primary odor groups. It follows that, to be perceived and properly classified by our olfactory apparatus, the molecular configuration of the presented material must be such that it will accurately "fit" a particular one of the seven receptor site designs. Certain physical and chemical requirements must also be met by any odor-bearing substance in order for it to correctly fit into any one of the seven receptor shapes. Perception and identification of multi-odor sensations is possible when each component of a composite odorous preparation correctly fits its respective receptor configuration. Similarly, by proper selection and blending of appropriate odiferous agents it is possible to obtain a composition in which masking or enhancement of one or more of the seven primary odors is achieved. Application of this ingenious theory, I believe, will ultimately make it possible to contribute patient-appeal features to even the most unappealing, though elegant, galenicals. This will be more surely accomplished if it can be established that the sense of taste also involves a series of specific receptor configurations. Seemingly, the latter is a logical expectation, in view of the fact that taste and smell are so intimately related.

The question comes to mind, then: Do you consider that the theory briefly outlined above may apply to animals and insects as well, and would, therefore, explain your findings which demonstrate that only compounds of a certain size act as pheromones and trigger a particular response in a particular species? In other words, in the animal and insect kingdoms is there a geometric configuration relationship dependency between receptor site and respective pheromone?

You showed, Dr. Wilson, how a pioneering ant lays down a trail as it returns to the colony to alert the whole group of ants to the fact that a food source has been found. The group can then find its way toward and reach the food supply by simply following the scent laid down by the trailblazer. We know that in the human the sense of smell is easily fatigued. Is this not also the case with the ants? If it were, it would seem that as the ants traveled toward the food, guided only by the trailblazed scent, their perception might gradually be dulled to the point that they ultimately would lose their way enroute and fail to reach their objective. Indeed, how do they manage to stay on the scent throughout a successful journey? Also, if each of three pioneers finds a different type food, is there a specific dissimilarity in the character of the three trails blazed back to the colony which subsequently enable each member of the colony to proceed toward any particular one of the three food sources according to individual preference? Or are the three trails completely identical, with the result that each of the colony sets out for one or another of the discovered food sources indiscriminately?

Finally, you have demonstrated that insect species characteristically respond to concentrations of certain odorous substances so infinitesimal as to make the extreme dilutions familiar to homeopathic medicine loom large in comparison. Does this capacity to perceive, properly identify, and respond to pheromones in remarkably small quantities gradually develop to a point of maximum sensitivity, development being timed carefully to coincide with a particular need of the insect to be attracted by a special pheromone? Or does the sensitivity "peak" at some stage in the organism's development and then simply remain in a stand-by status ready to function when proper conditions for reception prevail? For example, the "17-year locust"—more properly, cicada—slowly develops over that period of time. Yet, immediately upon completion of such a prolonged development process the cicada appears to be able correctly to sense the location of the most suitable spot to partake of a foliage feast. This would seem to indicate that the pheromone-perception capacity must have been fully mature before emergence of the adult form. If such is, indeed, the case, might it not be possible to artificially offer the respective pheromone in proper concentration at an appropriate time in the course of the 17-year development period, thus "falsely" communicating information to the organism? Premature reception of this fake information might then adversely interfere with the normal sequential development process of the insect, the ultimate result being practical control of the cicada's costly devastation of vegetation. Perhaps such an approach might effectively circumvent the need for the usual insecticidal techniques currently under fire due to publication of Rachel Carson's grossly controversial "Silent Spring." I will be pleased to have you candidly criticize the points I have presented.

E. O. Wilson: Once again, I confess my ignorance of the physicochemical basis of olfaction. I don't think anyone knows about the limitations of molecular size. There is no question, however, that the olfactory receptors of the insects are external, and are frequently adapted to respond to specific pheromones. I have mentioned the extreme example of the geometric isomers of bombykol. Butenandt found that one geometric isomer is 10^{12} or 10^{13} times more efficient than the least active geometric isomer. Schneider (100) got negative EAG's from saturniid antennae exposed to bombykol, so that the screening must be at the level of the receptor. These findings suggest that insect pheromones provide unusual opportunities for molecular biologists working on the physicochemical basis of olfaction.

In answer to your second question, we have no evidence that fire ant workers distinguish food finds, new nest sites, or recovered brood in the form or composition of the odor trails laid to these different classes of objects. Behavior seems to be identical in every case, and in experimental situations the responses did not differ even where there was a much larger quantity of food or a higher concentration of it. As long as the ant "chose" to lay a trail, it was essentially an all-or-none response. The quantitative communication involved was mass communication. That is to say, the number of ants laying the trail was determined by the number of ants who found the food acceptable. More than 90% found one mole of sucrose acceptable, less than 10% found pure water acceptable, and the number of ants laying trails was accordingly determined. The number of ants laying trails determined the number of ants responding. Further, the trail fades out in about 100 seconds, so that the number of responders at a given moment in time can be nicely regulated.

As to the ants' ability to follow the trail despite sensory adaptation, insects do adapt

714   EDWARD O. WILSON AND WILLIAM H. BOSSERT

to smell. Wharton and Wharton (114) demonstrated a typical adaptation curve with reference to the percentage of male *Periplaneta americana* responding to female sex attractant. My subjective experience is that the same is true of fire ants following odor trails. Yet the ants have no great difficulty, and they show only a weak tendency to turn back short of the end of the trail (117). This paradox has been commented on by authors in the past with respect to dogs following trails. We can only guess that the zigzag motion of the follower animals, causing them to move on and off the trail repeatedly, significantly reduces adaptation. Finally, with reference to timing of metamorphosis in cicadas, we may infer from information on better known insects that the process is primarily under endogenous endocrine control.

F. C. Bartter: It might be appropriate to mention very briefly some results relating hormonal activity to sensory threshold in man recently accumulated by Dr. Robert Henkin. The finding, in brief, is that in patients with Addison's disease taste sensitivity is much greater than that in people without Addison's disease. For example, median taste threshold in normal subjects for sodium chloride is 12 millimoles per liter, with a range of 6-60 millimoles per liter (Fig. A). The patients with Addison's disease, on the other hand, can taste solutions a hundred times more dilute. The same holds for potassium chloride, sodium bicarbonate, sucrose, and urea. For hydrochloric acid the difference is even greater, so that the Addisonian can taste about 10,000 times as well as the non-Addisonian when hydrochloric acid is the substance in the test water. Incidentally, this increased sensitivity is not changed in the least with DOC or sodium-retaining steroids, but is easily reversed with carbohydrate-active steroids like prednisolone. There is some evidence that this pertains to the sense of smell as well, as perhaps one would expect. Those studies are still in progress. It is difficult to bring this to as picturesque an example as those you use, but you might say that this means that a girl can attract a man with Addison's disease with only 1/10,000 as much Chanel No. 5 as she needs for a normal man.

L. Van Middlesworth: How do the pioneer ants find their way back to the nest? Have you studied the receptor organs?

E. O. Wilson: They find their way back to the nest by light-compass orientation. We did an experiment in which we kept the sole light source to one side of the feeding platform; we found, following the classical procedure, that when we suddenly switched the light source from one side to the other, trail-laying ants reversed their direction and headed away from the nest. The site of the receptors I have not investigated myself. This has not been investigated specifically in ants, but from what we know of other hymenopterans and on the basis of the early, crude experiments on ant olfaction by Adele Fielde in this country, olfaction is mediated primarily by antennal receptors. I think this is going to be found true generally of the insects. D. Schneider and J. Boeck [*Z. vergleich. Physiol.* **45**, 402 (1962)] have just published an interesting electrophysiological study of the antennae of moths and beetles, in which they suggest that qualitative and quantitative differences of response among the numerous antennal sensilla might provide the code allowing the brain to discriminate odors.

K. K. Carroll: I might mention an interesting example of chemical communication between hosts and its parasite which I worked on for some time [*Nematologica* **3, 154**, 197 (1958)]. This has to do with eel worms which will be familiar to people from Britain and Europe since they are a serious economic problem there, because they attack potatoes, sugar beets, and other crops. These microscopic organisms live in the roots of the host plant. The female, after fertilization, fills with eggs and the body hardens and forms a sac which drops off into the soil. The eggs are relatively resistant to any kind of chemical treatment as long as they are in the sac, and they emerge from the sac in response to a chemical excreted by the host plant. These chemicals are fairly specific. The potato eel worm emerges in response to something from the potato root, the beet eel worm in response to something that comes out of the sugar beet root, and

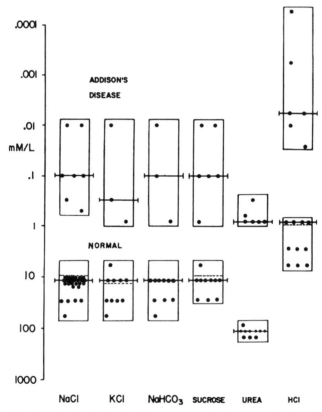

FIG. A. Detection thresholds for taste in normal subjects and patients with Addison's disease untreated for at least 3 days. Individual thresholds (dots), median values (bars) and range (rectangles) are shown. Dotted lines represent median values from literature.

so on. As you mentioned in respect to other chemical attractants, the hatching factors are active at very low concentrations. To my knowledge, the structure of these substances have still not been determined. I would have liked to continue the work on the problem, but since in America the potato eel is confined to Long Island, the authorities are not keen on having these organisms anywhere else on the continent.

H. L. Bradlow: Did I correctly understand you that this form of chemical communication in ants is about as efficient as the dancing system that has been described for bees?

716   EDWARD O. WILSON AND WILLIAM H. BOSSERT

In connection with this alarm mechanism, can you exhaust the ability of an insect to respond to the alarm chemical by repeated applications, so that it simply does not respond any more?

Finally, in connection with your information theory analogy, at these very low threshold levels is there a problem with a random noise, so that there is confusion in response to the chemical excitement?

E. O. Wilson: The dance of the honey bee probably contains more information in sum. My analysis extended purely to the arrival of both bees and ants with respect to the target and hence I had two information components—direction and distance. The honey bees are also able to communicate quality by what the Germans call "Tanzfreudigkeit." This is difficult to figure out. It has something to do with the liveliness of the dance, and I would judge from conversations with bee specialists that it really means amplitude of the abdomen waggle. Also, it has been pointed out that the bees sometimes bring in scents on their bodies which help the bees locate specific food sources. In addition, it is reported that when honey bees arrive at the food source, they extrude the Nassanoff gland of the abdomen and release a scent (geraniol) in the air at the target site which helps orient the workers more precisely when they arrive in the vicinity of the target.

Worker ants adapt to alarm substances as well as trail substances.

Noise undoubtedly occurs but is unmeasurable by present methods. It probably plays a key role in speciation of some taxa, as I suggested for the Saturniidae.

D. A. Denton: One of the problems which concerns us very much in Australia is that of shark attack. One of the theories which has been put forth is that the release of blood into the water may attract a shark. I am wondering whether you would care to comment on the feasibility of this from the point of view of diffusion, and, as a corollary, whether or not it could depend upon whether the release was deep in the water or on the surface, and the shark was cruising at a distance near the surface.

E. O. Wilson: It is certainly feasible that there could be some attractant present in blood to which the sharks are very sensitive, so that a high Q/K ratio obtains; and if this is true we can expect a long-range effect, perhaps over miles, occurring within a matter of hours or even minutes. Your hypothesis is a most intriguing one. It should be a straightforward procedure to test blood components in shark orientation experiments. I am reminded of some work reported by T. Hosoi [*J. Insect Physiol.* **3**, 191 (1959)] on the releaser of blood gorging by one of the mosquito species. Adenosine 5'-phosphates are evidently the main agents causing gorging. Possibly sharks orient over long distances by the Schwinck effect seen in moths, which permits an exceedingly low threshold. But it is difficult to see how sharks can judge currents in deep water in the absence of fixed reference points, if they rely on rheotaxis instead of chemical gradients.

Bird Migration: Influence of Physiological State upon Celestial Orientation

Abstract. *By means of photoperiod manipulation, the physiological states of spring and autumn migratory readiness were induced in indigo buntings. The orientational tendencies of these two groups of birds were tested simultaneously in May 1968, under an artificial, spring planetarium sky. Birds in spring condition oriented northward; those in autumnal condition, southward. These results suggest that changes in the internal physiological state of the bird rather than differences in the external stimulus situation are responsible for the seasonal reversal of preferred migration direction in this species.*

Migration, by definition, refers to a two-way journey. For most birds residing in North Temperate areas, this migration consists of a southward trip each autumn followed by a northward return the following spring. Distances covered in such round-trip journeys are considerable, frequently exceeding 4000 to 6000 miles (6400 to 9600 km) (*1*).

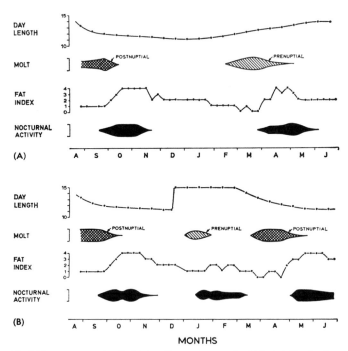

Fig. 1. The occurrence of molt, fat deposition, and nocturnal activity in indigo buntings (*7*). (A) Control birds on a natural photoperiod. (B) Experimental birds exposed to an accelerated photoperiod regimen.

Bird migrants have been the subjects of intensive study by both physiologists and ethologists. Physiologists have studied hormonal and biochemical changes associated with the physiological state of migratory readiness, whereas ethologists have concentrated upon the navigational problems posed by these long-distance flights. Neither group has focused its attention upon the proximate factors underlying the seasonal reversal of preferred migration direction.

The similarity in the physiological events leading up to migratory departure (particularly the deposition of large quantities of subcutaneous fat and the appearance of migratory activity) has suggested that the internal state of migratory readiness is a general one which recurs in similar form twice each year (2). Experiments involving the hormonal induction of fat deposition and nocturnal restlessness support this view. Injections of prolactin and adrenocortical hormone produced these two effects in white-crowned sparrows *Zonotrichia leucophrys* in both vernal (photostimulatory) and autumnal (photorefractory) conditions (3). This type of study has led to the hypothesis that changes in internal physiology cause a migrant to become highly receptive to external stimuli that can serve as orientational cues, and that the seasonal reversal of migration direction is not dependent upon internal changes in the physiological state of the bird but rather upon seasonal changes in the external stimuli themselves (4).

Many birds which migrate at night are able to determine their direction from the stars (5). Because of the inequality of the solar and sidereal day, the temporal positions of stars change with the seasons, with the result that very different stellar information is available from an autumn night sky in contrast to that of a spring night sky. This raises the possibility that migrants might possess a specific northward directional response to the stellar stimuli of the spring night sky, and a different, southerly, response to the different stellar stimuli present in the autumn sky. Indeed, experiments performed on European warblers supported this hypothesis (4).

Experiments on indigo buntings *Passerina cyanea* suggested that a very different mechanism might be operative in this species (6). Directional tendencies of caged buntings were re-

corded when the birds were exposed to various correct and "incorrect" planetarium skies during the spring of 1965. These birds, which oriented in a generally northward direction under an artificial spring sky, continued to display northerly preferences under a sky advanced 12 hours from local time —a sky that normally would be present in the autumn. Thus, the visual stimuli typical of the opposite migration season failed to evoke any change in directional response.

Experiments designed to determine which star patterns are essential for orientation in buntings suggested a second dichotomy with European warblers. Evidence obtained by blocking

various portions of the artificial sky from view suggested that the northern circumpolar area may be of special importance to orientation during both migration seasons (6). Although this finding requires confirmation, it raises the possibility that migrants may not use different cues during their northbound and southbound travels. This would indicate that it is not the external stimulus situation but rather some feature of the internal physiological state of the animal that dictates directional preferences.

To study this in further detail, I brought two groups of indigo buntings into spring and autumn migratory condition simultaneously by means of

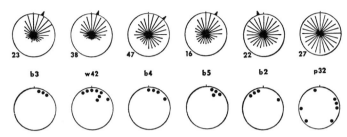

Fig. 2. Orientation of indigo buntings in spring migratory condition tested under a spring planetarium sky. (Upper) Vector summaries of total nocturnal activity. Arrows denote mean direction. (Lower) Mean directions adopted on different nights of testing. Vector diagrams are plotted such that the radius equals the greatest number of units of activity in any 15° sector. The number that this represents is presented at the lower left of each diagram. In all diagrams, 0° or 360° is north.

Table 1. Orientation of indigo buntings in spring (control) and autumn (experimental) migratory condition. Vector analysis of the total nocturnal activity (upper row) and of the nightly mean headings (lower row) are presented for each bird. All tests were conducted in May 1968 under a spring planetarium sky (30°N); ISS, insufficient sample size.

Bird	Activity (hr : min)	Nights active (No.)	Units of activity (total *N*)	*P*	Mean direction	Mean angular deviation
			Control			
b3	11 : 50		288	<< .001	23°	66°
		3			ISS	
w42	20 : 15		540	<< .001	16°	66°
		8		<< .001	12°	31°
b4	11 : 15		828	.019	45°	75°
		4			ISS	
b5	8 : 45		246	.016	29°	73°
		4			ISS	
b2	15 : 5		413	.026	342°	76°
		4			ISS	
p32	20 : 40		581	Random (.4)		
		7		Random (.4)		
			Experimental			
w95	20 : 0		594	<< .001	146°	55°
		6		.007	144°	27°
b62	10 : 40		311	<< .001	153°	68°
		5		.007	147°	19°
b63	17 : 30		882	<< .001	161°	72°
		6		.007	156°	36°
p38	15 : 40		561	<< .001	207°	71°
		6		.003	202°	26°
b64	8 : 35		337	Random (.4)		
		3			ISS	

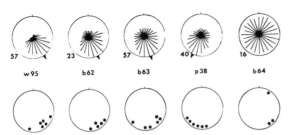

Fig. 3. Orientation of indigo buntings in autumn migratory condition tested under a spring planetarium sky. (Upper) Vector summaries of total nocturnal activity. (Lower) Nightly mean headings.

photoperiod manipulation. The directional preferences of both groups then were tested at the same time under identical planetarium skies.

Fifteen adult male indigo buntings were captured with mist nets in the vicinity of Ithaca, New York, during the autumn of 1967. They were housed in two identical 8 by 12 foot (2.4 by 3.6 m) flight rooms and given free access to food and water. Weight, fat level, and molt status were recorded weekly for each bird (7). Nocturnal activity was monitored continuously on an Esterline-Angus event recorder by means of microphone transducer.

As controls, eight birds were subjected to a photoperiod regimen simulating that encountered in nature. An astronomical-time clock maintained a day length equivalent to that on the wintering grounds at 18°N. These birds underwent normal, prenuptial molts between February and early April (Fig. 1). After this molt, large quantities of subcutaneous fat were deposited. Nocturnal activity first appeared in early April.

The experimental group of seven buntings was subjected to an accelerated photoperiod regimen. These birds were exposed to a long day length (15 hours light and 9 hours dark) after the termination of the refractory period in early December. This long day length was maintained until 1 March when the control of photoperiod was transferred to an astronomical timer set for 42°N and for a date 5 months advanced from local time. Hence, for example, the birds received the daylight equivalent of August on the breeding grounds during March. Each bird in this group initiated the prenuptial molt during the first 3 weeks of exposure to this long day length, but in all cases this molt was arrested

before completion (8). During March and April these birds underwent normal postnuptial molts, attaining the brown winter plumage during late April. After this, fat deposition occurred, and nocturnal activity commenced.

As a consequence of this photoperiod manipulation, both groups of birds were available for experimentation during May of 1968. Both were in a state of migratory readiness, the controls in preparation for a spring flight, the experimentals for an autumnal flight.

The orientational tendencies of these birds were tested under an artificial planetarium sky mimicking that present at 30°N on 1 May (9). Each bird was placed in an individual, circular, funnel-shaped cage, and its directional preference was recorded by the "footprint technique" (10) (Figs. 2 and 3, Table 1). For each data distribution the null hypothesis of randomness was tested by the "v" modification of the Rayleigh test (11), with the prediction that control birds would orient northward and experimentals, southward. Mean direction and mean angular deviation were calculated by vector analysis (12). Four buntings failed to display nocturnal activity and hence could not be tested. Of the six active control birds, one was random while the five others each demonstrated a significant north to northeasterly tendency that was fairly consistent from night to night. One of the five active birds in the experimental group also failed to show any directional preference. Four birds, however, did display significant directional tendencies which ranged from southeast to south-southwest.

In short, the behavior of these two groups of birds differed markedly when tested simultaneously under the same

planetarium sky. Individuals in spring condition oriented in a direction appropriate for spring migration, while those anticipating an autumnal migration took up directions appropriate for that season.

These findings indicate that important differences exist between the physiological states of migratory readiness in the spring and autumn. Results of planetarium blocking experiments mentioned earlier raised the possibility that the same celestial cues were used during both migration seasons. It thus seems plausible that differences in hormonal state may determine the manner in which an orientational cue will be used.

STEPHEN T. EMLEN
*Section of Neurobiology and Behavior,
Division of Biological Sciences,
Cornell University,
Ithaca, New York 14850*

References and Notes

1. See J. Dorst, *The Migrations of Birds* (Houghton Mifflin, Boston, 1962).
2. D. S. Farner, in *Recent Advances in Avian Biology*, A. Wolfson, Ed. (Univ. of Illinois Press, Urbana, 1955), p. 198; H. O. Wagner, *Z. Tierpsychol.* **18**, 302 (1961); J. R. King, *Proc. Int. Ornithol. Congr. 13th* **2**, 940 (1963); F. W. Merkel, *Ostrich* **6** (Suppl.), 239 (1964).
3. A. H. Meier and D. S. Farner, *Gen. Comp. Endocrinol.* **4**, 584 (1964); ———, J. R. King, *Animal Behav.* **13**, 453 (1965).
4. F. G. Sauer, *Z. Tierpsychol.* **14**, 29 (1957); ——— and E. M. Sauer, *Cold Spring Harbor Symp. Quant. Biol.* **25**, 463 (1960).
5. F. Sauer, *Z. Tierpsychol.* **14**, 29 (1957); ———, *Proc. Int. Ornithol. Congr. 13th* **2**, 454 (1963); ——— and E. M. Sauer, *Cold Spring Harbor Symp. Quant. Biol.* **25**, 463 (1960); W. J. Hamilton, *Auk* **79**, 208 (1962); L. R. Mewaldt, M. L. Morton, I. L. Brown, *Condor* **66**, 377 (1964); M. E. Shumakov, *Bionica* (1965), p. 371; S. T. Emlen, *Auk* **84**, 309 (1967).
6. S. T. Emlen, *Auk* **84**, 463 (1967).
7. Fat was recorded on an increasing 1 to 4 scale, according to C. M. Weise [*Ecology.* **37**, 274 (1956)]. Molt was assessed as absent, slight, moderate, or heavy, according to the index of M. L. Morton and L. R. Mewaldt [*Physiol. Zool.* **35**, 237 (1962)].
8. Possibly the arrest of molt was caused by the rapid increase in gonadal hormones, stimulated by the long days.
9. The Cornell research planetarium is a 30-foot (9 m) diameter, air-supported dome equipped with a Spitz A-3-P star projector. Experimental cages are placed 3 feet off-center to minimize distortion of the artificial sky.
10. S. T. Emlen and J. T. Emlen, Jr., *Auk* **83**, 361 (1966).
11. D. Durand and J. A. Greenwood, *J. Geol.* **66**, 229 (1958).
12. E. Batschelet, *Amer. Inst. Biol. Sci. Monogr.* *1* (1965). Sample sizes for the Rayleigh test were determined by dividing the total number of units of activity N by a correction factor. This divisor was determined empirically and represents the interval at which activity measures become independent of one another [see appendix 5 in S. T. Emlen, *The Living Bird* (Laboratory of Ornithology, Cornell University, Ithaca, N.Y., in press)].
13. Supported by an NIH health science advancement award to Cornell University. I thank W. Keeton, J. Planck, and A. van Tienhoven for commenting on the manuscript.

24 April 1969; revised 10 June 1969

J. Theoret. Biol. (1961) **1**, 295–317

The Logical Analysis of Animal Communication

PETER MARLER

University of California, Department of Zoology,
Berkeley, U.S.A.

(Received 16 January 1961)

An attempt has been made to describe some of the responses evoked by communication signals in certain animals and to infer the kind of information which the signals transmit. Using the methods developed by C. W. Morris (1946) for the logical analysis of human language, identifiors, designators, appraisors and prescriptors can be distinguished. Animal signals are rich in designative information, and five subcategories are distinguished: species-specific, sexual, individual, motivational and environmental information. The influence of natural selection upon the form of a signal will vary according to its information content. For example, the variable nature of some signals and the stereotypy of others can be related to the conveyance of different types of motivational information. A single signal often conveys several different items of information which are usually inherent in the whole signal and not represented by different parts of the signal. The form of some signals is arbitrary but the physical structure is often directly related to information content, in an iconic manner, or in other ways.

Introduction

By any reasonable definition of the term "communication" there can be no doubt that animals communicate with each other. Some authors even extend the term to include exchange of stimuli between organisms and their physical environment (Stevens, 1950), which is perhaps further than it is necessary to go. The position adopted in a recent book by C. Cherry (1957) serves very well to restrict the discussion to a social context. He defines communication as: "The establishment of a social unit from individuals by the use of language or signs". Inclusion of both signs and language in this definition ensures from the outset that studies of communication systems shall not be restricted to the languages of man. This simple step, which so many past authors have been reluctant to take leads Cherry into a lucid, illuminating account of the properties of communication systems and of the methodological problems which they pose. As a student of animal behavior who has been grappling with problems of animal communication the writer has been struck by the relevance to zoology of many of the ideas expressed in Cherry's book. This paper tries to apply some of them to animal communication and to show that they can open up new avenues to the understanding of the kind of evolutionary problems with which many zoologists are concerned.

THE ANTHROPOCENTRIC APPROACH

Comparative psychologists have neglected the subject of animal communication to a remarkable degree—remarkable, that is, until one reflects on the anthropocentric point of view of most psychologists. The strictures imposed by F. A. Beach (1959) on comparative psychology are nowhere more relevant than in the subject of animal communication. The main concern has been to differentiate man and the animals, rather than to determine the properties which their "languages" may have in common. Dozens of cases could be cited where this prejudice has influenced the questions that are asked, and therefore the answers obtained. Even in such distinguished contributions as the chapter on the social significance of animal studies by D. O. Hebb & W. R. Thompson (1954) in the "Handbook of Social Psychology" this bias is evident. After discussing the human capacity to combine and readily recombine sounds for different effects, they acknowledge that language has other distinctive characteristics but assert their belief that the above criteria "are enough to set it off fully from animal communication". A promising discussion thus terminates at the point where it is about to become productive. In the field of linguistics attempts to analyse animal communication have also been marred by anthropocentric viewpoints (e.g. Revesz, 1956), although it is also a linguist, C. F. Hockett (1961), who has succeeded in defining the properties of human language in a manner that permits us to test for their occurrence in animals. In doing so, he has omitted "purposiveness" as one of the criteria. This concept, which may also be associated with an anthropocentric viewpoint, has bedevilled investigations of animal behavior in the past (Thorpe, 1956).

PURPOSIVENESS

Hebb and Thompson (1954) question whether the waggle dance of the honey bee (von Frisch, 1954) is purposive, suggesting that it would be if:

(a) only the first of several returning bees made the waggle, since, if the message has already been conveyed to the colony by ten other bees, there is little sign of purpose in behavior that conveys it once more; and

(b) the worker still made the dance as though the audience was present even when it had been removed.

(a) seems to be based on the misconception that the entire contingent of perhaps ten thousand workers can perceive the performances of a dozen or so dancers. The solution to (*b*) is not certain, but personal observations suggest that an eager audience in the hive is certainly stimulating to a dancer. However, we may ask whether this is a sign of purposiveness, or whether the dancer is simply stimulated through palpation by the antennae of the audience.

If the concept of purposiveness has to be reduced to such a vague level before it can be tested, as Hebb and Thompson seem to imply, we may wonder whether it has not ceased to be valuable as a theoretical construct in the analysis of animal behavior. W. H. Thorpe (1956) has pointed out how difficult the subjective and objective aspects of purposiveness are to separate. It may be best to restrict the idea of purposiveness to a human context. Hebb & Thompson (1954) state that the essence of purposive communication is that "the sender remains sensitive to the receiver's responsiveness during sending, and by modification of his sending shows that his behavior is in fact guided by the intention of achieving a particular behavioral effect in the receiver". By this definition any dog-fight qualifies as purposive, as the authors admit. It is not clear what is gained by using a specialized and loaded term for a process which is basically a mutual communicatory exchange, unless to draw attention to possible subjective phenomena. If the latter, then we should recall Thorndike's (1911) still relevant warning about the dangers of the introspective method in animal studies, notwithstanding Tolman's (1932) demonstration that by placing special interpretations upon it, purposiveness can be given an objective basis.

AN OBJECTIVE APPROACH

The descriptive or taxonomic approach, which comes less readily to psychologists than to zoologists, has provided the bulk of our present knowledge about animal communication, as applied by such classic investigators as Charles Darwin and K. von Frisch. This in turn has led to new inductive generalizations by K. Z. Lorenz, N. Tinbergen, W. H. Thorpe and others which will provide the framework of future work for many years to come. Instead of approaching animal communication with anthropocentric preconceptions, they set out to describe the natural behavior in objective terms, seeking to derive conclusions about the evolutionary basis of behavior. Even such severe critics as D. S. Lehrman (1953) fully acknowledge the great value of the advances which this "ethological" school has achieved. Communicatory behavior has figured prominently in this work and provided the basis for much of the theoretical discussion in the early papers of Lorenz (1935) and Tinbergen (1940). The scope has subsequently been broadened to include other types of behavior, and the "ethological" school (Tinbergen, 1951; Thorpe, 1956) now provides a rationale for the analysis of animal behavior.

In proceeding thus far, it is the author's contention that some of the special circumstances surrounding communicatory behavior have been overlooked. Close attention has been given to the evolutionary basis of visual signals and the motivation which underlies them. Less attention has been given to the nature of the actual communicatory process; to the questions raised by the process of exchange of signals between one animal and another. The psychologists' concern with this aspect leads them to a consideration of purposiveness, but this does not prove to be a productive line of attack. A strictly objective approach is required which can be applied with equal efficacy to the communication of animals and of man. This paper seeks to show that the theoretical framework presented by Cherry (1957), building especially upon the ideas of Pierce (see Gallie, 1952) and Morris (1946), provides us with such an approach which can lead to advances in our understanding of animal communication.

Semiotic: The Theory of Signs

Dissatisfaction with the results of previous attempts to separate the subjective and objective aspects of human language led C. K. Ogden and I. A. Richards, in a book called "The Meaning of Meaning" (1923), to consider the implications of the theory of signs (or symbolism as they sometimes call it) as developed by the logician, C. S. Pierce. The relationship between a word or symbol and its external referent is shown to be elusive. Perception of external objects (referents) always involves sign situations. We respond only to a part of the whole object. That part comes to represent the whole object as a kind of symbol or sign. "If we realize that in *all* perception, as distinct from mere awareness, sign situations are involved we shall have a new method of approaching problems where a verbal deadlock seems to have arisen. Whenever we perceive what we name a chair we are interpreting a certain group of data, and treating them as signs of a referent." Narrowing down the discussion to the use of language they suggest that "when we consider the various kinds of sign situations . . . we find that these signs which men use to communicate with each other and as instruments of thought occupy a peculiar place". This comes to bear directly on our present problem with the statement that "the person actually interpreting a sign is not well placed to observe what is happening. We should develop our theory of signs from observations of other people, and only admit evidence from introspection when we know how to appraise it."

To explain the approach of C. S. Pierce to the problem of language analysis, W. B. Gallie (1952) gives the following example. "Suppose that in any particular case we are in doubt whether some sign made by an

THE LOGICAL ANALYSIS OF ANIMAL COMMUNICATION 299

individual A has been interpreted or understood by a second individual B. How should we set about trying to settle the question? Should we somehow or other try to discover directly what B's 'mental reaction' has been? It seems quite certain that we have no means whatever of doing this. What we would do, surely, is to try to discover whether B has made some overt response such as A's sign would justify." Cherry (1957) emphasizes the same point, that only a non-participant observer can make fully objective observations on communication systems.

The science of semiotic has arisen to deal with the kind of data that are obtained by direct, non-participant observation of communication systems. It is usually divided into three parts: *syntactics*, the formal study of signals as physical phenomena, and the laws relating to them; *semantics*, study of the "meaning" of signs; and *pragmatics*, the significance of signals to the communicants (Cherry, 1957). The application of syntactics to animal communication is clear, and great progress has been made by Tinbergen and others in this kind of analysis, especially in the sphere of visual communication (see Tinbergen, 1940, 1951, 1952, 1959). Semantics are of doubtful value in animal studies, and as Cherry points out there is considerable overlap with pragmatics, even in the sphere of human language. Pragmatics on the other hand forms the natural complement to syntactics, one defining the physical properties of signals, the other concerning itself with the role of those signals in the communicatory process, a role which we seek to establish by observing and interpreting the response which they evoke in other animals.

ANIMAL PRAGMATICS

The central problem is to determine the nature of the information content of communication signals. As Cherry points out "information content is not to be regarded as a commodity; it is more a property or potential". It cannot be discussed independently from the occurrences of responses to the signal in other organisms. We thus require a means of inferring information content from the nature of the response given. We may note in passing that the information theory developed by Wiener & Shannon (Shannon & Weaver, 1949) is of no help to us here since it operates only "at the syntactical level" (Cherry, 1957). The work of C. W. Morris (1946), however, is directly concerned with analysis of human language at the pragmatic level and can give us some clues as to how to proceed.

Morris seeks to distinguish between signals which function as *identifiors*, *designators*, *appraisors* and *prescriptors*. He emphasizes that this is not an exhaustive list, and elaborates some of them further to deal with special problems of human language. The four basic categories will suffice as a basis for further discussion. We can describe each of them as conveying a corresponding type of information, provided that we can discern an appropriate response from a communicant. The categories are not mutually exclusive, so that one signal might convey one or all of the different types of information.

Morris defines the four categories as follows: "In the case of *identifiors*, the interpreter is disposed to direct his responses to a certain spatio-temporal region; in the case of *designators* the interpreter is disposed toward response sequences which would be terminated by an object with certain characteristics; in the case of *appraisors* the interpreter is disposed to respond preferentially with respect to certain objects" as manifest in a choice situation; "in the case of *prescriptors*, the interpreter is disposed to perform certain response sequences rather than others." So identifiors may be said to signify (i.e. convey information about) location in space and time, designators to signify characteristics of the environment, appraisors to signify preferential status and prescriptors to signify that specific responses are required. This classification cuts across the division of language into emotive and referential (or symbolic) which received so much emphasis from Ogden & Richards (1923). Morris shows how his classification is subject to testing in a way that the other is not. Moreover we can see that while prescriptors and appraisors embody much of the quality of "emotive" language, and identifiors and designators are more obviously "referential"; in nature, the latter can be emotive in certain circumstances. Thus the new approach is more precise and should be regarded as replacing the older terminology, as Morris suggests.

We now have to demonstrate that this method of analysis can in fact be applied to animal communication systems. J. B. S. Haldane (1953; Haldane & Spurway, 1954) have already shown some ways in which this may be done, and the writer also made an attempt to analyse vocal communication in a small bird, the chaffinch (Marler, 1956) by a method similar to the one suggested here. A reinterpretation of those same data can serve as an illustration. In essence, given a knowledge of the response of other animals to the signal and of the other circumstances in which that same response is given, we can infer the nature of the "message" transmitted by the signal.

The song of the chaffinch is given only by the male. The species is normally monogamous, and the song is especially frequent in an unmated male, given only within his territory. An unmated female chaffinch in reproductive condition responds to repeated singing by persistently approaching the singing male, soliciting his courtship, and eventually establishing a pair bond with him. Circumstantial evidence suggests that some females learn the individual characteristics of their mates' song, and

THE LOGICAL ANALYSIS OF ANIMAL COMMUNICATION 301

subsequently respond to them in a preferential way. The behavioral exchanges consequent upon the female's response to the song are confined to a sexual context and are normally evoked by what we may describe as an "appropriate sexual partner". We may infer that frequent male singing conveys information about this particular class of objects which are the "designata" of the male's song, in this situation. What exactly is the information content which is implied?

An appropriate sexual partner for an unmated female chaffinch in reproductive condition is *an unmated male chaffinch in reproductive condition, in possession of a territory* (within which nesting will take place), *who is close to a location occupied by the female at the same time as she is there.* We are suggesting that all of these items of information are conveyed to her by the male's song. This does not imply that the song has any meaning for her, only that it performs selective actions upon her, appropriate to a certain input of information (Cherry, 1957). The male's individual identity may also be conveyed in some cases. To what extent can this be fitted into Morris's scheme?

"Identifiors" dispose the receiver to direct his responses to a certain spatio-temporal region. We can show that such identifying information is present in the male chaffinch's song which provides an abundance of clues for precise location of the singer in time and space (Marler, 1959). In some respects "locating" information might be a better description.

"Designators" dispose the receiver towards response sequences which would be terminated by an object with certain characteristics. Designative information is thus to be defined by the characteristics of the object normally evoking the response, in this case those of an appropriate sexual partner. This would encompass all of the items outlined above and we shall suggest in a moment that further sub-categories may be desirable.

"Prescriptors" dispose the receiver to perform certain response sequences rather than others. The response prescribed for the female chaffinch is to approach and to adopt certain postures which elicit male courtship. Prescriptors and designators may be confused in some cases because we need to know the kind of response prescribed before the object designated can be discovered. Circular reasoning can only be avoided when prescriptor and designator are contained in different signals. If they can be combined with other signals a different response can be prescribed with the same designator and the effects can be separated. When the same signal performs both functions, as seems to be common in animals, no logical separation between prescriptors and designators is possible.

Appraisors dispose the receiver to respond preferentially to certain objects. Although we have no quantitative information, the frequency with which a song is repeated probably conveys such appraisive information. Within the range of song frequencies that will evoke a response, a female confronted with two singing males may be most likely to choose the one who is singing most persistently.

A more detailed breakdown of the nature of designative (and therefore prescriptive) information is required if this system is to aid us in analysis of the evolution of animal communication systems. Most critical from the point of view of natural selection is the presence of the *species-specific* information—that the singer is a chaffinch. We can also separate *sexual* information—that the singer is a male; *individual* information—that the singer is a particular individual; *motivational* information—that the singer is in reproductive condition; *environmental* information—that he is within his territory and has no mate. The criteria by which these types of designative information may be identified are as follows.

Species-specific information and its evolutionary implications

If the response given to the signal is normally evoked only by members of one species we may infer that species-specific information is conveyed by the signal. Usually a member of the same species will be involved, since many animal communication signals play a role in reproductive isolation. Information about other species could come into this category, as for example in the signals exchanged between a commensal and its host. There are also mimics which emit signals with a false species specificity.

Some signals are lacking in such species-specific information. For example, in a situation involving acute danger male chaffinches have an alarm call consisting of a high thin squeak. It is typically given in response to a hawk flying overhead. It evokes the same response from other chaffinches as the stimulus provided by the hawk, namely, direct rapid flight to the nearest cover. However, several other small woodland birds have converged upon the same type of alarm call presumably because, as mentioned below, it is a difficult sound to locate, and so exposes the caller to a minimum of danger. Chaffinches will respond to the corresponding alarm calls of other species as promptly as to their own. Such cases of interspecific communication are very common in the woods in which chaffinches live. Thus species-specific information is not present in this call. Degrees of species specificity may be expected, decreasing to the extent that signals are of mutual value in communication within a group of different sympatric species.

A signal functioning to transmit species-specific information will be subject to certain evolutionary pressures, since there must be a minimum of confusion with signals used by other species at the same time and place. Circumstantial evidence suggests that many auditory and visual signals have been selected for specific distinctiveness (see Sibley, 1957). Con-

THE LOGICAL ANALYSIS OF ANIMAL COMMUNICATION 303

versely signals with an interspecific function may be subject to selection for convergence upon a common type—or at least to a minimum of selection for divergence. Where species specificity is required, it is desirable that, as well as being specifically distinct, the signal should also be biologically improbable and conspicuous for effective communication against a background of environmental "noise" (Lorenz, 1951). A relative lack of variability is also required among members of the same species, or at any rate of the same population, an important point when we compare signals which convey individual information.

Sexual information. Responses associated with reproduction are normally evoked by members of the opposite sex when in the appropriate physio-logical condition. A signal evoking such a response may be said to convey sexual information. There are, however, cases where such behavior patterns are also evoked by members of the same sex in what may be called homosexual or pseudo-sexual behavior (see Morris, 1955). The incidence of sexual information varies considerably as manifest in the extent to which the sets of communication signals of the male and female overlap in different species. The same principles often apply to visual and auditory signals, so that the more sexually dimorphic finches, for example, also show the greatest discrepancy between the repertoires of displays and vocalizations in the two sexes (Hinde 1955–6). The principles governing these variations in the prevalence of sexual information in the signals of different species have not yet been worked out.

In discussing differences between the signals produced by male and female animals, Hockett (1961) has elevated the principle of what he calls "interchangeability" to the level of a major criterion in the analysis of communication systems. He suggests that while it occurs in animals, it is especially characteristic of human language, implying that any person can theoretically reproduce sounds made by any other person. He makes a distinction between language and paralanguage (Trager, 1958) and applies the principle of interchangeability particularly to the former. However the same distinction, which seems to rest on an intuitive judgement with reference to human language, cannot be made with animals. If we regard the difference between the sexes as a means of conveying sexual informa-tion, this information is obviously present as a conspicuous and more or less consistent difference in frequency between the speech of men and women. While the auditory signals produced by women share many characteristics with the corresponding signals of men, there are also in Western Society certain unavoidable differences of pitch, unavoidable, that is, for *most* women (Potter, Kopp & Green, 1947). In this respect the lack of "interchangeability" in human speech is more striking than in some animals, since even strongly sexually dimorphic species often have some signals which are consistently identical in all respects in the two sexes.

Individual information. The transfer of individual information by a signal is implied whenever the response is normally only evoked, or most readily evoked, by the particular individual emitting the signal. The qualifications admit the possibility of appraisive information being included here, since the female chaffinch, for example, will respond to an unfamiliar chaffinch song, though she may choose a familiar song if given a choice. In many circumstances individual recognition of the signals of mates, rivals, young, and companions plays an important role in the social behavior of animals (Nice, 1943; Marler, in press).

A signal which transmits individual information is subject to selective influences different from those associated with species-specific information. The latter, as we have seen, is most readily transmitted by signals which show little variation, either in the individual or within a population of a given species. Individual information is again best conveyed by signals which vary little in the individual. But it is also a prerequisite that the signals emitted by individuals of the same species, especially within the same population, should differ from each other in a consistent manner. Circumstantial evidence suggests that there is an unduly high degree of intra-group variability in signals which are thought to be involved in individual recognition, such as visual signals originating from the head region of birds, and the songs of some species of birds (Marler, 1959, in press). Some bird songs appear to convey both species-specific and individual information by relegating the stereotyped and variable properties to different parameters of the song.

Motivational information. The last two categories of designative information, motivational and environmental, are the most difficult to define, the least understood, and perhaps ultimately the most important from an evolutionary point of view. The transmission of motivational information by a signal may be inferred if the response given is appropriate to a particular motivational state of the signaller. Such a signal conveys information about variations in the readiness of the signaller to engage in certain classes of activity, such as feeding, fighting or copulation and so on.

The male chaffinch's song evidently communicates to the female the fact that he is in a reproductive state. This condition usually lasts for about three or four months. Short-term changes in motivation may also be communicated by signals. When a mated female has built a nest and is preparing to ovulate she will allow the male to copulate at intervals for about four or five days. When actually ready for copulation she gives a special call which is restricted to this context. The male promptly

THE LOGICAL ANALYSIS OF ANIMAL COMMUNICATION   305

approaches and mounts. Similarly the calls given periodically by the young as they become hungry, cause the parents to bring food to them.

Information about still more subtle changes in motivation can also be transmitted. Here the best evidence comes from visual signals, and to discuss them we shall again have to anticipate consideration of the divisible parts of the signal and the information they convey. Many of the communication signals used by animals are subject to what Morris (1957) has called the "principle of typical intensity". This implies that the signal varies little or not at all, with variation in the level of motivation with which it is associated. Either it is given in "typical intensity" or it is not given. Such a signal can effectively communicate presence or absence of a certain type of motivational information but not variations in degree. For many purposes this appears to suffice. In general, a male chaffinch is either in reproductive condition or he is not, and an "all-or-none" type of signal can communicate this.

Other signals do not obey the principle of typical intensity, but vary widely in form, completeness and frequency with the intensity degree of motivation with which they are associated. Visual signals used in fighting behavior are particularly prone to vary in form with slight variations in the presumed balance between the tendencies to attack and withdraw. An opponent is often highly responsive to the slight shifts in motivation which these changes convey, advancing in response to signs of withdrawal, and vice versa, and the final outcome of the fight will normally be determined in this way. On the basis of his extensive studies of the behavior of cats, Leyhausen (1956) has been able to construct a Latin square of the changes in facial expression with changes in aggressiveness and readiness to flee, including all possible combinations between the two, a remarkable demonstration of the complex array of motivational information that such graded communication signals could convey. A function of this kind obviously has profound effects upon the way in which the signals will evolve.

The signals discussed above convey what we can describe as "positive" motivational information; they enable a receiver to "make a positive prediction" of the response which the signaller is likely to give when approached. The evolution of a second class of signals has been governed by a trend towards becoming the direct opposite of other signals, as Darwin (1872) pointed out with his principle of antithesis. His classical example is the behavior of a submissive dog which can only be described as the opposite in all respects of a dog which is fighting. Many other examples of such "antithetic" or "reversed" signals (Tinbergen, 1959) have been described, having the function of conveying something like "no offense meant", and so reducing the chance of an open conflict occurring (Tinbergen & Moynihan, 1952).

T.B.

306   PETER MARLER

In the light of the present analysis we can reinterpret this function as the conveyance of negative motivational information, making it possible for the receiver to predict that the signaller will *not* behave in a certain way when he is approached. All of the cases known so far occur in potentially aggressive situations and appear to function by reducing the chances of attack or flight, or both. Negative information about readiness to attack or to flee is conveyed in most cases. Once again there are evolu-

Increasing aggressiveness ————►

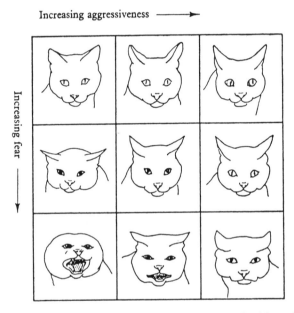

Increasing fear ————►

FIG. 1. Changes in the facial expressions of cats associated with variations in the intensity of aggressiveness and fear. (After Leyhausen, 1956.)

tionary implications which could be explored further. For example, aggressive displays usually have certain formal properties and also a certain orientation with respect to the opponent. A negative element can be introduced with respect to a certain receiver, both by reversing formal elements of the display and also by orienting away from the particular receiver. Both trends, in various combinations, can be traced in the examples given by Tinbergen (1959), ranging from simple reorientation of an aggressive display to the reversal of other aspects as well.

Environmental information. When Morris (1946) set up the category of designators he visualized their primary role as conveyors of environmental information, encompassing as they do the characteristically human tendency to give things names. In animals we may infer that a signal has transmitted environmental information if the response it evokes is appropriate to some characteristic of the environment of the signaller at the moment or in the immediate past. The exact temporal relationship will be discussed in a later section. If, for example, a particular sound is produced in the presence of food, as occurs in herring gulls, and if others respond by approaching and looking for food, as Frings, Frings, Cox & Peissner (1955) have shown to be the case, we may infer that information about the presence of food was conveyed by the signal.

We have inferred that the male chaffinch's song conveys two items of environmental information, one positive, that he is within his territory, and one negative, that he has no mate. Other examples are mainly concerned with what are perhaps the two most important aspects of the physical environment to animals, food and predators. The use of signals conveying the presence of food is probably widespread within family groups. The mock pecking movements, by which a domestic chicken attracts her chicks to a supply of grain, are a familiar example. Special calls are probably also used in this context, though no examples are known to the author. The gull call mentioned above certainly attracts adult gulls as well as young (Frings, Frings, Cox & Peissner, 1955).

The best known signals concerned with the communication of food are the dances of honey bees, analysed in detail by Karl von Frisch (1954). The round dance communicates the distance of a food source in the neighborhood of the hive and its richness. The waggle dance, when given in the hive, also communicates the direction, as well as the distance and richness of sources which are further from the hive. This is, in a sense, another case of communication within a family group, for the worker bees are all daughters of the same queen. The dances are also used in the swarm, to communicate the distance, direction and suitability of new home sites (Lindauer, 1957). Here the dances are given, not on the vertical combs in the darkness of the hive, as in the case with the food dances, but on the surface of the swarm. Clearly the context in which the dance is given affects its communicatory significance. In both cases the signal is a mechanical one, received particularly by the bee's antenna and the mechanoreceptors at its base. The different items of information are conveyed by different aspects of the dance: direction by the angle of the waggle run with respect to the vertical; distance by the tempo of the dance; richness of the food source by the persistence of the dancing. Information about the latter might also be placed under the heading of appraisive information, since it determines the choice made between different situations, particularly in the swarm, where departure from the temporary resting place does not occur until the dancers have reached a degree of unanimity. The final decision is achieved by the scout bees discovering the best site who, by their more persistent dancing, eventually sway those who have discovered alternative but less satisfactory sites (Lindauer, 1957).

If we place the environmental information conveyed by animal signals on a specificity-generality continuum it usually appears to be relatively unspecific in nature. Information is transmitted about food but not about which particular food. It is true that foraging honey bees may pick up the scent of flowers and so convey the identity of the nectar source to members of the hive (von Frisch, 1954), but the specificity of the signal has been evolved by the plant rather than by the bee. Similarly signals concerned with communicating danger usually seem to do little more than signify different degrees of danger without specifying which environmental agent is responsible. Many small birds have different vocal signals for sudden, acute danger, such as when a hawk appears overhead, and for less dangerous situations, such as when they discover a sleeping owl, or a cat on the ground beneath them. The responses which these stimuli evoke are quite different, sudden flight to cover and cryptic behavior on the one hand; approach to a safe distance and conspicuous "mobbing" behavior on the other. However, the circumstances may greatly modify the response to a given predator. In early spring a male chaffinch will normally give the call associated with acute danger only in response to a flying hawk. But later when he has nestlings, a variety of animals will evoke this call if they come near the nest. Thus we cannot say that the call communicates the presence of a hawk. There may be animals which convey more specific environmental information. Our knowledge is so fragmentary that we cannot begin to generalize. The European willow warbler is thought to have two mobbing calls, one given to a perched hawk, the other to a cuckoo, suggesting that more specific information may be conveyed in this case (Smith & Hosking, 1955).

Conclusions on information content. Although identifiors (or locators) and appraisors occur among the communication signals of animals, designators seem to be most richly represented. Five different categories of designative information have been described, all having particular implications for the evolution of animal communication systems. Perhaps the most prominent category in human language is environmental information, the one to which investigators most often turn when they wish to compare the animals with man. The basic capacity, to convey environmental information by signals, is present in both. However, the time element in this process is significant as several authors have pointed out (Haldane, 1956; Haldane & Spurway, 1954; Hockett, 1961). In animals the delay between perception of an object in the environment and emission of a signal conveying information about that object is usually a short one. In man the delay may be extended almost indefinitely, illustrating what Hockett

(1957) calls "displacement". The only well documented case involving a longer delay in animals comes again from the honey bee where the dance occurs after the forager has returned to the hive (von Frisch, 1954). Here there are finitè limits to the delay, which is short by human standards in any case, and it would hardly be useful to the honey bee if it were any longer since the food supply from a given plant varies from hour to hour. Human capacities in this direction are probably unique, although one may wonder if any but the most educated observer would be able to detect such extensive time delays in animals even if they occurred.

The context may have considerable significance to the animals themselves. Hockett (1961) has pointed out how the responses of honey bees to dancing differs when it takes place in the hive and on the swarm. In speaking of the communicatory process as though it were mediated by signals alone, we have thus been guilty of over-simplification. The response evoked by a signal—and therefore the information it conveys—may vary with changes in the circumstances both of the sender and the receiver. The song of a male chaffinch is seen in a different light if we observe the response of male chaffinches instead of females. A male chaffinch intruding into another's territory will flee if he hears the owner's song, implying reception of a further item of motivational information, that the owner is ready to attack male chaffinches found within the boundaries of his territory. The response of a male in an adjacent territory will be different again, and so on. The separation of all of the factors which bear on a given act of communication is thus an imposing task. The additional possibility always exists that the signaller may be emitting several different cues at the same time, as seems to be the case in rats, for example, where olfactory, tactile, visual and auditory signals may all play a part in the female's sexual responses to the male (Beach, 1942, 1947).

Divisible Elements of the Signal

In trying to determine the role of prescriptive information in animal signals we have been confronted by the dilemma that it cannot be distinguished from designative information in signals consisting of one indivisible unit. Only when prescriptors exist in physically separate parts of the signal can an unequivocal separation be made. It thus becomes important to transfer our attention from pragmatics to syntactics to consider the physical nature of some of the signals used in animal communication.

CONTINUITY VERSUS DISCRETENESS

Attention has been drawn to the fact that some signals vary to the extent that they sometimes grade continuously into other signals; others tend to appear in an all-or-none fashion, so that they are separate and discrete from all other signals. The degree of variation observed can be correlated to some extent with the information which the signal conveys. A degree of continuous variation may occur in at least three different circumstances.

First, appraisive information appears to be most commonly conveyed in animals by signal characteristics which vary in a continuous manner. The frequency with which the chaffinch song is repeated probably conveys appraisive information to the female about the male's relative suitability as a mate (cf. page 302). Similarly the persistence of dancing in the honey bee, as expressed by the number of dances given before the sequence is broken, conveys appraisive information about the richness of the food source or the suitability of a new nest site to other members of the hive. Cases may exist where appraisive information is conveyed by a discontinuous series of signals. For example, the remarkable series of postural displays given by the black-headed gull correlated with variations in the relative and absolute levels of tendencies to attack and to flee should come into this category, for while some intergrade, others are discrete, with a sudden switch from one to the other as the balance of motivation shifts (Moynihan, 1955; Tinbergen, 1959). However, it appears that this condition, which is characteristic of human language, is rare in animals.

Another function of continuously variable signals is the conveyance of subtle changes in motivational information. Some signals, as Morris (1957) has indicated, vary little with slight changes in the signaller's motivation, whereas others mirror the changes in motivation very closely (cf. page 308). Particularly in fighting behavior, where the communication of such subtle motivational information can assume great importance, such variable signals are often used. Human language is in some ways less well adapted to convey such continuously variable information because of the tendency to divide continuous phenomena into discrete classes, which is perhaps one of the reasons why animal signals of this type are difficult for zoologists to describe.

Continuously variable signals also occur as a means of conveying environmental information of a continuously variable nature. The best example is again from the honey bee dances, in which both the direction and distance of the food source are communicated. The former is conveyed by the angle of the waggle run with respect to the vertical, the latter by the tempo of the dance, both varying in a continuous manner. No doubt further examples will be discovered.

We may conclude that continuously variable signals have an important role to play in the communication systems of animals. More stereotyped, discrete signals are also common and, for example, make up the bulk of the vocal signals of such birds as the chaffinch. Continuously variable signals have certain disadvantages. Their interpretation may be slow, and

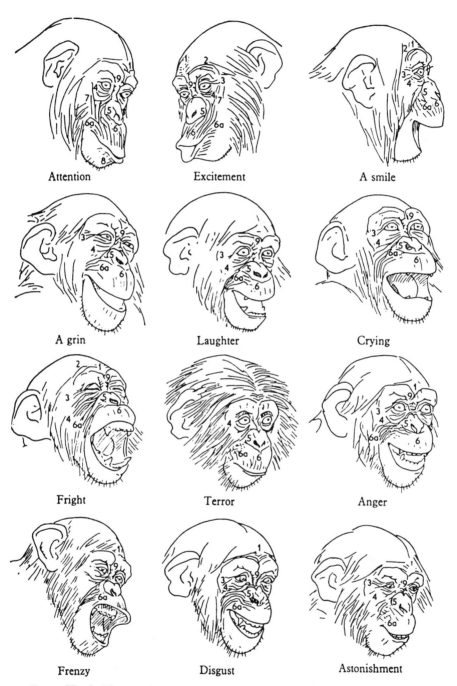

Attention Excitement A smile

A grin Laughter Crying

Fright Terror Anger

Frenzy Disgust Astonishment

FIG. 2. The facial expressions of a young chimpanzee in various moods. The creases on the face are numbered to emphasize that each one may be involved in several different expressions. (After Kohts, 1935.)

subject to error. Also appropriate inborn responsiveness to all properties of the signals, which characterizes the communication systems of many animals is more easy to visualize with discrete signals than with continuously variable signals. Finally there may be conflicts with other items of information conveyed by the same call. We have seen that the communication of species-specific and individual information both call for stereotyped signals, a requirement which may well override the need to convey subtle changes of motivation.

DIVISIBILITY

The impossibility of separating designative and prescriptive information hinges on the fact that in animal communication systems several items of information seem to be conveyed by one discrete, indivisible signal. We do not normally find the different items of information represented by different elements as is commonly the case in human language, where the component elements can be rearranged to create new "messages". However, this does seem to occur in some cases, particularly in visual communication signals. The facial expressions of chimpanzees probably serve as signals in intraspecific social behavior (Hebb, 1946). In describing them, Kohts (1935) took great care to point out that the same creases on the face may be involved in several different circumstances, the expressions as a whole presumably conveying different information. In her drawings, she even went so far as to number the facial lines to emphasize this point (Fig. 2). Assuming that all elements are necessary for the complete signal (which is difficult to test with visual signals), it appears that the divisible parts can be rearranged to create new "messages". In a similar way the sharing of components by different visual displays of such birds as the chaffinch may imply something similar.

Examples may also be found among vocal signals, but we have to proceed with care. Thus Hockett (1961) quotes Lanyon as presenting evidence that basic motifs in the songs of certain birds are rearranged in different ways to create new songs. A number of cases of this have been described, but there is no evidence that these recombined elements differ in any way in information content. Better examples are likely to be found in the alarm calls of certain birds. Some species have several discrete calls which are given, sometimes alone, sometimes together with other calls. The sequences of different signals may conceivably contribute to one overall signal whose information content varies with the constituents of which it is made up. If this proves to be the case, we are then approaching, at a very primitive level, the kind of lability in the manipulation of the information content of signals which is such a distinctive property of human language.

THE LOGICAL ANALYSIS OF ANIMAL COMMUNICATION   313

We must not assume that the lack of such lability among animals is simply a result of incapacity of the nervous system to handle such complex information. The way of life of most animals is so stringent and fraught with dangers that a high premium is placed upon quick production of brief signals, which can be accurately interpreted by receivers, often without the opportunity for previous practice. Given that the fight for survival is controlled by a limited number of factors, such as reproduction, fighting, food and predators which, as selective factors, dominate all others in their effect, there is little place in the biology of most animals for the kind of subtleties of communication which human language permits. Nor must we forget that communication is a social activity which often runs counter to the trend towards competition which characterizes most animal communities. A very elaborate social organization is required before the survival of an individual's genotype becomes so dependent upon survival of the group that natural selection will encourage individual sacrifice for the sake of the community. In most cases we may expect this to occur only within the family group which is one of the reasons for the strong emphasis on individual information in communication signals. With a more elaborate social organization and division of labor among its members, the immediate pressures upon individual survival are alleviated, and the stage is set for the exploitation of the more subtle gains arising from further elaboration of the systems of communication. The most elaborate communication system known in the animal kingdom occurs in the honey bee, whose social organization particularly from a genetic point of view begins to approach the ideal conditions we have postulated above.

RELATIONSHIPS BETWEEN INFORMATION CONTENT AND SIGNAL STRUCTURE

Human language is usually regarded as consisting of arbitrary symbols, bearing no direct physical relationship to the information which they carry. The communication system is thus based upon a convention. Some zoologists have asserted that the communication signals of animals are arbitrary in the same sense (Lorenz, 1935, 1951) and many of them seem to satisfy the criteria. However, in some cases physical structure is intimately related to the corresponding designata.

The conveyance of locating (i.e. identifying) information by sound signals is directly related to physical structure, since this controls the ease with which the sound source can be located. Vertebrate animals, for example, rely primarily upon differences of intensity, phase and time of arrival of sound at the two ears. The easiest sounds to locate are those providing all of these clues, the ideal being something like a repetitive click. This type of sound, used by many species of birds when they are mobbing an owl (cf. page 308), is a readily located call serving to attract the attention to the position of the owl. Conversely the calls given when a hawk flies overhead have a different structure which minimizes the clues available for location, making the source of the sound difficult to determine. Insects with different types of receptors which respond, not to pressure changes but to the actual displacement of molecules of the medium, are able to determine the direction of sound directly by reference to the vectorial properties of sound, so that their signals are not affected by the problems of location which confront vertebrates (Marler, 1959).

We have noted that appraisive information is sometimes conveyed by the frequency or length of time for which a signal is repeated. The honey bee dances longer for rich food sources than for poor ones, which implies an iconic relationship.

With the sub-categories of designative information we are on surer ground. Consider for example some of these signals in the light of the most commonly accepted alternative to "arbitrary" which is "iconic". A degree of direct physical correspondence between the signal and its referent is implied (see Cherry, 1957; Hockett, 1961) as with a picture for example. The portion of the honey bee's waggle dance that communicates the direction of the food source conveys this environmental information in an iconic manner by transposing directly from the direction with respect to the sun, to direction with respect to the vertical (Hockett, 1961).

Some other signals conveying environmental information appear to be non-iconic. Thus the various alarm calls of birds bear no relationship to the dangers which are their designata. However their physical structure is by no means arbitrary, in relationship to the locating information which they may convey. Thus the adjectives arbitrary and iconic cannot be applied to a signal as a whole, only to the relationship between signal structure and particular items of information which they communicate.

Signals conveying motivational information may be iconic or arbitrary. Most sound signals probably come into the latter category, although sounds used by many birds and mammals in fighting, having a grating, growling or rattling quality may be related in iconic fashion to the snapping of beaks or teeth which occurs in actual combat. Visual signals which are known to have originated as what zoologists call "intention movements" (Tinbergen, 1940, 1952; Daanje, 1950), which Darwin (1872) recognized as "serviceable associated habits" are more obvious illustrations. For example, many aggressive displays undoubtedly originated through emphasis of the actual physical preparations for attack—baring of the teeth, tensing of the muscles, and so on. More than one type of motivational information may be iconically represented in the same signal, conveying information about the existence of two or more types of motivation in the signaller at the same time. Many insights into the evolution of visual signals have arisen from the Tinbergen's discovery of this phenomenon of multiple motivation in communicatory behavior.

In the same way signals conveying *negative* motivational information are not arbitrary, since their physical structure is related in an inverse manner to the structure of other signals. These constitute a special class of iconic signals. Finally the need to communicate subtle changes in motivation has repercussions on signal structure, encouraging the use of signals which vary in a continuous manner instead of being discrete, in the appropriate circumstances.

Sexual information may equally well be arbitrary or iconic. The red breast of the reproductive male stickleback, which functions as a signal (Tinbergen, 1951) is arbitrary, whereas the swollen belly of a gravid female, also a signal, is iconic. Individual information also may be arbitrary or iconic. Arbitrariness becomes prominent with species-specific signals. It is no accident that Lorenz's (1951) emphasis on arbitrariness was largely derived from intensive study of the plumage and courtship behavior of ducks and other birds, as they play a role in reproductive isolation, all with a strong emphasis upon species-specific information.

The requirement here is that the signal should be readily distinguished from those of other species likely to be transmitted at the same time and place. The way in which they differ is arbitrary, as long as it is readily perceptible to members of the species. The evidence suggests that this has resulted in specific divergence in a wide variety of animal communication signals which function in reproductive isolation of the species. Even here the signals are not entirely arbitrary, since they are excluded from overlap with the signals of other species.

It will be clear from the above discussion that the classification of signals as either iconic or arbitrary is unsatisfactory. A signal may fail to be entirely arbitrary in several ways, which do not all conform closely to the usual definition of iconic. The structure can, however, be related in different ways to the different types of information being conveyed, be it locating, appraisive, species-specific, environmental, motivational and so on. It may be an aid to further progress if we treat signal structure from this point of view, instead of placing all non-arbitrary signals in the iconic category.

Conclusions on the Evolution of Signals

A detailed review of the evolution of the communication systems of animals is beyond the scope of this paper. We would need to present comparative data, on a much larger scale, and much of the evidence has been reviewed in recent papers (e.g. Tinbergen, 1952, 1959; Morris, 1956, 1957; Marler, 1959) together with discussion of the special problems which arise

316   PETER MARLER

with the different sensory modes. We may note that evolution from iconic to arbitrary signals is probably quite a common occurrence, as part of the process known as ritualization (Tinbergen, 1952; Blest, in press). The ontogenetic basis of sound signal systems has been considered in several recent papers (Sauer, 1954; Thorpe, 1958; Messmer & Messmer, 1956; Thielcke-Poltz & Thielcke, 1960; Lanyon, 1957) establishing that while the majority of signals are genetically controlled, some are passed on by the learning of traditions. In contrast we know almost nothing about the ontogenetic basis of responsiveness to signals. Learning probably plays an important role here, even in lower animals. All of these issues need to be considered in a complete analysis of the evolution of the communication systems of animals.

The aim of this essay is more restricted. It seeks only to demonstrate that by using the response evoked by signals as an index, we can derive a picture of the kind of information conveyed. An attempt is made to classify some of the types of information involved, and to show that the effects of natural selection upon the evolution of signals may be clarified by such an approach. The categories suggested are neither final nor exhaustive. The existing knowledge about animal communication is so scanty that we have little to use as a basis. Nevertheless we may make more rapid progress if we approach animal communication systems as a whole instead of treating each aspect as a separate issue. The problems occupy a unique position in the study of the evolution of behavior. It is a challenge for us to try to solve them, even at the most elementary level.

This paper results in part from a research program supported by a grant from the National Science Foundation. It has benefited greatly from the criticisms of Drs. F. A. Beach, C. Cherry, C. F. Hockett and O. P. Pearson.

REFERENCES

BEACH, F. A. (1942). *J. comp. Psychol.* **33,** 163.
BEACH, F. A. (1947). *Physiol. Rev.* **27,** 240.
BEACH, F. A. (1959). *Amer. Psychol.* **15,** 1.
BLEST, A. D. In "Modern Problems of the Behaviour of Man and Animals". Cambridge University Press, London. In press.
CHERRY, C. (1957). "On Human Communication". Wiley, New York.
DAANJE, A. (1950). *Behaviour,* **3,** 48.
DARWIN, C. (1872). "The Expression of the Emotions in Man and Animals". Murray, London.
FRINGS, H., FRINGS, M., COX, B. & PEISSNER, L. (1955). *Wilson Bull.* **67,** 155.
GALLIE, W. B. (1952). "Pierce and Pragmatism". Pelican Books, Harmondsworth, England.
HALDANE, J. B. S. (1953). *Diogenes,* **4,** 3.
HALDANE, J. B. S. (1956). *Science Progress,* **175,** 385.
HALDANE, J. B. S. & SPURWAY, H. (1954). *Insectes Sociaux,* **1,** 247.
HEBB, D. O. (1946). *Psychol. Bull.* **53,** 88.

THE LOGICAL ANALYSIS OF ANIMAL COMMUNICATION   317

HEBB, D. O. & THOMPSON, W. R. (1954). In "Handbook of Social Psychology", Vol. 1. (Lindzey, G., Ed.) Addison-Wesley, New York.

HINDE, R. A. (1955-6). *Ibis*, **97**, 706; **98**, 1.

HOCKETT, C. F. (1961). In "Animal Sounds and Communication". (Lanyon, W. E. & Tavolga, W. N., Eds.) A.I.B.S. Symposium Proceedings. A.I.B.S. Washington.

KOHTS, N. (1935). *Sci. Mem. Mus. Darwin.*, *Moscow*, **3**, 1. Russian with English summary.

LANYON, W. E. (1957). *Publ. Nuttall Ornithol. Cl.* **1**, 1.

LEHRMAN, D. S. (1953). *Quart. Rev. Biol.* **28**, 337.

LEYHAUSEN, P. (1956). *Handbuch Zool.*, *Berlin, Bd. VIII*, **10**(21), 1.

LINDAUER, M. (1957). *Nature, Lond.* **179**, 63.

LORENZ, K. Z. (1935). *J. Ornithol.* **83**, 137, 289.

LORENZ, K. Z. (1951). *Symp. Soc. exp. Biol.* **4**, 221.

MARLER, P. (1956). *Ibis*, **98**, 231.

MARLER, P. (1959). In "Darwin's Biological Work". (Bell, P. R., Ed.) Cambridge University Press, London.

MARLER, P. In "Vertebrate Speciation". (Blair, W. F., Ed.) University of Texas Press. In press.

MESSMER, E. & MESSMER, I. (1956). *Z. Tierpsychol.* **13**, 341.

MORRIS, C. W. (1946). "Signs, Language and Behavior". Prentice Hall, New York.

MORRIS, D. (1955). *Behaviour*, **8**, 46.

MORRIS, D. (1956). "The function and causation of courtship ceremonies". Fondation Singer Polignac Colloque Internationale sur L'Instinct, 1954.

MORRIS, D. (1957). *Behaviour*, **11**, 1.

MOYNIHAN, M. (1955). *Behaviour Suppl.* **4**, 1.

NICE, M. M. (1943). *Trans. Linn. Soc. N.Y.* **4**, 1.

OGDEN, C. K. & RICHARDS, I. A. (1923). "The Meaning of Meaning". Routledge and Kegan Paul, London.

POTTER, R. K., KOPP, G. A. & GREEN, H. C. (1947). "Visible Speech". van Nostrand, New York.

REVESZ, G. (1956). "The Origins and Prehistory of Language". Longmans, Green & Co., London.

SAUER, F. (1954). *Z. Tierpsychol.* **11**, 10.

SHANNON, C. E. & WEAVER, W. (1949). "The Mathematical Theory of Communication". University of Illinois Press, Urbana.

SIBLEY, C. R. (1957). *Condor*, **59**, 166.

SMITH, S. S. & HOSKING, E. (1955)."Birds Fighting". Faber & Faber, London.

STEVENS, S. S. (1950). *J. acoust. Soc. Amer.* **22**, 689.

THIELCKE-POLTZ, H. & THIELCKE, G. (1960). *Z. Tierpsychol.* **17**, 211.

THORNDIKE, E. L. (1911). "Animal Intelligence". Macmillan, New York.

THORPE, W. H. (1956). "Learning and Instinct in Animals". Methuen, London.

THORPE, W. H. (1958). *Ibis*, **100**, 535.

TINBERGEN, N. (1940). *Z. Tierpsychol.* **4**, 1.

TINBERGEN, N. (1951). "The Study of Instinct". Clarendon Press, Oxford.

TINBERGEN, N. (1952). *Quart. Rev. Biol.* **27**, 1.

TINBERGEN, H. (1959). *Behaviour*, **15**, 1.

TINBERGEN, N. & MOYNIHAN, M. (1952). *Brit. Birds*, **45**, 19.

TOLMAN, E. C. (1932). "Purposive Behavior in Animals and Man". Appleton-Century-Crofts, New York.

TRAGER, G. L. (1958). *Stud. Linguistics*, **13**, 1.

VON FRISCH, K. (1954). "The Dancing Bees". Methuen, London.

6

Evolution of Behavior

Approaches to Studying
Behavioral Change

Stevan J. Arnold and H. Jane Brockmann

Introduction

Understanding the evolution of behavior requires both a description of how behavior has changed over time and an analysis of why behavior has changed. Accordingly, we have grouped into two sections a selection of classic papers on the evolution of behavior. Papers in the first section document the phylogenetic history of behavior, and those in the second section provide analyses of the selective pressures that can produce new adaptations in behavior. Within this second section, we sorted papers into four approaches: comparative, correlational, experimental, and modeling. This categorization is not exhaustive. We have, for example, not included any studies that impose selection in experimental populations and document

evolutionary response. Likewise, the section on behavioral adaptation neglects papers that focus on inheritance, constraints, drift, and a host of other important topics.

In each section we have summarized primary research questions and considered the classic papers in the context of past and present conceptual developments. Our aim is not to produce a definitive history of science, but rather to guide the reader to related papers and developments. Early papers can be found in Burghardt (1985), and the early history of the field also is discussed by Klopfer and Hailman (1967). Richards (1987) discussed perspectives on behavioral evolution in the nineteenth century.

Phylogenetic History of Behavior

Studies of behavioral phylogeny ask two questions: (1) How do displays and other behavior patterns originate? and (2) What are the steps by which behavior and its underlying mechanisms evolve? Attention to these questions has a long history, beginning with Darwin (1859, 1871, 1872). Comparative studies addressing both issues blossomed during the period from 1940 to 1965 (e.g., Crane 1949; Daanje 1950; Baerends and Baerends-van Roon 1950; Tinbergen 1952; Jacobs 1953; Morris 1956, 1958; Moynihan 1962; Crook 1964; and van Tets 1965). For an overview of work during this period, see Hinde (1966). From about 1965 and until recently, comparative work languished, perhaps because the attention of ethologists was shifted by sociobiology and behavioral ecology. In recent years, however, comparative ethology has begun to flower again. The basic issues of historical transformation of behavior are the same, but methods are more explicit and rigorous (Brooks and McLennan 1991; Harvey and Pagel 1991). The new methods in comparative work are based on two principles. First, reconstruction of history requires a model of evolutionary process (Felsenstein 1973, 1985). In the past, such models were implicit and their connections to the inference of history were unstated. The contemporary challenge is to make explicit the models or assumptions about process that lurk behind studies of history. Second, the inference of phylogeny is a statistical undertaking (Felsenstein 1983, 1988). Although each group of species has only one actual phylogeny, that actual phylogeny is estimated by a set of more or less plausible alternative phylogenies. Statistical methods can be used to evaluate those alternatives. From this brief overview of past work and ongoing developments, we

now turn to the articles about behavioral phylogeny that are reproduced in this volume.

We chose papers by Konrad Lorenz (1941) and Howard Evans (1962) from a large field of candidate papers published during the heyday of ethology. These two papers are outstanding exemplars for two reasons. First, a large number of taxa are surveyed in each study. The size of each survey allows us to view many examples of behavioral differentiation. Second, results are displayed on a phylogeny. Such graphical portrayals are increasingly common and are likely to be the norm in the future (e.g., McLennan et al. 1988; Lauder 1990; Carpenter 1989; Proctor 1991; Langtimm and Dewsbury 1991). The reader may be surprised by the scant attention to methodology in these early papers by Lorenz and Evans. Until the advent of numerical taxonomy (Sokal and Sneath 1963) and its descendants (e.g., Farris 1988; Felsenstein 1991; Swofford 1991), it was normal practice to present phylogenetic conclusions with virtually no discussion of how those conclusions were reached. Phylogenetic conclusions were simply meant to represent the opinion of the author. That opinion was often offered without any justification of methods. The notion that methods of arriving at phylogenetic conclusions could be codified and compared did not appear until later.

Lorenz (1941, 1971) provides an important early illustration of the phylogenetic study of behavior. In addition to the fascinating discussion of display origins and modifications, this paper often is remembered for its "shaving brush" portrayal of a phylogeny (Lorenz 1971, p. 114), which enables the reader to see the evolution of behavioral traits on the tree. Limitations of space did not allow us to reproduce the entire paper. Instead, we have included sections from the beginning and end of the paper that discuss aims and methodology. To give the flavor of the ethograms and discussions that constitute the bulk of the paper, we have excerpted a section describing displays performed by the mallard drake and their possible origins. The interested reader may wish to compare Lorenz's phylogeny with those of Sibley and Ahlquist (1990) and Livezey (1991), and Lorenz's discussion of display evolution with McKinney's (1975).

Like Lorenz, Evans (1962) surveys a behaviorally diverse array of species and dissects the history of behavioral modification. Their approaches, however, are fundamentally different. Lorenz uses behavior to deduce phylogenetic relationships of his species and interprets the history of behavioral change in light of those relationships. Evans, in contrast, uses a phylogeny based on a separate (independent) set of characters, focusing on larval morphology (Evans 1959). He puts his behavioral traits on that phylogeny (fig. 6) and then discusses evolutionary trends in the behavior. This contrast in the methods used by Lorenz and Evans is mirrored by an ongoing debate in systematic biology. Proponents of the "total evidence" school argue that all available characters should be included in the construction of a phylogeny (Eernisse and Kluge 1993). Alternatively, one can argue that one set of characters should be used to construct the trees, so that the history of a second set of independent characters can be reconstructed on the tree, as in Evans (1962). By following this procedure, inferences about the evolution of the second set of characters are not confounded with the process of constructing the tree (Felsenstein 1988). Evans (1962) also uses the illustrative technique of having the phylogenetic tree grow up through a series of behavioral layers. The underlying concept was introduced by Simpson (1953), who referred to the layers as "adaptive zones." Phylogenetic studies of behavior recently have taken on new life and often are imbedded within the comparative approach, which we consider next.

Four Approaches to the Study of Adaptation

Four approaches have been used to study behavioral adaptation: comparative, correlational, experimental and modeling. One approach is not superior to the others. Rather, each asks different questions and should be viewed as complementary to the others (Endler 1986; Arnold 1988). In the future, the most illuminating studies of behavioral adaptation are likely to use two or more of these approaches in combination.

Comparative Approach

The comparative approach to adaptation attempts to identify the selective pressures responsible for evolutionary change in behavior or other traits. In modern practice, one begins with a phylogeny of taxa based on traits other than, and presumably independent of, those under scrutiny (e.g., molecular or morphological data) or with a phylogeny based on a combined behavior-nonbehavior data set (Grimaldi 1987; Sillen-Tullberg 1988; Arntzen and Sparreboom 1989; Emberton 1994; Titus and Larson 1995). Behavioral traits are placed on this phylogeny to reconstruct the probable history of behavioral change (procedures reviewed by Maddison and Maddison 1992). Next, aspects of the ecological or social environment that represent putative selective pressures are placed on the tree and their history is reconstructed. Finally, the statistical association between changes in behavior and changes in selective pressures is assessed (Coddington 1988; Harvey and Pagel 1991). The complete version of the comparative approach to adaptation goes beyond the phylogenetic studies of Lorenz (1941) and Evans (1962) by placing both behaviors and putative selection pressures on the tree and analyzing their statistical association. The approach is correlational (as is the method described next), but the time scale is commonly on the order of thousands or even millions of generations. Several early ethological papers use the spirit of the comparative approach to studying adaptation. Cullen's (1957) reprinted here, is a particularly fine example.

Cullen's (1957) comparison of a cliff-nesting gull, the kittiwake (*Rissa tridactyla*), with its ground-nesting relatives is a classic because of its success in identifying selective pressures. Cullen begins by arguing that the recent ancestor of the kittiwake and its congener, *Rissa brevirostris*, was a ground-nesting species. She reaches this conclusion because the most closely related genus to *Rissa* is the genus *Larus*, and nearly all of the many species in that genus nest on the ground. This form of argument is known today as an outgroup comparison (Brooks and McLennan 1991). (A modern practitioner also would ask whether the sister group to *Rissa* and *Larus* is a ground- or a cliff-nester.) Cullen then describes a large number of behavioral and morphological differences between the kittiwake and ground-nesting gulls. Most important, she relates nearly all of these differences to: (1) relaxation of selection due to the absence of predation, and (2) the special selective pressures that derive from cliff-nesting. Cullen's paper stands out be-

cause of the large number of traits that come under scrutiny and because of the plausibility of her selection arguments. Perhaps many niche shifts precipitate large suites of new adaptations. Cullen encourages us to look closely for interlocking sets of adaptations. Her predictions later were tested by others who worked on other cliff-nesting species. Indeed, other cliff-nesters like Galapagos Swallow-tailed gulls (Hailman 1965) or Noddy Terns (Cullen and Ashmole 1963) should show similar adaptations.

Correlational Approach

The correlational approach asks whether natural variation in behavior is statistically associated with differences in fitness within a population. In the most informative version of the approach, the behavioral traits of individuals are scored and then those individuals are followed through time so that their survivorship and/or reproduction can be assessed. In contrast to the comparative approach, the time scale for assessing selection with the correlational approach is typically a single generation or less and the focus is on a single population. Selection is measured in statistical terms that can be related to the selection coefficients that are used in equations for evolutionary change (Lande and Arnold 1983; Phillips and Arnold 1989). The correlational approach, with its emphasis on individual reproductive success, has its roots with Pearson at University College and Weldon at Cambridge (Provine 1971), in addition to the research groups of Lack and Tinbergen at Oxford. The recent surge of correlational studies of behavior has been reviewed by Fedigan (1983), Wiley (1991), Ryan and Keddy-Hector (1992), and Searcy (1992). A notable early paper using the correlational approach to behavior was Patterson (1965), who showed that gulls nesting on the periphery of a colony are less successful than those nesting in the middle, and that gulls nesting in synchrony were more successful than those that were asynchronous. Although this approach is useful for identifying possible selective pressures, one can not be sure of cause and effect. Gulls nesting in the middle of the colony may be more successful because they are older or in better condition, and not because they derive any protection from their colonial nesting behavior. For lack of space we have not reproduced any papers using the correlational approach.

Experimental Approach

The experimental approach to the study of adaptation asks whether a particular behavior or other trait is capable of affecting reproductive success. Usually the reproductive success of the two treatment groups is compared. In one group the expression of behavior is exaggerated, reduced or altogether eliminated. In the other group (the control), a sham operation of exaggeration, reduction or elimination is performed, but behavioral expression is unaltered and coincides with the average behavioral state in the population. The design and statistical analysis of such experiments is discussed in many textbooks (e.g., Cox 1958; Martin and Bateson 1968).

Behavioral experiments are particularly informative when they are performed in the field, for then they can tell us whether the normal array of predators and other selective agents are capable of exerting selection under natural circumstances. Successful field experiments are much admired because they are often difficult to do. The trick is to manipulate behavior while leaving everything else unchanged, have the selective agents do their work, and be able to score the re-

sults. This choreography is more difficult than it sounds! Selective agents such as free-ranging crows are especially prone to be uncooperative. Niko Tinbergen was one of the first masters of this approach.

In Tinbergen et al. (1962), the authors ask whether predators are the likely selective agents maintaining egg-shell removal behavior in a gull population. We have included only the parts of the paper that focus on the consequences of egg-shell removal. Due to space limitations we have not reproduced the middle section of the paper on the stimulus control of egg-shell recognition by parent gulls. It is worth noting, however, that this study, in its entirety, provides an excellent example of the value of studying both proximate and ultimate factors. The authors motivate their study in their opening paragraph by noting that egg-shell removal: (1) shows interesting evolutionary diversification (in particular, kittiwakes do not show egg-shell removal), and (2) is likely to be advantageous in some way because it is performed at apparent cost to the hatching nestling. The account of the experiments used to tackle the issue of survival value is illuminating because it takes us through the process of refining hypotheses. The authors systematically vary one characteristic at a time and maintain large sample sizes in each experiment. Finally, the discussion is worthy of special attention as the authors place egg-shell removal in the context of other traits that protect the nest from predators.

Models of Selection and Evolution

Models of selection and evolution ask whether it is logically feasible for a particular selective process to exert selection and cause evolutionary change. Models cannot tell us whether selection and evolution actually will occur in the natural world, but they can tell us whether our logic is sound or flawed. Models also can help us organize concepts so that our empirical work is focused on the right issues.

Hamilton's (1964) papers provided a new perspective on the evolution of social behavior. From the standpoint of style, the papers are unusual because they combine a novel theoretical viewpoint, the classical mechanics of population genetics, and a rich field of biological examples and applications. We have reproduced only the first of the pair of papers because of space constraint. The second, longer paper deals with applications of the model. The first paper is difficult to read, even for those trained in theoretical population genetics, but it develops the important concept of inclusive fitness and explains its utility in understanding the evolution of social behavior. It may help the reader to realize that the description of the model is directed at two issues. The first part (pp. 2–7) develops an equation for the change in frequency of a gene that affects social behavior; inclusive fitness plays a key role in this equation (eq. 2 on p. 6). The second part (pp. 7–8) is concerned with how inclusive fitness itself will change over time. In genetical models that do not involve social interactions, average fitness tends to show an increase over evolutionary time; Hamilton wants to know if inclusive fitness shows a similar property. Despite misleading claims in many articles and textbooks, average fitness (whether personal or inclusive) is not always maximized by evolution (Wright 1969; Lewontin 1978). The paper ends with a discussion of an inequality (p. 16) that has come to be known as Hamilton's Rule. In the second paper, Hamilton gives Haldane (1955) credit for expressing the crux of the idea: "Let us suppose that you carry a rare gene which affects your be-

havior so that you jump into a flooded river and save a child, but you have one chance in ten of being drowned. . . . If the child is your own child or your brother or sister, there is an even chance that the child will also have this gene, so five such genes will be saved in children for one lost in an adult. If you save a grandchild or a nephew the advantage is only two and a half to one" (Haldane 1955). Many papers have been devoted to defining the conditions under which Hamilton's Rule holds (Michod 1982; Grafen 1991).

Our next reprinted article is by Orians (1969), who deals not with social interactions between relatives but with interactions between unrelated mating partners. Orians' paper and its immediate ancestors (Verner 1964; Verner and Willson 1966) ushered in a new period of field research on the mating systems of birds and other animals. The Orians paper was (and is) influential because it presents a model. Today models are taken for granted, but in 1969 they were unusual. Orians' paper presents no equations. Instead the model is delivered as a graph that communicates the essentials of the argument. The Verner-Willson-Orians model, and especially the Orians graph, enabled workers to grasp the connections (including tradeoffs) between territory quality, polygyny, and reproductive success. These variables (and some others) became the dominant issues in field studies in the decades that followed. For a discussion of models closely related to the Verner-Willson-Orians model, see Altmann et al. (1977) and Lenington (1980).

In the opening page of his paper, Orians (1969) refers to the Wynne-Edwards (1962) theory of population regulation. Wynne-Edwards' book caused a sensation at the time of its publication by viewing a huge array of social phenomena as adaptations to regulate population size.

For example, Wynne-Edwards argued that nesting colonies of seabirds performed raucous displays at dusk, in which the entire colony took flight and circled the nesting area, in order to access and regulate numbers. If too many birds participated in the display, so the argument goes, then some of the birds would curtail their own reproduction. Williams (1966) argued against Wynne-Edwards by pointing out that social regulation of numbers would be opposed by individual selection. Selection would act against individuals that curtailed reproduction. The Wynne-Edwards mechanism required selection at the level of groups. Colonies that regulated their numbers would not overeat their food supply and so might persist longer than colonies that did not practice self-regulation. Group selection acting at the level of the colony, however, would be opposed by individual selection. Williams argued that individual selection would overwhelm group selection and that, consequently, the kinds of group adaptations proposed by Wynne-Edwards were not plausible. In the years that followed, Williams' view prevailed but, at the time of Orians' (1969) article, the arguments put forward by Wynne-Edwards still had some proponents.

In a different approach to modeling selection and evolution, the Trivers (1972) paper reprinted here pioneered a new perspective on sexual selection. The Trivers paper appeared in a symposium commemorating the centenary of Darwin's 1871 book *The Descent of Man and Selection in Relation to Sex*. (As a lesson to symposium organizers, the paper by Trivers is the only one in the symposium that is routinely cited today, despite the fact that in 1971 he was a graduate student among a distinguished company of senior scientists.) Trivers took a fresh perspective. He drew attention to a new set of variables (variation in reproductive success, parental investment, good genes) and used graphical models to portray his arguments. The connections that Trivers drew between his variables were outlined in an intuitively appealing but largely verbal model. The graphs in his paper were cost-benefit functions from econometrics and were only loosely related to the verbal arguments. Aside from bringing a modeling approach, however loosely constructed, to sexual selection studies, Trivers (1972) also resurrected a forgotten but insightful paper by Bateman (1948). One facet of Trivers' argument, the cost of parental investment, has recently been reformulated by Arnold and Duvall (1994).

In the final paper reprinted in this section, Maynard Smith and Price (1973) introduced the concept of the evolutionarily stable strategy (ESS) and showed how it can be used to analyze animal contests. The approach pioneered in this paper lead to a whole series of papers on animal contests, and to the application of the ESS concept to a broad range of evolutionary problems (Charnov 1982; Maynard Smith 1982). The ESS approach has been especially valuable for analyzing situations in which the fitness consequences of a particular action depend on the activities of other individuals in the population (e.g., sex ratio, sex allocation and sex change, sexual selection, parental investment and alternative strategies, sociality and cooperation). The paper opens with an easy-to-follow account of computer simulations and ends with the algebraic argument that became the standard in ESS models.

The models presented here on the evolution of social behavior, animal contests, mating systems, and sexual selection have utterly changed the way that we study these problems. Although the models admittedly were simplifications, they allowed us to formulate testable hypotheses, to quantify important variables, and to focus on the logic of selective processes.

Literature Cited

Altmann, S. A., S. S. Wagner, and S. Lenington. 1977. Two models for the evolution of polygyny. *Behavioral Ecology and Sociobiology* 2:397–410.

Arnold, S. J. 1988. Behavior, energy and fitness. *American Zoologist* 28:815–27.

Arnold, S. J., and D. Duvall. 1994. Animal mating systems: a synthesis based on selection theory. *American Naturalist* 143:317–48.

Arntzen, J. W., and M. Sparreboom. 1989. A phylogeny for the Old World newts, genus *Triturus*: biochemical and behavioural data. *Journal of Zoology*, London 219:645–64.

Baerends, G. P., and J. M. Baerends-van Roon. 1950. An introduction to the study of the ethology of cichlid fishes. *Behaviour*, Supplement 1:1–242.

Bateman, A. J. 1948. Intra-sexual selection in *Drosophila*. *Heredity* 2:349–68.

Brooks, D. R., and D. A. McLennan. 1991. *Phylogeny, Ecology, and Behavior*. Chicago: University of Chicago Press.

Burghardt, G., Ed. 1985. *Foundations of Comparative Ethology*. New York: van Nostrand Reinhold.

Carpenter, J. M. 1989. Testing scenarios: wasp social behavior. *Cladistics* 5:131–44.

Charnov, E. 1982. *The Theory of Sex Allocation*. Princeton: Princeton University Press.

Coddington, J. A. 1988. Cladistic tests of adaptational hypotheses. *Cladistics* 4:3–22.

Cox, D. R. 1958. *Planning of Experiments*. New York: John Wiley.

Crane, V. 1949. Comparative biology of salticid spiders at Rancho Grande, Venezuela: IV, An analysis of display. *Zoologica* 34:159–214.

Crook, J. 1964. The evolution of social organisation and visual communication in the weaver birds (Ploceinae). *Behaviour*, Supplement 10:1–178.

Cullen, E. E. 1957. Adaptations in the kittiwake to cliff-nesting. *Ibis* 99:275–302.

Cullen, J. M., and N. P. Ashmole. 1963. The black noddy, *Anous tenuirostris*, on Ascension Island. *Ibis* 103:423–46.

Daanje, A. 1950. On locomotory movements in birds and the intention movements derived from them. *Behaviour* 3:48–99.

Darwin, C. 1859. *On the Origin of Species by Means of Natural Selection or the Preservation of Favoured Races in the Struggle for Life*. London: Murray.

Darwin, C. 1871. *The Descent of Man and Selection in Relation to Sex*. London: Murray.

Darwin, C. 1872. *The Expression of the Emotions in Man and Animals*. London: D. Appleton and Co.

Eernisse, D. J., and A. G. Kluge. 1993. Taxonomic congruence versus total evidence, and amniote phylogeny inferred from fossils, molecules, and morphology. *Molecular Biology and Evolution* 10:1170–95.

Emberton, K. C. 1994. Polygyrid land snail phylogeny: external sperm exchange, early North American biogeography, iterative shell evolution. *Biological Journal of the Linnean Society* 52:241–71.

Endler, J. A. 1986. *Natural Selection in the Wild*. Princeton: Princeton University Press.

Evans, H. E. 1959. Studies on the larvae of digger wasps (Hymenoptera, Sphecidae). Part V: Conclusion. *Transactions of the American Entomological Society* 85:137–91.

Evans, H.E. 1962. The evolution of prey-carrying mechanisms in wasps. *Evolution* 16:468–83.

Farris, J. S. 1988. *HENNIG86, Version 1.5*. Distributed by the author, Port Jefferson Station, N.Y.

Fedigan, L. M. 1983. Dominance and reproductive success in primates. *Yearbook of Physical Anthropology* 26:91–129.

Felsenstein, J. 1973. Maximum-likelihood estimation of evolutionary trees from continuous characters. *American Journal of Human Genetics* 25:471–92.

Felsenstein, J. 1983. Parsimony in systematics: biological and statistical issues. *Annual Review of Ecology and Systematics* 14:313–33.

Felsenstein, J. 1985. Phylogenies and the comparative method. *American Naturalist* 125:1–15.

Felsenstein, J. 1988. Phylogenies and quantitative characters. *Annual Review of Ecology and Systematics* 19:445–71.

Felsenstein, J. 1991. *PHYLIP (Phylogeny Inference Package), Version 3.4*. Distributed by the author, University of Washington, Seattle.

Grafen, A. 1991. Modelling in behavioural ecology. Pp. 5–31 in *Behavioural Ecology: An Evolutionary Approach*, 3rd ed., J. R. Krebs and N. B. Davies, eds. London: Blackwell Scientific Publications.

Grimaldi, D. A. 1987. Phylogenetics and taxonomy of *Zygothrica* (Diptera: Drosophilidae). *Bulletin of the American Museum of Natural History* 186:103–268.

Hailman, J. P. 1965. Cliff-nesting adaptations of the

Galapagos swallow-tailed gull. *Wilson Bulletin* 77:346–62.

Haldane, J. B. S. 1955. Population genetics. *New Biology* 18:34–51.

Hamilton, W. D. 1964. The genetical evolution of social behavior I. and II. *Journal of Theoretical Biology* 7:1–16 and 17–52.

Harvey, P. H., and M. D. Pagel. 1991. *The Comparative Method in Evolutionary Biology.* Oxford: Oxford University Press.

Hinde, R. A. 1966. *Animal Behaviour, A Synthesis of Ethology and Comparative Psychology.* New York: McGraw-Hill.

Jacobs, W. 1953. Verhaltensbiologische Studien an Feldheuschrecken. *Zeitschrift für Tierpsychologie Beiheft* 1:1–228.

Klopfer, P. H., and J. P. Hailman. 1967. *An Introduction to Animal Behavior: Ethology's First Century.* New Jersey: Prentice-Hall, Englewood Cliffs.

Lande, R., and S. J. Arnold. 1983. The measurement of selection on correlated characters. *Evolution* 37:1210–26.

Langtimm, C. A., and D. A. Dewsbury. 1991. Phylogeny and evolution of rodent copulatory behaviour. *Animal Behaviour* 41:217–25.

Lauder, G. V. 1990. Functional morphology and systematics: studying functional patterns in an historical context. *Annual Review of Ecology and Systematics* 21:317–40.

Lenington, S. 1980. Female choice and polygyny in redwinged blackbirds. *Animal Behaviour* 28:347–61.

Lewontin, R. C. 1978. Fitness, survival, and optimality. Pp. 3–21 in *Analysis of Ecological Systems,* D. H. Horn, R. Mitchell, and G. R. Stairs, eds. Columbus: Ohio State University Press.

Livezey, B. C. 1991. A phylogenetic analysis and classification of recent dabbling ducks (tribe Anatini) based on comparative morphology. *Auk* 108:471–507.

Lorenz, K. 1941. Vergleichende Bewegungsstudien an Anatinen. *Journal für Ornithologie* 79: special volume.

Lorenz, K. 1971. Comparative studies of the motor patterns of Anatinae (1941), translated by Robert Martin. Pp. 14–18 and 106–14 in *Studies in Animal and Human Behavior,* vol. 2. Cambridge: Harvard University Press.

Maddison, W. P., and D. R. Maddison. 1992. *MacClade, Analysis of Phylogeny and Character Evolution,* Version 3. Sunderland, Mass.: Sinauer.

Martin, P., and P. Bateson. 1986. *Measuring Behaviour, An Introductory Guide.* Cambridge: Cambridge University Press.

Maynard Smith, J. 1982. *Evolution and the Theory of Games.* Cambridge University Press, Cambridge.

Maynard Smith, J., and G. R. Price. 1973. The logic of animal conflict. *Nature* 246:15–18.

McKinney, F. 1975. The evolution of duck displays. Pp. 331–357 in *Function and Evolution in Behaviour,* G. Baerends, C. Beer, and A. Manning, eds. Oxford: Clarendon Press.

McLennan, D. A., D. R. Brooks, and J. D. McPhail. 1988. The benefits of communication between comparative ethology and phylogenetic systematics: a case study using gasterosteid fishes. *Canadian Journal of Zoology* 66:2177–90.

Michod, R. E. 1982. The theory of kin selection. *Annual Reviews of Ecology and Systematics* 13:23–55.

Morris, D. 1956. The feather postures of birds and the problem of the origin of social signals. *Behaviour* 9:75–113.

Morris, D. 1958. The comparative ethology of grassfinches (Erythruae) and Mannikins (Amadinae). *Proceedings of the Zoological Society of London* 131:389–439.

Moynihan, M. 1962. Hostile and sexual behavior patterns of South American and Pacific Laridae. *Behaviour,* Supplement 8:1–365.

Orians, G. H. 1969. On the evolution of mating systems in birds and mammals. *American Naturalist* 103:589–603.

Patterson, I. J. 1965. Timing and spacing of broods in the black-headed full *Larus ridibundus. Ibis* 107:433–59.

Phillips, P. C., and S. J. Arnold. 1989. Visualizing multivariate selection. *Evolution* 43:1209–22.

Proctor, H. C. 1991. The evolution of copulation in water mites: a comparative test for nonreversing characters. *Evolution* 45:558–67.

Provine, W. B. 1971. *The Origins of Theoretical Population Genetics.* Chicago: University of Chicago Press.

Richards, R. J. 1987. *Darwin and the Emergence of Evolutionary Theories of Mind and Behavior.* Chicago: University of Chicago Press.

Ryan, M. J., and A. Keddy-Hector. 1992. Directional patterns of female mate choice and the role of sensory biases. *American Naturalist* 139:S4–S35.

Searcy, W. A. 1992. Song repertoire and mate choice in birds. *American Zoologist* 32:71–80.

Sibley, C. G., and J. E. Ahlquist. 1990. *Phylogeny and Classification of Birds. A Study in Molecular Evolution.* New Haven: Yale University Press.

Sillen-Tullberg, B. 1988. Evolution of gregariousness in aposematic butterfly larvae: a phylogenetic analysis. *Evolution* 42:293–305.

Simpson, G. G. 1953. *The Major Features of Evolution.* New York: Columbia University Press.

Sokal, R. R., and P. H. A. Sneath. 1963. *Numerical Taxonomy.* San Francisco: W. H. Freeman.

Swofford, D. L. 1991. *Phylogenetic Analysis Using*

Parsimony (PAUP), Version 3.0s. Champaign: Illinois Natural History Survey.

Tinbergen, N. 1952. Derived activities: their causation, biological significance, origin and emancipation during evolution. *Quarterly Review of Biology* 27:1–32.

Tinbergen, N., G. J. Broekhuysen, F. Feekes, J. C. W. Houghton, H. Kruuk, and E. Szulc. 1962. Egg shell removal by the black-headed gull, *Larus ridibundus* L.; a behavior component of camouflage. *Behaviour* 19:74–117.

Titus, T. A., and A. Larson. 1995. A molecular phylogenetic perspective on the evolutionary radiation of the salamander family Salmandridae. *Systematic Biology* 4:125:51.

Trivers, R. L. 1972. Parental investment and sexual selection. Pp. 136–79 in *Sexual Selection and the Descent of Man 1871–1971*, B. Campbell, ed. London: Heinemann.

van Tets, G. F. 1965. A comparative study of some social communication patterns in the Pelecaniformes. *Ornithological Monographs* 2:1–88.

Verner, J. 1964. Evolution of polygyny in the long-billed marsh wren. *Evolution* 18:252–61.

Verner, J., and M. F. Willson. 1966. The influence of habitat on mating systems of North American passerine birds. *Ecology* 47:143–47.

Wiley, R. H. 1991. Lekking in birds and mammals: behavioral and evolutionary issues. *Advances in the Study of Behavior* 20:201–91.

Williams, G. C. 1966. *Adaptation and Natural Selection*. Princeton: Princeton University Press.

Wright, S. 1969. *Evolution and the Genetics of Populations*. Vol. 2, *The Theory of Gene Frequencies*. Chicago: University of Chicago Press.

Wynne-Edwards, V. C. 1962. *Animal Disperson in Relation to Social Behavior*. Edinburgh: Oliver and Boyd.

Comparative Studies of the motor patterns
of Anatinae (1941)

I Introduction and aims

In zoological systematics, more than in any other branch of biological research, the success of the investigator is dependent upon a 'feel' for the subject matter. This property can be acquired, but it cannot be taught. In the introduction to his contribution (*Die Vögel*) to Bronn's *Klassen und Ordnungen des Tierreiches*, Gadow set out a neat theoretical experiment in the form of a 'thirty character classification'. He selected thirty generally applied characters of undoubted taxonomic importance and proceeded to compile a tabular classification of groups of birds based purely on the presence or aʋ⹀⹀ce of each character. The classification which emerged – although exhibiting extensive agreement in respect to some groups – in some sections showed amazingly crude departures from the 'obvious' and generally employed classification corresponding to the genuine relationships between the birds involved. Primarily, this is to be explained on the basis that so-called 'systematic intuition' is dependent upon subconscious evaluation of a *much larger number of characters*. These characters, which are extremely elusive on the conscious plane, are woven into the overall impression which an animal group makes upon the investigator. Such unanalysed complex qualities incorporate individual characters of the finest degree, which cannot be extracted from the overall impression even though they have a qualitative influence on the latter. This fact, which is self-evident to perceptual and Gestalt psychologists,[2] must be considered if one wishes to analyse 'systematic intuition' and to determine the reasons for one's own judgements regarding the degree of taxonomic affinity of different animal species.

The inadequacy of a classificatory system based on a limited number of specific, pre-determined characters is not only based on the fact that the *number* selected is too small. Far greater difficulties are presented by the fact that within the different subdivisions of a fairly wide systematic unit a given character *will by no means carry the same weight throughout.*

14

The rate of differentiation, the variability of a given character can differ even between two closely related species. The assumption that a specific character (e.g. the absence of the fifth pinion or the form of the furcula) exhibits the same taxonomic relevance in all of the Orders and Families of the entire Class Aves is wrong from the outset. The weight which must be attributed to a given character as an indicator of the degree of phylogenetic relationship must be determined in each individual case on the basis of its behaviour relative to the other characters of the group investigated. The statement that a given character is 'conservative' or 'variable' can only apply to a restricted selection of closely related species. This applies not only to fine systematic characters of restricted distribution but also, in many cases, to generally distributed characters. Even the characters of ontogeny (e.g. the juvenile plumage of many bird groups), which are usually very conservative, can be affected by caenogenetic changes within a narrowly defined taxonomic unit to such a degree that it would lead to great confusion if one were to attribute to these characters in the unit concerned the same taxonomic value as that applying to the same characters in general. For example, one can imagine the results if the details of the juvenile markings – which are so extremely conservative and taxonomically important for the Anatidae – were to be given the same importance in the classification of the rails, in which caenogenetic differentiation processes serving a releasing function overlay the basic features. Thus, the *number* of characters known to the investigator of phylogenetic relationships is not only operative through the determination of certain group-specific 'complex qualities'. Over and above this, the *relative taxonomic weight of an individual character* can be determined with greater exactitude as the number of characters known to the investigator (consciously or subconsciously) increases. All of this leads to 'systematic intuition' without necessitating analysis of this property by the possessor. However, the outcome of this property only becomes scientific when this analysis has been successfully conducted.

Even from this brief summary of the basis of conscious or intuitive estimation of phylogenetic relationships, it is possible to conclude that the most reliable assessment is produced not by investigators acquainted with *one* organ in all of its various manifestations within a large systematic unit (a goal often set by comparative anatomists), but by those who can survey a *small* systematic unit in consideration of a *maximum* number of characters. The opportunity for drawing reliable taxonomic conclusions increases in geometric (and not arithmetic) progression with the number of recognized characters, since each additional character investigated in all members of the group involved leads to greater accuracy of our estimation of the previously investigated characters. It is at once clear why zoologists who are acquainted (as zoo attendants or as amateur naturalists) with a large number of living representatives of a given group of animals manage to achieve outstanding 'systematic intuition' and critical illumination of phylogenetic relationships. Heinroth's Anatid studies and the Equid studies performed by Antonius provide two good examples. A zoo attendant who is acquainted with the anatomy, and possibly palaeontology, of a large number of representatives of an animal group is obviously provided with a significant advantage over any pure museum systematist through knowledge of *the characters of species-specific innate behaviour*. This undeniable fact is of significance and value not only for the systematist, but also – most particularly – for the psychologist. It has been emphasized for some time (since Wundt) that the theoretical approach of comparative phylogeny is just as indispensable in psychology and ethology for the understanding of the existing human and animal structures as it is in morphology. Even in the psychological field, living organisms are phylogenetically derived entities whose specific origin and form can only be interpreted in the light of their phylogentic history. Thus, comparative psychology (unfortunately little more than a programme as it now stands) is presented with the pressing primary task of performing purely descriptive behavioural research on a suitable group of animals, in order to incorporate the characters thus recognized, together with the maximum number of assessible morphological characters, into a fine classification of the group. Above all, correspondence with the related morphological characters would permit solid defence of the application of the phylogenetic *homology concept* to species-specific behaviour patterns, such that the pre-conditions for *comparative* psychology in the narrowest sense would be provided. Only such fine systematic studies of an exactly investigated group of animals is able to provide us with information about the manner in which phylogenetic alterations of instinctive motor patterns, taxes, innate schemata, and – later on – of all psychological mechanisms, occur. The basic importance of such information – which is entirely lacking at present – need not be emphasized here. The pathway along which research must proceed lies clearly before us, though it is extremely tedious and hazardous. This present paper represents a far from complete attempt to provide a fine systematic study of a single group, including behavioural characters, as a contribution to the described task.

II   Technical information

The task of establishing initially purely descriptive behavioural catalogues (ethograms) of a large number of animal species makes great demands upon the observational capacity of the investigator. In order to reach even an approximation to satisfactory catalogues, the investigator must live with the animals day after day, year after year. This alone would seem to exclude the possibility of acquiring the necessary information from field studies. Even after keeping the species studied for a period of years, conscientiously recording every observation in a diary, individual items of information which would be extremely important for comparison are still missing when the work is concluded. This paper unfortunately shows this only too well. The animal group investigated must therefore be one which can be maintained well in captivity. In addition, the group must contain a large number of comparable species and genera exhibiting the maximum possible range of degrees of relationship. The individual species themselves should exhibit a large complement of species-specific behaviour patterns, which exhibit interspecific resemblances and are yet distinct enough to present a test case for the application of the homology concept. All of these requirements are fulfilled in an ideal manner by two groups of animals, both of which I have employed as objects for the research purposes discussed here: the *Anatinae* among birds and the *Cichlidae* among fish. This paper is concerned with the former. The Anatinae have been subjected to particularly exact investigation by Heinroth, Delacour, von Boetticher and others. In addition, they present a special advantage in comparative phylogenetic studies in that *interspecific hybrids* can be obtained with particular ease. In very many cases, these hybrids are actually fertile so that they can be employed for investigation of the inheritance of species-specific behaviour patterns. This represents a fertile area for synthesising phylogeny and genetics. In many cases, hybrids can lead to phylogenetic conclusions regarding species-specific behaviour patterns because of a special peculiarity whereby they exhibit, morphologically and behaviourally, a condition which is distinct from that of the two parental species (rather than intermediate) and is in fact *more primitive*. The hybrids have a further significance in that the degree of fertility can be employed as a measure of the degree of relationship between the parental species, as Poll has shown.

I have been able to investigate the following species within the Family Anatinae and the two adjacent Families Cairininae and Casarcinae: ANATINI: mallard, *Anas platyrhynchos* L.; Meller's duck, *Anas melleri* Scl.; Japanese spot-billed duck, *Anas zonorhyncha zonorhyncha* Swinhoe; Indian spot-billed duck, *Anas z. poecilorhyncha* Forster; pintail, *Dafila acuta* L.; South-American pintail, *Dafila spinicauda* Viellot; Bahama pintail, *Poecilonetta bahamensis* L.; Red-billed pintail, *Poecilonetta erythrorhyncha* Gm.; common teal, *Nettion crecca* L.; South-American teal, *Nettion flavirostre* Viellot; gadwall, *Chaulelasmus strepera*; wigeon, *Mareca penelope*; South-American wigeon, *Mareca sibilatrix* Poepp.; garganey, *Querquedula querquedula* L.; shoveller, *Spatula clypeata* L. – CAIRININI : Muscovy duck, *Cairina moschata* L.; American wood-duck, *Lampronessa sponsa* L.; mandarin duck, *Aix galericulata* L. – TADORNINI : shelduck, *Tadorna tadorna* L.; ruddy shelduck, *Casarca ferruginea*, Pallas; Egyptian goose, *Alopochen aegyptiacus* L. The hybrids will be discussed elsewhere.

III   General discussion of motor display patterns

1. THE TAXONOMIC RELEVANCE OF MOTOR DISPLAY PATTERNS

In 1898, Whitman clearly emphasized that instinctive behaviour patterns require the same time intervals for phylogenetic development as morphological structures, and later Heinroth published his Anatid studies (1910) and his paper on the systematic distribution of certain motor patterns among vertebrates (1930). Since then, the applicability of innate, species-specific behaviour patterns as taxonomic criteria has gradually become a basic feature of zoological systematics (Stresemann on *Aves* in Kükenthal's *Handbuch*). Several years ago, I myself drew attention to the fact that certain instinctive behaviour patterns are quite particularly suited for taxonomic considerations and thus naturally suited for the analysis of the phylogenetic origin of innate motor patterns. Such instinctive behaviour patterns are those whose adaptive function is the *transmission of stimuli which evoke a specific response from conspecifics*, i.e. patterns referred to as 'releasers' (Lorenz 1935). Motor patterns of this kind are particularly common among birds, since responses to perception of movements play a very important part in their behaviour. In other groups of animals, the major rôle is played by chemical, acoustic or tactile releasers, which are by no means as favourable for comparative and experimental investigation by man – also an optically oriented animal. But quite apart from technical reasons of perceptibility, the optically operative motor display patterns of birds are particularly fertile objects for research (as is also the case with teleost fish). In the first place, there is a great abundance of characteristic and conspicuous individual characters, which provide a good foundation for inter-specific and inter-group comparisons. This abundance of characters is closely associated with the signal function involved, since animal releasers – just like human signals – must be *unmistakable* if they are to fully perform their functions. I have discussed this subject in detail in a special paper (Lorenz 1935), in which I demonstrated that the main characteristic of an unambiguous, easily perceived signal is a combination of maximum *simplicity* with maximum general *improbability*. Understandably, both

(g) Fighting between drakes

The fighting behaviour of the drakes provides a particularly interesting chapter in the ethology of *Aix*. The drakes, like many male birds with extreme social courtship and extreme developments of the 'Prachtkleid' (display plumage), no longer exhibit serious fights among themselves. Side-by-side darting, which is so characteristic of the fighting behaviour of the Carolina wood-duck, and which even in this species tends towards symbolic display and ritualization, is only present as a *pure symbolic pattern* in *Aix*, though as such it plays a large part in social courtship. Heinroth quite rightly compares the behaviour pattern of the wildly darting drakes in this ceremony with that of a flock of whirligigs (*Gyrinus*).

(h) Post-copulatory play

This consists merely of non-specific gestures of arousal, such as 'burping', demonstrative body-shaking and aiming movements of the head.

XXI Summary

If one wishes to carry out individual systematic study of a group, it is necessary to free oneself once and for all from the impression that a linear arrangement of the forms involved could ever reflect the genuine phylogenetic relationships prevailing between them. This of course applies to the 'approximate series' outlined for the ducks discussed so far. All animals alive today represent living branch-tips of the 'phylogenetic tree' and can thus *ipso facto* not be descendants of one another. Therefore, a comparison of their characters produces an arrangement which, taking the metaphor of the phylogenetic tree, permits spatial representation of relationships like those existing between the individual branch-tips of a rounded box-tree or yew-tree. All of the tips are located in *one* surface – a temporal cross-section of the expansively growing stock. Just as we can only estimate the probable grouping of the individual tips on common branches and the height of union and separation on the stock when looking through the thick, opaque leafy shroud of the tree, even the best systematic arrangement can only permit us probable estimations of the genuine phylogenetic relationships of a group of animals.

I should now like to attempt to give a graphic representation, in the form of a tabular schema, of the products of 'systematic intuition', which has been defined as simultaneous *survey* of a maximum number of characters. This itself provides the correct estimation of the taxonomic relevance of individual characters (p. 15).

motor patterns of Anatinae 107

In this connection, it is first necessary to conduct some fundamental consideration. Similarity in a series of forms need by no means correspond to a series of evolutionary *levels*, however clearly the series may be indicated by the distribution of characters. One can imagine that a given stock might have given rise to a number of forms which are all of the same age and all equally differentiated away from the basic

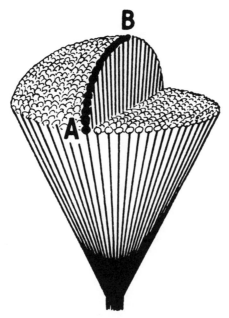

Fig. 51. *Schema of a series of similarities in recent animal species, where adjacent individuals are not linked by close phylogenetic relationship. Following loss of a proportion of the descendent lines, the remaining forms located on the line A–B can give the false impression of a phylogenetic series*

stock. This is symbolized in Fig. 51 as a kind of shaving brush. It can further be imagined that, as schematically shown in Fig. 52, a proportion of the hairs of the brush may have fallen out such that the remaining hairs are arranged in an approximate fan. The tips then represent a ladder which apparently speaks convincingly for the derivation of the forms involved 'from one another', particularly if the degree of differentiation at one edge of the fan is less than that at the other. Without doubt, it has already occurred very often that the terminal points of such 'phylogenetic fans' have been confused with phylogenetic series. Unfortunately, this has often pressed welcome weapons into the hands of opponents of the theory of evolution. On the other hand, we should not swing to the opposite extreme of this precipitate construction

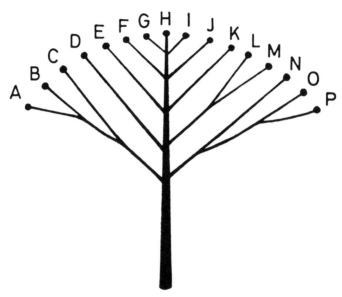

Fig. 52. *Schema of a similarity series of recent animal species based upon genuine phylogenetic relationship. Any two neighbouring forms in the series A–B owe their similarities to the common evolutionary pathway which they shared*

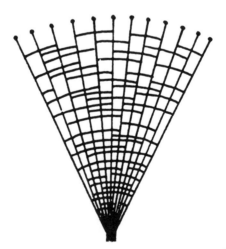

Fig. 53. *Schema of the expected character distribution among bush-like divergent and unbranching descendent lines. The cross-links represent common characters. Since similarities and differences can only be explained through greater or lesser divergence, the distributions of most characters overlap*

of 'phylogenetic' series and generalize the opinion that all similarity series among recent organisms can be explained through the principle of a fan-like arrangement of descendent lines. Without a trace of doubt, there are many cases where there has not only been monophyletic evolution of entire large animal stocks, with division into many individual forms occurring only at a later time, but also where the further evolution of the individual forms – at least in respect to individual characters – has proceeded at such different rates that similarity series rather like those illustrated in Fig. 53 have arisen. However, even when considering these genuinely phylogenetically graduated similarities *it must not be forgotten for an instant that the term 'primitive' can only be applied to one or more characters of a recent animal species and not to the entire animal.* Even *Sphenodon* or *Ornithorhynchus* are not 'primitive animals'. The fact that some, or even very many, characters of such a species are certainly primitive in phylogenetic terms by no means permits us to assume that all other characters are primitive too. Arrest in the evolution of one character tells us *nothing* about the further evolution of other characters.

The 'intuition' of a professional systematist mentioned at the beginning is generally quite adequate for distinction between similarity series based upon common origin of the type just described and those which originate from the previously discussed phenomenon of a fan-shaped arrangeme of descendent lines. However, in order to obtain an objectively tenable criterion for this distinction, I suggest the following probabilistic consideration: If one assumes that all representatives of an animal group (as symbolized in Fig. 51) are derived from one source divergently and without any close mutual relationships, it would be expected that the similarities between the characters decisive for the arrangement of the individual descendent lines would be fairly evenly distributed throughout the entire 'shaving brush'. If, for the sake of simpler graphic representation, we take a longitudinal section through the brush (i.e. a number of fan-like diverging descendent lines), the similarities linking each form to its systematic neighbour would necessarily be homogeneously distributed through the whole fan of lines. Above all, there would be links to both sides (i.e. on all sides in the spatial brush schema) equally binding species to species. If common characters are represented as cross-links and the more general, older and widely distributed common characters are placed towards the origin, whilst the others are placed more towards the periphery following a lesser distribution, greater specialization and thus more recent origin, we obtain in the ideal case of radiating, divergent speciation the arrangement represented in Fig. 53.

If it is now assumed that not every species in the group of forms investigated has proceeded along an individual evolutionary pathway from the origin, but that sub-units of the group have diverged from a common origin only at a later date, then it is to be expected that one or other character would be common to forms belonging to the sub-group concerned and exclusive to them. These characters would be products of the evolution of the common phylogenetic sub-stock in isolation from neighbouring forms. If two such sub-stocks separate from one another

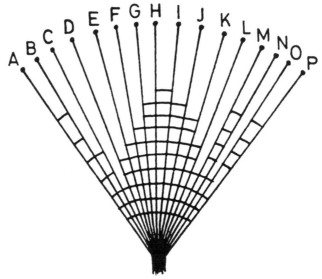

Fig. 54. *Schema of the expected character distribution with tree-like divergence of descendent lines. Since the connecting characters result from common evolutionary pathways, they are (excluding convergence) distributed according to common origin and do not overlap*

at a relatively early stage, we would expect that they would only be linked by very old characters common to larger group categories. Overlapping in the distribution of characters, as seen in Fig. 53, would not be expected for obvious reasons, as long as the possibility of convergence is excluded. Fig. 54 represents a character distribution of this type.

We shall now attempt to graphically represent the group of the Anatinae in the manner described without any prior assumptions and employing as many systematically applicable characters as possible. This will provide us with a judgement as to how far the representatives of the group can be classified into genuine phylogenetic groups and how far speciation has followed the type of straight-line divergent, brush-like evolution represented in Fig. 51. Since the brush of descendent lines can only be represented spatially, several plane projections would have to be made. Anybody who attaches great weight to illustrative techniques can draw up such projections and stick them together to give the requisite spatial pattern. I must admit that I myself used a bundle of stiff wires for arranging the species, united into sub-groups with thin wires representing 'common characters'. In order to summarize the character comparison conducted in this paper to give a single table, I am compelled to indicate specifically species which do *not* exhibit common characters indicated by crosslines. This is done with a heavy black dot. With species for which there is reason to believe that the lack of a character is not primitive but due to secondary loss, this is indicated by a cross at the point of intersection of the character cross-link and the descendent line. As can clearly be seen from the table, grouping into common stocks becomes more probable as we move down towards the origin, whereas with very many more recent characters obvious overlapping of character distribution occurs as in the schema of Fig. 53 (e.g. examine the distribution of the gruntwhistle, head-up-tail-up and hind-head display).

The few *morphological* characters distributed in the table are intended to show how similar their distribution is in many cases to that of the *innate behaviour patterns*. After filling in the gaps which are, above all, evident in the *list of species investigated*, I plan to set out a much larger table established on the same principle, including all possible morphological and behavioural characters as well as the fertility of the hybrids. However, before anything else the publication of this table must be preceded by publication of the comparative studies which Heinroth has carried out on the bony tracheal drums of the drakes, which are extremely rich in comparable characters. Even this provisional, incomplete table shows clearly the applicability of the phylogenetic homology concept to characters of innate behaviour. This fact, the demonstration of which was a major aim of my investigation, is of the greatest significance for *comparative psychology*.

Table: see overleaf

112   *Comparative studies of the*

Table

The *vertical* lines represent species; the horizontal lines characters common among them. A *cross* indicates the absence of a character in a species crossed at the point concerned by a character cross-line. A *circle* indicates special emphasis and differentiation of the character. A *question-mark* indicates the author's uncertainty.

SPECIES LIST

1. *Cairina moschata,* Muscovy duck
2. *Lampronessa sponsa,* Carolina wood-duck
3. *Aix galericulata,* mandarin duck
4. *Mareca sibilatrix,* Chiloë wigeon
5. *Mareca penelope,* wigeon
6. *Chaulelasmus streptera,* gadwall
7. *Nettion crecca,* teal
8. *Nettion flavirostre,* South American teal
9. *Virago castanea,* chestnut-breasted teal
10. *Anas* as genus including mallard, spot-billed duck, Meller's duck, etc.
11. *Dafila spinicauda,* South American pintail
12. *Dafila acuta,* pintail
13. *Poecilonetta bahamensis,* Bahama duck
14. *Poecilonetta*(?) *erythrorhyncha,* red-billed duck
15. *Querquedula querquedula,* garganey
16. *Spatula clypeata,* shoveller
17. *Tadorna tadorna,* shelduck
18. *Casarca ferruginea,* Ruddy shelduck
19. *Anser* as genus
20. *Branta* as genus

motor patterns of Anatinae   113

CHARACTERS

Mlp	monosyllabic 'lost-piping'	Bgsp	black-gold-green teal speculum
Dd	display drinking	Trc	chin-raising reminiscent of the triumph ceremony
Bdr	bony drum on the drake's trachea	Ibr	isolated bridling not coupled to head-up-tail-up
Adpl	Anatine duckling plumage	Kr	'Krick'-whistle
Wsp	wing speculum	Kd	'Koo-dick' of the true teals
Sbl	Sieve bill with horny lamellae	Pc	post-copulatory play with bridling and nod-swimming
Ddsc	disyllabic duckling social contact call	Ns	nod-swimming by the female
I	incitement by the female	Gg	*Geeeeegeeeee*-call of the true pintail drakes
Bs	body-shaking as a courtship or demonstrative gesture	Px	Pintail-like extension of the median tail-feathers
Ahm	aiming head-movements as a mating prelude	Rc	R-calls of the female in incitement and as social contact call
Sp	sham-preening of the drake, performed behind the wings	Iar	incitement with anterior of body raised
Scd	Social courtship of the drakes	Gt	graduated tail
B	'burping'	Bm	bill markings with spot and light-coloured sides
Lhm	lateral head movement of the inciting female	Dlw	drake lacks whistle
Spf	specific feather specializations serving sham-preening	Lsf	lancet-shaped shoulder feathers
Ibs	introductory body-shaking	Bws	blue wing secondaries
P	pumping as prelude to mating	Pi	pumping as incitement
Dc	decrescendo call of the female	Dw	drake whistle
Br	Bridling	Bwd	black-and-white duckling plumage
Cr	chin-raising		
Hhd	hind-head display of the drake	Psc	polysyllabic gosling social contact call of Anserinae
Gw	grunt-whistle	Udp	uniform duckling plumage
Dum	down-up movement	Nmp	neck-dipping as mating prelude
Hutu	head-up-tail-up		
Ssp	speculum same in both sexes		
Wm	black-and-white and red-brown wing markings of Casarcinae		

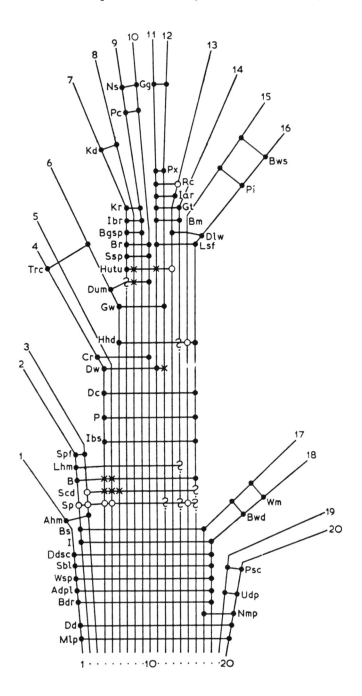

THE EVOLUTION OF PREY-CARRYING MECHANISMS IN WASPS[1]

Howard E. Evans

Museum of Comparative Zoology, Cambridge, Mass.

Received January 25, 1962

The elaborate behavior patterns of wasps provide almost unlimited possibilities for comparative ethological studies. That few such studies have actually been made is in large part a consequence of the fact that the behavior of very few species has been described in adequate detail. The available descriptions are widely scattered in the literature and often fragmentary and poorly documented. Observers have often been unaware of the importance of certain components of the behavior, just as taxonomists often overlook structural details that later prove essential in classification. For this reason, there may be justification for studies which focus attention upon one particular facet of behavior and attempt to trace its modifications in various taxa. However preliminary, attempts to discern trends in the evolution of some aspects of behavior may provide direction for further studies. In an early and now classic paper, Ducke (1913) outlined the evolution of nest building in the Vespidae. Wheeler (1928) and others have considered the matter of the origin of sociality among wasps. Leclercq (1954) discussed the probable phylogeny of the Sphecidae as suggested by some aspects of structure and nesting behavior. The present paper is an attempt to outline the different ways in which wasps carry their prey to the nest and to draw certain conclusions regarding the probable phylogeny of prey-carrying mechanisms. Such mechanisms may or may not have obvious morphological components; in some cases, knowledge of the behavior may help explain the significance of structures which might otherwise be difficult to understand.

Such an undertaking does not seem premature at this time for several reasons. The behavior patterns involved are simple, unambiguous, and subject to little intraspecific variation: indeed, they are often constant throughout major taxa. One needs to maintain the usual healthy scepticism of published observations, but in fact so much has been published on this aspect of behavior that one can usually identify reports which are inconsistent with the general picture. Information on this subject is so widely scattered in the literature that it is impractical to cite all original references in this review. Rather, I shall refer mainly to three general sources, each of which provides fuller documentation as well as bibliographic references. I shall make frequent reference to my own papers, since these papers are recent and pay particular attention to prey carriage. I shall also refer many times to two comprehensive studies of the behavior of solitary wasps. These are Iwata's "Comparative Studies on the Habits of Solitary Wasps" (1942) and Olberg's "Das Verhalten der Solitären Wespen Mitteleuropas" (1959). Although Olberg's book deals with a limited fauna, his remarkable photographs provide irrefutable documentation of the method of prey carriage in certain species. Iwata has reviewed the world literature and has presented an outline of types of prey carriage. He recognizes twelve types, arranged under three major headings. Certain of Iwata's types seem to me poorly documented and possibly incorrect, and his three major groupings seem to me in need of re-evaluation. Nevertheless Iwata's paper is an important pioneering work in this field.

THE ANCESTRY OF WASPS

There has been no recent reconsideration of the systematics of the order Hymenoptera as a whole. The classification and

[1] These studies have been supported by grants from the National Science Foundation, nos. G1794 and G17497.

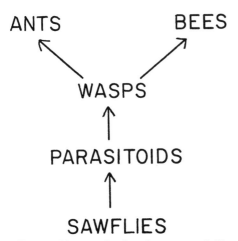

Fig. 1. Diagram showing the most probable relationships of the major groups of Hymenoptera.

arrangement of the families and super-families presented in Imms' textbook, as revised by Richards and Davies (1957), is accepted here. Wheeler (1928), although employing an older classification, discussed convincingly some aspects of the evolution of the Hymenoptera. There has evidently been much extinction in the past history of the Hymenoptera, leaving us with a diverse array of forms which may never be classified to everyone's satisfaction. Nevertheless the broad pattern of evolution seems reasonably clear (fig. 1). The wasps occupy a central position in the order, having evolved originally from the sawflies (Symphyta) and later from a common ancestor with an early parasitoid. The ants are generally regarded as having arisen from an unknown stock of early scolioid wasps, the bees from a now extinct stock of sphecoid wasps (see also fig. 6).

The sawflies, parasitoids, and wasps appear in the fossil record in that order. Clearly the wasps cannot have evolved from an extant (or known fossil) group of parasitoids, since all parasitoids exhibit structural simplifications and modifications not found in wasps. Nevertheless, it seems a safe assumption that the ancestral wasps behaved very much as do some of the more generalized Ichneumonoidea today. That

is, the female laid her egg directly on the host insect *in situ*, the larva developing upon the host while the latter continued its feeding, being killed only when the parasitoid had nearly completed its development. Quieting of the host by paralyzing substances produced by accessory glands and injected via the ovipositor may have at first functioned to permit deposition of the egg on a more specific part of the body of the host, as well as to permit the female to feed on the body fluids of the host. Temporary paralysis of the host occurs in some Ichneumonoidea and in some primitive, non-nest-building wasps, such as many Tiphiidae. With the development of the first simple nests, paralysis of the prey served to permit safe carriage to the nest and to prevent escape of the prey from the nest. Selection therefore favored more profound and lasting paralysis of the prey. Once set in motion, nest building and prey carriage both tended to become more complex and efficient, but independently of one another. That is, complex nests sometimes evolved in wasps exhibiting simple types of prey carriage—the social Vespidae, for example—and advanced types of prey carriage sometimes evolved in wasps making very simple nests—the solitary wasp *Oxybelus*, for example. These two facets of behavior bear no correlation whatever except that they necessarily had their inception simultaneously. These ideas are summarized in fig. 2.

It is important to remember that primitive wasps, having been derived from parasitoids, utilize a single host specimen per offspring. In the parasitoids there has been an important trend toward what is called "multiple parasitism," which means simply that several offspring develop at the expense of a single host. This trend reaches its ultimate in polyembryonic forms such at *Litomastix* (Encyrtidae), where as many as 3,000 parasitoids may develop in a single *Phytometra* caterpillar. Needless to say, such parasitoids are very much smaller than the host species.

In wasps (other than Bethyloidea), the trend has been in the opposite direction,

470   HOWARD E. EVANS

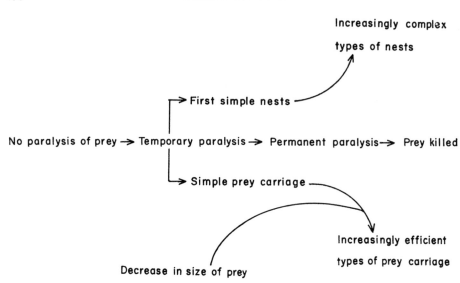

FIG. 2. Schematic representation of some of the factors involved in the evolution of prey-carrying mechanisms.

toward the utilization of several or many host specimens to feed a single larva (fig. 3). Thus the majority of wasps (except in primitive families such as the Tiphiidae, Bethylidae, and Pompilidae) are larger than the arthropods they prey upon. The highest number of prey recorded for a single nest-cell of a wasp is 104, for an aphid-storing species of *Passaloecus* (Sphecidae) (Iwata, 1942). Obviously, the smaller the prey the less of a burden it provides for the wasp carrying it, and it is not surprising that more advanced methods of prey carriage occur in wasps utilizing several prey per nest-cell. To say that in primitive wasps and parasitoids the hymenopteron and its host are equal in size is, of course, not quite correct, even as an approximation. Since the host inevitably contains certain parts which are not eaten or not digested, it follows that it has to be, on the average, larger than the predator. In other words, it is not quite correct to say that $P = H$ in fig. 3; using actual figures compiled by Iwata, H may weigh 0.9 to $8.0 \times P$. To a wasp dragging a paralyzed arthropod several times its own

weight, the mere struggle with the force of gravity may preclude the attainment of any appreciable speed or any noteworthy protection against factors which may injure or destroy the prey during transport.

Thus there is an important correlation between prey size and type of prey transport. There is, however, no particular correlation between the taxon of the prey and the type of prey transport, as will be discussed further in a later section. It is true that primitive wasps tend to prey upon phylogenetically earlier types of arthropods (e.g., Tiphiidae on beetle larvae, Pompilidae on spiders, Ampulicidae and Sphecini on Orthoptera). Predatism on such things as adult flies and bees is confined to more advanced groups of Sphecidae, which may also have evolved more advanced types of prey carriage. However, one finds little to support the contention that evolutionary changes in the kind of prey utilized have been closely accompanied by or dependent upon changes in type of prey carriage.

The most important factor to be considered with reference to prey carriage is the

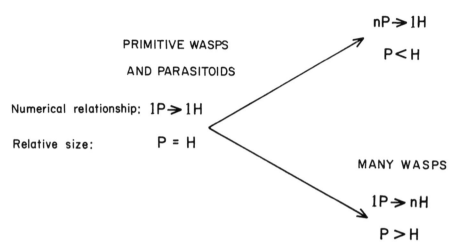

Fig. 3. Trends in the evolution of host size and number. P, predator or parasitoid; H, host.

manner in which the prey is grasped by the wasp. In general terms, the primitive condition is for the wasp to hold the prey in its mandibles; at a more advanced level it is held well back beneath the body by the legs, and finally it is held at and by the posterior extremity of the body, leaving the mandibles and legs free for other functions. The many different methods of prey carriage may be grouped under these three progressively more advanced types, which may be termed respectively mandibular, pedal, and abdominal mechanisms. In the following paragraphs I shall consider the various types and subtypes in turn.

MANDIBULAR MECHANISMS:
TYPE ONE (M1)

Certain members of the Scolioidea, finding the subterranean grubs on which they prey lying on top of the ground or in other unsuitable places, simply grasp the grub with their mandibles on any convenient part of the body and drag it backwards into a hole. Behavior of this type can be observed, for example, in the tiphiid genus *Methocha*. The female *Methocha* attacks the larvae of tiger beetles in their vertical burrows in the soil, stings them, oviposits,

then leaves the paralyzed grub in its own burrow. When the wasp encounters a tiger beetle outside its burrow, as sometimes occurs, she stings it, grasps it with her mandibles, and proceeds backwards until a suitable hole is located. The prey is generally much heavier than the wasp, and transport is slow and fraught with difficulties. I have observed this behavior twice in the North American *M. stygia*. Iwata records a *Methocha* dragging a larva twenty times its own weight. In most Scolioidea prey transport is facultative, if it occurs at all.

Wasps in which prey carriage is a fixed part of the behavior, that is, those in which there is a definite nest, however crude it may be, typically grasp the prey on some specific part of its body. The vast majority of spider wasps (Pompilidae) grasp the spider by the base of the hind legs and proceed backward; in a few cases the mouthparts or spinnerets are grasped (for specific examples, see Evans, 1953, and Evans and Yoshimoto, 1962). Similar prey transport is exhibited by members of the small, primitive sphecoid family Ampulicidae. As reported by Williams (1919a) and others, these wasps move the cockroaches on which they prey by simply grasping one of the antennae and proceeding backward. The practice of dragging the prey backward over the substrate also occurs in the family Bethylidae, but I am not aware that it occurs in the higher families of wasps, the Sphecidae and Vespidae.

This type of prey transport is essentially "blind," for the important sense organs of the wasp are on the end of the wasp which is headed away from the direction of travel. Many wasps deposit their prey on the ground for varying lengths of time while they either rest, groom, or explore ahead. When the prey is left alone, even for short periods, it is subject to attack by ants, tiger beetles, or wasps of the same or related species (I have observed, among Pompilidae, attacks on the prey from all these sources; see also examples in Evans and Yoshimoto, 1962). In describing this type of prey transport, one finds various authors using such adjectives as "awkward" and "inefficient." It is hard to think of such wasps as having a "prey-carrying mechanism"; rather they are characterized by the lack of any real "mechanism," by the lack of any structural or behavioral specializations which would shorten the time required to get the prey to the nest or reduce the hazards en route.

Some Pompilidae exhibit certain variations on this general theme; some of these variations appear to represent transitions to more advanced types of prey transport. In the genus Dipogon, the wasps typically walk sideways when dragging a spider, in this way perhaps making better use of their eyes and antennae than they would if walking backward (which they do when ascending a branch or tree trunk). There are also certain spider wasps that normally proceed backward but that now and then— especially when handling a spider of small size—turn around and walk forward (for example, Anoplius apiculatus, as reported by Evans, Lin, and Yoshimoto, 1953). Clearly such wasps may be thought of as representing a transition to the next type of prey transport to be discussed, which is forward transport. There are also some

spider wasps which, although normally proceeding backward over the ground, are able in some circumstances to take flight with the spider dangling from the mandibles. Although they of course fly forward, when they land they immediately turn and proceed backward. Particularly good examples of this are to be found in the related genera Sericopompilus, Episyron, and Poecilopompilus. It is not uncommon for wasps of these genera to climb backward up an herb or tree and then take flight, gradually losing altitude but gaining much distance by the procedure. That few or no Pompilidae have made the full transition to aerial transport is a consequence of the fact that all Pompilidae use a single spider per cell; therefore they must take large spiders, which can rarely be lifted from the substrate or flown great distances.

MANDIBULAR MECHANISMS:
TYPE TWO (M2)

Carriage of the prey forward is an obvious improvement over dragging it backward, and the step to forward transport was made by several stocks of wasps independently. The difficulties in forward transport are several: the prey may be difficult to lift for extended periods of time, it may block the view of the wasp, or it may impede walking, especially if the appendages of the prey are long. Consequently most wasps falling in this category exhibit structural or behavioral modifications, usually of a rather simple sort. One notes particularly the long legs and high stance of many wasps which carry their prey forward over the ground.[2] Spider wasps which drag the prey backward characteristically assume a very low stance, the

---

[2] An apparently unique method of prey transport has been described for the bethylid wasp, Epyris extraneus Bridwell. This wasp is depressed and relatively short-legged, but carries a beetle larva larger than itself forward over the ground by grasping a palpus with the mandibles and "slinging it over her back." The prey hides the wasp from view and makes it appear as if the paralyzed beetle larva were "making headway under its own steam" (Williams, 1919b).

body close to the ground and the legs spread widely. Species in which forward transport is fixed, such as the Palaearctic *Pompilus plumbeus*, hold the body far above the ground, walking on the extremities of their legs, so that they can partially straddle the spider (this is shown very well in the photographs in Olberg, 1959, pp. 198, 229). *P. plumbeus* holds the spider by the base of the hind legs, its anterior end forward; apparently the legs of the spider do not provide a serious impediment to the wasp. The North American spider wasp *Priocnessus nebulosus* also proceeds forward, straddling its spider, but in this case the spider is held by the spinnerets.

One major stock of Pompilidae, the Auplopodini, is characterized by forward prey carriage in which the spider is grasped by the spinnerets (less commonly the mouthparts), the wasp straddling the spider. These wasps amputate the legs of the spider shortly after it is captured. This remarkable behavior may have evolved from a simple malaxation of the prey for feeding purposes, as occurs in many wasps; removal of the legs may have been selected for because it improved the efficiency of prey transport. The Auplopodini may have evolved from a wasp not unlike *Priocnessus*, which exhibits essentially the same type of prey carriage but does not amputate the spider's legs.

Some Auplopodini fly with the prey to a considerable extent, but generally as a series of short flights, often starting from some high perch. Certain other Pompilidae fly with their prey with the aid of a "prop": they fly close to the substrate, dragging the prey over the substrate, which bears much of the weight of the spider. The best known of these is *Anoplius depressipes*, a wasp that has attained a considerable notoriety for its practice of towing large *Dolomedes* spiders over the surface of quiet waters (Evans and Yoshimoto, 1962).

With these examples we may leave the family Pompilidae, few if any members of which exhibit sustained flight with the prey or grasp the prey other than with the mandibles. Some of the more primitive genera of true digger wasps (Sphecidae) carry their prey forward over the ground in a manner similar to that of some Pompilidae. The genus *Priononyx* provides an excellent example (Evans, 1958a). These wasps prey on short-horned grasshoppers and use only one hopper per cell. The grasshoppers are often much larger than the wasp, but the wasp carries them rapidly over the ground to the nest, straddling them and holding their antennae with the mandibles. The wasps often vibrate their wings rapidly and thereby gain additional momentum, but they do not ordinarily lift the prey off the ground. These wasps hold their elongate bodies far above the ground, the femora extending out laterally, the tibiae almost perpendicular to them, thus forming a large space beneath the body to accommodate the prey. One of the diagnostic features of the genus *Priononyx* is the notch on the apical margin of the clypeus of the female. The thick antennae of the grasshopper fit into this notch and are supported beneath by the mandibles. An exceedingly tight grasp is doubtless a necessity for moving large grasshoppers effectively.

The related genus *Palmodes* is very similar in its behavior, but wasps of this genus prey upon long-horned grasshoppers, the antennae of which are much more slender and flexible. The clypeus of the female *Palmodes* lacks a notch. Most species of *Palmodes* use a single hopper per cell and therefore use very large hoppers which, however, are propelled over the ground very rapidly with much buzzing of the wings. LaRivers (1945) found that *P. laeviventris*, a predator on the Mormon cricket, used two somewhat smaller hoppers per cell in over half the nests he dug. He also noted some use of the forelegs in supporting the prey during transport. The related genus *Sphex* characteristically uses two or more long-horned grasshoppers per cell; these are carried in flight, held with the mandibles in the usual way but also supported by all the legs. Thus in this one complex of genera (the tribe Sphecini) one observes a change in prey carriage closely correlated with decreasing relative size of the prey.

Excellent transitions are also exhibited within the large genus *Ammophila* (tribe Ammophilini, subfamily Sphecinae). *A. procera* uses a single large caterpillar per cell; the caterpillar is carried over the ground venter-up, head-forward, the wasp grasping it with her mandibles a short distance behind the thoracic legs and also grasping it somewhat farther back with the front legs. This species, when handling a very large caterpillar, moves its long abdomen up and down rhythmically as it moves along, presumably gaining some mechanical advantage thereby. Some species of *Ammophila* use two or three caterpillars per cell, and these wasps take smaller caterpillars which can be carried over the ground more rapidly and without the up-and-down movements of the abdomen. Some species carry the prey short distances in flight, and species such as *harti*, which use many small caterpillars per cell, carry the prey considerable distances in flight. (For further discussion of *Ammophila*, see Evans, 1959a; also Olberg, 1959, the latter with excellent photographs of prey carriage in two species.)

In the sphecid subfamily Larrinae many forms carry their orthopterous prey forward over the ground: for example, *Larropsis*, *Motes*, *Lyroda*, *Dinetus*, *Tachysphex*, and other genera (Iwata, 1942); Evans, 1958c; Olberg, 1959). In each case one or both antennae of the hopper are held in the wasp's mandibles, and in many cases the front legs of the wasp also embrace the thorax of the prey. The use of the front legs is shown clearly in Olberg's photographs of *Dinetus pictus* (p. 271) and *Tachysphex helveticus* (p. 259). Most of these wasps are capable of carrying the prey short distances in flight (all use more than one hopper per cell).

MANDIBULAR MECHANISMS:
TYPE THREE (M3)

This type includes species which exhibit full aerial transport. Clearly there is no sharp distinction between this type and those members of the preceding type which fly with the prey in a series of short hops. In several genera one finds species which prey upon large insects and fly not at all, others which prey upon slightly smaller insects and fly for short stretches, and still others which take still smaller prey and fly all or most of the way to the nest: *Ammophila* and *Tachysphex* provide good examples. There are also many genera of Sphecidae which fall entirely within this category; most of these either prey on very small insects (for example, *Pemphredon* and *Xylocelia* and their aphids) or are unusually powerful fliers (for example, *Tachytes* and *Sphex* and their grasshoppers).

In the simplest situation, the prey is held with the mandibles alone. This occurs in *Pemphredon* and a number of related genera, also in the spider-hunting genera *Trypoxylon* and *Sceliphron*. Even in these genera, there is evidence that the front legs sometimes help support the prey during flight. Wasps which prey on larger insects generally support the prey in flight with all the legs. This is true of *Sphex*, *Tachytes*, *Astata*, *Mellinus*, and several other genera. When these wasps land at the nest entrance or elsewhere they hold the prey with the mandibles alone, standing on all three pairs of legs (see, for example, photographs in Evans, 1958b, and Olberg, 1959, p. 335).

An interesting and important variation on this theme is provided by certain species of *Aphilanthops* (Evans, 1962) and *Cerceris* (Olberg, 1959, p. 365) (both genera belong to the Philanthinae). These wasps hold their prey with the mandibles and support it in flight with the legs, but it is the middle legs that provide the major support. When *Cerceris arenaria* lands, she stands on all her legs and holds the prey only with the mandibles. *Aphilanthops frigidus* normally continues to hold the prey with the middle legs as well as with the mandibles unless the wasp has some occasion to walk about, in which case the middle legs release the prey. Olberg reports that *Philanthus triangulum* occasionally grasps the antenna of the prey with its mandibles, although the species of *Philanthus* normally use only the middle legs. Thus the Philanthinae show several transitional stages from mandibular to pedal prey carriage.

Another major family of wasps, the Vespidae, apparently had their beginnings as solitary digger wasps but have undergone a remarkable evolution in nesting behavior and in social behavior. Yet prey-carrying behavior is monotonously uniform throughout the Vespidae; all species fly with the prey and carry it in the mandibles (the social forms macerate the prey first and carry it as a ball in the mandibles). In many Vespidae the front legs assist the mandibles, and in Eumeninae such as *Odynerus* and *Eumenes*, which carry whole caterpillars, all the legs support the prey in flight (Olberg, 1959, fig. on pp. 132, 146). Iwata (1942) lists many vespids as using only the mandibles and many as using the mandibles assisted by the legs, but he lists no vespids under any other type of prey carriage. Cooper (1953), in his intensive studies of *Ancistrocerus antilope*, reports that when these wasps walk with their caterpillars, the prey is held by the mandibles alone, although some support is provided by the legs when the wasp is in flight. Cooper found that *antilope* occasionally takes caterpillars too large to lift from the ground, in which case the burden is carried in short hops, the wasp ascending vertical surfaces on foot. Presumably the Vespidae were derived from a stock which carried the prey over the ground, later achieved partial transport in flight (like some of the Pompilidae), and finally full aerial carriage. Doubtless a good many vespids return to the ground to some extent when handling large prey, as also occurs in Sphecidae such as the cicada killer, *Sphecius*.

In general, few structural modifications are associated with this type of prey carriage. There is a general trend toward more compact body form and shorter legs in the higher Sphecidae (also the Vespidae) as compared to the Sphecinae, the Ampuli-

cidae, and the Pompilidae. The spheroidal thorax suggests a stronger flight mechanism, and the shorter legs may be better adapted for holding the prey tightly beneath the body in flight. However, the correlation of body form and type of prey carriage is at most a vague one. Doubtless some of the modifications of the mandibles and clypeus in various stocks of Sphecidae represent devices for better grasping the prey, but I can cite no well-documented specific examples of this.

PEDAL MECHANISMS: TYPE ONE (P1)

Pedal mechanisms involve the use of the legs, unassisted by the mandibles. All wasps employing pedal mechanisms carry the prey in flight. Iwata (1942) lists one wasp, the Australian *Exeirus lateritus*, as carrying the prey over the ground holding it only with the hind legs, but I believe this record to be erroneous. McCulloch (1923) and Musgrave (1925) both state that the middle legs are employed, not the hind legs, and McCulloch mentions that there is much use of the wings and cites one author who states that the wasp "rides the cicada to the nest." I suspect that prey carriage in *Exeirus* is no different from that in the American cicada killer, *Sphecius speciosus*. These wasps are gorytines, and like other members of that tribe they typically carry the prey in flight by the middle legs. But because of the great weight of their prey, the cicada killers have secondarily returned to partial ground transport. *Sphecius* typically takes flight only from some high object and often can be found carrying its prey considerable distances over the ground.

Pedal prey carriage releases the mandibles for other functions, for example, removing impediments from the nest entrance or driving away potential predators. But the major advantage is more subtle than this. True pedal prey carriage involves only the middle or hind legs or both; there are no wasps that carry their prey by the front legs alone (the two examples cited by Iwata require confirma-

476   HOWARD E. EVANS

tion). When these wasps land at the nest entrance not only are the mandibles free but also the front legs, the major digging devices of wasps. The vast majority of these wasps close the nest entrance with soil when they leave to hunt prey; when they return they scrape it open with the front legs, the soil being thrown beneath the body and behind. They then enter quickly with the prey still clutched beneath them. Closure of the nest surely prevents various parasites and predators (such as hole-searching miltogrammine flies) from finding and entering the nest. The vast majority of wasps employing mandibular prey transport leave the nest open. In certain genera which employ the mandibles (*Ammophila* is a good example) the nest entrance is closed, but in this case the wasp has to put the prey down while the entrance is cleared. It is mechanically impossible to dig open a nest entrance while holding prey with the mandibles, for the front legs are unable to perform their digging movements.

Under pedal mechanisms of type one, I include the many Sphecidae in which the middle legs provide the major support. During flight, the prey may also be supported by the front and hind legs (if the prey is large; see Olberg's photographs of *Philanthus triangulum* carrying a honeybee, p. 354); or only the hind legs may assist the middle legs (if the prey is small; see Olberg's photographs of *Mimesa equestris*, p. 274, and *Lindenius pygmaeus*, p. 375). Although both Iwata and Olberg make a distinction between these two types, the difference seems to me unimportant. In either case the prey is held only by the middle legs when the wasp lands at the nest entrance, the wasp standing on the hind legs and opening the nest entrance with the front legs. Then, as the wasp enters the burrow, the prey is slipped backward and grasped by the hind legs, so that the prey follows the wasp down the small bore of the burrow. This type of prey transport seems characteristic of all Gorytini, Stizini, and Bembicini, also of most Psenini, Crabronini, and Philanthini.

Some Crabronini are reported to hold the prey with only one middle leg rather than both (Hamm and Richards, 1926), but others clearly use both middle legs.

One would expect various modifications of the legs which would enable the wasp to obtain a firmer grasp on the prey. Actually there seem to be no modifications in the legs of the females as striking as those which occur in some males and serve to hold the female during copulation. A careful survey of differences in leg structures of female Sphecidae would doubtless reveal that certain of these are associated with differences in type of prey or type of prey carriage. For example, those Bembicini which prey upon adult Lepidoptera exhibit marked reduction in the pretarsal arolia. In this instance there is no proof that the reduction in the arolia is of positive value in carrying moths or butterflies to the nest. In the case of the cicada killer, *Sphecius speciosus*, it has been shown that the very large, hooked tibial spurs actually play an important role in supporting the cicada in flight. When Howes (1919) removed the hind tibial spurs from a female cicada killer, the wasp continued to bring in cicadas, but they were held "suspended, tail down, in a line perpendicular to the wasp's body, the two insects forming the letter T while in the air." In normal prey carriage, the cicada is held parallel to the wasp's body, and it may be surmised that the hind tibial spurs hook onto some part of the cicada's body.

PEDAL MECHANISMS: TYPE TWO (P2)

A few wasps hold their prey only with the hind legs. In this case the prey is held far back, actually behind the wasp, so that the wasp and its prey are in tandem. The only wasps which without question exhibit this type of prey transport are certain members of the crabronine genus *Oxybelus*. As mentioned above, wasps which carry the prey with the middle legs generally shift it to the hind legs as they enter the burrow. By using the hind legs from the beginning, *Oxybelus* is able to avoid this shifting of its load; this behavior may in fact have evolved by a simple shifting forward in time of this behavioral component. Most other Crabroninae carry the prey with the middle legs.

There has been some dispute as to whether the species of *Oxybelus* actually hold the prey with the hind legs or impaled on the sting. These are among the smallest of digger wasps, and it is not easy to be sure of this point except by very close and repeated observation. There is now no question that some species of *Oxybelus* do impale the fly on the sting (also embracing it loosely with the hind legs when in flight). On the other hand, several reputable observers report that the hind legs alone are used in certain species (for example, Bohart and Marsh, 1960, have recently reported this for *O. sericeum*).

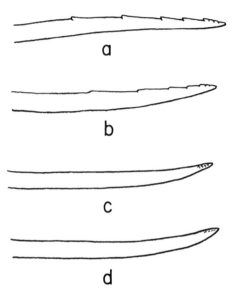

FIG. 4.   Stings of several Crabroninae, greatly enlarged.   a, *Oxybelus quadrinotatum*.   b, *O. sericeum*.   c, *Crossocerus elongatulus*.   d, *Crabro argus*.

ABDOMINAL MECHANISMS:
TYPE ONE (A1)

By holding the prey on the sting, some species of *Oxybelus* have effectively released all three pairs of legs for other purposes. Skeptics of this type of carriage should study Olberg's fine photographs of *O. uniglumis* carrying its fly (1959, pp. 379–381). During flight the hind tarsi of the wasp are also pressed against the thorax of the fly to give it additional support, but when the wasp lands the fly is held only by the sting. The tip of the wasp's abdomen turns down sharply, with the sting being inserted through the side of the anterior part of the thorax; the fly is upside down or more or less on its side during transport, extending straight out or somewhat obliquely behind the body of the wasp (see also Hamm and Richards, 1930, and the references cited therein).

I have observed *O. uniglumis quadrinotatum* and *O. bipunctatum* in some detail, and my observations agree closely with those of Olberg. Close study of the sting of *quadrinotatum* reveals that it possesses minute barbs (fig. 4a). So far as I know these barbs have not previously been described. In all probability they represent an adaptation for holding the prey more securely. Barbs are also present on the

sting of *O. sericeum*, a species that carries the prey with the hind legs (fig. 4b). However, while the barbs of *quadrinotatum* are clearly visible under a magnification of 40×, those of *sericeum* are barely discernible under twice that magnification. Whether the barbs of *sericeum* should be considered rudimentary or vestigial is a moot question. Clearly it would be worth studying the stings of other species of *Oxybelus* and attempting to correlate the strength of the barbs with the type of prey carriage.

Nielsen (1933) has reported that another crabronine wasp, *Crossocerus elongatulus*, carries its prey on the sting, and has provided a sketch of prey transport in this wasp. I have studied the sting of this species and found that it is not barbed (fig. 4c). Iwata also records *Crabro cingulatus* and *Aphilanthops quadrinotatus* as carrying the prey on the sting, but I believe these records to be erroneous. The latter species will be considered below, under abdominal mechanisms of type two. The record for *Crabro cingulatus* is based on

478   HOWARD E. EVANS

observations by the Raus (1918), but these
authors were not certain on this point.
They remark merely that the prey is carried beneath the abdomen, with the tip of
the abdomen curving forward beneath the
prey, the "sting holding the prey like a
hook." I have observed prey carriage in
several species of *Crabro* and found that
they often do, in fact, embrace the posterior end of the fly with the deflected tip
of the abdomen; however, the fly is held
principally with the middle legs and the
sting does not pierce the fly. This deflec-
tion of the tip of the abdomen may, however, represent a precursor of carriage on
the sting.

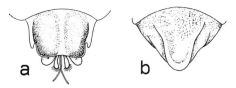

Fig. 5. Apical abdominal tergites of two species of *Aphilanthops* (Philanthinae). a, *A. (Clypeadon) haigi*, showing the biconcave tergite, and
protruding beyond it the bilobed sternite and the
sting and sting-sheaths. b, *A. (Aphilanthops)
frigidus*, showing a pygidial plate typical of many
fossorial wasps.

ABDOMINAL MECHANISMS:
TYPE TWO (A2)

Under this heading are included those
wasps that have structural modifications of
the apical abdominal segment for holding
the prey. This type was not discussed by
Iwata, since it has only recently been
described (Evans, 1962). It has been
established only in *Clypeadon*, a subgenus
of *Aphilanthops* (Philanthinae), but probably occurs also in the related subgenus
Listropygia. *Aphilanthops quadrinotatus*,
mentioned above, is a member of the subgenus *Clypeadon* (*laticinctus* is an earlier
name for the species). Although this species has been reported as carrying its prey
on its sting, the sting of these wasps is very
small, and it is doubtful that it pierces the
body during transport. The species of *Clypeadon* and *Listropygia* prey upon worker
ants of the genus *Pogonomyrmex*. My
observations indicate that females of at
least three species, after they sting the ant,
plunge the tip of their abdomen between
two pairs of coxae of the ant and fly off to
the nest. The ant is venter-up, its head
beneath the abdomen of the wasp, its
abdomen extending out behind. Originally
I believed that the tip of the wasp's abdomen was inserted between the middle and
hind coxae of the ant, but a careful study
of motion pictures taken in the summer of
1961 reveals that in *Clypeadon laticinctus*,

at least, the insertion is between the front
and middle coxae.

The apical abdominal segment of these
wasps is uniquely modified. The apical
tergite is expanded and biconcave (fig.
5a), the apical sternite bilobed and deeply
concave or biconcave. This double set of
concavities appears to embrace the coxae
or possibly portions of the mesothorax
grooved for reception of the coxae. Probably the wasp exerts pressure on the coxae
by forcing apart the tergite and sternite
slightly by muscular action. The result is
a highly efficient "ant-clamp," by means
of which the wasp carries the ant so far
behind that it is no impediment whatever
to the activities of the wasp. Prey transport in these wasps is very rapid, and I
have never seen a wasp drop its prey.
There is little doubt that the modifications
of the apical tergite evolved from the flattened pygidial plate present in many digger
wasps and used for packing soil in the
burrow (fig. 5b). Indeed, the related subgenus *Listropygia* was so named by Bohart
on the assumption that the elaborate apical
segment served as some sort of a scoop for
soil (*listron*, shovel, plus *pyge*, rump). I
have not observed prey transport in the
one known species of *Listropygia*, but I
have little doubt that the apical segment
functions as it does in *Clypeadon*.

DISCUSSION

The original method of prey carriage in
wasps was apparently simply to seize the
prey with the mandibles and drag it back-

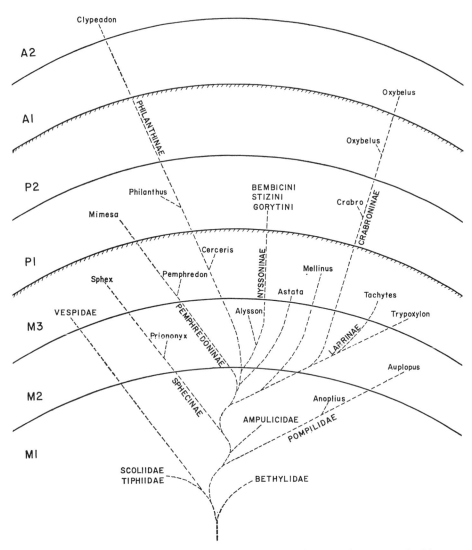

FIG. 6. Phylogenetic arrangement of major stocks and selected genera of wasps, emphasizing type of prey carriage achieved by each stock. M, mandibular mechanisms, P, pedal mechanisms, A, abdominal mechanisms; see text for significance of numbers. (Phylogenetic tree of Sphecidae based on Evans, 1959b).

ward over the ground, a method which is characteristic of the Bethylidae, Ampulicidae, many Pompilidae, and those Scolioidea which move their prey. Forward transport, still employing the mandibles, developed in several stocks of wasps independently. Use of several paralyzed arthropods per nest-cell in certain of these stocks involved

the use of smaller prey and permitted the carriage of the prey in flight. Thereafter the tendency was for the prey to be displaced progressively backward in the course of evolution. From its position in front of or beneath the wasp's head, the prey was moved to a position beneath the thorax and abdomen, finally to a position behind the body of the wasp. As one proceeds down the list to progressively more advanced types of carriage, fewer examples can be adduced for each type. Only four stocks of Sphecidae (and none of any other family) have developed pedal prey carriage, and only two of these stocks have progressed further to abdominal carriage (fig. 6).

The structural modifications accompanying these behavior patterns are often of a very generalized type: long legs and high stance in wasps that straddle their prey and carry it on the ground; a trend toward shorter legs and more compact body in wasps carrying their prey in flight. More specific adaptations for holding the prey appear to be uncommon among wasps employing the mandibles or legs, but more careful study will surely reveal more examples. At present one can cite such structures as the notched clypeus of *Priononyx*, apparently a device for holding the antennae of short-horned grasshoppers, and the large, hooked hind tibial spurs of *Sphecius*, which appear to play an important role in holding the cicada in flight. Wasps employing abdominal carriage exhibit structural modifications of the sting or apical abdominal segment, sometimes of unique form, for the abdomen is otherwise devoid of structures capable of holding the prey.

The selection pressures which have molded these behavior patterns have undoubtedly been complex, and different factors may have been of prime importance at different times and in different stocks of wasps. There is no question that wasps exhibiting more advanced types of prey carriage proceed much more rapidly to their nest. Many wasps which fly with their prey (but by no means all!) carry the prey far enough back so that its center of gravity is close to that of the wasp, in this way providing the least disturbance to the normal equilibrium of the wasp. But since these wasps employ several prey per cell, it may actually take a *Crabro*, for example, longer to provision its cell than a *Pompilus*. There is no evidence that any solitary wasp provisions more than about

one cell per day; thus greater speed in prey carriage appears to have no effect in increasing the number of progeny of wasps. It does, however, permit the wasps to obtain their prey at some distance from their nests. In *Sphex*, for example, the hunting grounds are often far from the nesting site, but in *Priononyx*, which drags its grasshoppers over the ground, the nests are always within the general area where the prey is captured. The very use of smaller prey, of course, requires greater speed in transport. A *Passaloecus* using 104 aphids per cell cannot afford to bring these in at a rate of one a day, the rate at which most pompilids procure spiders. The development of more advanced types of prey carriage appears to be part of the general picture of adaptive radiation in wasps, enabling them to take diverse types of prey, often at some distance from their nests.

It is also probable that the more advanced types of prey carriage serve to reduce attacks by predators and parasites. Unfortunately there are no quantitative data to support this, and I can only point to the innumerable observations by myself and others indicating that prey in transit may be attacked by miltogrammine flies, by cleptoparasitic wasps such as *Ceropales* (see Olberg, 1959, pp. 231–237), and by ants, tiger beetles, and other roving predators and scavengers. Here speed is unquestionably important, also the protection afforded by holding the prey tightly beneath the body. Still more important may be the avoidance of having to leave the prey on the ground, as pompilid wasps so commonly do. Furthermore, it is only those wasps that employ pedal and abdominal carriage that are able to close the nest entrance when they leave and reopen it without depositing their prey when they return. It is well known that hole-searching miltogrammine flies, which larviposit on the prey, and wasps of the families Chrysididae and Mutillidae, which attack the digger wasp larvae, readily find and enter open holes. That bombyliid flies, which are important parasites of many wasps and bees, deposit eggs indiscriminately in many types of open holes is well established (see, for example, a recent article by Frick, 1962). It is difficult to believe that nest closure is not a device for deterring various natural enemies. That the higher Sphecidae are appreciably more free of parasites than other wasps has not actually been demonstrated. In any event, one might expect some evolution in the behavior patterns of the parasites, adapting them to advances in the behavior of their hosts (e.g., the development of fossorial legs and digging behavior in the miltogrammine fly genus *Phrosinella*).

It is clear that there is no important correlation of type of prey carriage with kind of prey. For example, aphids may be carried in the mandibles (*Pemphredon*) or by the middle legs (*Diodontus*), while flies may be carried with the mandibles (*Mellinus*), the legs (*Bembix*), or the sting (*Oxybelus*). Although the philanthine subgenus *Clypeadon* has a special abdominal device for carrying worker ants, the genus *Tracheliodes* remains true to its subfamily (Crabroninae) and carries worker ants with its legs (Hicks, 1936). From another point of view, one finds that taxa restricted to one type of prey carriage often prey upon a variety of insects. The Bembicini, for example, prey upon stinkbugs, flies, butterflies, and moths, rarely even damselflies, yet they always carry their prey with their middle legs. A striking example is provided by *Microbembex*, a highly evolved genus of Bembicini which "preys" upon dead arthropods of two classes, including insects of at least ten orders. Yet these wasps, whether carrying a spider, a cricket, a caterpillar, or an ant, always employ the middle legs.

Nor is there any correlation with type of nest. The common mud dauber, *Sceliphron*, carries her prey no differently than her ground-nesting relatives. The twig-nesting genus *Ectemnius* exhibits exactly the same type of prey carriage as the related, ground-nesting genus *Crabro*.

Clearly type of prey carriage is "conserv-ative," that is, it is not readily modified even when major reorganizations occur in other, closely integrated aspects of the behavior. Prey carriage often provides excellent generic, tribal, and subfamilial characters. This being the case, one can predict with some confidence how various wasps of unknown ethology will be found to carry their prey. For example, the species of the genus *Bothynostethus* (Larrinae) must surely carry their prey with their mandibles, while the species of *Enoplolindenius* (Crabroninae) surely carry the prey in flight with the middle legs. Also, it seems evident that certain records in Iwata's compilation are in error, for example, the records for *Zyzzyx chilensis* and *Bembix hesione* carrying prey in their front legs. Curiously, in the few groups in which prey carriage is not conservative, the changes in prey carriage seem independent of other aspects of the behavior. The species of *Oxybelus* which employ the sting exhibit no other known behavioral differences from the species which employ the hind legs. Spooner (1948) reports that the nominate subgenus of *Mimesa* carries leafhoppers with its middle legs, while the very similar subgenus *Mimumesa* carries leafhoppers with its mandibles.[3]

Despite many unanswered questions, it is clear that a knowledge of methods of prey carriage is useful to taxonomists and essential to students of the bionomics of wasps. Attention should be focused upon more careful studies of the details of prey carriage, preferably documented with photographs such as those of Olberg. More data on the weight of prey and wasp, on the physics and physiology of carriage, and on the incidence of successful attacks by various parasites and predators, should do much to fill in the details of a picture which can only be sketched in a very preliminary way at this time.

---

[3] The generalization regarding *Mimumesa* is based solely on Adlerz' observations on *M. dahlbomi*. Although Adlerz is a very reliable authority, further verification of this point is needed.

482   HOWARD E. EVANS

SUMMARY

1. Wasps evolved from parasitoid Hymenoptera, and primitive wasps, like parasitoids, use a single host insect or spider for each offspring. Thus the prey is generally as large as or larger than the wasp.

2. Primitive wasps seize the prey with their mandibles and drag it backwards to the nest. Good examples of this can be found in the families Tiphiidae, Bethylidae, Ampulicidae, and Pompilidae.

3. At a more advanced stage, wasps acquired various mechanisms for straddling their prey and proceeding forward over the substrate. This occurs in many Pompilidae and in some Sphecidae.

4. Most Sphecidae, and all Vespidae, use more than one paralyzed insect or spider per cell; thus the prey is slightly to considerably smaller than the wasp. The prey is carried in flight, primitively held by the mandibles, often assisted by the legs.

5. Four stocks of Sphecidae have advanced to full pedal prey transport; that is, the prey is held by the middle or hind legs or both, unassisted by the mandibles.

6. Two stocks of Sphecidae have advanced still further to abdominal prey carriage. In one of these stocks (a portion of the subfamily Crabroninae), the prey is carried on the sting, which in some cases is barbed. In the other stock (two subgenera of the genus *Aphilanthops*, subfamily Philanthinae), the apical abdominal segment itself is greatly modified for clamping onto the prey.

7. The more advanced types of prey carriage permit more rapid provisioning of the nest and presumably provide fewer opportunities for predators and parasites to attack the prey in transit; they also enable the wasp to close the nest upon leaving and to reopen it upon returning without depositing the prey. The employment of rapid prey transport in flight also permits wasps to take their prey at a considerable distance from their nesting site.

LITERATURE CITED

BOHART, R. M., AND P. M. MARSH. 1960. Observations on the habits of *Oxybelus sericeum* Robertson (Hymenoptera: Sphecidae). Pan-Pac. Ent. **36**: 115–118.

COOPER, K. W. 1953. Biology of eumenine wasps. I. The ecology, predation and competition of *Ancistrocerus antilope* (Panzer). Trans. Amer. Ent. Soc., **79**: 13–35.

DUCKE, A. 1913. Ueber Phylogenie und Klassification der sozialen Vespiden. Zool. Jahrbücher, Syst. **36**: 303–330.

EVANS, H. E. 1953. Comparative ethology and the systematics of spider wasps. Syst. Zool., **2**: 155–172.

——. 1958a. Studies on the nesting behavior of digger wasps of the tribe Sphecini. Part I: Genus *Priononyx* Dahlbom. Ann. Ent. Soc. Amer., **51**: 177–186.

——. 1958b. Ethological studies on digger wasps of the genus *Astata* (Hymenoptera, Sphecidae). J. N. Y. Ent. Soc., **65**: 159–185.

——. 1958c. Observations on the nesting behavior of *Larropsis distincta* (Smith) (Hymenoptera, Sphecidae). Ent. News, **69**: 197–200.

——. 1959a. Observations on the nesting behavior of digger wasps of the genus *Ammophila*. Amer. Midl. Nat., **62**: 449–473.

——. 1959b. Studies on the larvae of digger wasps (Hymenoptera, Sphecidae). Part V: Conclusion. Trans. Amer. Ent. Soc., **85**: 137–191.

——. 1962. A review of the nesting behavior of digger wasps of the genus *Aphilanthops*, with special attention to the mechanics of prey carriage. Behaviour, **19**: 239–260.

——, C. S. LIN, AND C. M. YOSHIMOTO. 1953. A biological study of *Anoplius apiculatus autumnalis* (Banks) and its parasite, *Evagetes mohave* (Banks) (Hymenoptera, Pompilidae). J. N. Y. Ent. Soc., **61**: 61–78.

—— AND C. M. YOSHIMOTO. 1962. The ecology and nesting behavior of the Pompilidae (Hymenoptera) of the northeastern United States. Misc. Publ. Ent. Soc. Amer., **3**: 67–119.

FRICK, K. E. 1962. Ecological studies on the alkali bee, *Nomia melanderi*, and its bombyliid parasite, *Heterostylum robustum*, in Washington. Ann. Ent. Soc. Amer., **55**: 5–15.

HAMM, A. H., AND O. W. RICHARDS. 1926. The biology of the British Crabronidae. Trans. Ent. Soc. London, **74**: 297–331.

——. 1930. The biology of the British fossorial wasps of the families Mellinidae, Gorytidae, Philanthidae, Oxybelidae, and Trypoxylonidae. Trans. Ent. Soc. London, **78**: 95–132.

HICKS, C. H. 1936. *Tracheliodes hicksi* Sandhouse hunting ants (Hym. Sphecidae). Ent. News, **47**: 4–7.

Howes, P. G. 1919. Insect Behavior. Gorham Press, Boston. 176 p.

Imms, A. D. 1957. A General Textbook of Entomology. Ninth edition, revised by O. W. Richards and R. G. Davies. Methuen, London. 886 p.

Iwata, K. 1942. Comparative studies on the habits of solitary wasps. Tenthredo, 4: 1–146.

LaRivers, I. 1945. The wasp *Chlorion laeviventris* (Cresson) as a natural control of the Mormon cricket (*Anabrus simplex* Haldeman). Amer. Midl. Nat., 33: 743–763.

Leclercq, J. 1954. Monographie systématique, phylogénétique et zoogéographique des hyménoptères crabroniens. Lejeunia, Liege. 371 p.

McCulloch, A. R. 1923. War in the garden. Australian Mus. Mag., 1: 209.

Musgrave, A. 1925. The sand wasp's burrow. Australian Mus. Mag., 2: 243.

Nielsen, E. T. 1933. Sur les habitudes des Hyménoptères aculéates solitaires. III (Sphegidae). Ent. Meddel., 18: 259–348.

Olberg, G. 1959. Das Verhalten der solitären Wespen Mitteleuropas. Deutsch. Verlag Wissenshaften, Berlin. 402 p.

Rau, P., and N. Rau. 1918. Wasp Studies Afield. Princeton Univ. Press. 372 p.

Spooner, G. M. 1948. The British species of Psenine wasps (Hymenoptera: Sphecidae). Trans. R. Ent. Soc., London, 99: 129–172.

Wheeler, W. M. 1928. The Social Insects: their Origin and Development. Harcourt Brace, New York. 378 p.

Williams, F. X. 1919a. Philippine wasp studies. Bull. Hawaiian Sugar Planters' Assoc., Ent. Ser., no. 14. 186 p.

——. 1919b. *Epyris extraneus* Bridwell (Bethylidae), a fossorial wasp that preys on the larva of the tenebrionid beetle, *Gonocephalum seriatum* (Boisduval). Proc. Hawaiian Ent. Soc., 4: 55–63.

ADAPTATIONS IN THE KITTIWAKE TO CLIFF-NESTING.

By ESTHER CULLEN.

(Department of Zoology and Comparative Anatomy, University of Oxford.)

Received on 8 August 1956.

INTRODUCTION.

In 1952 I undertook a study of the breeding behaviour of the Kittiwake *Rissa tridactyla* in order to extend the scope of studies of various gulls, initiated by Dr. N. Tinbergen at Oxford. At that time accounts of the behaviour of this species were fragmentary (Selous 1905, Bent 1921, Perry 1940, Fisher & Lockley 1954), though recently the situation has been altered by a paper by Paludan (1955).

My observations were made while living on one of the Farne Islands off the Northumberland coast, where I was able to watch the birds through three seasons, 1953–55, and more briefly in 1952 and 1956. Although my observations were all made in these colonies there seems no reason to doubt that the behaviour described is typical for the species.

It is well known that the Kittiwake, together with its close relative *Rissa brevirostris*, differs from most gulls in two ways : outside the breeding season they live in the open sea rather than in the neighbourhood of land, and when they do visit the land to breed they nest on tiny cliff-ledges. Other gulls occasionally nest on ledges but these are far larger than the little shelves, sometimes only four inches wide, on which a Kittiwake can stick its nest. The present paper sets out some of the peculiarities of the Kittiwake's behaviour compared with the ground-nesting gulls and attempts to show that these peculiarities can be related to the cliff-nesting habit. A more detailed account of other aspects of the behaviour of the species is in preparation.

Since the great majority of gulls nest on the ground one may presume that this was also the ancestral breeding habitat of the Kittiwake. This view is supported by two facts : (1) Kittiwakes' eggs retain to some extent the cryptic pattern of blotching although this can be of little value, as every nest is marked conspicuously by a flag of white droppings (p. 288); (2) the young Kittiwakes are able to run under suitable conditions, though not quite as well as young ground-nesting gulls. This is an unusual feature for a species nesting in such precarious sites.

The advantage of cliff-nesting is certainly that it reduces predation. The nests seem fairly safe not only from ground-predators but also from such aerial ones as large gulls. I never saw a single Kittiwake chick or egg taken from the nest by the Herring Gulls *Larus argentatus* or Lesser Blackbacks *Larus fuscus*, which frequently preyed on the eggs and young of the Eider Ducks *Somateria mollissima* and terns nesting on the ground nearby; Nicholson (1930) remarks on a similar immunity even in the Kittiwakes nesting within a foot or so of the nests of Iceland Gulls *Larus leucopterus*.   Perry (1940) observed two pairs of Herring Gulls repeatedly robbing young Kittiwakes from their nests but such predation does not seem to be the rule. It is reported however that in some places in the Shetlands Great Skuas *Catharacta skua* are a greater danger to Kittiwakes' nests.   Lockie (1952) on Hermaness found them feeding largely on Kittiwakes' eggs and young and Venables & Venables (1955) also report nest-robbing by them in Foula. But heavy predation of nests seems restricted to few places (L. S. V. Venables, personal communication) and in the colony in Noss for instance the Skuas seem to specialise more on catching the Kittiwakes in the air (Perry 1948).

The Kittiwakes' security from large gulls probably results from the difficulty which they would often have in landing on the small perches, and this is enhanced by the complicated wind-eddies in such places.   In strong winds this turbulence is sometimes so awkward that even the agile Kittiwake has to circle again and again before it can alight where it wants to.

In addition to this protection from other species of gull the Kittiwake does not suffer from the depredations of its own species as do the ground-nesting gulls.   Unlike the other gulls, the Kittiwake does not take eggs or young of other birds but eats mainly fish and plankton.   This difference in diet is a subsidiary reason why the Kittiwake's eggs are safer than the ground-nesters'.   But even if Kittiwakes did prey on their own species they would suffer less than the other gulls because their nest-sites are much less accessible to ground-predators who might cause the parents to leave their nests exposed to be robbed by neighbours.

Throughout this paper I have contrasted the behaviour of the Kittiwake* with that of the ground-nesting gulls as a group.   Only two species, the Black-headed Gull *Larus ridibundus* and Herring Gull have been at all fully studied, but quite a lot is known about the other gulls in a more fragmentary way.   Although more studies are desirable, on the existing information there seems a good indication that in many respects the ground-nesters behave alike.   For information about the behaviour of the ground-nesters I have used published information (recent references summarised in Moynihan 1955 and Tinbergen 1956) and drawn extensively on personal communications from N. Tinbergen, M. Moynihan, R. & U. Weidmann.   I have also had the opportunity of seeing something myself of the behaviour of Herring and Lesser Black-backed Gulls on the Farne Islands.

---

* The behaviour of *Rissa brevirostris* is very little known but appears to be similar to that of *Rissa tridactyla* (Bent 1921).

FIGHTING METHODS AND RELATED BEHAVIOUR.

The ground-nesting gulls have various ways of attacking, the Kittiwake has specialized on one and this seems to have had a number of repercussions on the Kittiwake's behaviour.

According to Tinbergen (1953) and Moynihan (1955) an attacking Black-headed or Herring Gull tries to get above the enemy and peck down, or it may try to grasp any part of the other and then pull. If it happens to grasp the beak it sometimes tries to twist the other's head from side to side and thus may upset its balance. But this twisting seems rather rare : it is reported from the Common Gull *Larus canus* (Weidmann 1955), Tinbergen has seen it occasionally in the Herring Gull (personal communication) and I have observed it twice in the same species.

The Kittiwake on the other hand attacks by darting its head forward horizontally and always tries to grasp the opponent's beak. If it succeeds it twists the head from side to side. As most fights start on the ledges, which comprise the birds' whole territories, this method is very effective and often the opponent is twisted off the sill. It may however hang on to the other's beak and both tumble into the sea together, where they go on twisting with beaks still locked, ducking each other's heads under the water.

It would be difficult for the Kittiwake to use the other fighting methods of the ground-nesters; on their small perches they obviously cannot pull backwards and the structure of the cliffs often does not allow a bird to get above the other to peck downwards. On the water the ground-nesters' methods would probably be equally inefficient and it may be no mere coincidence that both times I saw the beak-twisting method used by Herring Gulls they were fighting on the water.

In connection with its specialized fighting method the Kittiwake seems to have lost a particular threat posture, which has been found in all the other 16 species of gull observed. In this posture, the " aggressive upright " (Moynihan 1955) or " upright threat " (Tinbergen 1953), the neck is stretched upwards and often a little forward and head and bill are pointed down. Tinbergen (1953) interprets the stretched neck and the attitude of head and beak as an intention-movement of pecking down at the opponent from above and later authors essentially agree with this interpretation in other species (Moynihan 1955, Weidmann 1955). As the Kittiwake does not fight in this way it is not surprising that it does not have the aggressive upright posture.

With the Kittiwake's specialization on one way of fighting the beak seems to have become an important stimulus in aggressive encounters. The fact that an attacking bird always goes for the beak of its opponent shows that the beak is important in directing the attacks. But it also appears to be a strong releasing stimulus. This can be concluded from observations in three situations : (1) when a female joins an aggressive male and tries to stay beside him and to prevent him attacking her, (2) when two females compete for one male, each trying to drive the other away, and (3) when one bird is trying to conquer a nest from another bird or from another pair. In each of these situations the bird who is attacked, whom I shall speak of as the visitor or intruder, may fight back, but often it seems too frightened to attack and turns its beak away hiding it in the breast-feathers, at the same time erecting the feathers of the neck. The attacker tries to get hold of the opponent's beak from either side, twisting and turning its neck over the other's; but if the beak is sufficiently hidden it does not succeed and usually after a while the attempts to attack stop, though the aggressor could easily peck other parts of the intruder's body. Should the intruder's posture relax so that it shows its beak again the owner of the nest starts to attack once more and tries to grab the beak and throw the other off the ledge.

It should be added that the beak-hiding does not always succeed in pacifying an opponent and a very aggressive bird may peck the neck or body of an intruder whose beak is out of reach. There is no doubt however that by tucking its beak out of sight a visitor can stay on a ledge beside the owner for much longer than it otherwise could. On one occasion I saw an intruder crouching on a ledge in this way beside the owner pair for a whole hour. A pair may ignore the presence of a stranger on their ledge to such an extent that they perform various friendly displays together such as mutual head-tossing, sometimes even standing on the intruder as if unaware of it, whereas a stranger with its beak visible would never be tolerated.

To test the importance of the beak experimentally I used models of Kitti-wake heads, which were cast of a plastic material (" Welvic " paste) from the same mould and then painted. I placed them on Kittiwakes' nests and counted the number of pecks delivered at them during a standard time after the bird had alighted on the nest. In some preliminary tests to determine the attack-releasing effect of the beak I presented two models in succession, one with a yellow beak like the adult's and one without a beak. Too few birds responded successively for a definite conclusion, but the results at least suggest that the head with the beak releases more attacks than the one without.

The experiments show more convincingly that the beak directs attacks. There were seven individuals who attacked the model with the beak, making 286 attacks in all. Of these attacks 98% (all but seven) were directed at the beak itself.

From general observations of fights it appears that the beak is not such an important stimulus in the ground-nesting gulls as it is in the Kittiwake. In particular the ground-nesters do not as a rule aim their attacks at the beak of the opponent (p. 277). Huxley & Fisher (1940) and R. Weidmann (personal communication) made some model-experiments with Black-headed Gulls which give information on the attack-directing effect of the beak in this species. Huxley and Fisher used stuffed birds and a corpse which they decapitated after a few experiments, presenting head and body separately. They found that the whole specimens were attacked and so was the head alone, but only a few pecks were aimed at the beak of the latter and none at all at the beaks of the whole birds. Weidmann repeated the experiments and also found that a far smaller percentage of the attacks were aimed at the beak of the head-model in the Black-headed Gull than I found in the Kittiwake though exact figures are not available. From general observations it seems probable that the beak directs attacks equally little in the other ground-nesters.

In releasing attacks the importance of the beak in the ground-nesters is more uncertain. I have said earlier that in the Kittiwake the effect of the beak-hiding demonstrated the attack-releasing effect of the beak. The ground-nesting gulls have a head-turning movement rather similar to that of the Kittiwake, the so-called " head-flagging " (named by Noble & Wurm 1943), but its effect is less conspicuous than in the Kittiwake and no experiments have been made to investigate its function. It has however been previously inferred without actual experiments that the beak had a certain attack-releasing effect in the Black-headed Gull (Tinbergen & Moynihan 1952).

None of the ground-nesting species shows such an elaborate hiding of the beak as the Kittiwake, they merely turn or jerk the face away from the opponent in the head-flagging. There cannot be any doubt that the movements are homologous, especially as the Kittiwake at a lower intensity also merely turns the face away, but one can understand why the Kittiwake has the elaborate beak-hiding : firstly because the beak may have a stronger attack-releasing value and secondly because the situation in which the Kittiwake does the elaborate beak-hiding does not occur in the ground-nesting gulls. There, a frightened bird which is attacked but wants to stay near the opponent or on the territory can always take a few steps away from the attacker to a place where it is not so frightened and does not provoke the other so much. But the Kittiwake has only the choice between staying very close to the other bird or leaving the territory altogether. Thus the intruding bird's greater fear and the attacker's increased aggressiveness might have led to the Kittiwake's exaggerated beak-hiding.

When young Kittiwakes are threatened they show the same head-turning movement as the adults and, as in the adults, it " appeases " the opponent, i.e. it tends to stop him from attacking. Young ground-nesters apparently do not head-flag at all. Adult gulls, Kittiwakes as well as ground-nesters, head-flag (or hide the beak) when they are pecked or threatened but want to stay. This situation hardly ever arises for young ground-nesters, for while they may be attacked by strangers, unlike young Kittiwakes, they can usually run away. This must be the reason that the Kittiwake is the only species known where the young shows the head-turning movement.

A young Kittiwake is not often pecked by a stranger.   In the first weeks after hatching the parents guard the nest so well that a strange bird is seldom able to land there.   But later the young are more often left alone and an adult who is searching for territory may land on their nest.   When this happens the adult may attack the young.   On the other hand fights between nest-mates over food are common and some kind of appeasement seems necessary then.   There are usually two young in a nest, one a day or two older than the other.   They may sit peacefully together for hours, but hostility flares up when the parent comes to feed them.   As soon as the younger starts to beg for food or tries to get the food, the older gives it a peck or prepares to do so and the smaller at once turns its head away.   This stops the older from attacking and while the smaller looks away the older often can get the food for itself.   Only when the older chick is satisfied is the younger one able to get food.

In the young ground-nesters fighting between brood-mates does not seem to occur in connection with food or for any other reason (personal communications, R. & U. Weidmann, N. Tinbergen).   The need for appeasement does not therefore normally arise.

I created this need artificially for two Herring Gulls and a Black-headed Gull which were placed in Kittiwakes' nests.   They were fiercely attacked by their Kittiwake foster-brothers and wire-netting fastened round the rim of the nest prevented them from escaping.   Even in these circumstances the young ground-nesters showed no trace of the head-turning.   As all three chicks had been hatched in Kittiwakes' nests this difference in behaviour must be innate.

The young ground-nesters were attacked more fiercely by the Kittiwake foster-brothers than young Kittiwakes would be in the same situation. Even two Kittiwake chicks which from hatching were always in the company of a young Herring Gull and Black-headed Gull respectively remained extremely hostile towards their companions whereas two young Kittiwakes settle down more quickly.   This continued hostility might well be due to the lack of head-turning in the ground-nester chicks.

It seems that the appeasement effect of the beak-hiding in the adult Kittiwake is mainly due to the removal of the beak from view.   But when a young Kittiwake turns its head away it not only conceals its beak, but it also shows off the black band across the nape, which is present only in this plumage (Fig. 1).   As the band is erected and displayed only in this situation one may infer that it is a special appeasement structure supporting the effect of hiding the beak.   Of all the 44 species of gull which Dwight (1925) described, only the young Kittiwake and the young of *Rissa brevirostris* have such a band.   This supports the idea of its function, since only the young of these two species must stay where they are however severely they may be attacked.

A slight difficulty remains. I have suggested that the head-turning of the young has been evolved in connection with fighting between the nest-mates, but this fighting is most fierce in the first days after hatching, before the neck-band develops. In these early fights the tiny chicks will hide their beaks just like the adults, squatting low in the nest-cup and tucking the beak under the breast. This gesture tends to appease the opponent although the displaying bird presents nothing but the pink skin of the neck shining through the white down. These early fights establish a kind of peck-order between the chicks and later when the black band has developed actual fights between the young are rarer. But the head-turning remains extremely common, especially at feeding times. There is however reason to think

FIGURE 1. Young Kittiwake beak-hiding.

that it is still important in preventing fights and an additional structure, the black band, may indeed be more necessary than at first, because the young are now much stronger.

We may conclude therefore that the young Kittiwake who has both the head-turning movement and the black neck-band has developed these characters as a consequence of cliff-nesting.

The head-turning in the Kittiwake seems to work because it conceals the weapon from the opponent and Tinbergen & Moynihan (1952) have suggested basically the same explanation for the head-flagging in the Black-headed Gull. But Lorenz has put forward another interpretation of similar movements, based mainly on his observations on dogs and wolves. He holds that the movements have appeasement effect because the inferior animal presents the most vulnerable part of its body to the enemy, who has a specific inhibition to attack the spot. This explanation cannot hold for the Kittiwake and probably not for the gulls in general, since in exceptional circumstances the neck may be pecked severely and the victim does not seem to take serious harm from it. It even appears uncertain at present whether Lorenz's explanation can hold for wolves and dogs, since, according to R. Schenkel (personal communication), Lorenz has misinterpreted the appeasement ceremony in these animals.

ADVERTISING DISPLAY OF THE MALE.

The Kittiwake differs from the ground-nesters in the way in which the new pairs form at the beginning of the breeding season.

In the Kittiwake the males occupy their territories, i.e. their nesting ledges, as soon as they arrive at their nesting cliffs from the winter quarters. They advertise themselves there with a special display which attracts females to them and it is on the nesting ledges that the pairs form. Unmated ground-nesting gulls usually do not seem to go straight to their breeding territories after they have arrived in the nesting area, but may spend the first few days either on neutral ground, the so-called " clubs " (Tinbergen 1953) or " pre- " or " pairing-territories " (Tinbergen & Moynihan 1952, Moynihan 1955). Here as a rule the pairs form and only later do they seem to occupy the actual breeding territories. In some species pairs may form outside the breeding season (Drost 1952) or perhaps even away from the breeding grounds (Moynihan, personal communication).

There may be several reasons why the Kittiwakes go straight to their nesting places while other species apparently do not. Firstly the Kittiwakes appear to be frightened of the land : they avoid sitting on the top of the nesting cliffs in the first few days after the occupation of the colony and it seems that the nesting ledges are the only places on the cliffs where the birds dare to land at this time. Secondly the Kittiwakes may have to stake their claims early because the number of suitable ledges is restricted and the competition for nest-sites is severe. On the other hand for the ground-nesters there is probably little shortage of nest-sites and they can therefore afford to wander about in clubs and pairing-territories.

The Kittiwake males advertise themselves to the females by a particular display, the " choking " (named by Noble & Wurm (1943) in the Laughing Gull *Larus atricilla*). Figure 2 shows the posture of the body in a choking bird and in this position head and neck are rhythmically jerked up and down with a nodding movement of a frequency of about 3 nods per second. Un-mated females are attracted by this display and land beside advertising males.

Choking occurs in other species of gulls, but the Kittiwake is the only species known regularly to use it when advertising for a mate. Unmated Black-headed Gulls use another display, the " oblique " posture with the " long call " (Moynihan 1955) and from recent observations it appears that unmated Herring Gulls also may use the long call as advertisement display (Tinbergen 1956).

On the other hand there are indications that both in the Herring Gull (Goethe 1937) and Black-headed Gull (R. & U. Weidmann, personal communication) an unmated male may occasionally choke in order to attract a mate. In these species and also in the Common Gull one can see much more often the attraction exerted by the choking display between a mated pair (Tinbergen 1953, Weidmann 1955). One bird, usually the male, walks away from its mate towards the nest or a prospective nest-site and chokes there; and this at once causes the female to approach and join in.

From these reports of the choking of the ground-nesters two points emerge. Firstly that choking attracts a mate and even though it may not be the usual display by which the unmated male first allures a female, it is not surprising that the Kittiwake has developed it for this purpose. Secondly it appears that the ground-nesting gulls tend to go to a nesting place in order to choke

FIGURE 2.   Kittiwake " choking ".

and this may be the reason why the Kittiwake males, whose pairing territory is the nesting ledge, have specialized more than any other gull in advertising themselves by choking.

The link between choking and the nest-site can be partly understood when the origin of the display is explained. The origin of choking has been discussed by several authors, most recently Moynihan (1955). Two different explanations have been put forward : one is that most if not all elements of the display are derived from nest-building movements; the other is that the pattern is derived from feeding the chicks. For various reasons, which will be published in a later paper, I believe the first explanation to be true and that choking behaviour is mainly derived from a displacement nest-building activity. It has been suggested (Armstrong 1950, Tinbergen 1952, Lorenz 1953 and more recent authors) that in situations where displacement activities might be expected external stimuli may play a part in determining which particular activity is chosen and it might well be that for this reason choking so often occurs at the nest-site.

It should however be stressed that this influence of the nest-site is not direct, because when a Kittiwake male advertises on the water or on the cliff-top, as happens occasionally, he still chokes as he would when advertising on the nest. The influence of the nest-site must be supposed rather to have acted phylogenetically in selecting for this particular type of advertisement display.

COPULATION.

Almost all the Kittiwake copulations take place on the nesting ledge. Only exceptionally are pairs seen to copulate elsewhere, for instance on the cliff-tops, and these are birds who do not own a ledge.

During copulation the female sits down; and Paludan (1955) has pointed out that this is different from the behaviour of a female ground-nester, who remains standing during the whole performance, often moving a little to and fro, apparently to keep her balance. Paludan has also pointed out that this method of copulating would be most unsuitable in the Kittiwake, as the female has only the tiny ledge to move on.

That this adaptation is not immediately dependent on the external situation is shown by the fact that the female sits down even during copulations on flatter places, where she would have plenty of room to move about.

NEST-BUILDING.

Nobody who has seen the loosely built nests of one of the ground-nesting gulls will expect that such a construction could be stuck onto the small shelves of the Kittiwake cliffs. And indeed the Kittiwake has a nest-building technique which is much more elaborate than the ground-nesters'.

To begin the nest the Kittiwakes mainly bring mud or soil, often mixed with roots or grass-tufts. Such trips for mud sometimes alternate with trips for grass, seaweed or other fibrous material. One bird usually deposits its material sideways over its shoulder with a rhythmic downward jerking movement of the head, performing about three to four jerks per second; at each downstroke the beak is opened and in this way the material, which often sticks to the beak, is jerked down piece by piece onto the nesting platform. Then the bird starts to trample on the ledge as if it were marking time. A bird trampling intensely does about six to seven steps a second and may go on trampling for up to half an hour with only short intervals. This behaviour is poorly directed as the bird often tramples a long time beside the nest-material on the bare rock, apparently without noticing the mistake. But as the pairs go on building in that way for several hours they succeed in the end in stamping the mud and the fibrous material to a firm platform, which sticks to the narrow ledges and enlarges them and produces horizontal shelves on amazingly steep slopes.   In the later stages of building the birds collect mainly material such a dry grasses, which they deal with in the same way as the mud : jerking it down and trampling on it, so as to form the nest itself on the reinforced mud-foundation.

In other gulls nest-building is simpler; they collect no mud but only fibrous material and they hardly trample on it at all.   The nests produced are piles of material loosely interwoven and this is all that is necessary on the ground.

Published descriptions (e.g. Moynihan 1953) of the depositing movements in ground-nesting gulls do not mention the downward jerking movement which I have described.   However the downward jerk sometimes occurs in the Black-headed Gull (R. & U. Weidmann, personal communication) and Herring Gull (own observation).   These species however do not seem to jerk as vigorously as the Kittiwake, usually performing only one movement, which is therefore easily overlooked, and only occasionally a few jerks in succession.   They apparently do not need to jerk as vigorously and as long as the Kittiwake, because the material they use does not stick to the beak while the Kittiwakes often have great difficulty in getting the sticky mud or humus onto the ledge.   I counted the number of jerks per depositing bout in eight individuals, which all brought material of both kinds (sticky, containing mud or soil and non-sticky, containing only grasses or seaweed), and found that 45 bouts with sticky material averaged 12·5 jerks, while 24 bouts with non-sticky material averaged only 2·6.

It is not known at present whether the differences found between species in the number of jerks are merely due to differences in the nest-material normally used or whether they are more intrinsic.   But there are strong indications for other nest-building activities that the specific differences depend not only on the immediate external stimuli : for instance once the Kittiwake has covered the mud with grass its nest is similar to that of a ground-nester, yet it still tramples intensely on the dry stuff, while the ground-nesters show only traces of the movement.   But this still does not decide whether the difference is innate or whether experience plays a part. On the other hand the fact that there is plenty of mud available to the ground-nesters and yet they do not collect it suggests that at least this difference between the Kittiwake and the other species is innate.

Another outcome of the different nest-building techniques is the different shape of the nest-cup; the Kittiwake's is relatively deeper than the ground-nester's, thus holding the eggs more safely.   This is necessary because if the Kittiwake's are accidentally kicked out, they almost always fall and are broken, whereas the ground-nesting gulls retrieve theirs in such an event (Kirkman 1937, Tinbergen 1953).

The nest-building behaviour of the Kittiwake has another peculiar feature : the birds collect nest-material in groups.   On the Inner Farne the collecting grounds were on the grassy top of the island and one would often see a single bird flying around hesitating to alight anywhere until it saw other birds alighting or already on the ground.   In this way parties of twenty birds or more would quickly assemble at one place.   A few birds after a while might detach themselves from the group and fly away with what they had gathered, while others joined the remaining body, so that there was a busy coming and going for a time, until suddenly the whole party would leave the place together.   Late-comers who either had no time to collect material or were just going to land flew off with the others empty-handed.   One could easily see that the birds were frightened on the flat ground even when they were in a flock.   One could not approach them within forty yards or so, whereas on the nest one could almost touch them.

This social collecting of nest-material is known only in the Kittiwake. The ground-nesters set off alone; they find their material either in the breeding colony itself or in places similar to where they nest, whereas the Kittiwake has to land on exposed flat ground and one can therefore understand its greater reluctance to collect alone.   As in many birds the flock seems to offer protection from predators and only the sense of security given by the flock seems to overcome the birds' reluctance to land on the ground.

There are other indications of the birds' fear of the land : on the Inner Farne they never flew over the island except when collecting nest-material and this avoidance of the land has struck other observers (Naumann 1820, Bent 1921).   Further, early in the season they will not assemble on the preening places on top of the island and away from the colony as they do later.   They restrict their visits to the ledges and rocks just above and below the nesting cliffs and only start to land on the preening places away from the colony when the gathering of nest-material has begun.   It seems thus that only the birds' strong urge to collect nest-material can overcome their fear of the unfamiliar land.   On the preening places, as on the collecting grounds, the birds always remain shy.

The birds do not gather nest-material all through the nest-building season. They have spells of building, during which collecting birds move to and fro between the colony and the collecting grounds and these active phases are interrupted by long spells when none or hardly any of the birds are building.   This synchronization is very remarkable and unusual for a gull-colony and might have the function of insuring that a bird finds companions on the collecting grounds.   It might be brought about by a strong mutual stimulation to build or by a certain stimulus to which the birds respond simultaneously, or by a combination of both.   While I have no evidence for or against the first possibility there is some evidence supporting the second. From my observations in the first two seasons I found that most of the big building outbreaks occurred during or after rain.   In 1955 I noted weather and building-activities in a colony of about fifty nests during five days, each divided into four roughly equal periods, at the height of the nest-building phase. In ten of the thirteen periods without rain there was little or no building (i.e. building at not more than three nests) and in three there was a lot (i.e. building at four nests and mostly more), whereas in all seven periods with rain a lot of building was recorded.

Although these figures are small they show that rain is correlated with the building activity in a Kittiwake colony (P< 1%). Apart from the function of the rain-stimulus in synchronizing the birds' building, collecting during and after rain might be advantageous to the Kittiwake, since during that time it probably is easier to pick up mud and humus from the softened ground.

The Kittiwakes are very ready to steal nest-material from unguarded nests, to such an extent that half- or almost finished nests may be dismantled completely. Most pairs however do not leave their nests undefended once they have started to build seriously; the mates take turns in guarding the nest, in contrast to their behaviour in earlier days when they frequently were both absent from the colony, sometimes for several days together.

Ground-nesting gulls often leave their nest alone until the first egg is laid and they steal material much more rarely (personal communication, N. Tinbergen, R. Weidmann). This difference again seems to be connected with cliff-nesting; it is obviously more convenient for a Kittiwake to collect from a nearby source on the cliff than from collecting grounds of which it is more frightened.

We can thus summarize the adaptations connected with nest-building : the Kittiwakes have specialized in collecting their material in groups as a consequence of their fear of open ground, which is itself apparently the result of cliff-nesting. They collect mud and soil in addition to fibrous material and work the material further with jerking and trampling. In this way they build a fairly safe nest on their small perches, and they guard it well from robbery by their neighbours.

CONCEALMENT AND DEFENCE OF BROOD.

I have mentioned earlier that the Kittiwakes' nests are much better protected from predators than the ground-nesters'. Because of this security the Kittiwake has been able to give up a number of behaviour patterns and morphological features which protect the eggs and young of the ground-nesting gulls.

When a predator approaches a colony of ground-nesting gulls the birds which can see it will start to give the alarm call and fly up when it is still at a distance; other birds join them even when they cannot see the enemy. (As Tinbergen (1953) has pointed out, this wariness at the nest must be correlated to the fact that the adult birds are very conspicuous whereas the eggs and young, being cryptically coloured and widely spaced, cannot be found easily.) Disturbed ground-nesters will moreover pursue an intruder and, particularly near hatching time and when they have young, they will swoop at it and may even strike it. This probably deters or at least distracts predators from searching diligently for the cryptic eggs or young (Tinbergen 1953).

In the Kittiwake it is more difficult to evoke these anti-predator reactions than in the ground-nesters. The birds allow an intruder to approach them far more closely before they react and very often they fly up without giving the alarm call at all. Some individuals can actually be lifted off the nest in broad daylight, even in the exceptional Kittiwake colony in Denmark where the birds nest on the ground (Paludan 1955). It is very striking how the alarm call is heard much less often in a Kittiwake colony than in a colony of ground-nesters, apparently because the Kittiwakes are safer and do not need to warn others so much.

The Kittiwakes also rarely attack predators at all. When I climbed among the nests I was hardly ever swooped at and never struck; and this applies also to the ground-nesting Kittiwakes studied by Paludan (1955). In one year on the Inner Farne a Herring Gull was seen several times to catch a young Kittiwake in the air. Even when it did so only a couple of feet from a nesting cliff the adult birds left their nests merely to hover in a completely silent cloud over the scene without interfering, while the chick was screaming and trying to defend itself against the powerful beak of its attacker. Such an assault in a ground-nesters' colony undoubtedly would have provoked violent attacks by the adult birds nesting nearby.

Although eggs and young ground-nesting gulls are cryptic they would be easily detected if the nest itself was not cryptic also. During incubation the ground-nesting gulls leave the nest in order to defaecate and when the young hatch the parents carry away the egg-shells. Apart from having any hygienic function, these actions are probably important in concealing the nest, as a collection of the white droppings or the white inside of an empty shell might easily betray the nest to a predator hunting by eye. A comparison of this behaviour in some species of terns points to the same conclusion (J. M. Cullen, personal communication).

On the other hand a Kittiwake while standing on the nest simply lets its droppings fall over the rim of the nest and a brooding bird merely gets up for defaecation. In this way soon a most conspicuous white flag forms below each nest. The egg-shells too are just left lying on the nest until they are accidentally knocked off. This usually happens soon after hatching, but may take several weeks.

Young ground-nesters are not only cryptically coloured but all the species which have been watched also show cryptic behaviour. When a predator approaches they run to hide under cover, and crouch. By contrast the young

Kittiwakes are most conspicuously coloured : in down they are white with a light grey back and in the juvenile plumage they are like the adults with the addition of some black bars.   In typical nest-sites it is impossible for them to run when danger approaches and Salomonsen (1941) reports that even in a Kittiwake colony on the ground the half-grown young remained on their nests when he came near (while the young Black-headed Gulls from the nests around all ran away).   Yet Kittiwake chicks will run when taken from the nest, put on the ground and frightened.

On the other hand two young Herring Gulls and one Black-headed Gull which were hatched and reared in Kittiwakes' cliff-nests would all have run over the edge of the nest had I not prevented them by fastening wire-netting round the nest-rim.   From this and Salomonsen's observations we can conclude that the difference in running between the young of the different species must be innate.

CLUTCH-SIZE.

Lack (1954 and previously) has suggested that in many birds the size of the clutch is adapted to correspond to the maximum number of young which the parents can, on the average, supply with food.

To test this hypothesis for the Kittiwake a number of nests were checked daily in a more or less random sample in the Kittiwake colony on the Inner Farne in 1953, 1954 and 1955 and the data were used to calculate clutch-size, nesting success etc.   In addition other nests were checked less often and might be expected to yield rather different results since eggs or young might disappear between two checks.   Clutch-size, hatching success and survival-rate were not significantly different for the two groups (P $<5\%$) and the data have been therefore combined in Tables 1 and 2.   A few young were not able to fly by the time I left the island, but in view of their age I have regarded them as surviving in compiling Tables 1 and 2.

The clutch-size distribution in 138 nests checked daily was 21 (15 %) with one egg, 104 (75%) with two and 13 (9%) with three.

Table 1 a shows that the hatching success increases with clutch-size. In order to discount this effect and to make comparison possible with the group of experimental nests mentioned below the nesting success of the Kittiwake was calculated on the basis of the number of young hatched rather than on the number of eggs laid (Table 1 b).   Since the numbers of families of one and three were small I adjusted the family size in some additional nests artificially, where necessary adding a young of a suitable age.   Table 1 c combines the results from these nests with those whose clutch-sizes were undisturbed and shows that families of three have a lower survival rate than families of one and two. (Survival of families of two is significantly higher than that of families of three (P $<3\%$).)   The results obtained so far from this single colony fit with one requirement of Lack's hypothesis but it should be noted that the number of young produced per nest is nevertheless greatest for the families of three.

Lack has emphasized the importance of food-shortage in determining the number of young which can be raised and there is some circumstantial evidence that food-shortage may account for mortality of young Kittiwakes. As noted above, young nest-mates regularly establish a peck-order which is important mainly at feeding times and when the parents have only a little food the superior chick may be the only one to be fed. This food-fighting suggests that food must at times be short. Furthermore the Kittiwakes' eggs hatch asynchronously, a device in birds which is thought to lessen the bad consequences of food-shortage (Lack 1954).

TABLE 1.   *Survival of Kittiwake eggs and young in relation to* (a) *clutch-size and* (b) *and* (c) *number of young hatched per nest. Only first clutches included.*

Clutch-size	No. of pairs	Young hatched		Young surviving	
		No.	Per cent hatching		
1	26	13	50	12	
2	130	182	70	164	(a)
3	18	40	74	32	

Size of family	No. of pairs	Young surviving			
		No.	Per cent	Average no. young per pair	
Undisturbed nests					
1 young	26	24	92	0·9	
2 young	91	164	90	1·8	(b)
3 young	9	20	74	2·2	
Both undisturbed and experimental nests					
1 young	31	28	90	0·9	
2 young	91	164	90	1·8	(c)
3 young	19	45	79	2·4	

In spite of this indirect evidence of food-shortage I was not able to see any direct signs of starvation except at a few nests and in some of these the parents seemed at fault, for nearby nests with the same number of young were being supplied adequately. Food-shortage might, however, be more evident in other colonies and under different conditions.

On the other hand food-shortage is probably not the only reason for a greater mortality in larger families and of the young whose deaths are recorded in Table 1 I judged that no more than half can have been due to starvation.

Chicks sometimes fall off their nests, or part of the nest ; or even the whole of it, may collapse, precipitating the young into the abyss.   One must suppose that the more young there are on a nest the more likely they are to fall, both because they have less room to stand and because their combined weight is greater.   I recovered some of the young which had fallen and found them perfectly healthy and they were in a number of cases successfully reared by foster-parents.   I seldom saw a chick actually fall from its nest, but when a chick was missing, one could get some idea of the likelihood that such an event had taken place from the condition of the nest.   From observations in a group of about twenty nests which I studied most intensely over three years, it would seem that the disappearance of young is just as often due to falling as to starvation.

It appears therefore that both shortage of food and the collapse of nests might account for the greater mortality in larger families in the Kittiwake.

The Kittiwake usually lays two eggs (p. 289) and so does its close relative *R. brevirostris* (Bent 1921), but the usual clutch-size of the gulls is three, except the Ivory Gull *Pagophila eburnea* which usually lays two (Buturlin 1906, Stark & Sclater 1906, Bent 1921, Baker 1935, Witherby *et al.* 1944).

Many young ground-nesting gulls are killed by predators, including members of the same species (Goethe 1937, Kirkman 1937, Darling 1938, Paynter 1949, Paludan 1951).   The Kittiwake young have a much lower mortality (Table 2) presumably thanks to their different nesting habit which protects them from predation and the disturbances which go with it.   The mortality of the young Kittiwakes, at least on the Inner Farne, is so low that in spite of their small clutch the birds produce on the average more young per nest than is in many cases reported of the other gulls with their larger clutches (Table 2).

Comparing my own data for the Kittiwake with Paynter's (1949) for the Herring Gull and Paludan's (1951) for the Herring and Lesser Black-backed Gull it appears that so many young ground-nesters die soon after hatching that within a week (Paludan) or from about three weeks (Paynter) there are fewer young per pair in these species than in the Kittiwake.   Other observers also remark on this high early mortality in the Herring Gull, Lesser Black-backed Gull and Black-headed Gull without giving quantitative evidence (Goethe 1937, Kirkman 1937, Darling 1938, Lockley 1947, Tinbergen 1953, R. & U. Weidmann, personal communication).

The available data thus suggest that through much of the nestling period the Kittiwake parents have as many or more young to feed than the ground-nesters.

Lack has repeatedly stressed that clutch-size in a species cannot be adapted to adult mortality, but he points out (1954) that mortality at an early age might be a subsidiary factor affecting the clutch-size.   Even if the number of young which could be raised by the Kittiwake and the ground-

TABLE 2. Nesting success in different species of gull. Different methods of sampling may account for some of the differences in the results.

	No. of		No. of young		Per cent hatched young surviving	Average young per pair surviving	Reference
	Pairs	Eggs	Hatching	Surviving			
Kittiwake 1953–55	179	348	240	211	88	1·18	this paper
Herring Gull							
1936	59	126	108	46	43	0·78	Darling (1938)
1937	68	198	189	77	41	1·13	ditto
1943	90	270	243	less than 131	<54	<1·46	Paludan (1951)
1944	87	371	206	41	20	0·47	ditto
1947	100				c. 51	c. 0·51	Paynter (1949)
Lesser Black-backed Gull							
1936	43	93	86	c. 50	c. 58	c. 1·16	Darling (1938)
1937	71	206	196	c. 104	c. 53	c. 1·47	ditto
1943	120	354	216	less than 32	<15	<0·27	Paludan (1951)
1944	112	362	220	12	5	0·11	ditto

nesting gulls were the same, one would expect that with relatively little early mortality, a consequence of cliff-nesting, the Kittiwake would evolve a smaller clutch-size than the other species.

FEEDING YOUNG.

The Kittiwake feeds its young in a different way from the ground-nesting gulls and this can be correlated with the different nesting habitats.

Young ground-nesters leave the nest a few days after hatching and may be fed at any place in the territory. Young Kittiwakes are confined to the nest until they can fly, that is, when they are about six weeks old, and have to be fed there. All gulls feed their young by regurgitation and in the ground-nesting species the parents very often drop the food to the ground and the young pick it up. But a young Kittiwake takes its food from the throat of the parent and food is rarely dropped. Gull's food is often half-digested and therefore cannot be picked up completely and it is very likely that soon a little heap of rotting food would collect in a Kittiwake's nest if the birds fed in the ground-nesters' way. This would not only smear the young, but it might also develop into a source of disease. For birds restricted to the nest it obviously is advantageous to have a more hygienic feeding method.

The Kittiwake keeps its nest clean in another way. It immediately picks up remains of food and other strange objects which may have fallen into the nest and either swallows them or, more frequently, flings them away with a vigorous head-shake. This habit has not been observed in the ground-nesters (N. Tinbergen, R. & U. Weidmann, personal communication), who do not seem to need it.

There are other behaviour patterns connected with the feeding of the young which seem to have changed in the Kittiwake. When a ground-nesting gull comes with food, its young are often hidden under cover, away from the place where the adult alights and if they are hungry they will run towards it with a characteristic food-begging movement : from a horizontal posture the so called " hunched posture " (Moynihan 1955) or " attitude of inferiority" (Tinbergen 1953) they quickly stretch their neck up vertically and withdraw it again, repeating these movements in succession so that head and neck perform a kind of pumping movement (Tinbergen 1953, Moynihan 1955, R. Weidmann personal communication). In the stretched phase they open the beak widely and give the food-begging call as they withdraw the neck again. One might guess that the function of this alternation and particularly the stretching of the neck is to make the young conspicuous to the parent who may alight some yards away. This explanation is confirmed by the behaviour of a young Black-headed Gull which I reared, for the pumping movement was much more pronounced when some distance away than when close to its " parent ".

By contrast the young Kittiwake lacks both the pumping movement and the need to make itself conspicuous for it is always on the nest.

It may be noted that in the food-begging of adult ground-nesting gulls, before courtship-feeding and copulation, a time when the pair is standing close together on the ground, the pumping is absent.

A related difference is that the Kittiwake lacks a special feeding call such as that which attracts the young ground-nester to the parent.

FLYING MOVEMENTS IN THE YOUNG.

Both the Kittiwakes and the ground-nesting gulls perform incipient flight movements when they are small, but this behaviour develops differently later in life.

From a very early age, before the wing-feathers have grown, the downy chicks of the ground-nesters flap their wings and at the same time jump up into the air. As the young grow these jumps get higher, the wing-flapping more vigorous and gradually it develops into proper flying.

The young Kittiwakes on the other hand, although they start wing-flapping from about the same age as the ground-nesting gulls (from the third day, Kittiwake, own observation ; from the fifth day, Black-headed Gull, personal communication R. Weidmann; from the seventh to ninth day, Herring Gull, Goethe 1955), do not perform the movement as vigorously as the other species and they do not start to jump into the air as early as the young ground-nesters. When eventually they start to do so, a few days before they fly, they never lift their feet for more than an inch or so from the nest, whereas much smaller ground-nesting chicks can be seen jumping a foot or two in the air.

This lack of vigorous flying movements in the young Kittiwake is certainly an adaptation connected with cliff-nesting for wing-flapping as vigorous as ground-nesters' would endanger their lives.

The orientation of the chicks when wing-flapping also differs with the species. The ground-nesters face into the wind, but the young Kittiwakes always face the wall. Kittiwake chicks spend much of their time orientated in this way, even when they are resting or sleeping and in the first days after hatching they hardly ever face outwards. This probably is a habit which prevents them from falling down. It is therefore not surprising that during a vigorous movement like the wing-flapping the young face the wall; moreover they often would not have room to spread their wings if they were facing the other way. When the young Kittiwakes prepare for their first real flight, however, they have to face outward. Thus their wing-flapping and jumping does not grade imperceptibly into proper flying as in the ground-nesters but their first flight is much more of a new achievement.

RECOGNITION OF YOUNG.

Several authors have claimed that ground-nesting gulls and terns are able to distinguish their own young from strangers (Watson & Lashley 1915, Culemann 1928, Dircksen 1932, Goethe 1937, Kirkman 1937, Palmer 1941, Tinbergen 1953, Pfeffer 1955). The experiments on which these claims are based are often unsatisfactory and in many cases do not distinguish between personal recognition by the parent of the young and other factors, such as the recognition by the young of the parent and possible differences in behaviour which this may lead to, which themselves may partly be the means by which the parent " recognizes " its young. Most of the experiments however show that if young more than a few days old are exchanged from different nests the parents drive them away. Tinbergen's experiments with Herring Gulls show that the parents begin to distinguish between their own young and strangers when their own young are about five days old and Pfeffer-Hülsemann (1955) claims the same of the Common Gull. Tinbergen stresses that this ability to recognize the young is very different from the ability to recognize the eggs, which depends almost entirely on location, and that this difference corresponds to the behaviour of eggs and young, as the eggs remain in the nest while the young soon start to move about. Furthermore it appears that the parents start to recognize their own young at the time when the young begin to move about actively. Thus

TABLE 3. *Exchange of young Kittiwakes from different nests. In each test, except the last, the whole brood was replaced by strangers, of the same number and age as the original young.*

Age in days	Age-group	No. of tests	Results
0–6	1	2	accepted
	1–2	4	,,
7–13	2	8	,,
	2–3	2	,,
14–20	3	12	,,
21–27	4	5	,,
28–34	5	1	pecked a little (see text)
Only one of two young exchanged	4	1	accepted

one may infer that the recognition enables the parents to attend to their own chicks by the time they are old enough to wander about. According to this explanation one would expect that the Kittiwakes, whose young, like the eggs, do not move from the nest, would also lack the ability to distinguish between their own young and those of strangers.

To examine this possibility I exchanged a number of young from different nests for five minutes (Tables 3, 4, 5) or longer (Table 6). Chicks of one nest were sometimes presented to several strange parents, and chicks which had been exchanged before were sometimes used again in an older age-group. In the first series I exchanged broods of the same age and the same number of chicks (Table 3). In another series I substituted young of different ages either for eggs or for young without altering the number of young (Table 4) and in some further experiments the number of individuals in the nest was altered as well as the nature of the contents (Table 5). In all the tests except

TABLE 4. *Young Kittiwakes of different age-groups exchanged. The whole brood was replaced by the same number of strangers. All accepted.*

Age-group of own brood	Age-group of replacing brood	Number of tests
2nd clutch ca. 1 week before hatch	3	1
near hatch	2	2
1	2	2
1–2	4	2
2	1	1
4	1–2	3
4	5	1
5	4	1
6	4	1

TABLE 5. *Number of individuals and nature of contents of nest changed in Kittiwakes' nests. All accepted.*

Owner's brood	Change made
1 young (age-group 1)	added 1 Kittiwake (age-group 1).
2 young (age-group 1)	added 1 Kittiwake (age-group 1).
1 young (age-group 4)	added 1 Kittiwake (age-group 4).
1 egg (pipped)	added 1 Shag (newly hatched).
1 young (2–4 days)	added 1 Shag (newly hatched).
2 young (1–3 days)	older replaced by a Herring Gull (newly hatched).
1 young (6 days)	added a Herring Gull (1 day).

the one in Table 3 with a young bird of the fifth age-group the parents showed the same behaviour to the strange young as they usually do to their own, which varies with the age of the chicks : small young are brooded, larger ones just guarded. Sometimes a young one was preened a little or even fed. The young Kittiwakes behaved towards the strange parents very much as they did towards their own; they usually stood beside them and might preen or peck at the adult's beak, begging for food, etc. It appears therefore that under the experimental conditions the parents do not recognize their offspring up to an age of at least four weeks either by number or the state of development, let alone by more subtle individual characteristics.

On the other hand Paludan (personal communication) found that the parents may attack their own young outside the nest, which confirms that they " recognize " their brood by location.

It is difficult to exchange young more than about four weeks old because from this age they are liable to jump off their ledges when approached. Of the three tests I made with young of this class the stranger was accepted in two. Once however the foreign chick was slightly pecked although its behaviour appeared not to differ from that of the parents' own young. These experiments are supported by other observations in which young who could

TABLE 6. *Strange young left in experimental nests to be reared after tests listed in Tables 3 and 4. (The whole brood was replaced without altering the number of young.)*

Number of exchanged young per nest	Age-group of own young	Age-group of strange young	Number of tests	Result
1	egg (ca. 1 week before hatch)	3	1	reared
1	1	1	2	ditto
2	1–2	1–2	2	ditto
2	3	3	2	3 reared, 1 died prematurely
1	6	4	1	reared

fly (and were therefore at least $5\frac{1}{2}$ weeks old) landed on strange ledges. Of twelve cases where a young bird landed on a nest with young, the stranger was pecked by the parent in six in a clearly hostile, although mostly inhibited way. In six cases the young was not molested and once it was even fed. In four of these intrusions the parent saw the chick landing and in three of the four it attacked; in four cases the young was on the nest first and in three of the four the parent accepted it. This and further observations suggest that a juvenile approaching the nest provokes attack more strongly than a juvenile already on the nest and this may apply even to the birds' own offspring. There are however signs that the parents can really distinguish their young from strangers at this age : an adult may threaten one of its own young who lands but does not actually attack it and I have seen a parent responding to the calls of its own young which landed on a strange nest. I suspect therefore that also in the cases when the strange fledglings were treated like their own young the birds might be aware of the change.

Summarizing these facts, it can be said that there is no evidence of personal recognition between parent and young in the Kittiwake up to four or five weeks after the young have hatched, but that the parents seem to discriminate later. Thus the behaviour of the Kittiwake agrees with what was expected

298   ESTHER CULLEN : ADAPTATIONS IN THE KITTIWAKE   Ibis 99

and confirms the idea about the function of recognition in the ground-nesters.

So far I have discussed mainly the recognition between adults and young, but the reaction of the chicks themselves to strange nest-companions is also interesting. Nest-mates know each other and fight strange young fiercely. Because of this hostility, it is difficult to foster a strange chick into a family with young, if at least one of the young is more than a few days old. Small strange young usually get used to each other after some fighting, but in larger ones the squabbling goes on and even after the loser has adopted the appeasement posture the other sometimes goes on pecking him fiercely for a long time.

This hostility of the young towards strangers is particularly surprising in view of the fact that the parents apparently do not distinguish their own young from strangers before the young fly, but the most probable explanation is that the young birds' hostility towards a stranger has another function from that of the parents'. In the ground-nesting gulls the parents' aggression is presumably a means of spacing out the families and ensuring that a stranger is not fed. Of this the Kittiwakes have no need. On the other hand the the fighting between the nest-mates is probably concerned with establishing the peck-order, already mentioned, whose function it is to determine the order in which the young are fed. In the normal course of events the peck-order is settled at an early age, when both the young are small and weak and easily tired, but if two strange young encounter one another on a nest (as happened during experiments) this may lead to a longer battle before one is subdued. It may be remarked that the outcome of these battles between strangers was a foregone conclusion as it always happens, in my experience, that the larger chick wins, whether or not it is the rightful owner of the nest.

To summarize this section we may say that the apparent inability of parents and young to recognize one another in the Kittiwake is another effect of cliff-nesting and the consequent isolation of the nests of different pairs from one another.

CONCLUSION AND SUMMARY.

The Kittiwake is probably derived from a ground-nesting gull and the change to cliff-nesting was presumably an anti-predator device. Because predation is less, the species seems to have lost a number of behaviour patterns and morphological features possessed by other gulls. On the other hand it has had to acquire a number of adaptations to suit its new life. Signal and non-signal movements have been altered : some have been modified, like the beak-hiding or the nest-building movements, others have been lost, like several anti-predator devices or the aggressive upright posture. There also seem to be a few new acquisitions like the black neck-band of the young and the collecting of mud.

Adaptations to fit an animal for a new kind of life are of two kinds. Some are inherited, others are acquired as a result of the individual's experience and environment. The chief morphological peculiarities of the Kittiwake, such as the black neck-band in the young, are certainly due to inherited differences. The adult's claws are sharper than those of other gulls, which helps the Kittiwake to hold on to small ledges and this may be partly because the tips are not abraded by walking as in the other gulls who spend the whole year close to land. On the other hand this difference is also apparent in the newly born young, so that here too it must have a genetic basis.

As I have shown, some of the differences in behaviour also are innate (the presence or absence of head-turning in the young, their readiness to run away when pecked). Others, for instance the use of mud as nest-material, are presumably also innate. Yet others, such as the social collecting of material, may be due to a complicated interaction of acquired and inherited factors. The existence of inherited factors leading to differences in behaviour should not blind one to the possibility that experience may be able to modify a pattern. This was shown in one of my experiments in which a newly-hatched Black-headed Gull learned after a day to take the food from the throat of its Kittiwake foster-parent in the way a young Kittiwake does from the first. Such modifiability is no objection to the existence of innate differences; it only shows that these differences may not be as rigid as has sometimes been supposed.

With the adaptations described in this paper, and which are summarized in the list below, I hope to have shown how this one change to nesting on tiny ledges on steep cliffs has had repercussions in many aspects of the life of the species and has led to morphological changes as well as a great many alterations in behaviour. In many animals adaptive differences between species have been described but I know of no other case where one relatively simple change can be shown to have been responsible for so many alterations.

GROUND-NESTING GULLS	KITTIWAKE
High predation-rate in nesting colonies.	*Predation pressure relaxed on cliffs.*
Alarm-call frequent.	Alarm-call rarer.
Adults leave nest when predator some way distant.	Remain on nest until predator very close.
Vigorously attacks predator intruding in colony.	Very weak attacks at most at intruding predator.
Brooding birds disperse droppings and carry egg-shells away from nest.	Neither droppings nor egg-shells dispersed.
Young cryptic in appearance and behaviour.	Young not cryptic either in appearance or behaviour.
Clutch-size normally three eggs.	Clutch-size normally two eggs.
Suited to life in colony on ground.	*Adapted to life on cliffs.*
A. Several fighting methods.	More specialized to fighting in one way (grabbing beak and twisting).

Upright threat posture occurs, derived from preparation to peck down at opponent.

Beak does not specially direct attacks. Not known if it is such a strong releasing stimulus as in the Kittiwake.

Beak turned away in appeasement but not elaborately hidden.

B. Young run away when attacked.
No head-flagging in young.

No neck-band.

C. Number of nest-sites probably less restricted and therefore probably less competition for nest-sites.
Often first occupy pairing territories before nesting territories and pairs form away from nest.
Choking not normally used by unmated males as advertisement display.

D. Copulation on the ground, female stands.

E. Nest-material collected near nest, building not synchronised, individual collecting.
Little stealing of nest-material.

Nests often unguarded before laying of first egg.
Nest-building technique relatively simple.
Mud not used.
Only one or, at most, very short series of depositing jerks.
Only traces of trampling on nest-material.
Nest has relatively shallow cup.

F. Young leave nest a few days after hatching.
Young fed by regurgitation on the ground.
Nest-cleaning absent or less conspicuous.
Parents have feeding call, probably to attract young.
Hungry young make themselves conspicuous to parents by head-pumping.
Parents learn to recognise own young in a few days.

G. Young face any direction. Vigorous wing-flapping in young.

H. Weaker claws, cannot hold on so well.

No upright threat.

Beak releases and directs attacks.

Beak turned away in appeasement and elaborately hidden.

Young do not run when attacked.
Head-turning and hiding of beak in young when pecked and appropriate behaviour in attacker.
Possess black neck-band.

Number of nest-sites restricted, probably more competition.

Occupy nesting ledges at arrival in breeding area and pairs form on the nest.
Choking normal advertisement display of unmated males.

Copulation on the tiny ledge or nest, female sits on tarsi.

Nest-material collected in unfamiliar places, synchronization of building and social collecting.
Birds very ready to steal nest-material.
Nests guarded.

Nest-building technique more elaborate.
Mud as nest-material.
Prolonged jerking of head when depositing nest-material.
Prolonged trampling on nest-material.

Nest has deeper cup.

Young have to stay on nest for long period.
Young fed from throat.

Young and adults pick up and throw away strange objects falling into nest.
Parents have no feeding call.

Head-pumping absent in young.

Parents do not recognize own chicks at least up to the age of four weeks.

Young face wall much of the time.
Flight movements much weaker.

Strongly developed claws and toe-musculature.

1957 ESTHER CULLEN : ADAPTATIONS IN THE KITTIWAKE 301

ACKNOWLEDGEMENTS.

I am grateful to the Janggen-Pöhn-Stiftung, St. Gallen, Switzerland, for a research grant which permitted this work and to the Nuffield Foundation for providing part of the equipment. Acknowledgements are also due to the Natural History Society of Northumberland, Durham and Newcastle-on-Tyne and to the National Trust for amenities for working on the islands. I would like to thank Dr. N. Tinbergen for his encouragement and most helpful advice and also Prof. A. C. Hardy for his hospitality in the Zoology Department in Oxford. Among others to whom I am grateful for helpful criticism are R. E. Moreau and my husband, who also helped in the field and revised the English text.

REFERENCES.

ARMSTRONG, E. A. 1950. The nature and function of displacement activities. Sympos. Soc. Exp. Biol. 4 : 361–384.

BAKER, E. C. S. 1935. The nidification of birds of the Indian Empire. 4. London.

BENT, A. C. 1921. Life histories of North American gulls and terns. U.S. Nat. Mus. Bull. 113.

BUTURLIN, S. A. 1906. The breeding-grounds of the Rosy Gull. Ibis (8) 6 : 131–139.

CULEMANN, H. W. 1928. Ornithologische Beobachtungen um und auf Mellum vom 13. Mai bis 5. September 1926. J. Orn. 76 : 609–653.

DARLING, F. F. 1938. Bird Flocks and the breeding Cycle. Cambridge.

DIRCKSEN, R. 1932. Die Biologie des Austernfischers, der Brandseeschwalbe und der Küstenseeschwalbe nach Beobachtungen und Untersuchungen auf Norderoog. J. Orn. 80 : 427–521.

DROST, R. 1952. Das Verhalten der männlichen und weiblichen Silbermöwen (*Larus a. argentatus* Pont.) ausserhalb der Brutzeit. Vogelwarte 16 : 108–116.

DWIGHT, J. 1925. The gulls (Laridae) of the world; their plumages, moults, variations, relationships and distribution. Bull. Amer. Mus. Nat. Hist. 52 : 63–408.

FISHER, J. & LOCKLEY, R. M. 1954. Sea-birds. London.

GOETHE, F. 1937. Beobachtungen und Untersuchungen zur Biologie der Silbermöwe (*Larus a. argentatus* Pontopp.) auf der Vogelinsel Memmertsand. J. Orn. 85 : 1–119.

GOETHE, F. 1955. Beobachtungen bei der Aufzucht junger Silbermöwen. Z. Tierpsych. 12 : 402–433.

HUXLEY, J. S. & FISHER, J. 1940. Hostility reactions in Black-headed Gulls. Proc. Zool. Soc. London (A) 110 : 1–10.

KIRKMAN, F. B. 1937. Bird Behaviour. London and Edinburgh.

LACK, D. 1954. The Natural Regulation of Animal Numbers. Oxford.

LOCKIE, J. D. 1952. The food of Great Skuas on Hermaness, Unst, Shetland. Scot. Nat. 64 : 158–162.

LOCKLEY, R. M. 1947. Letters from Skokholm. London.

LORENZ, K. 1953. Die Entwicklung der vergleichenden Verhaltensforschung in den letzten 12 Jahren. Zool. Anz. Suppl. 17 : 36–58.

MOYNIHAN, M. 1953. Some displacement activities of the Black-headed Gull. Behaviour 5 : 58–80.

MOYNIHAN, M. 1955. Some aspects of reproductive behaviour in the Black-headed Gull (*Larus ridibundus ridibundus* L.) and related species. Behaviour Suppl. 4 : 1–201.

NAUMANN, J. A. 1820. Naturgeschichte der Vögel Deutschlands. Leipzig.

NICHOLSON, E. M. 1930. Field-notes on Greenland birds. Ibis (12) 6 : 280–313, 395–428.

NOBLE, G. K. & WURM, M. 1943. The social behavior of the Laughing Gull. Ann. N.Y. Acad. Sci. 45 : 179–220.

PALMER, R. S. 1941. A behavior study of the Common Tern (*Sterna hirundo hirundo* L.). Proc. Bost. Soc. Nat. Hist. 42 : 1–119.

PALUDAN, K. 1951. Contributions to the breeding biology of *Larus argentatus* and *Larus fuscus*. Vidensk. Medd. Dansk. Naturh. Foren. 114 : 1–128.

PALUDAN, K. 1955. Some behaviour patterns of *Rissa tridactyla*. Vidensk. Medd. Dansk. Naturh. Foren. 117 : 1–21.

PAYNTER, R. A. 1949. Clutch-size and the egg and chick mortality of Kent Island Herring Gulls. Ecol. 30 : 146–166.

PERRY, R. 1940. Lundy, Isle of Puffins. London.

PERRY, R. 1948. Shetland Sanctuary. London.

PFEFFER-HÜLSEMANN, K. VON. 1955. Die angeborenen Verhaltensweisen der Sturm-möwe (*Larus c. canus* L.). Z. Tierpsych. 12 : 443-451.

SALOMONSEN, F. 1941. Tretaaet Maage (*Rissa tridactyla* (L.)) som Ynglefugl i Danmark. Dansk. Orn. Foren. Tidsskr. 35 : 159-179.

SELOUS, E. 1905. The bird watcher in the Shetlands. London.

STARK, A. & SCLATER, W. L. 1906. The birds of South Africa. 4. London.

TINBERGEN, N. 1952. " Derived " activities; their causation, biological significance, origin and emancipation during evolution. Q. Rev. Biol. 27 : 1-32.

TINBERGEN, N. 1953. The Herring Gull's World. London.

TINBERGEN, N. 1956. On the functions of territory in gulls. Ibis 98 : 401-411.

TINBERGEN, N. & MOYNIHAN, M. 1952. Head flagging in the Black-headed Gull; its function and origin. Brit. Birds 45 : 19-22.

VENABLES, L. S. V. & VENABLES, U. M. 1955. Birds and Mammals of Shetland. Edinburgh and London.

WATSON, J. B. & LASHLEY, K. S. 1915. Homing and related activities of birds. Pap. Dept. Mar. Biol. 7. Publ. Carnegie Inst. Washington 21 : 1-104.

WEIDMANN, U. 1955. Some reproductive activities of the Common Gull *Larus canus* L. Ardea 43 : 85-132.

WITHERBY, H. F. et al. 1944. The Handbook of British birds. 5. London.

EGG SHELL REMOVAL BY THE BLACK-HEADED GULL, *LARUS RIDIBUNDUS* L.; A BEHAVIOUR COMPONENT OF CAMOUFLAGE

by

N. TINBERGEN, G. J. BROEKHUYSEN [1]). F. FEEKES [2]),
J. C. W. HOUGHTON [3]), H. KRUUK [2]) and E. SZULC [4])

(Department of Zoology, University of Oxford) [5])

(With 11 Figs)
(Rec. 28-IV-1961)

I. INTRODUCTION

Many birds dispose in one way or another of the empty egg shell after the chick has hatched. A shell may be built in or trampled down; it may be broken up and eaten; or, more usually, it is picked up, carried away and dropped at some distance from the nest. C. and D. Nethersole Thompson (1942), who have given a detailed summary of our knowledge of egg shell disposal in birds, emphasise the inter- and even intra-specific variability of the responses involved. Since, in addition, the actual response is often over in a few seconds, and happens only once or twice for each egg, it is not surprising that our knowledge is still fragmentary. On the whole, the presence or absence of the response and its particular form seems to be typical of species or groups of species; for instance, it seems to be absent or nearly so in Anseres and in Gallinaceous birds; Accipitres often break up and eat the shell; Snipe are said to be "particularly lax" (Nethersole Thompson); Avocets, *Recurvirostra avosetta* L., remove discarded egg shells anywhere in the colony (Makkink, 1936). In the many species which carry the egg shell away, the response, occurring as it does just after hatching, when the young birds need warmth and protection from predators, must be supposed to have considerable survival value.

The Black-headed Gull invariably removes the egg shell in a matter of

1) Dept. of Zoology, University of Cape Town.
2) Dept. of Zoology, University of Utrecht.
3) City of Leeds Training College.
4) Nencki Institute of Biology, Warsaw.
5) We are indebted to Sir William Pennington Ramsden, Bart., and the Cumberland County Council for permission to camp and to carry out our work on the Drigg sand dunes; to the Nuffield Foundation and the Nature Conservancy for generous support; and Dr J. M. Cullen for constructive criticism.

hours after hatching (Pl. XI, fig. 1); it is extremely rare to find an egg shell in the nest once the chicks have dried. We have only a few direct observations on the time lapse between hatching and carrying in undisturbed gulls, but the 10 records we have (1′, 1′, 15′, 55′, 60′, 105′, 109′, 192′, 206′, 225′) suggest that the response is usually not very prompt. The carrying is done by the parent actually engaged at the nest, never (as far as we know) by the non-brooding partner which may be standing on the territory, even when it stands next to the sitting bird. At nest relief either the leaving partner, or, more often, the reliever carries the shell. Often however it is the sitting bird who starts looking at the shell, stretches its neck towards it, takes it in its bill and nibbles it (sometimes breaking off fragments while doing so, which then are swallowed), and finally rises and then either walks or flies away with the shell in its bill. The shell is dropped anywhere between a few inches and a hundred yards from the nest. We have also observed birds which flew off with the shell, made a wide loop in the air, and descended again at the nest with the shell still in their bill which they then either dropped on the nest or carried effectively straight away. There is no special place to which the shell is carried, though there may be a slight tendency to fly against the wind, or over an updraught, or where the carrier is less likely to be harrassed by other gulls; almost always the shell lands well beyond the territory's boundary. On rare occasions a shell may land in a neighbour's nest—where the latter then treats it as one of its own shells, *i.e.* removes it.

II. STATEMENT OF THE PROBLEMS

During our studies of the biology and the behaviour of gulls this response gradually began to intrigue us for a variety of reasons. (1) The shell does not differ strikingly from an egg, since it is only the small "lid" at the obtuse end which comes off during hatching; yet it is treated very differently from an egg, and eggs are never carried away. This raised the question of the stimuli by which the gulls recognise the shell. Systematic tests with egg shell dummies could provide the answer. (2) What could be the survival value of the response? (a) Would the sharp edges of the shell be likely to injure the chicks? NETHERSOLE THOMPSON raises this possibility, adding that poultry breeders know this danger well. (b) Would the shell tend to slip over an unhatched egg, thus trapping the chick in a double shell? (c) Would the shells interfere in some way with brooding? (d) Would the moist organic material left behind in the shell provide a breeding ground for bacteria or moulds? (e) Would egg shells, if left near the nest, perhaps attract the attention of predators and so endanger the brood?

76   N. TINBERGEN, ET AL.

The following facts, obtained earlier by our co-workers, seemed to give some clues.

(A) C. BEER (1960) found that Black-headed Gulls do not merely carry shells but a great variety of other objects as well if they happen to be found in the nest. Some of these objects are shown in Pl. XI, fig. 2. It seemed that the best characterisation of this class of objects would be: "Any object—perhaps below a certain size—which does not resemble an egg, or a chick, or nest material, or food"; in short: "any strange object". The very wide range of objects responded to suggests that the birds respond to very few sign stimuli; it might be that the response was adapted to deal with a much wider range of objects than just the egg shell.

(B) C. BEER (1960), testing the gulls' readiness to show this response at different times of the season, offered standard egg shell dummies (halved ping-pong balls painted egg shell colour outside, Pl. XI, fig. 2) to a large number of gulls once every day from the moment nest scrapes were formed (which is up to about three weeks before the laying of the first egg) till well beyond the hatching of the chicks. He found that under these conditions of standard (and near-optimal) stimulation the response could be elicited from at least 20 days before laying till 3 weeks after hatching. In this respect the response behaves rather like typical incubation responses such as sitting and egg retrieving which also develop gradually in the pre-egg period (BEER, 1960). In view of the heavy predation to which eggs are subjected (see below), and of the fact that the eggs are otherwise carefully guarded, this fact suggests that the response is important throughout the incubation period, and not merely during the few days when the chicks hatch.

(c) Finally, E. CULLEN (1957) found that the Kittiwake, Rissa tridactyla L. never carries the egg shell. The shells are just left in the nest until they are accidentally kicked off. It is true that this often happens in the first few days after hatching, but shells occasionally stay in the nest or on the rim for weeks, and at any rate they remain in or on the nest much longer than is the case with the Black-headed Gull. The Sandwich Tern, Sterna sandvicensis Lath. does not remove the egg shells either (J. M. CULLEN, 1960).

These observations combined suggest that neither the avoidance of injury, nor of parasitic infection, nor of interference with brooding are the main functions of egg shell removal — if this were so, then the Kittiwake as the most nidicolous species of gull would not lack the response. The most likely function seemed to be the maintenance of the camouflage of the brood — neither Kittiwake nor Sandwich Tern can be said to go in for camouflage to the extent of the other gulls and terns.

Thus these observations naturally led to an investigation into the function of the response and to a study of the stimuli eliciting it. In the following we shall deal with the problem of survival value first.

III. THE SURVIVAL VALUE OF EGG SHELL REMOVAL

The assumption that egg shell removal would serve to maintain the camouflage of the brood presupposes that the brood i s protected by camouflage. This basic assumption, usually taken for granted but — as far as we know — never really tested, was investigated in the following way.

First, we collected whatever observations we could about predation in the colony. While these observations are largely qualitative, they show convincingly that predation is severe throughout the season.

Very many eggs disappear in the course of spring. We did not make systematic counts but can give the following qualitative data. In the Ravenglass colony Carrion Crows, *Corvus corone* L., did not account for many egg losses, because, as we could observe time and again, they are easily chased away by the mass attack of the Black-headed Gulls. In fact we never saw a Crow alight in the colony. However, attacks by one, two or three gulls (which often occurred in our tests) did not deter Crows, and we must assume that nests on the fringe of the colony, and the dozens of nests we regularly find outside the colony and which do not survive, often fall victims to the Crows.

Egg predation within the colony was due to the following predators. Three pairs of Herring Gulls, *Larus argentatus* Pont., which bred in the gullery levied a constant toll of eggs, and later of chicks. Although the Black-headed Gulls attacked them, they could not altogether stop them from snatching eggs and chicks. We observed hundreds of occasions on which non-breeding Herring Gulls and Lesser Black-backed Gulls, *Larus fuscus* L., (many of them immature) passed near or over the colony or over our tests. On only one occasion did we see any of these taking an egg; they usually were totally uninterested in the colony.

Black-headed Gulls prey on each other's eggs to a certain extent. Most of those who visited our experiments did not attack an undamaged egg (the few exceptions are mentioned in our tables) but finished an egg once it had been broken by other predators. Later in the season individual Black-headed Gulls specialised on a diet of newly hatched chicks (see section VI).

Foxes, *Vulpes vulpes* L., which regularly visited the colony, and killed large numbers of adults early in the season and many half or fully grown chicks towards the end, did visit the gullery in the egg season, but we have no direct evidence of the amount of damage done by them in this part of the year. The gulls were greatly disturbed whenever a Fox entered the colony, but they did not attack him as fiercely as they attacked Herring Gulls and Crows. They flew over the Fox in a dense flock, calling the alarm, and made occasional swoops at it.

Stoats, *Mustela erminea* L., and Hedgehogs, *Erinaceus europaeus* L., visited the gullery and probably accounted for some losses. The gulls hovered over them in a low, dense flock but did not quite succeed in deterring them. In spite of several thousands of man-hours spent in hides in the gullery by us and by our colleagues Dr C. BEER and Dr G. MANLEY we never saw either a Fox, a Stoat or a Hedgehog actually taking an egg; they stayed in the dense cover, and though we have been able to read a great deal from their tracks whenever they had moved over bare sand, this method is of no avail in the egg season since the gull's nests are situated in vegetation.

Surprisingly, the Peregrine Falcons, *Falco peregrinus* L., which were often seen

Fig. 1. Black-headed Gull about to remove the empty egg-shell. (G. J. BROEKHUYSEN phot.)

Fig. 2. A sample collection of various objects which Black-headed Gulls remove from the nest. Top centre: BEER's standard model (a halved ping pong ball); bottom right: real egg-shell. After C. BEER, 1960. (N. TINBERGEN phot.)

Fig. 4. A small angle presented on a nest's rim. (N. TINBERGEN phot.)

Fig. 5. Some of the dummies used. Left, top to bottom: large angle, small angle, halved ping pong ball; centre top to bottom: khaki hens' shell, gulls' egg shell, black egg; right top to bottom: large ring, small ring. (N. TINBERGEN phot.)

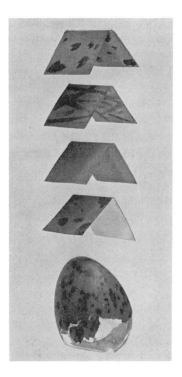

Fig. 10. Some of the small angles of Experiment 8, together with a gulls' egg shell. Top to bottom: "all egg", "cam", "sage green", and "shell". (J. Haywood phot.).

Fig. 11. Black-headed Gull trying to swallow rectangle measuring ½ × 18 cm. (G. J. Broekhuysen phot.).

near or over the gulleries left the Black-headed Gulls in peace, though they regularly took waders and carrier pigeons. The gulls panicked however when a Peregrine flew past.

In order to eliminate the effect of the gulls' social nest defence we put out eggs, singly and widely scattered (20 yards apart) in two wide valleys outside the gullery proper. These valleys had a close vegetation of grasses, sedges and other plants not exceeding 10 cm in height. Each egg was laid out in a small depression roughly the size of a Black-headed Gull's nest. Two categories of eggs were arranged in alternate pattern, and in successive tests exchanged position. While predators, particularly Carrion Crows, showed remarkable and quick conditioning to the general areas where we presented eggs, there was no indication that the exact spots were eggs had been found were remembered, and in any case such retention, if it would occur, would tend to reduce rather than enhance differential predation. We assumed (erroneously) that it might well be days before the first predator would discover the eggs, and in our first tests we therefore did not keep a continuous watch. We soon discovered however that the first eggs were taken within one or a few hours after laying out the test, and from then on we usually watched the test area from a hide put up in a commanding position, allowing a view of the entire valley. The tests were usually broken off as soon as approximately half the eggs had been taken.

Experiment 1.

Not wishing to take eggs of the Black-headed Gulls themselves (the colony is protected) we did our first test with hens' eggs, half of which were painted a matt white, the other half painted roughly like gulls' eggs. To the human eye the latter, though not quite similar to gulls' eggs, were relatively well camouflaged. It soon became clear that we had underrated the eye-sight of the predators, for the "artificially camouflaged" eggs were readily found.

TABLE I

Numbers of artificially camouflaged hens' eggs ("Artif. cam.") and white hens' eggs ("White") taken and not taken by predators.

	Artif.cam.		White	
	taken	not taken	taken	not taken
Carrion Crows	16		18	
Herring Gulls	0		1 (+3)	
Total	16	36	19 (+3)	33 (−3)

N presentations: 2 × 52. Difference between "artif.cam." and "white" not significant. [1]

---

1) The P-Values given in the tables were calculated by the χ^2-method, except where stated otherwise.

EGG SHELL REMOVAL BY THE BLACK-HEADED GULL 79

In 9 sessions, lasting from 20 minutes to 7½ hours, we saw Carrion Crows and Herring Gulls take the numbers of eggs (out of a total presented of 104) given in Table I.

We were astonished to see how easily particularly the Carrion Crows found even our "camouflaged" eggs. Each test area was usually visited every few hours by a pair of Crows [1]). They would fly in at a height of about 20 feet, looking down. Their sudden stalling and subsequent alighting near an egg were unmistakable signs that they had seen it; they usually discovered even camouflaged eggs from well over 10 metres distance. Often the Crows would discover every single egg, whether white or camouflaged, in the area over which they happened to fly. They either carried an egg away in their bills without damaging it to eat it elsewhere, or opened it on the spot and ate part of the contents, or (usually after they had first eaten four or five eggs) they carried it away and buried it. In some cases we saw Crows uncover these buried eggs one or more days after they had been cached.

Of the numerous Herring Gulls and Lesser Black-backed Gulls living in the general area, many of which flew over our test area every day, interest in eggs was shown only by three resident pairs of Herring Gulls. Their eyesight was undoubtedly less keen than that of the Crows (as expressed in the scores in table I, and particularly table II); further they were remarkably timid. For instance in experiment 1 we were certain on three separate occasions that a Herring Gull had discovered a white egg but did not dare approach it (the "3" in brackets in table I refers to these occasions). Yet as soon as Crows began to search the area, Herring Gulls would appear and attack them. Often Crows and Gulls attacked each other mutually, swooping down on their opponents from the air, for minutes before either of them alighted near an egg. Usually the Gulls succeeded in claiming an egg first. There were also occasions on which the Crows had the area to themselves.

It seemed obvious that our "camouflage" was not effective at all. Now the artificially camouflaged eggs differed from real gulls' eggs in four respects: (1) the ground colour, though to the human eye matching the overall colour of the rather brownish background very well, was slightly different from the ground colour of most of the gulls' eggs; (2) the dark grey dots which we painted on the eggs were more uniform in size and distribution than those on the real gulls' eggs; (3) unlike the natural dots they were all of one hue; and (4) hens' eggs are considerably larger than Black-headed Gulls' eggs, and hence probably more conspicuous.

---

1) We have good reasons to believe that each of our two test areas were visited by one pair.

Experiment 2.

Therefore, we next tested real, unchanged Black-headed Gulls' eggs against Black-headed Gulls' eggs painted a matt white. The tests were conducted in the same way as the previous ones. The results, obtained in 12 sessions lasting from 20 minutes to 4 hours, in which 137 eggs were presented, are summarised in table II.

TABLE II

Numbers of normal Black-headed Gulls' eggs ("Natural") and Black-headed Gulls' eggs painted white ("White") taken and not taken by predators.

	Natural		White	
	taken	not taken	taken	not taken
Carrion Crows	8		14	
Herring Gulls	1		19	
Black-headed Gulls	2		7	
Unknown	2		3	
Total	13	55	43	26

N presentations: 68 + 69: Difference between "Natural" and "White" significant at .1% level.

Experiment 3.

In order to get an impression of the parts played by size and by the nature of our "artificial camouflage", we painted Black-headed Gulls' eggs in the same way as the "camouflaged" hens' eggs, and compared their vulnerability with that of Black-headed Gulls' eggs painted white. The results of this test, to which we devoted only 5 sessions of from 1 to 3 hours, and in which 48 eggs were presented, are given in table III.

TABLE III

Numbers of artificially camouflaged Black-headed Gulls' eggs ("Artif. cam.") and Black-headed Gulls' eggs painted white ("White") taken and not taken by predators.

	Artif.cam.		White	
	taken	not taken	taken	not taken
Carrion Crows	4		5	
Herring Gulls	4		8	
Black-headed Gulls	1		1	
Total	9	15	14	10

N presentations: 24 + 24. Difference between "artif. cam." and "white" not significant (20 % < p < 30 %).

From these experiments, and particularly from experiment 2, we conclude that the natural egg colour of the Black-headed Gulls' eggs makes them less vulnerable to attack by predators hunting by sight than they would be if they were white; in other words that their colour acts as camouflage. The difference between the results of experiments 1 and 3 on the one hand and experiment 2 on the other indicates that we had underrated the eyesight of the predators, and also that their reactions to the different aspects of camouflage deserve a closer study: the parts played by over-all colour, by pattern and hue of the dotting, perhaps even by the texture of the eggs' surface we hope to investigate later — it was a pleasant surprise to discover that large scale experiments are possible. For our present purpose we consider it sufficient to know that painting the eggs white makes them more vulnerable.

Experiment 4.

We can now turn to the question whether or not the presence of an egg shell endangers an egg or a chick it is near. For obvious reasons we chose to investigate this for eggs rather than for chicks. The principle of this experiment was the same as that of the previous ones. We laid out, again avoiding site-conditioning in the predators, equal numbers of single Black-headed Gulls' eggs with and without an empty egg shell beside them. The shells used were such from which a chick had actually hatched the year before and which we had dried in the shade and kept in closed tins. The shells were put at about 5 cm. from the eggs which were again put in nest-shaped pits. Predators were watched during 5 sessions lasting from 45 minutes to 4½ hours. Sixty eggs were used. The results are given in table IV.

TABLE IV

Numbers of Black-headed Gulls' "eggs-with-shell" and "eggs-without-shell" taken and not taken by predators. Eggs not concealed.

	eggs-with-shell		eggs-without-shell	
	taken	not taken	taken	not taken
Carrion Crows	6		7	
Herring Gulls	9		7	
Total	15	15	14	16

N presentations: 30 + 30. Difference between egg with shell and egg without shell not significant.

We did not consider this result conclusive because the circumstances of the experiment differed from the natural situation in two respects. (1) Although a nest in which a chick has recently hatched may contain unhatched eggs, there are chicks equally or rather more often; and chicks, apart from having a less conspicuous shape than eggs, do at a quite early stage show a tendency to crouch at least half concealed in the vegetation when the parent gulls call the alarm. The Crows and Gulls might have less difficulty finding eggs than seeing chicks. (2) Both predators were always vigorously attacked by many Black-headed Gulls whenever they came in or near the gullery. In avoiding these attacks their attention is taken up for the greater part of the time (as judged by their head movements and evading action) by keeping an eye on the attackers; in the natural situation they never have the opportunity to look and search at their leisure. And to predators searching for camouflaged prey leisure means time for random scanning and opportunity for undivided attention — both probably factors enhancing discovery *i.e.* fixation of non-conspicuous objects. In other words our experiment had probably made things too easy for the predators, even though in the colony the nests themselves are often visible from a distance.

Experiment 5.

We therefore decided to repeat this experiment with slightly concealed eggs. This was done by covering each egg (whether or not accompanied by an egg shell) with two or three straws of dead Marram Grass, a very slight change which nevertheless made the situation far more similar to that offered by crouching chicks. Most tests of this experiment were done without watching from a hide; we knew by now who the main predators were, and for our main problem it was not really relevant to know the agent. In 8 tests, lasting from 2 hours 40 minutes to 4 hours 40 minutes, 120 eggs were offered, of which 60 with a shell at 5 cm. distance. The results are given in table V.

TABLE V.

Number of Black-headed Gulls' eggs-with-shell and eggs-without-shell taken and not taken by predators. Eggs slightly concealed.

eggs-with-shell		eggs-without-shell	
taken	not taken	taken	not taken
39	21	13	47

N presentations: 60 + 60. Difference between egg with shell and eggs without shell: $p < .1\%$.

The conclusion must be that the near presence of an egg shell helps Carrion Crows and Herring Gulls in finding a more or less concealed, camouflaged prey, and that therefore egg shells would endanger the brood if they were not carried away.

Experiment 6.

This was a series of pilot tests designed to examine whether the Carrion

EGG SHELL REMOVAL BY THE BLACK-HEADED GULL   83

Crows would be readily conditioned to shells once they had found eggs near shells, and whether, if they would find shells without eggs, they would lose interest in the shells. These tests began on April 21st, 1960, after the Crows operating in the valley had already gained experience with white and camouflaged eggs, and, since they had broken many of those while eating them, could have learned that egg shells meant food. In the very first test, egg shells were laid out in the Western area of the valley, where in previous tests with whole eggs the Crows had never been seen to alight. At their first visit the Crows alighted near these shells, pecked at them, and searched in the neighbourhood. They left after a few minutes whereupon we took the shells away. A few hours later we again laid out egg shells without eggs, this time scattered over the whole valley. When the Crows next returned they flew round over the valley, looking down as usual, but they did not alight. Next morning semi-concealed eggs were laid out over the whole valley, each with an egg shell at the usual distance of 5 cm. This time the Crows did alight and took a number of the eggs. A few hours later we once more laid out shells only, scattered over the valley, and at the next visit the Crows came down near several of them and searched. The morning after this we laid out just shells, this time in the Eastern part of the valley. This time the Crows did not alight. When next, between this and their next visit we gave each shell an egg at 5 cm., the Crows alighted again when they returned and took several eggs. Later that same day just shells were given in the Western part, which could not induce the Crows to alight. Such tests were continued until April 30, and while they did not give sufficient information, they strongly suggest: 1) that one experience with shells-without-eggs was sufficient to keep the Crows from alighting near shells the next time; and b) that renewed presentation of eggs with the shells attracted them again. Further, we had the impression (c) that the Crows later learned that egg shells in a part of the valley where shells only had been presented several times meant no food — in other words that they associated "shells only" with the locality. The full record of these experiments is given in table VI.

TABLE VI

Record of "conditioning test" explained in text.

Es + = eggs + shells offered, eggs taken; Es — = eggs + shells offered, eggs not taken; S + = shells alone offered, crows alighted and searched; S — = shells alone offered, crows did not alight.

Eastern area					S —	Es +		Es +			
Entire valley		S —	Es +	S +							S —
Western area	S +						S —		S —	S —	
Date (April)	21	21	22	22	23	23	23	24	24	25	25

84   N. TINBERGEN, ET AL.

	25	26	26	27	27	28	28	28	29	29	30
Eastern area	Es +	Es +			Es —	Es —	Es +		Es +		
Entire valley										S +	
Western area			S —	S —				S —			S +
Date (April)	25	26	26	27	27	28	28	28	29	29	30

Experiment 7.

We next investigated the effect of distance between egg and shell. This was done in a mass test without direct observation from the hide, and in the same valley where we had regularly seen Crows and Herring Gulls take eggs. Half concealed gulls' eggs were laid out; one third of them had a shell at 15 cm. distance, one third at 100 cm., and one third at 200 cm. A total of 450 eggs were presented in 15 tests, with the following results (table VII).

Table VII

Number of eggs taken out of the 450 offered with a shell 15, 100, and 200 cm. away.

15 cm.		100 cm.		200 cm.	
taken	not taken	taken	not taken	taken	not taken
63	87	48	102	32	118

N presentations: 3 × 150. Difference between 15 cm. and 200 cm. significant at .1% level. Significance of the total result: $p < 1\%$.

Part of each group of eggs may of course have been found without the aid of the shell (and this may explain why so many eggs were found even of the 200 cm.-group), but the figures show that the "betrayal effect" is reduced with increased distance.

A second, similar test was taken with the shells at 15 cm. and 200 cm.; altogether 60 eggs were presented in 3 tests; the results are given in table VIII and show a similar result.

TABLE VIII

Numbers of eggs-plus-shell taken | not taken for different distances (in cm.) between egg and shell. 60 eggs were offered.

15 cm.		200 cm.	
taken	not taken	taken	not taken
13	17	4	26

N presentations: 2 × 30. Difference between 15 cm. and 200 cm. significant at 2.5% level.

The following test, to be mentioned only briefly because it did not contribute to our results, was in fact done before any of the tests mentioned so far in an attempt to short-circuit our procedure. We put out 50 hens' eggs painted in what we then hoped would be a good camouflaged pattern (see above), and gave each egg an egg shell dummy at 15 cm. distance. These dummies were metal cylinders made by bending a strip of metal sheet measuring 2 × 10 cm. as used in our later tests with Black-headed Gulls, see experiments 11, 17 and 18 (Fig. 5). Half of these were painted like the eggs, inside and outside (they were in fact satisfactorily cryptic to the human eye); the other half were painted white. We assumed that predators would be slow to come, and that in the beginning one check per day would be sufficient. Upon our first check, about 24 hours after we laid out the eggs, we found that all 50 eggs had disappeared. Our first conclusion was that a whole horde of predators, such as a flock of Herring Gulls, had raided the valley, but we later saw that this valley was searched mainly by one pair of Carrion Crows and one, two, or sometimes three pairs of Herring Gulls. During the following weeks we discovered that the Crows kept digging up these eggs which they must have buried on the first day.

This failure forced us to check the effectiveness of our "camouflage" paint first (see experiments 1-3) and to test the effect of the natural egg shell (experiments 4-7); by the time these questions had been settled we had to start our tests on the stimuli eliciting egg shell removal in the Black-headed Gulls themselves, and so had to abandon for the moment any tests on the effect of the colour of the egg shell on the predators. However, the method having now been worked out, it is hoped to investigate this more fully.

The conclusion of this part of our study must therefore be that the eggs of the Black-headed Gulls are subject to predation; that in tests outside the colony the number of eggs found by Carrion Crows and Herring Gulls is lower than it would be if the eggs were white; that the proximity of an egg shell endangers the brood; and that this effect decreases with increasing distance. While it will now be worth investigating the predators' responsiveness to eggs and shells in more detail, the facts reported leave little room for doubt about the survival value of egg shell removal as an antipredator device. Whether or not the response has other functions is of course left undecided.

IV. THE STIMULI ELICITING EGG SHELL REMOVAL

Beer's demonstration that a great variety of objects can elicit the response naturally led to the question whether all these objects did so to exactly the same extent. This was investigated by presenting dummies of different types, one at a time at each nest, to hundreds of nests, and checking after a certain period (constant for each experiment but varying for different experiments according to the overall stimulating value of the dummies compared) whether or not the dummy had been removed. Nests were marked with numbered wooden pegs, and, unless otherwise mentioned each nest was used for only one experiment. Each experiment was arranged according to the latin square method, as set out in fig. 3. The nests were divided in as many equal groups as there were dummies, and each nest was offered each dummy once; the sequence of presentation to each group was arranged in such a way that each dummy was presented equally often in the first, second, third etc. position in the sequence.

Since each experiment involved repeated presentation of dummies, even

On various occasions the same model was given at different times of the season to different groups of new birds, and the scores therefore reflect their valence at different times without interference by waning. Because the duration of exposure to the models varied from one experiment to another comparison is possible only between tests of the same exposure time, that is, between figures in the same horizontal row. Further, the tests are arranged according to calendar date. While the colony as a whole is relatively well synchronised (most birds laying their first egg between April 12 and early May, with a peak in the last week of April — G. BEER, 1960), late birds do come in, and birds which have lost their first clutch do relay; this blurs the colony's calendar; yet on the average an arrangement of experiments according to calendar dates roughly reflects a classification according to age of eggs. In each horizontal row of table 30 the highest score is marked 1 and the next lower score is marked 2; it will be seen that while in two rows 1 precedes 2 in time, in the two others 2 precedes 1. Whatever differences in responsiveness there are between the different groups of birds within one horizontal series, they do not seem to be correlated with age of eggs; this, as far as it goes, is in accordance with BEER's findings.

It seems most likely therefore that the readiness to remove egg shells remains roughly constant through the incubation period, with perhaps a slight tendency to increase with time, which in BEER's figures might just have been offset by whatever slight waning may have occurred. Another possibility is that waning occurs only in the first few repeats, after which the response stays at a slightly lower but further constant level. If this were so, BEER's figures would mean that the build-up before egg laying is slightly steeper than apparent, and that the lack of increase in his figures after egg laying is real.

VI. THE LACK OF PROMPTNESS OF THE RESPONSE

The predator tests reported in section III demonstrated the intense pressure exerted at least by Carrion Crows and Herring Gulls against leaving the egg shell near the nest. Admittedly our observations refer to only one colony (which however contains a sizeable part of the British breeding population), but Carrion Crows are practically omnipresent and are notorious egg robbers. One should therefore expect that the Black-headed Gull would have developed a very prompt response and would remove the shell immediately after the chick has hatched. Yet, as we have seen, this is not usual. We cannot believe that the species has not been able to achieve promptness — egg shell removal is so widespread taxonomically that it must be an old response. The most likely reason is that there is a counteracting selection pressure — that too prompt removal is in some way penalised. At first we thought that the risk of carrying the chick with the shell before it had hatched completely might be responsible, but this risk is the same for gulls and other species. Yet we have observations (admittedly few in number) which show that Oystercatchers and Ringed Plovers carry the shell with far less delay — in spite of the fact that their chicks stay in the nest for a mere couple of hours, and shell removal therefore must be less urgent. In the course of 1959 and 1960 the reason for the delay became gradually clear: in both years there were a number of Black-headed gulls in the colony which preyed selectively on nearly hatched eggs and on wet chicks. Although we are certain that not all gulls engage in this "cannibalism", this type of predation is very common, particularly towards the end of the season. In fact many of our efforts to observe the development of the behaviour during the first few hours of a chick's life (which were usually done late in the season) were time and again frustrated by the wet chicks being snatched away by such robber gulls immediately after hatching. We have not made systematics notes on this, but twenty is a conservative estimate of the number of occasions on which we actually observed such chicks being taken in front of the hide, while dry chicks only hours older (equally or even more available) were left alone. The number of occasions on which we lost wet chicks without actually seeing it happen is much higher still. On only three occasions did we observe a Black-headed Gull trying to swallow a dry chick. While a wet chick is usually swallowed in a few seconds (often even in flight) dry chicks of less than a day old took approximately ten minutes to swallow. There can be no doubt that chicks are practically safe from predation by neighbours (though not from Herring Gulls) as soon as their plumage becomes dry and fluffy.

It is interesting to see the behaviour of parent gulls sitting on hatching chicks while there are robber gulls about. The parents are aware of the latters' intentions; they show signs of increased hostility whenever the robber comes near, and they are extremely loath to leave the nest. As soon as the gulls are disturbed by other predators or fly up for human visitors robbers snatch up the wet chicks in a fraction of a second. We had the impression that the robber gulls kept an eye on many nests and knew where chicks were hatching. On one occasion we saw a still wet chick being taken during the few seconds when the parent carried away the shell.

We feel justified therefore to ascribe the lack of promptness of the response to this tendency of some members of the colony to prey on wet chicks.

DISCUSSION

Removal of an egg shell lasts a few seconds. It is normally done three times in a year. Nothing would seem to be more trivial than this response, which at first glance might seem to be no more than fussy "tidying-up" by a "house proud" bird. Yet we have seen that it has considerable survival value, and that the behavioural organisation is complicated, and well adapted to the needs. In addition, our study has given us some insight in some more general problems of an ecological and evolutionary nature. The following discussion owes much to the stimulating studies of E. CULLEN (1956) and M. CULLEN (1960).

T e r r i t o r y . In 1956, N. TINBERGEN listed the alleged functions of territory in gulls as follows. One component, site attachment, assists in pair formation, in homing to the nest site, and in providing known shelter for the chicks. Inter-pair hostility, the other aspect of territoriality, prevents interference with pair formation, and, by forcing breeding pairs and thus nests apart, renders mass slaughter by predators less likely (see also TIN-BERGEN, 1952b). We believe that the facts mentioned in this paper show a hitherto unrecognised function of territorial hostility in the Black-headed Gull: reducing the likelihood of predation by neighbouring gulls. We distinguish this effect from the effect of spacing-out on inter-specific predation for the following reason. L. TINBERGEN (1960) has shown that some predatory birds can be guided for shorter or longer periods, by a "searching image": they can, through an unknown process, concentrate their attention on one particular type of prey while being less responsive to other types. The stability of this state of narrowed responsiveness seems to be controlled in part by the density of the prey species: a series of successes in a short time seems to strengthen or at least to maintain specialisation, but lack of success tends to widen their responsiveness again. This property of predators, which may well be wide-spread, puts a premium on spacing-out of potential prey animals, because it increases searching time between successes. In this connexion it is distance which counts.

We suggest that predation by neighbouring Black-headed Gulls is reduced by another aspect of territoriality. Each gull learns, during the prolonged period of territorial fighting in the early part of the season, not to intrude into the territories of its neighbours. The factor which reduces predation in this case is the existence of barriers, irrespective of distances between the broods (density).

Of course the inhibition of trespassing as a consequence of acquired knowledge of the boundaries does not totally prevent predation, but the fact that gulls do not trespass except very briefly and on rare moments (such as during a general alarm caused by another predator, *e.g.*, Man or Fox which require adult gulls to leave the ground) obviously reduces the amount of intra-species predation considerably.

The compromise character of colony density. As we have seen, the mass attack by Black-headed Gulls discourages at least one predator, the Carrion Crow, from penetrating into the colony. This demonstrates the advantage of colonial nesting — Crows were not deterred by attacks of one, two or even three Black-headed Gulls. On the other hand, spacing-out of breeding pairs within the colony and the establishment of knowledge of boundaries, both achieved by territorial hostility, have also distinct advantages. Thus the density of a Black-headed Gull colony has the character of a compromise between at least these two opposing demands. As E. CULLEN (1957) and J. M. GULLEN (1960) have shown, a study of the anti-predator devices of other species might help in elucidating the adaptedness of interspecific differences in colony density.

Synchronisation of the breeding calendar. Some of our data strengthen the conclusion that predator pressure may be an ultimate factor in the synchronisation of breeding. BEER's data (1960) show that the scatter in time of the appearance of fledged young on the beach does not differ strikingly from the scatter of egg laying. Yet there are every season a large number of late broods, partly those of birds arriving later than the main body, partly repeat clutches of birds who have lost their first clutch. Our observations show that at least intra-specific predation of wet chicks is particularly severe towards the end of the season, and it is striking how many of the late broods disappear. It is clear that most of the successful broods come from pairs which arrive early and which have not been forced to relay; failure to synchronise is heavily penalised.

The anti-predator system as a whole. At this stage of our studies it seems worthwhile to review what is now known of the ways in which the Black-headed Gull protects itself against predators. Some of these devices protect the individual— for brevity's sake we will call these "egoistic" devices, even though they may at the same time protect others. Others protect the brood, often at the cost of danger to the individual, and such devices we will call "altruistic". Naturally these terms refer to the function, not to motivation.

The most obvious, and seemingly trivial response is escape. Even this however takes different forms, dependent on the nature of the predator and on the age of the bird. A gull can simply fly away, as it does at the approach of a human being. The response to a Peregrine Falcon, which we have observed in detail several times, is different: the gulls fly up, form dense flocks and fly at great speed low over the ground or the water, executing quick zigzag manoeuvres. We believe that "erratic flights", in which individual gulls separate themselves from the flock and fly away, often downwards, at great speed and with very quick and sharp turns, are likely to be elicited by a Peregrine approaching gulls that are flying high. Chicks, in response to the alarmed behaviour of the adults in a colony, crouch, at first on the spot, but already after one day and occasionally even earlier they walk a little distance away from the nest and towards the surrounding vegetation. Each chick becomes soon conditioned to one or more individual hiding places; this conditioning allows them to reach safety more quickly than they would if they had to search for suitable cover.

Outside the breeding season Black-headed Gulls select wide open spaces: marshy meadows, open seashores, or water. This no doubt allows them to see an approaching predator in time.

In flocks of adult birds there operates at least one signaling system: the alarm call alerts other individuals.

The "altruistic" habitat selection in the breeding season shows signs of antipredator adaptedness: in the open sand dunes gulls avoid nesting on the bare sand, even though males may start by taking up a pairing territory there. Once paired however they select a nest site in moderately dense vegetation. Black-headed Gulls breeding on inland lakes usually select islands; either large ones which accomodate large numbers, or smaller islands such as individual *Molinia* bushes which offer space for one nest only. Where Black-headed Gulls nest on tidal saltings, this inclination to select islands can be the undoing of their broods in high spring tides (TINBERGEN, 1952a).

Several other behaviour patterns appear in the breeding season which, while endangering or at least not protecting the lives of the parents, do contribute to the safety of the brood. First, the scattering of the broods, which provides a certain degree of protection both from inter- and from intraspecific predation, is effected by the balanced attack-escape system, with its components of actual attack, withdrawal, and agonistic displays (TINBERGEN, 1959; MANLEY, 1960). Further, unlike camouflaged species such as ducks, curlews and several other waders, and pheasants, the incubating gull leaves the camouflaged brood at the first sign of danger. The camouflage of the eggs depends on a specialised pigmentation system in the upper reaches of the oviduct.

Parent gulls attack predators; the fierceness of the attack, and the degree to which it is counteracted by escape tendencies depends on the type of predator and the resultant seems highly adaptive. From the moment the first egg is laid, at least one parent stays on the territory and guards the brood. As we have seen, egg shell removal is also effective as an anti-predator device. There can finally be little doubt that the chick's plumage protects its bearer by being camouflaged.

Thus the picture that emerges is one of great complexity and beautiful adaptedness. It has further become clear that at least some of the different means of defence are not fully compatible with each other, and that the total system has the character of a compromise between various, in part directly conflicting demands. These conflicts are of different types. First, the safety requirements of the parents may differ from those of the brood. Thus the parent endangers itself by attacking predators. This is suggested by the fact that Foxes succeed in killing large numbers of adults in the colony. Though we have never seen a Fox killing an adult, their tracks in the sand cannot be misinterpreted. Often they kill many more birds than they eat. Some of these birds were "egg-bound" females (MANLEY, 1958), but in 1960, when we sexed 32 gulls killed by Foxes we found that 21 of these were males. Many of these gulls have their tails torn off and/or their legs broken. We believe that a Fox sometimes kills such birds by jumping at them when they "swoop". All this suggests that a certain balance between the tendency to attack a Fox and the tendency to flee from it is selected for.

The conflict between "egoistic" and "altruistic" behaviour is also very obvious in the time when the winter preference for wide, open spaces changes into the preference for the breeding habitat which, as we have just seen, is dangerous to the adults. The switch towards the breeding habitat selection is not sudden; there is a long period in which the birds show that they are afraid of it; even when, after long hesitation, they settle in the colony, there are frequent "dreads" when the birds suddenly fly off in panic; these dreads gradually subside (see also TINBERGEN, 1953 and CULLEN, 1956). Towards the end of the breeding season the adults begin to desert the colony in the evening to roost on the beach, leaving the chicks at the mercy of nocturnal predators.

Second, there are conflicts between two "altruistic" modes of defence, each of which has its advantages. Crowding, advantageous because it allows social attacks which are effective against Crows, has to compromise with spacing-out which also benefits the broods.

Finally there may be conflicts between the optimal ways of dealing with different predators. Herring Gulls and Crows might be prevented entirely from taking eggs and chicks if the gulls stayed on the nests, but this would expose them to the Foxes. While Herring Gulls and Crows exert pressure towards quick egg shell removal, neigbouring gulls exert an opposite pressure; the timing of the response is a compromise.

We cannot claim to have done more than demonstrate that egg shell disposal is a component of a larger system, nor are we forgetting that much in our functional interpretation requires further confirmation. It seems likely however that a more detailed study of all the elements of anti-predator systems of this and other species, and of the ways they are functionally interrelated, would throw light on the manifold ways in which natural selection has contributed towards interspecific diversity.

VIII. SUMMARY

The Black-headed Gull removes the empty egg shell shortly after the chick has hatched. The present paper describes some experiments on the function of this response, and on the stimuli eliciting it. Carrion Crows and Herring Gulls find white eggs more readily than normal gulls' eggs; it is concluded that the natural colours of the eggs afford a certain degree of cryptic protection. When normal eggs are given an egg shell at 15 cm. distance their vulnerability is greatly increased; this "betrayal effect" decreases rapidly with increased distance between egg and shell. We therefore conclude that egg shell removal helps to protect the brood from predators.

As reported by C. BEER (1960) the Black-headed Gull removes a surprisingly wide range of objects from the nest. Large scale tests with egg shell dummies in which colour, shape, size and distance from the nest were varied showed that objects of all colours are carried but that "khaki" (the normal ground colour of the egg) and white are particularly stimulating, while green elicits very few responses. Egg shells elicit more responses than cylindrical rings of the same colour, and these are responded to better than "angles". Size can be varied within wide limits; very large rings elicit fear which interferes with removal. Various other indications are mentioned which show that the score as obtained in the mass tests does not accurately reflect the responsiveness of the reaction itself but rather the result of its interaction with other behaviour tendencies. The eliciting effect decreases rapidly with increasing distance.

On the whole, the gulls' response is very well adapted to its main function of selectively removing the empty shell, but the relatively high scores for objects which have very little resemblance to egg shells suggest that it is adapted to the removal of any object which might make the brood more conspicuous.

A pilot test showed that gulls which have incubated black eggs respond better to black egg shell dummies than normal gulls.

The lack of promptness of the response as compared with non-colonial waders (Ringed Plover and Oystercatcher) is adaptive, since it tends to reduce predation by other Black-headed Gulls, which are shown to prey selectively on wet chicks. A hitherto unrecognised function of territory is suggested.

In a discussion of the entire anti-predator system of the Black-headed Gull its complexity and its compromise character are stressed: the safety demands of the individual clash with those of the brood; there are conflicts between the several safety devices which each benefit the brood; and there are clashes between the ideal safety measures required by each type of predator.

116   N. TINBERGEN, ET AL.

REFERENCES

BAERENDS, G. P. (1957). The ethological concept "Releasing Mechanism" illustrated by a study of the stimuli eliciting egg-retrieving in the Herring Gull. — Anatom. Rec. 128, p. 518-19.

—— (1959). The ethological analysis of incubation behaviour. — Ibis 101, p. 357-68.

BEER, C. (1960). Incubation and nest-building by the Black-headed Gull. — D. phil. thesis, Oxford.

CULLEN, E. (1957). Adaptations in the Kittiwake to cliff-nesting. — Ibis 99, p. 275-302.

CULLEN, J. M. (1960). Some adaptations in the nesting behaviour of terns. — Proc. 12th Internat. Ornithol. Congr. Helsinki 1958.

—— (1956). A study of the behaviour of the Arctic Tern, *Sterna paradisea*. — D. phil. thesis, Oxford.

HINDE, R. A. (1954). Factors governing the changes in strength of a partially inborn response, as shown by the mobbing behaviour of the Chaffinch (*Fringilla coelebs*). II. The Waning of the response. — Proc. Royal Soc. B 142, p. 331-58.

MAGNUS, D. (1958). Experimentelle Untersuchungen zur Bionomie und Ethologie des Kaisermantels *Argynnis paphia* L. (Lep. Nymph.). I. — Zs. Tierpsychol. 15, p. 397-426.

MAKKINK, G. F. (1936). An attempt at an ethogram of the European Avocet (*Recurvirostra avosetta* L.). — Ardea 25, p. 1-62.

MANLEY, G. H. (1957). Unconscious Black-headed Gulls. — Bird Study 4, p. 171-2.

—— (1960). The agonistic behaviour of the Black-headed Gull. — D.phil. thesis. Oxford.

NETHERSOLE THOMPSON, C. and D (1942). Egg-shell disposal by birds. — British Birds 35, p. 162-69, 190-200, 214-24, 241-50.

SCHWARTZKOPF, J. (1960). Physiologie der höheren Sinne bei Säugern und Vögeln. — Journ. f. Ornithol. 101, p. 61-92.

SIEGEL, S. (1956). Nonparametric Statistics for the Behavioral Sciences. — New York.

TINBERGEN, L. (1960). The natural control of insects in pine woods. I. Factors influencing the intensity of predation by songbirds. — Arch. néerl. Zool. 13, p. 265-336.

TINBERGEN, N. (1951). The Study of Instinct. — Oxford.

—— (1952a). When instinct fails. — Country Life, Feb. 15, p. 412-414.

—— (1952b). On the significance of territory in the Herring Gull. — Ibis 94, p. 158-59.

—— (1953). The Herring Gull's World. — London.

—— (1956). On the functions of territory in gulls. — Ibis 98, p. 408-11.

—— (1959). Comparative studies of the behaviour of gulls (Laridae); a progress Report. — Behaviour 15, p. 1-70.

VILLABOLOS, C. and J. VILLABOLOS (1947). *Colour Atlas.* — Buenos Aires.

WEIDMANN, U. (1956). Observations and experiments on egg-laying in the Black-headed Gull (*Larus ridibundus* L.) — Brit. Journ. Anim. Behav. 4, p. 150-62.

YTREBERG, N. J. (1956). Contribution to the breeding biology of the Black-headed Gull (*Larus ridibundus* L.) in Norway. — Nytt Magas. Zool. 4, p. 5-106.

ZUSAMMENFASSUNG

Die Lachmöwe trägt bald nach dem Schlüpfen eines Kückens die leere Eischale im Schnabel fort. Die vorliegende Arbeit beschreibt Versuche über die arterhaltende Funktion dieses Verhaltens und zur auslösenden Reizsituation. Aus der Tatsache, dass Rabenkrähen und Silbermöwen weiss gefärbte Eier häufiger finden als unveränderte, schliessen wir auf eine tatsächlich wirkungsvolle Tarnfärbung. Einzelne Eier, neben die wir in 15 cm. Abstand eine leere Eischale legten, fanden Krähen und Silbermöwen öfter als Eier ohne solche Nachbarschaft; mit zunehmendem Abstand

J. Theoret. Biol. (1964) **7**, 1–16

The Genetical Evolution of Social Behaviour. I

W. D. HAMILTON

The Galton Laboratory, University College, London, W.C.2

(*Received* 13 *May* 1963, *and in revised form* 24 *February* 1964)

A genetical mathematical model is described which allows for inter-actions between relatives on one another's fitness. Making use of Wright's Coefficient of Relationship as the measure of the proportion of replica genes in a relative, a quantity is found which incorporates the maximizing property of Darwinian fitness. This quantity is named "inclusive fitness". Species following the model should tend to evolve behaviour such that each organism appears to be attempting to maximize its inclusive fitness. This implies a limited restraint on selfish competitive behaviour and possibility of limited self-sacrifices.

Special cases of the model are used to show (a) that selection in the social situations newly covered tends to be slower than classical selection, (b) how in populations of rather non-dispersive organisms the model may apply to genes affecting dispersion, and (c) how it may apply approximately to competition between relatives, for example, within sibships. Some artificialities of the model are discussed.

1. Introduction

With very few exceptions, the only parts of the theory of natural selection which have been supported by mathematical models admit no possibility of the evolution of any characters which are on average to the disadvantage of the individuals possessing them. If natural selection followed the classical models exclusively, species would not show any behaviour more positively social than the coming together of the sexes and parental care.

Sacrifices involved in parental care are a possibility implicit in any model in which the definition of fitness is based, as it should be, on the number of adult offspring. In certain circumstances an individual may leave more adult offspring by expending care and materials on its offspring already born than by reserving them for its own survival and further fecundity. A gene causing its possessor to give parental care will then leave more replica genes in the next generation than an allele having the opposite tendency. The selective advantage may be seen to lie through benefits conferred in-differently on a set of relatives each of which has a half chance of carrying the gene in question.

T.B.   1

2 W. D. HAMILTON

From this point of view it is also seen, however, that there is nothing special about the parent-offspring relationship except its close degree and a certain fundamental asymmetry. The full-sib relationship is just as close. If an individual carries a certain gene the expectation that a random sib will carry a replica of it is again one-half. Similarly, the half-sib relationship is equivalent to that of grandparent and grandchild with the expectation of replica genes, or genes "identical by descent" as they are usually called, standing at one quarter; and so on.

Although it does not seem to have received very detailed attention the possibility of the evolution of characters benefitting descendants more remote than immediate offspring has often been noticed. Opportunities for benefitting relatives, remote or not, in the same or an adjacent generation (i.e. relatives like cousins and nephews) must be much more common than opportunities for benefitting grandchildren and further descendants. As a first step towards a general theory that would take into account all kinds of relatives this paper will describe a model which is particularly adapted to deal with interactions between relatives of the same generation. The model includes the classical model for "non-overlapping generations" as a special case. An excellent summary of the general properties of this classical model has been given by Kingman (1961b). It is quite beyond the author's power to give an equally extensive survey of the properties of the present model but certain approximate deterministic implications of biological interest will be pointed out.

As is already evident the essential idea which the model is going to use is quite simple. Thus although the following account is necessarily somewhat mathematical it is not surprising that eventually, allowing certain lapses from mathematical rigour, we are able to arrive at approximate principles which can also be expressed quite simply and in non-mathematical form. The most important principle, as it arises directly from the model, is outlined in the last section of this paper, but a fuller discussion together with some attempt to evaluate the theory as a whole in the light of biological evidence will be given in the sequel.

2. The Model

The model is restricted to the case of an organism which reproduces once and for all at the end of a fixed period. Survivorship and reproduction can both vary but it is only the consequent variations in their product, net reproduction, that are of concern here. All genotypic effects are conceived as increments and decrements to a basic unit of reproduction which, if possessed by all the individuals alike, would render the population both stationary and non-evolutionary. Thus the fitness $a^\bullet$ of an individual is treated as the sum of his basic unit, the effect δa of his personal genotype and the total e° of effects on him due to his neighbours which will depend on their genotypes:

$$a^\bullet = 1 + \delta a + e^\circ. \qquad (1)$$

The index symbol $\bullet$ in contrast to $\circ$ will be used consistently to denote the inclusion of the personal effect δa in the aggregate in question. Thus equation (1) could be rewritten

$$a^\bullet = 1 + e^\bullet.$$

In equation (1), however, the symbol $\bullet$ also serves to distinguish this neighbour modulated kind of fitness from the part of it

$$a = 1 + \delta a$$

which is equivalent to fitness in the classical sense of individual fitness.

The symbol δ preceding a letter will be used to indicate an effect or total of effects due to an individual treated as an addition to the basic unit, as typified in

$$a = 1 + \delta a.$$

The neighbours of an individual are considered to be affected differently according to their relationship with him.

Genetically two related persons differ from two unrelated members of the population in their tendency to carry replica genes which they have both inherited from the one or more ancestors they have in common. If we consider an autosomal locus, not subject to selection, in relative B with respect to the same locus in the other relative A, it is apparent that there are just three possible conditions of this locus in B, namely that both, one only, or neither of his genes are identical by descent with genes in A. We denote the respective probabilities of these conditions by c_2, c_1 and c_0. They are independent of the locus considered; and since

$$c_2 + c_1 + c_0 = 1,$$

the relationship is completely specified by giving any two of them. Li & Sacks (1954) have described methods of calculating these probabilities adequate for any relationship that does not involve inbreeding. The mean number of genes per locus i.b.d. (as from now on we abbreviate the phrase "identical by descent") with genes at the same locus in A for a hypothetical population of relatives like B is clearly $2c_2 + c_1$. One half of this number, $c_2 + \frac{1}{2}c_1$, may therefore be called the expected fraction of genes i.b.d. in a relative. It can be shown that it is equal to Sewall Wright's Coefficient of Relationship r (in a non-inbred population). The standard methods of calculating r without obtaining the complete distribution can be found in Kempthorne (1957). Tables of

$$f = \tfrac{1}{2}r = \tfrac{1}{2}(c_2 + \tfrac{1}{2}c_1) \quad \text{and} \quad F = c_2$$

4   W. D. HAMILTON

for a large class of relationships can be found in Haldane & Jayakar (1962).

Strictly, a more complicated metric of relationship taking into account the parameters of selection is necessary for a locus undergoing selection, but the following account based on use of the above coefficients must give a good approximation to the truth when selection is slow and may be hoped to give some guidance even when it is not.

Consider now how the effects which an arbitrary individual distributes to the population can be summarized. For convenience and generality we will include at this stage certain effects (such as effects on parents' fitness) which must be zero under the restrictions of this particular model, and also others (such as effects on offspring) which although not necessarily zero we will not attempt to treat accurately in the subsequent analysis.

The effect of A on specified B can be a variate. In the present deterministic treatment, however, we are concerned only with the means of such variates. Thus the effect which we may write $(\delta a_{\text{father}})'_A$ is really the expectation of the effect of A upon his father but for brevity we will refer to it as the effect on the father.

The full array of effects like $(\delta a_{\text{father}})_A$, $(\delta a_{\text{specified sister}})_A$, etc., we will denote

$$\{\delta a_{\text{rel.}}\}_A.$$

From this array we can construct the simpler array

$$\{\delta a_{r,\,c_2}\}_A$$

by adding together all effects to relatives who have the same values for the pair of coefficients (r, c_2). For example, the combined effect $\delta a_{\frac{1}{4},0}$ might contain effects actually occurring to grandparents, grandchildren, uncles, nephews and half-brothers. From what has been said above it is clear that as regards changes in autosomal gene-frequency by natural selection all the consequences of the full array are implied by this reduced array—at least, provided we ignore (a) the effect of previous generations of selection on the expected constitution of relatives, and (b) the one or more generations that must really occur before effects to children, nephews, grandchildren, etc., are manifested.

From this array we can construct a yet simpler array, or vector,

$$\{\delta a_r\}_A,$$

by adding together all effects with common r. Thus $\delta a_{\frac{1}{4}}$ would bring together effects to the above-mentioned set of relatives and effects to double-first cousins, for whom the pair of coefficients is $(\frac{1}{4}, \frac{1}{16})$.

Corresponding to the effect which A causes to B there will be an effect of similar type on A. This will either come from B himself or from a person who stands to A in the same relationship as A stands to B. Thus corresponding to an effect by A on his nephew there will be an effect on A by his uncle. The similarity between the effect which A dispenses and that which he receives is clearly an aspect of the problem of the correlation between relatives. Thus the term $e°$ in equation (1) is not a constant for any given genotype of A since it will depend on the genotypes of neighbours and therefore on the gene-frequencies and the mating system.

Consider a single locus. Let the series of allelomorphs be $G_1, G_2, G_3, ..., G_n$, and their gene-frequencies $p_1, p_2, p_3, ..., p_n$. With the genotype $G_i G_j$ associate the array $\{\delta a_{\text{rel.}}\}_{ij}$; within the limits of the above-mentioned approximations natural selection in the model is then defined.

If we were to follow the usual approach to the formulation of the progress due to natural selection in a generation, we should attempt to give formulae for the neighbour modulated fitnesses $a_{ij}^{\bullet}$. In order to formulate the expectation of that element of e_{ij}° which was due to the return effect of a relative B we would need to know the distribution of possible genotypes of B, and to obtain this we must use the double measure of B's relationship and the gene-frequencies just as in the problem of the correlation between relatives. Thus the formula for e_{ij}° will involve all the arrays $\{\delta a_{r,c_2}\}_{ij}$ and will be rather unwieldy (see Section 4).

An alternative approach, however, shows that the arrays $\{\delta a_r\}_{ij}$ are sufficient to define the selective effects. Every effect on reproduction which is due to A can be thought of as made up of two parts: an effect on the reproduction of genes i.b.d. with genes in A, and an effect on the reproduction of unrelated genes. Since the coefficient r measures the expected fraction of genes i.b.d. in a relative, for any particular degree of relationship this breakdown may be written quantitatively:

$$(\delta a_{\text{rel.}})_A = r(\delta a_{\text{rel.}})_A + (1-r)(\delta a_{\text{rel.}})_A.$$

The total of effects on reproduction which are due to A may be treated similarly:

$$\sum_{\text{rel.}} (\delta a_{\text{rel.}})_A = \sum_{\text{rel.}} r(\delta a_{\text{rel.}})_A + \sum_{\text{rel.}} (1-r)(\delta a_{\text{rel.}})_A,$$

or

$$\sum_r (\delta a_r)_A = \sum_r r(\delta a_r)_A + \sum_r (1-r)(\delta a_r)_A,$$

which we rewrite briefly as

$$\delta T_A^{\bullet} = \delta R_A^{\bullet} + \delta S_A,$$

where $\delta R_A^{\bullet}$ is accordingly the total effect on genes i.b.d. in relatives of A, and δS_A is the total effect on their other genes. The reason for the omission of an index symbol from the last term is that here there is, in effect, no question of whether or not the self-effect is to be in the summation, for if it is included it has to be multiplied by zero. If index symbols were used

6   W. D. HAMILTON

we should have $\delta S_A^\bullet = \delta S_A^\circ$, whatever the subscript; it therefore seems more explicit to omit them throughout.

If, therefore, all effects are accounted to the individuals that cause them, of the total effect $\delta T_{ij}^\bullet$ due to an individual of genotype G_iG_j a part $\delta R_{ij}^\bullet$ will involve a specific contribution to the gene-pool by this genotype, while the remaining part δS_{ij} will involve an unspecific contribution consisting of genes in the ratio in which the gene-pool already possesses them. It is clear that it is the matrix of effects $\delta R_{ij}^\bullet$ which determines the direction of selection progress in gene-frequencies; δS_{ij} only influences its magnitude. In view of this importance of the $\delta R_{ij}^\bullet$ it is convenient to give some name to the concept with which they are associated.

In accordance with our convention let

$$R_{ij}^\bullet = 1 + \delta R_{ij}^\bullet;$$

then $R_{ij}^\bullet$ will be called the *inclusive fitness*, $\delta R_{ij}^\bullet$ the *inclusive fitness effect* and δS_{ij} the *diluting effect*, of the genotype G_iG_j.

Let

$$T_{ij}^\bullet = 1 + \delta T_{ij}^\bullet.$$

So far our discussion is valid for non-random mating but from now on for simplicity we assume that it is random. Using a prime to distinguish the new gene-frequencies after one generation of selection we have

$$p_i' = \frac{\sum_j p_i p_j R_{ij}^\bullet + p_i \sum_{j,k} p_j p_k \delta S_{jk}}{\sum_{j,k} p_j p_k T_{jk}^\bullet} = p_i \frac{\sum_j p_j R_{ij}^\bullet + \sum_{j,k} p_j p_k \delta S_{jk}}{\sum_{j,k} p_j p_k T_{jk}^\bullet}.$$

The terms of this expression are clearly of the nature of averages over a part (genotypes containing G_i, homozygotes G_iG_i counted twice) and the whole of the existing set of genotypes in the population. Thus using a well known subscript notation we may rewrite the equation term by term as

$$p_i' = p_i \frac{R_{i.}^\bullet + \delta S_{..}}{T_{..}^\bullet}$$

$$\therefore \quad p_i' - p_i = \Delta p_i = \frac{p_i}{T_{..}^\bullet}(R_{i.}^\bullet + \delta S_{..} - T_{..}^\bullet)$$

or

$$\Delta p_i = \frac{p_i}{R_{..}^\bullet + \delta S_{..}}(R_{i.}^\bullet - R_{..}^\bullet). \tag{2}$$

This form clearly differentiates the roles of the $R_{ij}^\bullet$ and $\delta S_{ij}^\bullet$ in selective progress and shows the appropriateness of calling the latter diluting effects.

THE GENETICAL EVOLUTION OF SOCIAL BEHAVIOUR. I   7

For comparison with the account of the classical case given by Moran (1962), equation (2) may be put in the form

$$\Delta p_i = \frac{p_i}{T_{..}^{\bullet}} \left(\frac{1}{2} \frac{\partial R_{..}^{\bullet}}{\partial p_i} - R_{..}^{\bullet} \right)$$

where $\partial/\partial p_i$ denotes the usual partial derivative, written d/dp_i by Moran.

Whether the selective effect is reckoned by means of the $a_{ij}^{\bullet}$ or according to the method above, the denominator expression must take in all effects occurring during the generation. Hence $a_{..}^{\bullet} = T_{..}^{\bullet}$.

As might be expected from the greater generality of the present model the extension of the theorem of the increase of mean fitness (Scheuer & Mandel, 1959; Mulholland & Smith, 1959; a much shorter proof by Kingman, 1961a) presents certain difficulties. However, from the above equations it is clear that the quantity that will tend to maximize, if any, is $R_{..}^{\bullet}$, the mean inclusive fitness. The following brief discussion uses Kingman's approach.

The mean inclusive fitness in the succeeding generation is given by

$$R_{..}^{\bullet\prime} = \sum_{i,j} p_i' p_j' R_{ij}^{\bullet} = \frac{1}{T_{..}^{\bullet 2}} \sum_{i,j} p_i p_j R_{ij}^{\bullet}(R_{i.}^{\bullet}+\delta S_{..})(R_{.j}^{\bullet}+\delta S_{..}).$$

$$\therefore \quad R_{..}^{\bullet\prime} - R_{..}^{\bullet} = \Delta R_{..}^{\bullet} = \frac{1}{T_{..}^{\bullet 2}}\left\{ \sum_{i,j} p_i p_j R_{ij}^{\bullet} R_{i.}^{\bullet} R_{.j}^{\bullet} + 2\delta S_{..} \sum_{i,j} p_i p_j R_{ij}^{\bullet} R_{i.}^{\bullet} + \right.$$
$$\left. + R_{..}^{\bullet} \delta S_{..}^2 - R_{..}^{\bullet} T_{..}^{\bullet 2} \right\}.$$

Substituting $R_{..}^{\bullet} + \delta S_{..}$ for $T_{..}^{\bullet}$ in the numerator expression, expanding and rearranging:

$$\Delta R^{\bullet} = \frac{1}{T_{..}^{\bullet 2}}\left\{ \left(\sum_{i,j} p_i p_j R_{ij}^{\bullet} R_{i.}^{\bullet} R_{.j}^{\bullet} - R_{..}^{\bullet 3} \right) + \right.$$
$$\left. + 2\delta S_{..}\left(\sum_{i,j} p_i p_j R_{ij}^{\bullet} R_{i.}^{\bullet} - R_{..}^{\bullet 2} \right) \right\}.$$

We have ( ) $\geqslant 0$ in both cases. The first is the proven inequality of the classical model. The second follows from

$$\sum_{i,j} p_i p_j R_{ij}^{\bullet} R_{i.}^{\bullet} = \sum_i p_i R_{i.}^{\bullet 2} \geqslant \left(\sum_i p_i R_{i.}^{\bullet} \right)^2 = R_{..}^{\bullet 2}.$$

Thus a sufficient condition for $\Delta R_{..}^{\bullet} \geqslant 0$ is $\delta S_{..} \geqslant 0$. That $\Delta R_{..}^{\bullet} \geqslant 0$ for positive dilution is almost obvious if we compare the actual selective changes with those which would occur if $\{R_{ij}^{\bullet}\}$ were the fitness matrix in the classical model.

It follows that $R_{..}^{\bullet}$ certainly maximizes (in the sense of reaching a local maximum of $R_{..}^{\bullet}$) if it never occurs in the course of selective changes that $\delta S_{..} < 0$. Thus $R_{..}^{\bullet}$ certainly maximizes if all $\delta S_{ij} \geqslant 0$ and therefore also if all $(\delta a_{rel.})_{ij} \geqslant 0$. It still does so even if some or all δa_{ij} are negative, for, as we have seen δS_{ij} is independant of δa_{ij}.

8 W. D. HAMILTON

Here then we have discovered a quantity, inclusive fitness, which under the conditions of the model tends to maximize in much the same way that fitness tends to maximize in the simpler classical model. For an important class of genetic effects where the individual is supposed to dispense benefits to his neighbours, we have formally proved that the average inclusive fitness in the population will always increase. For cases where individuals may dispense harm to their neighbours we merely know, roughly speaking, that the change in gene frequency in each generation is aimed somewhere in the direction of a local maximum of average inclusive fitness,† but may, for all the present analysis has told us, overshoot it in such a way as to produce a lower value.

As to the nature of inclusive fitness it may perhaps help to clarify the notion if we now give a slightly different verbal presentation. Inclusive fitness may be imagined as the personal fitness which an individual actually expresses in its production of adult offspring as it becomes after it has been first stripped and then augmented in a certain way. It is stripped of all components which can be considered as due to the individual's social environment, leaving the fitness which he would express if not exposed to any of the harms or benefits of that environment. This quantity is then augmented by certain fractions of the quantities of harm and benefit which the individual himself causes to the fitnesses of his neighbours. The fractions in question are simply the coefficients of relationship appropriate to the neighbours whom he affects: unity for clonal individuals, one-half for sibs, one-quarter for half-sibs, one-eighth for cousins, ... and finally zero for all neighbours whose relationship can be considered negligibly small.

Actually, in the preceding mathematical account we were not concerned with the inclusive fitness of individuals as described here but rather with certain averages of them which we call the inclusive fitnesses of types. But the idea of the inclusive fitness of an individual is nevertheless a useful one. Just as in the sense of classical selection we may consider whether a given character expressed in an individual is adaptive in the sense of being in the interest of his personal fitness or not, so in the present sense of selection we may consider whether the character or trait of behaviour is or is not adaptive in the sense of being in the interest of his inclusive fitness.

3. Three Special Cases

Equation (2) may be written

$$\Delta p_i = p_i \frac{\delta R_{i.}^\bullet - \delta R_{..}^\bullet}{1 + \delta T_{..}^\bullet}.$$ (3)

† That is, it is aimed "uphill": that it need not be at all directly towards the local maximum is well shown in the classical example illustrated by Mulholland & Smith (1959).

Now $\delta T_{ij}^{\bullet} = \sum_{!r} (\delta a_r)_{ij}$ is the sum and $\delta R^{\bullet} = \sum_r r(\delta a_r)_{ij}$ is the first moment about $r = 0$ of the array of effects $\{\delta a_{rel.}\}_{ij}$ cause by the genotype $G_i G_j$; it appears that these two parameters are sufficient to fix the progress of the system under natural selection within our general approximation.

Let

$$r_{ij}^{\bullet} = \frac{\delta R_{ij}^{\bullet}}{\delta T_{ij}^{\bullet}}, \qquad (\delta T_{ij}^{\bullet} \neq 0); \tag{4}$$

and let

$$r_{ij}^{\circ} = \frac{\delta R_{ij}^{\circ}}{\delta T_{ij}^{\circ}}, \qquad (\delta T_{ij}^{\circ} \neq 0). \tag{5}$$

These quantities can be regarded as average relationships or as the first moments of reduced arrays, similar to the first moments of probability distributions.

We now consider three special cases which serve to bring out certain important features of selection in the model.

(a) The sums $\delta T_{ij}^{\bullet}$ differ between genotypes, the reduced first moment $r^{\bullet}$ being common to all. If all higher moments are equal between genotypes, that is, if all arrays are of the same "shape", this corresponds to the case where a stereotyped social action is performed with differing intensity or frequency according to genotype.

Whether or not this is so, we may, from equation (4), substitute $r^{\bullet} \delta T_{ij}^{\bullet}$ for $\delta R_{ij}^{\bullet}$ in equation (3) and have

$$\Delta p_i = p_i r^{\bullet} \frac{\delta T_{i.}^{\bullet} - \delta T_{..}^{\bullet}}{1 + \delta T_{..}^{\bullet}}.$$

Comparing this with the corresponding equation of the classical model,

$$\Delta p_i = p_i \frac{\delta a_{i.} - \delta a_{..}}{1 + \delta a_{..}}. \tag{6}$$

we see that placing genotypic effects on a relative of degree $r^{\bullet}$ instead of reserving them for personal fitness results in a slowing of selection progress according to the fractional factor $r^{\bullet}$.

If, for example, the advantages conferred by a "classical" gene to its carriers are such that the gene spreads at a certain rate the present result tells us that in exactly similar circumstances another gene which conferred similar advantages to the sibs of the carriers would progress at exactly half this rate.

In trying to imagine a realistic situation to fit this sort of case some concern may be felt about the occasions where through the probabilistic nature of things the gene-carrier happens not to have a sib, or not to have one suitably placed to receive the benefit. Such possibilities and their frequencies of reali-

10 W. D. HAMILTON

zation must, however, all be taken into account as the effects $(\delta a_{sibs})_A$, etc., are being evaluated for the model, very much as if in a classical case allowance were being made for some degree of failure of penetrance of a gene.

(b) The reduced first moments $r_{ij}^{\bullet}$ differ between genotypes, the sum $\delta T^{\bullet}$ being common to all. From equation (4), substituting $r_{ij}^{\bullet}\delta T^{\bullet}$ for $\delta R_{ij}^{\bullet}$ in equation (3) we have

$$\Delta p_i = p_i \frac{\delta T^{\bullet}}{T^{\bullet}}(r_{i.}^{\bullet} - r_{..}^{\bullet}).$$

But it is more interesting to assume δa is also common to all genotypes. If so it follows that we can replace $\bullet$ by $\circ$ in the numerator expression of equation (3). Then, from equation (5), substituting $r_{ij}^{\circ}\delta T^{\circ}$ for δR_{ij}°, we have

$$\Delta p_i = p_i \frac{\delta T^{\circ}}{T^{\bullet}}(r_{i.}^{\circ} - r_{..}^{\circ}).$$

Hence, if a giving-trait is in question (δT° positive), genes which restrict giving to the nearest relative ($r_{i.}^{\circ}$ greatest) tend to be favoured; if a taking-trait (δT° negative), genes which cause taking from the most distant relatives tend to be favoured.

If all higher reduced moments about $r = r_{ij}^{\circ}$ are equal between genotypes it is implied that the genotype merely determines whereabouts in the field of relationship that centres on an individual a stereotyped array of effects is placed.

With many natural populations it must happen that an individual forms the centre of an actual local concentration of his relatives which is due to a general inability or disinclination of the organisms to move far from their places of birth. In such a population, which we may provisionally term "viscous", the present form of selection may apply fairly accurately to genes which affect vagrancy. It follows from the statements of the last paragraph but one that over a range of different species we would expect to find giving-traits commonest and most highly developed in the species with the most viscous populations whereas uninhibited competition should characterize species with the most freely mixing populations.

In the viscous population, however, the assumption of random mating is very unlikely to hold perfectly, so that these indications are of a rough qualitative nature only.

(c) $\delta T_{ij}^{\bullet} = 0$ for all genotypes.

$$\therefore \quad \delta T_{ij}^{\circ} = -\delta a_{ij}$$

for all genotypes, and from equation (5)

$$\delta R_{ij}^{\circ} = -\delta a_{ij} r_{ij}^{\circ}.$$

Then, from equation (3), we have

$$\Delta p_i = p_i(\delta R_{i.}^{\bullet} - \delta R_{..}^{\bullet}) = p_i\{(\delta a_{i.} + \delta R_{i.}^{\circ}) - (\delta a_{..} + \delta R_{..}^{\circ})\}$$
$$= p_i\{\delta a_{i.}(1 - r_{i.}^{\circ}) - \delta a_{..}(1 - r_{..}^{\circ})\}.$$

Such cases may be described as involving transfers of reproductive potential. They are especially relevant to competition, in which the individual can be considered as endeavouring to transfer prerequisites of survival and reproduction from his competitors to himself. In particular, if $r_{ij}^{\circ} = r^{\circ}$ for all genotypes we have

$$\Delta p_i = p_i(1 - r^{\circ})(\delta a_{i.} - \delta a_{..}).$$

Comparing this to the corresponding equation of the classical model (equation (6)) we see that there is a reduction in the rate of progress when transfers are from a relative.

It is relevant to note that Haldane (1923) in his first paper on the mathematical theory of selection pointed out the special circumstances of competition in the cases of mammalian embryos in a single uterus and of seeds both while still being nourished by a single parent plant and after their germination if they were not very thoroughly dispersed. He gave a numerical example of competition between sibs showing that the progress of gene-frequency would be slower than normal.

In such situations as this, however, where the population may be considered as subdivided into more or less standard-sized batches each of which is allotted a local standard-sized pool of reproductive potential (which in Haldane's case would consist almost entirely of prerequisites for pre-adult survival), there is, in addition to a small correcting term which we mention in the short general discussion of competition in the next section, an extra overall slowing in selection progress. This may be thought of as due to the wasting of the powers of the more fit and the protection of the less fit when these types chance to occur positively assorted (beyond any mere effect of relationship) in a locality; its importance may be judged from the fact that it ranges from zero when the batches are indefinitely large to a halving of the rate of progress for competition in pairs.

4. Artificialities of the Model

When any of the effects is negative the restrictions laid upon the model hitherto do not preclude certain situations which are clearly impossible

12   W. D. HAMILTON

from the biological point of view. It is clearly absurd if for any possible set of gene-frequencies any $a_{ij}^{\bullet}$ turns out negative; and even if the magnitude of δa_{ij} is sufficient to make $a_{ij}^{\bullet}$ positive while $1 + e_{ij}^{\circ}$ is negative the situation is still highly artificial, since it implies the possibility of a sort of overdraft on the basic unit of an individual which has to be made good from his own takings. If we call this situation "improbable" we may specify two restrictions: a weaker, $e_{ij}^{\circ} > -1$, which precludes "improbable" situations; and a stronger, $e_{ij}^{\bullet} > -1$, which precludes even the impossible situations, both being required over the whole range of possible gene-frequencies as well as the whole range of genotypes.

As has been pointed out, a formula for $e_{ij}^{\bullet}$ can only be given if we have the arrays of effects according to a double coefficient of relationship. Choosing the double coefficient (c_2, c_1) such a formula is

$$e_{ij}^{\bullet} = \sum_{c_2, c_1}^{\bullet} [c_2 \operatorname{Dev}(\delta a_{c_2, c_1})_{ij} + \tfrac{1}{2} c_1 \{\operatorname{Dev}(\delta a_{c_2, c_1})_{i.} + \operatorname{Dev}(\delta a_{c_2, c_1})_{.j}\}] + \delta T_{..}^{\circ}$$

where

$$\operatorname{Dev}(\delta a_{c_2, c_1})_{ij} = (\delta a_{c_2, c_1})_{ij} - (\delta a_{c_2, c_1})_{..} \quad \text{etc.}$$

Similarly

$$e_{ij}^{\circ} = \sum^{\circ} ['']+ \delta T_{..}^{\circ},$$

the self-effect $(\delta a_{1,\,0})_{ij}$ being in this case omitted from the summations.

The following discussion is in terms of the stronger restriction but the argument holds also for the weaker; we need only replace $^\bullet$ by $^\circ$ throughout.

If there are no dominance deviations, i.e. if

$$(\delta a_{\text{rel.}})_{ij} = \tfrac{1}{2}\{(\delta a_{\text{rel.}})_{ii} + (\delta a_{\text{rel.}})_{jj}\} \quad \text{for all } ij \text{ and rel.,}$$

it follows that each ij deviation is the sum of the $i.$ and the $j.$ deviations. In this case we have

$$e_{ij}^{\bullet} = \sum^{\bullet} r \operatorname{Dev}(\delta a_r)_{ij} + \delta T_{..}^{\bullet}.$$

Since we must have $e_{..}^{\bullet} = \delta T_{..}^{\bullet}$, it is obvious that some of the deviations must be negative.

Therefore $\delta T_{..}^{\bullet} > -1$ is a necessary condition for $e_{ij}^{\bullet} > -1$. This is, in fact, obvious when we consider that $\delta T_{..}^{\bullet} = -1$ would mean that the aggregate of individual takings was just sufficient to eat up all basic units exactly. Considering that the present use of the coefficients of relationships is only valid when selection is slow, there seems little point in attempting to derive mathematically sufficient conditions for the restriction to hold;

intuitively however it would seem that if we exclude over- and under-dominance it should be sufficient to have no homozygote with a net taking greater than unity.

Even if we could ignore the breakdown of our use of the coefficient of relationship it is clear enough that if $\delta T_{\cdot}^{\bullet}$ approaches anywhere near -1 the model is highly artificial and implies a population in a state of catastrophic decline. This does not mean, of course, that mutations causing large selfish effects cannot receive positive selection; it means that their expression must moderate with increasing gene-frequency in a way that is inconsistent with our model. The "killer" trait of *Paramoecium* might be regarded as an example of a selfish trait with potentially large effects, but with its only partially genetic mode of inheritance and inevitable density dependance it obviously requires a selection model tailored to the case, and the same is doubtless true of most "social" traits which are as extreme as this.

Really the class of model situations with negative neighbour effects which are artificial according to a strict interpretation of the assumptions must be much wider than the class which we have chosen to call "improbable". The model assumes that the magnitude of an effect does not depend either on the genotype of the effectee or on his current state with respect to the pre-requisites of fitness at the time when the effect is caused. Where taking-traits are concerned it is just possible to imagine that this is true of some kinds of surreptitious theft but in general it is more reasonable to suppose that following some sort of an encounter the limited prerequisite is divided in the ratio of the competitive abilities. Provided competitive differentials are small however, the model will not be far from the truth; the correcting term that should be added to the expression for Δp_i can be shown to be small to the third order. With giving-traits it is more reasonable to suppose that if it is the nature of the prerequisite to be transferable the individual can give away whatever fraction of his own property that his instincts incline him to. The model was designed to illuminate altruistic behaviour; the classes of selfish and competitive behaviour which it can also usefully illuminate are more restricted, especially where selective differentials are potentially large.

For loci under selection the only relatives to which our metric of relationship is strictly applicable are ancestors. Thus the chance that an arbitrary parent carries a gene picked in an offspring is $\frac{1}{2}$, the chance that an arbitrary grandparent carries it is $\frac{1}{4}$, and so on. As regards descendants, it seems intuitively plausible that for a gene which is making steady progress in gene-frequency the true expectation of genes i.b.d. in a n-th generation descendant will exceed $\frac{1}{2}^n$, and similarly that for a gene that is steadily declining in frequency the reverse will hold. Since the path of genetic connection with a simple same-generation relative like a half-sib includes an "ascending part" and a "descending part" it is tempting to imagine that the ascending part can be treated with multipliers of exactly $\frac{1}{2}$ and the descending part by multipliers consistently more or less than $\frac{1}{2}$ according to which type of selection is in progress. However, a more rigorous attack on the problem shows that it is more difficult than the corresponding one for simple descendants, where the formulation of the factor which actually replaces $\frac{1}{2}$ is quite easy at least in the case of classical selection, and the author has so far failed to reach any definite general conclusions as to the nature and extent of the error in the foregoing account which his use of the ordinary co-efficients of relationship has actually involved.

Finally, it must be pointed out that the model is not applicable to the selection of new mutations. Sibs might or might not carry the mutation depending on the point in the germ-line of the parent at which it had occurred, but for relatives in general a definite number of generations must pass before the coefficients give the true—or, under selection, the approximate—expectations of replicas. This point is favourable to the establishment of taking-traits and slightly against giving-traits. A mutation can, however, be expected to overcome any such slight initial barrier before it has recurred many times.

5. The Model Limits to the Evolution of Altruistic and Selfish Behaviour

With classical selection a genotype may be regarded as positively selected if its fitness is above the average and as counter-selected if it is below. The environment usually forces the average fitness $a_{..}$ towards unity; thus for an arbitrary genotype the sign of δa_{ij} is an indication of the kind of selection. In the present case although it is $T_{..}^\bullet$ and not $R_{..}^\bullet$ that is forced towards unity, the analogous indication is given by the inclusive fitness effect $\delta R_{ij}^\bullet$, for the remaining part, the diluting effect δS_{ij}, of the total genotypic effect $\delta T_{ij}^\bullet$ has no influence on the kind of selection. In other words the kind of selection may be considered determined by whether the inclusive fitness of a genotype is above or below average.

We proceed, therefore, to consider certain elementary criteria which determine the sign of the inclusive fitness effect. The argument applies to any genotype and subscripts can be left out.

Let

$$\delta T^\circ = k \, \delta a. \tag{7}$$

According to the signs of δa and δT° we have four types of behaviour as set out in the following diagram:

THE GENETICAL EVOLUTION OF SOCIAL BEHAVIOUR. I   15

		Neighbours	
		gain; $\delta T° +$ve	lose; $\delta T° -$ve
gains;	$\delta a +$ve	$k +$ve *Selected*	$k -$ve Selfish behaviour ?
loses;	$\delta a -$ve	$k -$ve Altruistic behaviour ?	$k +$ve *Counter- selected*

(Individual — left margin label)

The classes for which k is negative are of the greatest interest, since for these it is less obvious what will happen under selection. Also, if we regard fitness as like a substance and tending to be conserved, which must be the case in so far as it depends on the possession of material prerequisites of survival and reproduction, k −ve is the more likely situation. Perfect conservation occurs if $k = -1$. Then $\delta T^* = 0$ and $T^* = 1$: the gene-pool maintains constant "volume" from generation to generation. This case has been discussed in Case (c) of section 3. In general the value of k indicates the nature of the departure from conservation. For instance, in the case of an altruistic action $|k|$ might be called the ratio of gain involved in the action: if its value is two, two units of fitness are received by neighbours for every one lost by an altruist. In the case of a selfish action, $|k|$ might be called the ratio of diminution: if its value is again two, two units of fitness are lost by neighbours for one unit gained by the taker.

The alarm call of a bird probably involves a small extra risk to the individual making it by rendering it more noticeable to the approaching predator but the consequent reduction of risk to a nearby bird previously unaware of danger must be much greater.† We need not discuss here just how risks are to be reckoned in terms of fitness: for the present illustration it is reasonable to guess that for the generality of alarm calls k is negative but $|k| > 1$. How large must $|k|$ be for the benefit to others to outweigh the risk to self in terms of inclusive fitness?

† The alarm call often warns more than one nearby bird of course—hundreds in the case of a flock—but since the predator would hardly succeed in surprising more than one in any case the total number warned must be comparatively unimportant.

16 W. D. HAMILTON

$$\delta R^{\bullet} = \delta R^{\circ} + \delta a$$
$$= r^{\circ}\, \delta T^{\circ} + \delta a \qquad \text{from (5)}$$
$$= \delta a(kr^{\circ} + 1) \qquad \text{from (7).}$$

Thus of actions which are detrimental to individual fitness (δa $-$ve) only those for which $-k > \dfrac{1}{r^{\circ}}$ will be beneficial to inclusive fitness ($\delta R^{\bullet}$ $+$ve).

This means that for a hereditary tendency to perform an action of this kind to evolve the benefit to a sib must average at least twice the loss to the individual, the benefit to a half-sib must be at least four times the loss, to a cousin eight times and so on. To express the matter more vividly, in the world of our model organisms, whose behaviour is determined strictly by genotype, we expect to find that no one is prepared to sacrifice his life for any single person but that everyone will sacrifice it when he can thereby save more than two brothers, or four half-brothers, or eight first cousins . . . Although according to the model a tendency to simple altruistic transfers ($k = -1$) will never be evolved by natural selection, such a tendency would, in fact, receive zero counter-selection when it concerned transfers between clonal individuals. Conversely selfish transfers are always selected except when from clonal individuals.

As regards selfish traits in general (δa $+$ve, k $-$ve) the condition for a benefit to inclusive fitness is $-k < \dfrac{1}{r^{\circ}}$. Behaviour that involves taking too much from close relatives will not evolve. In the model world of genetically controlled behaviour we expect to find that sibs deprive one another of reproductive prerequisites provided they can themselves make use of at least one half of what they take; individuals deprive half-sibs of four units of reproductive potential if they can get personal use of at least one of them; and so on. Clearly from a gene's point of view it is worthwhile to deprive a large number of distant relatives in order to extract a small reproductive advantage.

REFERENCES

HALDANE, J. B. S. (1923). *Trans. Camb. phil. Soc.* **23**, 19.
HALDANE, J. B. S. & JAYAKAR, S. D. (1962). *J. Genet.* **58**, 81.
KEMPTHORNE, O. (1957). "An Introduction to Genetical Statistics". New York: John Wiley & Sons, Inc.
KINGMAN, J. F. C. (1961a). *Quart. J. Math.* **12**, 78.
KINGMAN, J. F. C. (1961b). *Proc. Camb. phil. Soc.* **57**, 574.
LI, C. C. & SACKS, L. (1954). *Biometrics*, **10**, 347.
MORAN, P. A. P. (1962). *In* "The Statistical Processes of Evolutionary Theory", p. 54. Oxford: Clarendon Press.
MULHOLLAND, H. P. & SMITH, C. A. B. (1959). *Amer. math. Mon.* **66**, 673.
SCHEUER, P. A. G. & MANDEL, S. P. H. (1959). *Heredity*, **31**, 519.

Vol. 103, No. 934   The American Naturalist   November–December 1969

ON THE EVOLUTION OF MATING SYSTEMS
IN BIRDS AND MAMMALS

Gordon H. Orians

Department of Zoology, University of Washington, Seattle, Washington 98105

Mating systems and the selective factors that molded them have had an important place in the history of the theory of natural selection. Darwin (1871) himself gave considerable thought to the nature of sexual selection and its consequences for sexual dimorphism and mating patterns. He proposed two major forces in the evolution of sexual differences. First, that the fighting and display among males for the possession of females, which is especially prominent among mammals, accounted for the evolution of secondary sexual characteristics, such as horns and antlers, which are useful in battle. This aspect of sexual selection has been generally accepted. Second, Darwin suggested that the extreme development of plumage characters among males of some birds, such as pheasants and birds of paradise— features which did not seem of use in intermale combat—could be explained as being due to the cumulative effects of sexual preference exerted by the females at the time of mating. This aspect of his theory of sexual selection was challenged by a number of workers, but Fisher (1958) clearly showed that the notion of female choice is reasonable, notwithstanding the fact that direct evidence was then scarce for species other than man.

More recently, mating systems and sexual dimorphism have been assigned an important role in the theory of Wynne-Edwards (1962) as a device for regulating the reproductive output of populations. According to Wynne-Edwards, polygyny is one of a series of restrictive population adaptations arising through group selection which controls populations by reducing collective fecundity. He argues (1962, p. 515) that this restriction of breeding activity is possible because the territorial males of polygynous and promiscuous species can be fully informed about their own reproductive activity and, if the species engages in displays at communal mating grounds, of the total of matings performed by the group as well. These males could be conditioned to respond by becoming sexually inactive when an appropriate number of matings had taken place. The value of polygyny and group displays would lie in the fact that the assessment of total reproductive output by the population would be much easier than with a monogamous mating system in which the individuals are spaced out through the environment. He further suggested (1962, p. 525) that a balanced sex ratio would be maintained in nonmonogamous species because it would facilitate

589

more intensive intermale competition and thereby provide a more sensitive index of population density and total reproductive output.

This theory would best be tested by direct demonstration of the processes that are postulated to occur. For example, if females of polygynous species are unable to mate because males withhold coition after a certain number of copulations have been achieved, if low-ranking males do not attempt to solicit copulations after the quota has been reached, or if females are not receptive to their advances if they are made, then confidence in the theory would be strengthened.

Such evidence is extremely difficult to gather in the field, but there are now available data from a number of intensively studied polygynous species of birds, indicating that all females which appear in the breeding areas are able to obtain males and raise young. Jared Verner (personal communication) has not found any evidence for unmated females in the long-billed marsh wren (*Telmatodytes palustris*), and my own intensive work on red-winged blackbirds (*Agelaius phoeniceus*) and yellow-headed blackbirds (*Xanthocephalus xanthocephalus*) has failed to reveal a nonbreeding population of females in either of these species. In both species, there is a large and readily observable floating population of males. In the great-tailed grackle (*Quiscalus mexicanus*), a highly promiscuous species, adult males remain in full breeding condition and continue to attempt copulations after all females are nesting and are no longer receptive (Selander 1965). Therefore, in these species if there is some mechanism for limiting reproductive output, it must be due either to the failure of more females to present themselves at the breeding grounds or to a lowered effort per reproducing individual. At present, there is no evidence from any species that males withhold coition from receptive females.

It is the purpose of this paper to present an alternative theory of mating systems among birds and mammals which is based upon the assumption that the evolution of mate-selection behavior by individuals of both sexes has been influenced primarily by the consequences of these choices for individual fitness. The model is based upon mate-selection processes that can be observed directly in the field, and it is capable of generating a set of predictions which can be tested against the general mating patterns of broad groups of species for which there are no detailed observations of the factors influencing mating behavior.

A NATURAL SELECTION MODEL OF MATING SYSTEMS

This model is built upon the work of a number of people, especially Maynard Smith (1958), Verner (1964), Verner and Willson (1966), Lack (1968), and Willson and Pianka (1963), to which I have added some original ideas. Existing knowledge of the mating patterns of birds and mammals has been summarized by Lack (1968) and Eisenberg (1966), respectively. All theories of sexual selection involve an element of choice, and mine is no exception. In order for discrimination to be selected for, it is necessary that (*a*) the acceptance of one mate generally precludes the acceptance of another, and (*b*) the failure to accept one mate will be followed by an opportunity to mate with other individuals with such a high probability that the loss in reproductive output resulting from rejection of a potential mate is, on the average, less than the average gains that can be realized by obtaining a mate of superior fitness (Fisher 1958, p. 144).

The first condition is met by both sexes of many species and by females in virtually all species. Basically, a female produces gametes with a large amount of stored energy, while a male produces gametes with a complete set of genetic instructions but no significant amount of stored energy. Consequently, the number of gametes that can be produced by males is potentially, and in most cases actually, very large. On the other hand, the number of eggs produced by females is ultimately limited by the amount of energy that can be mobilized for their production or subsequent care. It follows that males can be expected to increase the number of offspring they produce by mating with more than one female, but females should not have more offspring by successive matings with more than one male unless one male were to provide insufficient gametes to fertilize all the eggs, an unlikely condition (Maynard Smith 1958, p. 146).

For the same reason, errors in mate selection are more serious for females than for males. An interspecific mating that produced inviable or sterile offspring might claim the entire season's gamete production for a female, while the male could have erred to the extent of no more than a few minutes and a few readily replaced gametes.

The inescapable conclusion is that mate selection will be practiced whenever sensory capabilities and locomotor abilities permit it and that females will, in the vast majority of cases, exercise a stronger preference. It is a well-known fact that males of many species court rather indiscriminately and can, especially when deprived of sexual activity for some time, be induced to mate with remarkably incomplete stimulus objects. Such behavior could not have evolved if errors were strongly selected against among males. For this reason, the following model assumes that females make a choice among available males. Since polygyny must always be advantageous to males, its presence or absence must depend primarily upon the advantages or disadvantages to the females.

I also assume that the environment inhabited by a species is variable and that mean reproductive success uncomplicated by density effects is correlated with this variation in quality of the habitat. For the purposes of graphic presentation, environments are treated as though they can be ordered linearly with respect to their intrinsic quality, as measured by reproductive success, but this is not essential to the argument. A model based upon these assumptions is presented in figure 1.

There are two bases upon which female choice could be made. The first, already mentioned, is the genetic quality of the male, that is, the nature of the genes that will be given to the offspring from a mating with that male. Given the existence of such differences, female choice must inevitably be

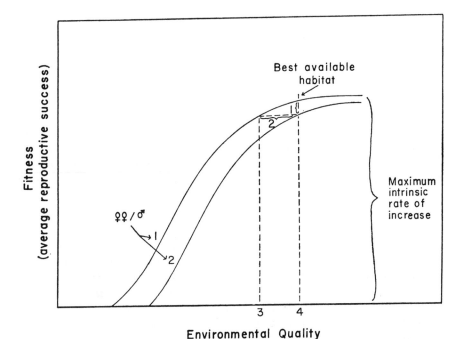

FIG. 1.—Graphic model of conditions necessary for the evolution of polygynous mating patterns. Average reproductive success is assumed to be correlated with environmental differences and females are assumed to choose their mates from available males. The distance *1* is the difference in fitness between females mated monogamously and females mated bigamously in the same environment; the distance *2* is the *polygyny threshold*, which is the minimum difference in quality of habitat held by males in the same general region sufficient to make bigamous matings by females favored by natural selection.

under strong selection, since those females mating with more fit males will thereby produce offspring that are more fit, on the average, than females mating with less fit males. Therefore, females should evolve to be especially responsive to those morphological and/or behavioral traits of males which reflect their fitness (Fisher 1958, p. 151).

In many species, however, the role of the male in reproduction is more extensive, involving provision of food for the offspring, possession of a territory within which resources can be gathered, protection against predation or inclement weather, and so forth. In these cases, I expect selection by the females to be influenced also by the quality of the territory and the probability that the male is capable of and disposed toward taking an active role in the care and defense of the offspring. The model accommodates both cases.

The model assumes that mean reproductive success is correlated with the quality of the environment in which the individuals are living. The exact shape of the function is immaterial as long as the slope is everywhere positive. Given this condition, the best strategy for a female is to mate with

EVOLUTION OF MATING SYSTEMS   593

a male on the best quality habitat and to rear her offspring, with or without his help, in that environment. However, as more individuals settle in these optimal environments, the average reproductive success will be expected to drop for three major reasons. First, the more individuals that are exploiting a given environment, the greater the likelihood that reduction of resources will adversely influence reproductive success. Second, the higher density of individuals may attract more predators, thereby increasing the probability of nest destruction. Third, if the male does play a role in the care of the offspring and females settle at a greater density than the males, then the aid of the male will have to be shared among the females, each getting less than if the male were able to devote his full attention to the offspring of one of them. For this reason, a curve representing the average reproductive success of females mated to males having more than one mate is drawn below the curve for females involved in monogamous matings.

Whatever the relative positions of these two curves, a situation may eventually be reached at which the quality of habitat on the territories of unmated males is such that the expected reproductive success of a newly arriving female is higher if she attempts to mate with a male already with one female but on a superior-quality habitat, rather than mating with an unmated male on poorer habitat. The difference in quality of habitats occupied by mated and unmated males required to make a bigamous mating advantageous for a female may be designated the *polygyny threshold* (Verner and Willson 1966), and polygyny is expected to evolve only when this situation regularly presents itself to the females of a species.

It follows that the likelihood that polygyny will evolve is influenced by all factors which determine how broad a range of environmental conditions will be occupied by the individuals of a species. For example, if the territories of the males of a species are nearly equal in quality and there are still unpaired males available, it is unlikely that the mean reproductive success of females attempting bigamous matings would be higher. Similarly, all factors that influence the amount of difference in mean reproductive success of females in monogamous and bigamous matings in equivalent environments will affect the likelihood that polygyny will evolve. More specifically, the greater the difference, that is, the farther apart the two curves are, the farther to the left will lie the polygyny threshold, and there must be greater differences in habitat quality for polygyny to pay for the females. Conversely, with less difference, that is, closer curves, a smaller difference in habitat will push part of the population to the left of the polygyny threshold.

a) Factors promoting occupancy of habitats differing widely in quality: The existence of individuals to the left of the polygyny threshold will be facilitated by (1) low mortality rates during the nonbreeding season, so that not all individuals can be accommodated in the better areas, and (2) the existence of large differences in the quality of habitats actually occupied by the species.

b) Factors influencing the differences in average reproductive success of monogamously and bigamously mated females: A major factor affecting the differences in reproductive success of females will be the role of the male in the care of the offspring. If his role is limited to the provision of genetic information, that is, if the male provides no food, territory, or protection, the success of the females will not be affected by the number of other females that have mated with the male unless his fertility declines with successive matings. Thus, in the limiting case of no male parental care, the two curves become congruent, and the mating choice by the females is based strictly upon the genotype of the male. A high degree of promiscuity is to be expected in such species.

Even if the male also cares for the offspring, in which case the number of other females he already has is a major factor in female choice, there are conditions which tend to minimize the reduction in reproductive success attendant upon polygynous matings. For example, if the successive females of a male are staggered in their breeding so that the periods of dependence of their offspring overlap little or not at all, more support from the male could be available (Verner 1964). This should give selective advantage to two different forms of female behavior. The first is the attempt to exclude additional females from the territory of the male until such time as the overlap in dependency periods of the young would be minimized, and second, the avoidance by newly arriving females of territories in which a prior female was just beginning to breed.

Another factor influencing the difference in reproductive success of monogamously and bigamously mated females is the nature of the food resources present in the area. If the food for the young is not being replenished during the breeding period, the individuals breeding earliest should experience the best foraging conditions, while later-breeding individuals are exploiting an already depressed supply. However, if the recruitment to the food supply is considerable during the breeding period, conditions for later breeders may be no worse (or may even be better) than conditions for earlier breeders. Therefore, other things being equal, species exploiting food supplies that are continually renewed are more likely to cross the polygyny threshold.

The above arguments all assume that the number of offspring being raised by a female (or pair) has evolved under the influence of natural selection to correspond to that which, on the average, is the largest number for which sufficient energy can be mobilized. The theoretical basis for this assumption and the supporting empirical data have been extensively summarized by Lack (1954 p. 21–44). However, in some species the number of offspring produced is strongly influenced by other factors. In such cases, the existence of male parental care may be of little consequence, and mate choice should be made primarily or strictly on the basis of phenotype and territory quality. Cases in which this situation may be operative will be discussed later.

The model implies the existence of several processes which can be directly observed in nature. For example, it should be true that females mate with already mated males when unmated males are readily available and perceived by the females. Evidence that this is true has been gathered for the long-billed marsh wren (Verner 1964), the red-winged blackbird, and the yellow-headed blackbird (Orians, unpublished observations). The great variations in number of females mated with different males in other polygynous species are suggestive of the widespread occurrence of this phenomenon. Verner has also shown that the number of females attracted to a male long-billed marsh wren is correlated with the features of his territory relatable to the available food.

The model also predicts that there should not be a negative correlation between average reproductive success per female and number of females mated with a given male, since females are assumed to enter polygynous matings only when it is advantageous for them to do so. This prediction has been verified for the red-winged blackbird (Haigh 1968).

The model does not require a skewed sex ratio in the breeding population for the initiation and evolution of polygynous mating patterns. This is important because there are theoretical reasons for expecting a sex ratio near equality when the young become independent in most species (Fisher 1958; Kolman 1960) and because sex ratios at the time of fledging are near equality in all polygynous species so far investigated (Haigh 1968; Selander 1960, 1961; Williams 1940).

Using the postulates upon which the polygyny model was erected, a series of seven predictions about mating patterns can be made. These predictions are subject to direct verification or falsification. Current knowledge of mating patterns among mammals, though by no means complete, is extensive enough that I can reasonably assess the goodness of the predictions. Moreover, the predictions from the model serve to draw attention to those cases which would be most rewarding of further study. Though predictions of wide application can be made, I restrict my present consideration to birds and mammals, the groups with which I am most familiar.

1. *Polyandry should be rare among all animal groups.* This prediction follows directly from the basic attributes of maleness and femaleness. A female could presumably increase her reproductive output if several males could be induced to care for her offspring, but such a situation would in most cases be sufficiently disadvantageous to the males to cause the evolution of male behavior patterns that would prevent the system from evolving. However, once a basic sexual role reversal had evolved in a species, males might profit by associating themselves with females on better territories, thus leading to polyandry.

The actual incidence of polyandry among birds and mammals is difficult to assess. Among mammals, Eisenberg (1966) reviewed several polygynous or promiscuous mating patterns but found no good case of a simultaneous association of a female with more than one male. Among birds the situation is confused by the fact that there has been a tendency to assume polyandry in all those species with a reversal of sexual roles. Most such species are tropical and subtropical in distribution and are not well known ecologically, making comparisons even more difficult. For example, it was formerly believed that the phalaropes (Phalaropodidae) were polyandrous, but recent data indicate that, though incubation and care of the young are exclusively the role of the males, the species are nonetheless monogamous (Höhn 1967). However, there is good evidence of polyandry for at least some species in five groups of birds, all with precocial young: the button quails (Turnicidae), painted snipe (Rostratulidae), jacanas (Jacanidae), tinamous (Tinamidae) and rails (Rallidae). Details in all these cases are summarized by Lack (1968).

Among rheas (Rheidae), emus (Dromiceidae), cassowaries (Casuariidae), kiwis (Apterygidae), tinamous and button quails, the males normally incubate the eggs and care for the young, but most species are apparently monogamous. In the best-studied species, the brushland tinamou (*Nothoprocta cinerascens*), the males defend territories while the females travel in small groups. Several females lay eggs in a single nest, but each female may mate with several males in rapid succession (Lancaster 1964). The classification of this type of pattern is somewhat ambiguous since the male may have several females at once, but each female may nevertheless mate with several different males during the laying of a single "clutch" of eggs. A clearer case of polyandry with a reversal of parental role is provided by the pheasant-tailed jacana (*Hydrophasianus chirurgus*) (Hoffmann 1949).

The evolution of sexual role reversal in birds may have had its origins in a monogamous system with equal sharing of parental care by the two sexes. In many such species, the males incubate first while the females recover the energy lost during egg production. Under such circumstances, especially among species with precocial young, if the females were able to obtain enough energy to produce more eggs, it would be advantageous to mate with another male were one available. It would also be advantageous for the incubating males to induce other females to mate with them and deposit their eggs in the nests, so that the tinamou type of system could readily evolve. A further advantage of such a system is that the length of time the nest is available to be destroyed by predators is reduced, since a full clutch of eggs is placed in the nest in a shorter period of time than if a single female were doing all the laying.

2. *Monogamy should be relatively rare among mammals but should be the predominant mating pattern among birds.* The physiology of mammalian reproduction dictates a minor role of the male in the care of the offspring, whereas among birds the only activity for which males are not equally adept as females is egg laying. This prediction is readily verified. In his extensive review of mammalian social organization, Eisenberg (1966) pointed out that there are very few cases of known monogamy among mammals, the apparent exceptions being the marmosets (Callithricidae), gibbons (Hylobatinae), beavers (Castoridae), the hooded seal (*Cystophora cristata*), the only pinniped known to form a stable family unit, and a number of terrestrial carnivores, such as foxes, badgers, and viverrids. The situation is more difficult to determine among large ungulates, but tem-

porary pair bonds that are apparently monogamous may form in hyraxes (*Dendrohyrax*), rhinoceroses, and some deer (*Capreolus capreolus, Odocoileus*).

The mammals among which monogamy is probably the most prevalent are the terrestrial carnivores, and they provide the most prominent exception to the generalization that the role of the male is limited or nonexistent in the care of offspring. For a carnivore, capturing food is a difficult task, and males can and do make kills and deliver the prey to either the female, who converts it to milk, or to the young once they are old enough to be able to ingest meat. It is difficult to imagine a comparable role for a male herbivore.

In contrast, monogamy is the prevalent mating pattern in the majority of bird species in virtually all families and orders. Assuming that all hummingbirds are promiscuous, Lack (1968) surmises that about 91% of all bird species are monogamous. Given the properties of the model, this is to be expected; but polygyny and promiscuity do exist among birds, and the model, if it is to be generally useful, must provide predictions capable of explaining those cases. Because polygyny should seldom evolve among birds, this group provides a particularly useful test case for the validity of the model. Fortunately, the mating patterns of birds are well enough known to allow tests of the predictions in most cases.

3. *Polygyny should evolve more readily among precocial birds than among altricial species.* This prediction follows from the ability of many precocial young to find their own food and be relatively independent of the provisioning activities of the adults. This decreases the potential role of the male. There are species with precocial young, such as gulls, in which the young, though able to run around actively at birth, are not able to forage for themselves. These species are not included in this prediction.

This prediction is only partially fulfilled in nature. Polygynous and promiscuous species are numerous among upland game birds (Phasianidae, Tetraonidae), and there are a few species among the shorebirds (ruff, *Philomachus pugnax;* buff-breasted sandpiper, *Tryngites subruficollis;* pectoral sandpiper, *Erolia melanotos;* and great snipe, *Capella media*), but most members of the Charadriidae (plovers) and Scolopacidae (sandpipers) are monogamous (Lack 1968, p. 116). Polygyny is also rare in the Anatidae (swans, geese, and ducks). In many of the monogamous species, both sexes take an active role in the care of the young, leading them to suitable foraging areas and keeping on the alert for predators. In geese the male defends the nest and young from predators, but in most species of ducks the female alone cares for the young, and yet monogamous pair bonding seems to be the rule. It is significant that the known exceptions among ducks are all tropical species (Lack 1968, p. 123), suggesting that perhaps the prevalence of monogamy among high latitude species may be the result of the advantage of pair formation on the wintering ground and rapid initiation of breeding which give a stronger advantage to monogamy for the males than would otherwise be the case.

4. *Polygyny is likely to evolve in species with altricial young that nest*

in marshes. The marsh environment possesses several features that make it more likely that the polygyny threshold will be crossed than in any other environment. First, the range in productivity of marshes greatly exceeds that found in upland habitats (Verner and Willson 1966). Differences of over tenfold are not unusual in aquatic environments, whereas the difference between the most productive and least productive woods is much less. Moreover, great differences in productivity in a terrestrial environment are likely to result in a sufficiently altered vegetational profile to cause a change in species rather than the occupation of a broad gradient by one species (L. L. Wolf, personal communication). In marshes, however, productivity differences are not necessarily associated with vegetation structure, and striking changes in vegetational features regularly occur within the span of a few years (Weller and Spatcher 1965), so that opportunities for evolving species that occupy only a small segment of the marsh vegetation pattern are more limited than in terrestrial environments.

The food supply for insectivorous birds in marshes is often rapidly renewed. Many of the breeding passerine birds of marshes exploit primarily the emerging individuals of insects with aquatic larval and terrestrial adult stages. These insects are vulnerable for only a few hours during their lives, and those not taken on the day they emerge are mostly unavailable on subsequent days. Therefore, the supply of food on a given day is not significantly affected by the number of insects removed from the system on previous days but rather upon those factors that regulate the size of the emergence on that particular day. In contrast, the supply of insects on the foliage of trees and shrubs may be seriously depleted by the foraging activities of birds, which lowers the expected reproductive success of later arriving individuals.

In a review of the mating systems of North American passerines, Verner and Willson (1966) demonstrated that, though marsh-nesting species constitute only a small fraction of the total species (about 5%), eight of the 15 polygynous species breed in marshes. Polygyny is also prevalent among the marsh-nesting weaverbirds (Ploceidae) in Africa (Crook 1963, 1964). Also, the only known nonpasserine species with altricial young that is regularly polygynous is the bittern (*Botaurus stellaris*), a marsh-nesting species (Gaukler and Kraus 1965).

5. *Polygyny should be more prevalent among species inhabiting early successional habitats.* Like marshes, early successional terrestrial vegetation changes rapidly, thus discouraging the evolution of species adapted specifically to minor variants of it. In addition, there are reasons for suspecting that variations in food supply in early successional sites might be considerable. Early successional plants are characterized by rapid growth and the apportionment of large amounts of energy to reproduction. They probably also devote less energy, on the average, to antiherbivore devices than plants of later successional stages. Accordingly, they should be vulnerable to insect attack when found, and may owe their success in part to the fact that many patches escape detection. If this is true, patches of early

successional vegetation should consist of some not yet found by insects, therefore containing relatively little food, and others supporting large populations of grazing insects. Five of the 15 regularly polygynous species of North American passerines breed in prairie or savannah habitats (Verner and Willson 1966; Zimmerman 1966), and some of them, notably the dick-cissel (*Spiza americana*) and the bobolink (*Dolichonyx oryzivorus*) are restricted to the very early successional stages of grassland vegetation (Zimmerman 1966). Differential food supply and its possible correlation with the number of females per male in different patches of early succes-sional vegetation has never been measured, but it should not be difficult to do so.

6. *Polygyny should be more prevalent among species in which feeding areas are widespread, but nesting sites are restricted.* If nest sites are re-stricted and a single male holds several of them, it should be advantageous for females to mate with such males even if they are already mated, par-ticularly if the alternative is accepting an inferior site or no site at all. Two of the polygynous passerines of North America, the house wren (*Troglodytes aedon*) and the winter wren (*T. troglodytes*), nest in cavities but are unable to excavate their own, and the same is true of the poly-gynous pied flycatcher (*Muscicapa hypoleuca*) of Europe (Curio 1959; von Haartman 1954). This may also be the explanation of the prevalence of polygyny among savannah species of weaverbirds in Africa and Asia (Crook 1962, 1963, 1964), since these are species that feed in grassland but require trees for their nests.

7. *Polygyny and promiscuity should be more prevalent among species in which clutch size is strongly influenced by factors other than the number of offspring that can be supported by the parents.* Clutches smaller than the number of young the parents can feed successfully might occur in species in which the adults feed primarily on low-energy food sources such as pulpy fruits and nectar which, though sufficient for maintenance energy, are not good for egg production. This supposition is supported by the fact that hummingbirds, unlike most avian species, lay no more than two eggs in all geographical areas and do not show the latitudinal gradient in clutch size characteristic of most birds. In addition, high predation rates may select against high feeding rates in tropical environments, reducing clutches below what the parents could feed (Skutch 1949). Finally, in stable en-vironments competitive ability may demand considerable time expenditure (Cody 1966), so that foraging time is reduced and males may spend all or most of their time at these activities. If any of these factors are operating, the contribution of the male to reproductive success by means of food de-livery to the nestlings would be decreased, and conditions favorable for polygyny would be created.

This prediction cannot be tested directly at present, but it is noteworthy, as pointed out by Snow (1963), that polygyny or promiscuity associated with lek displays in the tropics occur only among fruit- and nectar-eating birds such as hummingbirds, manakins (Snow 1962a, 1962b), cotingas

(Snow 1961; Gilliard 1962), birds of paradise (Iredale 1950) and bower-birds (Marshall 1954; Gilliard 1959a, 1959b), and not among insectivorous species, including insectivorous species in the same families, all of which are apparently monogamous. Only one of the lek species has been studied in detail (Snow 1962a, 1962b), and it was shown that fruits supporting the adults could not have been in short supply during the breeding season, but the effects of this kind of food source on egg production and nature of the food delivered to the young are completely unknown. There is evidence from the bowerbirds that frugivorous species feed their young on insects (Marshall 1954). Conversely, obligatory fruit eaters, such as parrots, have extremely long nestling periods, suggesting that rapid nestling growth and a fruit diet are mutually incompatible.

CONSEQUENCES OF THE EVOLUTION OF POLYGYNY

If polygynous mating systems evolve from monogamous ones as a result of the existence of choice situations in which it is advantageous for females to select mates already having at least one mate, the very existence of this choice system creates other selective forces that further influence the mating pattern and the morphology of the sexes.

First, in polygynous birds, there should be very keen competition among males for the better-quality territories, because possession of a high-quality territory is likely to result in the attraction of more than one female. The increased intermale competition for good areas should lead to stronger selection for secondary sexual characteristics useful in these contests. The existence of these characteristics, as indicated earlier, is well known among mammals, and Selander (1958) has demonstrated a strong correlation between the amount of sexual dimorphism in size and the degree of polygyny and promiscuity in mating pattern among American blackbirds (Icteridae). Great dimorphism in size is also characteristic of the polygynous marsh-nesting weaverbirds in Africa, the males even showing remarkable convergence toward the plumage patterns shown by males of polygynous species of icterids.

Nevertheless, unless the species are continually evolving to become more highly dimorphic, there must be counterselection against the more dimorphic individuals that stabilizes the degree of dimorphism at its present value. There are two obvious candidates for this counterselection. The first is predation, since the males are rendered exceedingly conspicuous by both their appearance and their behavior patterns. The second derives from the adverse ecological effects of larger size. In polygynous species, females are presumably not normally under selection for size other than that dictated by their basic ecological relationships with their environment (Amadon 1959). The greater the degree to which males depart from the presumably optimal size of the females, the more poorly adapted they should be for general existence, unless by their increased size they are able to exploit food resources not available to the smaller females. To date the only demonstration of a higher mortality rate during the nonbreeding season for males of a highly dimorphic species is that of Selander (1965) for the great-tailed grackle in south-central Texas. Sexing birds returning to communal roosts by examination of greatly enlarged photographs, Selander was able to show that males died at about twice the rate of the females during the winter. He also observed that the large tails of the males interfered with flying in strong winds, and that extremely strong winds completely prevented males from flying, while females were still able to navigate, though with difficulty.

Contrary to the theory of Wynne-Edwards, the model developed here predicts delayed maturation on the part of the males but not of females. Unless females are capable of preventing other females from settling in the area, all females should be able to obtain mates and reproduce. In fact, it should be extremely difficult for females to exclude other females from the territory of their mate. During the nest-building and egg-laying periods, defense of the area is easy; but once incubation has begun, eviction of a persistent intruder can only be accomplished at the expense of chilling and possible loss of the clutch of eggs. It is highly unlikely that the adverse effects of the second female could be so great as to select for such behavior. Moreover, by the time the first female is already incubating, the potential period of overlap in time when young are being fed is already minimized.

On the other hand, the strong competition among males for suitable territories and the failure of males with poor territories to obtain mates at all should produce a floating population of nonbreeding males. Such floating populations are known to be characteristic of a number of polygynous species of birds. Assuming that older and more experienced birds will be at an advantage in competition for territories, the chances of success for younger birds should be very low. If attempts to obtain territories result in higher mortality rates of the young birds and probability of success is sufficiently low, individuals making vigorous attempts might be selected against and delayed maturation would result. Selander (1965) gives a more detailed development of this argument. In the red-winged blackbird, first-year males do not acquire the full adult plumage, their testes develop later in the spring and do not reach the size characteristic of older males (Wright and Wright 1944), and they usually do not breed, though they may do so if the supply of adult males is reduced in some manner (Orians, unpublished field data).

SUMMARY

Predictions from a theory assuming mate selection on the part of females, which maximizes reproductive success of individuals, are found to accord closely, though not completely, with known mating patterns. These predictions are that (1) polyandry should be rare, (2) polygyny should be more common among mammals than among birds, (3) polygyny should be more prevalent among precocial than among altricial birds, (4) conditions for polygyny should be met in marshes more regularly than among terrestrial environments, (5) polygyny should be more prevalent among species of early successional habitats, (6) polygyny should be more prevalent among species in which feeding areas are widespread but nesting sites are restricted, and (7) polygyny should evolve more readily among species in which clutch size is strongly influenced by factors other than the ability of the adults to provide food for the young. Most cases of polygyny in birds, a group in which monogamy is the most common mating pattern, can be explained on the basis of the model, and those cases not apparently fitting into the predictions are clearly indicated. Thus, there is no need at present to invoke more complicated and restrictive mechanisms to explain the mating patterns known to exist.

ACKNOWLEDGMENTS

Valuable suggestions for improvement of the manuscript were provided by Henry S. Horn, Robert H. MacArthur, Dennis R. Paulson, Jared Verner, and Edwin O. Willis. My fieldwork on blackbird social organization has been supported by funds from the National Science Foundation.

LITERATURE CITED

Amadon, D. 1959. The significance of sexual differences in size among birds. Amer. Phil. Soc., Proc. 103:531–536.

Cody, M. L. 1966. A general theory of clutch size. Evolution 20:174–184.

Crook, J. H. 1962. The adaptive significance of pair formation types in weaver birds. Symposia (Zool. Soc. London) 8:57–70.

———. 1963. Monogamy, polygamy, and food supply. Discovery 24:35–41.

———. 1964. The evolution of social organization and visual communication in the weaver birds (Ploceinae). Behaviour (Suppl. 10):1–178.

Curio, E. 1959. Verhaltenstudien am Trauerschnäpper. Z. Tierpsychol. 3:1–118.

Darwin, C. 1871. The descent of man and selection in relation to sex. Appleton, New York. 2 Vol.

Eisenberg, J. F. 1966. The social organizations of mammals. Handbuch der Zoologie. Walter de Gruyter, Berlin. 8 (39):1–92.

Fisher, R. A. 1958. The genetical theory of natural selection. Dover, New York. 291 p.

Gaukler, A., and M. Kraus. 1965. Zur Brutbiologie der grossen Rohrdommel (Botaurus stellaris). Vogelwelt 86:129–146.

Gilliard, E. T. 1959a. Notes on the courtship behavior of the blue-backed manakin (Chiroxiphia pareola). Amer. Mus. Novitates. No. 1942:1–19.

———. 1959b. A comparative analysis of courtship movements in closely allied bowerbirds of the genus Chlamydera. Amer. Mus. Novitates. No. 1936:1–8.

———. 1962. On the breeding behavior of the cock-of-the-rock (Aves, Rupicola rupicola). Amer. Mus. Natur. Hist. Bull. 124:31–65.

Haartman, L. von. 1954. Der Trauerfliegenschnäpper. III. Die Nahrungsbiologie. Acta Zool. Fennica 83:1–96.

Haigh, C. R. 1968. Sexual dimorphism, sex ratios, and polygyny in the red-winged blackbird. Ph.D. thesis. Univ. Washington. Seattle. 116 p.

Hoffmann, A. 1949. Über die Brutpflege des polyandrischen Wasserfasans, Hydrophasianus chirurgus (Scop.). Zool. Jahrb. (Syst.), 78:367–403.

Höhn, E. O. 1967. Observations on the breeding biology of Wilson's phalarope (*Steganopus tricolor*) in central Alberta. Auk 84:220–244.

Iredale, T. 1950. Birds of paradise and bower birds. Georgian H., Melbourne. 239 p.

Kolman, W. A. 1960. The mechanism of natural selection for the sex ratio. Amer. Natur. 94:373–377.

Lack, D. 1954. The natural regulation of animal numbers. Clarendon, Oxford. 343 p.

———. 1968. Ecological adaptations for breeding in birds. Methuen, London. 409 p.

Lancaster, D. A. 1964. Biology of the brushland tinamou, *Nothoprocta cinerascens*. Amer. Mus. Natur. Hist. Bull. 127:270–314.

Marshall, A. J. 1954. Bower-birds. Clarendon, Oxford. 208 p.

Maynard Smith, J. 1958. The theory of evolution. Penguin, Harmondsworth. 320 p.

Selander, R. K. 1958. Age determination and molt in the boat-tailed grackle. Condor 60:355–376.

———. 1960. Sex ratio of nestling and clutch size in the boat-tailed grackle. Condor 62:34–44.

———. 1961. Supplemental data on the sex ratio of nestling boat-tailed grackles. Condor 63:504.

———. 1965. On mating systems and sexual selection. Amer. Natur. 99:129–141.

Skutch, A. F. 1949. Do tropical birds rear as many young as they can nourish? Ibis 91:430–455.

Snow, B. K. 1961. Notes on the behavior of three cotingidae. Auk 78:150–161.

Snow, D. W. 1962a. A field study of the black and white manakin, *Manacus manacus*, in Trinidad. Zoologica 47:60–104.

———. 1962b. A field study of the golden-headed manakin, *Pipra erythrocephala*, in Trinidad, W. I. Zoologica 47:183–198.

———. 1963. The evolution of manakin displays, p. 553–561. *In* 13th Int. Ornithological Congr., Proc., Ithaca, N.Y., 1962.

Verner, J. 1964. Evolution of polygamy in the long-billed marsh wren. Evolution 18:252–261.

Verner, J., and M. F. Willson. 1966. The influence of habitats on mating systems of North American passerine birds. Ecology 47:143–147.

Weller, M. W., and C. S. Spatcher. 1965. Role of habitat in the distribution and abundance of marsh birds. Iowa State Univ. Agr. Home Econ. Exp. Sta. Spec. Rep. No. 43.

Williams, J. F. 1940. The sex ratio in nestling eastern red-wings. Wilson Bull. 52:267–277.

Willson, M. F., and E. R. Pianka. 1963. Sexual selection, sex ratio, and mating system. Amer. Natur. 97:405–407.

Wright, P. L., and M. H. Wright. 1944. The reproductive cycle of the male red-winged blackbird. Condor 46:46–59.

Wynne-Edwards, V. C. 1962. Animal dispersion in relation to social behavior. Oliver & Boyd, Edinburgh. 653 p.

Zimmerman, J. L. 1966. Polygyny in the dickcissel. Auk 83:534–546.

7

ROBERT L. TRIVERS

HARVARD UNIVERSITY

Parental Investment and Sexual Selection

Introduction

Charles Darwin's (1871) treatment of the topic of sexual selection was sometimes confused because he lacked a general framework within which to relate the variables he perceived to be important: sex-linked inheritance, sex ratio at conception, differential mortality, parental care, and the form of the breeding system (monogamy, polygyny, polyandry, or promiscuity). This confusion permitted others to attempt to show that Darwin's terminology was imprecise, that he misinterpreted the function of some structures, and that the influence of sexual selection was greatly overrated. Huxley (1938), for example, dismisses the importance of female choice without evidence or theoretical argument, and he doubts the prevalence of adaptations in males that decrease their chances of surviving but are selected because they lead to high reproductive success. Some important advances, however, have been achieved since Darwin's work. The genetics of sex has now been clarified, and Fisher (1958) has produced a model to explain sex ratios at conception, a model recently extended to include special mechanisms that operate under inbreeding (Hamilton 1967). Data from the laboratory and the field have confirmed that females are capable of very subtle choices (for example, Petit & Ehrman 1969), and Bateman (1948) has suggested a general basis for female choice and male-male competition, and he has produced precise data on one species to support his argument.

I thank E. Mayr for providing me at an early date with the key reference. I thank J. Cohen, I. DeVore, W. H. Drury, M. Gadgil, W. D. Hamilton, J. Roughgarden, and T. Schoener for comment and discussion. I thank M. Sutherland (Harvard Statistics Department) for statistical work on my *A. garmani* data, H. Hare for help with references, and V. Hogan for expert typing of drafts of the paper. I thank especially E. E. Williams for comment, discussion and unfailing support throughout. The work was completed under a National Science Foundation predoctoral fellowship and partly supported by NSF Grant B019801 to E. E. Williams.

136

This paper presents a general framework within which to consider sexual selection. In it I attempt to define and interrelate the key variables. No attempt is made to review the large, scattered literature relevant to sexual selection. Instead, arguments are presented on how one might *expect* natural selection to act on the sexes, and some data are presented to support these arguments.

Variance in Reproductive Success

Darwin defined sexual selection as (1) competition within one sex for members of the opposite sex and (2) differential choice by members of one sex for members of the opposite sex, and he pointed out that this usually meant males competing with each other for females and females choosing some males rather than others. To study these phenomena one needs accurate data on differential reproductive success analysed by sex. Accurate data on female reproductive success are available for many species, but similar data on males are very difficult to gather, even in those species that tend toward monogamy. The human species illustrates this point. In any society it is relatively easy to assign accurately the children to their biological mothers, but an element of uncertainty attaches to the assignment of children to their biological fathers. For example, Henry Harpending (personal communication) has gathered biochemical data on the Kalahari Bushmen showing that about two per cent of the children in that society do not belong to the father to whom they are commonly attributed. Data on the human species are, of course, much more detailed than similar data on other species.

To gather precise data on both sexes Bateman (1948) studied a single species. *Drosophila melanogaster,* under laboratory conditions. By using a chromosomally marked individual in competition with individuals bearing different markers, and by searching for the markers in the offspring, he was able to measure the reproductive success of each individual, whether female or male. His method consisted of introducing five adult males to five adult female virgins, so that each female had a choice of five males and each male competed with four other males.

Data from numerous competition experiments with *Drosophila* revealed three important sexual differences: (1) Male reproductive success varied much more widely than female reproductive success. Only four per cent of the females failed to produce any surviving offspring, while 21 per cent of the males so failed. Some males, on the other hand, were phenomenally successful, producing nearly three times as many offspring as the most successful female. (2) Female reproductive success did not appear to be limited by ability to attract males. The four per cent who failed to copulate were apparently courted as vigorously as those who did copulate. On the other hand, male reproductive success was severely limited by ability to attract or arouse females. The 21 per cent who failed to reproduce showed no disinterest in trying to copulate, only an inability to be accepted. (3) A female's reproductive success did not increase much, if any, after the first copulation and not at all after the second; most females were uninterested in copulating more than once or twice. As shown by genetic markers in the offspring, males showed an almost linear increase in reproductive success with increased copulations. (A corollary of this finding is that males tended not to mate with the same female twice.) Although these results were obtained in the laboratory, they may apply with even greater force to the wild, where males are not limited to five females and where females have a wider range of males from which to choose.

Bateman argued that his results could be explained by reference to the energy investment of each sex in their sex cells. Since male *Drosophila* invest very little metabolic energy in the production of a given sex cell, whereas females invest considerable energy, a male's reproductive success is not limited by his ability to produce sex cells but by his ability to fertilize eggs with these cells. A female's reproductive success is not limited by her ability to have her eggs fertilized but by her ability to produce eggs. Since in almost all animal and plant species the male produces sex cells that are tiny by comparison to the female's sex cells, Bateman (1948) argued that his results should apply very widely, that is, to "all but a few very primitive organisms, and those in which monogamy combined with a sex ratio of unity eliminated all intra-sexual selection."

Good field data on reproductive success are difficult to find, but what data exist, in conjunction with the assumption that male reproductive success varies as a function of the number of copulations,[1] support the contention that in all species, except those mentioned below in which male parental care may be a limiting resource for females, male reproductive success varies more than female reproductive success. This is supported, for example, by data from dragonflies (Jacobs 1955), baboons (DeVore 1965), common frogs (Savage 1961), prairie chickens (Robel 1966), sage grouse (Scott 1942), black grouse (Koivisto 1965), elephant seals (LeBoeuf & Peterson, 1969), dung flies (Parker 1970a) and some anoline lizards (Rand 1967 and Trivers, in preparation, discussed below.) Circumstantial evidence exists for other lizards (for example, Blair 1960, Harris 1964) and for many mammals (see Eisenberg 1965). In monogamous species, male reproductive success would be expected to vary as female reproductive success, but there is always the possibility of adultery and differential female mortality (discussed below) and these factors should increase the

1. Selection should favor males producing such an abundance of sperm that they fertilize all a female's available eggs with a single copulation. Futhermore, to decrease competition among offspring, natural selection may favor females who prefer single paternity for each batch of eggs (see Hamilton 1964). The tendency for females to copulate only once or twice per batch of eggs is supported by data for many species (see, for example, Bateman 1948, Savage 1961, Burns 1968 but see also Parker 1970b).

variance of male reproductive success without significantly altering that
of the female.

Relative Parental Investment

Bateman's argument can be stated in a more precise and general form
such that the breeding system (for example, monogamy) as well as the
adult sex ratio become functions of a single variable controlling sexual selection. I first define parental investment as *any investment by the parent
in an individual offspring that increases the offspring's chance of surviving
(and hence reproductive success) at the cost of the parent's ability to invest in other offspring.* So defined, parental investment includes the metabolic investment in the primary sex cells but refers to any investment (such
as feeding or guarding the young) that benefits the young. It does not include effort expended in finding a member of the opposite sex or in subdu-
ing members of one's own sex in order to mate with a member of the opposite sex, since such effort (except in special cases) does not affect the
survival chances of the resulting offspring and is therefore not *parental*
investment.

Each offspring can be viewed as an investment independent of other offspring, increasing investment in one offspring tending to decrease investment in others. I measure the size of a parental investment by reference to
its negative effect on the parent's ability to invest in other offspring: a large
parental investment is one that strongly decreases the parent's ability to
produce other offspring. There is no necessary correlation between the size
of parental investment in an offspring and its benefit for the young. Indeed,
one can show that during a breeding season the benefit from a given
parental investment must decrease at some point or else species would not
tend to produce any fixed number of offspring per season. Decrease in reproductive success resulting from the negative effect of parental investment on *nonparental* forms of reproductive effort (such as sexual competition for mates) is excluded from the measurement of parental investment.
In effect, then, I am here considering reproductive success as if the only
relevant variable were parental investment.

For a given reproductive season one can define the total parental investment of an individual as the sum of its investments in each of its offspring
produced during that season, and one assumes that natural selections has
favored the total parental investment that leads to maximum net reproductive success. Dividing the total parental investment by the number of
individuals produced by the parent gives the typical parental investment
by an individual per offspring. Bateman's argument can now be reformulated as follows. Since the total number of offspring produced by one sex
of a sexually reproducing species must equal the total number produced
by the other (and assuming the sexes differ in no other way than in their

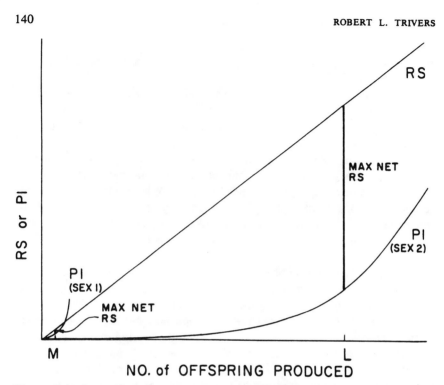

Figure 7.1. Reproductive success (RS) and decrease in future reproductive success resulting from parental investment (PI) are graphed as functions of the number of offspring produced by individuals of the two sexes. At M and L the net reproductive success reaches a maximum for sex 1 and sex 2 respectively. Sex 2 is limited by sex 1 (see text). The shape of the PI curves need not be specified exactly.

typical parental investment per offspring)[2] then the sex whose typical parental investment is greater than that of the opposite sex will become a limiting resource for that sex. Individuals of the sex investing less will compete among themselves to breed with members of the sex investing more, since an individual of the former can increase its reproductive success by investing successively in the offspring of several members of the limiting sex. By assuming a simple relationship between degree of parental investment and number of offspring produced, the argument can be presented graphically (Figure 7.1). The potential for sexual competition in the sex investing less can be measured by calcuiating the ratio of the number of offspring that sex optimally produces (as a function of parental invest-

2. In particular, I assume an approximately 50/50 sex ratio at conception (Fisher 1958) and no differential mortality by sex, because I later derive differential mortality as a function of reproductive strategies determined by sexual selection. (Differential maturation, which affects the adult sex ratio, can also be treated as a function of sexual selection.) For most species the disparity in parental investment between the sexes is so great that the assumptions here can be greatly relaxed.

ment alone, assuming the opposite sex's investment fixed at its optimal value) to the number of offspring the limiting sex optimally produces (L/M in Figure 7.1).

What governs the operation of sexual selection is the relative parental investment of the sexes in their offspring. Competition for mates usually characterizes males because males usually invest almost nothing in their offspring. Where male parental investment per offspring is comparable to female investment one would expect male and female reproductive success to vary in similar ways and for female choice to be no more discriminating than male choice (except as noted below). Where male parental investment strongly exceeds that of the female (regardless of which sex invests more in the sex cells) one would expect females to compete among themselves for males and for males to be selective about whom they accept as a mate.

Note that it may not be possible for an individual of one sex to invest in only part of the offspring of an individual of the opposite sex. When a male invests less per typical offspring than does a female but more than one-half what she invests (or vice-versa) then selection may not favor male competition to pair with more than one female, if the offspring of the second female cannot be parcelled out to more than one male. If the net reproductive success for a male investing in the offspring of one female is larger than that gained from investing in the offspring of two females, then the male will be selected to invest in the offspring of only one female. This argument is graphed in Figure 7.2 and may be important to understanding differential mortality in monogamous birds, as discussed below.

Fisher's (1958) sex ratio model compares the parental expenditure (undefined) in male offspring with that in female offspring and suggests energy and time as measures of expenditure. Restatements of Fisher's model (for example, Kolman 1960, Willson & Pianka 1963, T. Emlen 1968, Verner 1965, Leigh 1970) employ either the undefined term, parental expenditure, or the term energy investment. In either case the key concept is imprecise and the relevant one is parental investment, as defined above. Energy investment may often be a good approximation of parental investment, but it is clearly sometimes a poor one. An individual defending its brood from a predator may expend very little energy in the process but suffer a high chance of mortality; such behavior should be measured as a large investment, not a small one as suggested by the energy involved.

Parental Investment Patterns

Species can be classified according to the relative parental investment of the sexes in their young. In the vast majority of species, the male's only contribution to the survival of his offspring is his sex cells. In these species, female contribution clearly exceeds male and by a large ratio.

A male may invest in his offspring in several ways. He may provide his

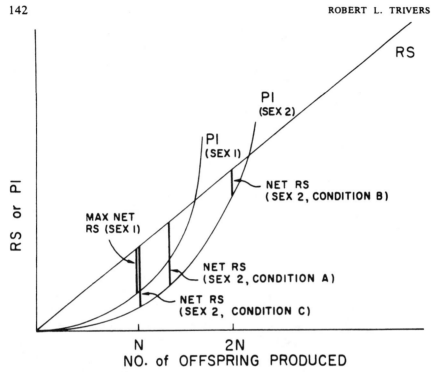

N   2N
NO. of OFFSPRING PRODUCED

Figure 7.2. RS and PI as functions of the number of offspring produced for two sexes. Sex 2 invests per typical offspring more than half of what sex 1 invests. Condition A: maximum net RS for a member of sex 2 assuming he can invest in any number of offspring between N and 2N. Condition B: net RS assuming member of sex 2 invests in 2N offspring. Condition C: net RS assuming member of sex 2 invests in N offspring. If member of sex 2 must invest in an integral multiple of N offspring, natural selection favors condition C.

mate with food as in baloon flies (Kessel 1955) and some other insects (Engelmann 1970), some spiders, and some birds (for example, Calder 1967, Royama 1966, Stokes & Williams, 1971). He may find and defend a good place for the female to feed, lay eggs or raise young, as in many birds. He may build a nest to receive the eggs, as in some fish (for example, Morris 1952). He may help the female lay the eggs, as in some parasitic birds (Lack 1968). The male may also defend the female. He may brood the eggs, as in some birds, fish, frogs, and salamanders. He may help feed the young, protect them, provide opportunities for learning, and so on, as in wolves and many monogamous birds. Finally, he may provide an indirect group benefit to the young (such as protection), as in many primates. All of these forms of male parental investment tend to decrease the disparity in investment between male and female resulting from the initial disparity in size of sex cells.

To test the importance of relative parental investment in controlling sexual selection one should search for species showing greater male than female parental investment (see Williams 1966, pp. 185–186). The best candidates include the Phalaropidae and the polyandrous bird species reviewed by Lack (1968). In these species, a female's parental investment ends when she lays her eggs; the male alone broods the eggs and cares for the young after hatching. No one has attempted to assess relative parental investment in these species, but they are striking in showing very high male parental investment correlating with strong sex role reversal: females tend to be more brightly colored, more aggressive and larger than the males, and tend to court them and fight over them. In the phalaropes there is no evidence that the females lay multiple broods (Höhn 1967, Johns 1969), but in some polyandrous species females apparently go from male to male laying successive broods (for example, Beebe 1925; see also Orians 1969). In these species the female may be limited by her ability to induce males to care for her broods, and female reproductive success may vary more than male. Likewise, high male parental investment in pipefish and seahorses (syngnathidae) correlates with female courtship and bright coloration (Fiedler 1954), and female reproductive success may be limited by male parental investment. Field data for other groups are so scanty that it is not possible to say whether there are any instances of sex role reversal among them, but available data for some dendrobatid frogs suggest at least the possibility. In these species, the male carries one or more young on his back for an unknown length of time (for example, Eaton 1941). Females tend to be more brightly colored than males (rare in frogs) and in at least one species, *Dendrobates aurata*, several females have been seen pursuing, and possibly courting, single males (Dunn 1941). In this species the male carries only one young on his back, until the tadpole is quite large, but females have been found with as many as six large eggs inside, and it is possible that females compete with each other for the backs of males. There are other frog families that show male parental care, but even less is known of their social behavior.

In most monogamous birds male and female parental investment is probably comparable. For some species there is evidence that the male invests somewhat less than the female. Kluijver (1933, cited in Coulson 1960) has shown that the male starling (*Sturnus vulgaris*) incubates the eggs less and feeds the young less often than the female, and similar data are available for other passerines (Verner & Willson, 1969). The fact that in many species males are facultative polygynists (von Haartman 1969) suggests that even when monogamous the males invest less in the young than their females. Because sex role reversal, correlating with evidence of greater male than female parental investment, is so rare in birds and because of certain theoretical considerations discussed below, I tentatively classify most monogamous bird species as showing somewhat greater female than male investment in the young.

A more precise classification of animals, and particularly of similar species, would be useful for the formulation and testing of more subtle hypotheses. Groups of birds would be ideal to classify in this way, because slight differences in relative parental investment may produce large differences in social behavior, sexual dimorphism and mortality rates by sex. It would be interesting to compare human societies that differ in relative parental investment and in the details of the form of the parental investment, but the specification of parental investment is complicated by the fact that humans often invest in kin other than their children. A wealthy man supporting brothers and sisters (and their children) can be viewed functionally as a polygynist if the contributions to his fitness made by kin are devalued appropriately by their degree of relationship to him (see Hamilton 1964). There is good evidence that premarital sexual permissiveness affecting females in human societies relates to the form of parental investment in a way that would, under normal conditions, tend to maximize female reproductive success (Goethals 1971).

The Evolution of Investment Patterns

The parental investment pattern that today governs the operation of sexual selection apparently resulted from an evolutionarily very early differentiation into relatively immobile sex cells (eggs) fertilized by mobile ones (spermatozoa). An undifferentiated system of sex cells seems highly unstable: competition to fertilize other sex cells should rapidly favor mobility in some sex cells, which in turn sets up selection pressures for immobility in the others. In any case, once the differentiation took place, sexual selection acting on spermatozoa favored mobility at the expense of investment (in the form of cytoplasm). This meant that as long as the spermatozoa of different males competed directly to fertilize eggs (as in oysters) natural selection favoring increased parental investment could act only on the female. Once females were able to control which male fertilized their eggs, female choice or mortality selection on the young could act to favor some new form of male investment in addition to spermatozoa. But there exist strong selection pressures against this. Since the female already invests more than the male, breeding failure for lack of an additional investment selects more strongly against her than against the male. In that sense, her initial very great investment commits her to additional investment more than the male's initial slight investment commits him. Furthermore, male-male competition will tend to operate against male parental investment, in that any male investment in one female's young should decrease the male's chances of inseminating other females. Sexual selection, then, is both controlled by the parental investment pattern and a force that tends to mold that pattern.

The conditions under which selection favors male parental investment have not been specified for any group of animals. Except for the case of polygyny in birds, the role of female choice has not been explored; instead, it is commonly assumed that, whenever two individuals can raise more in-dividuals together than one alone could, natural selection will favor male parental investment (Lack 1968, p. 149), an assumption that overlooks the effects of both male-male competition and female choice.

INITIAL PARENTAL INVESTMENT

An important consequence of the early evolutionary differentation of the sex cells and subsequent sperm competition is that male sex cells remain tiny compared to female sex cells, even when selection has favored a total male parental investment that equals or exceeds the female investment. The male's initial parental investment, that is, his investment at the mo-ment of fertilization, is much smaller than the female's, even if later, through parental care, he invests as much or more. Parental investment in the young can be viewed as a sequence of discrete investments by each sex. The relative investment may change as a function of time and each sex may be more or less free to terminate its investment at any time. In the hu-man species, for example, a copulation costing the male virtually nothing may trigger a nine-month investment by the female that is not trivial, fol-lowed, if she wishes, by a fifteen-year investment in the offspring that is considerable. Although the male may often contribute parental care during this period, he need not necessarily do so. After a nine-month pregnancy, a female is more or less free to terminate her investment at any moment but doing so wastes her investment up until then. Given the initial im-balance in investment the male may maximize his chances of leaving sur-viving offspring by copulating and abandoning many females, some of whom, alone or with the aid of others, will raise his offspring. In species where there has been strong selection for male parental care, it is more likely that a mixed strategy will be the optimal male course—to help a single female raise young, while not passing up opportunities to mate with other females whom he will not aid.

In many birds, males defend a territory which the female also uses for feeding prior to egg laying, but the cost of this investment by the male is difficult to evaluate. In some species, as outlined above, the male may pro-vision the female before she has produced the young, but this provisioning is usually small compared to the cost of the eggs. In any case, the cost of the copulation itself is always trivial to the male, and in theory the male need not invest anything else in order to copulate. If there is any chance the fe-male can raise the young, either alone or with the help of others, it would be to the male's advantage to copulate with her. By this reasoning one would expect males of monogamous species to retain some psychological traits consistent with promiscuous habits. A male would be selected to dif-ferentiate between a female he will only impregnate and a female with whom he will also raise young. Toward the former he should be more eager for sex and less discriminating in choice of sex partner than the female toward him, but toward the latter he should be about as discriminating as she toward him.

If males within a relatively monogamous species are, in fact, adapted to pursue a mixed strategy, the optimal is likely to differ for different males. I know of no attempt to document this possibility in humans, but psychology might well benefit from attempting to view human sexual plasticity as an adaptation to permit the individual to choose the mixed strategy best suited to local conditions and his own attributes. Elder (1969) shows that steady dating and sexual activity (coitus and petting) in adolescent human females correlate inversely with a tendency to marry up the socioeconomic scale as adults. Since females physically attractive as adolescents tend to marry up, it is possible that females adjust their reproductive strategies in adolescence to their own assets.

Desertion and Cuckoldry

There are a number of interesting consequences of the fact that the male and female of a monogamous couple invest parental care in their offspring at different rates. These can be studied by graphing and comparing the cumulative investment of each parent in their offspring, and this is done for two individuals of a hypothetical bird species in Figure 7.3. I have graphed no parental investment by the female in her young before copulation, even though she may be producing the eggs before then, because it is not until the act of copulation that she commits the eggs to a given male's genes. In effect, then, I have graphed the parental investment of each individual in the other individual's offspring. After copulation, this is the same as graphing investment in their own offspring, assuming, as I do here, that the male and female copulate with each other and each other only.

To discuss the problems that confront paired individuals ostensibly cooperating in a joint parental effort, I choose the language of strategy and decision, as if each individual contemplated in strategic terms the decisions it ought to make at each instant in order to maximize its reproductive success. This language is chosen purely for convenience to explore the adaptations one might expect natural selection to favor.

At any point in time the individual whose cumulative investment is exceeded by his partner's is theoretically tempted to desert, especially if the disparity is large. This temptation occurs because the deserter loses less than his partner if no offspring are raised and the partner would therefore be more strongly selected to stay with the young. Any success of the partner will, of course, benefit the deserter. In Figure 7.3, for example, desertion by the male right after copulation will cost him very little, if no offspring are raised, while the chances of the female raising some young alone may be great enough to make the desertion worthwhile. Other factors are

Parental Investment and Sexual Selection   147

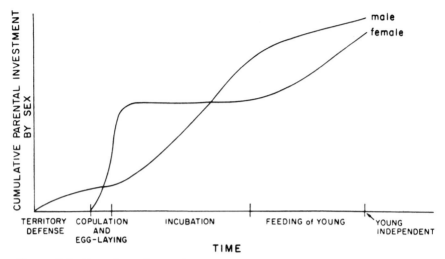

Figure 7.3. Hypothetical cumulative parental investment of a male and a female bird in their offspring as a function of time. Territory defense: *Male defends area for feeding and nest building.* Copulation and egg-laying: *Female commits her eggs to male who commits his defended nest to the female.* Incubation: *Male incubates eggs while female does nothing relevant to offspring.* Feeding of young: *Each parent feeds young but female does so at a more rapid rate.*

important in determining the adaptiveness of abandonment, factors such as the opportunities outside the pair for breeding and the expected shape of the deserter's investment curve if he does not desert. If the male's investment curve does not rise much after copulation, then the female's chances of raising the young alone will be greater and the time wasted by the male investing moderately in his offspring may be better spent starting a new brood.

What are the possible responses of the deserted individual? If the male is deserted before copulation, he has no choice but to attempt to start the process over again with a new female; whatever he has invested in that female is lost. If either partner is deserted after copulation, it has three choices. (1) It can desert the eggs (or eat them) and attempt to breed again with another mate, losing thereby all (or part of) the initial investment. (2) It can attempt to raise the young on its own, at the risk of overexertion and failure. Or, (3) it can attempt to induce another partner to help it raise the young. The third alternative, if successful, is the most adaptive for it, but this requires deceiving another organism into doing something contrary to its own interests, and adaptations should evolve to guard individuals from such tasks. It is difficult to see how a male could be successful in deceiving a new female, but if a female acts quickly, she might fool a male. As time goes on (for example, once the eggs are laid), it is unlikely that a male could easily be fooled. The female could thus be programmed to try the third strategy first, and if it failed, to revert to the first or second. The male deserter gains most if the female succeeds in the third strategy, nothing if she chooses the first strategy, and possibly an intermediate value if she chooses the second strategy.

If neither partner deserts at the beginning, then as time goes on, each invests more and more in the young. This trend has several consequences. On the one hand, the partner of a deserter is more capable of finishing the task alone and natural selection should favor its being more predisposed to try, because it has more to lose. On the other hand, the deserter has more to lose if the partner fails and less to gain if the partner succeeds. The balance between these opposing factors should depend on the exact form of the cumulative investment curves as well as the opportunities for further breeding outside the pair.

There is another effect with time of the increasing investment by both parents in the offspring. As the investments increase, natural selection may favor *either* partner deserting even if one has invested more in the young than the other. This is because the desertion may put the deserted partner in a cruel bind: he has invested so much that he loses considerably if he also deserts the young, even though, which should make no difference to him, the partner would lose even more. The possibility of such binds can be illustrated by an analogous situation described by Rowley (1965). Two neighboring pairs of wrens happened to fledge their young simultaneously and could not tell their young apart, so both pairs fed all six young indiscriminately, until one pair "deserted" to raise another brood, leaving their neighbors to feed all six young, which they did, even though this meant they were, in effect, being taken advantage of.

Birds should show adaptations to avoid being deserted. Females, in particular, should be able to guard against males who will only copulate and not invest subsequent parental effort. An instance of such an adaptation may be found in the red-necked phalarope, *Phalaropus lobatus*. In phalaropes the male incubates the eggs alone and alone cares for the young after hatching (Höhn 1967, Johns 1969), so that a graph of cumulative parental investment would show an intitial large female investment which then remains the same through time, whereas the initial male investment is nil and increases steadily, probably to surpass the female investment. Only the female is vulnerable to being deserted and this right after copulation, since any later desertion by the male costs him his investment in incubation, the young being almost certain to perish. Tinbergen (1935) observed a female vigorously courting a male and then flying away as soon as he responded to the courtship by attempting to copulate. This coy performance was repeated numerous times for several days. Tinbergen attributed it to the "waxing and waning of an instinct," but the behavior may have been a test

of the male's willingness to brood the female's eggs. The male under observation was, in fact, already brooding eggs and was courted when he left
the eggs to feed on a nearby pond. In order to view a complete egg-laying
sequence, Tinbergen destroyed the clutch the male was brooding. Within
a half day the female permitted the male sexual access, and he subsequently brooded her eggs. The important point is that the female could
apparently tell the difference between a free and an encumbered male, and
she withheld sex from the latter. Courtship alternating with flight may be
the test that reveals the male's true attachments: the test can show, for example, whether he is free to follow the female.

It is likely that many adaptations exist in monogamous species to guard
against desertion, but despite evidence that desertion can be common
(Rowley 1965) no one has attempted to analyze courtship with this danger
in mind. Von Haartman (1969) has reviewed some evidence for adaptations of females to avoid being mated to a polygynous male, and being so
mated is sometimes exactly equivalent to being deserted by the male (von
Haartman, 1951).

External fertilization requires a synchrony of behavior such that the
male can usually be certain he is not attempting to fertilize previously
fertilized eggs. With the evolution of internal fertilization the male cannot
be so certain. For many species (for example, most mammals), the distinction is not important because the male loses so little by attempting to fertilize previously fertilized eggs. Where male parental care is involved, however, the male runs the risk of being cuckolded, of raising another male's
offspring. For Figure 7.1 it was assumed that the pair copulated with each
other and each other only, but the male can usually not be sure that such
is the case and what is graphed in such a situation is the male's investment
in the *female's* offspring. Adaptations should evolve to help guarantee that
the female's offspring are also his own, but these can partly be countered
by the evolution of more sophisticated cuckolds.

One way a male can protect himself is to ensure that other males keep
their distance. That some territorial aggression of monogamous male birds
is devoted to protecting the sanctity of the pair bond seems certain, and human male aggression toward real or suspected adulterers is often extreme.
Lee (1969), for example, has shown that, when the cause is known, the
major cause of fatal Bushman fights is adultery or suspected adultery. In fact,
limited data on other hunter-gathering groups (including Eskimos and
Australian aborigines) indicate that, while fighting is relatively rare (in
that organized intergroup aggression is infrequent), the "murder rate" may
be relatively high. On examination, the murderer and his victim are usually
a husband and his wife's real or suspected lover. In pigeons (*Columba
livia*) a new male arriving alone at a nocturnal roosting place in the fall is
attacked day after day by one or more resident males. As soon as the same

male appears with a mate, the two are treated much more casually (Trivers, unpublished data), suggesting that an unpaired male is more threatening than a paired one.

I have argued above that a female deserted immediately after copulation may be adapted to try to induce another male to help raise her young. This factor implies adaptations on the part of the male to avoid such a fate. A simple method is to avoid mating with a female on first encounter, sequester her instead and mate with her only after a passage of time that reasonably excludes her prior impregnation by another male. Certainly males guard their females from other males, and there is a striking difference between the lack of preliminaries in promiscuous birds (Scott 1942, Kruijt & Hogan 1967) and the sometimes long lag between pair bonding and copulation in monogamous birds (Nevo 1956), a lag which usually seems to serve other functions as well.

Biologists have interpreted courtship in a limited way. Courtship is seen as allowing the individual to choose the correct species and sex, to overcome antagonistic urges and to arouse one's partner (Bastock 1967). The above analysis suggests that courtship should also be interpreted in terms of the need to guard oneself from the several possibilities of maltreatment at the hands of one's mate.

Differential Mortality and the Sex Ratio

Of special interest in understanding the effects of sexual selection are accurate data on differential mortality of the sexes, especially of immature individuals. Such data are, however, among the most difficult to gather, and the published data, although important, are scanty (for example, Emlen 1940, Hays 1947, Chapman, Casida, & Cote 1938, Robinette et al. 1957, Coulson 1960, Potts 1969, Darley 1971, Myers & Krebs 1971). As a substitute one can make use of data on sex ratios within given age classes or for all age classes taken together. By assuming that the sex ratio at conception (or, less precisely, at birth) is almost exactly 50/50, significant deviations from this ratio for any age class or for all taken together should imply differential mortality. Where data exist for the sex ratio at birth and where the sex ratio for the entire local population is unbalanced, the sex ratio at birth is usually about 50/50 (see above references, Selander 1965, Lack 1954). Furthermore, Fisher (1958) has shown, and others refined (Leigh 1970), that parents should invest roughly equal energy in each sex. Since parents usually invest roughly equal energy in each individual of each sex, natural selection, in the absence of unusual circumstances (see Hamilton 1967), should favor approximately a 50/50 sex ratio at conception.

It is difficult to determine accurately the sex ratio for any species. The most serious source of bias is that males and females often make themselves differentially available to the observer. For example, in small mammals sexual selection seems to have favored male attributes, such as high mobility, that tend to result in their differential capture (Beer, Frenzel, & MacLeod 1958; Myers & Krebs, 1971). If one views one's capture techniques as randomly sampling the existing population, one will conclude that males are more numerous. If one views one's capture techniques as randomly sampling the effects of mortality on the population, then one will conclude that males are more prone to mortality (they are captured more often) and therefore are less numerous. Neither assumption is likely to be true, but authors routinely choose the former. Furthermore, it is often not appreciated what a large sample is required in order to show significant deviations from a 50/50 ratio. A sample of 400 animals showing a 44/56 sex ratio, for example. does not deviate significantly from a 50/50 ratio. (Nor, although this is almost never pointed out, does it differ significantly from a 38/62 ratio.)

Mayr (1939) has pointed out that there are numerous deviations from a 50/50 sex ratio in birds and I believe it is likely that, if data were sufficiently precise, most species of vertebrates would show a significant deviation from a 50/50 sex ratio. Males and females differ in numerous characteristics relevant to their diff   nt reproductive strategies and these characters are unlikely to have equivalent effects on survival. Since it is not advantageous for the adults of each sex to have available the same number of adults of the opposite sex, there will be no automatic selective agent for keeping deviations from a 50/50 ratio small.

A review of the useful literature on sex ratios suggests that (except for birds) when the sex ratio is unbalanced it is usually unbalanced by there being more females than males. Put another way, males apparently have a tendency to suffer higher mortality rates than females. This is true for those dragonflies for which there are data (Corbet, Longfield, & Moore 1960), for the house fly (Rockstein 1959), for most fish (Beverton & Holt 1959), for several lizards (Tinkle 1967, Harris 1964, Hirth 1963, Blair 1960, Trivers, discussed below) and for many mammals (Bouliere & Verschuren 1960. Cowan 1950. Eisenberg 1965, Robinette et al. 1957, Beer, Frenzel, & MacLeod 1958, Stephens 1952, Tyndale-Biscoe & Smith, 1969, Myers & Krebs, 1971, Wood 1970). Hamilton (1948) and Lack (1954) have reviewed studies on other animals suggesting a similar trend. Mayr (1939) points out that where the sex ratio can be shown to be unbalanced in monogamous birds there are usually fewer females, but in polygynous or promiscuous birds there are fewer males. Data since his paper confirm this finding. This result is particularly interesting since in all other groups in which males tend to be less numerous monogamy is rare or nonexistent.

THE CHROMOSOMAL HYPOTHESIS

There is a tendency among biologists studying social behavior to regard the adult sex ratio as an independent variable to which the species reacts with appropriate adaptations. Lack (1968) often interprets social behavior as an adaptation in part to an unbalanced (or balanced) sex ratio, and Verner (1964) has summarized other instances of this tendency. The only mechanism that will generate differential mortality independent of sexual differences clearly related to parental investment and sexual selection is the chromosomal mechanism, applied especially to humans and other mammals: the unguarded X chromosome of the male is presumed to predispose him to higher mortality. This mechanism is inadequate as an explanation of differential mortality for three reasons.

1. The distribution of differential mortality by sex is not predicted by a knowledge of the distribution of sex determining mechanisms. Both sexes of fish are usually homogametic, yet males suffer higher mortality. Female birds are heterogametic but suffer higher mortality only in monogamous species. Homogametic male meal moths are outsurvived by their heterogametic female counterparts under laboratory conditions (Hamilton & Johansson 1965).

2. Theoretical predictions of the degree of differential mortality expected by males due to their unguarded X chromosome are far lower than those observed in such mammals as dogs, cattle and humans (Ludwig & Boost 1951). It is possible to imagine natural selection favoring the heterogametic sex determining mechanism if the associated differential mortality is slight and balanced by some advantage in differentiation or in the homogametic sex, but a large mortality associated with heterogamy should be counteracted by a tendency toward both sexes becoming homogametic.

3. Careful data for humans demonstrate that castrate males (who remain of course heterogametic) strongly outsurvive a control group of males similar in all other respects and the earlier in life the castration, the greater the increase in survival. (Hamilton & Mestler 1969). The same is true of domestic cats (Hamilton, Hamilton & Mestler 1969), but not of a species (meal moths) for which there is no evidence that the gonads are implicated in sexual differentiation (Hamilton & Johansson 1965).

An Adaptive Model of Differential Mortality

To interpret the meaning of balanced or unbalanced sex ratios one needs a comprehensive framework within which to view life historical phenomena. Gadgil & Bossert (1970) have presented a model for the adaptive interpretation of differences between species' life histories; for example, in the age of first breeding and in the growth and survival curves. Although they did not apply this model to sexual differences in these parameters, their model is precisely suited for such differences. One can, in effect, treat the sexes as if they were different species, the opposite sex being a resource relevant to producing maximum surviving offspring. Put this way, female "species" usually differ from male species in that females compete among themselves for such resources as food but not for members of the opposite sex, whereas males ultimately compete only for members of the opposite sex, all other forms of competition being important only insofar as they affect this ultimate competition.

To analyze differential mortality by sex one needs to correlate different reproductive strategies with mortality, that is, one must show how a given reproductive strategy entails a given risk of mortality. One can do this by graphing reproductive success (RS) for the first breeding season as a function of reproductive effort expended during that season, and by graphing the diminution in future reproductive success (D) in units of first breeding season reproductive success. (Gadgil and Bossert show that the reproductive value of a given effort declines with age, hence the need to convert future reproductive success to comparable units.) For simplicity I assume that the diminution, D, results entirely from mortality between the first and second breeding seasons. The diminution could result from mortality in a later year (induced by reproductive effort in the first breeding season) which would not change the form of the analysis, or it could result from decreased ability to breed in the second (or still later) breeding season, which sometimes occurs but which is probably minor compared to the diminution due to mortality, and which does not change the analysis as long as one assumes that males and females do not differ appreciably in the extent to which they suffer this form of diminution.

Natural selection favors an individual expending in the first breeding season the reproductive effort (RE) that results in a maximum net reproductive success (RS—D). The value of D at this RE gives the degree of expected mortality between the first and second breeding seasons (see Figures 7.4 and 7.5). Differences between the sexes in D will give the expected differential mortality. The same analysis can be applied to the nth breeding season to predict mortality between it and the nth + 1 breeding season. Likewise, by a trivial modification, the analysis can be used to generate differences in juvenile mortality: let D represent the diminution in chances of surviving to the first breeding season as a function of RE at first breeding. Seen this way, one is measuring the cost in survival of developing during the juvenile period attributes relevant to adult reproductive success.

SPECIES WITH LITTLE OR NO MALE PARENTAL INVESTMENT

In Figure 7.4, I have graphed RS and D as functions of reproductive effort in the first breeding season for females of a hypothetical species in which males invest very little parental care. The RS function is given a sigmoidal shape for the following reasons. I assume that at low values of RE, RS increases only very gradually because some investment is necessary just to initiate reproduction (for example, enlarging the reproductive organs). RS then increases more rapidly as a function of RE but without achieving a very steep slope. RS finally levels off at high values of RE because of increased inefficiencies there (for example, inefficiencies in foraging; see Schoener 1971). I have graphed the value, f, at which net reproductive success for the female reaches a maximum. Technically, due to competition, the shape of the RS function for any given female will depend partly on the reproductive effort devoted by other females; the graph therefore assumes that other females tend to invest near the optimal value, f, but an important feature of a female's RS is that it is *not* strongly dependent on the RE devoted by other females: the curve would not greatly differ if all other females invested much more or less. I have graphed D as a linear function of RE. So doing amounts to a definition of reproductive effort, that is, a given increment in reproductive effort during the first breeding season can be detected as a proportionately increased chance of dying be-

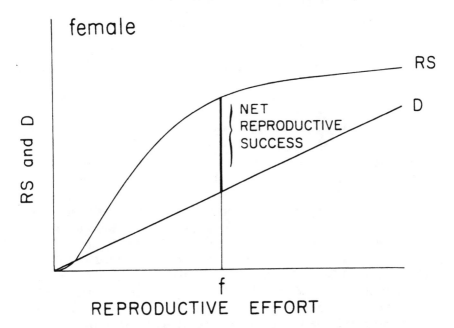

Figure 7.4. *Female reproductive success during the first breeding season (RS) and diminution of future reproductive success (D) as functions of reproductive effort during first breeding. D is measured in units of first breeding (see text). At f the net reproductive success reaches a maximum. Species is one in which there is very little male parental investment.*

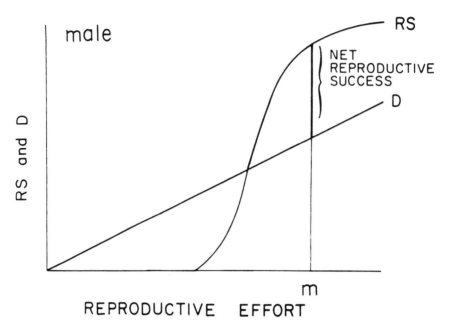

Figure 7.5. Same as Figure 7.4 except that it is drawn for the male instead of the female. At m the net reproductive success reaches a maximum.

tween the first and second breeding seasons. Note that reproductive effort for the female is essentially synonymous with parental investment.

Male RS differs from female RS in two important ways, both of which stem from sexual selection. (1) A male's RS is highly dependent on the RE of other males. When other males invest heavily, an individual male will usually not outcompete them unless he invests as much or more. A considerable investment that is slightly below that of other males may result in zero RS. (2) A male's RS is potentially very high, much higher than that of a conspecific female, but only if he outcompetes other males. There should exist some factor or set of factors (such as size, aggressiveness, mobility) that correlates with high male RS. The effect of competition between males for females is selection for increased male RE, and this selection will continue until greater male than female RE is selected as long as the higher associated D is offset by the potentially very high RS. This argument is graphed in Figure 7.5, where the steep slope of RS reflects the high interaction between one male's RS and the RE of the other males. Note that the argument here depends on the existence of a set of factors correlated with high male reproductive success. If these factors exist, natural selection will predispose the male to higher mortality rates than the female. Where a male can achieve very high RS in a breeding season (as in land-breeding seals, Bartholemew 1970), differential mortality will be correspondingly high.

SPECIES WITH APPRECIABLE MALE PARENTAL INVESTMENT

The analysis here applies to species in which males invest less parental care than, but probably more than one-half, what females invest. I assume that most monogamous birds are so characterized, and I have listed reasons and some data above supporting this assumption. The reasons can be summarized by saying that because of their initial large investment, females appear to be caught in a situation in which they are unable to force greater parental investment out of the males and would be strongly selected against if they unilaterally reduced their own parental investment.

Functions relating RS to parental investment are graphed for males and females in Figures 7.6 and 7.7, assuming for each sex that the opposite sex shows the parental investment that results for it in a maximum net reproductive success. The female curve is given a sigmoidal shape for the reasons that apply to Figure 7.4; in birds the female's initial investment in the eggs will go for nothing if more is not invested in brooding the eggs and feeding the young, while beyond a certain high RE further increments do not greatly affect RS. Assuming the female invests the value, f, male RS will vary as a function of male parental investment in a way similar to female RS, except the function will be displaced to the left (Figure 7.7) and some RS will be lost due to the effects of the cuckoldry graphed in Figure 7.8.

Because males invest in parental care more than one-half what females

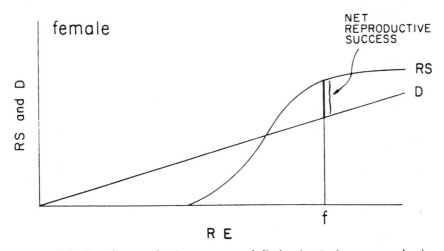

Figure 7.6. Female reproductive success and diminution in future reproductive success as functions of reproductive effort (RE) assuming male reproductive effort of m_1. Species is a hypothetical monogamous bird in which males invest somewhat less than females in parental care (see Figure 7.7 and 7.8).

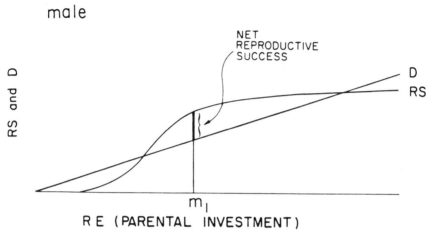

male

Figure 7.7. *Male reproductive success and diminution in future reproductive success as functions of reproductive effort, assuming female reproductive effort of* f. *Species is same as in Figure 7.6. Reproductive effort of male is invested as parental care in one female's offspring. Net reproductive success is a maximum at* m_1.

invest and because the offspring of a given female tend to be inseminated by a single male, selection does not favor males competing with each other to invest in the offspring of more than one female. Rather, sexual selection only operates on the male to inseminate females whose offspring he will not raise, especially if another male will raise them instead. Since selection presumably does not strongly favor female adultery and may oppose it (if, for example, detection leads to desertion by the mate), the opportunities for cuckoldry are limited: high investment in promiscuous activity will bring only limited RS. This argument is graphed in Figure 7.8. The predicted differential mortality by sex can be had by comparing D (f) with D $(m_1 + m_2)$.

It may seem ironic, but in moving from a promiscuous to a monogamous life, that is, in moving toward *greater* parental investment in his young, the male tends to *increase* his chances of surviving relative to the female. This tendency occurs because the increased parental investment disproportionately decreases the male's RE invested in male-male competition to inseminate females.

Note that in both cases above differential mortality tends to be self-limiting. By altering the ratio of possible sexual partners to sexual competitors differential mortality sets up forces that tend to keep the differential mortality low. In species showing little male parental investment differential male mortality increases the average number of females available for those males who survive. Other things being equal, this increase tends to make it more difficult for the most successful males to maintain their relative ad-

158   ROBERT L. TRIVERS

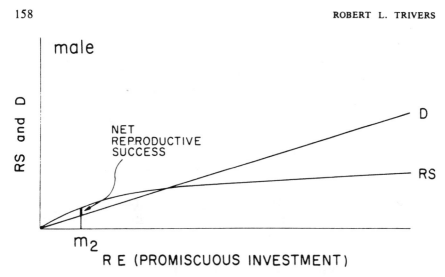

Figure 7.8. *Male reproductive success and diminution of future reproductive success as a function of reproductive effort solely devoted to promiscuous behavior. Net reproductive success at* m_2 *is a maximum. Same species as in Figures 7.6 and 7.7.*

vantage. In monogamous birds differential female mortality induces competition among males to secure at least one mate, thereby tending to increase male mortality. Such competition presumably also increases the variance in male reproductive success above the sexual differential expected from cuckoldry.

SPECIES WITH GREATER MALE THAN FEMALE PARENTAL INVESTMENT

Since the above arguments were made with reference to relative parental investment and not sex, they apply to species in which males invest more parental effort than females, except that there is never apt to be a female advantage to cuckolding other females, and this advantage is always alive with males. Where females invest more than one-half what males invest, one would predict differential female mortality. Where females invest less than one-half what males invest, one would predict competition, and a resulting differential female mortality.

Male–Male Competition

Competition between males does not necessarily end with the release of sperm. Even in species with internal fertilization, competition between sperm of different males can be an important component of male-male competition (see the excellent review by Parker 1970b). In rare cases, competition between males may continue after eggs are fertilized. For example an adult male langur (*Presbytis entellus*) who ousts the adult male of a group may systematically kill the infants of that group (presumably fathered by the ousted male) thereby bringing most of the adult females quickly into estrus again (Sugiyama 1967). While clearly disadvantageous for the killed infants and their mothers, such behavior, benefiting the new male, may be an extreme product of sexual selection. Female mice spontaneously abort during the first four days of pregnancy when exposed to the smell of a strange male (Bruce 1960, reviewed in Sadleir 1967), a situation subject to several interpretations including one based on male-male competition.

Sperm competition may have important effects on competition between males prior to release of sperm. In those insects in which later-arriving sperm take precedence in fertilizing eggs, selection favors mating with a female just prior to release of eggs, thereby increasing competition at ovulation sites and intensifying selection for a postovulatory guarding phase by the male (see Parker 1970bcd, Jacobs 1955). I here concentrate on male-male competition prior to the release of sperm in species showing very little male parental investment.

The form of male-male competition should be strongly influenced by the distribution in space and time of the ultimate resource affecting male reproductive success, namely, conspecific breeding females. The distribution can be des¬¬ ¡hed in terms of three parameters: the extent to which females are clumped or dispersed in space, the extent to which they are clumped or dispersed in time, and the extent to which their exact position in space and time is predictable. I here treat females as if they are a passive resource for which males compete, but female choice may strongly influence the form of male-male competition, as, for example, when it favors males clumping together on display grounds (for example, S. Emlen 1968) which females then search out (see below under "Female Choice").

DISTRIBUTION IN SPACE

Cervids differ in the extent to which females are clumped in space or randomly dispersed (deVos, Broky & Geist 1967) as do antelopes (Eisenberg 1965), and these differences correlate in a predictable way with differences in male attributes. Generally male-male aggression will be the more severe the greater the number of females two males are fighting over at any given moment. Searching behavior should be more important in highly dispersed species especially if the dispersal is combined with unpredictability.

DISTRIBUTION IN TIME

Clumped in time refers to highly seasonal breeders in which many females become sexually available for a short period at the same moment (for example, explosive breeding frogs; Bragg 1965, Rivero & Estevez 1969), while highly dispersed breeders (in time) are species (such as chimpanzees; Van Lawick-Goodall 1968) in which females breed more or less randomly throughout the year. One effect of extreme clumping is that it becomes more difficult for any one male to be extremely successful: while he is copulating with one female, hundreds of other females are simultaneously being inseminated. Dispersal in time, at least when combined with clumping in space, as in many primates, permits each male to compete for each newly available female and the same small number of males tend repeatedly to inseminate the receptive females (DeVore 1965).

PREDICTABILITY

One reason males in some dragonflies (Jacobs 1955) may compete with each other for female oviposition sites is that those are highly predictable places at which to find receptive females. Indeed, males display several behaviors, such as testing the water with the tips of their abdomen, that apparently aid them in predicting especially good oviposition sites, and such sites can permit very high male reproductive success (Jacobs 1955). In the cicada killer wasp (*Sphecius spheciosus*) males establish mating territories around colony emergency holes, presumably because this is the most predictable place at which to find receptive females (Lin 1963).

The three parameters outlined interact strongly, of course, as when very strong clumping in time may strongly reduce the predicted effects of strong clumping in space. A much more detailed classification of species with nonobvious predictions would be welcome. In the absence of such models I present a partial list of factors that should affect male reproductive success and that may correlate with high male mortality.

SIZE

There are very few data showing the relationship between male size and reproductive success but abundant data showing the relationship between male dominance and reproductive success: for example, in elephant seals (LeBoeuf & Peterson 1969), black grouse (Koivisto 1965, Scott 1942), baboons (DeVore 1965) and rainbow lizards (Harris 1964). Since dominance is largely established through aggression and larger size is usually helpful in aggressive encounters, it is likely that these data partly reveal the relationship between size and reproductive success. (It is also likely that they reflect the relationship between experience and reproductive success.)

Circumstantial evidence for the importance of size in aggressive encounters can be found in the distribution of sexual size dimorphism and aggressive tendencies among tetrapods. In birds and mammals males are generally larger than females and much more aggressive. Where females are known to be more aggressive (that is, birds showing reversal in sex roles) they are also larger. In frogs and salamanders females are usually larger than males, and aggressive behavior has only very rarely been recorded. In snakes, females are usually larger than males (Kopstein 1941) and aggression is almost unreported. Aggression has frequently been observed between sexually active crocodiles and males tend to be larger (Allen Greer, personal communication). In lizards males are often larger than females, and aggression is common in some families (Carpenter 1967). Male aggressiveness is also common, however, in some species in which females are larger, for example, *Sceloporus,* (Blair 1960). There is a trivial reason for the lack of evidence of aggressiveness in most amphibians and reptiles: the species are difficult to observe and few behavioral data of any sort have been recorded. It is possible, however, that this correlation between human ignorance and species in which females are larger is not accidental. Humans tend to be more knowledgeable about those species that are also active diurnally and strongly dependent on vision, for example, birds and large mammals. It may be that male aggressiveness is more strongly selected in visually oriented animals because vision provides long-range information on the behavior of competitors. The male can, for example, easily observe another male beginning to copulate and can often quickly attempt to intervene (for example, baboons, DeVore 1965 and sage grouse, Scott 1942).

Mammals and birds also tend towards low, fixed clutch sizes and this may favor relatively smaller females, since large female size may be relatively unimportant in reproductive success. In many fish, lizards and salamanders female reproductive success as measured by clutch size is known to correlate strongly within species with size (Tinkle, Wilbur & Tilley 1970, Tilley 1968).

Measuring reproductive success by frequency of copulation, I have analyzed male and female reproductive success as a function of size in *Anolis garmani* (Figures 7.9 and 7.10). Both sexes show a significant positive correlation between size and reproductive success, but the trend in males is significantly stronger than the trend in females ($p < .01$). Consistent with this tendency, males grow faster at all sizes than females (Figure 7.11) and reach an adult weight two and one-half times that of adult females. The sex ratio of all animals is unbalanced in favor of females, which would seem to indicate differential mortality, but the factors that might produce the difference are not known. Males are highly aggressive and territorial, and large males defend correspondingly large territories with many resident females. No data are available on size and success in aggressive encounters, but in the closely related (and behaviorally very similar) *A. lineatopus,* 85 per cent of 182 disputes observed in the field were won by the larger animal (Rand 1967). Females lay only one egg at a time, but it is likely that larger adult females lay eggs slightly more often

Figure 7.9. Male and female Anolis garmani *copulating face down four feet up the trunk of a cocoanut tree.* Photo by Joseph K. Long.

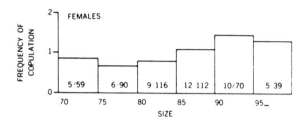

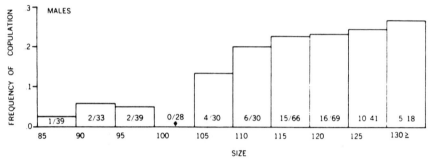

Figure 7.10. Reproductive success in male and female A. garmani *as a function of size. Reproductive success is measured by the number of copulations observed per number of individuals (male or female) in each nonoverlapping 5 mm size category. Data combined from five separate visits to study area between summer 1969 and summer 1971.*

than smaller ones, and this may partly be due to advantages in feeding through size-dependent aggressiveness, since larger females wander significantly more widely than smaller adult ones. An alternate interpretation (based on ecological competition between the sexes) has been proposed for sexual dimorphism in size among animals (Selander 1966), and the interpretation may apply to *Anolis* (Schoener 1967).

METABOLIC RATE

Certainly more is involved in differential male mortality than size, even in species in which males grow to a larger size than females. Although data show convincingly that nutritional factors strongly affect human male survival *in utero*, a sexual difference in size among humans is not detected until the twenty-fourth week after conception whereas differences in mortality appear as soon as the twelfth week. Sellers et al. (1950) have shown that male rats excrete four times the protein females do; the difference is removed by castration. Since males suffer more from protein-deficient diets than females (they gain less weight and survive less well) the sex-linked proteinuria, apparently unrelated to size, may be a factor in causing lower male survival in wild rats (Schein 1950). (The connection between

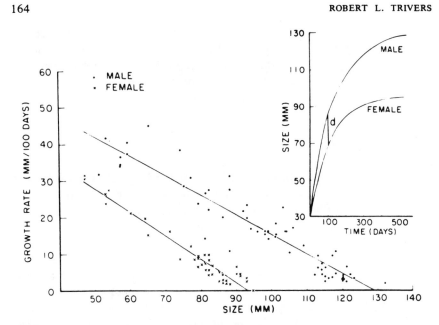

Figure 7.11. Male and female growth rates in A. garmani *as a function of initial size based on summer 1970 recaptures of animals marked 3 to 4 months before. A line has been fitted to each set of data; d indicates how much larger a male is when a similar aged female reaches sexual maturity.*

proteinuria and male reproductive success is obscure.) Again, although human male survival is more adversely affected by poor nutritional conditions than female survival, Hamilton (1948) presents evidence that the higher metabolic rate of the male is an important factor increasing his vulnerability to many diseases which strike males more heavily than females. Likewise, Taber & Dasmann (1954) argue that greater male mortality in the deer, *Odocoileus hemionus,* results from a higher metabolic rate. High metabolic rate could relate to both aggressiveness and searching behavior.

EXPERIENCE

If reproductive success increases more rapidly in one sex than the other as a function of age alone (for example, through age-dependent experience), then one would expect a postponement of sexual maturity in that sex and a greater chance of surviving through a unit of time than in the opposite sex. Thus, the adult sex ratio might be biased in favor of the earlier maturing sex but the sex ratio for all ages taken together should be biased in favor of the later maturing sex. Of course, if reproductive success for one sex increases strongly as a function of experience and experience only partly correlates with age, then the sex may be willing to suffer increased mortality if this mortality is sufficiently offset by increases in experience. Selander (1965) has suggested that the tendency of immature male blackbirds to exhibit some mature characteristics may be adaptive in that it increases the male's experience, although it also presumably increases his risk of mortality.

MOBILITY

Data from mammals (reviewed by Eisenberg 1965 and Brown 1966) and from some salamanders (Madison & Shoop 1970) and numerous lizards (Tinkle 1967 and Blair 1960) suggest that males often occupy larger home ranges and wander more widely than females *even when males are smaller* (Blair 1965). Parker (1970a) has quantified the importance of mobility and searching behavior in dung flies. If females are a dispersed resource, then male mobility may be crucial in exposing the male to a large number of available females. Again, males may be willing to incur greater mortality if this is sufficiently offset by increases in reproductive success. This factor should only affect the male during the breeding season (Kikkawa 1964) unless factors relevant to mobility (such as speed, agility or knowledge of the environment) need to be developed prior to the reproductive season. Lindburg (1969) has shown that macaque males, but not females, change troops more frequently during the reproductive season than otherwise and that this mobility increases male reproductive success as measured by frequency of copulation, suggesting that at least in this species, greater mobility can be confined to the reproductive season (see also Miller 1958). On the other hand, Taber & Dasmann (1954) present evidence that as early as six months of age male deer wander more widely from their mothers than females—a difference whose function, of course, is not known. Similar very early differences in mobility have been demonstrated for a lizard (Blair 1960) and for several primates, including man (Jensen, Bobbitt & Gordon 1968).

Female Choice

Although Darwin (1871) thought female choice an important evolutionary force, most writers since him have relegated it to a trivial role (Huxley 1938, Lack 1968; but see Fisher 1958, and Orians 1969). With notable exceptions the study of female choice has limited itself to showing that females are selected to decide whether a potential partner is of the right species, of the right sex and sexually mature. While the adaptive value of such choices is obvious, the adaptive value of subtler discriminations among broadly appropriate males is much more difficult to visualize or document. One needs both theoretical arguments for the adaptive value of such female choice and detailed data on how females choose. Neither of these criteria is met by those who casually ascribe to female (or male) choice the evolution of such traits as the relative hairlessness of both human sexes (Hershkovitz 1966) or the large size of human female breasts (Morris 1967). I review here theoretical considerations of how females might be expected to choose among the available males, along with some data on how females do choose.

SELECTION FOR OTHERWISE NEUTRAL OR DISFUNCTIONAL MALE ATTRIBUTES

The effects of female choice will depend on the way females choose. If some females exercise a preference for one type of male (genotype) while others mate at random, then other things being equal, selection will rapidly favor the preferred male type and the females with the preference (O'Donald 1962). If each female has a specific image of the male with whom she prefers to mate and if there is a decreasing probability of a female mating with a male as a function of his increasing deviation from her preferred image, then it is trivial to show that selection will favor distributions of female preferences and male attributes that coincide. Female choice can generate continuous male change only if females choose by a relative rather than an absolute criterion. That is, if there is a tendency for females to sample the male distribution and to prefer one extreme (for example, the more brightly colored males), then selection will move the male distribution toward the favored extreme. After a one generation lag, the distribution of female preferences will also move toward a greater percentage of females with extreme desires, because the granddaughters of females preferring the favored extreme will be more numerous than the granddaughters of females favoring other male attributes. Until countervailing selection intervenes, this female preference will, as first pointed out by Fisher (1958), move both male attributes and female preferences with increasing rapidity in the same direction. The female preference is capable of overcoming some countervailing selection on the male's ability to survive to reproduce, if the increased reproductive success of the favored males when mature offsets their chances of surviving to reproduce.

There are at least two conditions under which one might expect females to have been selected to prefer the extreme male of a sample. When two species, recently speciated, come together, selection rapidly favors females who can discriminate the two species of males. This selection may favor females who prefer the appropriate extreme of an available sample, since such a mechanism would minimize mating mistakes. The natural selection of females with such a mechanism of choice would then initiate sexual selection in the same direction, which in the absence of countervailing selection would move the two male phenotypes further apart than necessary to avoid mating error.

Natural selection will always favor female ability to discriminate male sexual competence, and the safest way to do this is to take the extreme of a sample, which would lead to runaway selection for male display. This case is discussed in more detail below.

SELECTION FOR OTHERWISE FUNCTIONAL MALE ATTRIBUTES

As in other aspects of sexual selection, the degree of male investment in the offspring is important and should affect the criteria of female choice. Where the male invests little or nothing beyond his sex cells, the female has only to decide which male offers the ideal genetic material for her off-spring, assuming that male is willing and capable of offering it. This question can be broken down to that of which genes will promote the survival of her offspring and which will lead to reproductive success, assuming the offspring survive to adulthood. Implicit in these questions may be the relation between her genes and those of her mate: do they complement each other?

Where the male invests parental care, female choice may still involve the above questions of the male's genetic contribution but should also involve, perhaps primarily involve, questions of the male's willingness and ability to be a good parent. Will he invest in the offspring? If willing, does he have the ability to contribute much? Again, natural selection may favor female attentiveness to complementarity: do the male's parental abilities complement her own? Can the two parents work together smoothly? Where males invest considerable parental care, most of the same considerations that apply to female choice also apply to male choice. The alternate criteria for female choice are summarized in Table 7.1.

SEXUAL COMPETENCE

Even in males selected for rapid, repeated copulations the ability to do so is not unlimited. After three or four successive ejaculations, for example, the concentration of spermatozoa is very low in some male chickens (Parker, McKenzie & Kempster 1940), yet males may copulate as often as 30 times in an hour (Guhl 1951). Likewise, sperm is completely depleted in male *Drosophila melanogaster* after the fifth consecutive mating on the same day (Demerec & Kaufmann 1941, Kaufmann & Demerec 1942). Duration of copulation is cut in half by the third copulation of a male dung fly on the same day and duration of copulation probably correlates with sperm transferred (Parker 1970a). In some species females may be able to judge whether additional sperm are needed (for example, house flies; Riemann, Moen & Thorson 1967) or whether a copulation is at least behaviorally successful (for example, sea lions; Peterson & Bartholomew 1967), but in many species females may guarantee reproductive success by mat-

Table 7.1. Theoretical criteria for female choice of males

I.   All species, but especially those showing little or no male parental investment
  A.   Ability to fertilize eggs
     (1)   correct species
     (2)   correct sex
     (3)   mature
     (4)   sexually competent
  B.   Quality of genes
     (1)   ability of genes to survive
     (2)   reproductive ability of genes
     (3)   complementarity of genes
II. Only those species showing male parental investment
  C.   Quality of parental care
     (1)   willingness of male to invest
     (2)   ability of male to invest
     (3)   complementarity of parental attributes

---

ing with those males who are most vigorous in courtship, since this vigor may correlate with an adequate supply of sperm and a willingness to transfer it.

When the male is completely depleted, there is no advantage in his copulating but selection against the male doing so should be much weaker than selection against the female who accepts him. At intermediate sperm levels, the male may gain something from copulation, but the female should again be selected to avoid him. Since there is little advantage to the male in concealing low reproductive powers, a correlation between vigor of courtship and sperm level would not be surprising. Females would then be selected to be aroused by vigorous courtship. If secondary structures used in display, such as bright feathers, heighten the appearance of vigorousness, then selection may rapidly accentuate such structures. Ironically, the male who has been sexually most successful may not be ideal to mate with if this success has temporarily depleted his sperm supply. Males should not only be selected to recover rapidly from copulations but to give convincing evidence that they have recovered. It is not absurd to suppose that in some highly promiscuous species the most attractive males may be those who, having already been observed to mate with several females, are still capable of vigorous display toward a female in the process of choosing.

GOOD GENES

Maynard Smith (1956) has presented evidence that, given a choice, female *Drosophila subobscura* discriminate against inbred males of that species and that this behavior is adaptive: females who do not so discriminate leave about ¼ as many viable offspring as those who do. Females may choose on the basis of courtship behavior: inbred males are apparently unable to perform a step of the typical courtship as rapidly as outbred males. The work is particularly interesting in revealing that details of courtship behavior may reveal a genetic trait, such as being inbred, but it suffers from an artificiality. If inbred males produce mostly inviable offspring, then, even in the absence of female discrimination, one would expect very few, if any, inbred males to be available in the adult population. Only because such males were artificially selected were there large numbers to expose to females in choice experiments. Had that selection continued one generation further, females who chose inbred males would have been the successful females.

Maynard Smith's study highlights the problem of analyzing the potential for survival of one's partner's genes: one knows of the adult males one meets that they have survived to adulthood; by what criterion does one decide who has survived better? If the female can judge age, then all other things being equal, she should choose older males, as they have demonstrated their capacity for long survival. All other things may not be equal, however, if old age correlates with lowered reproductive success, as it does in some ungulates (Fraser 1968) through reduced ability to impregnate. If the female can judge the physical condition of males she encounters, then she can discriminate against undernourished or sickly individuals, since they will be unlikely to survive long, but discrimination against such individuals may occur for other reasons, such as the presumed lowered ability of such males to impregnate successfully due to the weakened condition.

In some very restricted ways it may be possible to second-guess the future action of natural selection. For example, stabilizing selection has been demonstrated to be a common form of natural selection (see Mayr 1963) and under this form of selection females may be selected to exercise their own discrimination against extreme types, thereby augmenting the effects of any stabilizing selection that has occurred prior to reproduction. Mason (1969) has demonstrated that females of the California Oak Moth discriminate against males extreme in some traits, but no one has shown independent stabilizing selection for the same traits. Discrimination against extreme types may run counter to selection for diversity; the possible role of female choice in increasing or decreasing diversity is discussed below as a form of complementarity.

Reproductive success, independent of ability to survive is easier for the female to gauge because she can directly observe differences in reproductive success before she chooses. A striking feature of data on lek behavior of birds is the tendency for females to choose males who, through competition with other males, have already increased their likelihood of mating. Female choice then greatly augments the effects of male–male competition. On the lek grounds there is an obvious reason why this may be adaptive.

By mating with the most dominant male a female can usually mate more quickly, and hence more safely, than if she chooses a less dominant individual whose attempts at mating often result in interference from more dominant males. Scott (1942) has shown that many matings with less dominant individuals occur precisely when the more dominant individuals are unable, either because of sexual exhaustion or a long waiting line, to quickly service the female. Likewise, Robel (1970) has shown that a dominant female prevents less dominant individuals from mating until she has mated, presumably to shorten her stay and to copulate while the dominant male still can. A second reason why choosing to mate with more dominant males may be adaptive is that the female allies her genes with those of a male who, by his ability to dominate other males, has demonstrated his reproductive capacity. It is a common observation in cervids that females placidly await the outcome of male strife to go with the victor. DeVore (1965) has quantified the importance of dominance in male baboon sexual success, emphasizing the high frequency of interference by other males in copulation and the tendency for female choice, when it is apparent, to be exercised in favor of dominant males. That previous success may increase the skill with which males court females is suggested by work on the black grouse (Kruijt, Bossema and deVos, *in press*), and females may prefer males skillful at courting in part because their skill correlates with previous success.

In many species the ability of the male to find receptive females quickly may be more important than any ability to dominate other males. If this is so, then female choice may be considerably simplified: the first male to reach her establishes thereby a *prima facie* case for his reproductive abilities. In dung flies, in which females must mate quickly while the dung is fresh, male courtship behavior is virtually nonexistent (Parker 1970a). The male who first leaps on top of a newly arrived female copulates with her. This lack of female choice may also result from the *prima facie* case the first male establishes for his sound reproductive abilities. Such a mechanism of choice may of course conflict with other criteria requiring a sampling of the male population, but in some species this sampling could be carried out prior to becoming sexually receptive.

There are good data supporting the importance of complementarity of genes to female choice. Assortative mating in the wild has been demonstrated for several bird species (Cooch & Beardmore 1959, O'Donald 1959) and disassortative mating for a bird species and a moth species (Lowther 1961, Sheppard 1952). Petit & Ehrman (1969) have demonstrated the tendency in several *Drosophila* species for females to prefer mating with the rare type in choice experiments, a tendency which in the wild leads to a form of complementarity, since the female is presumably usually of the common type. These studies can all be explained plausibly in terms of selection for greater or lesser genetic diversity, the female choosing a male whose genes complement her own, producing an "optimal" diversity in the offspring.

GOOD PARENT

Where male parental care is involved, females certainly sometimes choose males on the basis of their ability to contribute parental care. Orians (1969), for example, has recently reviewed arguments and data suggesting that polygyny evolves in birds when becoming the second mate of an already mated male provides a female with greater male parental contribution than becoming the first mate of an unmated male would. This will be so, for example, if the already mated male defends a territory considerably superior to the unmated male's. Variability in territory quality certainly occurs in most territorial species, even in those in which territories are not used for feeding. Tinbergen (1967), for example, has documented the tendency for central territories in the black-headed gull to be less vulnerable to predation. If females compete among themselves for males with good territories, or if males exercise choice as well, then female choice for parental abilities will again tend to augment intra-male competition for the relevant resources (such as territories). The most obvious form of this selection is the inability of a nonterritory holding male to attract a female.

Female choice may play a role in selecting for increased male parental investment. In the roadrunner, for example, food caught by a male seems to act on him as an aphrodisiac: he runs to a female and courts her with the food, suggesting that the female would not usually mate without such a gift (Calder 1967). Male parental care invested after copulation is presumably not a result of female choice after copulation, since she no longer has anything to bargain with. In most birds, however, males defend territories which initially attract the females (Lack 1940). Since males without suitable territories are unable to attract a mate, female choice may play a role in maintaining male territorial behavior. Once a male has invested in a territory in order to attract a mate his options after copulating with her may be severely limited. Driving the female out of his territory would almost certainly result in the loss of his investment up until then. He could establish another territory, and in some species some males do this (von Haartman 1951), but in many species this may be difficult, leaving him with the option of aiding, more or less, the female he has already mated. Female choice, then, exercised *before* copulation, may indirectly force the male to increase his parental investment *after* copulation.

There is no reason to suppose that males do not compete with each other to pair with those females whose breeding potential appears to be high. Darwin (1871) argued that females within a species breeding early for nongenetic reasons (such as being in excellent physical condition) would produce more offspring than later breeders. Sexual selection, he argued, would favor males competing with each other to pair with such females. Fisher (1958) has nicely summarized this argument, but Lack (1968, p. 157) dismisses it as being "not very cogent," since "the date of breeding in birds has been evolved primarily in relation to two different factors, namely the food supply for the young and the capacity of the female to form eggs." These facts are, of course, fully consistent with Darwin's argument, since Darwin is merely supposing a developmental plasticity that allows females to breed earlier if they are capable of forming the eggs, and data presented elsewhere in Lack (1968) support the argument that females breeding earlier for nongenetic reasons (such as age or duration of pair bond) are more successful than those breeding later (see also, for example, Fisher 1969, and Coulson 1966). Goforth & Baskett (1971) have recently shown that dominant males in a penned Mourning Dove population preferentially pair with dominant females; such pairs breed earlier and produce more surviving young than less dominant pairs. It would be interesting to have detailed data from other species on the extent to which males do compete for females with higher breeding potential. Males are certainly often initially aggressive to females intruding in their territories, and this aggressiveness may act as a sieve, admitting only those females whose high motivation correlates with early egg laying and high reproductive potential. There is good evidence that American women tend to marry up the socioeconomic scale, and physical attractiveness during adolescence facilitates such movement (Elder 1969). Until recently such a bias in female choice presumably correlated with increased reproductive success, but the value, if any, of female beauty for male reproductive success is obscure.

The importance of choice by both female and male for a mate who will not desert nor participate in sex outside the pair bond has been emphasized in an earlier section ("Desertion and cuckoldry"). The importance of complementarity is documented in a study by Coulson (1966).

CRITERIA OTHER THAN MALE CHARACTERS

In many species male–male competition combined with the importance of some resource in theory unrelated to males, such as oviposition sites may mitigate against female choice for male characters. In the dragonfly *Parthemis tenera* males compete with each other to control territories containing good oviposition sites, probably because such sites are a predictable place at which to find receptive females and because sperm competition in insects usually favors the last male to copulate prior to oviposition (Parker 1970b). It is clear that the females choose the oviposition site and not the male (Jacobs 1955), and male courtship is geared to advertise good oviposition sites. A male maintaining a territory containing a good oviposition site is *not* thereby contributing parental investment unless that maintenance benefits the resulting young.

Female choice for oviposition sites may be an especially important determinant of male competition in those species, such as frogs and salamanders. showing external fertilization. Such female choice almost certainly predisposed these species to the evolution of male parental investment. Female choice for good oviposition sites would tend to favor any male investment in improving the site, and if attached to the site to attract other females the male would have the option of caring more or less for those eggs already laid. A similar argument was advanced above for birds. Internal fertilization and development mitigate against evolution of male parental care in mammals, since female choice can then usually only operate to favor male courtship feeding, which in herbivores would be nearly valueless. Female choice may also favor males who mate away from oviposition sites if so doing reduced the probability of predation.

Where females are clumped in space the effects of male competition may render female choice almost impossible. In a monkey troop a female preference for a less dominant male may never lead to sexual congress if the pair are quickly broken up and attacked by more dominant males. Apparent female acquiescence in the results of male–male competition may reflect this factor as much as the plausible female preference for the male victor outlined above.

Summary

The relative parental investment of the sexes in their young is the key variable controlling the operation of sexual selection. Where one sex invests considerably more than the other, members of the latter will compete among themselves to mate with members of the former. Where investment is equal, sexual selection should operate similarly on the two sexes. The pattern of relative parental investment in species today seems strongly influenced by the early evolutionary difference into mobile sex cells fertilizing immobile ones, and sexual selection acts to mold the pattern of relative parental investment. The time sequence of parental investment analyzed by sex is an important parameter affecting species in which both sexes invest considerable parental care: the individual initially investing more (usually the female) is vulnerable to desertion. On the other hand, in species with internal fertilization and strong male parental investment, the male is always vulnerable to cuckoldry. Each vulnerability has led to the evolution of adaptations to decrease the vulnerability and to counter-adaptations.

Females usually suffer higher mortality rates than males in monogamous birds. but in nonmonogamous birds and all other groups, males usually suffer higher rates. The chromosomal hypothesis is unable to account for the data. Instead, an adaptive interpretation can be advanced based on the relative parental investment of the sexes. In species with little or no male parental investment, selection usually favors male adaptations that lead to high reproductive success in one or more breeding seasons at the cost of increased mortality. Male competition in such species can only be analyzed in detail when the distribution of females in space and time is properly described. Data from field studies suggest that in some species, size, mobility, experience and metabolic rate are important to male reproductive success.

Female choice can augment or oppose mortality selection. Female choice can only lead to runaway change in male morphology when females choose by a relative rather than absolute standard, and it is probably sometimes adaptive for females to so choose. The relative parental investment of the sexes affects the criteria of female choice (and of male choice). Throughout, I emphasize that sexual selection favors different male and female reproductive strategies and that even when ostensibly cooperating in a joint task male and female interests are rarely identical.

REFERENCES

Bartholomew, G. A. 1970. A model for the evolution of pinniped polygyny. *Evolution* 24: 546–559.

Bastock, M. 1967. *Courtship: An ethological study*. Chicago: Aldine.

Bateman, A. J. 1948. Intrasexual selection in Drosophila. *Heredity* 2: 349–368.

Beebe, W. 1925. The variegated Tinamou *Crypturus variegatus variegatus* (Gmelin). *Zoologica* 6: 195–227.

Beer, J. R., L. D. Frenzel, & C. F. MacLeod. 1958. Sex ratios of some Minnesota rodents. *American Midland Naturalist* 59: 518–524.

Beverton, J. M., & S. J. Holt. 1959. A review of the lifespan and mortality rates of fish in nature and their relation to growth and other physiological characteristics. In *The lifespan of animals*, ed. G. Wolstenhome & M. O'Connor, pp. 142–177. London: J. & A. Churchill.

Blair, W. F. 1960. *The Rusty Lizard*. Austin: University of Texas.

Bouliere, Z. F., & Verschuren, J. 1960. *Introduction a l'ecologie des ongules du Parc National Albert*. Bruxelles: Institut des Parcs Nationaux du Congo Belge.

Bragg, A. N. 1965. *Gnomes of the night*. Philadelphia: University of Pennsylvania Press.

Brown, L. E. 1966. Home range and movement of small mammals. *Symposium of the Zoological Society of London* 18: 111–142.

Bruce, H. 1960. A block to pregnancy in the mouse caused by the proximity of strange males. *Journal of Reproduction and Fertility* 1: 96–103.

Burns, J. M. 1968. Mating frequency in natural populations of skippers and butterflies as determined by spermatophore counts. *Proceedings of the National Academy of Sciences* 61: 852–859.

Calder, W. A. 1967. Breeding behavior of the Roadrunner, *Geococcyx californianus*. *Auk* 84: 597–598.

Carpenter, C. 1967. Aggression and social structure in Iguanid lizards. In *Lizard ecology*, ed. W. Milstead. Columbia, Mo.: University of Missouri.

Parental Investment and Sexual Selection   175

Chapman, A. B., L. E. Casida, & A. Cote. 1938. Sex ratios of fetal calves. *Proceedings of the American Society of Animal Production* 1938, pp. 303–304.

Cooch, F. G., & M. A. Beardmore. 1959. Assortative mating and reciprocal difference in the Blue-Snow Goose complex. *Nature* 183: 1833–1834.

Corbet, P., C. Longfield, & W. Moore. 1960. *Dragonflies.* London: Collins.

Coulson, J. C. 1960. A study of the mortality of the starling based on ringing recoveries. *Journal of Animal Ecology* 29: 251–271.

————. 1966. The influence of the pair-bond and age on the breeding biology of the kittiwake gull *Rissa tridactyla. Journal of Animal Ecology* 35: 269–279.

Cowan, I. M. 1950. Some vital statistics of big game on overstocked mountain range. *Transactions of North American Wildlife Conference* 15: 581–588.

Darley, J. 1971. Sex ratio and mortality in the brown-headed cowbird. *Auk* 88: 560–566.

Darwin, C. 1871. *The descent of man, and selection in relation to sex.* London: John Murray.

Demerec, M., & Kaufmann, B. P. 1941. Time required for *Drosophila* males to exhaust the supply of mature sperm. *American Naturalist* 75: 366–379.

DeVore, I. 1965. Male dominance and mating behavior in baboons. In *Sex and behavior,* ed. Frank Beach. New York: John Wiley and Sons.

deVos, A., P. Broky, & V. Geist. 1967. A review of social behavior of the North American Cervids during the reproductive period. *American Midland Naturalist* 77: 390–417.

Dunn, E. R. 1941. Notes on *Dendrobates auratus. Copeia* 1941, pp. 88–93.

Eaton, T. H. 1941. Notes on the life history of *Dendrobates auratus. Copeia* 1941, pp. 93–95.

Eisenberg, J. F. 1965. The social organizations of mammals. *Handbuch der Zoologie* 10 (7): 1–92.

Elder, G. 1969. Appearance and education in marriage mobility. *American Sociological Review* 34: 519–533.

Emlen, J. M. 1968. A note on natural selection and the sex-ratio. *American Naturalist* 102: 94–95.

Emlen, J. T. 1940. Sex and age ratios in the survival of the California Quail. *Journal of Wildlife Management* 4: 91–99.

Emlen, S. T. 1968. Territoriality in the bullfrog, *Rana catesbeiana. Copeia* 1968, pp. 240–243.

Engelmann, F. 1970. *The physiology of insect reproduction.* Oxford: Pergamon Press.

Fiedler, K. 1954. Vergleichende Verhaltensstudien an Seenadeln, Schlangennadeln und Seepferdchen (Syngnathidae). *Zeitsch. Tierpsych.* 11: 358–416. 358–416.

Fisher, H. 1969. Eggs and egg-laying in the Laysan Albatross, *Diomedea immutabilis. Condor* 71: 102–112.

Fisher, R. A. 1958. *The genetical theory of natural selection.* New York: Dover Publications.

Fraser, A. F. 1968. *Reproductive behavior in Ungulates.* London and New York Academic Press.

Gadgil, M., & W. H. Bossert. 1970. Life historical consequences of natural selection. *American Naturalist* 104: 1–24.

Goethals, G. W. 1971. Factors affecting permissive and nonpermissive rules regarding premarital sex. In *Studies in the sociology of sex: a book of readings,* ed. J. M. Henslin. New York: Appleton-Century-Croft.

Goforth, W., & T. Baskett. 1971. Social organization of penned Mourning Doves. *Auk* 88: 528–542.

176   ROBERT L. TRIVERS

Guhl, A. M. 1951. Measurable differences in mating behavior of cocks. *Poultry Science* 30: 687.

Haartman, L. von. 1951. Successive polygamy. *Behavior* 3: 256–274.

————. 1969. Nest-site and evolution of polygamy in European Passerine birds. *Ornis Fennica* 46: 1–12.

Hamilton, J. B. 1948. The role of testicular secretions as indicated by the effects of castration in man and by studies of pathological conditions and the short lifespan associated with maleness. *Recent Progress in Hormone Research* 3: 257–322.

Hamilton, J. B., & M. Johansson. 1965. Influence of sex chromosomes and castration upon lifespan: studies of meal moths, a species in which sex chromosomes are homogenous in males and heterogenous in females. *Anatomical Record* 24: 565–578.

Hamilton, J. B., & G. E. Mestler. 1969. Mortality and survival: comparison of eunuchs with intact men and women in a mentally retarded population. *Journal of Gerontology* 24: 395–411.

Hamilton, J. B., R. S. Hamilton, & G. E. Mestler. 1969. Duration of life and causes of death in domestic cats: influence of sex, gonadectomy and inbreeding. *Journal of Gerontology* 24: 427–437.

Hamilton, W. D. 1964. The genetical evolution of social behavior. *Journal of Theoretical Biology* 7: 1–52.

————. 1967. Extraordinary sex ratios. *Science* 156: 477–488.

Harris, V. A. 1964. *The life of the Rainbow Lizard.* Hutchinson Tropical Monographs. London.

Hays, F. A. 1947. Mortality studies in Rhode Island Reds II. *Massachusetts Agricultural Experiment Station Bulletin* 442: 1–8.

Hershkovitz, P. 1966. Letter to *Science* 153: 362.

Hirth, H. F. 1963. The ecology of two lizards on a tropical beach. *Ecological Monographs* 33: 83–112.

Höhn, E. O. 1967. Observations on the breeding biology of Wilson's Phalarope (*Steganopus tricolor*) in Central Alberta. *Auk* 84: 220–244.

Huxley, J. S. 1938. The present standing of the theory of sexual selection. In *Evolution*, ed. G. DeBeer. New York: Oxford Univ. Press.

Jacobs, M. 1955. Studies in territorialism and sexual selection in dragonflies. *Ecology* 36: 566–586.

Jensen, G. D., Bobbitt, R. A. & Gordon, B. N. 1968. Sex differences in the development of independence of infant monkeys. *Behavior* 30: 1–14.

Johns, J. E. 1969. Field studies of Wilson's Phalarope. *Auk* 86: 660–670.

Kaufmann, B. P., & Demerec, M. 1942. Utilization of sperm by the female *Drosophila melanogaster. American Naturalist* 76: 445–469.

Kessel, E. L. 1955. The mating activities of baloon flies. *Systematic Zoology* 4: 97–104.

Kikkawa, J. 1964. Movement, activity and distribution of small rodents *Clethrionomys glareolus* and *Apodemus sylvaticus* in woodland. *Journal of Animal Ecology* 33: 259–299.

Kluijver, H. N. 1933. Bijrage tot de biologie en de ecologie van den spreeuw (*Sturnus vulgaris* L.) gedurende zijn voortplantingstijd. *Versl. Plantenziektenkundigen dienst, Wageningen* 69: 1–145.

Koivisto, I. 1965. Behaviour of the black grouse during the spring display. *Finnish Game Research* 26: 1–60.

Kolman, W. 1960. The mechanism of natural selection for the sex ratio. *American Naturalist* 94: 373–377.

Parental Investment and Sexual Selection   177

Kopstein, F. 1941. Über Sexualdimorphismus bei Malaiischen Schlangen. *Temminckia*, 6: 109–185.
Kruijt, J. P, I. Bossema, & G. J. deVos. *In Press*. Factors underlying choice of mate in Black Grouse. *15th Congr. Intern. Ornith.*, The Hague, 1970.
Kruijt, J. P., & J. A. Hogan. 1967. Social behavior on the lek in Black Grouse, *Lyrurus tetrix tetrix* (L.) *Ardea* 55: 203–239.
Lack, D. 1940. Pair-formation in birds. *Condor* 42: 269–286.
————. 1954. *The natural regulation of animal numbers.* New York: Oxford University Press.
————. 1968. *Ecological adaptations for breeding in birds.* London: Methuen.
LeBoeuf, B. J., & R. S. Peterson. 1969. Social status and mating activity in elephant seals. *Science* 163: 91–93.
Lee, R., 1969, King Bushman violence. Paper presented at meeting of American Anthropological Association, November, 1969.
Leigh, E. G. 1970. Sex ratio and differential mortality between the sexes. *American Naturalist* 104: 205–210.
Lin, N. 1963. Territorial behavior in the Cicada killer wasp *Sphecius spheciosus* (Drury) (Hymenoptera: Sphecidae.) I. *Behaviour* 20: 115–133.
Lindburg, D. G. 1969. Rhesus monkeys: mating season mobility of adult males. *Science* 166: 1176–1178.
Lowther, J. K. 1961. Polymorphism in the white-throated sparrow, *Zonotrichia albicollis* (Gmelin). *Canadian Journal of Zoology* 39: 281–292.
Ludwig, W., & C. Boost. 1951. Über Beziehungen zwischen Elternalter, Wurfgrösse und Geschlechtsverhältnis bei Hunden. *Zeitschrift fur indukt. Abstammungs und Vererbungslehre* 83: 383–391.
Madison, D. M., & Shoop, C. R. 1970. Homing behavior, orientation, and home range of salamanders tagged with tantalum–182. *Science* 168: 1484–1487.
Mason. L. G. 1969. Mating selection in the California Oak Moth (Lepidoptera, Droptidae). *Evolution* 23: 55–58.
Maynard Smith, J. 1956. Fertility, mating behaviour and sexual selection in *Drosophila subobscura. Journal of Genetics* 54: 261–279.
Mayr, E. 1939. The sex ratio in wild birds. *American Naturalist* 73: 156–179.
————. 1963. *Animal species and evolution.* Cambridge: Harvard University Press.
Miller, R. S. 1958. A study of a wood mouse population in Wytham Woods, Berkshire. *Journal of Mammalogy* 39: 477–493.
Morris, D. 1952. Homosexuality in the Ten-spined Stickleback (*Pygosteus pungitius*). *Behaviour* 4: 233–261.
————. 1967. *The naked ape.* New York: McGraw Hill.
Myers. J., & C. Krebs. 1971. Sex ratios in open and closed vole populations: demographic implications. *American Naturalist* 105: 325–344.
Nevo, R. W. 1956. A behavior study of the red-winged blackbird. 1. Mating and nesting activities. *Wilson Bulletin* 68: 5–37.
O'Donald, P. 1959. Possibility of assortative mating in the Arctic Skua. *Nature* 183: 1210.
————. 1962. The theory of sexual selection. *Heredity* 17: 541–552.
Orians, G. H. 1969. On the evolution of mating systems in birds and mammals. *American Naturalist* 103: 589–604.
Parker. G. A. 1970a. The reproductive behavior and the nature of sexual selection in *Scatophaga stercoraria* L. (Diptera: Scatophagidae) 2. The fertilization rate and the spatial and temporal relationships of each sex around the site of mating and oviposition. *Journal of Animal Ecology* 39: 205–228.

178   ROBERT L. TRIVERS

————. 1970b. Sperm competition and its evolutionary consequences in the insects. *Biological Reviews* 45: 525–568.

————.1970c. The reproductive behaviour and the nature of sexual selection in *Scatophaga stercoraria* L. (Diptera: Scatophagidae) VI. The adaptive significance of emigration from the oviposition site during the phase of genital contact. *Journal of Animal Ecology* 40: 215–233.

————. 1970d. The reproductive behaviour and the nature of sexual selection in *Scatophaga stercoraria* L. (Diptera: Scatophagidae). VI. The adaptive sig-evolution of the passive phase. *Evolution* 24: 774–788.

Parker, J. E., F. F. McKenzie, & H. L. Kempster. 1940. Observations on the sexual behavior of New Hampshire males. *Poultry Science* 19: 191–197.

Peterson, R. S., & G. A. Bartholomew. 1967. *The natural history and behavior of the California Sea Lion.* Special Publications #1, American Society of Mammalogists.

Petit, C., & L. Ehrman. 1969. Sexual selection in *Drosophila*. In *Evolutionary biology*, vol. 5, ed. T. Dobzhansky, M. K. Hecht, W. C. Steere. New York: Appleton-Century-Crofts.

Potts, G. R. 1969. The influence of eruptive movements, age, population size and other factors on the survival of the Shag (*Phalacrocorax aristotelis* L.). *Journal of Animal Ecology* 38: 53–102.

Rand, A. S. 1967. Ecology and social organization in the Iguanid lizard *Anolis lineatopus. Proc. U.S. Nat. Mus.* 122: 1–79.

Riemann, J. G., D. J. Moen, & B. J. Thorson. 1967. Female monogamy and its control in house flies. *Insect Physiology* 13: 407–418.

Rivero, J. A., & A. E. Estevez. 1969. Observations on the agonistic and breeding behavior of *Leptodactylus pentadactylus* and other amphibian species in Venezuela. *Breviora No.* 321: 1–14.

Robel, R. J. 1966. Booming territory size and mating success of the Greater Prairie Chicken (*Tympanuchus cupido pinnatus*). *Animal Behaviour* 14: 328–331.

Robel, R. J. 1970. Possible role of behavior in regulating greater prairie chicken populations. *Journal of Wildlife Management* 34: 306–312.

Robinette, W. L., J. S. Gashwiler, J. B. Low, & D. A. Jones. 1957. Differential mortality by sex and age among mule deer. *Journal of Wildlife Management* 21: 1–16.

Rockstein, M. 1959. The biology of ageing insects. In *The lifespan of animals*, ed. G. Wolstenhome & M. O'Connor, pp. 247–264. London: J. A. Churchill.

Rowley, I. 1965. The life history of the Superb Blue Wren *Malarus cyaneus. Emu* 64: 251–297.

Royama, T. 1966. A re-interpretation of courtship feeding. *Bird Study* 13: 116–129.

Sadleir, R. 1967. *The ecology of reproduction in wild and domestic mammals.* London: Methuen.

Savage, R. M. 1961. *The ecology and life history of the common frog.* London: Sir Isaac Pitman and Sons.

Schein, M. W. 1950. The relation of sex ratio to physiological age in the wild brown rat. *American Naturalist* 84: 489–496.

Schoener, T. W. 1967. The ecological significance of sexual dimorphism in size in the lizard *Anolis conspersus. Science* 155: 474–477.

————. 1971. Theory of feeding strategies. *Annual Review of Ecology and Systematics,* 2: 369–404.

Scott, J. W. 1942. Mating behavior of the Sage Grouse. *Auk* 59: 477–498.

Parental Investment and Sexual Selection 179

Selander, R. K. 1965. On mating systems and sexual selection. *American Naturalist* 99: 129–141.

————. 1966. Sexual dimorphism and differential niche utilization in birds. *Condor* 68: 113–151.

Sellers, A., H. Goodman, J. Marmorston, & M. Smith. 1950. Sex differences in proteinuria in the rat. *American Journal of Physiology* 163: 662–667.

Sheppard, P. M. 1952. A note on non-random mating in the moth *Panaxia dominula*. (L.) *Heredity* 6: 239–241.

Stephens, M. N. 1952. Seasonal observations on the Wild Rabbit (*Oryctolagus cuniculus cuniculus* L.) in West Wales. *Proceedings of the Zoological Society of London* 122: 417–434.

Stokes, A., & H. Williams. 1971. Courtship feeding in gallinaceous birds. *Auk* 88: 543–559.

Sugiyama, U. 1967. Social organization of Hanuman langurs. In *Social communication among primates*, ed. S. Altmann. Chicago: University of Chicago Press.

Taber, R. D., & R. F. Dasmann. 1954. A sex difference in mortality in young Columbian Black-tailed Deer. *Journal of Wildlife Management* 18: 309–315.

Tilley, S. 1968. Size-fecundity relationships and their evolutionary implications in five Desmognathine salamanders. *Evolution* 22: 806–816.

Tinbergen, N. 1935. Field observations of East Greenland birds. 1. The behavior of the Red-necked Phalarope (*Phalaropus lobatus* L.) in Spring. *Ardea* 24: 1–42.

————. 1967. Adaptive features of the Black-headed Gull *Larus ridibundus* L. *Proceedings of the International Ornithological Congress* 14: 43–59.

Tinkle, D. W. 1967. The life and demography of the Side-blotched Lizard *Uta stansburiana*. *Miscellaneous Publications of the Museum of Zoology, University of Michigan* 132: 1–182.

Tinkle, D., H. Wilbur, & S. Tilley. 1970. Evolutionary strategies in lizard reproduction. *Evolution* 24: 55–74.

Tyndale-Biscoe, C. H. and R. F. C. Smith. 1969. Studies on the marsupial glider, *Schoinobates volans* (Kerr). 2. Population structure and regulatory mechanisms. *Journal of Animal Ecology* 38: 637–650.

Van Lawick-Goodall, J. 1968. The behavior of free-living chimpanzees in the Gombe Stream Reserve. *Animal Behaviour Monographs* 1: 161–311.

Verner, J. 1964. Evolution of polygamy in the long-billed marsh wren. *Evolution* 18: 252–261.

————. 1965. Selection for sex ratio. *American Naturalist* 99: 419–421.

Verner, J., & M. Willson. 1969. Mating systems, sexual dimorphism, and the role of male North American passerine birds in the nesting cycle. *Ornithological Monographs* 9: 1–76.

Williams, G. C. 1966. *Adaptation and natural selection*. Princeton: Princeton University Press.

Willson, M., & E. Pianka. 1963. Sexual selection, sex ratio, and mating system. *American Naturalist* 97: 405–406.

Wood, D. H. 1970. An ecological study of *Antechinus stuartii* (Marsupialia) in a Southeast Queensland rain forest. *Australian Journal of Zoology* 18: 185–207.

The Logic of Animal Conflict

J. MAYNARD SMITH
School of Biological Sciences, University of Sussex, Falmer, Sussex BN1 9QG

G. R. PRICE
Galton Laboratory, University College London, 4 Stephenson Way, London NW1 2HE

Conflicts between animals of the same species usually are of "limited war" type, not causing serious injury. This is often explained as due to group or species selection for behaviour benefiting the species rather than individuals. Game theory and computer simulation analyses show, however, that a "limited war" strategy benefits individual animals as well as the species.

In a typical combat between two male animals of the same species, the winner gains mates, dominance rights, desirable territory, or other advantages that will tend toward transmitting its genes to future generations at higher frequencies than the loser's genes. Consequently, one might expect that natural selection would develop maximally effective weapons and fighting styles for a "total war" strategy of battles between males to the death. But instead, intraspecific conflicts are usually of a "limited war" type, involving inefficient weapons or ritualized tactics that seldom cause serious injury to either contestant. For example, in many snake species the males fight each other by wrestling without using their fangs[1,2]. In mule deer (*Odocoileus hemionus*) the bucks fight furiously but harmlessly by crashing or pushing antlers against antlers, while they refrain from attacking when an opponent turns away, exposing the unprotected side of its body[3]. And in the Arabian oryx (*Oryx leucoryx*) the extremely long, backward pointing horns are so inefficient for combat that in order for two males to fight they are forced to kneel down with their heads between their knees to direct their horns forward[4]. (For additional examples, see Collins[5], Darwin[6], Hingston[6], Huxley *et al.*[7], Lorenz[8] and Wynne-Edwards[9].)

How can one explain such oddities as snakes that wrestle with each other, deer that refuse to strike "foul blows", and antelope that kneel down to fight?

The accepted explanation for the conventional nature of contests is that if no conventional methods existed, many individuals would be injured, and this would militate against the survival of the species (see, for example, Huxley[7]). The difficulty with this type of explanation is that it appears to assume the operation of "group selection". Although one cannot rule out group selection as an agent producing adaptations, it is only likely to be effective in rather special circumstances[10-12]. Consequently it seems to us that group selection cannot by itself account for the complex anatomical and behavioural adaptations for limited conflict found in so many species, but there must also be individual selection for these, which means that a "limited war" strategy must be differentially advantageous for individuals.

We consider simple formal models of conflict situations, and ask what strategy will be favoured under individual selection. We first consider conflict in species possessing offensive weapons capable of inflicting serious injury on other members of the species. Then we consider conflict in species where serious injury is impossible, so that victory goes to the contestant who fights longest. For each model, we seek a strategy that will be stable under natural selection; that is, we seek an "evolutionarily stable strategy" or ESS. The concept of an ESS is fundamental to our argument; it has been derived in part from the theory of games, and in part from the work of MacArthur[13] and of Hamilton[14] on the evolution of the sex ratio. Roughly, an ESS is a strategy such that, if most of the members of a population adopt it, there is no "mutant" strategy that would give higher reproductive fitness.

A Computer Model

A main reason for using computer simulation was to test whether it is possible even in theory for individual selection to account for "limited war" behaviour.

We consider a species that possesses offensive weapons capable of inflicting serious injuries. We assume that there are two categories of conflict tactics: "conventional" tactics, C, which are unlikely to cause serious injury, and "dangerous" tactics, D, which are likely to injure the opponent seriously if they are employed for long. (Thus in the snake example, wrestling involves C tactics and use of fangs would be D tactics. In many species, C tactics are limited to threat displays at a distance, without any physical fighting. We consider a conflict between two individuals to consist of a series of alternate "moves". At each move, a contestant can employ C or D tactics, or retreat, R. If a contestant employs D tactics, there is a fixed probability that his opponent will be seriously injured: a contestant who is seriously injured always retreats. If a contestant retreats, the contest is at an end and his opponent is the winner. A possible conflict between contestants A and B can be represented in this way:

A's move $C\ C\ C\ C\ C\ C\ C\ C\ C\ C\ C\ C\ D\ C\ C\ C\ C\ C\ C\ D$

B's move $C\ C\ C\ C\ C\ C\ C\ C\ C\ C\ C\ C\ D\ C\ C\ C\ C\ C\ C\ R$

If a contestant plays D on the first move of a contest, or plays D in response to C by his opponent, this is called a "probe" or a "provocation". A probe made after the opening move is said to "escalate" a contest from C to D level. A contestant who plays D in reply to a probe is said to "retaliate". In the example shown above, A probes on his twelfth and twentieth moves; B retaliates after the first probe, but retreats after the second, leaving A the winner. At the end of a contest there are "pay-offs" to each contestant. The pay-offs are taken as measures of the contribution the contest has made to the reproductive success of the individual. They take account of three factors: the advantages of winning as compared with losing, the disadvantage of being seriously injured, and the disadvantage of wasting time and energy in the contest.

A "strategy" for a contestant is a set of rules which ascribe probabilities to the C, D, and R plays, as functions of what has previously happened in the course of the current contest. (No memory of what has happened in previous contests with the same or other opponents is assumed.) For computer simulation we programmed five possible strategies, each of which might be thought on *a priori* grounds to be optimal in certain circumstances. The strategies considered were as follows:

(1) "Mouse". Never plays D. If receives D, retreats at once before there is any possibility of receiving a serious injury. Otherwise plays C until the contest has lasted a preassigned number of moves.

(2) "Hawk". Always plays D. Continues the contest until he is seriously injured or his opponent retreats.

(3) "Bully". Plays D if making the first move. Plays D in response to C. Plays C in response to D. Retreats if opponent plays D a second time.

(4) "Retaliator". Plays C if making the first move. If opponent plays C, plays C (but plays R if contest has lasted a preassigned number of moves). If opponent plays D, with a high probability retaliates by playing D.

(5) "Prober-Retaliator". If making the first move, or after opponent has played C, with high probability plays C and with low probability plays D (but plays R if contest has lasted a preassigned number of moves). After giving a probe, reverts to C if opponent retaliates, but "takes advantage" by continuing to play D if opponent plays C. After receiving a probe, with high probability plays D.

The contestants were programmed as having identical fighting prowess, so that they differed only in the strategies they followed. The five strategies represent extremes, but from results with these it is possible to estimate the results likely to be found with intermediate types. The Hawk strategy is a "total war" strategy; Mouse, Retaliator, and Prober-Retaliator are "limited war" strategies. The question of main interest is whether individual selection will favour the former or one of the latter types.

The Simulation Test

The five strategies determine fifteen types of two-opponent contests. Two thousand contests of each type were simulated by computer, using pseudo-random numbers generated by an algorithm to vary the contests. The following probabilities were used: Probability of serious injury from a single D play$=0.10$. Probability that a Prober-Retaliator will probe on the opening move or after opponent has played $C=0.05$. Probability that Retaliator or Prober-Retaliator will retaliate against a probe (if not injured) by opponent$=1.0$. Pay-offs were calculated as follows: Pay-off for winning$=+60$. Pay-off for receiving serious injury$=-100$. Pay-off for each D received that does not cause serious injury (a "scratch")$=-2$. Pay-off for saving time and energy (awarded to each contestant not seriously injured) varied from 0 for a contest of maximum length, to $+20$ for a very short contest. The contest example shown earlier was one of the 2,000 Prober-Retaliator versus Prober-Retaliator contests.

Table 1 shows the average pay-off to each contestant in each type of contest. The number in a given row and

column is the pay-off gained by the row strategy when the opponent uses the column strategy. For example, in contests between Mouse and Hawk, the average pay-offs are 19.5 to Mouse and 80.0 to Hawk.

To tell whether a strategy is evolutionarily stable against the other four strategies, we examine the corresponding column in Table 1. For Hawk to be an ESS, it is necessary that it be the most profitable strategy in a population almost entirely of Hawks. In such a population, a given animal of any type will almost always have a Hawk as opponent. Therefore the pay-offs in the "Hawk" column apply. These show that Mouse and Bully are both more successful than Hawk. Therefore natural selection will cause alleles for Mouse and Bully behaviour to increase in frequency, and alleles giving Hawk behaviour to decrease. Thus Hawk is not an ESS.

Examining the other columns, we see that Mouse is not an ESS because Hawk, Bully, and Prober-Retaliator average higher pay-offs in a population almost entirely of Mouse. Nor is Bully an ESS. However, Retaliator is an ESS since no other strategy does better, though Mouse does equally well. And the last column shows that Prober-Retaliator is almost an ESS.

How would we expect such a population to evolve? It will come to consist mainly of Retaliators or Prober-Retaliators, with the other strategies maintained at a low frequency by mutation. The balance between the two main types will depend on the frequency of Mouse, since the habit of probing is only an advantage against Mouse. For the particular values in Table 1, it can be shown that if the frequency of Mouse is greater than 7%, Prober-Retaliator will replace Retaliator as the predominant type. It is worth noting that a real population would contain young, senile, diseased and injured individuals adopting the strategy Mouse for non-genetic reasons.

Thus the simulation shows emphatically the superiority, under individual selection, of "limited war" strategies in comparison with the Hawk strategy.

Briefly, the reason that conflict limitation increases individual fitness is that retaliation behaviour decreases the fitness of Hawks, while the existence of possible future mating opportunities reduces the loss from retreating uninjured.

This general result will not be altered by moderate changes in the program parameters, though very large changes will alter it. One way would be by changing the probability of serious injury from a single D from 0.10 to 0.90. This would give advantage to "Pre-emptive Strike" policies, making Hawk an ESS. (Such species are probably rare, because excessively dangerous weapons would be opposed by kin selection.) Another way to make selection favour "total war" behaviour would be by giving the same pay-off penalty for retreating uninjured as for serious injury. This would correspond to a species where an individual fights only a single battle in its lifetime, on which its reproductive success entirely depends. Our choice of $+60$ for winning, 0 for retreating uninjured, and -100 for serious injury represents a species where males have more than one opportunity to gain a mate. Changing these values to $+60$, -100, and -100 respectively, would make Hawk the optimal strategy. Conversely, $+60$, 0,

Table 1 Average Pay-offs in Simulated Intraspecific Contests for Five Different Strategies

		Opponent				
		"Mouse"	"Hawk"	"Bully"	"Retaliator"	"Prober-Retaliator"
Contestant receiving the pay-off	"Mouse"	29.0	19.5	19.5	29.0	17.2
	"Hawk"	80.0	−19.5	74.6	−18.1	−18.9
	"Bully"	80.0	4.9	41.5	11.9	11.2
	"Retaliator"	29.0	−22.3	57.1	29.0	23.1
	"Prober-Retaliator"	56.7	−20.1	59.4	26.9	21.9

−500 would represent a long-lived species with numerous opportunities to gain mates, where individual selection would still more strongly favour cautious strategies.

Real Animals

Real animal conflicts are vastly more complex than our simulated conflicts. (An interesting study by Dingle[15] shows that this holds true even at the lowly level of the mantis shrimp.) Probably our models are true to nature in emphasising a category distinction rather than an intensity distinction between "conventional" and "dangerous" tactics. In real animals, however, there exist not only the category distinction, but also individual differences in the intensity and skill with which each kind of tactic is employed. Also, in many species there are several categories of increasingly dangerous tactics, instead of only one.

The advantage from making a category distinction is that this simplifies behavioural requirements for limited conflict. It is probably easier for genetics to program a snake not to use fangs at all in certain situations than to program it to use fangs as intensively as possible up to intensity k, but not at intensities greater than k. Similarly, fair and foul blows are distinguished in boxing, and conventional and nuclear weapons in war.

Under the condition that any act of physical aggression is treated as a D act, the theoretical model will result in symbolic fighting by threat from a distance. This would be advantageous for a species that has an inherent difficulty in fighting physically at a safe level. For example, domestic and wild cattle, which have very dangerous horns and are somewhat clumsy in their charges, make much use of threat displays (stomping, pawing, bellowing). The model will not, however, give rise to conflict behaviour that is wholly symbolic and never backed up by physical aggression or other sanctions, since such behaviour would not be evolutionarily stable without some mechanism reducing the reproductive success of mutant individuals deficient in responding to the symbols. An interesting problem is how the felids, with their dangerous teeth and claws, limit their physical combats to non-fatal levels. Probably the explanation is that they have a hierarchy of many conflict categories and limit their probing to small escalations. Consequently, it takes repeated escalations to raise the conflict to the most dangerous level.

In most animal species there is probably a high correlation between prowess in C tactics and in D tactics. This means that C level conflict provides information to each animal about how its opponent is likely to perform if the conflict is escalated. This permits improvement in strategies over those used in the computer model. Instead of probing at random, an animal will be more likely to probe if its opponent is inferior in conventional fighting. On the other hand, if its opponent is very superior in conventional tactics, an animal will frequently retreat without waiting for its opponent to try a probe. Thus actual animals may combine Prober-Retaliator and Mouse capabilities.

If animals can adopt different strategies according to the opponent that confronts them, then an interesting possibility appears. The "Hawk" column of Table 1 shows that the best strategy against a Hawk is Mouse: that is, retreat immediately. If a species includes deviant individuals who follow the Hawk strategy and fight recklessly against every opponent, then it will be advantageous for ordinary members of the species to be able to estimate recklessness and avoid combat with Hawks. But if this happens, then it will be advantageous to simulate wild, incontrollable rage. And in fact the threat displays of some species do have an appearance of maniacal fury, hence there probably is some advantage in acting this way. However, if most species members simulate insane rage when actually their fighting is limited and controlled, then selection will favour indi-

viduals who partly discount the threat displays, and "call the bluff" of the pseudo-Hawks.

This leads to the suggestion that it might be advantageous for an individual animal to be maniacal in an easily recognisable way that could not be counterfeited. A possible instance of this is the phenomenon of going "on musth", which occurs periodically in adult male elephants[16,17]. The temporal glands secrete a dark brown fluid that runs down the face, giving a visual and olfactory sign that cannot be counterfeited. The madness of the animal "on musth" causes other elephants to avoid him, and this may give an increase in dominance status that persists for a time after the musth period is over.

Conflict in which Injury is Impossible

The previous section offers an explanation of why, in a species with offensive weapons capable of inflicting serious injury, escalated fighting may be rare or absent. In doing so, it raises a second problem. In a contest between opponents who are unable to inflict serious injury, victory goes to the one who is prepared to continue for a longer time. How are such contests decided?

Suppose that the pay-off to the victor is v. If a contest is ever to be settled, there must also be some disadvantage to the contestants in a long contest. If so, the only choice of strategy open to a contestant is of the period for which he is prepared to continue, and hence of the pay-off, say $-m$, he is prepared to accept. Thus if two contestants adopt strategies m_1 and m_2, where $m_1 > m_2$, the pay-off to the first is $v - m_2$ and to the second is $-m_2$. Our problem then is how a contestant should choose a value of m, or, more precisely, whether there is a method of choosing m which is an ESS.

To answer this question, we need a more precise definition of an ESS. We define $E_J(I)$ as the expected pay-off to I played against J. Then I is an ESS if, for all J, $E_I(I) > E_I(J)$; if for any strategy J, $E_I(I) = E_I(J)$, then evolutionary stability requires that $E_J(I) > E_J(J)$. The relevance of the latter condition is as follows. If in a population adopting strategy I a mutant J arises whose expectation against I is the same as I's expectation against itself, then J will increase by genetic drift until meetings between two J's becomes a common event.

It is easy to show that no "pure" strategy (that is, no fixed value of m) is an ESS. Thus in a population adopting strategy m, a mutant adopting $m + \epsilon$ would always do better (and if $m > v$, a mutant adopting a zero strategy would also do better). It is, however, possible to find a mixed strategy which is an ESS. Let strategy I be a mixed strategy which selects a value of m between x and $x + \delta x$ with probability $p(x)\delta x$. Then if

$$p(x) = (1/\nu)\exp(-x/\nu) \qquad (1)$$

it can be shown that I is an ESS.

We conclude that an evolutionary stable population is either genetically polymorphic, the strategies of individuals being distributed as in equation (1), or that it consists of individuals whose behaviour differs from contest to contest as in (1). There is no stable pure strategy, and hence no behaviourally uniform population can be stable.

Conclusions

There are many complications left out of these simple models. The analysis is, however, sufficient to show that individual selection can explain why potentially dangerous offensive weapons are rarely used in intraspecific contests; a stable strategy does, however, require that contestants should respond to an "escalated" attack by escalating in return. Also, if contests are settled by a process of attrition, then evolutionary stability requires that the population be genetically polymorphic, or that individuals vary their behaviour from contest to contest.

A more detailed analysis will be published elsewhere.

Ideas similar to those described here have been applied to human neurotic behaviour by J. S. Price[18].

For suggestions, we thank Professors Hans Kalmus and R. C. Lewontin, and Drs W. D. Hamilton, Gerald Lincoln, T. B. Poole and M. J. A. Simpson. We thank the Science Research Council for support.

[1] Shaw, C. E., *Herpetologica*, **4**, 137 (1948).
[2] Shaw, C. E., *Herpetologica*, 7, 149 (1951).
[3] Linsdale, J. M., and Tomich, P. Q., *A Herd of Mule Deer*, 511f (Univ. of California Press, Berkeley and Los Angeles, 1953).
[4] Darwin, C., *The Descent of Man and Selection in Relation to Sex*, chap. 17 (Murray, London, 1882).
[5] Collins, N. E., *Physiol. Zool.*, **17**, 83 (1944).
[6] Hingston, R. W. G., *Character Person.*, **2**, 3 (1933).
[7] Huxley, J. S., *Phil. Trans. R. Soc.*, **251B**, 249 (1956).
[8] Lorenz, K., *On Aggression* (Methuen, London, 1966).
[9] Wynne-Edwards, V. C., *Animal Dispersion in Relation to Social Behaviour*, chap. 8 (Oliver and Boyd, Edinburgh and London, 1962).
[10] Maynard Smith, J., *Nature*, **201**, 1145 (1964).
[11] Levins, R., in *Some Mathematical Questions in Biology* (American Mathematical Society, 1970).
[12] Price, G. R., *Ann. hum. Genet.*, **35**, 485 (1972).
[13] MacArthur, R. H., in *Theoretical and Mathematical Biology* (edit. by Waterman, T., and Horowitz, H.) (Blaisdell, New York, 1965).
[14] Hamilton, W. D., *Science, N.Y.*, **156**, 477 (1967).
[15] Dingle, H., *Anim. Behav.*, **17**, 561 (1969).
[16] West, L. J., Pierce, C. M., and Thomas, W. D., *Science, N.Y.*, **138**, 1100 (1962).
[17] Eisenberg, J. F., McKay, G. M., and Jainudeen, M. R., *Behaviour*, **38**, 193 (1971).
[18] Price, J. S., *Proc. R. Soc. Med.*, **62**, 1107 (1969).

Contributors

Elizabeth Adkins-Regan
Department of Psychology
Cornell University
Ithaca, New York 14853

Stevan J. Arnold
Department of Ecology and Evolution
University of Chicago
Chicago, Illinois 60637

H. Jane Brockmann
Department of Zoology
University of Florida
Gainesville, Florida 32611

Lee C. Drickamer
Department of Zoology
Southern Illinois University
Carbondale, Illinois 62901-6501

Fred C. Dyer
Department of Zoology
Michigan State University
East Lansing, Michigan 48824

Bennett G. Galef
Department of Psychology
McMaster University
Hamilton, Ontario L8S 4K1
Canada

Lynne D. Houck
Department of Ecology and Evolution
University of Chicago
Chicago, Illinois 60637

Charles Snowdon
Department of Psychology
University of Wisconsin
Madison, Wisconsin 53706